ANIMAL FEEDING AND NUTRITION

Seventh Edition

Marshall H. Jurgens
Iowa State University

KENDALL/HUNT PUBLISHING COMPANY
2460 Kerper Boulevard P.O. Box 539 Dubuque, Iowa 52004-0539

Contents

List of Tables of Animal Nutrient Requirements and Feedstuff Composition

List of Tables of Suggested Animal Diet Formulations

Preface

This text provides the reader with a basic understanding of animal nutrition including feedstuff characteristics and principles for formulating nutritionally balanced diets. The book should be particularly useful to students in the animal sciences and preveterinary medicine. It is also intended to be a suitable information source for livestock producers, veterinarians, extention agents and professional animal nutritionists.

This seventh edition of *Animal Feeding and Nutrition* contains an updated table of feedstuff nutrient composition. Also included are tables showing nutrient requirements for various types of animals quoted from current National Research Council publications prepared by the National Academy of Sciences. Several sections of this edition have been revised and expanded in an effort to make a more complete text for a course in animal nutrition. In addition, over 170 suggested ration formulations for various stages of animal and poultry life-cycle nutrition are available to the reader of this seventh edition. To guide students in learning essential principles and facts concerning subjects presented, study questions and problems follow each chapter. An extensive glossary of animal nutrition and a detailed table of equivalents section provides the reader information and access to terminology, weights and measures, conversion factors and feed storage capacity calculations which students and participants of animal nutrition are likely to encounter.

In order to reduce the vast amount of reading necessary for a complete coverage of feeding practices, the author has selected and condensed subject matter into an outline of four basic topics: (1) Nutrients and digestive systems; (2) Feedstuffs and formulations; (3) Commercial feeds and additives; and (4) Feeding guides and recommendations: swine, beef, dairy, sheep, horses, poultry, dogs and cats. To provide the reader with the most current information, contents of this text have been adapted from data of agricultural universities, experiment stations, U.S.D.A. and industry together with the author's personal thoughts and observations. It is hoped that this text can be used by students in combination with additional information obtained from classroom lecture as well as the several textbooks and periodicals available for supplemental reading. In the large and rapidly changing field of livestock nutrition and feeding practices students should be directed by their instructor to pertinent current scientific literature to amplify the subject matter of this text.

Students using this text are expected to have a basic knowledge of general chemistry and preferably some exposure to organic and biochemistry. In addition, the student is expected to have some understanding of animal anatomy, physiology and husbandry.

Grateful appreciation is expressed to my colleagues and students who have offered numerous helpful suggestions in preparing this latest edition.

Marshall H. Jurgens

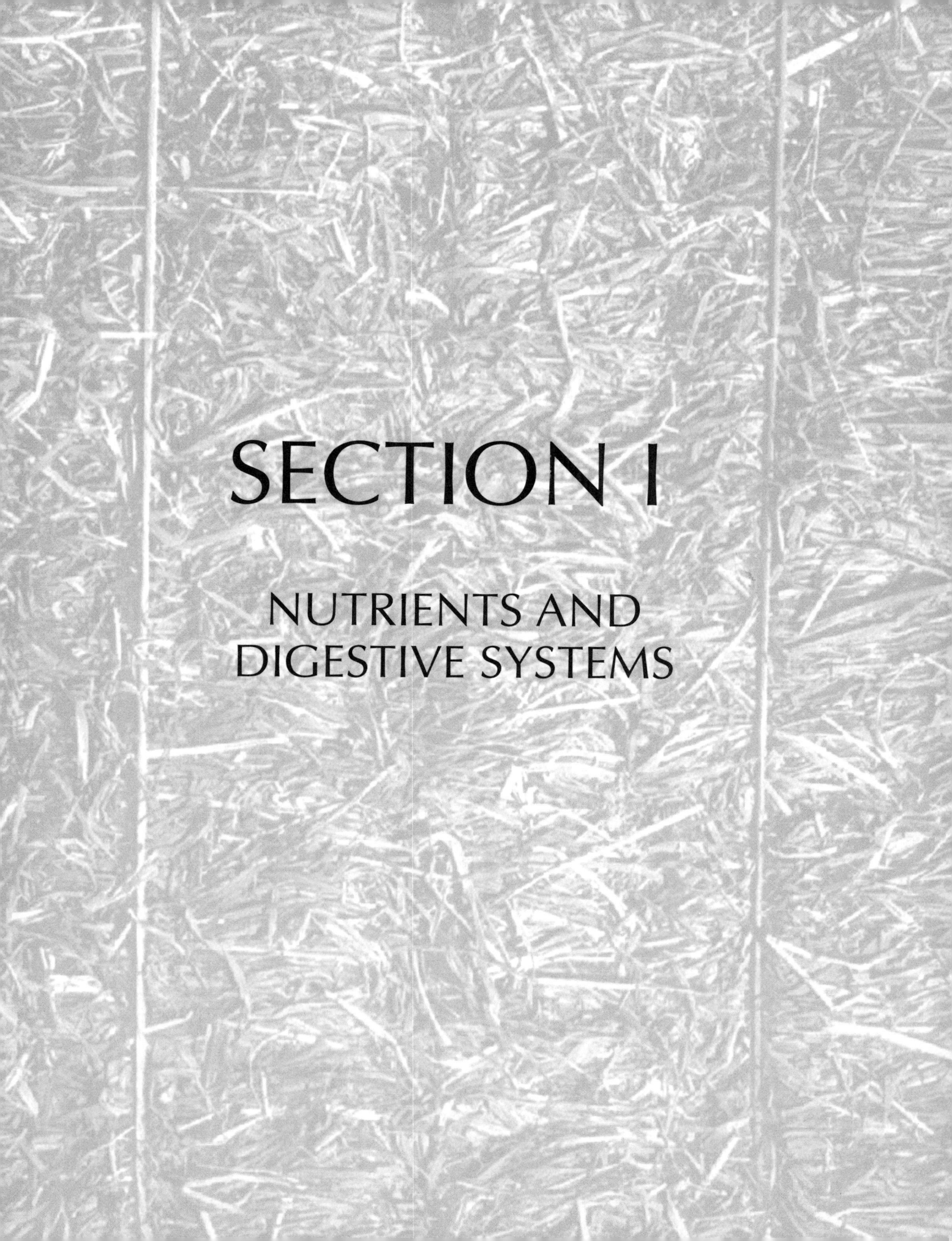

SECTION I

NUTRIENTS AND DIGESTIVE SYSTEMS

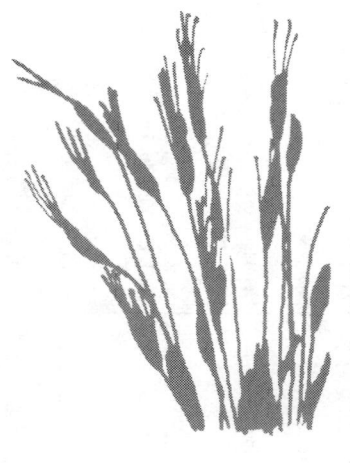

1

Review of Nutrients and Digestion

Chapter Goals

- Define the nutrient needs of animals.
- Explain the differences in the six basic nutrients.
- Identify symptoms of nutrient deficiency or toxicity.
- Identify parts and functions of animal digestive systems.
- Compare differences and similarities of the digestive systems.
- Explain feedstuff digestion and nutrient absorption.

A nutrient is defined by Morrison[1] as any feed constituent or group of feed constituents of the same general chemical composition that aids in the support of animal life. We must interpret this definition somewhat more broadly than was originally intended, because we now have to include substances that are not of feed origin. Although feeds are parcels of nutrients usually mixed with nonnutrient material, a complete ration may be more than a combination of feeds. It may include synthetically produced vitamins, chemically prepared inorganic salts or biogenically synthesized amino acids. What animals eat in terms of the products actually consumed is fundamentally of much less importance than the quantity and assortment of the nutrients furnished by the rations made available to them.

One cannot be sure today that a list of the presently recognized nutrients found in feeds, or in animal tissues, represents a complete list of the operating needs of the body. For practical purposes, however, we do know what most of these nutrients are. We understand the general functions of most nutrients and, consequently, why they must be supplied to the animal.

The description of nutrients and their requirements that have single entities (or that have physiologically specific functions) is quite simple. On the other hand, any description of nutrients that serve to supply energy (which cannot within itself be classified as a nutrient) or that serve multiple functions in the metabolic processes of releasing and using energy, may become confusing in the learning process. Thus, any scheme of brief classification of nutrient makeup of feed must be oversimplified. We therefore generally offer chemical classification of feeds rather than nutrient classification. Figure 1–1 is a chemical analysis scheme.

The following outline briefly discusses the six basic nutrients.

I. Water

  A. *General*

    1. Cheapest and most abundant nutrient.

    2. Makes up 65% to 85% of animal body weight at birth and 45% to 60% of body weight at maturity.

    3. Percentage of body water decreases with animal age and has an inverse relationship with body fat.

1. Morrison FB, *Feeds and Feeding,* 22nd ed, Clinton, Iowa: Morrison Publishing Co, 1959.

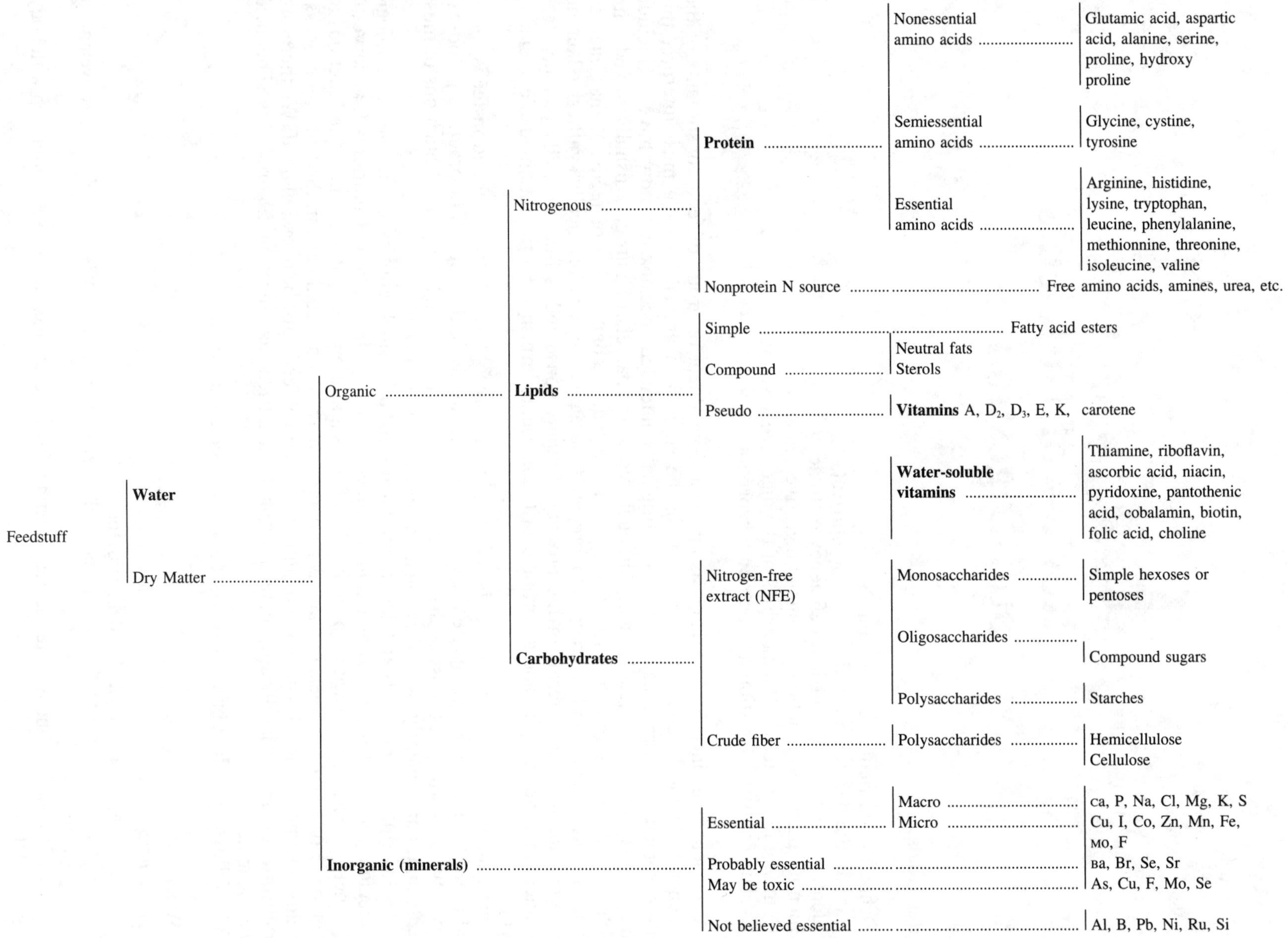

Feedstuff

- **Water**
- Dry Matter
  - Organic
    - Nitrogenous
      - **Protein**
        - Nonessential amino acids .... Glutamic acid, aspartic acid, alanine, serine, proline, hydroxy proline
        - Semiessential amino acids .... Glycine, cystine, tyrosine
        - Essential amino acids .... Arginine, histidine, lysine, tryptophan, leucine, phenylalanine, methionine, threonine, isoleucine, valine
      - Nonprotein N source .... Free amino acids, amines, urea, etc.
    - **Lipids**
      - Simple .... Neutral fats
      - Compound .... Fatty acid esters / Sterols
      - Pseudo .... **Vitamins** A, D_2, D_3, E, K, carotene
      - **Water-soluble vitamins** .... Thiamine, riboflavin, ascorbic acid, niacin, pyridoxine, pantothenic acid, cobalamin, biotin, folic acid, choline
    - **Carbohydrates**
      - Nitrogen-free extract (NFE)
        - Monosaccharides .... Simple hexoses or pentoses
        - Oligosaccharides .... Compound sugars
        - Polysaccharides .... Starches
      - Crude fiber
        - Polysaccharides .... Hemicellulose / Cellulose
  - **Inorganic (minerals)**
    - Essential
      - Macro .... Ca, P, Na, Cl, Mg, K, S
      - Micro .... Cu, I, Co, Zn, Mn, Fe, Mo, F
    - Probably essential .... Ba, Br, Se, Sr
    - May be toxic .... As, Cu, F, Mo, Se
    - Not believed essential .... Al, B, Pb, Ni, Ru, Si

FIGURE 1–1. Chemical analysis scheme.

2

4. Accounts for 90% to 95% of blood. Many tissues contain 70% to 90% water.
5. Found in the animal body as:
   a. Intracellular water—mainly muscle and skin.
   b. Extracellular water—mainly interstitial fluids, blood plasma, lymph, synovial and cerebro-spinal fluids.
   c. Water present in urinary and gastrointestinal tract.

B. *Function and Deficiencies of Water in the Animal*
   1. Functions
      a. Transportation of nutrients and excretions.
      b. Chemical reactions and solvent properties.
      c. Body temperature regulation.
      d. Maintains shape of body cells.
      e. Lubricates and cushions joints and organs in the body cavity.
   2. Deficiencies or restrictions
      a. Reduced feed intake and reduced productivity.
      b. Weight loss due to dehydration.
      c. Increased excretion of nitrogen and electrolytes such as sodium and potassium.

C. *Sources of Water to the Animal*
   1. Drinking water
      a. Factors that affect drinking water consumption
         (1) Environmental temperature and humidity (increased water consumption at heat stress).
         (2) Dry matter consumption (water consumption directly correlated at moderate temperatures).
         (3) Dietary factors
            (a) High water content of feed reduces drinking.
            (b) High fiber, salt or protein content of diet increases drinking.
         (4) Function of the animal (lactating cow vs dry cow).
         (5) Type of urinary system (mammal vs avian).
         (6) Water quality
            (a) Good water should have less than 2500 mg/liter of total dissolved solids.
            (b) Water containing over 1 g/liter sulfates may cause diarrhea.
            (c) Levels above 100–200 ppm nitrates are potentially toxic.
      b. Approximate water consumption (mature, nonstressed animal)
         (1) Swine—1½–3 gal/hd/day.
         (2) Sheep—1–3 gal/hd/day.
         (3) Cattle—10–14 gal/hd/day.

Water may be the most overlooked but most important of all nutrients. Allow animals free access to plenty of clean, fresh water at all times.

3

(4) Horses—10–14 gal/hd/day.
(5) Poultry—2 parts water for each 1 part dry feed.
2. Water contained in or on feed
    a. Highly variable in feedstuffs.
      (1) Grains may range from below 8% to more than 30% water.
      (2) Forages may range from below 5% in a dry hay to more than 90% water in a lush young grass.
    b. Precipitation or dew on feeds may decrease water consumption.
3. Metabolic water
    a. Results from oxidation of organic nutrients in the tissues.
    b. May account for 5% to 10% of total water intake.
D. *Water Losses from the Animal Body*
  1. Urine.
  2. Feces.
  3. Vaporization from the lungs and dissipation through the skin (insensible losses).
  4. Sweat from the sweat glands.

II. Carbohydrates (CHO)

A. *General*
  1. Made up (by % mol wt) of C (40%), H (7%), and O (53%) with H and O having a chemical similarity of H_2O.
  2. Includes sugar, starch, cellulose and gums.
  3. Very little occurs as such in animal body.
  4. CHO serve as structural components in the plant.
  5. CHO make up approximately ¾ of plant dry weight and thus form the largest part of animal's food supply.
  6. Formed by photosynthesis in plants:

$$6\,CO_2 \;+\; 6\,H_2O \;+\; 686\ kcal \;\xrightarrow[\text{chlorophyll}]{\text{Plant}}\; C_6H_{12}O_6 \;+\; 6\,O_2$$

carbon dioxide + water + solar energy → glucose + oxygen

B. *Structure*
  1. Structure consists of C atoms arranged in chains to which are attached H and O.
  2. Carbohydrates are characterized by having a ketone $(C{-}\overset{\displaystyle O}{\overset{\|}{C}}{-}C)$ or an aldehyde $({-}\overset{\displaystyle O}{\overset{\|}{C}}{-}H)$ group in their structure.

C. *Classification (by Number of Sugar Molecules)*
  1. Monosaccharides (1 sugar molecule)
    a. Hexoses (6 C)
      (1) Glucose.

(2) Fructose.

HOH₂C, O, CH₂OH / C, C / H H HO OH / C C / OH H

(3) Galactose.

CH₂OH / HO C O OH / C C / H OH H H / C C / H OH

(4) Mannose.

CH₂OH / H C O H / H C / HO C OH HO OH / C C / H H

b. Pentoses (5 C)
   (1) Arabinose.
   (2) Xylose.
   (3) Ribose.
2. Disaccharides (2 sugar molecules)
  a. Sucrose (glucose + fructose).

CH₂OH / H C O H HOH₂C O H / C C O C C / HO OH H H HO CH₂OH / C C C C / H OH OH H

b. Maltose (glucose + glucose).

c. Lactose (galactose + glucose).

d. Cellobiose (glucose + glucose).

3. Polysaccharides (many sugar molecules)
   a. Starch—α (alpha) linkage of glucose.

    (1) Amylose—unbranched-chain plant starch.
    (2) Amylopectin—branched-chain plant starch.
    (3) Glycogen—branched-chain animal starch.

b. Cellulose—β (beta) linkage of glucose.

c. Mixed polysaccharides
   (1) Hemicellulose—mixture of hexoses and pentoses.
   (2) Pectins—hexoses and pentoses mixed with salts of complex acids.
   (3) Gums—mixtures of hexoses and pentoses.
d. Lignin
   (1) Not considered a true carbohydrate as the proportions of C, H and O are different from other polysaccharides.
   (2) Essentially indigestible by all animals and may reduce the digestibility of other nutrients.
   (3) Found most commonly in forages of poor quality and tends to be higher in legumes than in grasses.
   (4) Lignification increases with plant age.
D. *Function and Deficiencies*
   1. Functions in animal body
      a. Source of energy.
      b. Source of heat.
      c. Building blocks for other nutrients.
      d. Stored in animal body by converting to fats.
   2. Deficiencies or abnormal metabolism
      a. Ketosis
         (1) Caused by a disorder in carbohydrate or lipid metabolism creating an excess of ketones (acetone, acetoacetic acid and β-hydroxybutyric acid) to accumulate in blood and tissues.
         (2) Common in animals requiring high amounts of energy such as in cattle at the peak of lactation or in sheep in late pregnancy.
         (3) Symptoms may include
            (a) Increased breakdown of tissue protein for energy.
            (b) Loss of body weight.
            (c) Increased water consumption.
            (d) Decreased milk production in lactating animals.
            (e) Abortion in pregnant animals.
            (f) Acetone smell about the animal's breath.
      b. Diabetes mellitus
         (1) More common in humans than farm animals.
         (2) Results from insufficient insulin production by the pancreas.
         (3) Generally has a genetic basis; may be induced by overfeeding and obesity.
         (4) Symptoms include
            (a) High blood glucose level.
            (b) Excess urinary loss of glucose.
            (c) Elevated mobilization of adipose tissue lipids.
            (d) Increased production of ketones.
E. *Natural Sources*
   Most feedstuffs of plant origin are high in carbohydrate content, especially cereal grains.

F. *Digestion and Metabolism*
  1. Digestion

$$CHO \begin{cases} \text{Crude fiber (cellulose, hemicellulose, lignin)—poorly digested} \\ \text{Nitrogen-free extract (soluble sugars and starches)—readily digested} \end{cases}$$

Carbohydrates must be broken down to monosaccharides for absorption from the digestive system. This requires digestive enzymes elaborated by the host or by microflora inhabiting the digestive system of the host. A high proportion of these monosaccharides will be converted to glucose in the wall of the small intestine. The remainder is modified by the liver or metabolized in other ways. The carbohydrate-splitting enzymes are effective in hydrolyzing most complex carbohydrates to monosaccharides except for those with the β (beta) linkage such as cellulose. Microflora of the rumen of ruminants and the cecum of some nonruminants, such as the horse and rabbit, produce cellulase, so that these species can utilize large quantities of cellulose. Anaerobic fermentation of carbohydrates by these microflora results in the production of large quantities of volatile fatty acids (VFA), mainly acetic, propionic and butyric acids. These acids provide the host animal a large proportion of the total energy supply.

  2. Metabolism
   a. The complexities of the metabolism of energy within animal cells may be centered on a few common cycles termed "the final oxidative pathways" (Figure 1–2).

   Glucose, for example, is fragmented into two three-carbon units (pyruvate) via the aerobic pathway of glycolysis (Embden-Myerhoff pathway). Pyruvate eventually enters the tricarboxylic acid cycle (TCA cycle, Krebs cycle) as acetyl-CoA. Also oxidized by the Krebs cycle are the long-chain fatty acids and the carbon skeletons of amino acids.

   The end result of the Krebs cycle is the oxidation of carbon and hydrogen to form carbon dioxide, water, energy stored as adenosine triphosphate (ATP) and heat. The energy contained within the high energy phosphate bonds of ATP is used as the "energy currency" of the body to drive various physiological processes that require energy, such as muscle contraction and protein synthesis. Carbon dioxide is eventually exhaled by the lungs. The animal dissipates the heat resulting from oxidation.

   Complete breakdown of 1 mole of glucose by the final oxidative pathways yields the following amounts of ATP trapped at each stage of oxidation to CO_2 and H_2O:

E-M pathway	8 moles of ATP
Pyruvic acid to acetyl CoA (2)	6 moles of ATP
Krebs cycle (2)	24 moles of ATP
Total	38 moles of ATP

   One mole of ATP has an energy value of approximately 8 kcal/mole.
   b. Carbohydrates in the feed are also broken down by rumen microorganisms to form glucose (Figure 1–3). Glucose is converted by the rumen microorganisms via glycolysis (Embden-Myerhoff pathway) to pyruvic acid. Two molecules of pyruvic acid, each containing three carbon atoms, are formed from one molecule of glucose. Pyruvic acid is a key intermediate in rumen VFA formation because it can be metabolized to any of the VFAs. Fermentation of glucose to pyruvic acid and of pyruvic acid to the VFAs is very rapid.

   The efficiencies associated with different VFA productions are summarized in Table 1–1. The efficiency of conversion values shown represent that portion of VFA energy available to the animal for metabolism relative to the gross energy value of 1 mole of glucose (686.0 kcal/mole).

   The production of propionic acid results in the most efficient conversion of feed energy to energy usable by the ruminant animal. The production of acetic or butyric acids is less efficient because the waste gases carbon dioxide and methane are generated, and lost in the process of eructation. These gases are of no benefit to the host animal, representing a significant loss of dietary energy. The efficiency greater than 100% shown for propionic acid in Table 1–1 is possible due to the addition of energy from hydrogen.

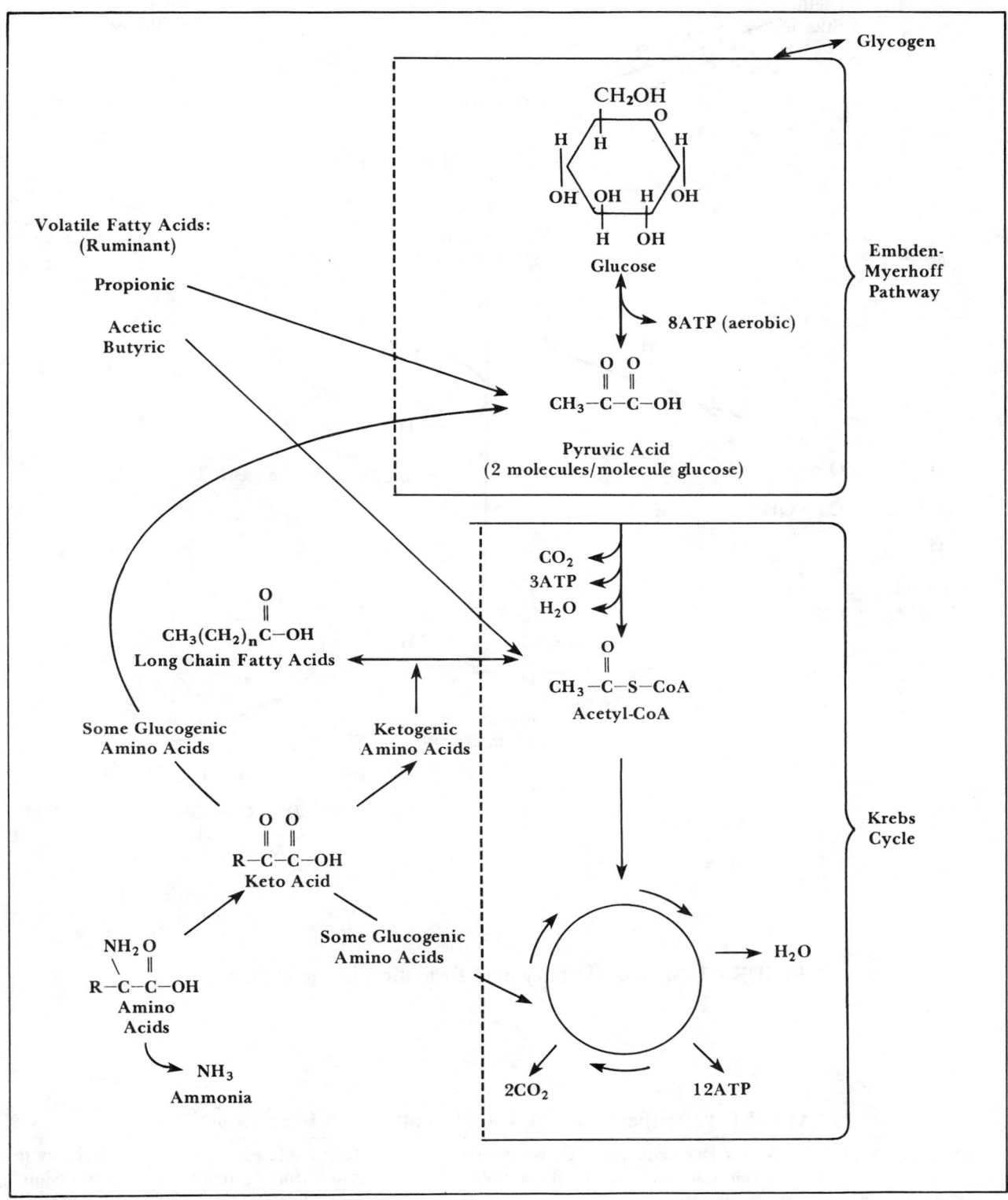

FIGURE 1–2. Final oxidative pathways and relationships of carbohydrates, fats and proteins.

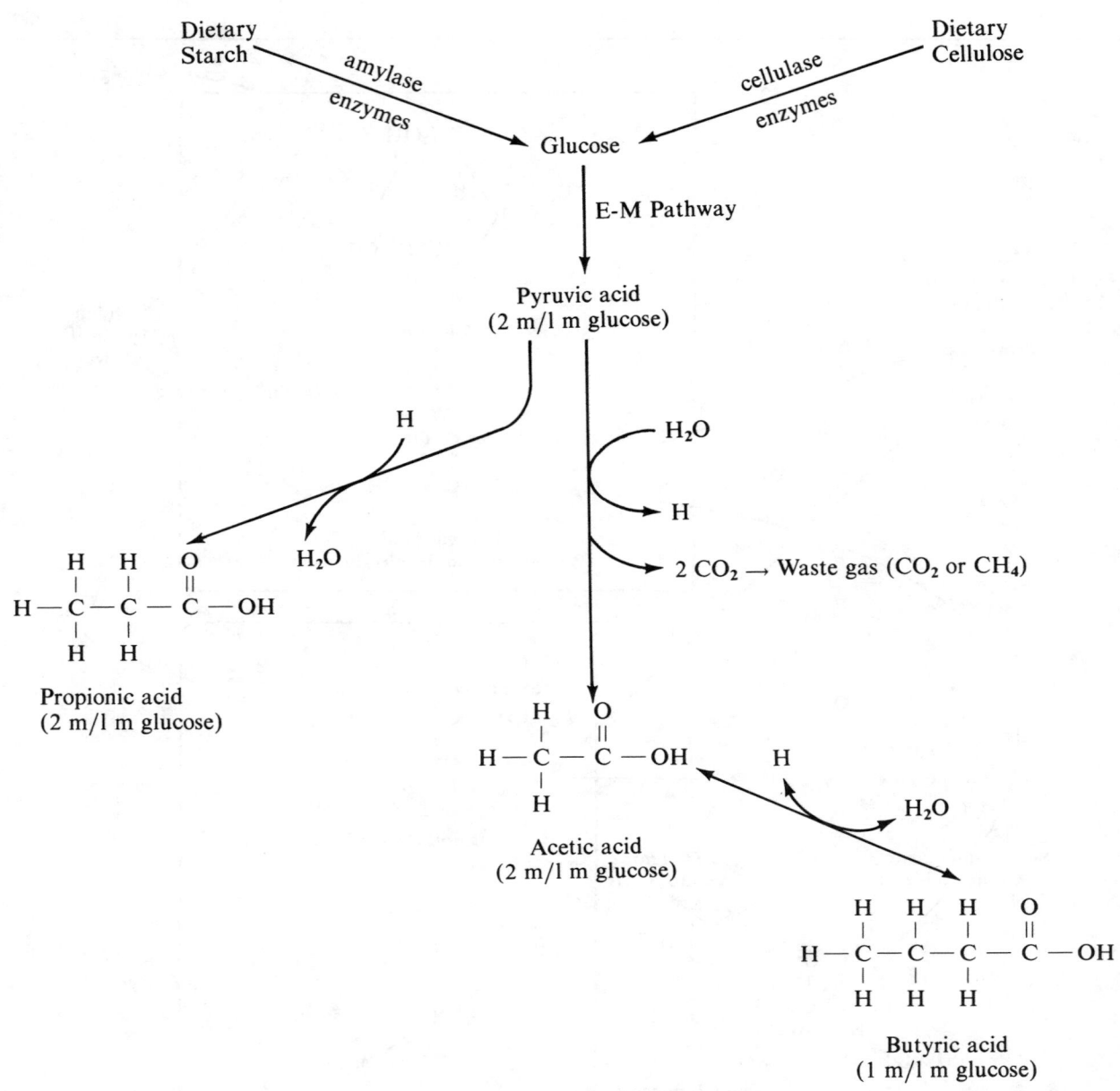

FIGURE 1–3. Volatile fatty acid formation in the rumen.

TABLE 1–1. Efficiencies of Volatile Fatty Acid Formation*

Volatile Fatty Acid	Moles Derived/ Mole Glucose	Gross Energy (kcal/mole)	Total Usable Energy Derived/Mole Glucose (kcal)	Efficiency of Conversion† (%)
Acetic	2	209.4	418.8	61.0
Propionic	2	367.2	734.4	107.1
Butyric	1	524.3	524.3	76.4

*Gross energy values obtained from *Handbook of Chemistry and Physics*.
†Relative to glucose, glucose = 686.0 kcal/mole.

c. The three primary VFAs produced in the rumen are metabolized differently following absorption:
  (1) Acetic acid is not significantly metabolized by the rumen mucosa or the liver, but rather by peripheral tissue (e.g., muscle and adipose tissue). Acetic acid eventually is oxidized by the Krebs cycle, entering as acetyl-CoA. At maintenance levels, 10 moles of ATP are produced per mole of acetic acid. Acetic acid is also utilized in long-chain fatty acid synthesis.
  (2) Nearly all of the propionic acid is metabolized in the liver where it either gives rise to glucose or is metabolized via the Krebs cycle to yield carbon dioxide, water and 18 moles of ATP per mole of acid oxidized. Glucose can be used to synthesize long-chain fatty acids. Therefore, propionic acid can indirectly contribute to fat synthesis.
  (3) The rumen mucosa converts approximately 50% of the absorbed butyric acid to β-hydroxybutyric acid, an intermediate in fatty acid oxidation that is eventually used in the Krebs cycle as acetyl-CoA. The remainder of the butyric acid is utilized by the liver. Again, it is converted to acetyl-CoA and enters the Krebs cycle, yielding 27 moles of ATP per mole of acid metabolized. Since very little butyric acid is present in the bloodstream, butyric acid is not highly involved in fat synthesis.
  G. *Units of Measure*—refer to Measures of Feedstuff Energy in Chapter 2 of this text.

III. Lipids (Fat or Ether Extract)
  A. *General*
    1. Made up (by % mol wt) of C (77%), H (12%) and O (11%).
    2. Insoluble in water but soluble in organic solvents.
    3. One gram of a typical fat yields about 9.45 kcal gross energy (heat) when completely combusted, compared with about 4.2 kcal for a typical carbohydrate; thus, fats produce approximately 2.25 times more energy than carbohydrates.
  B. *Classification*
    1. Simple lipids
      a. True fat—esters of fatty acids with glycerol.
      b. Waxes—esters of fatty acids with alcohol other than glycerol.
    2. Compound lipids—esters containing groups in addition to an alcohol and fatty acid
      a. Phospholipids—contain phosphoric acid and nitrogen.
        (1) Lecithin.
        (2) Cephalin.
        (3) Sphingomyelin.
      b. Glycolipids—contain a carbohydrate and nitrogen but no phosphoric acid.
      c. Lipoproteins—lipids bound to proteins in blood and other tissues.
    3. Derived lipids—substances derived from the above groups by hydrolysis
      a. Fatty acids.
      b. Sterols—large molecular alcohols.
  C. *Structure of Fat*

$$\begin{array}{l}
\text{H}_2\text{C}-\text{OH} \quad \text{HO}-\overset{\overset{\text{O}}{\|}}{\text{C}}-\text{R}_1 \qquad\qquad \text{H}_2\text{C}-\text{O}-\overset{\overset{\text{O}}{\|}}{\text{C}}-\text{R}_1 \\[4pt]
\text{HC}-\text{OH} + \text{HO}-\overset{\overset{\text{O}}{\|}}{\text{C}}-\text{R}_2 \;\rightleftharpoons\; \text{HC}-\text{O}-\overset{\overset{\text{O}}{\|}}{\text{C}}-\text{R}_2 + 3\text{H}_2\text{O} \\[4pt]
\text{H}_2\text{C}-\text{OH} \quad \text{HO}-\overset{\overset{\text{O}}{\|}}{\text{C}}-\text{R}_3 \qquad\qquad \text{H}_2\text{C}-\text{O}-\overset{\overset{\text{O}}{\|}}{\text{C}}-\text{R}_3
\end{array}$$

Glycerol + Fatty acids $\rightleftharpoons$ Fat (triglyceride) + Water

  1. The R groups represent the various possible fatty acids that could be attached to the glycerol molecule by an ester linkage. It is possible to have a mono-, di- or triglyceride, that is, an ester linkage of a fatty acid with one, two or three of the hydroxyl groups of glycerol, respectively.

2. The fatty acids consist of chains of C atoms ranging from 2 to 24 or more Cs in length and characterized by a carboxyl group on the end.
   a. Saturated fatty acids are those in which each C atom in the chain (except the carboxyl group) has two H atoms attached to it (three Hs at the terminal C).

$$-\overset{|}{\underset{|}{C}}-\overset{|}{\underset{|}{C}}-\overset{|}{\underset{|}{C}}-\overset{|}{\underset{|}{C}}-$$

   b. Unsaturated fatty acids have one or more pairs of C atoms in their chain attached by a double bond, and H has been removed.

$$-\overset{|}{C}=\overset{|}{C}-\overset{|}{C}=\overset{|}{C}-$$

3. Examples of structural formulas
   a. Saturated.

$$\text{H}\overset{\overset{\textstyle H}{|}}{\underset{\underset{\textstyle H}{|}}{C}}-\overset{\overset{\textstyle H}{|}}{\underset{\underset{\textstyle H}{|}}{C}}-\overset{\overset{\textstyle H}{|}}{\underset{\underset{\textstyle H}{|}}{C}}-\overset{\overset{\textstyle O}{\|}}{C}-\text{OH}$$

Butyric acid—$C_4H_8O_2$

Myristic acid—$C_{14}H_{28}O_2$

Stearic acid—$C_{18}H_{36}O_2$

b. Unsaturated.

H H H H H H H H H   H H H H H H H H O
H C—C—C—C—C—C—C—C—C=C—C—C—C—C—C—C—C—C—OH
H H H H H H H H   H H H H H H H

Oleic acid—$C_{18}H_{34}O_2$ (C18:1)

H H H H H H H H H H H H H H H H H O
H C—C—C—C—C—C=C—C—C=C—C—C—C—C—C—C—C—C—OH
H H H H H   H   H H H H H H H

Linoleic acid—$C_{18}H_{32}O_2$ (C18:2)

H H H H H H H H H H H H H H H H H H H O
H C—C—C—C—C—C=C—C—C=C—C—C=C—C—C—C=C—C—C—C—OH
H H H H H   H   H   H   H H H

Arachidonic acid—$C_{20}H_{32}O_2$ (C20:4)

D. Constants for Measuring Chemical Properties

1. Iodine number—denotes the degree of unsaturation of a fat or fatty acid. An unsaturated fat unites easily with iodine, two atoms of this element being added for each double bond. Thus, the iodine number is the amount of iodine in grams that can be taken up by 100 g of fat.
2. Saponification number—gives a measure of the average chain length of the three fatty acids in a fat. When fat is boiled with alkali (NaOH or KOH), it is split into glycerol and the alkali salt (called soaps) of the fatty acid. Thus, the saponification number is the number of milligrams of alkali required for the saponification (hydrolysis) of 1 g of fat. The smaller the fatty acid molecules (short-chain fatty acids), the greater is the number of these molecules per gram of fat and, thus, the larger the amount of alkali required for saponification or the larger the number.
3. Reichert-Meissl (RM) number—measures the amount of water soluble, steam-volatile fatty acids (short-chain) present in a fat. Thus, the RM number is the number of milliliters of 0.1 N KOH solution required to neutralize these volatile acids obtained by hydrolysis of 5 g of fat. Lard and other high molecular weight fats contain practically no VFAs (RM numbers near 0), but butterfat contains a higher proportion of VFAs (RM number 20–33).
4. Melting point—the temperature at which a fat changes from a solid to liquid state is referred to as its melting point. This is dependent on chain length and degree of unsaturation of the molecule. Short fatty acids of low molecular weight tend to be more volatile and have a lower melting point. As chain length increases, melting point increases. As the number of double bonds increases (more unsaturated) the melting point decreases.

E. Function and Deficiencies

1. Function
   a. Dietary energy supply.
   b. Source of heat, insulation and protection for the animal body.
   c. Source of essential fatty acids.
      (1) Linoleic acid (C18:2) and linolenic acid (C18:3) apparently cannot be synthesized by animal tissues, or at least not in sufficient amounts to prevent pathological changes, and so must be supplied in the diet.
      (2) Arachidonic acid (C20:4) can be synthesized from linoleic, and therefore is required in the diet only if linoleic is absent.
   d. Serve as a carrier for absorption of fat-soluble vitamins.

2. Deficiency or abnormal metabolism
   a. Skin lesions, hair loss, poor feathering in chicks and reduced growth or reproductive rate may result from a deficiency of certain essential fatty acids in monogastric diets.
   b. Ketosis—insufficient dietary energy intake in high producing animals may cause catabolism of body reserves for needed energy. If these steps do not occur normally, the two carbon fragments from fat catabolism accumulate and produce toxic blood levels of ketone bodies (acetone, acetoacetic acid and β-hydroxybutyric acid). This physiological disorder is most common in dairy cattle during early lactation and sheep during late pregnancy.
   c. Fatty livers—a possible result of abnormal liver function is the accumulation of lipids in the liver. Fatty liver may arise from high dietary fat or cholesterol intake, increased liver lipogenesis, increased mobilization of lipids from adipose tissue or various other deficiencies or abnormalities.

F. *Location and Natural Sources of Fat*
   1. Location
      a. Animal body.
         (1) Subcutaneous.
         (2) Surrounding internal organs.
         (3) Marbling and milk.
      b. Plant—seed germ or embryo.
   2. Natural sources—most feeds contain low levels of fat, generally less than 10% (cereal grains, forages, animal products etc.). Unprocessed "oil seeds" may contain up to 20% fat (soybean, cottonseed, sunflower etc.).

G. *Digestion and Metabolism*
   1. The digestibility of fat is quite high (usually exceeding 80%). The triglyceride molecule undergoes hydrolysis in the small intestine to monoglyceride and free fatty acids. These small units rarely pass into the blood circulatory system directly but normally are absorbed into the lymph system following hydrolysis and resynthesized during the absorption process.
   2. Breakdown of long-chain fatty acids proceeds by stepwise removal of two carbons at a time beginning at the carboxyl end. Acetyl-CoA is the form in which the C2 fragments are removed (see Figure 1–2). The acetyl-CoA released in oxidation is available for resynthesis of fatty acids, for synthesis of steroids or ketones or for entry into the Krebs cycle (Figure 1–2).

H. *Units of Measure*
   1. Amount per animal daily (g, kg, lb).
   2. Percent of animal diet.

IV. Protein

A. *General*
   1. Made up (by % mol wt) of C (53%), H (7%), O (23%), N (16%) and S and P (<1%).
   2. Protein is the principal constituent of the organs and soft structures of the animal body.
   3. Dietary requirement (%) is highest in the young growing animal and declines gradually to maturity.
   4. Proteins are large molecules with molecular weights ranging from approximately 5000 to many millions. Because of their large size, they vary widely in chemical composition, physical properties, shape, solubility and biological function.

B. *Structure of Proteins*
   1. All proteins have one common property: their basic structure is made up of simple units, amino acids. The general structure representing all amino acids (except proline and hydroxyproline) is shown below:

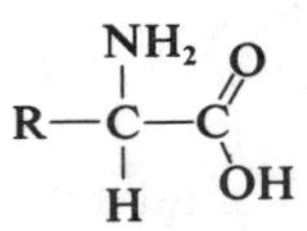

$$R-\underset{\underset{H}{|}}{\overset{\overset{NH_2}{|}}{C}}-\underset{OH}{\overset{O}{C}}$$

14

The essential component is an α-amino group (NH_2) on the C atom adjacent to the carboxyl group (—COOH). The R represents the remainder of the molecule attached to the C atom associated with the α-amino group of the amino acid.

2. Twenty-two amino acids are commonly found in proteins: they are linked together by peptide bonds that couple the α-carboxyl group of one amino acid residue to the α-amino group of another residue, illustrated as follows:

3. The arrangement of amino acids in the chain and the length of the chain are two factors that determine the composition of the protein.

4. The usual classification of amino acids depends on the number of acidic and basic groups present. Thus, the neutral amino acids contain one amino and one carboxyl group. The acidic amino acids have an excess of carboxyl over amino groups. The basic amino acids possess an excess of basic groups. Two amino acids are imino acids rather than primary α-amino acids.

5. The chemical structure of the amino acids follows:
   a. Neutral amino acids
      (1) Aliphatic amino acids
         (a) Glycine.

      (b) Alanine.

      (c) Valine.

      (d) Leucine.

(e) Isoleucine.

$$H_3C-\overset{\overset{\displaystyle H}{|}}{C}-\overset{\overset{\displaystyle H}{|}}{\underset{\underset{\displaystyle H_3C}{|}}{C}}-\overset{\overset{\displaystyle NH_2}{|}}{\underset{\underset{\displaystyle H}{|}}{C}}-\overset{O}{\underset{OH}{\diagup}}C$$

(f) Serine.

$$HO-\overset{\overset{\displaystyle H}{|}}{\underset{\underset{\displaystyle H}{|}}{C}}-\overset{\overset{\displaystyle NH_2}{|}}{\underset{\underset{\displaystyle H}{|}}{C}}-\overset{O}{\underset{OH}{\diagup}}C$$

(g) Threonine.

$$H_3C-\overset{\overset{\displaystyle H}{|}}{\underset{\underset{\displaystyle OH}{|}}{C}}-\overset{\overset{\displaystyle NH_2}{|}}{\underset{\underset{\displaystyle H}{|}}{C}}-\overset{O}{\underset{OH}{\diagup}}C$$

(2) Aromatic amino acids
   (a) Phenylalanine.

$$\text{C}_6\text{H}_5-\overset{\overset{\displaystyle H}{|}}{\underset{\underset{\displaystyle H}{|}}{C}}-\overset{\overset{\displaystyle NH_2}{|}}{\underset{\underset{\displaystyle H}{|}}{C}}-\overset{O}{\underset{OH}{\diagup}}C$$

(b) Tyrosine.

$$HO-\text{C}_6\text{H}_4-\overset{\overset{\displaystyle H}{|}}{\underset{\underset{\displaystyle H}{|}}{C}}-\overset{\overset{\displaystyle NH_2}{|}}{\underset{\underset{\displaystyle H}{|}}{C}}-\overset{O}{\underset{OH}{\diagup}}C$$

(c) Tryptophan.

$$-\overset{\overset{\displaystyle H}{|}}{\underset{\underset{\displaystyle H}{|}}{C}}-\overset{\overset{\displaystyle NH_2}{|}}{\underset{\underset{\displaystyle H}{|}}{C}}-\overset{O}{\underset{OH}{\diagup}}C$$

(3) Sulfur-containing amino acids
   (a) Cystine.

$$\begin{array}{ccccc} & & \text{H} & \text{NH}_2 & \text{O} \\ & & | & | & \| \\ \text{S} & - & \text{C} & - & \text{C} & - & \text{C} \\ & & | & | & \\ & & \text{H} & \text{H} & \text{OH} \end{array}$$

$$\begin{array}{ccccc} & & \text{H} & \text{NH}_2 & \text{O} \\ & & | & | & \| \\ \text{S} & - & \text{C} & - & \text{C} & - & \text{C} \\ & & | & | & \\ & & \text{H} & \text{H} & \text{OH} \end{array}$$

   (b) Cysteine.

$$\text{HS} - \overset{\text{H}}{\underset{\text{H}}{\text{C}}} - \overset{\text{NH}_2}{\underset{\text{H}}{\text{C}}} - \overset{\text{O}}{\text{C}} \diagdown \text{OH}$$

   (c) Methionine.

$$\text{H}_3\text{C} - \text{S} - \overset{\text{H}}{\underset{\text{H}}{\text{C}}} - \overset{\text{H}}{\underset{\text{H}}{\text{C}}} - \overset{\text{NH}_2}{\underset{\text{H}}{\text{C}}} - \overset{\text{O}}{\text{C}} \diagdown \text{OH}$$

b. Acidic amino acids
   (1) Aspartic acid.

$$\underset{\text{HO}}{\overset{\text{O}}{\text{C}}} - \overset{\text{H}}{\underset{\text{H}}{\text{C}}} - \overset{\text{NH}_2}{\underset{\text{H}}{\text{C}}} - \overset{\text{O}}{\text{C}} \diagdown \text{OH}$$

   (2) Asparagine.

$$\underset{\text{H}_2\text{N}}{\overset{\text{O}}{\text{C}}} - \overset{\text{H}}{\underset{\text{H}}{\text{C}}} - \overset{\text{NH}_2}{\underset{\text{H}}{\text{C}}} - \overset{\text{O}}{\text{C}} \diagdown \text{OH}$$

   (3) Glutamic acid.

$$\underset{\text{HO}}{\overset{\text{O}}{\text{C}}} - \overset{\text{H}}{\underset{\text{H}}{\text{C}}} - \overset{\text{H}}{\underset{\text{H}}{\text{C}}} - \overset{\text{NH}_2}{\underset{\text{H}}{\text{C}}} - \overset{\text{O}}{\text{C}} \diagdown \text{OH}$$

(4) Glutamine.

$$\underset{H_2N}{\overset{O}{\Vert}}\!C\!-\!\overset{\overset{H}{\vert}}{\underset{\underset{H}{\vert}}{C}}\!-\!\overset{\overset{H}{\vert}}{\underset{\underset{H}{\vert}}{C}}\!-\!\overset{\overset{NH_2}{\vert}}{\underset{\underset{H}{\vert}}{C}}\!-\!\overset{O}{\underset{OH}{\overset{\Vert}{C}}}$$

c. Basic amino acids
   (1) Histidine.

d. Imino acids
   (1) Proline.

(2) Arginine.

(3) Lysine.

18

(2) Hydroxyproline.

C. *Classification of Proteins*
1. Simple (globular) proteins—those yielding only amino acids or their derivatives on hydrolysis; abundant in nature
   a. Albumins, histones, protamines—soluble in water.
   b. Globulins—soluble in dilute neutral solutions of salts.
   c. Glutelins—soluble in dilute acids or bases.
   d. Prolamins—soluble in 80% ethanol.
2. Fibrous proteins—constitutes about 30% of total protein in animal body; connective tissue
   a. Collagens—insoluble in water; become digestible after conversion to gelatin in dilute acids or bases.
   b. Elastins—similar to collagens but cannot be converted to gelatin.
   c. Keratins—highly insoluble and indigestible.
3. Conjugated proteins—those in which simple proteins are combined with a nonprotein radical
   a. Nucleoproteins—contain nucleic acid.
   b. Glycoproteins (mucoproteins)—contain carbohydrate.
   c. Phosphoproteins—contain phosphorus.
   d. Hemoproteins—contain hematin.
   e. Lecithoproteins—contain lecithin.
   f. Lipoprotein—contain lipids.
   g. Metalloproteins—complexed with metals.

D. *Protein Terminology*
1. True protein—protein composed only of amino acids.
2. Nonprotein nitrogen (NPN)—compounds that are not true protein in nature but contain N and can be converted to protein by bacterial action (example: urea).
3. Crude protein (or total protein)—protein composed of true protein and any other nitrogenous product.

   Crude protein = % N × 6.25

4. Digestible protein—that portion of the crude protein which the animal can digest; represented by the difference between what is present in the feed and what appears in the feces.
5. Essential amino acid (EAA)—those amino acids that are essential to the animal and must be supplied in the diet because the animal body cannot synthesize them fast enough to meet its requirement. The following are considered essential amino acids for normal growth. They may be remembered by taking the first letter of each amino acid to spell PVT TIM HALL.

Phenylalanine*	Tryptophan	Histidine‡	(for poultry)
Valine	Isoleucine	Arginine‡	Glycine
Threonine	Methionine†	Leucine‡	Proline
		Lysine	
PVT	TIM	HALL	

*About ⅓ to ½ of the requirement can be met (or spared) by tyrosine.
†About ½ of the requirement can be met (or spared) by cystine.
‡The mature pig does not need arginine, histidine or leucine for maintenance as measured by nitrogen balance.

6. Nonessential amino acid—those amino acids that are essential to the animal but are normally synthesized or sufficient in the diet and need not be supplemented:

Alanine	Cystine	Hydroxyproline
Asparagine	Glutamic acid	Proline
Aspartic acid	Glutamine	Serine
Cysteine	Glycine	Tyrosine

Note: Ruminant animals may not require dietary amino acids to the same extent as monogastric animals. This is so because the rumen and intestinal microorganisms are capable of synthesizing the EAA (and non-EAA) from feedstuff sources of nitrogen and carbohydrates.

7. Protein quality—refers to the amount and ratio of essential amino acids present in a protein. Evaluation of protein quality may be made on the basis of:

   a. Biological value (BV)—a measure of the relationship of protein (or N) retention to protein (or N) absorption. May also be defined as the percent of true absorbed protein (or N) from the intestinal tract available for productive body functions.

   $$BV\ (\%) = \frac{N\ intake - (Urinary\ N + Fecal\ N)}{N\ intake - Fecal\ N} \times 100$$

   Egg protein is considered to have the highest BV of natural sources (94%+). In general, BV is higher from proteins of animal origin (60% to 80%+) than from proteins of plant origin (40% to 65%). When combined in correct proportions with other proteins, those that individually have very poor BV may yield a BV similar to that of a single high quality protein. A protein with a BV of 70 or more (70% of the intake of N is retained) is considered capable of supporting growth, assuming that the calorie value of the diet is adequate. Proteins with BV of less than 70 are less capable of supporting growth through the use of more of that protein.

   b. Net protein value (NPV)—since the BV is based only on the amount of N absorbed, it does not take into account differences in digestibility from one protein to another. When digestibility as well as BV data are used, a NPV can be computed, which is simply the product of the BV × the digestion coefficient.

   NPV = BV × digestion coefficient

   This value is corrected for either very low or very high digestibility and is a more useful value than BV.

   c. Net protein utilization (NPU)—measures efficiency of growth by comparing body N content resulting from feeding a test protein with that resulting from feeding a comparable group of animals a protein-free diet for the same length of time. Thus,

   $$NPU = \frac{(Body\ N\ with\ test\ protein) - (Body\ N\ with\ protein\text{-}free\ diet)}{Total\ N\ intake}$$

   An advantage of this method is the large number of values that can be obtained over a brief test period (7 to 10 days), with a minimum of measurements.

   d. Protein efficiency ratio (PER)—procedure involving a feeding trial in which protein sources are compared in terms of gain in animal body weight per gram of protein or nitrogen fed. Thus,

   $$PER = \frac{body\ wt\ gain,\ g}{protein\ consumed,\ g}$$

   To compare specific proteins or protein sources, a nitrogen-free, otherwise adequate basal diet is used in which the protein sources to be compared are included for different groups of young animals, and records are kept of growth and feed consumption.

8. Protein quality for ruminants

   a. For many years ruminant nutritionists have assumed that protein leaving the rumen and entering the lower digestive tract was of good quality and adequate to meet the essential amino acid needs of the animal. More recent studies have suggested that a feedstuff protein of good quality may actually be lowered in quality through rumen microbial protein degradation and microbial protein synthesis. In some instances the amount and quality of the protein leaving the rumen could be inadequate to meet the essential amino acid needs of the high producing ruminant.

Several theoretical protein systems have been proposed that have potential application to feeding ruminants. (*Ruminant Nitrogen Usage*, NAS/NRC, 1985). These systems assume that dietary intake crude protein is either degraded in the rumen, with partial or total conversion to bacterial and protozoal crude protein, or passed from the rumen as undegraded intake protein. The undegraded dietary intake protein that passes to the omasum is often called "bypass" or "undegraded" protein to differentiate it from protein synthesized by microbes (bacterial protein) in the rumen.

Several factors influence the rate of protein degradation in the rumen, but two appear to be most important: solubility of feedstuff protein and rate of passage through the rumen.

Protein utilization by the ruminant is depicted in Figure 1–4. Approximately 60% of the various feed proteins are broken down (degraded) in the rumen by bacteria and other microorganisms to their component amino acids and then to ammonia (NH_3). The other 40% is not broken down and simply passes on through the rumen (bypass or escape protein) and moves on down the digestive tract. Nonprotein nitrogen sources, such as urea and anhydrous ammonia, are very soluble and are rapidly converted to ammonia. The insoluble protein fractions are also degraded in the rumen but at a slower rate than are the soluble fractions. Typical values for many common feedstuffs are included in Table 1–2. The rumen microorganisms utilize ammonia plus carbon skeletons and energy to reproduce themselves. The various bacteria and other microorganisms are subsequently washed down the tract with the rumen fluid and the nondegraded or bypass protein for digestion and absorption. Thus a ruminant has the ability to convert low quality feed protein into higher quality microbial protein that better meets the animal's needs.

When excessive amounts of soluble protein or NPN are fed, more ammonia is produced than the bacteria can utilize. The excess ammonia is absorbed from the rumen, converted to urea by the liver and then is excreted by the kidneys. In extreme cases of overconsumption of urea or NPN, the ammonia buildup can make cattle sick or result in their death.

Several methods of "protecting" high quality feedstuff protein or critical essential amino acids from excessive ruminal degradation are currently being investigated. Among them are:

(1) Heat treatment.
(2) Treatment with formaldehyde or tannins.
(3) Encapsulation of amino acids.
(4) Use of amino acid analogs.
(5) Control of microbial metabolism in the rumen.

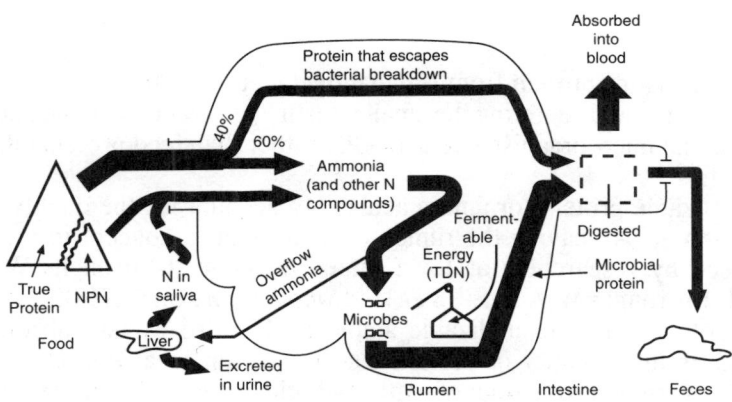

FIGURE 1–4. Fate of certain nitrogen compounds in the ruminant and synthesis of microbial protein.

TABLE 1–2. Ruminal Degradability of Protein in Selected Feeds

Feed	CP, % d.m. basis	Degradability of CP, %	Undegradability of CP, %
Grains			
Barley	13.0	73	27
Beet pulp	9.8	55	45
Citrus pulp	6.7	80	20
Corn, shelled	10.1	48	52
Corn, dry-rolled	10.1	40	60
Corn, high-moisture	10.1	44	56
Oats	13.3	83	17
Sorghum grain	11.3	46	54
Wheat	14.7	78	22
Intermediate proteins			
Alfalfa meal, dehydrated	18.9	41	59
Brewer's dried grains	29.5	51	49
Corn gluten feed	25.5	78	22
Cottonseed, whole	23.6	55	45
High proteins			
Blood meal	94.7	18	82
Cottonseed meal, solvent	45.4	57	43
Corn gluten meal	47.3	45	55
Fish meal, Menhaden	67.9	40	60
Linseed meal, solvent	38.0	65	35
Meat meal	54.0	24	76
Peanut meal	52.9	75	25
Soybean meal, solvent	51.0	65	35
Soybeans, raw	42.5	74	26
Soybeans, heated	39.5	51	49
Sunflower meal	48.9	74	26
Urea	281.0	100	0
Forages			
Alfalfa hay	18.7	72	28
Alfalfa silage	18.4	77	23
Clover hay, red	14.7	69	31
Corn silage	8.1	69	31
Grass hay	9.5	63	37
Average		60.3	39.7

Protecting feedstuff protein from ruminal degradation would not only increase the quantity of protein or amino acids entering the small intestine but could have the added benefit of encouraging the ruminal microorganisms to use NPN rather than feed protein nitrogen in their fermentation activities.

b. Metabolizable protein (or amino acids)—to account for the quality and quantity of protein (or amino acids) leaving the rumen compartment, a concept in protein nutrition has been proposed by researchers at Iowa State University (Burroughs W.A., et al, *J Dairy Sci* 58:611; Burroughs W.A., et al, *Vet Med Small Anim Clin* 69:713). The expression "metabolizable protein" or "metabolizable amino acid" (also called "absorbable protein") can be defined as the quantity of protein digested or amino acid(s) absorbed in the postruminal portion of the digestive tract of cattle and other ruminants. In nonruminants, it has the same meaning as apparent digestible protein or absorbable amino acids. In ruminants, however, metabolizable protein differs from digestible protein in that metabolizable protein reflects the quantity of feed protein consumed that escapes degradation in the rumen. Secondly, metabolizable protein reflects the quantity of degraded protein reformed into rumen microbial protein. Consideration is finally given to digestibility and the absorbable quantity of amino acids arising from the two sources of protein from each feedstuff consumed.

The chief advantage for the metabolizable protein system is its greater accuracy in predicting protein needs of cattle. This greater accuracy is accounted for by the variable destructive action of rumen degradation of different dietary proteins coupled with a restrictive limitation of the maximum amount of microbial protein that can be synthesized in the rumen based on dietary energy intake. A second major advantage of the metabolizable protein system is that it gives a better insight into the individual amino acid needs of cattle and how these needs can best be supplied by a combination of performed protein and NPN. The system of establishing requirements of feedlot cattle in terms of metabolizable protein or metabolizable amino acids is believed to be superior to current methods, even though amino acid composition data and data on rumen degradation of feed proteins are not yet complete and require the establishment of additional values for greatest accuracy. Additional information and example formulations are presented in Chapter 8 of this text.

E. *Functions and Deficiencies*
  1. Function
    a. Basic structural unit of the animal body
       (1) Collagen.
       (2) Elastin.
       (3) Contractile proteins.
       (4) Keratin proteins.
       (5) Blood proteins.
    b. Body metabolism
       (1) Enzymes.
           (a) Digestion process.
           (b) Degradative process.
           (c) Synthesis process.
       (2) Hormones.
       (3) Immune antibodies.
       (4) Hereditary transmission.
    c. Source of energy after deamination
  2. Deficiencies and abnormalities
    a. Symptoms of protein deficiency include reduced growth rate and feed efficiency, anorexia, reduced nitrogen balance and serum protein concentration, anemia, fatty liver, infertility, reduced birth weight of newborn, reduced milk production and reduced synthesis of certain hormones and enzymes.
    b. Amino acid deficiency—a deficiency of an individual amino acid results in the animal body deaminating the other amino acids and using the carbon chains for energy. Symptoms of a deficiency may resemble those described for protein deficiency with certain amino acid deficiencies producing specific lesions.
    c. Amino acid imbalance—equally important with the presence of all essential amino acids for body protein synthesis is the balance of amino acids available to the cell. The closer the yields of amino acids from dietary protein approximate the proportions required by the animal, the more efficiently the dietary protein will be utilized for the synthesis of new protein. Under conditions of low dietary protein, when an amino acid that is limiting to growth is added in excess, a growth depression may occur, which can be overcome by addition of the next most limiting amino acid(s). The growth depression caused by an "imbalanced" diet is not well understood but seems to be related in some manner to depressed feed intake.
    d. Amino acid antagonisms—amino acid antagonisms are somewhat different from a simple imbalance. Whereas the growth depression characteristic of an amino acid imbalance may be alleviated by supplementing another growth-limiting amino acid into the diet, in an antagonism the growth depression caused by a single amino acid may be reversed by adding a structurally related amino acid.

    e. Amino acid toxicity—some amino acids seem to be flatly toxic when fed at high levels. This toxicity may be partially alleviated by other amino acids but not completely. Methionine is particularly depressing at high levels.

  F. *Natural Sources*

    Most common feeds contain protein. However, the quantity and quality of protein will vary and may be limiting to the animal's growth and production. In order to properly feed protein, one must know minimum protein need plus the minimum need for each EAA and supply them in adequate amounts and in correct balance. This may be done by selecting the protein, combining proteins or supplementing protein with individual amino acids in such a way as to meet the need. A detailed discussion of feedstuff sources is presented in Chapter 3 of this text.

  G. *Digestion and Metabolism*

    1. Dietary proteins must be broken down to amino acid fragments for absorption from the digestive system. An exception to this is in early postnatal life of mammals (in most species during the first 24 to 48 hours), when intact protein is absorbed across the intestinal epithelium.

    Hydrolysis of dietary proteins is accomplished by proteolytic enzymes elaborated by epithelial cells lining the lumen of the intestinal tract and by the pancreas. The efficiency with which hydrolysis occurs determines the degree of absorption of individual amino acids and thus contributes to the nutritional value of the dietary protein. Another important factor contributing to nutritional value is the balance of essential amino acids. Even proteins easily hydrolyzed in the intestinal tract do not have a high nutritional value if they have a deficiency or an imbalance of one or more amino acids.

    2. The fate of amino acids after absorption can be divided broadly into three categories:
      a. Tissue protein synthesis.
      b. Synthesis of enzymes, hormones, and other metabolites.
      c. Deamination and use of the carbon skeleton for energy. This involves the release of the amino group from the carbon skeleton and entrance of the carbon skeleton into the Krebs cycle (Figure 1–2).

    3. Metabolism of nitrogen in ruminants
      a. Microorganisms in the rumen hydrolyze most feed protein to peptides and amino acids, many of which will be degraded further to ammonia, organic acids and carbon dioxide.
      b. Ammonia released on microbial degradation of feed protein or NPN sources is removed from the rumen by:
        (1) Absorption through the rumen wall, particularly at a relatively high rumen pH. Much of the absorbed ammonia is recycled back to the rumen as urea via the blood or saliva.
        (2) Utilization by the microorganisms together with a readily available carbohydrate for synthesis of microbial protein.
      c. Microbial protein resynthesized by the microorganisms results in a supply of fairly constant quality protein to the lower digestive tract. The quality of protein from moderate to poor feeds will usually be improved by rumen metabolism, whereas the opposite may occur with high quality feed proteins.

  H. *Units of Measure*
    1. Amount per animal daily (g, kg, lb).
    2. Percent of the animal diet.

V. Minerals

  A. *General*
    1. Inorganic, solid, crystalline chemical elements.
    2. The total mineral content of plants or animals is often called ash.
    3. Make up 3% to 5% of animal body dry weight, depending on species; calcium accounts for nearly one-half of the total body mineral; phosphorus, approximately one-fourth; and all other minerals approximately one-fourth.

B. *Classification*
   1. Major minerals (macrominerals)
      a. Minerals normally present at greater levels in animal body or needed in relatively large amounts in the diet.
      b. Include Ca, P, Na, Cl, Mg, K, S.
   2. Trace minerals (microminerals)
      a. Minerals normally present at low levels in animal body or needed in very small amounts in the diet.
      b. Include Co, Cu, F, I, Fe, Mn, Mo, Se, Zn.
      c. Small amounts of F and Se are considered beneficial in certain geographical areas, but toxic if fed in excess.
C. *Functions, Deficiencies and Toxicities*
   1. A summary of individual mineral functions, deficiencies, toxicities and interrelationships is shown in Table 1–3.
   2. General functions of minerals
      a. Skeletal formation and maintenance—Ca, P, Mg, Cu, Mn.
      b. Function in protein synthesis—P, S, Zn.
      c. Oxygen transport—Fe, Cu.
      d. Fluid balance (osmotic pressure and excretion)—Na, Cl, K.
      e. Regulating acid-base balance of the entire system—Na, Cl, K.
      f. Activators and/or components of enzyme systems—Ca, P, K, Mg, Fe, Cu, Mn, Zn.
      g. Mineral-vitamin relationship—Ca, P, Co, Se.
   3. In the case of several of the essential minerals, the initial knowledge of their need and the symptoms resulting from the deficiency in the ration may be gained by mineral analysis of the soil. Many studies have reported the occurrence of diseases resulting from deficiencies or excesses of minerals in the soil.
D. *Natural Sources of Minerals*

   Forage plants are generally considered good sources of most minerals; grains are a fair source of phosphorus but low in many of the other minerals. Chapter 3 provides a more complete comparison of feedstuff mineral content.
E. *Digestion and Metabolism*
   1. Mineral elements are absorbed primarily from the small intestine in the ionic form.
   2. Absorption occurs as the result of active absorption (Ca, P and Na) or diffusion (all other minerals). Refer to Nutrient Absorption at the end of this chapter.
   3. Factors affecting mineral absorption
      a. Age of animal (young more efficient than old in absorbing minerals).
      b. Form of the element (organic vs inorganic)—inorganic generally more available.
      c. pH of intestinal tract—lower pH enhances absorption.
      d. Binding or chelating components (oxalates, phytates, fats etc.) may reduce absorption.
      e. Excesses or interactions within minerals.
F. *Units of Measurement*
   1. Amount required per animal daily (g, mg, mcg).
   2. Percent or amount per weight unit of diet.
      a. Major minerals generally expressed as percentage.
      b. Trace minerals generally expressed as milligrams or micrograms per kilogram or pound.

VI. Vitamins

A. *History*

   Vitamins were the last group of dietary essentials to be recognized. In 1897, the Dutch physician Eijkmann showed that beriberi, which resulted from the prolonged consumption of polished rice, could be cured by adding the rice polishings back to the diet. Those observations suggested the presence of substances in natural foods that were indeed indispensable for health, but did not fall into any of the familiar categories of carbohydrate, fat or protein. Frederick Hopkins, in 1906, called these substances "accessory food factors"; in 1912 Casimir Funk introduced the term "vitamine" (*vita* = life; *amine* = contains N). Subsequent work showed that many "vitamines" existed,

25

TABLE 1–3. Summary of Individual Mineral Functions, Deficiencies, Toxicities and Interrelationships

Mineral	Major Functions	Specified Deficiency Symptoms	Major Interrelationships and Toxicities
Calcium (Ca)	Bone and teeth formation; blood coagulation; muscle contraction; nerve function; cell permeability; milk production, egg shell formation	Rickets (young); osteomalacia (adult); tetany; thin-shelled eggs; reduced egg production; hypocalcemia or milk fever in dairy cattle	Vitamin D involved in absorption and bone deposition; excess PO_4 decreases absorption; excess Mg decreases absorption, replaces Ca in bone and increases Ca excretion; Ca:P ratio should be 1:1 to 2:1
Phosphorus (P)	Bone and teeth formation; phosphorylation; high-energy phosphate bonds; PO_4 chief anion radical of intracellular fluid; PO_4 important in acid-base balance	Rickets (young), osteomalacia (adult); reduced egg production	Vitamin D involved in renal reabsorption and bone deposition; excess Ca and Mg causes decrease in absorption; Ca:P ratio should be 1:1 to 2:1; in male ruminants, excess P may cause urinary calculi
Sodium (Na)	Major cation of extracellular fluid where it is involved in osmotic pressure and acid-base equilibrium; preservation of normal muscle cell irritability; cell permeability	Reduced growth; eye disturbances with corneal lesions; reproduction impairment (infertility in males, delayed sexual maturity in females)	Salt toxicity readily occurs in nonruminants with levels above 8% in diet; staggering gait, blindness, nervous disorders and hypertension
Chlorine (Cl)	Major anion involved in osmotic pressure and acid-base balance (chlorine shift); hydrochloric acid in digestion	Hypochloremic alkalosis (usually due to physiological disturbance such as vomiting rather than deficiency); reduced growth	Toxicity unlikely
Magnesium (Mg)	Enzyme activator primarily in glycolytic system; bone formation	Vasodilation; hyperirritability with convulsions, loss of equilibrium and trembling; tetany	Excess upsets Ca and P metabolism; toxicity not likely
Potassium (K)	Major cation of intracellular fluid where it is involved in osmotic pressure and acid-base balance; muscle activity	Hypokalemia; lethargic condition with high incidence of comas and death; diarrhea, distended abdomen and untidy appearance	Excess reduces Mg absorption; Mg deficiency reduces K retention leading to K deficiency
Sulfur (S)	Sulfur-containing amino acids; SH group function in tissue respiration; component of biotin and thiamine	Primarily reduced growth effect due to sulfur amino acid requirement for protein synthesis	Toxicity unlikely
Iron (Fe)	Cellular respiration (hemoglobin, cytochromes, myoglobin)	Hypochromic-microcytic anemia (less than normal amount of hemoglobin and fewer red cells); anemia may be common in baby pigs unless Fe is supplied	Ca-P ratio influences absorption; Cu required for proper metabolism; pyridoxine deficiency decreases absorption

TABLE 1–3 (continued). Summary of Individual Mineral Functions, Deficiencies, Toxicities and Interrelationships

Mineral	Major Functions	Specified Deficiency Symptoms	Major Interrelationships and Toxicities
Copper (Cu)	Cofactor in several oxidation-reduction enzyme systems; hemoglobin synthesis; bone formation; maintenance of myelin of nerves; hair pigmentation	Fading hair coat or lack of wool; nervous symptoms or ataxia; lameness, swelling of joints and fragility of bones; anemia	Excess Mo, Zn inhibit its utilization and storage; toxicity occurs at levels above 250 ppm with much the same symptoms as deficiency
Zinc (Zn)	Component or cofactor of several enzyme systems including peptidases and carbonic anhydrase; needed for bone and feather development	Poor hair or feather development and slipping of wool; rough and thickened skin or parakeratosis in swine	High Ca or phytate ties up Zn; excess Zn interferes with Cu metabolism and may cause anemia
Manganese (Mn)	Thought to be an activator of enzyme systems involved in oxidative phyosphorylation, amino acid metabolism, fatty acid synthesis and cholesterol metabolism; bone formation (organic matrix); growth and reproduction	Poor growth; shortened long bones; impaired reproduction (testicular degeneration of males, defective ovulation of females); perosis or slipped tendon in poultry	Excess Ca and P decreases absorption; toxicity unlikely
Cobalt (Co)	Component of vitamin B_{12}; needed by rumen bacteria for growth and B_{12} synthesis	Anemia (varies from normocytic-normochromic to megaloblastic or macrocytic; deficiency in ruminants causes reduced appetite, reduced growth and body weight and eventually death	Related in vitamin B_{12}; toxicity unlikely
Iodine (I)	Thyroxine formation	Goiter; stillbirths; hairless pigs or wool-less lambs at birth	Long-term intake of high amounts of I reduces thyroid uptake of I
Selenium (Se)	Not completely known but thought to be involved in vitamin E absorption and/or retention	Mortality in poultry; if E is also deficient or suboptimal: exudative diathesis (chicks), muscle dystrophy (lambs) or liver necrosis (pigs)	Chronic toxicity yields blind staggers at 10–20 ppm or alkali disease at 5–10 ppm; acute toxicity occurs at 20 ppm and above; sudden death; SO_4 protector against toxicity
Molybdenum (Mo)	Purine metabolism; stimulates microbial activity in rumen	Lack of conversion of xanthine to uric acid but not likely to be deficient in natural diet	Excess interferes with Cu activation of enzymes; causes anemia and diarrhea; SO_4 protects against toxicity
Fluorine (F)	Traces protect against teeth decay	Excesses of F are of more concern than deficiencies in livestock production	Levels above 5–10 ppm block vital oxidative enzymes by interfering with Mn; causes bone deformities, enamel defects and organ degeneration; Ca and Al salts protect against toxicity; F is a cumulative poison so toxicity may not be noted for some time

but only a few were "amine" in nature, and so the final *e* was dropped to give the familiar name *vitamin*.

Soon thereafter, Osborne and Mendel and McCollum and Davis, two teams of United States researchers working independently, reported an elusive, unidentified substance in fat that was necessary for growth and reproduction in animals. This they designated fat-soluble A to differentiate it from the water-soluble B, believed to be responsible for preventing and curing beriberi.

Subsequently, other studies of Holst and Frolich suggested that there must be a third factor since there seemed to be something in fresh fruits and vegetables that prevented scurvy in humans. It was believed to be different from either of those discovered earlier since it was water soluble but did not contain nitrogen. Consequently, it was designated vitamin C.

From this simple classification of vitamins has grown a list of four completely different fat-soluble vitamins and twelve water-soluble substances that have been grouped together as the vitamin B complex and vitamin C.

B. *General*
1. Organic components of natural food but distinct from carbohydrates, fat, protein, and water.
2. Present in food in minute amounts and effective in the animal body in small amounts.
3. Essential for development of normal tissue; necessary for metabolic activity but do not enter into the structural portion of the body.
4. When absent from the diet or not properly absorbed or utilized, results in a specific deficiency disease or syndrome.
5. Cannot be synthesized by the animal and therefore must be obtained from the diet (or microbial synthesis in digestive tract).
6. Related substances
   a. Provitamins or precursors—substances that are chemically related to the biologically active form of the vitamin but have no vitamin activity until the body converts it into the active form (e.g., carotene → vitamin A).
   b. Antivitamins, vitamin antagonists or pseudovitamins—substances that are usually chemically related to the biologically active vitamin. The body does not discriminate between the useful form of the vitamin and the antagonist. The antagonist not only does not function as the active vitamin, but also stubbornly refuses to be replaced by the proper substance, which would allow the reaction to proceed.
7. General comparison of fat- versus water-soluble vitamins:
   a. Chemical composition—fat-soluble vitamins contain only carbon, hydrogen and oxygen, whereas the water-soluble B vitamins contain these three elements plus either nitrogen, sulfur or cobalt.
   b. Occurrence—fat-soluble vitamins differ from the water-soluble B vitamins in that they may occur in plant tissues in the form of provitamin (a precursor of vitamin), which can be converted into a vitamin in the animal's body. No provitamins are known for any water-soluble vitamins. Tryptophan, which can be converted into niacin, has never been considered a provitamin.
   c. Physiological action—the water-soluble B vitamins almost collectively are concerned with the transfer of energy. Fat-soluble vitamins are required for the regulation of the metabolism of structural units.
   d. Absorption—fat-soluble vitamins are absorbed from the intestinal tract in the presence of fat. The absorption of water-soluble vitamins is a simpler process because water from the intestine is constantly resorbed into the bloodstream.
   e. Storage—any of the fat-soluble vitamins may be stored whenever fat is deposited; the storage increases with intake. The water-soluble B vitamins are not stored in the same way or to the same extent.
   f. Excretion—fat-soluble vitamins are excreted wholly in the feces. Water-soluble B vitamins may be present in the feces, but their chief pathway of excretion following metabolic use is through urine.
   g. Synthesis—Rumen microorganisms can generally synthesize the water-soluble vitamins, while vitamin K is the only fat-soluble vitamin readily synthesized by microorganisms.

Therefore, the supplementation of water-soluble vitamins is not necessary in rations for ruminants.

C. *Classification and Structure*

   1. Fat-soluble vitamins

      a. Vitamin A.

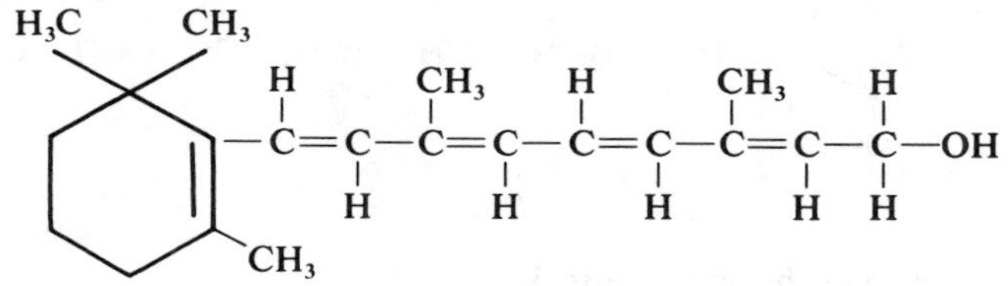

$$H_3C \quad CH_3$$

$$-C=C-C=C-C=C-C=C-C-OH$$

(with CH_3, H substituents along the chain and CH_3 on the ring)

      b. Vitamin D.

         (1) Ergocalciferol (D_2).

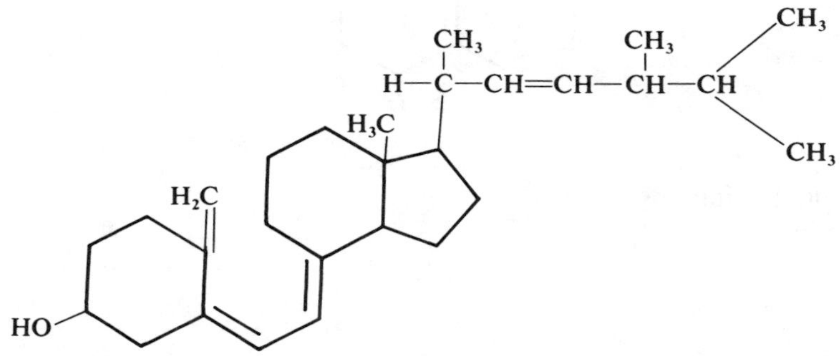

Side chain: $H-\overset{CH_3}{\underset{|}{C}}-CH=CH-CH-CH\begin{smallmatrix}CH_3\\CH_3\end{smallmatrix}$

Ring system with H_3C, H_2C, HO.

         (2) Cholecalciferol (D_3).

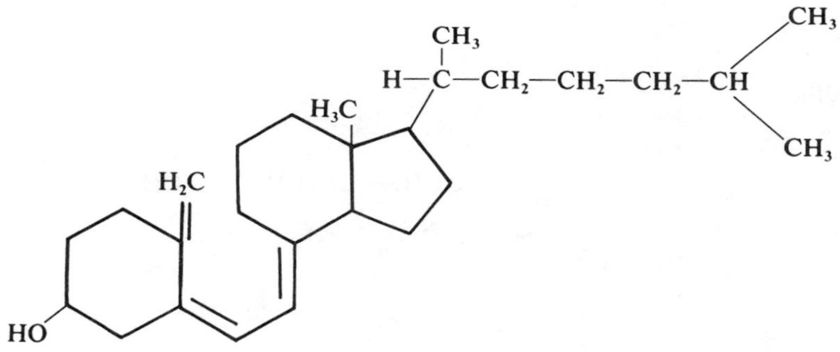

Side chain: $H-\overset{CH_3}{\underset{|}{C}}-CH_2-CH_2-CH_2-CH\begin{smallmatrix}CH_3\\CH_3\end{smallmatrix}$

Ring system with H_3C, H_2C, HO.

      c. Vitamin E (α-tocopherol).

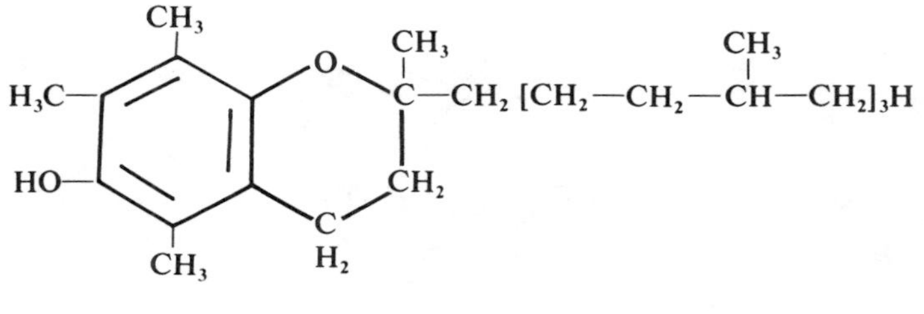

$$H_3C- \quad CH_3 \quad O \quad CH_3 \quad CH_3$$

$$\overset{CH_3}{\underset{|}{C}}-CH_2\ [CH_2-CH_2-\overset{CH_3}{\underset{|}{CH}}-CH_2]_3H$$

with $HO-$, CH_3, CH_2, $\overset{C}{\underset{H_2}{}}$.

29

d. Vitamin K.
   (1) Phylloquinone or K_1 (natural green plant source).

   (2) Menadione (synthetic vitamin K_3).

2. Water-soluble vitamins
   a. Thiamine.

   b. Riboflavin.

   c. Niacin (nicotinic acid).

d. Pyridoxine.

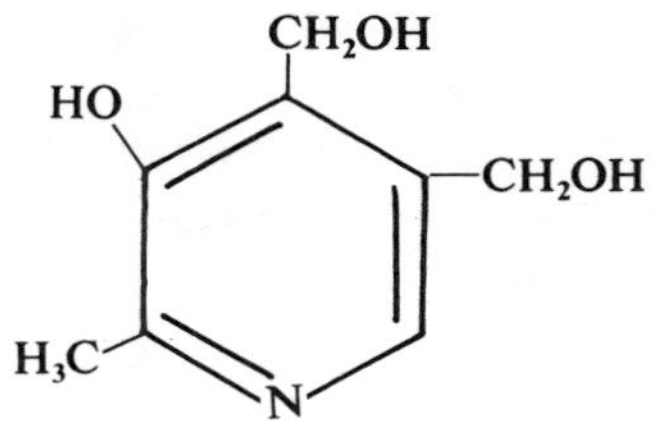

CH_2OH
HO
CH_2OH
H_3C
N

e. Pantothenic acid.

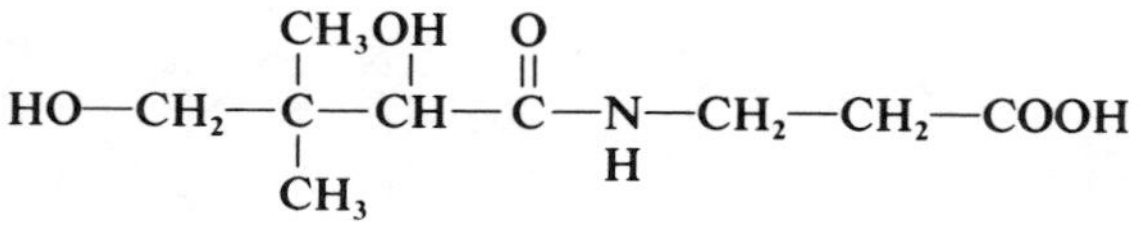

CH_2OH O
HO—CH_2—C—CH—C—N—CH_2—CH_2—COOH
H
CH_3

f. Biotin.

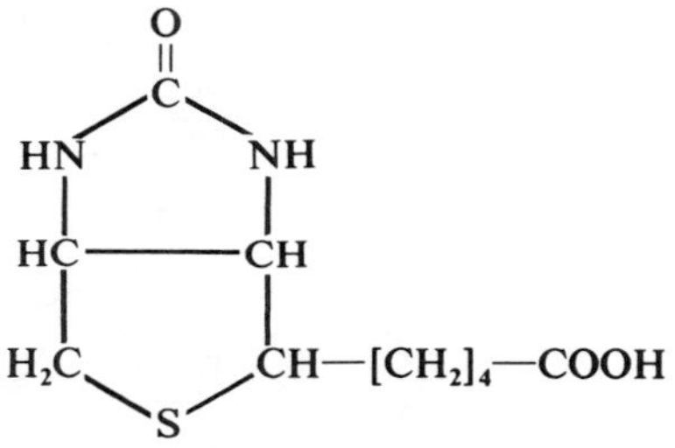

O
C
HN
NH
HC
CH
H_2C
CH—$[CH_2]_4$—COOH
S

g. Choline.

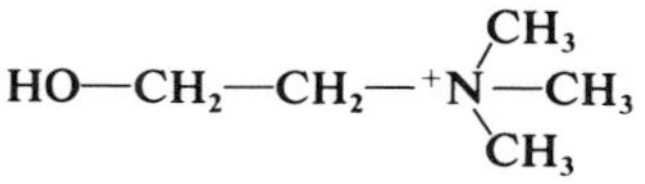

CH_3
HO—CH_2—CH_2—$^+$N—CH_3
CH_3

h. Folic acid.

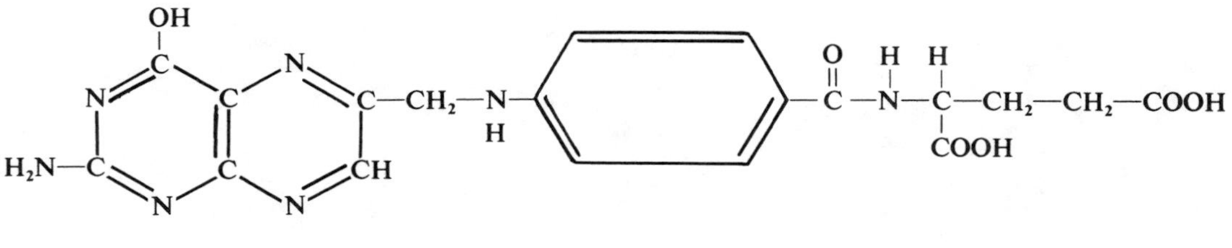

OH
C
N
N
C
C—CH_2—N
H
H_2N—C
C
CH
N
N
O H H
C—N—C—CH_2—CH_2—COOH
COOH

i. Vitamin B_{12} (cobalamin).

3. Other water-soluble vitamins
   a. Vitamin C (ascorbic acid).

b. Inositol.

c. Paraaminobenzoic acid (PABA).

D. *Functions and Deficiencies*
    1. Since vitamins play various roles as regulators of metabolism, they are necessary for growth and maintenance of life. Therefore, proper vitamin levels in the diet are not only important from the standpoint of preventing specific deficiency symptoms, but also to promote general health, vigor and the ability to combat stress and disease. For example, most vitamins are apparently involved in antibody synthesis whereby animals acquire immunity to specific infections.
    2. Vitamin requirements may also increase in old age due to difficulties in absorption and utilization. Today, suboptimal vitamin levels in the rations fed livestock and laboratory animals that cause undetected reductions in performance are probably of greater importance than gross deficiencies with the resulting typical deficiency symptom. On the other hand, vitamin supplementation should not be used as a ''cure-all'' treatment.
    3. A summary of vitamin functions and deficiency or toxicity symptoms is presented in Table 1–4.
E. *Natural Sources*
    1. Vitamin A—feeds rich in carotene such as green, leafy forages from pastures, hays or silages; dehydrated alfalfa meal; yellow corn; whole milk; fish oils.
    2. Vitamin D—certain fish liver oils; sun-cured hay; irradiated yeast; (sunlight).
    3. Vitamin E—seed germ or germ oils from plants; green forage or hay.
    4. Vitamin K—green forage or well-cured hays; fish meal; menadione (synthetic vitamin K_3).
    5. Thiamine—green forage or well-cured hays; cereal grains (especially seed coat or bran); brewer's yeast.
    6. Riboflavin—green forages, hays or silages; milk and milk products; meat or fish meal; distiller's or brewer's byproducts.
    7. Pantothenic acid—brewer's yeast; liver meal; dehydrated alfalfa meal; fish solubles; most feedstuffs fairly good sources.
    8. Niacin—animal byproducts; brewer's yeast; green alfalfa; some present in most feeds, but that in grains is largely unavailable to nonruminants.
    9. Pyridoxine—most feedstuffs are fair to good sources; cereal grains and their byproducts; rice and rice bran; green forages and alfalfa hay; yeast.
   10. Biotin—widely distributed; especially rich in egg yolk, liver, kidney, milk and yeast.
   11. Folacin—green forages and alfalfa meal; some animal proteins.
   12. Choline—most commonly used feeds are fair to good sources.
   13. Vitamin B_{12}—protein feeds of animal origin; fermentation products.
   14. Inositol—all feeds.
   15. Paraaminobenzoic acid—not well known.
   16. Vitamin C—citrus fruits; green, leafy forages; well-cured hays.

TABLE 1–4. Summary of Vitamin Functions and Deficiency Symptoms

Vitamin	Main Functions	Deficiency Symptoms	Comments
A	Bone formation (mucopolysaccharide release); vision (rhodopsin); epithelial tissue maintenance; glucose synthesis (adrenocortical hormones); growth	Xerophthalmia; night blindness; hyperkeratosis; skeletal lesions and bone remodeling; poor growth; reproductive failures; reduced egg production and hatchability	Hypervitaminosis may cause hyperostosis or hyperkeratosis; most of the same symptoms that occur with deficiency; both carotene and vitamin A readily destroyed by oxidation
D	Bone formation (Ca absorption, P absorption from renal tubules; osteoblast formation and calcification); CHO metabolism (phosphorylation); growth (related to bone formation)	Rickets (growing period); osteomalacia (adults); soft egg shells and reduced egg production and hatchability	Hypervitaminosis may cause decalcification of skeletal and calcification of soft tissue; most mammals can use either D_2 or D_3, but poultry require D_3
E	Antioxidant; muscle structure (muscle dystrophy); reproduction (seminiferous epithelium)	Muscle dystrophy; encephalomalacia; exudative diathesis; reproductive failures; steatitis	Relatively nontoxic; utilization dependent on adequate Se
K	Prothrombin formation and blood clotting	Spontaneous hemorrhages and increased blood clotting time with lowered prothrombin levels	Relatively nontoxic; antagonists of K include dicoumarol and warfarin
Thiamine	Coenzyme, thiamine pyrophosphate; decarboxylation of α-keto acids; transketolase reactions	Polyneuritis and convulsions (head retraction in chickens); cardiovascular disturbances; beriberi (humans); anorexia and emaciation	Relatively nontoxic; seldom deficient in livestock; antagonist of thiamine is an enzyme called thiaminase found in some fish feeds
Riboflavin	Coenzyme, flavin mononucleotide (FMN) and flavin adenine dinucleotide (FAD); dehydrogenase (hydrogen acceptance); important in CHO and protein metabolism	Ecodermal lesions; dermatitis and hair loss; curled toe paralysis in birds; moon blindness in horses; leg troubles in pigs	Nontoxic; common swine or poultry rations low or deficient
Pantothenic acid	Coenzyme A; acyl transfer	Dermatitis, loss of hair and greying of hair; spastic gait, goose-stepping or posterior incoordination and paralysis; enteritis; poor growth and reproduction	Relatively nontoxic; low content in cereal grains; commonly deficient for swine or poultry
Niacin	Constituent of coenzymes, nicotinamide adenine dinucleotide (NAD) and nicotinamide adenine dinucleotide phosphate (NADP)	3 D's—dermatitis, diarrhea and dementia; pellagra in humans; irritability; inflamation and ulceration of mouth, tongue and digestive tract (black tongue in dog)	Vasodilation with itching and burning of skin; fatty liver; not available from grains to the pig; can be synthesized in body tissue from surplus tryptophan

TABLE 1–4 (continued).　Summary of Vitamin Functions and Deficiency Symptoms

Vitamin	Main Functions	Deficiency Symptoms	Comments
Pyridoxine	Coenzyme, pyridoxal phosphate; amino acid decarboxylation, transamination and removal of sulfhydryl gorups; red blood cell formation	Convulsions; neuritis and hyperirritability; hypochromic-microcytic anemia; increased excretion of xanthurenic acid	Convulsion and death occur in hypervitaminosis; normally adequate in livestock rations
Biotin	Coenzyme, carboxylase; carboxylation; β-decarboxylation; CO_2 fixation	Dermatitis and loss of hair (spectacle eye in rats and mice); reduced growth; perosis in chicks	Nontoxic; rendered unavailable by raw egg white; normally adequate from diet or intestinal synthesis
Folacin (Folic acid)	One carbon carrier; related to B_{12} metabolism	Macrocytic anemia and leukopenia; cervical paralysis in turkeys; poor growth	Nontoxic; unlikely to be deficient for livestock
Choline	Methyl donor; lipotropic substance; constituent of acetylcholine and phospholipids (nerve impulses)	Fatty liver and kidney degeneration; poor reproduction and lactation in swine; perosis in chicks	Persistent diarrhea occurs with hypervitaminosis; may be synthesized in body, especially with a high-protein diet
B_{12}	Labile methyl group metabolism; isomerization reactions; closely linked with folacin	Macrocytic anemia with megaloblastic marrow; neurological disturbances; hatching problems in chicks; reduced growth	Nontoxic; not available from plant sources; usually deficient in swine diets based on grains
Inosital and Paraaminobenzoic acid	Not clearly established	None demonstrated in livestock	Unlikely to be deficient for livestock
Vitamin C	Collagen formation; hydrogen transport (activation of folic acid)	Scurvy—slow wound healing, spongy gums, swollen joints, hemorrhaging and anemia	Nontoxic; synthesized in body tissue of farm livestock

F. *Digestion and Metabolism*
　　1. Little is known about the digestion of vitamins, but they probably can be used as such within the body without conversion to simpler compounds.
　　2. Most of the vitamins are absorbed in the upper portion of the intestine, with the exception of vitamin B_{12}, which is absorbed in the ileum. Water-soluble vitamins are rapidly absorbed, but the absorption of fat-soluble vitamins relies heavily on the fat absorption mechanisms, which generally are slow.

G. *Units of Measure*
　　1. International unit (IU)—a standard unit potency as defined by the International Conference for Unification of Formulae. The unit of potency is based on bioassay that produces a particular effect agreed on internationally. May also be termed as USP unit (United States Pharmacopoeia).
　　　　a. Vitamin A
　　　　　　(1) One IU (or USP) of vitamin A = 0.3 mcg vitamin A alcohol (retinol) = 0.344 mcg vitamin A acetate = 0.55 mcg vitamin A palmitate.
　　　　　　(2) One IU of vitamin A equivalent from carotene = 0.6 mcg beta carotene.

(3) The conversion of feedstuff beta carotene to vitamin A activity by animals will differ among species as well as among breeds and individuals within a species.

    (a) Rats and poultry—1 mg beta carotene = approximately 1667 IU vitamin A.

    (b) Cattle, sheep and horses—1 mg beta carotene = approximately 400 IU vitamin A.

    (c) Swine—1 mg beta carotene = approximately 200 to 500 IU vitamin A.

  b. Vitamin D

    (1) One IU of vitamin D = 0.025 mcg of cholecalciferol (D_3) or ergocalciferol (D_2); 1 mg of either source contains 40,000 IU activity.

    (2) International chick unit (ICU)—often used to express the vitamin D requirements of poultry based on using the chick as the assay animal because of the unequal activity of vitamin D_2 and vitamin D_3 for this species. Vitamin D_2 is only about ⅓₀ to ¼₀ as effective for birds as vitamin D_3.

  c. Vitamin E

    (1) One mg DL-α-tocopheryl acetate = 1 IU.

    (2) One mg DL-α-tocopherol (alcohol) = 1.1 IU.

    (3) One mg D-α-tocopheryl acetate = 1.36 IU.

    (4) One mg D-α-tocopherol (alcohol) = 1.49 IU.

  d. Vitamin C—the vitamin C value of a food is commonly expressed in milligrams; however, 1 IU is the activity of 0.05 mg of ascorbic acid or 1 mg = 20 IU.

2. All the other vitamins are measured in weight units (mcg, mg, g etc.)

  a. Vitamin K.

    (1) 0.3 mg menadione = 1 mg vitamin K_1 (naturally occurring source).

    (2) 0.5 mg menadione sodium bisulfite = 1 mg K_1.

  b. Pantothenic acid.

    (1) DL-calcium pantothenate has 47% of the activity of the naturally occurring pantothenic acid.

    (2) DL-calcium pantothenate—$CaCl_2$ has 37% of the activity of the naturally occurring pantothenic acid.

H. *Stability and Solubility of Vitamins*

Several factors can influence vitamin stability during storage, including time, temperature, mineral content of the mixture, and presence of ultraviolet light. In general, water-soluble vitamins are less susceptible to destruction than the fat-soluble vitamins. The stability of vitamins in feeds and premixes has been improved tremendously in recent years by chemical stabilization as an ester and by physical protection using antioxidants, emulsifying agents, gelatin and sugar in spray-dried, beaded or prilled products.

1. Relatively stable vitamins

Heat	Oxidation	Minerals	Light
Vitamin E	Vitamin E	Riboflavin	Vitamin E
Riboflavin	Vitamin K (MSB)	Ca pantothenate	Thiamine HCl
Niacin	Riboflavin	Biotin	Niacin
Pyridoxine	Niacin	Choline chloride	Pyridoxine
Choline chloride	Pyridoxine		Ca pantothenate
Vitamin C	Ca pantothenate		Biotin
	Biotin		Choline chloride
	Choline chloride		
	Folic acid		

2. Least stable vitamins

Heat	Oxidation	Minerals	Light
Vitamin A	Vitamin A	Vitamin A	Vitamin A
Vitamin D	Thiamine HCl	Vitamin D	Vitamin D
Vitamin K (MSB)	Vitamin C	Vitamin K (MSB)	Riboflavin
Thiamine HCl		Thiamine HCl	Folic acid
		Niacin	Vitamin C
		Pyridoxine	
		Vitamin C	

3. Ranked in order of stability under conditions usually found in complete mixed diets

a. Choline chloride
b. Biotin
c. Vitamin B_{12}
d. Ca pantothenate
e. Pyridoxine

f. Vitamin E
g. Niacin
h. Riboflavin
i. Folic acid
j. Thiamine HCl

k. Vitamin K (MSB)
l. Vitamin C
m. Vitamin D
n. Vitamin A

4. Water solubility. Ranked in order (most soluble first)

a. Choline chloride
b. Thiamine HCl
c. Vitamin K (MSB)
d. Calcium pantothenate

e. Vitamin C
f. Biotin
g. Niacin

h. Vitamin B_{12}
i. Riboflavin
j. Folic acid

VII. Nutrients in Metabolism

A. *Requirements for the various nutrients in any feeding program are usually composite because the program may be meeting needs for two or more of the following:*
1. Maintenance—necessary to prevent any loss or gain in weight.
   a. Heat for body temperature.
   b. Energy for internal work such as heart rate and breathing.
   c. Minimum movement.
   d. Repair of body tissue.
   e. Energy is the greatest need for maintenance but other nutrients are also required.
2. Growth.
   a. Protein and major minerals are important because they are part of the body structure, but energy and other nutrients are needed to build the structure.
   b. Especially critical when carrying on another function such as work or lactation because growth will be sacrificed at the expense of the others.
3. Work or activity—mainly energy but some need for other nutrients such as water due to perspiration.
4. Reproduction.
   a. Requirements for pregnancy are minimal, particularly the first third.
      (1) A 70-lb calf × 30% d.m. = 21 lb dry matter over 9 months or 8 pigs × 3 lb = 24 lb × 30% d.m. = approximately 7 lb d.m. over almost 4 months.
      (2) Have traditionally overfed gestating and underfed lactating animals.
   b. Over- or underfatness reduces reproductive efficiency.
5. Lactation.
   a. Demand very great in comparison to pregnancy, e.g., a dairy cow may produce d.m. equivalent to 2.5 times her body weight during a 10-month lactation.
   b. More efficient to feed for lactation requirement rather than build up body stores (fat) during pregnancy to be used during lactation.
   c. Also more efficient to feed the offspring animal itself rather than feed mother and have her produce milk to feed young.
6. Fattening.
   a. High energy (CHO and fat) to protein ratio in ration with high level of intake needed.
   b. Body will meet all other needs first.
7. Other needs.
   a. Wool growth.
   b. Egg production.

B. *Requirements of nutrients for the above needs will vary with many factors, including*
1. State of growth or age—younger animals usually grow faster in proportion to their metabolic size and total nutrient intake.
2. Size of the animal—primarily relates to capacity of feedstuff intake.
3. Environment.
   a. Cold environment requires more energy for maintenance.
   b. Hot environment requires more water for perspiration and other water losses.

4. Heredity.
5. Disease.
6. Activity—15% larger energy requirement when standing compared to lying down.
7. Degree of condition—may be able to use some body reserves.
8. Antagonistic factors or ration imbalances.
   a. High level of fat or minerals may destroy some vitamins in feedstuff storage.
   b. An excess or deficiency of one nutrient may impair absorption or cause destruction of others.
9. Species.

VIII. Digestive Systems

A. *General*

The gastrointestinal tract (GIT), sometimes referred to as the alimentary tract, is the passage from the mouth to the anus through which feed passes following consumption as it is subjected to various digestive processes. The various organs, glands and other structures involved are concerned with procuring, chewing and swallowing feed and with the digestion and absorption of nutrients as well as some excretory functions.

The following terms are associated with the study of animal digestion:
1. Prehension—taking in of feed or water.
2. Mastication—reduction of feed particle size; generally by chewing.
3. Deglutition—act of swallowing.
4. Regurgitation—casting up of undigested material.
5. Digestion—breakdown of feed particles into suitable products for absorption. May include
   a. Mechanical forces.
   b. Chemical action.
   c. Enzymic activity.
6. Absorption—transfer of substances from GIT to the circulating blood or lymph system.
7. Anabolism—growth or building process.
8. Catabolism—breakdown or destruction reactions.
9. Metabolism—combination of anabolic and catabolic reations occurring in the body with the liberation of energy.
10. Excretion—removal of wastes.
11. Herbivores—species of animals that depend entirely on plants for food.
12. Carnivores—species of animals that feed almost entirely on flesh of other animals.
13. Omnivores—species of animals that consume both plants and flesh.

B. *Monogastric Digestive System (Nonruminant)*

These are animals, such as the pig, horse, dog, cat or avian, which have a single-compartment stomach. Most monogastric animals make very poor use of fibrous feeds. However, by means of the microbial fermentation process of the cecum and large intestine, the horse is able to utilize such feeds effectively.
1. Swine
   A schematic illustration of the swine digestive system is shown in Figure 1–5.
   a. Mouth—proximal organ of the GIT containing three accessory organs:
      (1) Tongue—prehension, mixing and deglutition.
      (2) Teeth—prehension and mastication.
      (3) Salivary glands—three paired glands that secrete fluid called saliva; components of saliva include:
         (a) Water—moistens consumed feed and aids in the taste mechanisms.
         (b) Mucin—lubrication aid for swallowing.
         (c) Bicarbonate salts—act as a buffer to regulate pH of stomach.
         (d) Enzyme—salivery amylase initiates carbohydrate breakdown; not present in all species.
   b. Esophagus—hollow muscular tube that transports ingesta from the mouth to the stomach; ingesta material is moved by a series of muscular contractions referred to as peristaltic waves. A valve (cardiac sphincter) is at the junction of the stomach and esophagus.

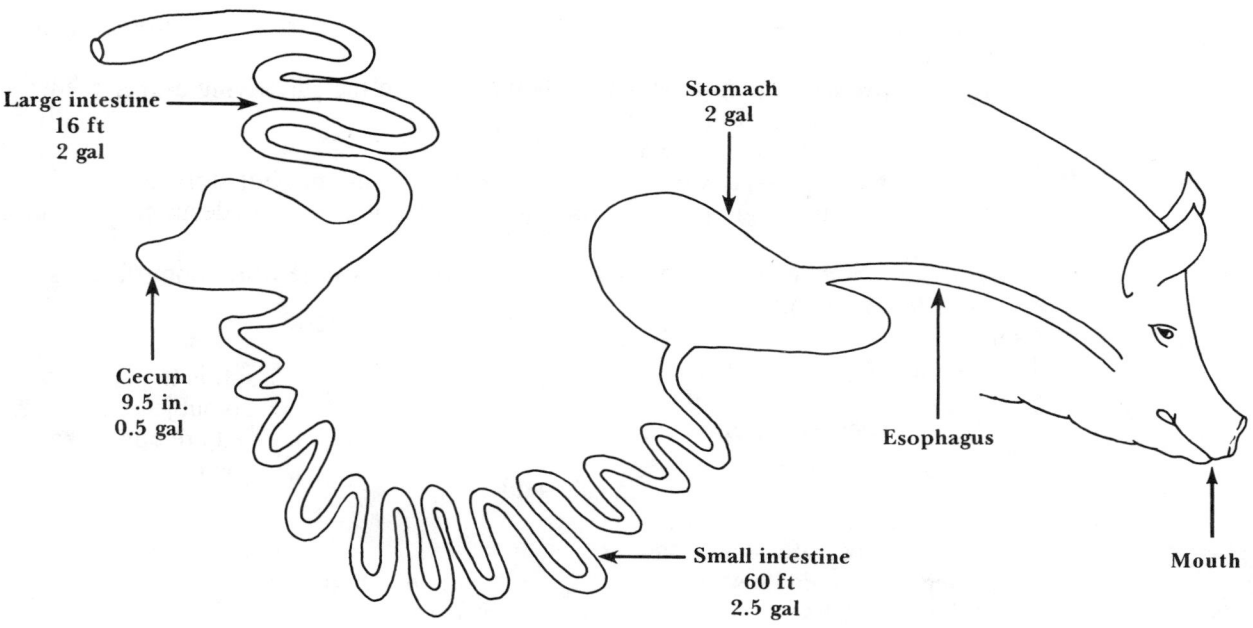

Simple digestive tract—14 × body length. Total capacity 7 gal. 44 teeth.

FIGURE 1–5. Digestive system of the pig.

  c. Stomach—a hollow, pear-shaped, muscular digestive organ.
    (1) Functions
      (a) Storage of ingested feed.
      (b) Muscular movements causing physical breakdown.
      (c) Secretes digestive juices.
        (1′) Hydrochloric acid.
        (2′) Pepsin.
        (3′) Rennin.
    (2) Stomach contents approximately pH 2 (bactericidal effect).
    (3) Material leaving stomach is called chyme.
    (4) Parts of the stomach (Figure 1–6).
      (a) Cardia—sphincter at the junction of the esophagus and stomach, which controls passage of ingesta into the stomach, or out.
      (b) Esophageal region—nonglandular area surrounding the cardia.

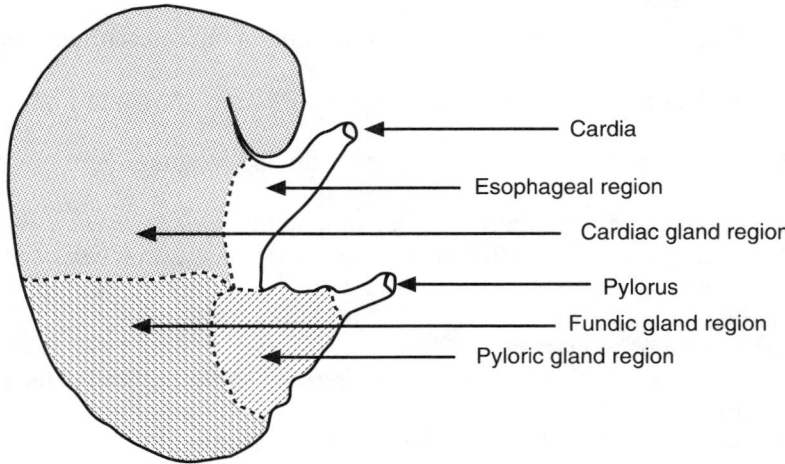

FIGURE 1–6. Swine stomach.

       (c) Cardiac gland region—contains cells that produce primarily mucus (protects stomach lining).

       (d) Fundic gland region—contains cells that provide the gastric secretions needed for the initial stages of digestion.

          (1') Parietal cells—produce hydrochloric acid.

          (2') Chief cells—produce enzymes or precursors of enzymes.

       (e) Pyloric gland region—contains cells that produce mucus and some proteolytic enzymes.

       (f) Pylorus—sphincter at the beginning of the small intestine which controls passage of material (chyme) out of the stomach.

   d. Small intestine

     (1) Divided into three sections:

       (a) Duodenum—first section.

          (1') Receives secretions from

             (a') Pancreas.

             (b') Liver (bile stored in gallbladder).

             (c') Intestinal walls.

          (2') Active site of digestion.

       (b) Jejunum—middle section; active in nutrient absorption.

       (c) Ileum—last section; active in nutrient absorption.

     (2) Walls of the small intestine are lined with a series of fingerlike projections called villi, which in turn have minute projections called microvilli that increase the nutrient absorption area. Each villus contains an arteriole and venule, together with a drainage tube of the lymphatic system, a lacteal. The venules ultimately drain into the portal blood system, which goes directly to the liver; the lymph system empties via the thoracic duct into the vena cava.

     (3) Small intestine contents approximately pH 6 to 7.

   e. Large intestine

     (1) Divided into three sections

       (a) Cecum—first section; size varies considerably in different species; little functional significance in the pig.

       (b) Colon—middle section; largest part of large intestine.

       (c) Rectum—last section.

     (2) Function of large intestine

       (a) Site of water resorption.

       (b) Secretion of some mineral elements such as calcium.

       (c) Storage reservoir of undigested GIT contents.

       (d) Bacterial fermentation:

          (1') Synthesis of some water-soluble vitamins and vitamin K.

          (2') Some bacterial breakdown of fibrous ingredients.

          (3') Synthesis of some protein.

       (e) Limited absorption of nutrients (feedstuff or microbial origin) from the large intestine.

2. Horse

A schematic illustration of the horse digestive system is shown in Figure 1–7.

   a. Mouth

     (1) Prehensive agents include teeth, upper lip and tongue.

     (2) Teeth.

       (a) Jaw movement is both vertical and lateral.

       (b) Upper jaw is wider than lower jaw; thus, mastication can occur on only one side of mouth at a time.

     (3) Saliva.

       (a) Contains no enzyme.

       (b) Secretion is stimulated by scratching (mechanical action) of feed on mucus membrane of inner cheeks.

       (c) May secrete up to 10 gal/day (mature horse).

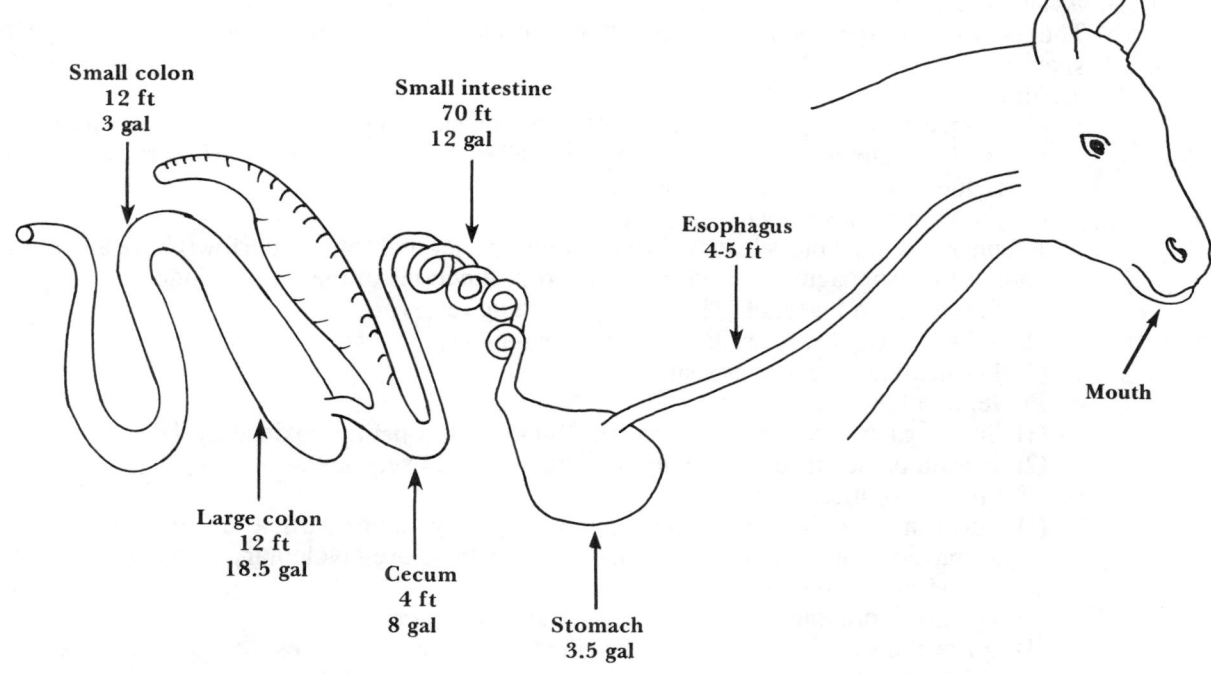

Small colon
12 ft
3 gal

Small intestine
70 ft
12 gal

Esophagus
4-5 ft

Mouth

Large colon
12 ft
18.5 gal

Cecum
4 ft
8 gal

Stomach
3.5 gal

More complex digestive tract—12 X body length. Total capacity 45 gal. 40 or 42 teeth.

FIGURE 1–7. Digestive system of the horse.

   b. Esophagus
     (1) 50- to 60-inch–long tube from mouth to stomach on the left side of the neck.
     (2) Only one-way peristaltic movements, thus it is very difficult for horse to regurgitate.
   c. Stomach
     (1) Horse has smaller stomach compared to other species, thus, must be fed small portions of ration several times daily.
     (2) Stomach does not have as extensive muscular movement activity as other species, and ingesta tends to arrange itself in layers. This can result in the horse being prone to greater digestive disorders originating in the stomach.
   d. Small intestine—same as the pig, except horse does not have a gallbladder; there is a direct secretion of bile into duodenum.
   e. Large intestine
     (1) Accounts for large part of GIT capacity (over 60%).
     (2) Divided into cecum, large colon, small colon and rectum.
     (3) Cecum and large colon.
       (a) Contain an active flora of bacteria similar to the microbial population in rumen compartment of ruminants.
         (1') Bacterial breakdown of cellulose and other carbohydrate material to produce VFAs (acetic, propionic, butyric) thus, the horse can utilize fibrous feeds.
         (2') Bacterial synthesis of water-soluble vitamins.
         (3') Bacterial synthesis of protein.
       (b) Some absorption of VFA occurs from the cecum; because proteins and other large molecules originating in the cecum and large colon are not subject to action of the digestive juices, they are assumed to be of limited use to the horse. Further data are required to clearly understand the overall digestive and absorbative function of the cecum and large colon.
     (4) Small colon—primary area of water resorption from intestinal contents.
     (5) Because the large intestine is usually expanded with ingesta material, impaction may easily occur.

3. Avian species

The GIT of avian species (Figure 1–8) differs considerably in anatomy from typical monogastric species.

a. Mouth
   (1) No teeth; does contain rigid and attached tongue and poorly developed salivary glands.
   (2) Beak is adapted for rapidly picking up small particles of feed; beak also serves to partially reduce feed to a size that may be swallowed.
   (3) Saliva contains salivary amylase.

b. Esophagus—most birds (except insect-eating species and some waterfowl) have an enlarged area in the esophagus referred to as the crop; functions of the crop include
   (1) Serves as an ingesta holding and moistening reservoir.
   (2) Allows breakdown reaction by salivary amylase.
   (3) Fermentation occurs in some species.

c. Proventriculus
   (1) Site of gastric juice production (HCl and pepsin); pH approximately 4.
   (2) Ingesta passes through rapidly (approximately 14 seconds).

d. Gizzard (or ventriculus)
   (1) Thick, muscular walled area acting to physically reduce particle size of ingesta (similar to mastication in mammals); muscle contractions are involuntary occurring at the rate of one every 20 to 30 seconds.
   (2) The gizzard lining is a hardened glandular secretion.
   (3) Gizzard normally contains grit (small stones or hard particles) that aids in the grinding of ingested seeds and grains.
   (4) Gizzard does not secrete any enzymes; however, HCl and pepsin from the proventriculus work in the gizzard.

e. Small intestine
   (1) Most enzymes found in mammalian species are present, with the exception of lactase.
   (2) The pH of the small intestine is slightly acid.
   (3) Absorption is similar to mammalian species except that the hormone enterogastrone, which affects fat absorption, is not present in birds.

f. Ceca and large intestine
   (1) The avian GIT contains 2 "blind pouches" (ceca), compared with one in mammals (cecum).
   (2) The ceca and large intestine are site of water resorption; some fiber digestion and water-soluble vitamin synthesis occurs in the ceca because of bacterial fermentation, but at much lower levels than in most mammals.
   (3) Large intestine is very short (2 to 4 in.) and empties into the cloaca where fecal material will be voided via the vent.

C. *Ruminant Digestive System*

These are animals such as the cow and sheep, which have a multicompartment stomach as schematically shown in Figure 1–9. By means of the microbial fermentation processes of the rumen compartment, such animals are able to make effective use of high fiber feeds and as a result are frequently fed rations containing high levels of fibrous feeds. These animals can also ruminate or chew a cud. An animal's cud consists of boluses of feed consumed earlier, regurgitated up the esophagus, remasticated and reswallowed.

1. Mouth
   a. No upper incisors; only an upper dental pad and lower incisors which act in conjunction with lips and tongue for prehension of feedstuffs.
   b. Molar teeth (both upper and lower) are so shaped and spaced that the animal can chew only on one side of the jaw at one time. Lateral jaw movements aid in shredding tough plant fibers.
   c. Saliva
      (1) Production is relatively continuous, although greater quantities are produced when eating and ruminating than when resting. Quantity produced may reach more than 12 gal/day in adult cows and 2 gal or more daily in sheep.

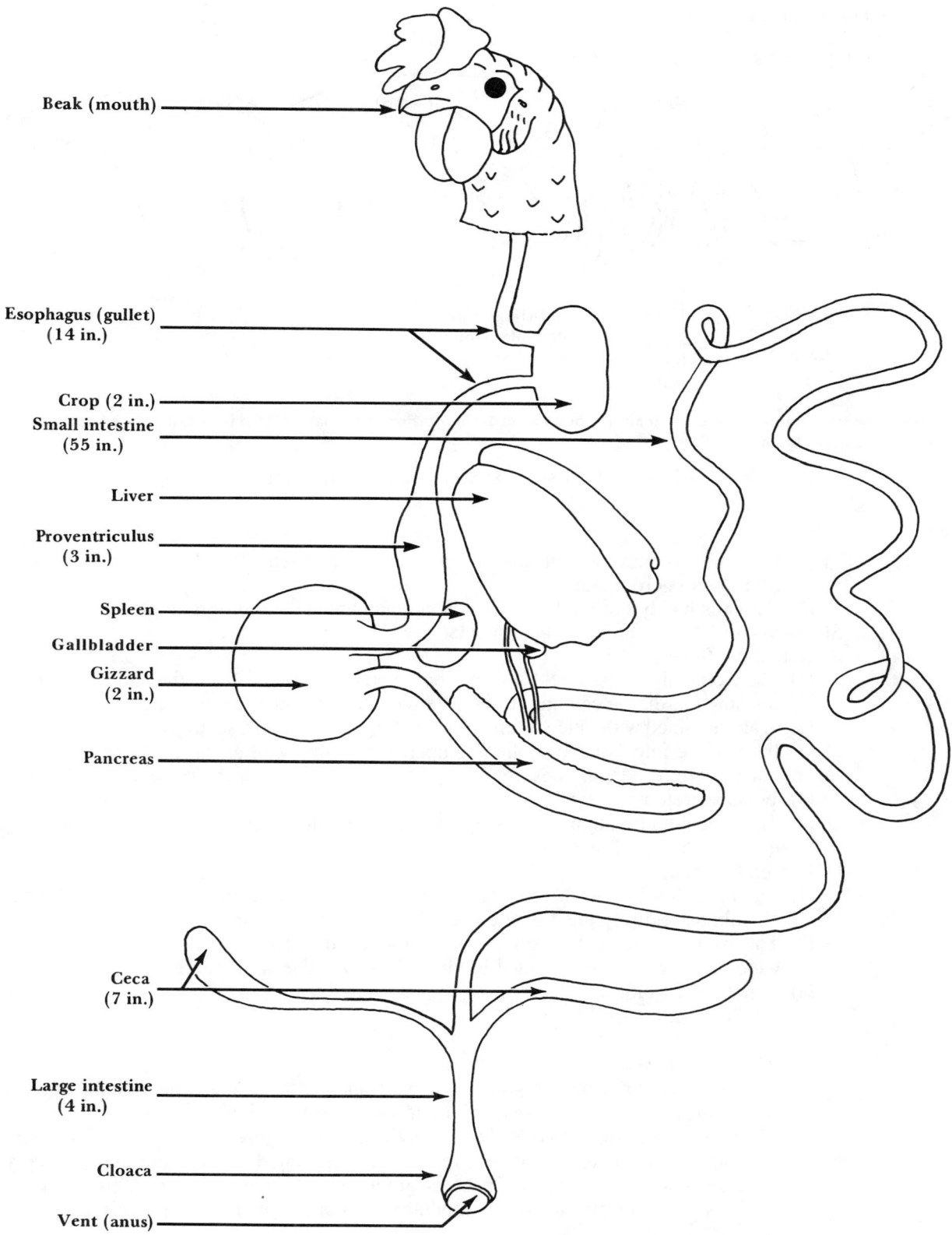

Beak (mouth)

Esophagus (gullet)
(14 in.)

Crop (2 in.)

Small intestine
(55 in.)

Liver

Proventriculus
(3 in.)

Spleen

Gallbladder

Gizzard
(2 in.)

Pancreas

Ceca
(7 in.)

Large intestine
(4 in.)

Cloaca

Vent (anus)

Total length 80 inches. Total capacity 85 grams. No teeth.

FIGURE 1–8. Digestive system of the avian.

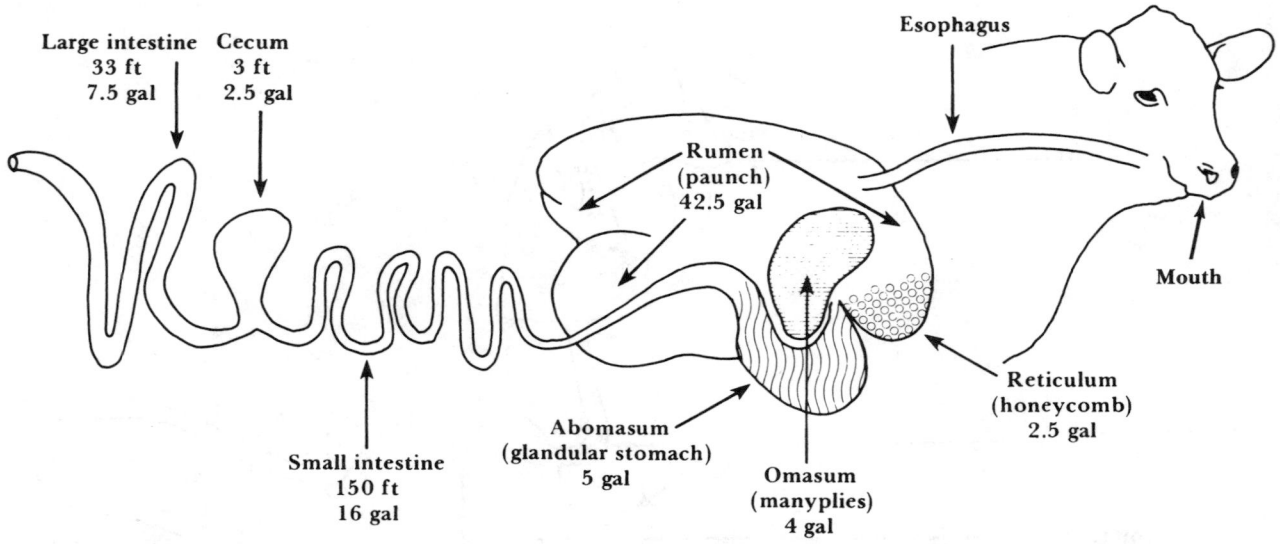

Complex digestive tract—20 X body length. Chews after fermentation (ruminates). Stomach capacity 54 gal. Total capacity 80 gal. 32 teeth.

FIGURE 1–9. Digestive system of the ruminant (cow).

(2) Contains no enzymes but provides a source of N (urea), P and Na, which are utilized by rumen microorganisms.

(3) Saliva is highly buffered, which aids in maintaining an appropriate pH in the rumen.

2. Stomach—divided into four compartments

a. Reticulum (honeycomb)

(1) Most cranial compartment and not completely separated from the rumen; esophagus opening (cardia) is common to both reticulum and rumen compartments.

(2) Walls are lined with mucus membrane containing many intersecting ridges that subdivide the surface into honeycomblike compartments; this wall arrangement traps hardware (nails, wire, etc.) and does not allow it to proceed through the remainder of the GIT.

(3) Walls secrete no enzymes.

(4) The reticulum functions in moving ingested feed into the rumen or into the omasum and in regurgitation of ingesta during rumination.

b. Rumen (paunch)

(1) Large, hollow, muscular compartment that extends from the diaphragm to the pelvis and almost entirely fills the left side of the abdominal cavity.

(2) The walls of the mature rumen contain small tonguelike projections called papillae, which can be readily identified by the naked eye. The walls secrete to enzymes.

(3) Functions include

(a) Storage.

(b) Soaking.

(c) Physical mixing and breakdown.

(d) Fermentation chamber—the reticulum-rumen compartments provide an ideal environment for microbial (bacterial and protozoal) survival and activity—it is moist, warm, anaerobic, desirable in pH and there is an irregular introduction of new ingesta and a more or less continual removal of fermented digesta and end products of digestion. Various types of bacteria are found with typical counts approaching numbers of 25 to 50 billion per milliliter of ruminal fluid. This extensive pregastric fermentation results in:

(1′) Bacterial synthesis of water-soluble vitamins and vitamin K

(2′) Bacterial synthesis of amino acids and protein. Bacteria will combine N from dietary protein or NPN sources with a C skeleton from carbohydrate sources to form their own body protein. As the bacteria move out of the rumen, they

44

will be degraded in the lower portion of the host animal's GIT and serve as a source of amino acids to the host animal.

   (3') Breakdown of fibrous feeds (high in cellulose). Bacteria contain enzymes to rupture cellulose bonding as well as starch bonding. In the breakdown of dietary carbohydrate, bacteria release VFAs (acetic, propionic, butyric). These VFAs are absorbed through the rumen wall and serve as a source of energy to the host animal (refer to the previous final oxidative pathway).

   (4) Rumen compartment is quite undeveloped at birth but may be functional by 6 to 8 weeks of age.

c. Omasum (manyplies)

   (1) The omasum is a spherical organ filled with muscular laminae studded with short, blunt papillae. The walls secrete no enzymes.

   (2) The omasum is located to the right of the rumen and reticulum.

   (3) Function of omasum appears to be in reducing particle size of ingesta before it enters the abomasum and some absorption of water.

d. Abomasum (true or glandular stomach)

   (1) First glandular portion of the ruminant GIT (walls secrete enzymes).

   (2) Located ventral to the omasum and extends caudally on the right side of the rumen.

   (3) In general, the gland regions of the abomasum correspond to the gland regions in the simple stomach of the nonruminant.

3. Small and large intestine.

   These areas of the ruminant GIT are similar in structure and function to those of the pig.

4. Additional peculiarities of the ruminant digestive system

   a. Esophageal (or reticular) groove

   (1) A passageway that extends from the cardia to the omasum, formed by two heavy muscular folds or lips, which can close to direct ingesta from the esophagus into the omasum directly, or open and permit the ingesta to enter the rumen and reticulum.

   (2) Functions to allow milk consumed by the suckling animal to bypass the reticulorumen and escape bacterial fermentation.

   (3) Does not appear to remain functional in older animals.

   b. Rumination

   (1) A process that permits an animal to forage and ingest feed rapidly, then complete the chewing at a later time; steps include regurgitation of the feed, remastication, resalivation and finally reswallowing.

   (2) The regurgitation step is preceded by contraction of the reticulum; probably a reverse peristalsis in the esophagus is the major factor in moving the material up to the mouth, where excess liquid is squeezed out and swallowed.

   (3) The regurgitated material consists largely of roughage and fluid with little if any concentrate.

   (4) Cattle average about 8 hours per day ruminating. One rumination cycle requires about 1 minute, of which 3 to 4 seconds is utilized for both regurgitation and reswallowing.

   c. Eructation (belching of gas)

   (1) Microbial fermentation in the rumen results in production of large amounts of gases (primarily carbon dioxide and methane), which must be eliminated.

   (2) Contractions of the upper sacs of the rumen force gases forward and down; the esophagus then dilates and allows the gases to escape.

   (3) Bloat is a common problem in ruminants in which gas cannot escape. This creates a distention of the rumen, which can be seen on the left side of the animal; in most cases of bloat, a stable froth or foam is produced in the rumen. This interferes with normal belching and gas accumulates. Therefore, control measures must prevent foam or break it rapidly after it has formed.

D. *Nutrient Digestion*

Nutrient digestion refers to the processes involved in the conversion of various feed nutrients into forms (called end products) that can be absorbed from the digestive tract. Most of these processes are accomplished through the action of various enzymes found in the different digestive juices secreted into the digestive tract. Enzymes are organic catalysts that produce changes in the feed without being destroyed themselves. Most enzyme names end in "ase." A few early named enzymes end in "in."

Five major areas where digestive fluids are secreted into the digestive tract by glands or tissues and the end products of digestion are as outlined below.

1. Saliva (digestion in the mouth)
   a. Salivary amylase (ptyalin)
      (1) Starch to maltose.
      (2) None in ruminants; of minor importance in other species.
2. Gastric (digestion in the stomach)
   a. Pepsin—protein to polypeptides.
   b. Rennin—(proteinase)—milk curdling or protein denaturing enzyme.
   c. Lipase—fat splitting (a minor activity compared to pancreatic digestion).
3. Pancreas (digestion in the duodenum and upper small intestine)
   a. Trypsin—protein to peptides and amino acids.
   b. Chymotrypsin—protein to peptides and amino acids.
   c. Carboxypeptidase—protein to peptides and amino acids.
   d. Amylase—undigested carbohydrates to sugars.
   e. Lipase—fat splitting; fats to fatty acids and glycerol.
   f. Pancreatic secretions contain buffers (e.g., bicarbonates) to neutralize stomach acids.
4. Liver
   Does not produce enzymes but produces bile salts, which are emulsifiers that separate fat globules and give the lipase enzyme greater access or more surface area upon which it may react. Bile salts also aid in maintaining an alkaline pH in the small intestine.
5. Intestine (digestion in small intestine)
   a. Aminopeptidase—protein to peptides and amino acids.
   b. Dipeptidase—peptides to amino acids.
   c. Nucleases (several types)—nucleoproteins to purine and pyrimidine bases, phosphoric acid and pentose sugars.
   d. Maltase—maltose to glucose.
   e. Lactase—lactose to glucose and galactose.
   f. Sucrase—sucrose to glucose and fructose.

E. *Nutrient Absorption*
   1. Nutrient absorption involves the movement of the end products of digestion from the digestive tract into the blood system, the lymph system, or both. It is accomplished by the process of osmosis through the semipermeable membranes that line much of the digestive tract.
   2. In broad terms, absorption occurs as the result of diffusion (passive) or active transport.
      a. Diffusion involves the movement of molecules from a region of high concentration of those molecules to a region of low concentration; only physical forces are involved, i.e., the migration of molecules from one place to another.
      b. Active transport involves the movement of molecules from one region to another with the expenditure of energy required.
   3. Most nutrient absorption takes place in the small and large intestines. The villi of the small intestine especially facilitate absorption at this location. Considerable VFA absorption also takes place through the rumen wall in ruminants. All of the end products of digestion, except those of fat digestion and possibly the fat-soluble vitamins, are absorbed directly into the bloodstream. The end products of fat digestion are absorbed into the lymph system through the lacteals of the villi in the form of chyle. The latter subsequently enters the blood through the thoracic duct and undergoes certain metabolic changes in the liver before being used in the tissues.
   4. The schematic summary of nutrient digestion and absorption is shown in Figure 1–10.

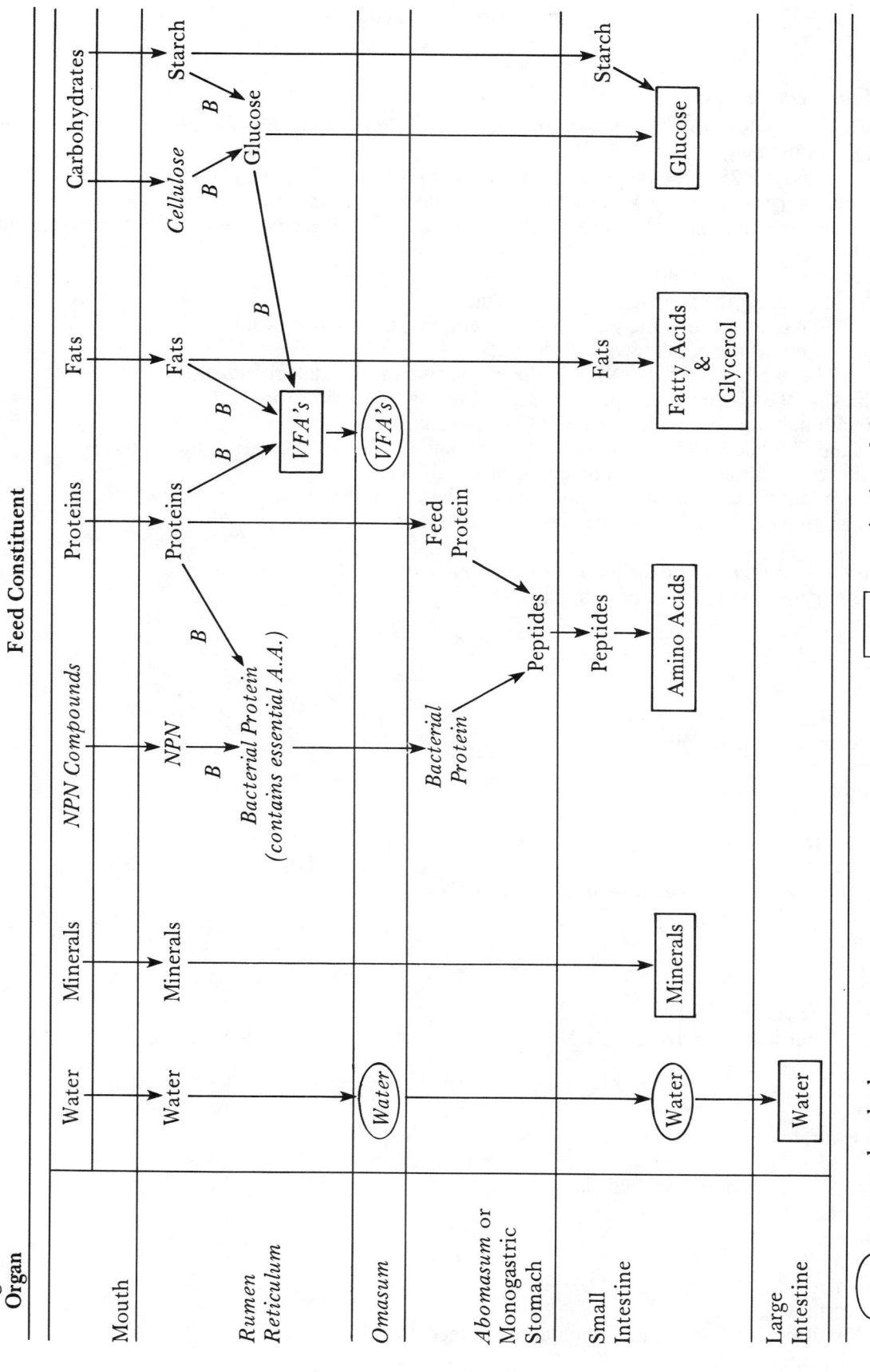

FIGURE 1–10. Summary of digestion and absorption.

47

Study Questions and Problems

Chapter 1

I. True or False Questions

1. Essential amino acids (EAAs) are not synthesized in sufficient amounts by livestock and must therefore be provided in the diet.
2. Fats liberate about 2.25 times as much energy as carbohydrates when metabolized.
3. Minerals play very important roles as activators in enzymatic reactions.
4. Urea is an example of nonprotein nitrogen (NPN) and contributes the carbon chain backbone for amino acid synthesis.
5. Glucose is an example of a disaccharide.
6. The material entering the stomach is called chyme.
7. In birds, food passes through the gizzard before entering the proventriculus.
8. Deglutition refers to the breakdown of feedstuffs.
9. Protein may be synthesized from NPN in the abomasum of the adult ruminant.
10. In the chicken, the gizzard corresponds to the glandular stomach of the pig.
11. Animals with only one cecum are referred to as monogastrics.
12. Gastric glands secrete sodium bicarbonate to maintain a stable level of pH in the stomach.
13. Anabolism refers to the breakdown reactions of metabolism.
14. Most absorption of digested nutrients into the circulatory system occurs in the small intestine.
15. Pepsin is an enzyme found in the intestinal juice that acts on protein.

II. Multiple Choice (select the single best answer)

1. The trace mineral that is a part of vitamin B_{12} is
   a. Iodine.
   b. Iron.
   c. Copper.
   d. Cobalt.
   e. Magnesium.
2. Fats are composed of various fatty acids plus
   a. Glucose.
   b. Glycerol.
   c. Inositol.
   d. Insoleucine.
   e. Glycogen.
3. Proteins are made up of carbon, hydrogen, oxygen and
   a. Calcium.
   b. Sodium.
   c. Potassium.
   d. Nitrogen.
   e. None of these.
4. The class of nutrients supplying the main source of energy in usual livestock rations is
   a. Fat.
   b. Protein.
   c. Water.
   d. Carbohydrates.
   e. Minerals.
5. The precursor of vitamin A in feeds is
   a. Carnosine.
   b. Carotene.
   c. Cholesterol.
   d. Vitamin A.
   e. None of these.
6. The readily digestible fraction of carbohydrate in feedstuffs is
   a. Essential amino acid.
   b. Nonprotein nitrogen.

c. Phenylalanine.
d. Nitrogen-free extract.
e. Crude fiber.
7. The compartment in the ruminant that is most like a monogastric stomach is the
   a. Rumen.
   b. Abomasum.
   c. Reticulum.
   d. Omasum.
   e. None of the above.
8. The most active portion of the pig's digestive tract for digestion and absorption of nutrients is in the
   a. Stomach.
   b. Small intestine.
   c. Cecum.
   d. Large intestine.
   e. Colon.
9. Which of the following is much more developed in the horse than in the pig?
   a. Rumen.
   b. Omasum.
   c. Ileum.
   d. Duodenum.
   e. Cecum.
10. Which of these is not in the proper sequence (duodenum-caecum-jejunum-ileum-colon) starting from the anterior end of the digestive tract?
    a. Duodenum.
    b. Cecum.
    c. Jejunum.
    d. Ileum.
    e. Colon.
11. Which compartment of the ruminant stomach absorbs large quantities of water?
    a. Reticulum.
    b. Rumen.
    c. Omasum.
    d. Abomasum.
12. The following nutrients are synthesized by the rumen bacteria:
    a. Vitamins and minerals.
    b. Proteins and carbohydrates.
    c. Vitamins and water.
    d. Proteins and vitamins.
    e. All of the above.
13. Which of the following is the major site of roughage fermentation in the horse?
    a. Cecum.
    b. Omasum.
    c. Rumen.
    d. Small intestine.
    e. Stomach.
14. The enzyme most active in fat degradation in the pig is
    a. Lipase.
    b. Rennin.
    c. Lactase.
    d. Amylase.
    e. Pepsin.
15. Which of the following digestive fluids has no degradation effect on protein?
    a. Renin.
    b. Bile.
    c. Trypsin.
    d. Chymotrypsin.
    e. Carboxypeptidase.

III. Matching and Identification

A. Match the following concepts with the correct nutrient.

a.	Pantothenic acid	f.	Selenium
b.	Zinc	g.	Vitamin K
c.	Manganese	h.	Vitamin A
d.	Vitamin E	i.	Iron
e.	Magnesium	j.	Vitamin D

1. Fat-soluble vitamin synthesized by rumen microorganism.
2. Prevents night blindness in cattle.
3. Prevents parakeratosis in pigs.
4. Antioxidant properties.
5. Prevents perosis in birds.
6. Deficiency causes anemia in baby pigs.
7. Involved in prothrombin formation and blood clotting.
8. Mineral found in hemoglobin.
9. Deficiency causes goose-stepping in pigs.
10. Produced in the body by sunlight.
11. Mineral associated with grass tetany in cattle.
12. Has carotene as a precursor.
13. Vitamin involved in Ca absorption.
14. Mineral that functions similarly to vitamin E.
15. Chemical form is DL-α-tocopheryl acetate.
16. Mineral that is easily tied up by excess Ca or phytate.
17. Mineral that prevents muscle dystrophy (white muscle) in lambs.
18. Antagonists include dicoumarol and warfarin.

B. Classify the following items within these eight categories.

a.	Digestive enzyme	f.	Carbohydrate
b.	Digestive tract part	g.	Amino acid
c.	Major mineral	h.	Vitamin
d.	Trace mineral		
e.	Lipid		

1. Thiamine
2. Selenium
3. Pepsin
4. Calcium
5. Rumen
6. Zinc
7. Trypsin
8. Tyrosine
9. Manganese
10. Stearic acid
11. Fructose
12. Maltase
13. Cecum
14. Biotin
15. Sodium
16. Cellulose
17. Pantothenic acid
18. Linoleic acid

IV. Short Answer Questions

1. Describe the formation of carbohydrates in plants.
2. What is meant by the term "unsaturated fat"?
3. Of what units are proteins composed?
4. Explain what is meant by an essential amino acid (EAA).

5. List one or more examples of the interrelationships among minerals.
6. What is the structural difference between amylopectin and amylose?
7. Diagram a peptide linkage.
8. How do fats differ in their chemical composition from carbohydrates?
9. What is meant by the biological value of a protein?
10. What is the function of rumination?

V. Problems

1. The complete metabolic breakdown of three units of maltose by the final oxidative pathways will yield how many ATPs?
2. In a protein quality trial, an animal consumed 1224 g of crude protein. The animal excreted 331 g of fecal protein and 252 g of urinary protein. Calculate the biological value (BV) of this protein.
3. Assume there are 28 mg of beta carotene in 1 lb of bromegrass hay. If a 1200-lb pregnant cow retired 28,000 IU of vitamin A daily, how many pounds of hay would be needed to meet the vitamin A requirement?

Notes

SECTION II

FEEDSTUFFS
AND FORMULATIONS

2

Evaluating Feedstuffs for Farm Livestock

Chapter Goals

- Discuss the various analytical methods used to determine the nutrient composition of feedstuffs.
- Explain how feed samples should be collected for analysis and how nutrient composition is reported (dry matter, as-fed, air-dry).
- Describe procedures in determining the apparent digestibility of feedstuffs.
- Describe the various energy measurements and explain their usage in diet formulation or evaluation.
- Describe how feeds can be physically and economically evaluated.

Evaluating feedstuffs for use in livestock diets may employ a number of procedures. These procedures should identify nutrient composition, palatability, digestibility, productive value, physical or handling characteristics, and provide economic comparisons.

I. Analytical Methods for Nutrient Composition

For many nutrients required by animals there are direct analytical methods by which we can establish the potency of feedstuffs in these nutrients. There are three general types of analytical methods: (1) chemical procedures (gravimetric procedures, titration, colorimetry, chromatography etc.); (2) biological procedures (employ animals such as chick or rat to give a much more accurate estimate of animal utilization but which makes procedure more tedious and expensive); and (3) microbiological procedures (similar to biological procedures but employ isolated bacteria).

A. *Obtaining Samples for Analysis*

The key to reliable feed nutrient evaluation is obtaining a sample that represents the feed the animals will eat. A good sample of each lot of feed is essential because this small amount, usually less than a quart, will probably represent several tons of feed.

1. Identification

The first step in obtaining samples is recording information about the feed. Label each container with your name, address, sample number, feed name and type (including variety name), stage of maturity and harvest date. It may also be useful to note where the sample is stored and any special conditions such as color, odor, mold etc. This information will be useful when formulating rations, and it may aid in predicting value of future samples.

2. Sampling

Outlined below are recommendations for obtaining representative samples.

a. Grains or mixed feeds

(1) Sacked feeds should be sampled by taking two samples (approximately a handful) each from five to seven different sacks. If a sack probe is available, it should be used. If not, it is best to sample sacked feeds when they are being emptied.

(2) Bulk feeds or grain in bins should also have 12 to 15 samples taken from a given lot. These samples should be as widely separated as possible. It may be easiest to sample this material while it is being delivered or fed, so a representative sample is obtained. If a grain probe is available, it will aid in obtaining a good random sample.

(3) The random samples should be mixed in a clean pail and a 1- to 2-lb (approximately 1-qt) sample sent to the laboratory. When mixing these samples, care should be taken that the feed is not segregated by particle size.

b. Hay

(1) For maximum reliability, use a drill-type core sampler such as the Penn State Forage Sampler (this type of sampler is usually available through the local extension service).

(2) For stacked or chopped hay, take samples along the sides where it is sufficiently packed to obtain a good sample or stand on top of the stack and take core samples between your feet. Insert the sampler 12 to 15 in. into the end of square-baled hay, across the center of round bales or full depth in stacked hay. This allows a sample to be taken that will contain both stems and leaves, giving a representative sample.

(3) Take 12 to 15 widely separated samples from each lot, taking only one sample per bale if hay is baled. In stacked hay try to obtain some samples from the middle of the stack—by using an extension for the core sampler.

(4) If a core sampler is not available, at least ten "grab" samples should be taken. In baled hay the bales should be broken, and a "handful" of sample should be taken from a central slab in the bale without separating stems and leaves (this can be done at feeding time when bales are broken for feeding).

(5) Mix core samples or "grab" samples (which have been cut to 1- to 2-in. lengths using heavy scissors or lawn shears) in a clean pail. Mix carefully so that stems and leaves are not separated. Take a random sampling of this mixed material and place in container for shipping to the laboratory.

c. Haylage or silage

(1) In upright silos materials can be collected during the entire feeding period while the unloader is in operation. If hand feeding, collect the sample after material is removed from the silo into the cart or similar feeding unit (throwing material into the cart will aid in mixing for a reliable sample).

(2) In pit or bunker silos, four to five grab samples should be taken from the freshly opened face of the silage daily.

(3) Do not collect spoiled material unless you mix it so thoroughly that animals do not separate it. Two or three feet of silage should be removed from upright silos before sampling and no sample should be taken within 12 to 18 in. of the edge of a bunker silo. Remember, you are trying to get a representative sample of the material the animal will consume.

(4) Approximately 1 to 2 lb of material should be collected each day for 2 to 4 days.

(5) The samples should be frozen immediately after collection, adding each day's collection into the same bag. Freezing wet samples prevents bacterial fermentation and moisture loss, both of which are important to accurate analysis.

(6) The samples taken should be mixed thoroughly in a clean pail, and a random sampling of this material should be analyzed. The material should be packed tightly in the sample container to exclude air and then frozen for shipment to the laboratory.

d. Harvest sampling

Many times it is more convenient and more reliable to obtain samples during harvest. In addition, harvest sampling and analysis allows easier identification of specific forage lots and analytical results that are available by the time of feeding. Most forages, except those below 30% dry matter, can be sampled at harvest. If any changes are noticed during storage (such as seepage) or feeding (such as drastic change in color, odor etc.), a new sample should be taken.

The procedure for sampling at harvest is the same as listed above for the feed type. However, at harvest the 12 to 15 random samplings are divided into two to three samplings per load from five to seven different loads per day. With hay it is important to note where the material is stacked and the field from which it was taken. For silages the sampled material can be identified at feeding by using punch holes or notches from colored plastic or by adding 1 to 2 bushels of oats to the silo before starting a new lot of forage.

3. General

A representative sample is as important as the accuracy of analysis in obtaining reliable results. A carefully taken sample when one is mindful of why it is taken, the use of the results and the cost involved, yields the information needed to improve the feeding program. To be beneficial, however, feed analysis must be used in designing the feeding program. Properly balancing the protein and energy of the ration can increase profits three ways. First, if protein is higher than needed, then it can be reduced and supplement costs decreased. Second, if underfeeding is the problem, then adding additional protein or energy may increase production with only a slight increase in supplemental cost. Third, if protein and energy meet requirements, yet production is below expectation, feed analysis can confirm that some other factor is limiting production.

Feed analysis also may provide information useful in improving feed harvesting programs to obtain the highest quality feed for the animal that will yield the most profit to the producer.

B. *Proximate Analysis*

Proximate analysis is a combination of analytical procedures developed in Weende, Germany more than a century ago. The different fractions that result from the proximate analysis include water, ash, crude protein, ether extract, crude fiber and nitrogen-free extract (NFE). This procedure is probably the most generally used chemical scheme for describing feedstuffs in spite of the fact that the information it gives often is of uncertain nutritional significance, or may even be misleading. We should, therefore, consider in some detail the nature, peculiarities and limitations of proximate analysis as a description of the nutritional properties of feedstuffs.

Determination by proximate analysis:

1. Dry matter

Heat sample to a constant weight at a temperature above the boiling point of water (100–105° C). This removes the water, so loss in weight equals water. (100% − H_2O = d.m.%)

Sources of error: Any materials that volatilize at the temperature are lost. (Silages or other fermented products.) Some liquids oxidize when heated and therefore increase in weight. Other methods for determination of water in feeds include toluene distillation, drying under vacuum and freeze drying.

2. Ash (minerals)

Burn sample by placing a weighed amount in a muffle furnace for 2 hours at 600°C. Ash is considered as the dry inorganic residue remaining. The water, fat, protein and carbohydrates have been removed by the process. This high a temperature may alter forms of some minerals and may even volatilize some such as chlorine, zinc, selenium and iodine.

3. Crude protein (Kjeldahl process)

Digest small dried sample in concentrated sulfuric acid until all organic matter is destroyed. Nitrogen from feed is now in the form of ammonium sulfate. The digest is neutralized with sodium hydroxide, distilled, driving the ammonia over into standard acid and titrated. This determines the amount of nitrogen in the sample.

Protein contains an average of 16% nitrogen; therefore, total nitrogen × 6.25 equals the crude protein in the sample. The analysis does not distinguish one form of N from another. Thus, we cannot tell if a feed mixture has true protein (amino acids) or other NPN sources. In addition, nitrate N is not converted to ammonium salts by this procedure.

4. Ether extract (fat)

Extract dry sample with ether for a period of 4 hours or more. This removes the fat, so again the loss in weight after drying (evaporation of ether) equals the fat. This procedure includes as "fat" any ether-soluble compounds including some nonnutrient compounds such as chlorophyll, volatile oils, resins, pigments and plant waxes, which are of little value to animals.

5. Carbohydrates (CHO)

Not determined by analysis as such, CHO = CF + NFE

a. Crude fiber (CF)

After removal of water and ether extract from a sample of feed, the sample is boiled in weak acid (0.255 N H_2SO_4), then in weak alkali (0.312 N NaOH). This removes the proteins, sugars and starches, which are discarded. Cellulose, lignin and mineral matter are left in the feed residue. This material is dried and weighed, then burned in a muffle furnace at

600°C. Loss in weight is reported as crude fiber. This fraction consists primarily of hemicellulose, cellulose and some insoluble lignin.

    b. Nitrogen-free extract (NFE)

Nitrogen-free extract is found by difference—not by actual analysis. The percentages of water, ash, protein, fiber and fat are merely added together and subtracted from 100. NFE is made up primarily of readily available carbohydrates, such as the sugars and starches, but it may also contain some hemicellulose and lignin, particularly in such feedstuffs as forages.

C. *The Van Soest (Detergent) Method of Forage Evaluation*

The proximate analysis system of feed analysis has served for many years and continues to serve a very useful purpose in predicting the nutritive value of feeds. However, there are some definite limitations of this system, especially with respect to both the crude fiber and NFE fractions. The material that was dissolved by the solvents (NFE) was assumed to be digestible and the residue (crude fiber) was assumed to be indigestible. Later studies showed that in some cases crude fiber was more digestible than was NFE. Low digestibility of the NFE fraction arises from the fact that nearly all of the fibrous hemicellulose and a variable amount of the lignin are dissolved in the weak acid and alkali. Both fractions have been shown to be relatively indigestible. Thus, predicting nutritive value from the proximate analysis figures became a major problem, especially with the more fibrous feeds. As a result, numerous workers over the past several years have tested various procedures that might provide a more definitive separation of feed carbohydrate than does the proximate analysis.

To develop a chemical procedure that fractionates forages into relatively digestible and indigestible portions, Van Soest[1] proposed the extraction scheme shown below.

The portion of the plant material that is soluble in neutral detergent, cell contents, is virtually completely digested by the ruminant. The residue from this fraction, cell wall constituents, is of low but variable availability depending upon the species of plant and its stage of maturity. A similar statement can be made about acid detergent fiber and cellulose. Lignin is nearly indigestible.

    1. Detergent feed analysis scheme (Figure 2–1)

    2. Uses for the Van Soest system

        a. Predict intake—NDF is used as an index of gut fill to predict voluntary feed intake.

        b. Predict digestibility—ADF is used as an indicator of forage digestibility.

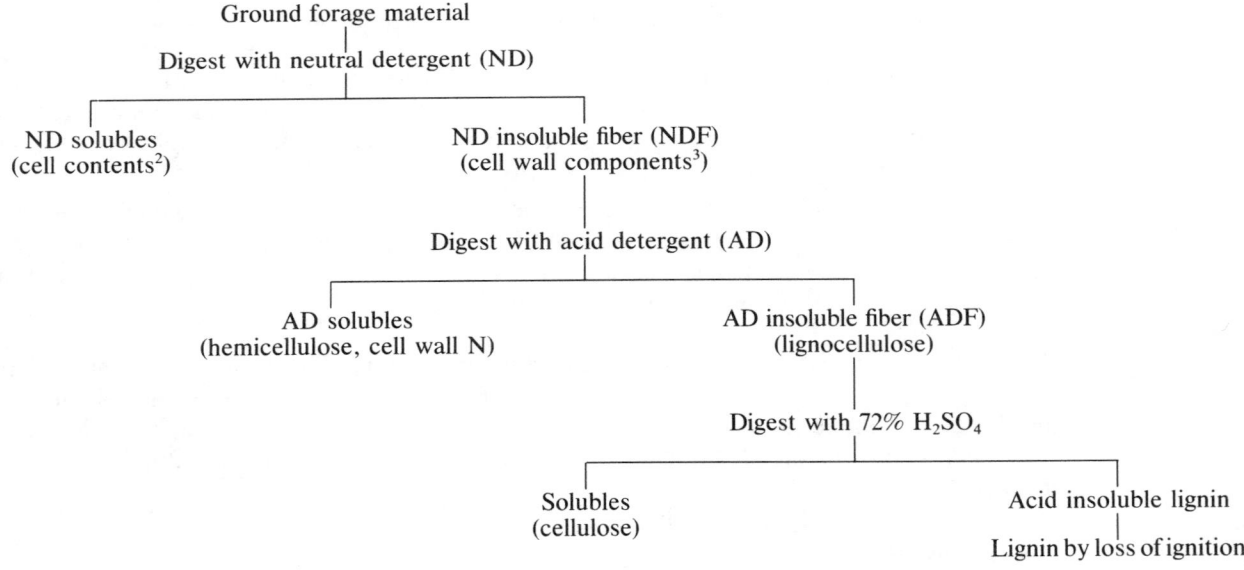

FIGURE 2–1. Detergent feed analysis scheme.

1. Van Soest PJ, 1967, J Animal Sci, 26:119. Also see Goering HK and PJ Van Soest, 1970, *Forage Fiber Analysis,* USDA, AR Handbook No. 379.
2. Cell contents include sugar, starch, soluble carbohydrates, pectin, protein, NPN, lipids, miscellaneous vitamins etc.
3. Cell wall components include cellulose, hemicellulose, lignin, silica, heat-damaged protein.

A Micro-Kjeldahl apparatus for making nitrogen determinations.

c. Heat-damaged forages—heat damage is caused by ensiling of forages, particularly alfalfa, at low moisture levels, allowing air into the ensiled mass. This results in aerobic instead of anaerobic fermentation. During aerobic fermentation, the temperature in the silo increases and the forage protein and carbohydrate combine, making the protein unavailable to animals. Other forage or concentrate feeds exposed to heat may also be susceptible to this reaction. The amount of heat damage or decrease in protein availability can be determined by the acid detergent insoluble nitrogen (ADIN) method. This test involves analysis of the forage for ADF, total N, and N in the ADF (both by Kjeldahl). The unavailable or indigestible protein content of feed (ADIN) is subtracted from the total protein content (total N) to determine the amount of protein actually available to the animal. Example:

(1) Alfalfa silage analysis (dry matter basis):
 Total N, % = 3.05%
  ADIN, % = 0.25%

(2) 3.05 − 0.25 = 2.80% N available

(3) 2.80 divided by 3.05 (100) = 91.8% of the total N (or protein) is available to the animal (or 8.2% is unavailable).

(4) As a general guideline:

Unavailable protein	Level of heat damage
0–10%	None or very slight
11–25%	Moderate to severe
> 26%	Severe

d. Determining net energy (NE) and total digestible nutrient (TDN) levels of forages—the following equations can be used to calculate the energy content of forages (all values are on a dry matter basis):

(1) Legumes
 $NEm(Mcal/lb) = 1.1698 - 0.0111 \times NDF$
 $NEg(Mcal/lb) = 0.9095 - 0.0122 \times NDF$
 $NE_L(Mcal/lb) = 1.0440 - (0.0119 \times ADF)$
 $TDN (\%) = 4.8980 + (NE_L \times 89.796)$

(2) Grasses
 $NEm(Mcal/lb) = 1.3875 - 0.0125 \times NDF$
 $NEg(Mcal/lb) = 1.2596 - 0.0148 \times NDF$
 $NE_L(Mcal/lb) = 1.0850 - (0.0124 \times ADF)$
 $TDN (\%) = 4.8980 + (NE_L \times 89.796)$

 Note: Unfamiliar acronyms are defined in Section IV, below.

D. *Near Infrared Reflectance Spectroscopy*

Development of near infrared reflectance spectroscopy (NIRS) technology has been a real boon for assessing forage quality. This procedure involves an instrumental method for rapidly and reproducibly measuring the chemical composition of samples with little or no sample preparation. It is based on the fact that each of the major chemical components of a sample has near infrared absorption properties that can be used to distinguish one component from the others. In the near infrared range, absorption occurs primarily as a result of vibrations of light-weight atoms that have strong molecular bonds. When there are weak chemical bonds, or heavy atoms, the vibration frequency is low and will not be detected in the near infrared range. Thus, NIR is primarily limited to chemical bonds containing hydrogen attached to atoms such as nitrogen, carbon or oxygen. Consequently, the detection of minerals is poor unless the mineral exists in association with some organic constituent.

The NIRS method of analysis has four main advantages over the more traditional wet-lab procedures: speed, simplicity of sample preparation and ability to analyze multiple components in one operation. No portion of the sample is consumed by the procedure. The NIR measurements can be made in less than 1 second, although typical times range from 30 seconds to 3 minutes, and the only sample preparation usually required is to grind the sample. In contrast, some wet-lab procedures take several hours to perform and require more extensive sample preparation. NIRS has its disadvantages too: the requirement of a high-precision instrument, dependence on calibration procedures and inability to measure minor constituents.

The NIRS method is still a developing technology. It is rapidly gaining popularity in the feed industry because of its speed in obtaining analysis data. When utilized in the field, such as at a hay auction, it can provide valuable information quickly. As the instrumentation continues to improve, new applications, such as amino acid analysis and the detection of molds and mycotoxins, may become possible.

E. *Determination of Vitamins*

Because of the diversity of compounds in this class, there is no routine analysis for vitamins. However, methods are available for assaying individual vitamins. Biological assays are used for some vitamins, whereas others are determined strictly on the basis of chemical analysis.

F. *Determination of Energy*

The bomb calorimeter is an instrument used for determining the gross energy content of a material (solid, liquid or gas). The energy value of a given sample is determined by burning it in an atmosphere of oxygen. When the sample is burned, the heat liberated raises the temperature of water surrounding the container in which the sample is enclosed, and the temperature increase provides the basis for calculating the energy value.

The energy value is expressed in units called calories where 1 calorie is the amount of heat required to raise the temperature of 1 g water from 14.5° to 15.5° C.

G. *The nutrient or energy content of a feed may be expressed as a percentage or quantity per unit of weight (mg/kg, g/lb, etc.) on one of the following bases:*
1. Dry matter basis—amount contained in only the dry matter fraction, without water.
2. As-fed basis—amount contained in the feed as it would be fed to the animal, including water.
3. Air-dry basis—generally assumed to be approximately 90% dry matter. Most feeds will equilibrate to about 90% dry matter after prolonged storage (aerobic). Air-dry and as-fed basis may be the same for many common feeds.
4. Since feeds contain varying amounts of dry matter, it would be much simpler, and more accurate, if both feed composition and nutrient requirement values were on a dry matter basis.
5. Conversion of feed nutrients from an as-fed to a dry matter basis.
   a. Assume alfalfa silage analyzed 7% crude protein on an as-fed basis and contained 40% dry matter. What percent crude protein would the alfalfa silage contain when expressed on a dry matter basis?
   The solution for this example can be obtained by the following equation:

   $$\frac{\% \text{ nutrient (as-fed basis)}}{\% \text{ feed dry matter}} = \frac{\% \text{ nutrient (dry matter basis)}}{100\% \text{ dry matter}}$$

thus:

(1) $\dfrac{7}{40} = \dfrac{X}{100}$

(2) $40X = 700$ (values obtained by cross-multiplying)

(3) $X = \dfrac{700}{40} = 17.5$

(4) The alfalfa silage contains 17.5% crude protein on a dry matter basis.

b. Two samples of shelled corn were sent to a laboratory for analysis of crude protein. One sample was "dry" corn and the other "high-moisture" corn. The laboratory sent back the following analysis:

"Dry" corn		"High-moisture" corn
89.0	% dry matter	75.0
8.8	% crude protein (as-fed basis)	7.4

To compare crude protein content to the two samples, calculate the composition on a dry matter basis.

(1) "Dry" corn:

(a) $\dfrac{8.8}{89} = \dfrac{X}{100}$

(b) $89X = 880$

(c) $X = \dfrac{880}{89} = 9.89$

(2) "High-moisture" corn

(a) $\dfrac{7.4}{75} = \dfrac{X}{100}$

(b) $75X = 740$

(c) $X = \dfrac{740}{75} = 9.87$

(3) Thus, the two samples of corn contain the same percent crude protein when expressed on a dry matter basis.

6. Conversion of feed nutrients from a dry matter to an as-fed basis.

a. Assume a feed nutrient composition table lists linseed meal as containing 10% crude fiber on a dry matter basis. If the linseed meal contains 91% dry matter, what is the percent crude fiber expressed on an as-fed basis? The solution for this problem can be obtained by the same equation given in the previous example.

(1) $\dfrac{X}{91} = \dfrac{10}{100}$

(2) $100X = 910$

(3) $X = \dfrac{910}{100} = 9.1$

(4) The linseed meal contains 9.1% crude fiber on an as-fed basis.

b. Three samples of corn silage contain the following digestible energy (dry matter basis) and % dry matter levels:

Sample	DE, kcal/kg	% dry matter
A	1230	40
B	1225	36
C	1237	33

Calculate digestible energy (DE) kcal/kg, on an as-fed basis. The same equation is followed with nutrient concentration per weight unit (kcal/kg) used rather than percent. (See Section IV for a definition of DE.)

(1) Sample A:

(a) $\dfrac{X}{40} = \dfrac{1230}{100}$

(b) $100X = 49{,}200$

(c) $X = \dfrac{49{,}200}{100} = 492$

(2) Sample B:

(a) $\dfrac{X}{36} = \dfrac{1225}{100}$

(b) $100X = 44{,}100$

(c) $X = \dfrac{44{,}100}{100} = 441$

(3) Sample C:

(a) $\dfrac{X}{33} = \dfrac{1237}{100}$

(b) $100X = 40{,}821$

(c) $X = \dfrac{40{,}821}{100} = 408$

(4) Digestible energy levels for the three samples on an as-fed basis:

A = 492 kcal/kg

B = 441 kcal/kg

C = 408 kcal/kg

7. Convert the weight of ration ingredients from an as-fed to a dry matter basis.

   Problem: How many kilograms of ration dry matter are consumed daily if a steer is being fed the following amounts of as-fed feeds?

Corn silage	10.0 kg	(40% dry matter)
Corn grain	4.0 kg	(89% dry matter)
Supplement	0.5 kg	(92% dry matter)
Total	14.5 kg	as-fed ration

   The solution is obtained with the following equation:

   Parts dry matter feed = Parts as-fed feed × % Dry matter in feed
   (kg, lb, g etc.)

   Thus:

Feed	kg, as-fed	×	% dry matter	=	kg dry matter
Corn silage	10.0	×	.40	=	4.00
Corn grain	4.0	×	.89	=	3.56
Supplement	0.5	×	.92	=	0.46
	14.5 kg				8.02 kg

8. Convert the weight of ration ingredients from a dry matter basis to an as-fed basis.

   Problem: The following concentrate mixture is being fed to yearling horses. Feeds are presented as pounds of dry matter. Calculate the pounds of as-fed feeds in this diet.

Rolled oats	1045
Cracked corn	425
Soybean meal	182
Molasses (liquid)	80
Dicalcium phosphate	23
Vitamin-mineral premix	10
	1765 lb dry matter

   The solution for the above example can be obtained with the following equation:

   Parts (kg, lb, g etc.) as-fed feed = Parts dry matter feed ÷ % dry matter in feed thus:

Feed	lb, dry matter	÷	% dry matter	=	lb, as-fed
Rolled oats	1045	÷	.87	=	1201.1
Cracked corn	425	÷	.90	=	477.5
Soybean meal	182	÷	.91	=	200.0
Molasses (liquid)	80	÷	.75	=	106.7
Dicalcium phosphate	23	÷	.96	=	24.0
Vitamin-mineral premix	10	÷	1.00	=	10.0
	1765				2019.3

9. "Thumb rules" for converting to and from dry matter and as-fed
   a. When converting as-fed to dry matter
      (1) Nutrient concentration will increase.
      (2) Weight will decrease.
   b. When converting dry matter to as-fed
      (1) Nutrient concentration will decrease.
      (2) Weight will increase.

H. *A chemical or proximate analysis fails to give adequate information regarding digestibility, palatability, toxicity or nutritional adequacy. Thus, further steps need to be taken to evaluate a feed.*

II. Feeding Trial

A feeding trial simply gives an indication as to whether the animal will accept the feedstuff and the performance obtained from the feedstuff as compared to others. It tells nothing of why different results were obtained.

III. Digestion or Metabolism Trial

A. *Chemical analysis is the starting point for determining the nutritive value of feeds.*

But the actual value of ingested nutrients depends on the use the body can make of them. The first consideration here is digestibility, since undigested nutrients do not get into the body proper.

B. *A Digestion Trial Consists Chiefly of*
   1. Running a proximate analysis of feed.
   2. Feeding an animal a given amount of feed, or feeding at a constant rate.
   3. Collecting feces from given amount by use of a marker or collecting feces at a given time on a constant rate feeding.
   4. Running a proximate analysis of feces.
   5. The difference is the apparent digestible portion of the feed.
   6. Computed as follows

$$\text{Apparent digestibility (\%)} = \frac{\text{Nutrient intake} - \text{Nutrient in feces}}{\text{Nutrient intake}} \times 100$$

C. *Methods Employed for Fecal Collection*
   1. Marker-fed with the ration at beginning and end of the collection period
      a. Desirable properties of markers
         (1) Physiologically inert.
         (2) Contain no element under investigation.
         (3) Will not diffuse.
      b. Types of markers
         (1) Carmine.
         (2) Ferric oxide.
         (3) Chromic oxide.
         (4) Soot.
      c. Use of markers is not desirable in animals with larger and more complicated digestive tracts (ruminants).
      d. Using the marker method requires accurate measurement of the total amount of feed.

2. Indicator method—involves the use of an "inert reference substance" as an indicator
   a. Ideal specifications of indicators
      (1) Totally indigestible and unabsorbable.
      (2) Have no pharmacological action on the digestive tract.
      (3) Pass through the tract at a uniform rate.
      (4) Are readily determined chemically.
      (5) Preferably a natural constituent of the feed under test.
   b. By determining the ratio of the concentration of the reference substance to that of a given nutrient in the feed and the same ratio in the feces resulting from the feed, the apparent digestibility of the nutrient can be obtained without measuring either the feed intake or feces output.
   c. The calculation is made as follows

$$\text{Apparent digestibility} = 100 - \left[100 \times \frac{\% \text{ indicator in feed}}{\% \text{ indicator in feces}} \times \frac{\% \text{ nutrient in feces}}{\% \text{ nutrient in feed}} \right]$$

   d. Indicator materials
      (1) Chromic oxide.
      (2) Lignin.
      (3) Various naturally occurring "chromogen" compounds.
3. Metabolism or digestion stalls—used to confine the animal for quantitative collection of the feces uncontaminated by urine. An essential feature of these stalls is that the animal must have freedom of movement, particularly as regards lying down or getting up. Can also be designed to collect urine separately.
4. Feces collection bags—a bag on a harness arrangement attached to the animal in which feces and/or urine can be collected from animals grazing on pasture.
5. Digestion trials generally consist of two periods:
   a. Preliminary or adjustment period—to free digestive tract of any prior undigested feed and accustom animal to the facility.
   b. Collection period—time period in which measurement of feed and collection of feces occurs.
   c. For pigs, preliminary and collection periods of 3 to 5 days each are commonly used. For ruminants, these periods must be extended to 8 or 10 days.

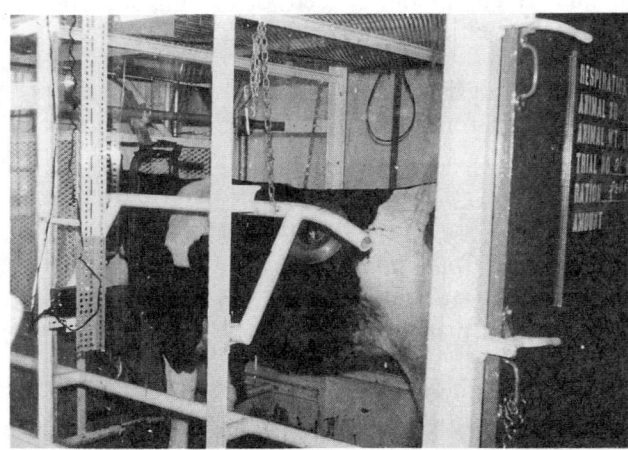

A holstein cow in a metabolism stall used for quantitative measurement of feed intake and collection of feces and urine. This cow is also fitted with a rumen fistula (opening) which can be opened and closed. This allows the collection of rumen fluid and the study of rumen microbe digestion.

D. *A Digestion Trial*
1. An example of the data obtained in a digestion trial is presented in Table 2–1.
2. In obtaining the data for intake of nutrients, the feed intake was multiplied by figures for its percentage composition as determined by chemical analysis. Similarly, the data for excreted nutrients were calculated and the digested nutrients obtained by subtraction. The final figures, expressed as percentages, are called digestion coefficients. The general formula for determination of the digestion coefficient may be written as:

Digestion coefficient =

$$\frac{(\text{Wt of feed} \times \% \text{ nutrient in feed}) - (\text{Wt of feces} \times \% \text{ nutrient in feces})}{\text{Wt of feed} \times \% \text{ nutrient in feed}} \times 100$$

In such a trial, several animals must be used and the results averaged to minimize the factor of individual variability.

3. The apparent digestibility of individual feeds may be determined insofar as they provide a satisfactory ration for the period of the test, when fed alone. The apparent digestibility of concentrates by ruminants cannot be determined in this way because they do not furnish sufficient bulk; their coefficients can be obtained only by difference. In this procedure the apparent digestibility of a roughage as a basal ration is first determined, and then the concentrate is added to the roughage for a second test. By a consideration of differences between the figures obtained from the roughage alone and for the combination, coefficients for digestibility of the concentrate are calculated.

4. Errors in a digestibility trial
   a. The nutrient digestibility is only apparent because the feces may contain portions of nutrients from sources other than from the consumed feed such as
      (1) Enzymes excreted into the digestive tract.
      (2) Nutrients of bacterial origin.
      (3) Nutrients (primarily N) from abraded intestinal mucosa.
      (4) Some mineral excretion into the digestive tract.
   b. Feed or fecal spillage and wastage.
   c. Errors in proximate analysis.

E. *Factors Affecting Digestibility Within Species*
   1. Age—small influence; could be associated with teeth etc.
   2. Disease and/or parasites (worms).

TABLE 2–1. Digestibility of Dried Grass by a Dairy Cow

Given:

50,800 g dry matter grass consumed
11,609 g dry matter feces excreted
Proximate analyses (dry matter basis):

	Grass	Feces		
Crude protein (CP), %	20.11	22.04		
Fiber, %	16.25	18.59		
NFE, %	40.99	34.82		
Ether extract, %	3.34	6.74		

Calculate:	CP	Fbr.	NFE	EE
Nutrients consumed (grass cons. × % anal.), g	10216	8255	20823	1697
Nutrients excreted (feces excr. × % anal.), g	2559	2158	4042	782
Nutrients digested (cons. - excr.), g	7657	6097	16781	915
Digestion coefficient (dig. ÷ cons. × 100), %	75.0	73.9	80.6	53.9

Source: JA Newlander and CH Jones, The digestibility of artificially dried grass, *Vermont Agr Expt Sta Bull, 348, 1932.*

3. Feed source and composition—feeds may vary in the type of fiber and/or ether extract material present and thus influence the digestibility.
4. Level of feed intake—as the level of feed intake increases above a certain value, the digestibility of all nutrients tends to decrease.
5. Rate of passage through intestinal tract
   a. Rapid—lack of time for complete digestive action and complete absorption.
   b. Slow—may be excessively subject to wasteful fermentations.
   c. Highly influenced by particle size and degree of feedstuff preparation.
6. Nutrient excess or deficiency—may cause poorer absorption and utilization of other nutrients.
7. Digestibility of a mixture is not necessarily the average of values for its constituents determined separately or indirectly.

IV. Measures of Feedstuff Energy

The term *feedstuff energy* is used to denote the value of feed for its primary function—to furnish energy for body processes and to form the nonnitrogenous, organic matter of tissues and secretions, functions in which all organic nutrients can take part. Total digestible nutrients (TDN), gross energy, digestible energy, metabolizable energy, net energy and roughage/concentrate ratio are all different measures of feed energy value.

Because these measures differ with the actual feeding represented, one should clearly understand their significance and know something of their advantages and limitations in practice.

A. *Total Digestible Nutrients (TDN)*
1. As a general measure of the nutritive value of a feed, digestion coefficients are used to compute its content of total digestible nutrients. The dried grass used in the digestion trial presented previously had the following composition by chemical analysis:
   Crude protein—20.11%
   Crude fiber—16.25%
   Nitrogen-free extract—40.99%
   Ether extract—3.34%
2. The digestible nutrients are obtained from these data by multiplying them by the digestion coefficients given in Table 2–2. The digestible fat is multiplied by the factor 2.25 in arriving at the figure 4.05 because of its higher energy value.
3. The principal limitation of the usefulness of TDN as a measure of feed energy is that it does not account for other important losses, such as urine, combustible gases in the case of herbivore and, most important, heat loss. These losses are relatively larger for roughages than for concentrates, and thus a pound of TDN in roughage has considerably less value for productive purposes than a pound in concentrates.

   The following illustrates the approximate relationships between TDN and net energy (NE):

   1 lb TDN from 1.2 lb corn $\cong$ 1.2 Mcal NE
   1 lb TDN from 2.1 lb better hay $\cong$ 1.0 Mcal NE
   1 lb TDN from 2.4 lb poor hay $\cong$ 0.8 Mcal NE

   In general, the TDN values for roughages consistently overestimate the usable energy of such feeds by ruminant animals.

TABLE 2–2. Calculating Total Digestible Nutrient Value of Dried Grass Fed to a Dairy Cow

Nutrient	Chemical Analysis of Nutrients, %	Digestion × Coefficients, % =	Digestible Nutrients, %	Calculated TDN, %
Protein	20.11	75.0	15.08	15.08
Fiber	16.25	73.9	12.01	12.01
Nitrogen-free extract	40.99	80.6	33.04	33.04
Ether extract	3.34	53.9	1.80 (×2.25) =	4.05
			Total digestible nutrients =	64.18

4. Total digestible nutrients is expressed as a percentage of the ration or in units of weight (lb or kg) and not as an actual caloric figure. Conversion may be made for comparison assuming:

    1 lb TDN = 2000 kcal digestible energy or 1640 kcal metabolizable energy

B. *Partition of Feedstuff Energy in Digestion and Metabolism*

   1. The energy unit
      a. Calorie (cal)—that amount of heat required to raise the temperature of 1 g of water from 14.5 to 15.5°C.
      b. Kilocalorie (kcal)—1000 small calories.
      c. Megacalorie (Mcal)—1000 kcal or 1,000,000 cal; sometimes called a therm.
   2. Figure 2–2 shows the interrelationship among the fractions of feed energy during their utilization by the animal.
      a. Gross energy (GE)—the total potential energy of a feedstuff consumed. It is determined in a bomb calorimeter, which determines the amount of heat (measured in calories) that is released when a feedstuff is completely oxidized. A known amount of substance is sealed in the bomb compartment and 20 to 30 atmospheres of oxygen are added. The contents are ignited with an electrical spark and the temperature rise of water bathing the bomb is measured.
      b. Fecal energy (FE)—during digestion and absorption, gross energy is broken down. Part of it escapes the body via the large intestine in the form of undigested food residues and the energy-yielding metabolic products that have originated in the digestive system. These latter include abraded and sloughed mucosa cells, principally the worn-out absorptive cells of the villi of the intestine, the microflora residue that has escaped digestion by the host, and digestive enzymes that have escaped reabsorption following their excretion into the lumen of the gut, where they have helped prepare the food residues for absorption. Fecal energy is readily determined by collection of the solid fecal residue and use of the bomb calorimeter, which measures its gross energy.
      c. Digestible energy (DE)—computed as GE − FE. The measurement of DE takes account of digestible losses and is subject to the same variables that affect digestion and the same additional losses in metabolism as have been mentioned for TDN.

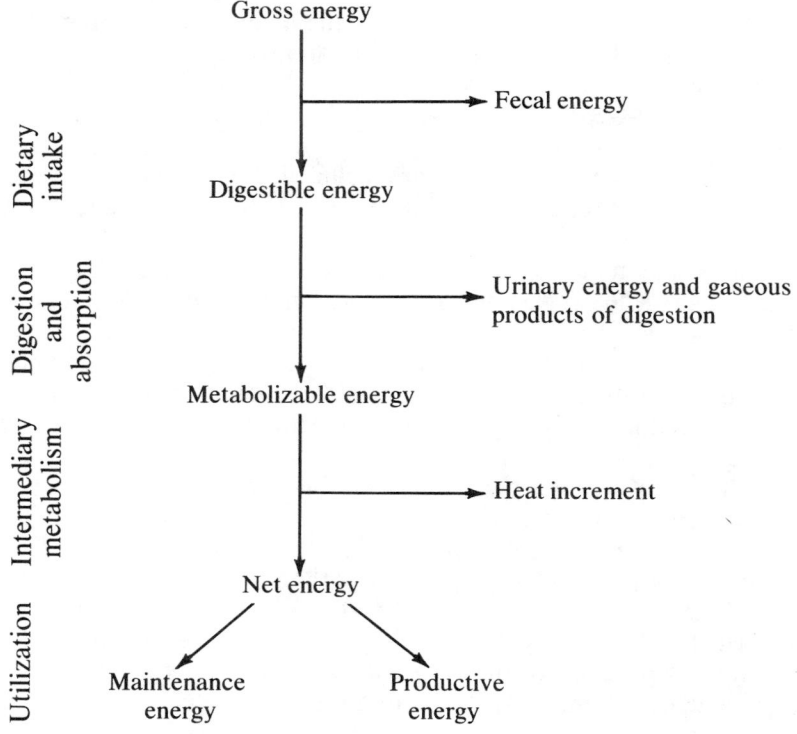

FIGURE 2–2. Feed energy fractions.

d. Gasseous products of digestion (GPD)—these include combustible gases that escape the body during the digestion and absorption process. The gases mainly consist of methane with traces of H, CO, H_2S, acetone and ethane. While fecal residues are fairly simply recovered, these gases frequently escape and not all the unused energy is recorded. Gaseous products of digestion are important primarily in herbivore or ruminant animals as very few are produced in swine. The measurement of GPD is difficult and is expensive in both equipment and technical labor.

e. Urinary energy (UE)—intermediary metabolism can be roughly defined as the series of changes that absorbed materials undergo in the course of the synthesis of tissues needed by the body, including enzymes, and the molecular changes necessary to release the potential energy of the molecules and capture it in forms that can be used to "drive" the metabolic machinery. Those portions of the absorbed nutrients that have served their useful purpose must be discarded, and the exit channel is the urine. Some of this material is directly of food origin, such as the nitrogenous components of amino acids; some of it is of endogenous origin, arising from the catabolism of some of the tissues in the body.

f. Metabolizable energy (ME)—sometimes referred to as the usable portion of the ingested energy, normally computed as DE − UE − GPD.

In accounting for losses in the urine and, in the case of ruminants, the combustible gases, as well as fecal losses, ME represents a step beyond TDN or DE as a measure of nutritive value. In poultry ME is easier to determine than DE because the feces and urine are voided together, so these losses can be determined in one step. In pigs, the feces and urine are determined separately. The same procedure is required for ruminants, plus the measurement of the combustible gases. The additional losses in arriving at ME are smaller than those in digestion. This is particularly true in the pig, where the urine loss is the only additional one. In ruminants, the urine losses are somewhat greater for roughages than for concentrates, but in both cases they are exceeded by losses of the combustible gases, the combined loss thus greatly exceeding that in nonruminants.

g. Heat increment (HI)—the increase in heat production following consumption of feed when the animal is in a thermally neutral environment. It consists of increased heats of fermentation (HF is the heat produced in the digestive tract as a result of microbial action) and heats of nutrient metabolism (HNM is the heat produced in intermediary metabolism as a result of using absorbed nutrients). The energy of the heat increment is wasted except when the temperature of the environment is below the animal's critical temperature. This heat may then be used to help keep the body warm, and so becomes part of the net energy required for maintenance.

h. Net energy (NE)—computed as NE = ME − HI. It includes the amount of energy used either for maintenance only or for maintenance plus production. When reporting NE, one must state clearly which functions are included. For example, there may be NE values for maintenance plus production (NEm + p), NE for maintenance only (NEm) or NE for production only (NEp).

(1) NEm—fraction of total NE expended to keep the animal in energy equilibrium. In this state, there is no net gain or loss of energy in the body tissue. Net energy for maintenance might include

(a) Basal metabolism—BM is the chemical change that takes place in the cells of an animal in the fasting and resting state when it uses just enough energy to maintain vital cellular activity, respiration and circulation as measured by the basal metabolic rate.

(b) Energy of voluntary activity—VAE is the amount of energy needed in getting up, standing, moving about to obtain food, grazing, drinking, lying down etc.

(c) Heat to keep body warm—HBW is the additional heat needed to keep the animal's body warm when the temperature of the environment is below its critical temperature. The HI and HF, in total or in part can be used for keeping the animal warm.

(d) Heat to keep body cool—HBC is the extra energy expended by the animal when the temperature of the environment is above its zone of thermal neutrality. This includes panting, respiration rate, heart rate etc.

(2) NEp—fraction of total NE required (in addition to that needed for body maintenance) for involuntary work, for tissue gain (growth or fat production) or for the synthesis of a fetus, milk, eggs, wool, fur, feathers etc.

3. Energy in beef cattle nutrition

One of the problems facing nutritionists has been to find a system that accurately expresses the energy requirements of beef cattle. Various systems have been proposed and utilized, but after a period of usage the shortcomings have become obvious.

A system utilizing net energy developed by Lofgreen and Garrett (1968, *J of Anim Science,* 27:793) has been recognized as a milestone in beef nutrition. The system can be used to determine the expected daily body weight gain in cattle fed a given ration under average conditions or to determine the amount of feed needed to produce a desired amount of gain. Keep in mind that these values were developed for medium frame cattle in a thermoneutral environment. Adjustments for adverse weather are warranted under certain conditions. Presently, various agricultural experimental stations are determining data that can be used to calculate rations in their areas. Larger-framed cattle at a given weight require less energy for a given amount of gain than smaller-framed cattle. This is because muscle is a more efficient tissue for growth than fat. Tables 2–3 and 2–4 assume medium frame (70% Choice at 1100 lb for steers; 950 lbs for heifers) cattle. A simple thumb rule for adjusting for frame size is to decrease the NEg necessary for a given gain by the ratio of

Estimated weight at 70% Choice/1100 lb, steers or 950 lb, heifers

For example, the NEg required for a 900-lb medium frame steer to gain 3 lbs per day is about 7 Mcals. A larger-framed group of steers (70% Choice at 1200 lbs) would require

$(1100/1200 = .92)$, $.92 \times 7 = 6.44$ Mcals

to gain 3 lbs/day. More specific guidelines for adjusting larger-framed beef cattle are given in the beef NRC (1984) publication.

Some net energy requirements and the net energy values for various feedstuffs are stated in Tables 2–3, 2–4, and 2–5. A more complete listing of feed values is found in Table 3–2, and a complete beef requirement listing is found in the tables of Chapter 8. The use of the net energy procedure is based on the fact that feeds given feedlot beef cattle have different fuel values

TABLE 2–3. NE Requirements for Growing and Finishing Steers (Medium Frame)*

Daily Gain lb	Body Weight, lb							
	300	400	500	600	700	800	900	1000
				NEm required, Mcal/day				
0.0	3.07	3.81	4.50	5.16	5.80	6.41	7.00	7.58
				NEg required, Mcal/day				
0.6	0.53	0.66	0.78	0.90	1.00	1.11	1.21	1.31
0.8	0.73	0.91	1.07	1.23	1.38	1.52	1.66	1.80
1.0	0.93	1.16	1.37	1.57	1.76	1.94	2.12	2.30
1.2	1.14	1.41	1.67	1.91	2.15	2.38	2.59	2.81
1.4	1.35	1.67	1.98	2.27	2.55	2.81	3.07	3.33
1.6	1.56	1.94	2.29	2.62	2.95	3.26	3.56	3.85
1.8	1.78	2.20	2.61	2.99	3.35	3.71	4.05	4.38
2.0	1.99	2.47	2.92	3.35	3.76	4.16	4.54	4.92
2.2	2.12	2.75	3.25	3.72	4.18	4.62	5.05	5.46
2.4	2.44	3.02	3.57	4.10	4.60	5.08	5.55	6.01
2.6	2.66	3.30	3.90	4.47	5.02	5.55	6.06	6.56
2.8	2.88	3.58	4.23	4.85	5.44	6.02	6.57	7.11
3.0	3.11	3.86	4.56	5.23	5.87	6.49	7.09	7.67
3.2	3.34	4.14	4.90	5.61	6.30	6.97	7.61	8.24
3.4	3.57	4.43	5.23	6.00	6.74	7.45	8.13	8.80
3.6	3.80	4.71	5.57	6.39	7.17	7.93	8.66	9.37

*Adapted from NRC *Nutrient Requirements of Beef Cattle* (1984) tables. Medium frame cattle are assumed 70% Choice at approximately 1100 lb for steers and 950 lb for heifers.

TABLE 2–4. NE Requirements for Growing and Finishing Heifers (Medium Frame)*

Daily Gain lb	Body Weight, lb							
	300	400	500	600	700	800	900	1000
	NEm required, Mcal/day							
0.0	3.07	3.81	4.50	5.16	5.80	6.41	7.00	7.58
	NEg required, Mcal/day							
0.6	0.64	0.79	0.93	1.07	1.20	1.33	1.45	1.57
0.8	0.88	1.09	1.29	1.48	1.66	1.83	2.00	2.17
1.0	1.13	1.40	1.65	1.90	2.13	2.35	2.57	2.78
1.2	1.38	1.72	2.03	2.33	2.61	2.89	3.15	3.41
1.4	1.64	2.04	2.41	2.76	3.10	3.43	3.75	4.06
1.6	1.91	2.37	2.80	3.21	3.60	3.98	4.35	4.71
1.8	2.18	2.70	3.19	3.66	4.11	4.54	4.96	5.37
2.0	2.45	3.04	3.59	4.12	4.63	5.11	5.59	6.04
2.2	2.73	3.38	4.00	4.58	5.15	5.69	6.21	6.72
2.4	3.00	3.73	4.41	5.05	5.67	6.27	6.85	7.41
2.6	3.29	4.08	4.82	5.53	6.20	6.86	7.49	8.11
2.8	3.57	4.43	5.24	6.00	6.74	7.45	8.14	8.81
3.0	3.86	4.79	5.66	6.49	7.28	8.05	8.79	9.51
3.2	4.15	5.14	6.08	6.97	7.83	8.65	9.45	10.23
3.4	4.44	5.51	6.51	7.46	8.38	9.26	10.11	10.95
3.6	4.73	5.87	6.94	7.95	8.93	9.87	10.78	11.67

*Adapted from NRC *Nutrient Requirements of Beef Cattle* (1984) tables. Medium frame cattle are assumed 70% Choice at approximately 1100 lb for steers and 950 lb for heifers.

TABLE 2–5. NE Content of Certain Feeds Expressed as Mcal/lb on an As-fed Basis

Feed (IFN)	For Growing Beef Cattle		For Lactating Dairy Cattle
	Maintenance NEm	Gain NEg	NF$_L$
Alfalfa, aerial part, dehydrated meal, min 17% protein (1-00-023)	0.58	0.33	0.57
Alfalfa, hay, sun-cured, prebloom (1-00-054)	0.57	0.34	0.58
Alfalfa, hay sun-cured, midbloom (1-00-063)	0.53	0.30	0.52
Alfalfa, aerial part, grazed (2-00-196)	0.16	0.09	0.16
Alfalfa, ensiled, early bloom, min 30% max 50% dry matter (3-08-150)	0.21	0.10	0.23
Alfalfa-Brome, Smooth, aerial part, grazed, early bloom (2-00-262)	0.14	0.08	0.14
Alfalfa-Brome, Smooth, aerial part, ensiled, min 30% max 50% d.m. (3-08-147)	0.20	0.10	0.21
Alfalfa-Orchardgrass, aerial part, ensiled, min 30% max 50% d.m. (3-08-144)	0.18	0.09	0.20
Animal-Poultry, fat, heat rendered, min 90% FA (4-00-409)	2.66	1.99	2.04
Barley, hay, sun-cured (1-00-495)	0.45	0.22	0.45
Barley, straw (1-00-498)	0.29	0.07	0.41
Barley, grain (4-00-549)	0.79	0.53	0.82
Beet, Sugar, aerial part with crowns, ensiled (3-00-660)	0.13	0.06	0.14
Beet, Sugar, molasses (4-00-668)	0.70	0.47	0.65
Beet, Sugar, pulp, dehydrated (4-00-669)	0.73	0.47	0.70
Bermudagrass, Coastal, hay, sun-cured (1-00-716)	0.31	0.09	0.53
Bluegrass, Kentucky, aerial part, grazed, early vegetative (2-00-777)	0.24	0.15	0.23
Brome, hay, sun-cured (1-00-947)	0.54	0.30	0.53
Brome, aerial part, grazed, early vegetative (2-00-963)	0.17	0.10	0.17
Canarygrass, Reed, hay, sun-cured (1-01-104)	0.46	0.24	0.50
Canarygrass, Reed, aerial part, grazed (2-01-113)	0.15	0.09	0.15
Cattle, milk, skimmed dehydrated (5-01-175)	0.69	0.43	0.83
Citrus, pulp without fines (4-01-237)	0.74	0.48	0.78
Clover, Alsike, hay, sun-cured (1-01-313)	0.49	0.26	0.51
Clover, Ladino, hay, sun-cured (1-01-378)	0.58	0.34	0.58

TABLE 2–5 (continued). NE Content of Certain Feeds Expressed as Mcal/lb on an As-fed Basis

Feed (IFN)	For Growing Beef Cattle		For Lactating Dairy Cattle
	Maintenance NEm	Gain NEg	NF$_L$
Clover, Red, hay, sun-cured (1-01-415)	0.50	0.28	0.53
Clover, Red, aerial part, grazed, early bloom (2-01-428)	0.15	0.09	0.14
Corn, aerial part, with ears and husks, sun-cured (1-02-772)	0.59	0.37	0.58
Corn, aerial part without ears or husks, sun-cured, mature (1-02-776)	0.51	0.29	0.52
Corn, cobs, meal (1-02-782)	0.39	0.17	0.44
Corn, aerial part, with ears and husks, ensiled, mature, well-eared, max 50% min 30% dry matter (3-02-823)	0.25	0.16	0.24
Corn, aerial part without ears or husks, ensiled (3-02-836)	0.15	0.08	0.15
Corn, ears, meal (4-02-849)	0.86	0.60	0.76
Corn Grain, gluten, meal (5-02-900)	0.84	0.57	0.80
Corn Grain, gluten with bran, dehydrated (5-02-903)	0.82	0.55	0.78
Corn, Dent Yellow, grain, gr 2 US (4-02-931)	0.89	0.61	0.84
Cotton, hulls (1-01-599)	0.25	0.04	0.39
Cotton, seeds, meal, mechanical-extracted, min 41% protein (5-01-617)	0.73	0.46	0.76
Cotton, seeds, meal, solvent-extracted, min 41% protein (5-01-621)	0.74	0.48	0.73
Cowpea, hay, sun-cured (1-01-645)	0.58	0.34	0.56
Fescue, Meadow, hay sun-cured (1-01-912)	0.54	0.32	0.55
Fish, Menhaden, meal, extracted (5-02-009)	0.75	0.49	0.72
Flax, seeds, meal, mechanical-extracted (5-02-045)	0.81	0.54	0.78
Flax, seeds, meal, solvent-extracted (5-02-048)	0.73	0.49	0.72
Grains Screening's, brewer's grains, dehydrated (5-02-141)	0.64	0.39	0.67
Lespedeza, hay, sun-cured, early bloom (1-02-607)	0.50	0.26	0.52
Native Plants, Midwest, hay, sun-cured, midbloom (1-07-956)	0.47	0.13	0.44
Oats, hay, sun-cured (1-03-280)	0.57	0.33	0.55
Oats, straw (1-03-283)	0.43	0.20	0.48
Oats, aerial part, ensiled (3-03-298)	0.17	0.09	0.18
Oats, cereal byproduct (4-03-303)	0.95	0.67	0.89
Oats, grain (4-03-309)	0.80	0.54	0.70
Oats, groats (4-03-331)	1.02	0.72	0.88
Orchardgrass, hay, sun-cured (1-03-438)	0.50	0.27	0.52
Orchardgrass, aerial part, grazed, early bloom (2-03-442)	0.16	0.10	0.16
Peanut, kernels, meal, solvent-extracted (5-03-650)	0.78	0.51	0.74
Rape, seeds, meal, solvent-extracted (5-03-871)	0.66	0.41	0.65
Redtop, aerial part, grazed, full bloom (2-03-891)	0.16	0.09	0.17
Rice, bran with germ (4-03-928)	0.62	0.38	0.54
Rye, straw (1-04-007)	0.25	0.03	0.35
Rye, grain (4-04-047)	0.80	0.54	0.74
Ryegrass, Italian, aerial part, grazed (2-04-073)	0.14	0.08	0.15
Safflower, seeds, meal, mechanical-extracted (5-04-109)	0.50	0.27	0.50
Safflower, seeds, meal, solvent-extracted (5-04-110)	0.47	0.24	0.53
Safflower, seeds without hulls, meal, solvent-extracted (5-07-959)	0.75	0.49	0.72
Sesame, seeds, meal, mechanical-extracted (5-04-220)	0.75	0.49	0.72
Sorghum, Grain Variety, aerial part without heads, sun-cured (1-07-961)	0.38	0.16	0.44
Sorghum, Grain Variety, grain (4-04-383)	0.54	0.30	0.55
Sorghum, Johnsongrass, hay, sun-cured (1-04-407)	0.48	0.25	0.51
Sorghum, Sudangrass, hay, sun-cured (1-04-480)	0.49	0.25	0.52
Sorghum, Sudangrass, aerial part, grazed, early vegetative (2-04-484)	0.13	0.08	0.13
Sorghum, Sudangrass, aerial part, ensiled (3-04-499)	0.13	0.07	0.14
Soybean, hay, sun-cured (1-04-558)	0.52	0.29	0.55
Soybean, straw (1-04-567)	0.21	0.00	0.32

TABLE 2–5 (continued). NE Content of Certain Feeds Expressed as
Mcal/lb on an As-fed Basis

Feed (IFN)	For Growing Beef Cattle		For Lactating Dairy Cattle
	Maintenance NEm	Gain NEg	NF$_L$
Soybean, aerial part, ensiled (3-04-581)	0.15	0.07	0.16
Soybean, seeds, meal, mechanical-extracted (5-04-600)	0.86	0.59	0.81
Soybean, seeds, meal, solvent-extracted (5-04-604)	0.84	0.57	0.79
Sugarcane, molasses, dehydrated (4-04-695)	0.70	0.44	0.69
Sugarcane, molasses (4-04-696)	0.77	0.54	0.64
Sunflower, seeds without hulls, meal, solvent-extracted (5-04-739)	0.62	0.37	0.62
Sweet Clover, hay, sun-cured (1-04-754)	0.49	0.26	0.51
Timothy, hay, sun-cured, early bloom (1-04-882)	0.52	0.29	0.54
Timothy, hay, sun-cured, late bloom (1-04-885)	0.49	0.26	0.52
Timothy, aerial part, grazed, late vegetative (2-04-903)	0.19	0.11	0.19
Timothy, aerial part, ensiled (3-04-922)	0.20	0.11	0.20
Trefoil, Birdsfoot, hay, sun-cured (1-05-044)	0.55	0.31	0.57
Trefoil, Birdsfoot, aerial part, grazed (2-20-786)	0.16	0.10	0.15
Vetch, hay, sun-cured (1-05-106)	0.55	0.32	0.56
Wheat, straw (1-05-175)	0.36	0.14	0.40
Wheat, aerial part, grazed, early vegetative (2-05-176)	0.19	0.13	0.18
Wheat, bran (4-05-190)	0.65	0.41	0.64
Wheat, middlings (4-05-205)	0.79	0.52	0.79
Wheat, grain (4-05-211)	0.88	0.60	0.81
Wheatgrass, Crested, hay, sun-cured (1-05-418)	0.48	0.24	0.52
Wheatgrass, Crested, aerial part, grazed, early vegetative (2-05-429)	0.29	0.17	0.29
Yeast, brewer's, dehydrated (7-05-527)	0.81	0.54	0.77
Yeast, Torula, dehydrated (7-05-534)	0.82	0.55	0.78

depending on whether they are being used for the maintenance component (NEm) or the production (gain) component (NEg) of the total energy requirement. Further, a given feed has a much higher value in a maintenance ration than in a finishing ration. Therefore, evaluating feed on this value for a given purpose is of great economic importance.

The procedure also considers the relationship between the type of tissue being produced (relative fat and lean production) and the energy requirement.

The following examples are developed to illustrate how this technique can be used by the cattle feeder.

a. Table 2–6 shows a typical beef finishing diet with calculations to show the development of net energy values for maintenance (NEm) and gain (NEg).

b. After the NEm and NEg values have been determined for a diet, the information can be used to estimate the amount of feed required to support a given level of daily gain. In Table 2–7 the objective is to predict the amount of feed required to produce a 2.8-lb daily gain for an 800-lb medium frame steer being finished in drylot.

c. This procedure is also used to predict rate of gain and cost of gain at different levels of ration intake. Table 2–8 illustrates an example predicting rate and cost of gain for 800-lb steers consuming 15, 20 or 25 pounds of the diet shown in Table 2–6.

4. Energy in dairy cattle nutrition (See also Chapter 9)

Both TDN and DE have been criticized as measures of the useful energy value of forages and concentrates for dairy cattle. As previously described for beef cattle, net energy (NE) is a more scientifically sound system for expressing energy requirements and the energy value of feeds. The NE value of a feed, however, depends on whether it is used for maintenance, fattening, growth or milk production.

Two systems of expressing net energy have been used. The first system (Lofgreen and Garrett, 1968, *J Anim Sci,* 27:793), previously described for beef cattle, expresses the requirements for

TABLE 2–6. Beef Finishing diet with Calculations for NE

Ingredients	% of lb/100 lb	NEm Mcal*	NEg Mcal
Rolled corn (4-02-931)	60.85	54.16	37.12
Cottonseed meal (5-01-621)	8.00	5.92	3.84
Dehydrated alfalfa meal (1-00-023)	6.00	3.48	1.98
Molasses (4-04-695)	3.00	2.10	1.32
Urea	1.00	—	—
Cottonseed hulls (1-01-599)	20.00	5.00	0.80
Salt	0.50	—	—
Limestone (6-02-632)	0.60	—	—
Vitamin A premix	0.03	—	—
Trace mineral premix	0.02	—	—
TOTAL	100.00	70.66	45.06
Mcal/100 lb		70.66	45.06
Mcal/lb		0.71	0.45

*These values are derived by multiplying the level of ingredient times the respective NEm or NEg values for that ingredient. For example, the diet contains 60.85 lb of corn per 100 lb. The NEm value for corn is 0.89 Mcal/lb and the NEg value for corn is 0.61 Mcal/lb. Thus, 60.85 × 0.61 = 37.12. Values used are from Table 2–5.

TABLE 2–7. Calculation of Daily Feed Required for NE Needs of a Finishing Steer (Medium Frame)

Item	Maintenance Component		Gain Component
Net energy required, Mcal/day	6.41[1]		6.02[2]
Quantity of ration required for maintenance (6.41/.71), lb	9.03[3]		—
Quantity of ration required for gain (6.02/.45), lb	—		13.38[3]
Total daily ration, lb (maintenance and gain)		22.41	
Feed/pound of gain, lb (22.41/2.8)		8.00	
Feed cost/100 lb of gain[4]		40.00	

1. Obtained from Table 2–3 listing NEm requirement of steers with body weight of 800 lb.
2. Obtained from Table 2–3 listing NEg requirements for cattle of different weights.
3. The example diet in Table 2–6 is used in this example.
4. Assume feed cost of 5.0 cents per pound (800.0 lb feed × $.05/lb = $40.00)

TABLE 2–8. Predicting Rate and Cost of Gain in Beef Steers (Medium Frame) Fed at Various Levels of Intake

Daily Ration Intake (lb)	Quantity of Ration Required for Maint. (lb)	Quantity of Ration Remaining for Gain (lb)	NE Available for Gain (Mcal)	Predicted Daily Gain (lb)	Feed per Pound of Gain (lb)	Feed Cost per 100 lb of Gain
15	9.03[1]	5.97[2]	2.69[3]	1.31[4]	11.45[5]	57.25[6]
20	9.03	10.97	4.94	2.31	8.66	43.30
25	9.03	15.97	7.19	3.30	7.58	37.90

1. NEm requirement/NEm value of ration or 6.41/.71.
2. Daily feed intake minus daily feed required for maintenance.
3. Ration remaining for gain × NEg value of ration.
4. Obtained from Table 2–3 listing NE requirements for cattle of different weights.
5. Daily feed/daily gain.
6. Feed/100 lb gain × 5.0 cents per pound.

specific physiological functions, such as net energy for maintenance (NEm) and net energy for gain (NEg). With this system, there are two values in table 2–3 and 2–4 (or tables 8–4 and 9–4) on requirements and two values in tables 2–5 and 3–2 on feed composition. Calculating a diet by this system is similar to that outlined for beef cattle.

The second system (Moe and Flatt, 1969, *J Dairy Sci,* 52:928) adjusts the net energy requirements for the various physiological functions by a regression equation that equates net energy for other functions to that for milk production. Thus, one feed value is obtained that can be used for maintenance, pregnancy and milk production. It is designated as net energy for lactating dairy cows (NE$_L$) in Tables 2–5, 3–2, and 9–5 since the value is used for all physiological functions.

Calorimetric studies (Armstrong et al., 1964, *J Agr Sci,* 62:417; Flatt et al., 1965, *Proc Third Symposium on Energy Metabolism,* Troon, Scotland, European Assoc Anim Prod Publ, 11:121; Moe and Flatt, 1969, *J Dairy Sci,* 52:928) have shown that digestible or metabolizable energy is used with different degrees of efficiency for maintenance and body gain in nonlactating animals but is used with similar degrees of efficiency for maintenance and milk production in lactating animals. For this reason, three net energy values are included in the tables on feed composition (2–5 and 3–2). The net energy value for maintenance (NEm) is the net energy value of feeds for the maintenance of nonlactating animals. The net energy for gain (NEg) is the net energy value of feeds for the deposition of body gain in nonlactating animals. The net energy for lactation (NE$_L$) is the net energy value of feeds for maintenance, gain, pregnancy and milk production of lactating cows.

Since energy is used with different degrees of efficiency for maintenance and body gain in nonlactating animals, both NEm and NEg are used to express the total energy needs for growing dairy heifers and bulls (Table 9–4) or for finishing steers or heifers (tables 2–3, 2–4, and 8–4). The amount of NEm required for maintenance is based on body size of the animal; the amount of NEg required is based on the desired rate of gain.

Since energy is used with similar degrees of efficiency for maintenance and milk production in lactating animals, a single net energy value of feeds (NE$_L$) is adequate to calculate rations for both maintenance and milk production. An example of how to calculate a diet by this system is outlined in Chapter 9.

5. Net energy for sheep (See also Chapter 10)

   A comparative slaughter technique similar to that used by Lofgreen and Garrett (1968) for growing and finishing beef cattle was used by Rattray et al. (1973) to establish the net energy requirements for growing and finishing lambs and yearlings. The NEm and NEg requirements for different classes of sheep are presented in Table 10–3. The net energy values of feeds presented for cattle in Table 3–2 can also be used for sheep.

6. Energy for swine and poultry

   a. Swine

      (1) At present, NE requirements for maintenance and production are not available, and, therefore, the energy requirements of swine and the energy values of feedstuffs commonly fed to swine are presented as DE and ME values.

      (2) The ME value of a swine feed, diet concentration or daily allowance can be estimated from the DE value using the following formula

      $$ME = \frac{DE \times 96 - (0.202 \times \text{crude protein \%})}{100}$$

   b. Poultry excrete feces and urine via a cloaca, thus making it difficult to measure digestibility. As a consequence, DE values are not generally employed in poultry feed formulation. For poultry the gaseous products are usually negligible so that ME represents the gross energy of a feed minus the gross energy of the excreta. A correction for nitrogen retained in the body is frequently applied to yield an MEn value. No absolute NE values are available.

   c. Growing swine and poultry tend to eat to satisfy their energy requirements if fed ad libitum. It is accordingly possible, within limits, to regulate the intake of all nutrients, except water, by including them in the diet in a specific ratio to the amount of available energy in the diet. The importance of considering the energy content of the diet in order to promote the desired intake of all nutrients is therefore of obvious importance. This expression of nutrients as weight per unit of energy is called the nutrient/calorie ratio. For example, it is suggested that the protein/calorie ratio should be expressed as grams of protein per 1000 kcal metabolizable energy (g protein/1000 kcal ME or mg protein/kcal ME).

Example:

Calculate the protein/calorie ratio of a poultry diet containing 20% CP and 3200 kcal/kg ME.

(1) 1 kg diet = 1000 g @ 20% CP and 3200 kcal ME

(2) 1000 g diet × .20 (%CP) = 200.0 g CP

(3) Thus: $\dfrac{200.0 \text{ g CP}}{3200 \text{ kcal}} = \dfrac{X \text{ g CP}}{1000 \text{ kcal ME}}$

$$3200 \text{ X} = 200,000$$
$$\text{X} = 62.5$$

(4) Answer: 62.5 g CP/1000 kcal ME

C. *Roughage/Concentrate Ratio*

1. This is still another way of expressing the approximate energy level of a ration. Approximate energy levels at different dietary roughage and concentrate percents are

	Low Energy		High Energy	
% roughage	100	65	35	0
% concentrate	0	35	65	10

2. An example of how to calculate R/C ratio is shown below. The procedure expresses ratio of high and low energy feeds—assuming concentrates are high energy and roughages are low energy.

Daily Beef Cattle Ration		Concentrate	Roughage
Corn	12.0 lb	12.0	—
Supplement	1.5 lb	1.5	—
Hay	5.0 lb	—	5.0
Total	18.5 lb	13.5	5.0

Thus: % C = 13.5/18.5 = 73
% R = 5.0/18.5 = 27
R/C Ratio = 27/73

V. Physical Evaluation of Feedstuffs

A. *Eye Appraisal of Feeds—Used Mostly with Forages*
  1. Type of forage—% legume, etc.
  2. Color of forage.
  3. Percent leaves on forage.
  4. Contamination with weeds etc.
  5. Spoilage.

B. *Palatability—Based on Eye Appraisal, Smell etc.*

C. *Factors That Might Influence the Value of a Feedstuff*
  1. Soil fertility.
  2. Growing conditions.
  3. Harvesting.
    a. Stage of maturity.
    b. Losses.
  4. Processing and storage.

VI. Evaluating Feeds on the Basis of Cost per Unit of Nutrient

A. *General Formula*

$$\text{Cost per unit of nutrient} = \frac{\text{Cost of feed per unit of wt}}{\text{Unit of wt} \times \% \text{ nutrient concentration}}$$

B. *Cost per Pound of Crude Protein*
   Examples:
   1. Soybean meal @ $180/ton (44% CP)

   $$\frac{\$180}{2000 \times .44} = \frac{\$180}{880} = \$0.204/\text{lb of CP}$$

   2. Cottonseed meal @ $175/ton (41% CP)

   $$\frac{\$175}{2000 \times .41} = \frac{\$175}{820} = \$0.213/\text{lb of CP}$$

C. *Cost per Pound of TDN*
   Examples:
   1. Alfalfa hay @ $70/ton (51% TDN)

   $$\frac{\$70}{2000 \times .51} = \frac{\$70}{1020} = \$0.069/\text{lb TDN}$$

   2. Milo grain @ $4.00/100 (77% TDN)

   $$\frac{\$4.00}{100 \times .77} = \frac{\$4.00}{77} = \$0.052/\text{lb TDN}$$

D. *Cost per Unit of Net Energy for Lactation (NE$_L$)*
   Examples:
   1. Corn @ $3.50/100 lb (.84 Mcal NE$_L$/lb)

   $$\frac{\$3.50}{100 \times .84} = \frac{\$3.50}{84} = \$0.042/\text{Mcal}$$

   2. Alfalfa hay @ $70/ton (.58 Mcal NE$_L$/lb)

   $$\frac{\$70}{2000 \times .58} = \frac{\$70}{1160} = \$0.060 \text{ Mcal}$$

   3. Oat straw @ $50/ton (.48 Mcal NE$_L$/lb)

   $$\frac{\$50}{2000 \times .48} = \frac{\$50}{960} = \$0.052/\text{Mcal}$$

E. This same approach can be used to calculate the cost per unit of elements in mineral sources, cost per unit of vitamins in vitamin concentrates, etc. In some cases the cost per unit of available nutrient is more appropriate for comparison if an attempt is being made to purchase the nutrient at lowest cost.

Study Questions and Problems
Chapter 2

I. True or False Questions

1. To determine the digestibility of protein in a feedstuff fed an animal, you must measure the protein in the urine.
2. One kcal is equivalent to the amount of heat required to raise the temperature of 1000 g of water from 14.5 to 15.5°C.
3. In feedlot cattle, heifers are less efficient at utilizing net energy for maintenance (NE_m) than are steers of the same body weight.
4. Metabolizable energy (ME) is used with similar degrees of efficiency for maintenance and milk production in lactating cattle.
5. Total digestible nutrients (TDN) tends to underestimate the usable energy value of forages, especially low-grade forages.
6. The gross energy value of a feed is a good indication of its TDN value.
7. The detergent analysis scheme allows the separate determination of lignin.
8. The term nitrogen-free extract (NFE) refers to the portion of carbohydrate that is poorly digested.
9. The heat increment (HI) of a feed serves no worthwhile function in livestock production.
10. It is possible for a feed to contain more than 100% TDN.

II. Multiple Choice (Select the single best answer)

1. In the proximate analysis of a feedstuff, the protein content is calculated
   a. From amino acids.
   b. By direct protein determination.
   c. From nitrogen content.
   d. From ether extract (EE).
   e. From nitrogen-free extract (NFE).
2. A feed source containing 4% nitrogen would contain ___?___ % protein by the proximate analysis system.
   a. 4
   b. 21.5
   c. 25
   d. 26
   e. Cannot be calculated without further data
3. Which of the following terms equals the calories of heat released when a substance is completely oxidized in a bomb calorimeter?
   a. Digestible energy (DE)
   b. Gross energy (GE)
   c. Heat increment (HI)
   d. Metabolizable energy (ME)
   e. Net energy (NE)
4. Assume that you fed a steer 20 lb of ground corn (corn is 9% crude protein) and later recovered 1.62 lb of protein in the feces. What is the digestion coefficient for crude protein in corn?
   a. 90%
   b. 55%
   c. 38%
   d. 10%
   e. None of these
5. Net energy for maintenance (NE_m) does *not* include which function:
   a. Basal metabolism (BM)
   b. Synthesis of milk
   c. Energy of voluntary activity (VAE)
   d. Heat to keep body warm (HBW)
   e. Heat to keep body cool (HBC)

6. A cattle ration having a roughage/concentrate (R/C) ratio of 22/78 would be considered which of the following:
   a. Low energy
   b. Medium energy
   c. High energy
   d. Toxic
   e. None of the above

7. In calculating the TDN value of a feedstuff, which of the following nutrients is not used?
   a. % digestible protein
   b. % digestible fiber
   c. % digestible ash
   d. % digestible NFE
   e. % digestible ether extract

8. If a commercial trace mineral premix contained 454 ppm (parts per million) of $ZnCO_3$, how many milligrams per kilogram (mg/kg) would this represent?
   a. 0.00045
   b. 0.454
   c. 454.0
   d. 454,000.0
   e. 454,000,000.0

9. Which of the following feeds should be frozen before sending to a laboratory for analysis?
   a. Alfalfa hay
   b. Barley grain
   c. Trace mineral premix
   d. Corn silage
   e. Commercial protein supplement

10. The portion of forage material that is soluble in neutral detergent is
    a. Cell wall.
    b. Cell contents.
    c. Both of the above.
    d. None of the above.

III. Matching and Identification

A. Indicate into which nutrient class from the proximate analysis the following words or statements would best fit:
   a. Ash
   b. Crude protein (CP)
   c. Crude fat
   d. Crude fiber (CF)
   e. Nitrogen-free extract (NFE)

1. Cellulose
2. Arachidonic acid
3. Salt
4. Amino acids
5. Sucrose
6. Vitamin C
7. Plant waxes
8. Nonprotein nitrogen
9. Potassium
10. Amylose
11. Lysine
12. Forms the major constituents of enzymes
13. The major form of stored energy in an animal body
14. Provides the highest concentration of energy for swine

15. Contains an average of 16% nitrogen
16. Provides the most total energy in poultry diets
17. Not included in TDN
18. % N × 6.25

IV. Short Answer Questions

1. Define total digestible nutrients (TDN), and describe how the amount is computed.
2. Name five considerations used to physically evaluate a forage to be used for livestock feed.
3. Describe a recommended procedure for obtaining representative grain samples for chemical analysis.
4. What determinations are made by the proximate analysis procedure?
5. Briefly describe the procedure followed in a digestion trial.

V. Problems

1. A feed analysis report shows a feedstuff to contain 2.2% nitrogen and 86% dry matter (as-fed basis). Calculate the percent crude protein on
   a. As-fed basis
   b. Dry matter basis
   c. Air-dry basis
2. A cattle feed has the following composition on as as-fed basis

Ash	5.1%
Crude protein	7.5%
Crude fat	2.4%
Crude fiber	29.6%
NFE	43.8%
Calcium	0.36%
Carotene	47.2 mg/kg

   a. Calculate the percentage of dry matter.
   b. Calculate the percentage of CP on an air-dry basis.
   c. Calculate the percentage of fiber on a dry matter basis.
   d. What is the percentage of carbohydrate on a dry matter basis?
   e. How many ppm of calcium are in the feed (as-fed basis)?
   f. What is the percentage of carotene in the feed (as-fed basis)?
   g. How many international units (IU) of vitamin A equivalent are in each kilogram (kg) (as-fed basis)?
3. Assume the following information on a feedstuff (as-fed basis)

Nutrient Content		Digestion Coefficient, %
Moisture, %	11.0	—
Ash, %	7.6	95
Protein, %	15.2	70
Fat, %	1.8	80
Fiber, %	27.4	60
NFE, %	37.0	90
NE_m, Mcal/lb	0.55	—
NE_g, Mcal/lb	0.30	—

   a. Calculate the percentage of TDN (as-fed basis).
   b. A 600-lb steer requires the following net energy:

   NE_m – 5.21 Mcal/day
   NE_g – 2.43 Mcal/day (to gain 1.4 lb)

   How many pounds of the above feed must be given to meet energy needs?

4. Calculate the roughage/concentrate ratio of the following ration:

Ingredients	lb (dry matter)
Barley grain	6
Corn silage*	10
Bromegrass hay	6
Supplement	1

*Assume the corn silage was harvested from a field that yielded 15 tons/acre (40% dry matter). Part of the same field was harvested for grain and yielded 110 bu/acre (90% dry matter).

5. A 600-kg lactating dairy cow producing 25 kg of milk daily requires 20 Mcal of net energy for lactation (NE_L). The following diet is being prepared by the dairy producer (as-fed basis)

   195 kg Alfalfa hay (1.11 Mcal/kg NE_L; 90% d.m.)
   440 kg Corn silage (0.65 Mcal/kg NE_L; 40% d.m.)
   365 kg Concentrate mix (1.50 Mcal/kg NE_L; 88% d.m.)
1000 kg TOTAL

  a. What is the percentage of dry matter in the above diet?
  b. What is the NE_L/kg of diet (as-fed)?
  c. How many as-fed kg of diet must be fed daily to meet energy needs?
  d. Calculate the dry matter intake as a percentage of body weight.

3

Feedstuffs Used in Livestock Diets

Chapter Goals

- Describe feedstuff nomenclature and explain how the table of feed composition is used.
- Classify feeds into the various categories for discussion
  —concentrate feeds.
  —roughage feed.
  —additives (minerals and vitamins).
- Identify various feeds by nutrient content (high or low).
- Identify desirable and undesirable characteristics of feeds.
- Highlight certain milling or industrial processes used to inactivate deleterious factors in feeds or improve palatability and feeding value.
- Discuss how stage of plant maturity influences nutritional value to the animal.
- Discuss harvesting, storing and processing techniques available for various feeds.
- Identify important factors regulating feed intake by animals.

Many different feedstuffs are used at one place or another in the feeding of livestock. Morrison lists analyses for 173 dry roughages (not including grades of the same feed), 143 green roughages and roots, 46 silages and 432 concentrates. In 1971, the Subcommittee on Feed Composition of the NRC Committee on Animal Nutrition of the US National Research Council published an *Atlas* on 6152 feeds, with the available data on their composition. In this tabulation, varieties and grades of feeds of common plant or animal origin are listed as separate feedstuffs, whereas in most tabulations all analyses of feeds of one common name are averaged under one heading, such as "Barley, grain, all analyses."

With such reference data readily available, there is little reason to attempt here any extensive discussion except for a few key feeds. The plan of this section is rather to discuss first the general nutritional properties of feed by groups (i.e., dry roughages, green forages, silages, energy feeds, protein supplements). We shall augment this discussion by considering in detail the characteristics of one or more key feeds in each group. Finally, we shall consider the unique properties of some common feeds within subgroups that can replace each other, and the consequences of such substitutions.

With a sound understanding of the general nutritional nature of the feeds making up the several groups, together with enough description of a particular product to assign it correctly to its proper category in the classification and to disclose any pronounced peculiarity it may have, we can incorporate feedstuffs that are not well known into rations with reasonable assurance that we will make the most of their potential feeding values, and avoid major undesirable consequences of improper use.

I. International Nomenclature[1]

International Feed Names are used in the tables of feed composition. The nomenclature of the feeds under which the analytical data are shown primarily follows the "International Feed Vocabulary" as

---

1. This section was prepared by Dr. Lorin E. Harris, International Feedstuffs Institute, Utah State University, Logan, Utah.

outlined by Harris et al.[2] Many feeds in the United States have official names and definitions designated by the Association of American Feed Control Officials (AAFCO). Frequently, these names are common or trade names, and the origin of the feed name does not follow a standardized naming system. A name should clearly state the source of the material and describe any process, alteration or special circumstance which affects the nutritional value of that feed.

The International Feed Vocabulary is designed to give a comprehensive name to each feed as concisely as possible. Each feed name was coined by using descriptors taken from one or more of six facets.

- Origin, consisting of scientific names (genus, species, variety); common name (generic, kind or breed, strain), and chemical formula as appropriate.
- Part fed to animals as affected by process(es).
- Process(es) and treatment(s) to which the part has been subjected.
- Stage of maturity or development (applicable to forages and animals).
- Cutting (primarily applicable to forages).
- Grade (official grades with guarantees).

A. *Feed Classes*

Feeds of the same origin (and of the same species, variety or kind, if one of these is stated) have been grouped into eight classes, each of which is designated by a number in parentheses. The numbers and classes they designate are as follows: (1) dry forages and roughages; (2) pasture, range plants and forages fed fresh; (3) silages; (4) energy feeds; (5) protein supplements; (6) mineral supplements; (7) vitamin supplements; and (8) additives. The following is a brief explanation of the physical and chemical characteristics that distinguish the feed classes.

1. Dry forages and roughages
   All forages and roughages cut and cured and other products with more than 18% crude fiber or containing more than 35% cell wall (dry basis). Forages and roughages are low in net energy per unit weight usually because of the high cell-wall content.
   a. Carbonaceous roughages (low protein)
      (1) Straws.
      (2) Stalks.
      (3) Weathered grass.
   b. Proteinaceous roughages
      (1) Legume hays.
      (2) Grass or legume hays.
2. Pasture, range plants and forages fed fresh
   Included in this group are all forage feeds either not cut (including feeds cured on the stem) or cut and fed fresh.
   a. Carbonaceous (low protein)
      (1) Grama grass, fresh.
      (2) Wheatgrass fresh, postripe.
   b. Proteinaceous (high protein)
      (1) Wheatgrass, fresh, early vegetative.
3. Silages
   This class includes only ensiled forages (maize, alfalfa, grass etc.), but not ensiled fish, grain, roots and *tubers*.
   a. Carbonaceous (low protein)
      (1) Corn silages.
      (2) Grass silage.
   b. Proteinaceous
      (1) Alfalfa silage.
      (2) Clover silage.

2. Harris LE, F Jager, TF Leche, H Mayr and LC Kearl, 1980, International Network of Feed Information Centres, *Publication 5*. Prepared on behalf of INFIC by the International Feedstuffs Institute, Utah State University, Logan, Utah, 84322 USA. Harris LE, LC Kearl and PV Fonnesbeck, 1981, *A rationale for naming feeds*. Utah Agr Expt Sta Bull 501.

4. Energy feeds

   Products with less than 20% protein and less than 18% crude fiber or less than 35% cell wall (dry basis), as for example grain, mill byproducts, fruit, nuts, roots, and tubers. Also, when these feeds are ensiled they are classified as energy feeds.

   a. Carbonaceous concentrates (low protein)

      (1) Cereal grains—corn, oats, barley, rye and wheat.

      (2) Sorghums—kafir, milo and hybrids.

      (3) Byproduct feeds—bran, middlings, shorts, cobs, molasses.

5. Protein supplements

   Products that contain 20% or more of protein (dry basis) from animal origin (including ensiled products) as well as oil meals.

   a. Supplements of vegetable origin

      (1) Soybean meal.

      (2) Flaxseed meal (linseed meal).

      (3) Cottonseed meal.

      (4) Peanut meal.

      (5) Corn gluten meal.

      (6) Sorghum gluten meal.

      (7) Brewer's dried grains.

      (8) Sesame meal.

   b. Supplements of animal origin

      (1) Animal tissue

         (a) Tankage.

         (b) Tankage with bone.

         (c) Meat scraps.

         (d) Meat and bone scraps.

         (e) Blood meal.

         (f) Meat meal.

      (2) Fish products

         (a) Fish meal.

         (b) Dried fish solubles.

         (c) Condensed fish solubles.

      (3) Milk products

         (a) Dried skim milk.

         (b) Dried whole milk.

         (c) Dried butter milk.

         (d) Condensed butter milk.

         (e) Dried whey.

6. Mineral supplements

   a. Bone meal, steamed.

   b. Calcium carbonate.

   c. Limestone.

7. Vitamin supplements

   a. Ensiled yeast.

   b. Carotene.

   c. Fish, salmon, oil.

   d. Wheat germ oil.

8. Additives

   Feed supplements such as antibiotics, coloring material, flavors, hormones and medicants.

   a. Aluminum sulfate (antigelling agent for molasses).

   b. Monosodium glutamate (seasoning spice).

   c. Propylene glycol (emulsifying agent).

   d. Titanium dioxide (color additive).

B. *International Feed Number*

Each International Feed Name is assigned a six-digit International Feed Number (IFN) for its identification. These numbers are the link between the Feed Names and chemical and biological data in the U.S. and world databanks. The numbers are particularly useful as tags when calculations involve various feeds to make up diets or feed mixtures. In the tables of feed composition, the Feed Class Number has been put in front of the International Feed Number.

The system has been revised. In some cases, the class number has been changed to conform to the composition of the feed. In other cases, two feeds have been combined.

C. *The International Feed Name*

As explained above, the Feed Names are made up of six facets. Each facet has its own parameters to which descriptors are fixed. Also, the information provided by each entry within a facet differs from that of another facet.

1. Facet 1: Original Material (Origin)

Facet 1 has many options because an immense number of plants, animals or other original materials (minerals, chemicals etc.) can be used as feed.

For ease of use, the descriptors of this facet were split into two parts. One is the scientific name (Latin) of biological subjects (or an adequate quasiscientific name in cases where no real scientific name exists); the other is a common term chosen to include the material in question. Thus, the original material may be one of three types (descriptors are capped).

plants	with scientific name CLOVER, WHITE, (TRIFOLIUM PRATENSE), SOYBEAN (GLYCINE MAX), COCONUT (COCOSNUCIFERA) without scientific name (CEREALS, GRASS, MEADOW PLANTS)
animals	with scientific name CATTLE (BOS TAURUS), CHICKENS (GALLUS DOMESTICUS), SWINE (SUSSCROFA) without scientific name (ANIMAL, POULTRY)
minerals	chemical products and others (without scientific name)

The scientific name (Latin) has three elements

Genus
Species
Variety or cultivar

Usually, common terms follow the same principle. Thus, the preferred terms have three elements

General or common name
Breed or kind
Strain or chemical formula

Minerals, drugs and chemicals are listed according to the nomenclature of the Chemical Rubber Company (CRC).[3] The chemical formulae are designated where applicable. Table 3–1A gives examples of Facet 1 with and without scientific names.

2. Facet 2: Parts of the Material Used as Feed as Affected by Process(es)

This facet of the feed description represents the actual part of the parent material fed. In the past, the edible parts of plants and animals were obvious such as leaves, stems, seeds, meat trimmings or bones. Today, due to the extensive fractionation of plant seeds and the reconstitution of many of the parts into new processed foods, many byproducts are available for animal feeding. Also, there are those byproducts from the preparation of meat and fish for human consumption.

---

3. Chemical Rubber Company, 1977, *Handbook of Biochemistry* selected data for molecular biology, The Chemical Rubber Company.

TABLE 3–1A. International Feed Names: Original Material + Part + Process + Maturity + Cut + Grade (Examples)

With scientific name

	Example 1	Example 2	Example 3	Example 4	Example 5	Example 6
Original material						
Genus	MELILOTUS	SEBASTES	VICIA	MEDICAGO	GLYCINE	ZEA
Species	INDICA	MARINUS	FABA	SATIVA	MAX	MAYS
Variety	—	—	MINOR	—	—	INDENTATA
Generic name	SWEET CLOVER	FISH	HORSEBEAN	ALFALFA	SOYBEAN	MAIZE
Breed or kind	YELLOW	HADDOCK	SMALL	—	—	DENT
Strain	ANNUAL	BIG	—	—	—	YELLOW
Part	—	LIVERS	STEMS	STEMS	SEEDS	EARS WITH HUSKS
Process	FRESH	FRESH	FRESH	SUN-CURED	MEAL SOLVENT EXTRACTED	SILAGE
Maturity	—	—	EARLY VEGETATIVE	EARLY BLOOM	—	—
Cut	CUT 1	—	—	CUT 1	43% PROTEIN	—
Grade	—	—	—	—	—	—

Without scientific name

	Example 1	Example 2	Example 3	Example 4	Example 5	Example 6
Original material						
Genus	CALCIUM	MARSH PLANTS	LEGUME-GRASS	ANIMAL	POULTRY	GRASS-LEGUME-FORB
Species	PHOSPHATE	IN SEA WATER	—	—	—	—
Variety	DIBASIC	—	—	—	—	—
Generic name	CALCIUM PHOSPHATE DIBASIC	MARSH PLANTS	LEGUME-GRASS	ANIMAL	POULTRY	GRASS-LEGUME-FORB
Breed or kind	—	IN SEA WATER	—	—	—	—
Strain	—	—	—	—	—	—
Part	—	—	—	BONE	FEATHERS	—
Process	—	HAY SUN-CURED	SILAGE	FRESH GROUND	MEAL	SILAGE
Maturity	—	LATEBLOOM	LATE VEGETATIVE	—	—	—
Cut	—	CUT 1	CUT 1	—	—	CUT 1
Grade	CHEMICALLY PURE	—	—	—	—	—

Each part has to be described precisely by a descriptor, the use of which is clearly defined. Examples of definitions are

BRAN: The pericarp of cereal grains

COB: The fibrous inner portion of the ear of maize (corn) from which the kernels have been removed.

GERM: The embryo found in seeds and frequently separated from the starch endosperm during milling.

Examples of feed names with parts are given in Table 3–1A.

3. Facet 3: Process(es) and Treatment(s)
Many processes are used in the preparation of animal feeds, and some of these may significantly alter their nutritional value. Heat may damage some nutrients and conversely, it may make other nutrients more available.

Pelleting increases consumption, while grinding may affect digestibility of protein and carbohydrates. Also, such treatments alter the proportions of the lower fatty acids produced by rumen microflora of lactating cows, thus decreasing the concentration of fat in the milk.

It is important, then, that a feeder be aware of the processes to which a feed has been subjected, whether intended for preservation, purification, concentration or to improve nutritive value; or as uncontrollable factors that reduce nutritive value. Therefore, origin and part terms are followed by those distinguishing the different methods of processing, such as preservative processes, separating, reducing particle size or thermal treatments.

Examples of definitions for processes are

DRY-RENDERED: Residues of animal tissues cooked in open steamjacketed vessels until the water has evaporated; fat removed by draining and processing the solid residue.

FRESH: Recently produced or gathered; not stored, cured or preserved.

HYDROLYZED: Subjected to hydrolysis, a process by which complex molecules (e.g., those in proteins) are split into simpler units by chemical reaction with water molecules. (The reaction may be produced by an enzyme, catalyst or acid, or by heat and pressure.)

MECHANICAL-EXTRACTED: Extracted by heat and mechanical pressure. Refers to removal of fat or oil from the feeds. Synonyms: expeller-extracted, hydraulically extracted, old process.

Examples of feed names with processes are given in Table 3–1A.

4. Facet 4: Stage of Maturity for Plants and Animals
Stage of maturity is an important factor influencing the nutritive value of forages, silages, and some animal products. There is an optimal stage of maturity for forage crops beyond which the chemical composition, ratio of leaf to stem or portion of seeds or grain greatly influences the nutritive value. When plants bloom intermittently (usually in the tropics), the stage of growth is described by the length of time the plant has been growing (Table 3–1B). In some cases, the stage of development in an animal is extremely important. For example, when comparing the nutritive value of day-old chicks to aged laying hens (Table 3–1C). Examples of feed names with stage of maturity are given in Table 3–1A.

5. Facet 5: Cutting
Many forage crops are cut and harvested several times during a year. Each cutting has a unique nutrient content as well as characteristic physical properties. The description for cutting refers to the sequence of cutting from the first to the last during a year (cut 1, cut 2, etc.). The maturity terms refer to stage of growth or of regrowth within the limits of each cutting.

Since stage of maturity is more important than cutting date, the various cuts for forages are sometimes combined with the stage of maturity when data are summarized for feed composition tables. Examples of feed names with cuttings are given in Table 3–1A.

6. Facet 6: Grade
Some commercial feeds and feed ingredients are given official grades on the basis of their composition and other quality characteristics. Such feeds are sold on a quality description basis in accordance with their official grading. Thus, these grades and quality designations must be included as a definitive component in the description of the feed. The sixth facet makes available descriptors that characterize grades of quality as sometimes used in the feed trade. Descriptors

TABLE 3–1B. Alphabetic List of Facet 4 Stage of Maturity Descriptors for Plants with Definitions on Their Use in the International Feed Name

	Preferred Term Definition	Comparable Terms
For plants that bloom		
GERMINATED	Stage in which the embryo in a seed resumes growth after a dormant period	Sprouted
EARLY VEGETATIVE	Stage at which the plant is vegetative and before the stems elongate	Fresh new growth, before heading out, before inflorescence emergence, immature, prebud stage, very immature, young
LATE VEGETATIVE	Stage at which stems are beginning to elongate to just before blooming; first bud	Before bloom, bud stage, budding plants heading to bloom, heads just showing, jointing and boot (grasses), prebloom, preflowering, stems elongated
EARLY BLOOM	Stage between initiation of bloom and state in which 1/10 of the plants are in bloom; some grass heads are in anthesis	Early anthesis, first flower, headed out in head, up to 1/10 bloom
MIDBLOOM	Stage in which 1/10 to 2/3 of the plants are in bloom, most grass heads are in anthesis	Bloom, flowering, flowering plants, half bloom, in bloom, mid anthesis
FULL BLOOM	Stage in which 2/3 or more of the plants are in bloom	3/4 to full bloom, late anthesis
LATE BLOOM	Stage in which blossoms begin to dry and fall and seeds begin to form	15 days after silking, before milk, in bloom to early pod, late to past anthesis
MILK STAGE	Stage in which seeds are well formed but soft and immature	After anthesis, early seed, fruiting, in tassel, late bloom to early seed, past bloom, pode stage, postanthesis, postbloom, seed developing, seed-forming, soft, soft immature
DOUGH STAGE	Stage in which the seeds are of doughlike consistency	Dough stage, nearly mature, seeds dough, seeds well developed, soft, dent
MATURE	Stage in which plants are normally harvested for seed	Dent, dough to glazing, fruiting plants, in seed, kernels ripe, ripe seed
POSTRIPE	Stage that follows maturity; some seeds cast and plants have begun to weather (applies mostly to range plants)	Late seed, overripe, very mature
STEM-CURED	Stage in which plants are cured on the stem; seeds have been cast and weathering has taken place (applies mostly to range plants)	Dormant, mature and weathered, seeds cast
REGROWTH EARLY VEGETATIVE	Stage in which regrowth occurs without flowering activity; vegetative crop aftermath; regrowth in stubble (applies primarily to fall regrowth in temperate climates); early dry-season regrowth	Vegetative recovery growth
REGROWTH LATE VEGETATIVE	Stage in which stems begin to elongate to just before blooming; first bud to first flowers; regrowth in stubble with stem elongation (applies primarily to fall regrowth in temperate climates)	Recovery growth, stems elongating jointing and boot (grasses)
*For plants that do not bloom**		
1 TO 14 DAYS' GROWTH	A specified length of time after plants have started to grow	2 weeks' growth
15 TO 28 DAYS' GROWTH	do	4 weeks' growth
29 TO 42 DAYS' GROWTH ETC.	do	6 weeks' growth

*These classes are for species that remain vegetative for long periods and apply primarily to the tropics. When the name of a feed is developed, the age classes form part of the name (e.g., pangola grass, 15 to 28 days' growth). Do not use terms which apply to plants that bloom and those that do not bloom in the same name.

TABLE 3–1C. Maturity Terms for Animals, Facet 4

Ruminants and Nonruminants	Poultry	Fish
DAY-OLD	DAY-OLD	LARVAL
SUCKLING	CHICK	FRY
GROWER	BROILER	FINGERLING
ADULT	ADULT	GROWER
AGED	AGED	ADULT
		AGED

of this kind are often expressed in terms of "more than" (minimum) and "less than" (maximum) or even "from—to" of designated contents of crude fiber, protein, fat etc. Generally, artificial grades should not be used unless they are used in the trade.

Examples of descriptors of this facet are given in Table 3–1A.

D. *How to Find International Feed Names*

The International Feed Names begin with the name of the plant (Barley; Wheat) and the name of the animal (Cattle; Chicken; Sheep; Fish, salmon). If the name of the plant is not known, the name begins with Vegetable; Cereal; Marsh plants; etc. If the name of the animal is not known, the name begins with Animal or Poultry. Animal products are listed under the name of the animal (Cattle, milk; Cattle, fat) or Animal blood meal. Molasses is listed under Beet or Sugarcane. Cross references are used, i.e., Whey see Cattle, whey; Hominy feed see Corn.

E. *Composition of Feeds*

Table 3–2 presents the nutrient composition of nearly 300 important U.S. and Canadian feeds. Nutrient concentrations are organized as follows:

Table 3–2A Proximate Analysis and General Nutrients
Table 3–2B Energy Values
Table 3–2C Mineral Values
Table 3–2D Vitamin Values
Table 3–2E Amino Acid Values
Table 3–2F Mineral Supplements

The analytical data in Table 3–2 is expressed in the metric system and is shown on an as-fed and dry matter basis. In some instances, instead of numerical values, dashes (—) appear in the feed composition table. Dashes indicate that composition values are not available or are insignificant in formulating animal diets.

The data were compiled from the Feed Composition Data Bank of the USDA's National Agriculture Library, Beltsville, MD.

Energy values in Table 3–2 were calculated using the procedures outlined in the NRC United States–Canadian Tables of Feed Composition.[4]

Feedstuffs are not of constant composition. Individual feed samples may vary widely from the values set forth in these tables. Variation is caused by many factors such as variety, climate, soil and length of storage. Actual analysis should be obtained and used whenever possible. Often, however, if it is either difficult to determine actual composition or there is insufficient time to obtain such analysis, tabulated data is the next best source of information.

For more detail on the International Nomenclature and definitions of part, process and grade descriptors see Harris et al.[5]

II. International Composition of Feeds (Instructions for Using Table 3–2)

The International Feed Names begin with the name of the plant (Barley; Wheat) or the name of the animal (Cattle, Chickens; Sheep; Fish, salmon). If the name of the plant is not known, the name begins

4. National Research Council, 1982, *United States—Canadian Tables of Feed Composition,* Washington, DC: National Academy Press.
5. Harris LE, LC Kearl and PV Fonnesbeck, 1981. *A rationale for naming feeds,* Utah Agr Expt Sta Bull 501. Harris LE, F Jager, TF Leche, H Mayr and LC Kearl, 1980, International Network of Feed Information Centres, *Publication 5.* Prepared on behalf of INFIC by the International Feedstuffs Institute, Utah State University, Logan, Utah, 84322 USA.

with Vegetable; Marsh plants; Cereal, etc. If the name of the animal is not known, the name begins with Animal or Poultry.

Animal products begin with the name of the animal, i.e., Cattle, milk not Milk, cattle. Molasses is listed under the plant name, i.e., Beet, sugar, molasses or Sugarcane, molasses. Fats and oils are listed under the name of the animal, i.e., Cattle, fat not Fat, cattle; or plant, i.e., Soybean, oil, not Oil, soybean.

Also, feed names are arranged alphabetically by feed classes within origin (except names listed in order of ascending stage of maturity). The feed class is the first digit of the International Feed Number (IFN). The feed classes are: (1) dry forages and roughages; (2) pasture, range plants, and forages fed fresh; (3) silages; (4) energy feeds; (5) protein supplements; (6) mineral supplements; (7) Vitamin supplements and (8) additives.

For example

Alfalfa, fresh	IFN 2-10-196
Alfalfa, hay, sun-cured, late vegetative	IFN 1-00-054
Alfalfa, hay, sun-cured, early bloom	IFN 1-00-059
Alfalfa, meal dehydrated	IFN 1-00-022
Alfalfa, silage, 30–50% dry matter	IFN 3-08-150

TABLE 3–2A. Composition of Feeds—Proximate Analysis and General Nutrients

		IFN	DRY MATT %	PRO TEIN %	CELL ULOS %	CRUID FIBR %	ETHR EXTR %	ASH %	ACID DETR %	NEUT DETR %
	ALFALFA. MEDICAGO SATIVA									
001	Fresh	2-00-196	26.0	5.3	—	6.0	1.0	2.5	—	—
002			100.0	20.5	—	23.0	3.8	9.5	—	—
003	Hay, sun-cured, late vegetative	1-00-054	89.7	20.3	27.3	21.5	3.8	8.2	28.2	36
004			100.0	22.6	30.5	24.0	4.2	9.1	31.5	40
005	Hay, sun-cured, early bloom	1-00-059	90.5	18.0	31.3	25.8	2.6	8.4	29.1	38
			100.0	19.9	34.6	28.5	2.9	9.2	32.1	42
007	Hay, sun-cured, midbloom	1-00-063	91.0	17.0	23.1	25.5	3.3	7.8	33.6	42
			100.0	18.7	25.4	28.0	3.6	8.5	36.9	46
009	Hay, sun-cured, full bloom	1-00-068	90.9	15.5	32.9	27.3	3.1	7.1	35.2	45
010			100.0	17.0	36.2	30.1	3.4	7.8	38.7	50
011	Hay, sun-cured, mature	1-00-071	91.2	15.2	—	29.3	2.9	6.7	40.1	—
012			100.0	16.7	—	32.1	3.2	7.4	44.0	—
013	Meal, dehydrated, 15% protein	1-00-022	90.3	15.6	25.0	26.2	2.2	8.9	33.8	46
014			100.0	17.3	27.7	29.0	2.4	9.9	37.5	51
015	Meal, dehydrated, 17% protein	1-00-023	91.8	17.4	—	24.0	2.8	9.8	31.5	41
016			100.0	18.9	—	26.2	3.0	10.6	34.3	45
017	Meal, dehydrated, 20% protein	1-00-024	91.6	20.3	—	20.8	3.3	10.2	27.0	38
018			100.0	22.1	—	22.7	3.6	11.2	29.4	42
019	Silage, 30–50% dry matter	3-08-150	43.4	8.0	11.5	14.5	1.1	3.8	15.1	—
			100.0	18.4	26.6	33.5	2.6	8.9	34.7	—
	ALFALFA-BROME, SMOOTH. MEDICAGO SATIVA-BROMUS INFERMIS									
021	Fresh	2-00-262	21.3	3.9	—	5.4	0.8	2.1	6.8	—
022			100.0	18.6	—	25.3	3.6	9.8	32.2	—
023	Silage, 30–50% dry matter	3-08-147	39.8	6.2	—	13.1	0.9	—	—	—
			100.0	15.5	—	33.0	2.3	—	—	—
	ALFALFA-GRASS. MEDICAGO SATIVA-GRASS									
025	Hay, sun-cured	1-08-331	91.4	14.5	—	30.3	2.1	6.6	37.5	—
026			100.0	15.9	—	33.2	2.3	7.2	41.0	—
	ALFALFA-ORCHARDGRASS. MEDICAGO SATIVA-DACTYLIS GLOMERATA									
027	Silage, 30–50% dry matter	3-08-144	36.4	6.3	12.9	11.0	1.4	4.0	15.5	—
028			100.0	17.2	35.5	30.2	4.0	10.9	42.5	—
	ANIMAL									
029	Blood, spray dehydrated	5-00-381	92.6	87.7	—	1.0	0.7	2.3	—	—
030			100.0	94.7	—	1.1	0.8	2.5	—	—
031	Bone, meal, steamed	6-00-400	96.1	10.9	—	2.0	3.0	71.3	—	—
032			100.0	11.4	—	2.1	3.1	74.2	—	—
033	Meal, meal, rendered	5-00-385	93.8	50.6	—	2.7	8.9	28.1	—	—
034			100.0	54.0	—	2.9	9.5	29.9	—	—
035	Meat with bone, meal, rendered	5-00-388	93.3	50.2	—	2.3	10.4	27.8	—	—
036			100.0	53.8	—	2.4	11.2	29.7	—	—

	IFN	DRY MATT %	PRO TEIN %	CELL ULOS %	CRUID FIBR %	ETHR EXTR %	ASH %	ACID DETR %	NEUT DETR %
ANIMAL—Continued									
037 Tankage, meal, rendered	5-00-386	92.0	59.2	—	1.8	7.7	21.1	—	—
038		100.0	64.4	—	2.0	8.3	22.9	—	—
039 Tankage with bone, meal, rendered	5-00-387	92.9	47.2	—	2.2	12.7	28.2	—	—
040		100.0	50.8	—	2.4	13.7	30.4	—	—
ANIMAL-POULTRY									
041 Fat	4-00-409	99.1	—	—	—	99.1	—	—	—
042		100.0	—	—	—	100.0	—	—	—
043 Tankage, meal, rendered, 50% protein	5-00-410	94.2	51.4	—	2.2	10.2	29.3	—	—
044		100.0	54.6	—	2.4	10.8	31.1	—	—
APPLE. MALUS SPP									
045 Pomace, dehydrated	4-00-423	89.6	4.5	—	16.3	4.3	3.0	—	—
046		100.0	5.0	—	18.2	4.8	3.4	—	—
BAHIAGRASS. PASPALUM NOTATUM									
047 Fresh	2-00-464	28.7	3.6	8.1	8.7	0.5	3.2	10.2	20
048		100.0	12.6	28.2	30.4	1.6	11.1	35.5	68
049 Hay, sun-cured	1-00-462	90.0	8.5	—	28.1	1.8	5.7	33.9	66
050		100.0	9.5	—	31.2	2.0	6.3	37.7	73
BAKERY									
051 Waste, dehydrated (dried bakery product)	4-00-466	91.2	10.1	—	1.3	10.9	3.7	1.6	16
052		100.0	11.1	—	1.4	11.9	4.0	1.8	18
BARLEY. HORDEUM VULGARE									
053 Grain	4-00-549	88.6	11.5	5.2	4.9	1.8	2.4	7.7	17
054		100.0	13.0	5.9	5.6	2.0	2.7	8.7	19
055 Grain, Pacific coast	4-07-939	88.6	9.7	—	6.0	2.0	2.4	6.3	19
		100.0	11.0	—	6.8	2.2	2.7	7.1	21
057 Grain screenings	4-00-542	88.9	11.5	—	7.5	2.3	3.1	9.8	—
058		100.0	13.0	—	8.4	2.6	3.4	11.0	—
059 Hay, sun-cured	1-00-495	88.4	7.8	—	23.6	1.9	6.6	—	—
060		100.0	8.8	—	26.7	2.1	7.5	—	—
061 Straw	1-00-498	91.4	4.0	34.1	37.9	1.7	6.7	42.2	73
062		100.0	4.4	37.3	41.5	1.9	7.3	46.2	80
BEAN, NAVY. PHASEOLUS VULGARIS									
063 Seeds	5-00-623	89.8	22.2	2.9	4.3	1.5	4.0	5.6	—
064		100.0	24.8	3.2	4.8	1.7	4.5	6.2	—
BEET, MANGEL. BETA VULGARIS MACRORRHIZA									
065 Roots, fresh	4-00-637	11.0	1.3	—	0.8	0.1	1.1	—	—
066		100.0	11.8	—	7.4	0.7	9.6	—	—
BEET, SUGAR. BETA VULGARIS ALTISSIMA									
067 Aerial part with crowns, fresh	2-00-649	16.5	2.5	—	1.8	0.3	3.4	—	—
068		100.0	15.0	—	10.7	2.1	20.4	—	—
069 Aerial part with crowns, silage	3-00-660	25.2	3.4	—	3.2	0.7	8.6	—	—
070		100.0	13.4	—	12.8	2.9	34.0	—	—
071 Molasses, more than 48% invert sugar more than 79.5 degrees	4-00-668	77.9	6.6	—	—	0.2	8.9	—	—
072 brix		100.0	8.5	—	—	0.2	11.4	—	—
073 Pulp, dehydrated	4-00-669	91.0	8.9	18.3	18.2	0.5	4.9	25.0	49
074		100.0	9.8	20.1	20.0	0.6	5.3	27.5	54
075 Pulp, wet	4-00-671	11.0	1.2	—	2.4	0.2	0.5	3.3	6
076		100.0	10.8	—	21.7	2.1	4.9	30.5	54
077 Pulp with molasses, dehydrated	4-00-672	91.8	9.3	—	15.2	0.6	5.7	24.5	40
078		100.0	10.1	—	16.6	0.6	6.2	26.6	44
BERMUDAGRASS. CYNODON DACTYLON									
079 Fresh	2-00-712	28.9	4.2	8.1	7.7	0.6	3.3	9.1	—
080		100.0	14.6	28.0	26.6	2.0	11.4	31.4	—
081 Hay, sun-cured	1-00-703	91.2	9.4	29.4	28.5	2.0	7.9	35.7	69
082		100.0	10.3	32.3	31.3	2.1	8.7	39.1	76
BERMUDAGRASS, COASTAL. CYNODON DACTYLON									
083 Fresh	2-00-719	30.3	3.8	8.7	8.6	1.1	2.4	11.1	—
084		100.0	12.5	28.8	28.4	3.7	8.1	36.8	—
085 Hay, sun-cured	1-00-716	91.6	11.2	27.8	27.6	2.1	6.2	34.4	70
086		100.0	12.3	30.3	30.2	2.2	6.8	37.6	76
BLOOD (SEE ANIMAL)									

		IFN	DRY MATT %	PRO TEIN %	CELL ULOS %	CRUID FIBR %	ETHR EXTR %	ASH %	ACID DETR %	NEUT DETR %
	BLUEGRASS, CANADA. POA COMPRESSA									
087	Fresh	2-00-764	31.4	5.3	—	8.3	1.2	2.8	—	—
088			100.0	17.0	—	26.4	3.7	8.9	—	—
089	Hay, sun-cured	1-00-762	91.7	9.5	—	27.6	2.4	6.5	—	—
090			100.0	10.3	—	30.1	2.6	7.0	—	—
	BLUEGRASS, KENTUCKY. POA PRATENSIS									
091	Fresh, early vegetative	2-00-777	30.8	5.4	—	7.8	1.1	2.9	9	17
092			100.0	17.4	—	25.2	3.5	9.4	29	55
093	Fresh, early bloom	2-00-779	35.0	5.8	—	9.6	1.4	2.5	11	23
094			100.0	16.6	—	27.4	3.9	7.1	32	65
095	Hay, sun-cured	1-00-776	88.9	9.3	26.1	26.7	2.6	6.0	—	—
096			100.0	10.5	29.4	30.0	2.9	6.7	—	—
	BLUESTEM. ANDROPOGON SPP									
097	Fresh, early vegetative	2-00-821	26.8	3.6	8.3	6.7	0.7	2.4	—	—
098			100.0	13.6	30.9	24.9	2.8	8.9	—	—
099	Fresh, mature	2-00-825	59.0	3.4	—	20.2	1.4	3.3	—	—
100			100.0	5.8	—	34.2	2.4	5.6	—	—
	BONES (SEE ANIMAL)									
	BREWER'S GRAINS (SEE CEREALS)									
	BROME. BROMUS SPP									
101	Fresh	2-00-900	36.0	4.2	—	11.0	1.0	3.4	—	—
102			100.0	11.8	—	30.6	2.8	9.3	—	—
103	Hay, sun-cured	1-00-890	90.8	8.7	—	29.8	1.9	7.1	31.7	—
104			100.0	9.5	—	32.9	2.1	7.8	34.9	—
	BROME, SMOOTH. BROMUS INFERMIS									
105	Fresh	2-00-963	27.0	4.1	—	7.5	0.8	—	9.5	16
106			100.0	15.2	—	28.0	3.0	—	35.2	58
107	Hay, sun-cured	1-00-947	89.6	12.4	—	29.0	2.6	7.1	33.4	58
108			100.0	13.9	—	32.4	2.9	8.0	37.3	65
	BUCKWHEAT, COMMON. FAGOPYRUM SAGITTATUM									
109	Grain	4-00-994	88.4	11.3	—	10.8	2.3	2.1	11.6	—
110			100.0	12.8	—	12.2	2.6	2.4	13.1	—
	BUFFALOGRASS. BUCHLOE DACTYLOIDES									
111	Fresh	2-01-010	45.8	4.7	—	12.7	0.9	5.1	16.7	—
112			100.0	10.2	—	27.7	1.9	11.1	36.5	—
	BUTTERMILK (SEE CATTLE)									
	CANARY GRASS, REED. PHALARIS ARUNDINACEA									
113	Fresh	2-01-113	22.8	3.9	4.9	5.6	0.9	2.3	6.5	—
114			100.0	17.0	21.6	24.4	4.1	10.2	28.3	—
115	Hay, sun-cured	1-01-104	89.3	9.1	—	30.2	2.7	7.3	32.7	—
116			100.0	10.2	—	33.9	3.0	8.1	36.6	—
	CARROT. DAUCUS SPP									
117	Roots, fresh	4-01-145	11.5	1.2	—	1.1	0.2	1.0	1	1
118			100.0	10.0	—	9.5	1.3	8.4	8	9
	CASEIN									
119	Acid precipitated dehydrated	5-01-162	91.6	85.6	—	0.2	0.5	2.2	0.0	0.0
120			100.0	93.5	—	0.2	0.5	2.4	0.0	0.0
	CATTLE. BOS TAURUS									
121	Buttermilk, dehydrated	5-01-160	92.3	31.7	—	0.3	5.2	9.1	—	—
122			100.0	34.4	—	0.4	5.6	9.9	—	—
123	Manure, dehydrated	1-01-190	93.0	14.5	—	27.9	2.0	14.2	43.7	—
124			100.0	15.6	—	30.1	2.1	15.3	47.0	—
125	Milk, dehydrated	5-01-167	95.5	25.4	—	0.2	26.4	5.4	—	—
126			100.0	26.6	—	0.2	27.7	5.7	—	—
127	Milk, fresh	5-01-168	12.4	3.3	—	—	3.6	0.8	—	—
128			100.0	26.7	—	—	29.5	6.2	—	—
129	Skimmilk, dehydrated	5-01-175	94.1	33.4	0.0	0.2	1.0	7.9	0.0	0.0
130			100.0	35.5	0.0	0.2	1.0	8.4	0.0	0.0
131	Skimmilk, fresh	5-01-170	9.6	3.0	—	—	0.1	0.7	—	—
132			100.0	31.2	—	—	1.0	6.8	—	—
133	Whey, dehydrated	4-01-182	93.3	13.1	—	0.2	0.7	8.7	0.2	—
134			100.0	14.0	—	0.2	0.8	9.4	0.2	—
135	Whey, fresh	4-08-134	6.9	0.9	—	—	0.3	0.7	—	—
136			100.0	13.1	—	—	4.3	9.6	—	—

		IFN	DRY MATT %	PRO TEIN %	CELL ULOS %	CRUID FIBR %	ETHR EXTR %	ASH %	ACID DETR %	NEUT DETR %
	CATTLE. BOS TAURUS—Continued									
137	Whey low lactose, dehydrated (dried whey product)	4-01-186	93.7	16.6	—	0.2	1.0	16.0	—	—
138			100.0	17.8	—	0.2	1.1	17.1	—	—
	CEREALS									
139	Brewer's grains, dehydrated	5-02-141	92.2	27.1	12.0	13.2	6.5	3.7	21.1	42
140			100.0	29.5	13.0	14.3	7.1	4.0	22.9	46
141	Grain screenings, uncleaned	4-02-153	92.1	13.7	—	17.1	5.8	10.2	—	—
142			100.0	14.9	—	18.6	6.3	11.1	—	—
	CITRUS. CITRUS SPP									
143	Pomace without fines, dehydrated (dried citrus pulp)	4-01-237	91.1	6.1	—	11.6	3.4	6.0	21.0	21
144			100.0	6.7	—	12.8	3.7	6.6	23.0	23
145	Pomace, silage (pulp)	4-01-234	20.8	1.5	—	3.3	2.1	1.2	4.2	—
146			100.0	7.3	—	15.7	9.9	5.6	20.0	—
147	Syrup (molasses)	4-01-241	66.9	5.7	—	—	0.2	5.1	—	—
148			100.0	8.5	—	—	0.3	7.7	—	—
	CLOVER, ALSIKE. TRIFOLIUM HYBRIDUM									
149	Fresh	2-01-316	22.5	4.1	—	5.2	0.8	2.1	—	—
150			100.0	18.1	—	23.3	3.6	9.3	—	—
151	Hay, sun-cured	1-01-313	87.7	12.4	—	26.2	2.4	7.6	—	—
152			100.0	14.2	—	29.9	2.8	8.7	—	—
	CLOVER, CRIMSON. TRIFOLIUM INCARNATUM									
153	Fresh	2-01-336	17.5	3.0	—	4.9	0.6	1.7	—	—
154			100.0	17.0	—	27.7	3.3	9.5	—	—
155	Hay, sun-cured	1-01-328	87.8	14.7	—	28.1	2.0	7.8	—	—
156			100.0	16.8	—	31.9	2.3	8.9	—	—
	CLOVER, LADINO. TRIFOLIUM REPENS									
157	Fresh	2-01-383	17.7	4.4	—	2.5	0.9	1.9	—	—
158			100.0	24.7	—	14.2	4.8	10.5	—	—
159	Hay, sun-cured	1-01-378	89.1	20.0	—	18.5	2.4	8.4	28.5	32
160			100.0	22.4	—	20.8	2.7	9.4	32.0	36
	CLOVER, RED. TRIFOLIUM PRATENSE									
161	Fresh, early bloom	2-01-428	19.6	4.1	—	4.6	1.0	2.0	6	8
162			100.0	20.8	—	23.2	5.0	10.2	31	40
163	Fresh, full bloom	2-01-429	26.2	3.8	—	6.8	0.8	2.0	9	11
164			100.0	14.6	—	26.1	2.9	7.8	35	43
165	Hay, sun-cured	1-01-415	88.4	13.0	—	27.1	2.5	6.7	36.2	41
166			100.0	14.7	—	30.7	2.8	7.5	41.0	46
	COCONUT. COCOS NUCIFERA									
167	Kernels with coats, meal, mechanical-extracted (copra meal)	5-01-572	91.6	21.1	—	12.0	6.7	6.4	18.3	—
168			100.0	23.1	—	13.1	7.3	7.0	20.0	—
169	Kernels with coats, meal, solvent-extracted (copra meal)	5-01-573	91.1	21.3	—	14.4	2.1	6.1	21.9	—
170			100.0	23.4	—	15.8	2.3	6.7	24.0	—
	CORN (SEE MAIZE)									
	COTTON. GOSSYPIUM SPP									
171	Bolls, sun-cured	1-01-596	91.4	9.0	—	34.5	2.8	6.2	—	—
172			100.0	9.8	—	37.8	3.0	6.8	—	—
173	Hulls	1-01-599	90.4	3.8	—	43.2	1.5	2.6	59.0	81
174			100.0	4.2	—	47.8	1.7	2.9	65.3	90
175	Seeds, ground	5-01-608	92.2	21.7	—	18.2	22.4	3.5	24	34
176			100.0	23.6	—	19.8	24.3	3.8	26	37
177	Seeds, meal, mechanical-extracted	5-01-609	92.7	37.4	—	14.5	5.3	6.1	—	—
178			100.0	40.3	—	15.6	5.7	6.6	—	—
179	Seeds, meal, mechanical-extracted, 41% protein	5-01-617	92.6	41.0	9.1	11.9	4.7	6.2	16.7	26
180			100.0	44.3	9.8	12.9	5.0	6.6	18.0	28
181	Seeds, meal, solvent-extracted, 41% protein	5-01-621	91.0	41.3	12.0	12.2	1.5	6.5	18.2	24
182			100.0	45.4	13.2	13.4	1.7	7.1	20.0	26
183	Seeds without hulls, meal, prepressed solvent-extracted, 50%	5-07-874	93.0	50.3	—	8.2	1.3	6.6	—	—
184	protein		100.0	54.0	—	8.8	1.4	7.1	—	—
	COWPEA, COMMON. VIGNA SINENSIS									
185	Hay, sun-cured	1-01-645	90.4	17.7	—	24.4	2.6	10.5	—	—
186			100.0	19.6	—	27.1	2.9	11.7	—	—
	DROPSEED, SAND. SPOROBOLUS CRYPTANDRUS									
187	Fresh, stem cured	2-05-596	88.0	5.4	—	31.6	1.1	7.0	—	—
188			100.0	6.1	—	35.9	1.2	7.9	—	—
	FAT (SEE ANIMAL-POULTRY)									

TABLE 3–2A (continued). Composition of Feeds—Proximate Analysis and General Nutrients

		IFN	DRY MATT %	PRO TEIN %	CELL ULOS %	CRUID FIBR %	ETHR EXTR %	ASH %	ACID DETR %	NEUT DETR %
	FESCUE, MEADOW. FESTUCA ELATIOR									
189	Fresh	2-01-920	28.4	3.5	—	8.6	1.0	2.4	—	—
190			100.0	12.5	—	30.1	3.7	8.4	—	—
191	Hay, sun-cured	1-01-912	87.5	8.2	—	28.0	2.4	7.9	43.8	63
192			100.0	9.4	—	32.0	2.7	9.0	50.0	72
	FISH									
193	Solubles, condensed	5-01-969	50.4	31.5	—	0.5	6.0	10.1	—	—
194			100.0	62.5	—	1.0	11.9	20.1	—	—
195	Solubles, dehydrated	5-01-971	92.8	60.3	—	2.0	9.2	13.1	—	—
196			100.0	65.0	—	2.1	9.9	14.1	—	—
	FISH, ANCHOVETA. ENGRAULIS RINGEN									
197	Meal mechanical extracted	5-01-985	92.0	65.5	—	1.0	4.1	14.7	—	—
198			100.0	71.2	—	1.1	4.5	16.0	—	—
	FISH, HERRING. CLUPEA HARENGUS									
199	Meal mechanical extracted	5-02-000	91.8	71.1	—	1.8	8.4	10.7	—	—
200			100.0	77.4	—	1.9	9.1	11.7	—	—
	FISH, MENHADEN. BREVOORTIA TYRANNUS									
201	Meal mechanical extracted	5-02-009	91.7	62.2	—	0.7	9.8	18.9	—	—
202			100.0	67.9	—	0.8	10.7	20.6	—	—
	FISH, SARDINE. CLUPEA SPP-SARDINOPS SPP									
203	Meal, mechanical-extracted	5-01-015	92.9	65.1	—	1.0	5.0	15.8	—	—
204			100.0	70.0	—	1.1	5.4	17.0	—	—
	FISH, WHITE. GADIDAE (FAMILY)—LOPHIIDAE (FAMILY)—RAJIDAE (FAMILY)									
205	Meal, mechanical-extracted	5-02-025	91.2	62.9	—	0.5	4.6	22.9	—	—
206			100.0	68.9	—	0.6	5.1	25.1	—	—
	FLAX, COMMON. LINUM USITATISSIMUM									
207	Seed screenings	4-02-056	91.2	16.6	—	12.1	9.3	6.2	—	—
208			100.0	18.2	—	13.2	10.2	6.8	—	—
209	Seeds, meal, mechanical-extracted	5-02-045	90.8	34.5	—	8.9	4.8	5.7	15.4	23
210			100.0	38.0	—	9.8	5.3	6.2	17	25
	GALLETA. HILARIA JAMESII									
211	Fresh, stem-cured	2-05-594	86.0	4.3	—	28.4	1.4	13.3	—	—
212			100.0	5.0	—	33.0	1.7	15.5	—	—
	GAMAGRASS, EASTERN. TRIPSACUM DACTYLOIDES									
213	Fresh, full bloom	2-02-084	25.0	1.9	—	7.4	0.5	2.6	—	—
214			100.0	7.7	—	29.5	2.0	10.6	—	—
	GAMAGRASS, FLORIDA. TRIPSACUM FLORIDANUM									
215	Hay, sun-cured	1-02-087	92.3	8.4	—	26.0	1.8	6.8	—	—
216			100.0	9.1	—	28.2	1.9	7.4	—	—
	GARBAGE, HOTEL AND RESTAURANT									
217	Boiled	4-07-865	22.9	3.6	—	0.7	5.4	1.3	—	—
218			100.0	15.8	—	3.0	23.7	5.5	—	—
	GARBAGE, INSTITUTIONAL									
219	Boiled	4-07-867	17.9	2.7	—	0.5	2.8	1.0	—	—
220			100.0	15.0	—	2.8	15.8	5.5	—	—
	GRAMA. BOUTELOUA SPP									
221	Fresh, early vegetative	2-02-163	41.0	5.4	—	11.2	0.8	4.6	—	—
222			100.0	13.1	—	27.2	2.0	11.3	—	—
223	Fresh, mature	2-02-166	63.4	4.1	—	20.7	1.1	7.2	—	—
224			100.0	6.5	—	32.7	1.7	11.4	—	—
	GRAPE. VITIS SPP									
225	Pomace, dehydrated (marc)	1-02-208	90.4	12.1	—	27.9	7.6	7.5	49.2	50
226			100.0	13.4	—	30.9	8.4	8.3	54.4	55
	GRASS									
227	Hay, sun-cured, full bloom	1-02-244	89.3	8.5	—	28.7	2.1	6.6	—	—
228			100.0	9.5	—	32.1	2.3	7.4	—	—
	GRASS-LEGUME									
229	Silage	3-02-303	31.1	3.7	9.4	10.1	1.1	2.5	12.4	—
230			100.0	11.9	30.1	32.6	3.6	8.0	40.0	—
	HOMINY (SEE MAIZE)									
	LEGUME (SEE GRASS-LEGUME)									
	LESPEDEZA, CHINESE. LESPEDEZA CUNEATA									
231	Hay, sun-cured	1-02-607	90.4	10.7	—	29.9	2.0	4.8	—	—
232			100.0	11.8	—	33.1	2.2	5.3	—	—

		IFN	DRY MATT %	PRO TEIN %	CELL ULOS %	CRUID FIBR %	ETHR EXTR %	ASH %	ACID DETR %	NEUT DETR %
	LESPEDEZA, COMMON-LESPEDEZA, KOREAN. LESPEDEZA STRIATA-LESPEDEZA STIPULACEA									
233	Fresh	2-26-029	—	—	—	—	—	—	—	—
234			100.0*	—	—	—	—	—	—	—
235	Hay, sun-cured	1-26-035	—	—	—	—	—	—	—	—
236			100.0*	—	—	—	—	—	—	—
	LINSEED (SEE FLAX)									
	LUCERNE (SEE ALFALFA)									
	MAIZE. ZEA MAYS									
237	Aerial part with ears, sun-cured, mature (fodder)	1-02-772	82.1	6.6	—	18.6	1.9	4.4	—	—
238			100.0	8.0	—	22.6	2.3	5.4	—	—
239	Aerial part without ears without husks, sun-cured (stover) (straw)	1-02-776	85.2	5.4	—	29.3	1.1	6.1	33.2	57
240			100.0	6.4	—	34.4	1.3	7.2	39	67
241	Cobs, ground	1-02-782	89.8	2.8	—	32.2	0.6	1.6	39.5	80
242			100.0	3.1	—	35.8	0.7	1.8	44.0	89
243	Distiller's grains, dehydrated	5-02-842	93.5	27.8	—	11.5	8.9	2.2	15.9	40
244			100.0	29.7	—	12.3	9.5	2.4	17	43
245	Distiller's grains with solubles, dehydrated	5-02-843	91.8	27.1	—	9.1	9.2	4.5	16.5	40
246			100.0	29.5	—	9.9	10.1	4.9	18	44
247	Distiller's solubles, dehydrated	5-02-844	92.9	27.4	—	4.6	8.6	7.2	6.5	21
248			100.0	29.5	—	4.9	9.3	7.7	7	23
249	Ears with husks, silage	4-02-839	43.4	4.0	—	4.8	1.6	1.5	6.1	—
250			100.0	9.2	—	11.1	3.7	3.3	14	—
251	Ears, ground (maize and cob meal)	4-02-849	86.5	7.8	—	8.2	3.2	1.7	9.5	24
252			100.0	9.0	—	9.4	3.7	1.9	11	28
253	Gluten, meal	5-02-900	91.3	43.2	—	4.5	2.2	3.1	8.2	34
254			100.0	47.3	—	4.9	2.4	3.4	9	37
255	Gluten with bran (maize bluten feed)	5-02-903	89.9	22.9	—	8.7	2.2	6.7	10.8	40
256			100.0	25.5	—	9.7	2.5	7.4	12	45
257	Grits byproduct (hominy feed)	4-02-887	90.2	10.3	—	4.8	6.5	2.8	11.7	50
258			100.0	11.4	—	5.3	7.2	3.1	13	55
259	Silage, well-eared	3-02-823	34.1	2.8	—	8.1	1.0	1.5	9.7	17
260			100.0	8.1	—	23.7	3.1	4.5	28.3	51
261	Silage, 30–50% dry matter	3-20-506	36.9	3.1	8.3	8.1	1.4	2.1	9.8	—
262			100.0	8.3	22.4	21.9	3.8	5.8	26.7	—
263	Silage, aerial part without ears without husks (stalklage) (stover)	3-02-836	30.7	1.9	—	9.6	0.8	2.4	17.0	20
264			100.0	6.3	—	31.2	2.6	7.8	55.4	67
	MAIZE, DENT WHITE. ZEA MAYS INDENTATA									
265	Grits byproduct (hominy feed)	4-02-990	90.0	10.6	—	5.1	7.9	3.0	—	—
266			100.0	11.8	—	5.7	8.8	3.3	—	—
	MAIZE, DENT YELLOW. ZEA MAYS INDENTATA									
267	Grain	4-02-935	88.0	9.1	2.1	2.2	3.6	1.3	3.8	8
268			100.0	10.4	2.4	2.5	4.1	1.5	4.3	9
269	Grain, grade 2 69.5 kg/hl (54 lb/bushel)	4-02-931	87.3	8.8	4.1	2.2	3.9	1.2	5.3	8
270			100.0	10.1	4.7	2.5	4.5	1.4	6.1	9
271	Grain, grade 3 66.9 kg/hl (52 lb/bushel)	4-02-932	86.5	8.6	—	2.1	3.9	1.2	3.3	—
272			100.0	10.0	—	2.4	4.5	1.4	3.8	—
	MAIZE, DENT YELLOW, OPAQUE 2. ZEA MAYS INDENTATA									
273	Grain	4-28-253	91.0	9.6	—	2.3	4.9	1.6	—	—
274			100.0	10.6	—	2.6	5.3	1.8	—	—
	MAIZE, SWEET. ZEA MAYS SACCHARATA									
275	Process residue, silage	3-07-955	31.0	2.5	—	10.1	1.3	1.7	10.5	—
276			100.0	8.0	—	32.7	4.3	5.4	34.0	—
277	Process residue, wet (cannery residue)	2-02-975	76.8	6.8	—	17.0	1.9	3.5	22.3	—
278			100.0	8.8	—	22.2	2.4	4.6	29.0	—
	MANURE (SEE CATTLE; POULTRY)									
	MEAT (SEE ANIMAL)									
	MILK (SEE CATTLE)									
	MILLET. SETARIA SPP									
279	Grain	4-03-098	89.9	11.9	—	6.2	3.9	2.8	15.3	—
280			100.0	13.2	—	6.9	4.3	3.1	17	—

		IFN	DRY MATT %	PRO TEIN %	CELL ULOS %	CRUID FIBR %	ETHR EXTR %	ASH %	ACID DETR %	NEUT DETR %
	MILLET, PEARL. PENNISETUM GLAUCUM									
281	Fresh	2-03-115	20.7	2.1	—	6.5	0.6	1.9	—	—
282			100.0	10.1	—	31.1	2.9	9.2	—	—
283	Silage	3-20-903	28.3	3.6	9.8	9.2	—	4.2	11.5	—
284			100.0	12.6	34.8	32.5	—	14.9	40.8	—
	MOLASSES (SEE BEET; SUGARCANE)									
	NAPIERGRASS. PENNISETUM PURPUREUM									
285	Fresh, late vegetative	2-03-158	20.2	1.8	—	6.7	0.6	1.7	9.1	14
286			100.0	8.7	—	33.0	3.0	8.6	45	70
287	Fresh, late bloom	2-03-162	23.0	1.8	—	9.0	0.3	1.2	10.8	17
288			100.0	7.8	—	39.0	1.1	5.3	47	75
	NEEDLE AND THREAD. STIPA COMATA									
289	Fresh, stem-cured	2-07-989	92.0	3.7	33.4	—	5.0	19.4	39.7	—
290			100.0	4.1	36.3	—	5.4	21.1	43.2	—
	OATS. AVENA SATIVA									
291	Cereal byproduct, less than 4% fiber (feeding oat meal (oat	4-03-303	90.7	14.8	—	3.6	6.4	2.3	—	—
292	middlings)		100.0	16.3	—	4.0	7.0	2.5	—	—
293	Grain	4-03-309	89.2	11.8	—	10.7	4.6	3.1	14.2	28
294			100.0	13.3	—	12.0	5.2	3.4	15.9	32
295	Grain, Pacific coast	4-07-999	90.9	9.1	—	11.2	5.0	3.8	—	—
296			100.0	10.0	—	12.3	5.5	4.2	—	—
297	Groats	4-03-331	89.6	15.5	—	2.5	6.1	2.0	—	—
298			100.0	17.3	—	2.8	6.8	2.3	—	—
299	Hay, sun-cured	1-03-280	90.7	8.6	—	29.1	2.2	7.2	34.8	56
300			100.0	9.5	—	32.0	2.4	7.9	38.4	62
301	Hulls	1-03-281	92.4	3.8	27.3	30.6	1.4	6.1	36.5	72
302			100.0	4.1	29.5	33.2	1.5	6.6	39.6	78
303	Silage	3-03-298	30.5	2.9	—	9.6	1.0	2.6	11.6	—
304			100.0	9.5	—	31.4	3.3	8.5	38.1	—
305	Straw	1-03-283	92.2	4.1	39.5	37.2	2.0	7.2	44.2	64
306			100.0	4.4	42.8	40.4	2.2	7.8	47.9	70
	ORANGE. CITRUS SINENSIS									
307	Pomace without fines, dehydrated (dried orange pulp)	4-01-254	88.1	7.5	—	8.4	1.7	3.7	14.1	18
308			100.0	8.5	—	9.6	1.9	4.2	16	21
	ORCHARDGRASS. DACTYLIS GLOMERATA									
309	Fresh, early vegetative	2-03-439	22.6	5.7	4.0	5.7	1.1	2.6	5.2	12
310			100.0	25.1	17.8	25.2	5.0	11.4	22.8	55
311	Fresh, early bloom	2-03-442	23.5	3.0	6.0	7.5	0.9	1.9	7.2	—
312			100.0	12.8	25.6	32.0	3.7	8.1	30.7	—
313	Hay, sun-cured	1-03-438	89.6	10.5	26.4	31.1	2.8	6.5	34.0	56
314			100.0	11.8	29.4	34.7	3.1	7.3	37.9	63
	OYSTER. CRASSOSTREA SPP-OSTREA SPP									
315	Shells, fine ground (oyster shell flour)	6-03-481	99.1	1.0	—	—	—	81.3	—	—
316			100.0	1.0	—	—	—	82.1	—	—
	PANGOLAGRASS. DIGITARIA DECUMBENS									
317	Fresh	2-03-493	20.2	1.8	7.0	6.6	0.5	1.5	7.5	—
318			100.0	9.1	34.5	32.6	2.3	7.6	36.9	—
319	Hay, sun-cured	1-09-459	88.8	5.0	—	30.9	—	—	36.4	62
320			100.0	5.6	—	34.8	—	—	41	70
	PAPER									
321	Waste, corrugated	1-28-257	90.0	0.5	63.0	—	—	3.6	73.8	83
322			100.0	0.5	70.0	—	—	4.0	82.0	92
	PEA. PISUM SPP									
323	Haulm with pods, silage	3-03-596	24.5	3.2	—	7.3	0.8	2.2	12.0	14
324			100.0	13.1	—	29.8	3.3	9.0	49	59
325	Seeds	5-03-600	89.1	23.4	—	5.6	0.9	2.8	—	—
326			100.0	26.3	—	6.3	1.0	3.2	—	—
	PEANUT. ARACHIS HYPOGAEA									
327	Hay, sun-cured	1-03-619	90.7	9.9	—	30.3	3.3	8.2	37.2	—
328			100.0	10.9	—	33.4	3.6	9.0	41.0	—
329	Seeds without coats, meal, mechanical-extracted (peanut meal)	5-03-649	92.6	49.2	4.2	6.2	5.6	5.0	5.6	13
330			100.0	53.1	4.5	6.7	6.0	5.4	6.1	14
331	Seeds without coats, meal, solvent-extracted (peanut meal)	5-03-650	92.4	48.9	—	7.7	2.1	5.8	—	—
332			100.0	52.9	—	8.4	2.3	6.3	—	—
	PEARL MILLET (SEE MILLET, PEARL)									

TABLE 3–2A (continued). Composition of Feeds—Proximate Analysis and General Nutrients

		IFN	DRY MATT %	PRO TEIN %	CELL ULOS %	CRUID FIBR %	ETHR EXTR %	ASH %	ACID DETR %	NEUT DETR %
	PINEAPPLE. ANANAS COMOSUS									
333	Process residue, dehydrated (pineapple bran)	4-03-722	87.1	4.0	—	18.2	1.3	3.0	32	63
334			100.0	4.6	—	20.9	1.5	3.5	37	73
	POTATO. SOLANUM TUBEROSUM									
335	Process residue, dehydrated	4-03-775	90.0	7.3	—	5.4	2.6	3.3	6.8	—
336			100.0	8.1	—	6.0	2.9	3.6	7.6	—
337	Tubers, dehydrated	4-07-850	91.1	8.0	—	2.1	0.5	7.2	2.7	—
338			100.0	8.8	—	2.3	0.5	7.9	3.0	—
339	Tubers, fresh	4-03-787	21.9	2.1	—	0.5	0.1	1.0	—	—
340			100.0	9.7	—	2.2	0.4	4.4	—	—
341	Tubers, silage	4-03-768	24.7	1.8	—	0.8	0.1	1.4	1.2	—
342			100.0	7.3	—	3.2	0.4	5.5	5.0	—
	POULTRY									
343	Byproduct, meal, rendered (viscera with feet with heads)	5-03-798	93.8	61.1	—	2.1	13.1	14.6	—	—
344			100.0	65.1	—	2.3	13.9	15.6	—	—
345	Feathers, meal, hydrolyzed	5-03-795	92.9	83.4	—	1.3	5.5	2.9	5.7	—
346			100.0	89.8	—	1.4	5.9	3.1	6.2	—
347	Manure, dehydrated	5-14-015	90.6	27.2	—	12.2	2.1	25.0	14.5	—
348			100.0	30.0	—	13.5	2.3	27.5	16.0	—
	PRAIRIE PLANTS, MIDWEST									
349	Hay, sun-cured, midbloom	1-07-956	94.1	5.6	—	33.7	2.1	5.8	—	—
350			100.0	6.0	—	35.9	2.2	6.1	—	—
351	Hay, sun-cured, late bloom	1-07-957	91.0	—	—	—	—	—	—	—
352			100.0	—	—	—	—	—	—	—
353	Hay, sun-cured, milk stage	1-03-185	89.6	5.9	—	28.9	2.4	8.0	—	—
354			100.0	6.6	—	32.2	2.7	8.9	—	—
	PRICKLY PEAR. OPUNTIA SPP									
355	Fresh	2-01-061	16.4	0.9	—	2.1	0.3	3.2	—	—
356			100.0	5.5	—	12.7	2.0	19.8	—	—
	RAPE. BRASSICA NAPUS									
357	Fresh	2-03-867	16.7	2.9	—	2.4	0.6	2.1	—	—
358			100.0	17.6	—	14.7	3.8	12.6	—	—
359	Seeds, meal, mechanical-extracted	5-03-870	92.0	35.4	—	11.9	7.4	6.9	—	—
360			100.0	38.5	—	12.9	8.1	7.5	—	—
361	Seeds, meal, solvent-extracted	5-03-871	91.2	36.9	—	11.9	1.7	6.8	—	—
362			100.0	40.5	—	13.1	1.9	7.5	—	—
	RAPE, SUMMER. BRASSICA NAPUS ANNUA									
363	Seeds, meal, boiled mechanical-extracted	5-08-136	94.0	35.2	—	15.5	7.0	6.8	15.0	32
364			100.0	37.4	—	16.5	7.4	7.2	16	34
365	Seeds, meal, prepressed solvent-extracted	5-08-135	91.6	39.2	—	10.2	2.5	7.2	16.5	33
366			100.0	42.8	—	11.1	2.7	7.9	18	36
	RAPE, TURNIP. BRASSICA CAMPESTRIS									
367	Seeds, meal, mechanical-extracted	5-07-871	93.7	34.5	—	13.4	7.6	6.8	—	—
368			100.0	36.9	—	14.3	8.2	7.2	—	—
369	Seeds, meal, solvent-extracted	5-07-870	90.9	37.1	—	12.2	2.3	6.4	—	—
370			100.0	40.8	—	13.5	2.6	7.0	—	—
	REDTOP. AGROSTIS ALBA									
371	Fresh, full bloom	2-03-891	26.3	2.1	—	6.6	0.9	1.8	—	—
372			100.0	8.1	—	25.1	3.5	7.0	—	—
373	Hay, sun-cured, midbloom	1-03-886	92.8	11.1	—	29.0	2.4	6.0	—	—
374			100.0	12.0	—	31.2	2.6	6.5	—	—
	RICE. ORYZA SATIVA									
375	Bran with germs (rice, bran)	4-03-928	90.5	13.1	—	11.7	13.6	10.4	25.7	30
376			100.0	14.4	—	12.9	15.0	11.5	28.4	33
377	Grain, ground	4-03-938	89.0	7.5	—	8.6	1.6	5.3	—	—
378			100.0	8.4	—	9.7	1.8	6.0	—	—
379	Groats, polished	4-03-942	88.6	6.8	—	0.4	0.5	0.5	—	—
380			100.0	7.7	—	0.4	0.5	0.6	—	—
381	Groats, polished broken	4-03-932	88.5	8.0	0.4	0.8	0.8	0.9	1.2	14
382			100.0	9.0	0.5	0.9	0.9	1.0	1.4	16
383	Hulls	1-08-075	91.9	2.8	34.2	39.2	1.0	19.0	63.1	75
384			100.0	3.1	37.2	42.7	1.1	20.6	68.7	82
385	Polishings	4-03-943	90.2	11.9	—	3.3	12.5	7.5	3.6	—
386			100.0	13.2	—	3.6	13.9	8.3	4.0	—

		IFN	DRY MATT %	PRO TEIN %	CELL ULOS %	CRUID FIBR %	ETHR EXTR %	ASH %	ACID DETR %	NEUT DETR %
	RUSSIAN THISTLE, TUMBLING. SALSOLA KALI TENUIFOLIA									
387	Fresh, stem-cured	2-08-000	88.0	8.1	—	—	—	10.7	38.8	—
388			100.0	9.2	—	—	—	12.2	44.1	—
389	Hay, sun-cured	1-03-988	86.4	10.2	—	25.1	1.7	13.1	—	—
390			100.0	11.8	—	29.0	2.0	15.2	—	—
	RYE. SECALE CEREALE									
391	Distiller's grains, dehydrated	5-04-023	91.9	23.0	—	12.3	6.0	2.3	—	—
392			100.0	25.1	—	13.4	6.5	2.5	—	—
393	Flour byproduct, less than 8.5% fiber (rye middlings)	4-04-031	89.5	16.4	—	5.2	3.4	3.4	6.3	—
394			100.0	18.3	—	5.8	3.8	3.8	7.0	—
395	Fresh, early vegetative	2-04-013	15.5	—	—	—	—	—	—	—
396			100.0	—	—	—	—	—	—	—
397	Grain	4-04-047	87.5	12.0	2.1	2.2	1.5	1.6	3.7	—
398			100.0	13.7	2.4	2.5	1.7	1.9	4.2	—
399	Mill run, less than 9.5% fiber	4-04-034	88.8	15.6	—	4.9	3.2	4.5	—	—
400			100.0	17.6	—	5.5	3.7	5.0	—	—
401	Silage	3-04-020	35.0	4.2	—	12.1	1.2	2.8	13.6	—
402			100.0	12.1	—	34.6	3.3	7.9	38.9	—
403	Straw	1-04-007	90.9	2.8	—	38.3	1.4	3.8	—	—
404			100.0	3.0	—	42.1	1.5	4.2	—	—
	RYEGRASS, ITALIAN. LOLIUM MULTIFLORUM									
405	Fresh	2-04-073	22.6	4.0	—	4.7	0.9	3.9	7.5	9
406			100.0	17.9	—	20.9	4.1	17.4	33	38
	RYEGRASS, PERENNIAL. LOLIUM PERENNE									
407	Fresh	2-04-086	26.6	3.0	—	6.7	1.3	2.4	—	—
408			100.0	11.3	—	25.2	4.9	9.0	—	—
409	Hay, sun-cured	1-04-077	86.6	8.4	—	25.1	2.4	7.9	26.0	35
410			100.0	9.7	—	28.9	2.8	9.1	30	41
	SACATON, ALKALI. SPOROBOLUS AIROIDES									
411	Fresh, stem-cured	2-05-599	76.0	3.1	—	28.3	1.1	7.7	—	—
412			100.0	4.1	—	37.3	1.5	10.1	—	—
	SAFFLOWER. CARTHAMUS TINCTORIUS									
413	Seeds	4-07-958	93.1	14.9	—	23.6	30.8	3.0	37.2	—
414			100.0	16.0	—	25.3	33.1	3.2	40.0	—
415	Seeds, meal, mechanical-extracted	5-04-109	91.6	21.3	—	31.3	6.6	4.4	40.0	54
416			100.0	23.3	—	34.2	7.2	4.8	43.7	59
417	Seeds, meal, solvent-extracted	5-04-110	91.7	22.9	—	29.9	1.2	5.3	37.6	53
418			100.0	25.0	—	32.6	1.3	5.8	41	58
419	Seeds without hulls, meal, solvent-extracted	5-07-959	91.0	42.7	—	13.4	1.2	7.8	—	—
420			100.0	46.9	—	14.7	1.3	8.5	—	—
	SAGE, BLACK. SALVIA MELLIFERA									
421	Browse, fresh, stem-cured	2-05-564	65.0	5.5	—	—	7.0	3.6	—	—
422			100.0	8.5	—	—	10.7	5.5	—	—
	SAGEBRUSH, BIG. ARTEMISIA TRIDENTATA									
423	Browse, fresh, stem-cured	2-07-992	65.0	6.1	—	—	6.4	2.8	—	—
424			100.0	9.3	—	—	9.8	4.3	—	—
	SAGEBRUSH, BUD. ARTEMISIA SPINESCENS									
425	Browse, fresh, early vegetative	2-07-991	23.0	4.0	—	—	1.1	4.9	—	—
426			100.0	17.3	—	—	4.9	21.4	—	—
427	Browse, fresh, late vegetative	2-04-124	32.0	5.6	—	7.3	0.8	6.9	—	—
428			100.0	17.5	—	22.7	2.5	21.6	—	—
	SAGEBRUSH, FRINGED. ARTEMISIA FRIGIDA									
429	Browse, fresh, midbloom	2-04-129	43.0	4.0	—	14.3	0.9	2.8	—	—
430			100.0	9.4	—	33.2	2.0	6.5	—	—
431	Browse, fresh, mature	2-04-130	60.0	4.3	—	19.1	2.0	10.3	21.1	—
432			100.0	7.1	—	31.8	3.4	17.1	35.1	—
	SALTBUSH, NUTTALL. A TRIPLEX NUTTALLI									
433	Browse, fresh, stem-cured	2-07-993	55.0	4.0	—	—	1.2	11.8	—	—
434			100.0	7.2	—	—	2.2	21.5	—	—
	SALTBUSH, SHADSCALE. A TRIPLEX CONFERTIFOLIA									
435	Browse, fresh, stem-cured	2-05-565	80.0	6.1	—	—	2.0	18.9	—	—
436			100.0	7.7	—	—	2.5	23.6	—	—

		IFN	DRY MATT %	PRO TEIN %	CELL ULOS %	CRUID FIBR %	ETHR EXTR %	ASH %	ACID DETR %	NEUT DETR %
	SALT GRASS. DISTICHLIS SPP									
437	Fresh	2-04-170	74.4	4.8	—	22.5	1.3	5.6	—	—
438			100.0	6.5	—	30.3	1.8	7.5	—	—
439	Hay, sun-cured	1-04-168	89.5	8.0	—	28.3	1.8	11.4	—	—
440			100.0	8.9	—	31.6	2.0	12.7	—	—
	SALTGRASS, DESERT. DISTICHLIS STRICTA									
441	Fresh	2-04-171	29.0	1.6	—	8.6	0.5	2.0	—	—
442			100.0	5.7	—	29.7	1.7	6.8	—	—
	SCREENINGS (SEE BARLEY; CEREALS; WHEAT)									
	SEAWEED, KELP. LAMINARIALES (ORDER)-FUCALES (ORDER)									
443	Whole, meal	1-08-073	90.9	6.5	—	6.7	0.5	35.0	—	—
444			100.0	7.1	—	7.4	0.5	38.6	—	—
	SEDGE. CAREX SPP									
445	Hay, sun-cured	1-04-193	89.3	8.6	—	27.0	2.1	6.4	33.6	—
446			100.0	9.7	—	30.2	2.4	7.2	37.6	—
	SESAME. SESAMUM INDICUM									
447	Seeds, meal, mechanical-extracted	5-04-220	92.7	45.6	—	5.7	6.4	11.4	15.7	16
448			100.0	49.2	—	6.2	6.9	12.3	17	17
	SKIM MILK (SEE CATTLE)									
	SORGHUM. SORGHUM BICOLOR									
449	Aerial part without heads, sun-cured, mature	1-07-961	88.1	3.9	30.9	28.4	1.2	—	37.7	—
450			100.0	4.4	35.1	32.3	1.4	—	42.8	—
451	Distiller's grains, dehydrated	5-04-374	93.8	30.8	—	12.1	8.3	4.3	—	—
452			100.0	32.9	—	12.9	8.8	4.6	—	—
453	Grain	4-04-383	90.1	11.5	5.0	2.6	2.7	1.7	8.3	—
454			100.0	12.7	5.6	2.8	3.0	1.9	9.3	—
455	Grain, 9–12% protein	4-08-139	89.4	10.1	—	2.4	2.6	1.8	—	—
456			100.0	11.3	—	2.7	2.9	2.0	—	—
	SORGHUM, JOHNSONGRASS. SORGHUM HALEPENSE									
457	Hay, sun-cured	1-04-407	90.5	6.7	—	30.4	3.0	7.7	—	—
458			100.0	7.5	—	33.6	2.2	8.6	—	—
	SORGHUM, KAFIR. SORGHUM BICOLOR CAFFROFUM									
459	Grain	4-04-428	89.1	10.8	—	2.0	2.8	1.5	1.8	—
460			100.0	12.1	—	2.2	3.1	1.7	2.0	—
	SORGHUM, MILO. SORGHUM BICOLOR SUBGLABRESCENS									
461	Grain	4-04-444	88.5	10.0	2.7	2.2	2.8	1.6	3.5	16
462			100.0	11.3	3.0	2.5	3.2	1.8	4.0	18
463	Heads	4-04-446	89.9	8.6	—	8.9	2.6	4.8	—	—
464			100.0	9.6	—	9.9	2.9	5.4	—	—
	SORGHUM, SORGO. SORGHUM BICOLOR SACCHARATUM									
465	Silage	3-04-468	28.8	1.9	—	7.0	0.7	2.4	11.0	—
466			100.0	6.7	—	24.4	2.5	8.3	38.0	—
	SORGHUM, SUDANGRASS. SORGHUM BICOLOR SUDANENSE									
467	Fresh, early vegetative	2-04-484	17.8	3.0	—	5.5	0.7	1.6	5.2	10
468			100.0	16.8	—	30.9	3.9	9.0	29	55
469	Fresh, midbloom	2-04-485	22.8	2.0	—	8.2	0.4	2.4	9	15
470			100.0	8.8	—	36.1	1.8	10.5	40	65
471	Hay, sun-cured	1-04-480	91.1	10.9	29.9	26.2	1.6	10.7	37.2	62
472			100.0	12.0	32.8	28.7	1.7	11.8	40.8	68
473	Silage	3-04-499	23.3	2.5	—	7.9	0.7	2.1	5.3	16
474			100.0	10.9	—	34.0	2.9	9.2	22.6	68

		IFN	DRY MATT %	PRO TEIN %	CELL ULOS %	CRUID FIBR %	ETHR EXTR %	ASH %	ACID DETR %	NEUT DETR %
	SOYBEAN. GLYCINE MAX									
475	Hay, sun-cured	1-04-558	89.1	14.1	—	30.6	2.3	7.2	35.7	—
476			100.0	15.8	—	34.3	2.5	8.0	40.0	—
477	Seed coats (hulls)	1-04-560	90.3	11.2	41.6	35.5	1.9	4.4	44.2	60
478			100.0	12.4	46.1	39.3	2.1	4.9	49.0	67
479	Seeds	5-04-610	92.0	39.2	—	5.3	17.1	5.1	8.2	—
480			100.0	42.5	—	5.7	18.6	5.5	8.9	—
481	Seeds, heat-processed	5-04-597	92.6	36.6	—	5.1	18.7	4.7	—	—
482			100.0	39.5	—	5.6	20.2	5.1	—	—
483	Seeds, meal, mechanical-extracted	5-04-600	90.0	43.3	—	5.8	4.7	6.0	—	—
484			100.0	48.1	—	6.5	5.2	6.6	—	—
485	Seeds, meal, solvent-extracted	5-04-604	89.6	45.7	—	5.8	1.2	6.2	—	—
486			100.0	51.0	—	6.5	1.3	6.9	—	—
487	Seeds without hulls, meal, solvent-extracted	5-04-612	89.9	49.3	4.1	3.5	1.0	6.0	5.5	7
488			100.0	54.8	4.5	3.8	1.1	6.7	6.1	8
489	Silage	3-04-581	30.2	5.2	—	9.0	0.8	3.0	—	—
490			100.0	17.1	—	29.9	2.6	9.9	—	—
491	Straw	1-04-567	87.7	4.6	—	38.9	1.3	5.6	—	—
492			100.0	5.2	—	44.3	1.4	6.4	—	—
	SQUIRREL TAIL. SITANION SPP									
493	Fresh, stem-cured	2-05-566	50.0	1.5	—	—	1.1	8.5	—	—
494			100.0	3.1	—	—	2.2	17.0	—	—
	SUDANGRASS (SEE SORGHUM)									
	SUGARCANE. SACCHARUM OFFICINARUM									
495	Molasses, dehydrated	4-04-695	94.4	9.0	—	7.1	0.8	12.0	—	—
496			100.0	9.5	—	7.5	0.9	12.7	—	—
497	Molasses, more than 46% invert sugars more than 79.5 degrees	4-04-696	74.3	4.3	—	0.4	0.2	9.9	0.3	—
498	brix		100.0	5.8	—	0.5	0.2	13.3	0.4	—
	SUNFLOWER, COMMON. HELIANTHUS ANNUUS									
499	Seeds without hulls, meal, mechanical-extracted	5-04-738	92.7	40.5	—	12.7	8.5	6.7	—	—
500			100.0	43.7	—	13.7	9.2	7.2	—	—
501	Seeds without hulls, meal, solvent-extracted	5-04-739	92.5	45.2	—	11.7	2.7	7.5	—	—
502			100.0	48.9	—	12.7	2.9	8.1	—	—
	SWEET CLOVER, YELLOW. MELILOTUS OFFICINALIS									
503	Hay, sun-cured	1-04-754	88.5	13.7	—	28.9	1.9	7.6	—	—
504			100.0	15.4	—	32.6	2.2	8.6	—	—
	TANKAGE—SEE ANIMAL; ANIMAL-POULTRY									
	TIMOTHY. PHLEUM PRATENSE									
505	Fresh, late vegetative	2-04-903	26.7	3.3	—	8.6	1.0	2.0	—	—
506			100.0	12.2	—	32.1	3.8	7.5	—	—
507	Fresh, midbloom	2-04-905	29.2	2.7	—	9.8	0.9	1.9	—	—
508			100.0	9.1	—	33.5	3.0	6.6	—	—
509	Hay, sun-cured, late vegetative	1-04-881	89.3	12.3	—	27.9	2.6	6.2	29.4	49
510			100.0	13.8	—	31.3	2.9	7.0	32.9	55
511	Hay, sun-cured, early bloom	1-04-882	89.1	9.6	31.2	30.0	2.5	5.1	31.4	54
512			100.0	10.8	35.0	33.6	2.8	5.7	35.2	61
513	Hay, sun-cured, midbloom	1-04-883	88.9	8.6	30.1	30.0	2.3	5.4	32.3	60
514			100.0	9.7	33.9	33.8	2.6	6.1	36.4	67
515	Hay, sun-cured, late bloom	1-04-885	88.3	6.9	31.0	28.8	2.4	4.8	36.9	62
516			100.0	7.8	35.1	32.6	2.7	5.4	41.8	70
517	Silage	3-04-922	33.5	3.9	—	11.7	1.1	2.9	12.9	—
518			100.0	11.8	—	35.0	3.4	8.6	38.6	—

TABLE 3-2A (continued). Composition of Feeds—Proximate Analysis and General Nutrients

		IFN	DRY MATT %	PRO TEIN %	CELL ULOS %	CRUID FIBR %	ETHR EXTR %	ASH %	ACID DETR %	NEUT DETR %
	TOMATO. LYCOPERSICON ESCULENTUM									
519	Pomace, dehydrated	5-05-041	92.2	21.0	18.7	25.6	9.9	6.4	46.9	51
520			100.0	22.8	20.3	27.7	10.8	7.0	50.9	55
	TREFOIL, BIRDSFOOT. LOTUS CORNICULATUS									
521	Fresh	2-20-786	19.3	4.0	—	4.1	0.8	2.2	—	—
522			100.0	20.6	—	21.2	4.0	11.2	—	—
523	Hay, sun-cured	1-05-044	90.6	13.9	—	29.3	1.9	6.7	32.6	43
524			100.0	15.3	—	32.3	2.1	7.4	36	47
	TRITICALE. TRITICALE HEXAPLOIDE									
525	Grain	4-20-362	89.2	14.7	—	2.9	1.5	1.8	—	—
526			100.0	16.5	—	3.3	1.6	2.0	—	—
	TURNIP. BRASSICA RAPA RAPA									
527	Roots, fresh	4-05-067	9.2	1.2	—	1.1	0.2	0.8	3.1	4
528			100.0	13.2	—	11.5	1.8	8.7	34	44
	UREA									
529	45% nitrogen 281% protein equivalent	5-05-070	97.0	276.9	—	—	—	1.5	—	—
530			100.0	285.4	—	—	—	1.5	—	—
	VEGETABLE									
531	Oil	4-05-077	99.8	—	—	—	99.7	—	—	—
532			100.0	—	—	—	99.9	—	—	—
	VETCH. VICIA SPP									
533	Hay, sun-cured	1-05-106	88.7	18.4	—	24.8	2.7	7.8	29.3	43
534			100.0	20.7	—	27.9	3.0	8.8	33	48
	WHEAT. TRITICUM AESTIVUM									
535	Bran	4-05-190	89.0	15.4	9.5	10.0	3.8	5.9	12.5	45
536			100.0	17.4	10.7	11.3	4.3	6.6	14.0	51
537	Flour byproduct, less than 4% fiber (wheat red dog)	4-05-203	88.3	15.7	4.9	2.9	3.4	2.5	8.1	—
538			100.0	17.7	5.6	3.3	3.9	2.8	9.2	—
539	Flour byproduct, less than 7% fiber (wheat shorts)	4-05-201	88.4	16.7	—	6.3	4.3	4.4	—	—
540			100.0	18.9	—	7.2	4.8	5.0	—	—
541	Flour byproduct, less than 9.5% fiber (wheat middlings)	4-05-205	88.9	16.4	7.5	7.8	4.2	4.6	10.5	33
			100.0	18.4	8.4	8.8	4.7	5.2	11.8	37
543	Fresh, early vegetative	2-05-176	22.2	6.1	5.3	3.9	1.0	3.0	6.3	11
544			100.0	27.4	23.9	17.4	4.4	13.3	28.4	52
545	Germ, ground	5-05-218	88.4	24.5	—	3.1	8.5	4.3	4.4	—
546			100.0	27.8	—	3.5	9.7	4.9	5.0	—
547	Grain	4-05-211	89.0	13.0	2.1	2.5	1.7	1.7	3.3	—
548			100.0	14.7	2.4	2.8	2.0	2.0	3.7	—
549	Grain screenings	4-05-216	88.5	13.3	—	5.0	2.9	2.9	—	—
550			100.0	15.0	—	5.6	3.2	3.3	—	—
551	Hay, sun-cured	1-05-172	88.7	7.7	—	25.7	2.0	7.0	36.4	60
552			100.0	8.7	—	29.0	2.2	7.9	41	68
553	Mill run, less than 9.5% fiber	4-05-206	89.9	15.6	—	8.2	4.1	5.1	9.9	—
554			100.0	17.3	—	9.1	4.6	5.7	11.0	—
555	Silage, early vegetative	3-05-184	29.9	3.5	—	8.4	1.0	1.9	9.0	—
556			100.0	11.7	—	28.1	3.4	6.4	30.0	—
557	Straw	1-05-175	91.1	3.3	40.0	38.0	1.8	7.1	49.9	77
558			100.0	3.6	43.9	41.7	2.0	7.8	54.8	85

		IFN	DRY MATT %	PRO TEIN %	CELL ULOS %	CRUID FIBR %	ETHR EXTR %	ASH %	ACID DETR %	NEUT DETR %
	WHEAT, DURUM, TRITICUM DURUM									
559	Grain	4-05-224	87.6	13.7	—	2.2	1.8	1.6	—	—
560			100.0	15.6	—	2.6	2.0	1.9	—	—
	WHEAT, HARD RED SPRING, TRITICUM AESTIVUM									
561	Grain	4-05-258	87.6	14.9	7.2	2.5	1.8	1.6	11.0	—
561			100.0	17.1	8.2	2.8	2.0	1.8	12.6	
	WHEAT, HARD RED WINTER, TRITICUM AESTIVUM									
563	Grain	4-05-268	88.8	12.8	—	2.6	1.6	1.8	3.9	—
564			100.0	14.4	—	2.9	1.8	2.0	4.4	—
	WHEAT, SOFT RED WINTER, TRITICUM AESTIVUM									
565	Grain	4-05-294	88.4	11.4	—	2.3	1.6	1.9	—	—
566			100.0	12.9	—	2.6	1.8	2.1	—	—
	WHEAT, SOFT WHITE WINTER, TRITICUM AESTIVIUM									
567	Grain	4-05-337	90.2	10.4	—	2.3	1.5	1.5	3.6	13
568			100.0	11.5	—	2.6	1.7	1.7	4	14
	WHEATGRASS, GRESTED, AGROPRON DESERTORUM									
569	Fresh	2-05-429	45.0	6.3	—	13.4	1.0	3.4	16.1	—
570			100.0	13.9	—	29.7	2.2	7.6	35.6	
571	Hay, sun-cured	1-05-418	91.8	10.3	—	30.8	2.1	6.4	33.2	—
572			100.0	11.2	—	33.6	2.3	7.0	36.2	—
	WHEY—SEE CATTLE									
	YEAST, BREWER'S, SACCHAROMYCES CEREVISIAE									
573	Dehydrated	7-05-527	93.1	43.4	—	3.2	1.0	6.7	3.7	—
574			100.0	46.6	—	3.5	1.1	7.2	4.0	—
	YEAST, IRRADIATED, SACCHAROMYCES CEREVISIAE									
575	Dehydrated	7-05-529	93.9	48.1	—	6.2	1.1	6.2	—	—
576			100.0	51.2	—	6.5	1.2	6.6	—	—
	YEAST, TORULA, TORULOPSIS UTILIS									
577	Dehydrated	7-05-534	93.0	49.5	—	2.5	1.6	8.6	4.0	—
578			100.0	53.2	—	2.7	1.7	8.6	4.0	

TABLE 3–2B. Composition of Feeds—Energy Values

		IFN	DRY MATT %	TDN CTTL %	TDN SHEP %	DE CTTL Mcal /kg	DE SHEP Mcal /kg
	ALFALFA, MEDICAGO SATIVA						
001	Fresh	2-00-196	26.0	15.9	15.8	0.70*	0.70*
002			100.0	61.0	60.7	2.69*	2.67*
003	Hay, sun-cured, late vegetative	1-00-054	89.7	56.4	51.6	2.49*	2.75
004			100.0	62.9	57.6	2.78*	3.07
005	Hay, sun-cured, early bloom	1-00-059	90.5	51.8	53.2	2.46	2.61
006			100.0	57.2	58.8	2.72	2.89
007	Hay, sun-cured, midbloom	1-00-063	91.0	52.1	52.2	2.38	2.14
008			100.0	57.2	57.4	2.61	2.35
009	Hay, sun-cured, full bloom	1-00-068	90.9	51.4	50.0	2.27*	2.39
010			100.0	56.6	55.0	2.50*	2.63
011	Hay, sun-cured, mature	1-00-071	91.2	50.2	60.8	2.21*	2.68
012			100.0	55.0	66.7	2.42*	2.94
013	Meal, dehydrated, 15% protein	1-00-022	90.3	54.6	51.9	2.41*	2.29*
014			100.0	60.5	57.4	2.67*	2.53*
015	Meal, dehydrated, 17% protein	1-00-023	91.8	55.6	55.0	2.45*	2.28
016			100.0	60.6	60.0	2.67*	2.48
017	Meal, dehydrated, 20% protein	1-00-024	91.6	57.4	55.7	2.53*	2.38
018			100.0	62.7	60.8	2.77*	2.59
019	Silage, 30–50% dry matter	3-08-150	43.4	23.0	25.4*	1.01*	1.12*
020			100.0	53.0	58.6*	2.34*	2.58*
	ALFALFA-BROME, SMOOTH. MEDICAGO SATIVA-BROMUS INERMIS						
021	Fresh	2-00-262	21.3	13.4	14.0*	0.59*	0.62*
022			100.0	63.0	65.9*	2.78*	2.90*
023	Silage, 30–50% dry matter	3-08-147	39.8	21.1	—	0.93*	—
024			100.0	53.0	—	2.34*	—
	ALFALFA-GRASS. MEDICAGO SATIVA-GRASS						
025	Hay, sun-cured	1-08-331	91.4	51.8	49.3	2.11	2.10
026			100.0	56.7	53.9	2.31*	2.30
	ALFALFA-ORCHARDGRASS. MEDICAGO SATIVA-DACTYLIS GLOMERATA						
027	Silage, 30–50% dry matter	3-08-144	36.4	19.7	20.7*	0.87*	0.91
028			100.0	54.0	56.9*	2.38*	2.51
	ANIMAL						
029	Blood, spray dehydrated	5-00-381	92.6	85.4*	84.3	3.76*	3.71*
030			100.0	92.2*	91.0	4.06*	4.01*
031	Bone, meal, steamed	6-00-400	96.1	15.4	—	0.68*	—
032			100.0	16.0	—	0.71*	—
033	Meat, meal, rendered	5-00-385	93.8	66.9	66.3	2.35*	2.92*
034			100.0	71.3	70.7	2.50*	3.12*
035	Meat with bone, meal, rendered	5-00-388	93.3	66.8	65.0	2.94*	2.87*
036			100.0	71.5	69.7	3.15*	3.07*
037	Tankage, meal, rendered	5-00-386	92.0	68.1	66.0	3.00*	2.91*
038			100.0	74.0	71.7	3.26*	3.16*
039	Tankage with bone, meal, rendered	5-00-387	92.9	80.6*	63.4	3.55*	2.80*
040			100.0	86.8*	68.3	3.83*	3.01*
	ANIMAL-POULTRY						
041	Fat	4-00-409	99.1	187.7	—	7.87*	—
042			100.0	189.4	—	7.94*	—
043	Tankage, meal, rendered, 50% protein	5-00-410	94.2	79.2*	56.0*	3.49*	2.47*
044			100.0	84.0*	59.4*	3.70*	2.62*
	APPLE. MALUS SPP						
045	Pomace, dehydrated	4-00-423	89.6	62.7	58.6	2.76*	2.58*
046			100.0	69.9	65.3	3.08*	2.88*
	BAHIAGRASS. PASPALUM NOTATUM						
047	Fresh	2-00-464	28.7	15.7	15.2	0.69*	0.67*
048			100.0	54.7	53.0	2.41*	2.34*
049	Hay, sun-cured	1-00-462	90.0	45.8	50.6*	2.02*	2.23*
050			100.0	50.8	56.2*	2.24*	2.48*
	BAKERY						
051	Waste, dehydrated (dried bakery product)	4-00-466	91.2	81.3	81.8	3.53*	3.60*
052			100.0	89.1	89.6	3.87*	3.95*

	ME CTTL Mcal/kg	ME SHEP Mcal/kg	NE-M CTTL Mcal/kg	NE-G CTTL Mcal/kg	NE-L CTTL Mcal/kg	ME-N CHKN kcal/kg	TDN HORS %	DE HORS Mcal/kg	DE SWIN kcal/kg	ME SWIN kcal/kg
001	0.59*	0.59*	0.35*	0.20*	0.36*	—	14.1*	0.58*	—	—
002	2.27*	2.25*	1.34*	0.77*	1.37*	—	54.0*	2.23*	—	—
003	2.11*	2.37*	1.26*	0.74*	1.28*	—	48.7*	2.01*	—	—
004	2.35*	2.65*	1.41*	0.82*	1.42*	—	54.3*	2.24*	—	—
005	2.12	2.23*	1.33*	0.80*	1.21*	—	48.1*	2.24*	—	—
006	2.35	2.47*	1.47*	0.88*	1.33*	—	53.1*	2.48	—	—
007	1.99*	1.75*	1.17*	0.65*	1.15*	—	45.5*	2.07*	—	—
008	2.19*	1.92*	1.28*	0.71*	1.27*	—	50.0*	2.28*	—	—
009	1.88*	2.01*	1.09*	0.57*	1.15*	—	43.6*	1.97*	—	—
010	2.07*	2.21*	1.20*	0.63*	1.27*	—	48.0*	2.17*	—	—
011	1.82*	2.30*	1.04*	0.53*	1.12*	—	43.1*	1.81*	—	—
012	2.00*	2.52*	1.14*	0.58*	1.23*	—	47.2*	1.98*	—	—
013	2.03*	1.90*	1.20*	0.68*	1.23*	1533.	45.9*	2.00*	—	1291.
014	2.24*	2.11*	1.33*	0.75*	1.36*	1698.	50.8*	2.21*	—	1429.
015	2.06	1.88*	1.27*	0.73*	1.25*	1508.	47.2*	2.16*	1417.	1322.
016	2.25	2.05*	1.38*	0.80*	1.37*	1643.	51.4*	2.36*	1544.	1441.
017	2.15*	1.99*	1.28*	0.75*	1.30*	1625.	50.3*	2.07*	2079.	1923.
018	2.34*	2.17*	1.40*	0.82*	1.42*	1774.	54.9*	2.27*	2270.	2099.
019	0.83*	0.94*	0.46*	0.22*	0.51*	—	—	—	—	—
020	1.91*	2.16*	1.07*	0.52*	1.18*	—	—	—	—	—
021	0.50*	0.53*	0.30*	0.18*	0.30*	—	10.7*	0.45*	—	—
022	2.36*	2.48*	1.41*	0.83*	1.42*	—	50.4*	2.10*	—	—
023	0.76*	—	0.43*	0.21*	0.47*	—	—	—	—	—
024	1.91*	—	1.07*	0.52*	1.18*	—	—	—	—	—
025	1.72*	1.71*	0.96*	0.46*	0.99*	—	45.5*	1.90*	—	—
026	1.88*	1.87*	1.05*	0.50*	1.08*	—	49.8*	2.08*	—	—
027	0.71*	0.76*	0.40*	0.20*	0.44*	—	—	—	—	—
028	1.95*	2.08*	1.11*	0.55*	1.20*	—	—	—	—	—
029	3.38*	3.34*	2.13*	1.48*	1.98*	2770.	—	—	2701.	1942.
030	3.66*	3.60*	2.30*	1.60*	2.14*	2991.	—	—	2916.	2097.
031	0.25*	—	-0.36*	-0.85*	0.26*	1003.	—	—	—	—
032	0.26*	—	-0.37*	-0.88*	0.27*	1044.	—	—	—	—
033	1.95	2.53*	1.15*	0.62*	1.53*	2087.	—	—	2599.*	2224.
034	2.08	2.70*	1.23*	0.66*	1.63*	2225.	—	—	2771.*	2371.
035	2.55*	2.48*	1.57*	1.00*	1.52*	2017.	—	—	2367.	2266.
036	2.74*	2.65*	1.68*	1.07*	1.63*	2161.	—	—	2536.	2428.
037	2.62*	2.52*	1.62*	1.05*	1.56*	2556.	—	—	2448.*	2087.
038	2.85*	2.74*	1.76*	1.14*	1.69*	2778.	—	—	2660.*	2268.
039	3.17*	2.41*	1.99*	1.36*	1.86*	2618.	—	—	3052.*	2644.
040	3.41*	2.59*	2.14*	1.47*	2.01*	2818.	—	—	3285.*	2846.
041	7.51	—	5.85*	4.38*	4.48*	7677.	—	7.94*	8259.*	7973.
042	7.57	—	5.90*	4.42*	4.52*	7744.	—	8.00*	8332.*	8044.
043	3.10*	2.07*	1.94*	1.32*	1.83*	—	—	—	174.*	-19.*
044	3.29*	2.19*	2.06*	1.40*	1.94*	—	—	—	184.*	-20.*
045	2.39*	2.20*	1.46*	0.92*	1.43*	—	—	—	3059.*	2883.*
046	2.66*	2.46*	1.63*	1.03*	1.59*	—	—	—	3413.*	3216.*
047	0.57*	0.55*	0.32*	0.16*	0.35*	—	13.2*	0.61*	—	—
048	1.98*	1.91*	1.13*	0.57*	1.22*	—	45.8*	2.03*	—	—
049	1.63*	1.85*	0.90*	0.40*	1.01*	—	42.3*	1.75*	—	—
050	1.81*	2.05*	1.00*	0.45*	1.13*	—	47.0*	1.94*	—	—
051	3.16	3.23*	2.19*	1.54*	1.88*	3835.	—	—	3955.	3711.
052	3.46	3.54*	2.41*	1.69*	2.06*	4203.	—	—	4335.	4068.

TABLE 3–2B (continued). Composition of Feeds—Energy Values

		IFN	DRY MATT %	TDN CTTL %	TDN SHEP %	DE CTTL Mcal /kg	DE SHEP Mcal /kg
	BARLEY, HORDEUM VULGARE						
053	Grain	4-00-549	88.6	73.7	76.0	3.42	3.35*
054			100.0	83.2	85.9	3.86	3.79*
055	Grain, Pacific coast	4-07-939	88.6	72.7	78.1	3.20*	3.44*
056			100.0	82.0	88.1	3.62*	3.89*
057	Grain screenings	4-00-542	88.9	71.1	71.4	3.14*	3.15*
058			100.0	80.0	80.3	3.53*	3.54*
059	Hay, sun-cured	1-00-495	88.4	49.3	50.7	2.11*	2.23*
060			100.0	55.7	57.3	2.39*	2.53*
061	Straw	1-00-498	91.4	41.8	43.6	1.80*	1.92*
062			100.0	45.7	47.7	1.97*	2.10*
	BEAN, NAVY. PHASEOLUS VULGARIS						
063	Seeds	5-00-623	89.8	75.0	78.5	3.30*	3.46*
064			100.0	83.5	87.4	3.68*	3.85*
	BEET, MANGEL. BETA VULGARIS MACRORRHIZA						
065	Roots, fresh	4-00-637	11.0	8.5	8.9	0.38*	0.39*
066			100.0	77.8	81.4	3.43*	3.59*
	BEET, SUGAR. BETA VULGARIS ALTISSIMA						
067	Aerial part with crowns, fresh	2-00-649	16.5	10.0*	10.7	0.44*	0.47*
068			100.0	60.7*	64.6	2.68*	2.85*
069	Aerial part with crowns, silage	3-00-660	25.2	13.6	12.5	0.60*	0.55*
070			100.0	54.0	49.8	2.38*	2.19*
071	Molasses, more than 48% invert sugar more than 79.5 degrees	4-00-668	77.9	62.0	60.3	2.61*	2.66*
072	brix		100.0	79.6	77.4	3.36*	3.41*
073	Pulp, dehydrated	4-00-669	91.0	67.7	67.2	2.82*	2.96*
074			100.0	74.4	73.8	3.10*	3.26*
075	Pulp, wet	4-00-671	11.0	7.5	8.3	0.33*	0.37*
076			100.0	68.0	76.1	3.00*	3.35*
077	Pulp with molasses, dehydrated	4-00-672	91.8	67.9	70.6	2.99*	3.11*
078			100.0	74.0	76.9	3.26*	3.39*
	BERMUDAGRASS. CYNODON DACTYLON						
079	Fresh	2-00-712	28.9	17.8	17.4	0.78*	0.77*
080			100.0	61.5	60.0	2.71*	2.65*
081	Hay, sun-cured	1-00-703	91.2	41.1	44.5	1.93	1.96*
082			100.0	45.1	48.8	2.12	2.15*
	BERMUDAGRASS, COASTAL. CYNODON DACTYLON						
083	Fresh	2-00-719	30.3	18.8	20.0	0.83*	0.88*
084			100.0	62.3	66.0	2.75*	2.91*
085	Hay, sun-cured	1-00-716	91.6	49.9	49.0	2.40	2.23
086			100.0	54.5	53.5	2.62	2.43
	BLOOD-SEE ANIMAL						
	BLUEGRASS, CANADA. POA COMPRESSA						
087	Fresh	2-00-764	31.4	21.9	18.8	0.97*	0.83*
088			100.0	70.0	59.9	3.09*	2.64*
089	Hay, sun-cured	1-00-762	91.7	59.6	54.7	2.63*	2.41*
090			100.0	65.0	59.7	2.87*	2.63*
	BLUEGRASS, KENTUCKY. POA PRATENSIS						
091	Fresh, early vegetative	2-00-777	30.8	22.2	20.2*	0.98*	0.89*
092			100.0	72.0	65.5*	3.17*	2.89*
093	Fresh, early bloom	2-00-779	35.0	24.3	24.3	1.07*	1.07*
094			100.0	69.5	69.4	3.06*	3.06*
095	Hay, sun-cured	1-00-776	88.9	51.4*	53.9	2.26*	2.38*
096			100.0	57.8*	60.6	2.55*	2.67*
	BLUESTEM. ANDROPOGON SPP						
097	Fresh, early vegetative	2-00-821	26.8	17.2*	18.1	0.76*	0.80*
098			100.0	64.4*	67.7	2.84*	2.99*
099	Fresh, mature	2-00-825	59.0	35.1*	31.4	1.55*	1.38*
100			100.0	59.5*	53.1	2.63*	2.34*
	BONES (SEE ANIMAL)						
	BREWERS GRAINS (SEE CEREALS)						

	ME CTTL Mcal /kg	ME SHEP Mcal /kg	NE-M CTTL Mcal /kg	NE-G CTTL Mcal /kg	NE-L CTTL Mcal /kg	ME-N CHKN kcal /kg	TDN HORS %	DE HORS Mcal /kg	DE SWIN kcal /kg	ME SWIN kcal /kg
053	2.57	2.99*	1.73*	1.16*	1.81*	2621.	72.6	3.26*	3081.	2937.
054	2.91	3.37*	1.95*	1.31*	2.05*	2959.	82.0	3.68*	3478.	3316.
055	2.84*	3.08*	1.77*	1.20*	1.67*	2583.	—	3.17*	3095.*	2919.
056	3.20*	3.47*	2.00*	1.35*	1.89*	2914.	—	3.58*	3492.*	3293.
057	2.77*	2.78*	1.73*	1.15*	1.64*	1797.	—	—	2495.*	2330.
058	3.11*	3.13*	1.94*	1.30*	1.84*	2021.	—	—	2806.*	2620.
059	1.73	1.86*	0.99*	0.49*	1.10*	—	42.3*	1.77*	—	—
060	1.96	2.10*	1.11*	0.56*	1.25*	—	47.8*	2.01*	—	—
061	1.40	1.53*	0.64*	0.15*	0.91*	—	28.1*	1.47*	—	—
062	1.54	1.67*	0.70*	0.16*	1.00*	—	30.7*	1.62*	—	—
063	2.93*	3.09*	1.84*	1.24*	1.73*	2327.	—	—	3849.*	3386.
064	3.27*	3.44*	2.05*	1.39*	1.93*	2593.	—	—	4288.*	3772.
065	0.33*	0.35*	0.21*	0.14*	0.20*	—	—	—	399.*	377.*
066	3.02*	3.17*	1.88*	1.24*	1.79*	—	—	—	3641.*	3439.*
067	0.37*	0.40*	0.22*	0.13*	0.23*	—	7.9*	0.32*	—	—
068	2.25*	2.43*	1.33*	0.76*	1.37*	—	72.2*	2.11*	—	—
069	0.49*	0.44*	0.28*	0.14*	0.30*	—	—	—	—	—
070	1.95*	1.77*	1.11*	0.55*	1.20*	—	—	—	—	—
071	2.29	2.33*	1.54*	1.04*	1.42*	1929.	—	—	2489.*	2338.
072	2.94	2.99*	1.98*	1.33*	1.83*	2476.	—	—	3195.*	3002.
073	2.44	2.58*	1.60*	1.04*	1.55*	649.	—	2.33*	2879.*	2704.
074	2.68	2.84*	1.76*	1.14*	1.70*	713.	—	2.56*	3163.*	2971.
075	0.28*	0.32*	0.17*	0.11*	0.17*	—	—	—	307.*	287.*
076	2.58*	2.94*	1.57*	0.97*	1.55*	—	—	—	2798.*	2612.*
077	2.61*	2.73*	1.61*	1.04*	1.55*	660.	—	—	3061.*	2881.
078	2.85*	2.98*	1.76*	1.14*	1.69*	719.	—	—	3335.*	3139.
079	0.66*	0.64*	0.39*	0.23*	0.40*	—	14.3*	0.60*	—	—
080	2.29*	2.22*	1.36*	0.78*	1.39*	—	49.3*	2.06*	—	—
081	1.41	1.57*	0.65*	0.16*	0.88*	—	39.8*	1.69*	—	—
082	1.55	1.72*	0.71*	0.18*	0.96*	—	43.7*	1.85*	—	—
083	0.70*	0.75*	0.42*	0.24*	0.43*	—	12.9*	0.72*	—	—
084	2.32*	2.49*	1.39*	0.80*	1.41*	—	42.6*	2.38	—	—
085	1.46	1.84*	0.69*	0.20*	1.17*	—	45.1*	1.89*	—	—
086	1.59	2.01*	0.75*	0.22*	1.28*	—	49.3*	2.06*	—	—
087	0.84*	0.70*	0.51*	0.32*	0.50*	—	15.2*	0.64*	—	—
088	2.67*	2.22*	1.63*	1.03*	1.60*	—	48.5*	2.03*	—	—
089	2.24*	2.03*	1.35*	0.81*	1.35*	—	41.3*	1.74*	—	—
090	2.44*	2.21*	1.47*	0.88*	1.47*	—	45.0*	1.90*	—	—
091	0.85*	0.76*	0.52*	0.33*	0.51*	—	15.4*	0.64*	—	—
092	2.76*	2.47*	1.70*	1.08*	1.64*	—	49.9*	2.08*	—	—
093	0.93*	0.92*	0.57*	0.35*	0.55*	—	17.1*	0.71*	—	—
094	2.64*	2.64*	1.62*	1.01*	1.58*	—	48.8*	2.04*	—	—
095	1.89*	2.00*	1.10*	0.59*	1.15*	—	39.4*	1.52*	—	—
096	2.12*	2.25*	1.23*	0.67*	1.30*	—	44.3*	1.71*	—	—
097	0.65*	0.69*	0.39*	0.23*	0.39*	—	13.4*	0.56*	—	—
098	2.42*	2.57*	1.45*	0.87*	1.46*	—	49.9*	2.08*	—	—
099	1.30*	1.13*	0.76*	0.43*	0.79*	—	23.1*	0.99*	—	—
100	2.20*	1.92*	1.29*	0.72*	1.34*	—	39.1*	1.69*	—	—

TABLE 3–2B (continued). Composition of Feeds—Energy Values

	IFN	DRY MATT %	TDN CTTL %	TDN SHEP %	DE CTTL Mcal /kg	DE SHEP Mcal /kg
BROME. BROMUS SPP						
101 Fresh	2-00-900	36.0	21.7*	23.5	0.96*	1.03*
102		100.0	60.2*	65.1	2.66*	2.87*
103 Hay, sun-cured	1-00-890	90.8	49.0	60.2*	2.30	2.66
104		100.0	54.0	66.3*	2.53	2.92
BROME, SMOOTH. BROMUS INERMIS						
105 Fresh	2-00-963	27.0	16.9*	13.9	0.75*	0.61*
106		100.0	62.8*	51.4	2.77*	2.27*
107 Hay, sun-cured	1-00-947	89.6	53.5	53.4	2.39	2.55
108		100.0	59.7	59.6	2.67	2.84
BUCKWHEAT, COMMON. FAGOPYRUM SAGITTATUM						
109 Grain	4-00-994	88.4	61.9	64.6	2.73*	2.85*
110		100.0	70.0	73.1	3.09*	3.22*
BUFFALOGRASS. BUCHLOE DACTYLOIDES						
111 Fresh	2-01-010	45.8	27.1*	25.6	1.19*	1.13*
112		100.0	59.1*	55.9	2.61*	2.46*
BUTTERMILK (SEE CATTLE)						
CANARYGRASS, REED. PHALARIS ARUNDINACEA						
113 Fresh	2-01-113	22.8	14.8*	14.0	0.65*	0.62*
114		100.0	65.0*	61.2	2.87*	2.70*
115 Hay, sun-cured	1-01-104	89.3	49.2*	44.1	2.17*	1.95*
116		100.0	55.1*	49.5	2.43*	2.18*
CARROT. DAUCUS SPP						
117 Roots, fresh	4-01-145	11.5	9.4	10.0	0.42*	0.44*
118		100.0	82.0	86.5	3.62*	3.82*
CASEIN						
119 Acid precipitated, dehydrated	5-01-162	91.6	75.3	87.6	3.09*	3.86*
120		100.0	82.2	95.7	3.38*	4.22*
CATTLE. BOS TAURUS						
121 Buttermilk, dehydrated	5-01-160	92.3	81.0	82.9	3.23*	3.66*
122		100.0	87.7	89.8	3.50*	3.96*
123 Manure, dehydrated	1-01-190	93.0	47.6*	49.0*	2.10*	2.16*
124		100.0	51.2*	52.7*	2.26*	2.32*
125 Milk, dehydrated	5-01-167	95.5	109.8	117.1	4.84*	5.16*
126		100.0	115.0	122.6	5.07*	5.41*
127 Milk, fresh	5-01-168	12.4	15.9	15.7	0.71	0.69*
128		100.0	128.6	127.3	5.75	5.61*
129 Skim milk, dehydrated	5-01-175	94.1	79.0	80.3	2.76*	3.54*
130		100.0	84.0	85.3	2.93*	3.76*
131 Skim milk, fresh	5-01-170	9.6	8.8	8.8	0.39	0.39*
132		100.0	91.6	91.6	4.11	4.04*
133 Whey, dehydrated	4-01-182	93.3	74.4	78.0	3.22*	3.44*
134		100.0	79.7	83.6	3.45*	3.69*
135 Whey, fresh	4-08-134	6.9	—	6.5	—	0.29*
136		100.0	—	94.2	—	4.15*
137 Whey low lactose, dehydrated (dried whey product)	4-01-186	93.7	73.6	74.5	2.92*	3.28*
138		100.0	78.6	79.5	3.12*	3.51*
CEREALS						
139 Brewers grains, dehydrated	5-02-141	92.2	65.1	65.9	2.61*	2.91*
140		100.0	70.6	71.5	2.83*	3.15*
141 Grain screenings, uncleaned	4-02-153	92.1	59.8	58.9*	2.64*	2.60*
142		100.0	65.0	64.0*	2.87*	2.82*
CITRUS. CITRUS SPP						
143 Pomace without fines, dehydrated (dried citrus pulp)	4-01-237	91.1	74.1	76.4	2.85*	3.37*
144		100.0	81.4	83.8	3.13*	3.70*
145 Pomace, silage (pulp)	4-01-234	20.8	18.3	18.2	0.81*	0.80*
146		100.0	88.0	87.7	3.88*	3.87*
147 Syrup (molasses)	4-01-241	66.9	50.0	50.9	2.16*	2.24*
148		100.0	74.8	76.1	3.22*	3.36*
CLOVER, ALSIKE. TRIFOLIUM HYBRIDUM						
149 Fresh	2-01-316	22.5	16.0	16.0	0.70*	0.71*
150		100.0	71.0	71.3	3.13*	3.15*
151 Hay, sun-cured	1-01-313	87.7	50.5	51.0	2.22*	2.25*
152		100.0	57.5	58.1	2.54*	2.56*

	ME CTTL Mcal /kg	ME SHEP Mcal /kg	NE-M CTTL Mcal /kg	NE-G CTTL Mcal /kg	NE-L CTTL Mcal /kg	ME-N CHKN kcal /kg	TDN HORS %	DE HORS Mcal /kg	DE SWIN kcal /kg	ME SWIN kcal /kg
101	0.80*	0.88*	0.47*	0.27*	0.49*	—	15.1*	0.65*	—	—
102	2.23*	2.45*	1.32*	0.74*	1.36*	—	42.0*	1.79*	—	—
103	1.91*	2.27*	1.11*	0.60*	1.10*	—	38.8*	1.65*	—	—
104	2.11*	2.50*	1.22*	0.66*	1.22*	—	42.7*	1.82*	—	—
105	0.63*	0.50*	0.38*	0.22*	0.38*	—	12.7*	0.54*	—	—
106	2.35*	1.84*	1.40*	0.82*	1.42*	—	47.2*	1.98*	—	—
107	2.01*	2.17*	1.19*	0.67*	1.17*	—	47.0*	1.71*	—	—
108	2.24*	2.42*	1.33*	0.75*	1.30*	—	52.5*	1.91*	—	—
109	2.36*	2.48*	1.44*	0.91*	1.41*	2668.	62.6	—	3059.*	2884.
110	2.67*	2.80*	1.63*	1.03*	1.60*	3016.	70.8	—	3459.*	3261.
111	1.00*	0.93*	0.59*	0.32*	0.61*	—	20.4*	0.86*	—	—
112	2.18*	2.04*	1.28*	0.71*	1.33*	—	44.6*	1.89*	—	—
113	0.56*	0.52*	0.34*	0.20*	0.34*	—	10.7*	0.58*	—	—
114	2.44*	2.28*	1.47*	0.88*	1.47*	—	46.8*	2.54*	—	—
115	1.79*	1.56*	1.02*	0.52*	1.10*	—	42.8*	1.78*	—	—
116	2.00*	1.75*	1.14*	0.58*	1.23*	—	48.0*	2.00*	—	—
117	0.37*	0.39*	0.23*	0.16*	0.22*	458.	—	0.43*	458.	430.
118	3.20*	3.40*	2.00*	1.35*	1.89*	3979.	67.2*	3.78*	3979.	3736.
119	2.71	3.49*	1.83*	1.23*	1.73*	4160.	—	—	3543.*	3100.*
120	2.96	3.81*	2.00*	1.35*	1.89*	4543.	—	—	3869.*	3385.*
121	2.84	3.28*	1.94*	1.32*	1.87*	2752.	—	—	3485.*	3046.
122	3.08	3.55*	2.10*	1.43*	2.03*	2982.	—	—	3776.*	3300.
123	1.70*	1.76*	0.94*	0.43*	1.05*	—	38.4*	1.64*	2816.	2687.
124	1.83*	1.90*	1.01*	0.46*	1.13*	—	41.3*	1.77*	3029.	2890.
125	4.46*	4.78*	2.81*	2.03*	2.58*	—	—	—	4817.*	4270.*
126	4.67*	5.01*	2.95*	2.12*	2.70*	—	—	—	5044.*	4471.*
127	0.66*	0.65*	0.42*	0.30*	0.40*	—	—	—	681.	606.*
128	5.36*	5.22*	3.37*	2.46*	3.22*	—	—	—	5510.	4902.*
129	2.37	3.15*	1.52*	0.95*	1.82*	2537.	—	3.81*	3874.	3591.
130	2.51	3.35*	1.62*	1.01*	1.94*	2695.	—	4.05*	4115.	3815.
131	0.35*	0.35*	0.22*	0.16*	0.21*	—	—	—	415.	365.*
132	3.70*	3.63*	2.33*	1.63*	2.20*	—	—	—	4339.	3819.*
133	2.83	3.05*	1.92*	1.30*	1.71*	1941.	—	3.79*	3185.	3104.
134	3.03	3.27*	2.06*	1.40*	1.83*	2081.	—	4.06*	3415.	3328.
135	—	0.26*	—	—	—	—	—	—	—	—
136	—	3.74*	—	—	—	—	—	—	—	—
137	2.53	2.89*	1.66*	1.08*	1.69*	2117.	—	3.37*	2933.*	2753.
138	2.70	3.09*	1.78*	1.15*	1.81*	2260.	—	3.61*	3131.*	2939.
139	2.22	2.52*	1.41*	0.86*	1.48*	2312.	47.9	2.53*	2181.	2120.
140	2.41	2.73*	1.53*	0.93*	1.61*	2509.	52.0	2.75*	2367.	2301.
141	2.25*	2.21*	1.36*	0.81*	1.36*	—	—	—	2858.*	2682.*
142	2.44*	2.40*	1.47*	0.88*	1.47*	—	—	—	3104.*	2913.*
143	2.47	2.99*	1.63*	1.06*	1.71*	1337.	—	2.56*	3059.	2423.
144	2.71	3.28*	1.79*	1.16*	1.87*	1468.	—	2.81*	3357.	2659.
145	0.72*	0.72*	0.45*	0.31*	0.42*	—	—	—	781.*	739.*
146	3.47*	3.46*	2.18*	1.50*	2.04*	—	—	—	3762.*	3558.*
147	1.88	1.97*	1.25*	0.82*	1.15*	—	—	—	2369.*	2235.
148	2.81	2.94*	1.87*	1.23*	1.71*	—	—	—	3543.*	3344.
149	0.61*	0.61*	0.37*	0.24*	0.36*	—	11.8*	0.49*	—	—
150	2.71*	2.73*	1.67*	1.05*	1.62*	—	52.3*	2.17*	—	—
151	1.85*	1.88*	1.08*	0.58*	1.13*	—	40.8*	1.72*	—	—
152	2.11*	2.14*	1.23*	0.66*	1.29*	—	46.5*	1.96*	—	—

TABLE 3–2B (continued).   Composition of Feeds—Energy Values

		IFN	DRY MATT %	TDN CTTL %	TDN SHEP %	DE CTTL Mcal /kg	DE SHEP Mcal /kg
	CLOVER, CRIMSON. TRIFOLIUM INCARNATUM						
153	Fresh	2-01-336	17.5	11.2*	11.4	0.49*	0.50*
154			100.0	63.7*	64.9	2.81*	2.86*
155	Hay, sun-cured	1-01-328	87.8	53.7	46.7	2.37*	2.06*
156			100.0	61.2	53.2	2.70*	2.34*
	CLOVER, LADINO. TRIFOLIUM REPENS						
157	Fresh	2-01-383	17.7	13.8	13.2	0.61*	0.58*
158			100.0	78.2	74.7	3.45*	3.29*
159	Hay, sun-cured	1-01-378	89.1	56.6	59.2	2.50*	2.61*
160			100.0	63.5	66.5	2.80*	2.93*
	CLOVER, RED. TRIFOLIUM PRATENSE						
161	Fresh, early bloom	2-01-428	19.6	13.6	654.8*	0.60*	17.31*
162			100.0	69.3	464.7*	3.06*	29.16*
163	Fresh, full bloom	2-01-429	26.2	16.8	17.2*	0.74*	0.76*
164			100.0	64.0	65.7*	2.82*	2.90*
165	Hay, sun-cured	1-01-415	88.4	51.6	52.0	2.28*	2.67
166			100.0	58.4	58.9	2.57*	3.02
	COCONUT. COCOS NUCIFERA						
167	Kernels with coats, meal, mechanical-extracted (copra meal)	5-01-572	91.6	73.5	74.9	3.34	3.30*
168			100.0	80.3	81.8	3.65	3.61*
169	Kernels with coats, meal, solvent-extracted (copra meal)	5-01-573	91.1	68.3*	68.6	3.01	3.02*
170			100.0	75.0*	75.3	3.31	3.32*
	CORN (SEE MAIZE)						
	COTTON. GOSSYPIUM SPP						
171	Bolls, sun-cured	1-01-596	91.4	38.4	41.8	1.69*	1.84*
172			100.0	42.0	45.7	1.85*	2.02*
173	Hulls	1-01-599	90.4	40.6	43.8	1.90	1.93*
174			100.0	44.9	48.5	2.10	2.14*
175	Seeds, ground	5-01-608	92.2	90.3	90.3	3.98*	3.98*
176			100.0	98.0	98.0	4.32*	4.32*
177	Seeds, meal, mechanical-extracted	5-01-609	92.7	51.9	62.2	2.29*	2.74*
178			100.0	56.0	67.1	2.47*	2.96*
179	Seeds, meal, mechanical-extracted, 41% protein	5-01-617	92.6	73.6	69.7	3.22	3.37
180			100.0	79.4	75.2	3.48	3.64
181	Seeds, meal, solvent-extracted, 41% protein	5-01-621	91.0	71.0	64.9	3.11	3.45
182			100.0	78.0	71.3	3.41	3.79
183	Seeds without hulls, meal, prepressed solvent-extracted, 50%	5-07-874	93.0	69.8	73.0*	3.08*	3.22*
184	protein		100.0	75.0	78.5*	3.31*	3.46*
	COWPEA, COMMON. VIGNA SINENSIS						
185	Hay, sun-cured	1-01-645	90.4	56.9	52.0	2.52	2.29*
186			100.0	63.0	57.6	2.78	2.54*
	DROPSEED, SAND. SPOROBOLUS CRYPTANDRUS						
187	Hay, sun-cured	2-05-596	88.0	49.4*	51.9	2.18*	2.12
188			100.0	56.2*	59.0	2.48*	2.41
	FAT (SEE ANIMAL-POULTRY)						
	FESCUE, MEADOW. FESTUCA ELATIOR						
189	Fresh	2-01-920	28.4	17.6*	17.2	0.77*	0.76*
190			100.0	61.8*	60.7	2.72*	2.67*
191	Hay, sun-cured	1-01-912	87.5	54.3	52.1	2.39*	2.30*
192			100.0	62.0	59.6	2.73*	2.63*
	FISH						
193	Solubles, condensed	5-01-969	50.4	42.0	40.9	1.93*	1.80*
194			100.0	83.3	81.1	3.83*	3.58*
195	Solubles, dehydrated	5-01-971	92.8	73.8	80.6	3.05*	3.55*
196			100.0	79.6	86.8	3.29*	3.83*
	FISH, ANCHOVETA. ENGRAULIS RINGEN						
197	Meal, mechanical-extracted	5-01-985	92.0	71.7	73.0	3.16*	3.22*
198			100.0	78.0	79.3	3.44*	3.50*
	FISH, HERRING. CLUPEA HARENGOS						
199	Meal, mechancial-extracted	5-02-000	91.8	72.6	75.2	3.20*	3.32*
200			100.0	79.0	81.9	3.48*	3.61*
	FISH, MENHADEN, BREVOORTIA TYRANNUS						
201	Meal, mechanical-extracted	5-02-009	91.7	69.2	64.0	3.05*	2.82*
202			100.0	75.5	69.9	3.33*	3.08*

	ME CTTL Mcal /kg	ME SHEP Mcal /kg	NE-M CTTL Mcal /kg	NE-G CTTL Mcal /kg	NE-L CTTL Mcal /kg	ME-N CHKN kcal /kg	TDN HORS %	DE HORS Mcal /kg	DE SWIN kcal /kg	ME SWIN kcal /kg
153	0.42*	0.43*	0.25*	0.15*	0.25*	—	8.5*	0.35*	—	—
154	2.39*	2.44*	1.43*	0.85*	1.44*	—	48.3*	2.02*	—	—
155	2.00*	1.68*	1.19*	0.68*	1.21*	—	43.4*	1.81*	—	—
156	2.28*	1.92*	1.35*	0.77*	1.38*	—	49.4*	2.06*	—	—
157	0.54*	0.51*	0.33*	0.22*	0.32*	—	10.7*	0.44*	—	—
158	3.03*	2.88*	1.89*	1.25*	1.80*	—	60.8*	2.48*	—	—
159	2.12*	2.24*	1.27*	0.75*	1.28*	—	56.8*	1.96*	—	—
160	2.38*	2.51*	1.42*	0.84*	1.44*	—	63.7*	2.20*	—	—
161	0.52*	21.39	0.32*	0.20*	0.31*	—	9.4*	0.50*	—	—
162	2.64*	50.00	1.61*	1.01*	1.58*	—	48.1*	2.53*	—	—
163	0.63*	0.65*	0.38*	0.22*	0.38*	—	13.4*	0.66*	—	—
164	2.40*	2.48*	1.44*	0.86*	1.45*	—	51.3*	2.25*	—	—
165	1.90*	2.30*	1.11*	0.61*	1.16*	—	42.3*	1.96*	—	—
166	2.15*	2.60*	1.26*	0.69*	1.31*	—	47.9*	2.22*	—	—
167	2.96*	2.92*	1.86*	1.25*	1.75*	1497.	—	—	3308.	3393.
168	3.24*	3.19*	2.03*	1.37*	1.91*	1635.	—	—	3612.	3705.
169	2.63*	2.64*	1.63*	1.06*	1.55*	1525.	—	—	3218.	3055.
170	2.89*	2.90*	1.79*	1.16*	1.70*	1674.	—	—	3532.	3353.
171	1.30*	1.45*	0.62*	0.14*	0.83*	—	33.0*	1.44*	—	—
172	1.42*	1.59*	0.68*	0.15*	0.91*	—	36.1*	1.58*	—	—
173	1.31	1.55*	0.55*	0.08*	0.86*	—	29.4*	1.71*	—	—
174	1.45	1.71*	0.61*	0.08*	0.95*	—	32.5*	1.89*	—	—
175	3.61*	3.61*	2.28*	1.60*	2.10*	—	—	—	3570.*	3123.*
176	3.91*	3.91*	2.47*	1.74*	2.28*	—	—	—	3873.*	3389.*
177	1.90*	2.35*	1.09*	0.57*	1.16*	—	—	—	2773.*	2385.*
178	2.04*	2.54*	1.18*	0.61*	1.25*	—	—	—	2989.*	2572.*
179	2.44	2.51	1.60*	1.02*	1.67*	2236.	—	—	2880.	2641.
180	2.63	2.71	1.72*	1.11*	1.81*	2414.	—	—	3109.	2851.
181	2.48	2.67	1.64*	1.07*	1.61*	1950.	—	2.74*	2669.	2359.
182	2.73	2.93	1.80*	1.17*	1.77*	2142.	—	3.01*	2933.	2592.
183	2.69*	2.83*	1.66*	1.08*	1.60*	2141.	—	—	2854.	2613.
184	2.89*	3.04*	1.79*	1.16*	1.72*	2301.	—	—	3068.	2809.
185	2.13*	1.91*	1.28*	0.75*	1.24*	—	45.5*	1.90*	—	—
186	2.36*	2.11*	1.41*	0.83*	1.37*	—	50.4*	2.10*	—	—
187	1.80*	1.82	1.04*	0.54*	1.11*	—	35.2*	1.51*	—	—
188	2.05*	2.07	1.18*	0.62*	1.26*	—	40.0*	1.72*	—	—
189	0.65*	0.64*	0.39*	0.22*	0.40*	—	11.6*	0.63*	—	—
190	2.30*	2.25*	1.37*	0.79*	1.39*	—	40.6*	2.22*	—	—
191	2.02*	1.93*	1.20*	0.70*	1.22*	—	37.1*	1.75*	—	—
192	2.31*	2.20*	1.38*	0.80*	1.40*	—	42.4*	*2.00	—	—
193	1.73	1.60*	1.20*	0.84*	0.97*	1664.	—	—	1908.	1622.
194	3.42	3.16*	2.37*	1.66*	1.92*	3300.	—	—	3784.	3217.
195	2.66	3.17*	1.78*	1.19*	1.70*	2915.	—	—	3240.*	2818.
196	2.87	3.42*	1.92*	1.28*	1.83*	3140.	—	—	3491.*	3036.
197	2.78*	2.84*	1.73*	1.14*	1.65*	2745.	—	2.76*	3020.	2478.
198	3.02*	3.08*	1.88*	1.24*	1.79*	2985.	—	3.00*	3283.	2695.
199	2.82*	2.94*	1.76*	1.17*	1.67*	3257.	—	—	3930.	2779.
200	3.07*	3.20*	1.91*	1.27*	1.82*	3546.	—	—	4280.	3026.
201	2.67*	2.44*	1.65*	1.08*	1.59*	2851.	—	2.93*	3482.	2602.
202	2.91*	2.66*	1.80*	1.18*	1.73*	3110.	—	3.20*	3799.	2839.

TABLE 3–2B (continued). Composition of Feeds—Energy Values

		IFN	DRY MATT %	TDN CTTL %	TDN SHEP %	DE CTTL Mcal /kg	DE SHEP Mcal /kg
	FISH, SARDINE. CLUPEA SPP-SARDINOPS SPP						
203	Meal, mechanical-extracted	5-02-015	92.9	69.7	70.3	3.07*	3.10*
204			100.0	75.0	75.6	3.31*	3.33*
	FISH, WHITE. GADIDAE (FAMILY)—LOPHIIDAE (FAMILY)—RAJIDAE (FAMILY)						
205	Meal, mechanical-extracted	5-02-025	91.2	71.2	69.2	3.14*	3.05*
206			100.0	78.0	75.8	3.44*	3.34*
	FLAX, COMMON. LINUM USITA TISSIMUM						
207	Seed screenings	4-02-056	91.2	58.3	58.1	2.57*	2.56*
208			100.0	64.0	63.8	2.82*	2.81*
209	Seeds, meal, mechanical-extracted	5-02-045	90.8	74.3	75.6	3.04*	3.33*
210			100.0	81.8	83.2	3.35*	3.67*
	GALLETA. HILARIA JAMESII						
211	Fresh, stem-cured	2-05-594	86.0	51.6	36.7	2.28*	1.58
212			100.0	60.0	42.7	2.65*	1.83
	GAMAGRASS, EASTERN. TRIPSACUM DACTYLOIDES						
213	Fresh, full bloom	2-02-084	25.0	14.4*	14.8*	0.63*	0.65*
214			100.0	57.7*	59.3*	2.54*	2.61*
	GAMAGRASS, FLORIDA. TRIPSACUM FLORIDANUM						
215	Hay, sun-cured	1-02-087	92.3	51.4*	48.5	2.27*	2.14*
216			100.0	55.7*	52.5	2.45*	2.31*
	GARBAGE, HOTEL AND RESTAURANT						
217	Boiled	4-07-865	22.9	21.3*	20.4	0.94*	0.90*
218			100.0	93.1*	89.4	4.11*	3.94*
	GARBAGE, INSTITUTIONAL						
219	Boiled	4-07-867	17.9	15.7*	15.8*	0.69*	0.70*
220			100.0	87.4*	88.0*	3.85*	3.88*
	GRAMA. BOUTELOUA SPP						
221	Fresh, early vegetative	2-02-163	41.0	24.7*	25.9*	1.09*	1.14*
222			100.0	60.3*	63.1*	2.66*	2.78*
223	Fresh, mature	2-02-166	63.4	34.6*	36.3*	1.53*	1.60*
224			100.0	54.6*	57.2*	2.41*	2.52*
	GRAPE. VITIS SPP						
225	Pomace, dehydrated (marc)	1-02-208	90.4	—	24.0	—	1.06*
226			100.0	—	26.5	—	1.17*
	GRASS						
227	Hay, sun-cured, full bloom	1-02-244	89.3	49.1*	49.2*	2.16*	2.17*
228			100.0	55.0*	55.1*	2.42*	2.43*
	GRASS-LEGUME						
229	Silage	3-02-303	31.1	18.1	17.6	0.84	0.77*
230			100.0	58.4	56.5	2.71	2.49*
	HOMINY (SEE MAIZE)						
	LEGUME (SEE GRASS-LEGUME)						
	LESPEDEZA, CHINESE. LESPEDEZA CUNEATA						
231	Hay, sun-cured	1-02-607	90.4	51.5*	41.7	2.27*	1.84*
232			100.0	57.0*	46.2	2.51*	2.04*
	LESPEDEZA, COMMON-LESPEDEZA, KOREAN. LESPEDEZA STRIATA-LESPEDEZA STIPULACEA						
233	Fresh	2-26-029	—	—	—	—	—
234			100.0*	—	—	—	—
235	Hay, sun-cured	1-26-035	—	—	—	—	—
236			100.0*	—	—	—	—
	LINSEED (SEE FLAX)						
	LUCERNE (SEE ALFALFA)						
	MAIZE. ZEA MAYS						
237	Aerial part with ears, sun-cured, mature (fodder)	1-02-772	82.1	56.2	48.1*	2.48*	2.12*
238			100.0	68.5	58.5*	3.02*	2.58*
239	Aerial part without ears without husks, sun-cured (stover)	1-02-776	85.2	51.3	49.9	2.26*	2.20*
240	(straw)		100.0	60.2	58.5	2.65*	2.58*
241	Cobs. ground	1-02-782	89.8	43.6	45.4	2.00*	2.00*
242			100.0	48.5	50.5	2.23*	2.23*
243	Distillers' grains, dehydrated	5-02-842	93.5	80.4	81.6	3.21*	3.60*
244			100.0	86.0	87.3	3.44*	3.85*
245	Distillers' grains with solubles, dehydrated	5-02-843	91.8	81.7	79.9	3.33*	3.52*
246			100.0	89.1	87.1	3.63*	3.84*

	ME CTTL Mcal /kg	ME SHEP Mcal /kg	NE-M CTTL Mcal /kg	NE-G CTTL Mcal /kg	NE-L CTTL Mcal /kg	ME-N CHKN kcal /kg	TDN HORS %	DE HORS Mcal /kg	DE SWIN kcal /kg	ME SWIN kcal /kg
203	2.69*	2.71*	1.66*	1.08*	1.60*	2889.	—	—	2923.*	2525.
204	2.89*	2.92*	1.79*	1.16*	1.72*	3109.	—	—	3146.*	2717.
205	2.76*	2.67*	1.72*	1.14*	1.63*	2594.	—	—	3062.*	2657.
206	3.02*	2.93*	1.88*	1.24*	1.79*	2843.	—	—	3356.*	2912.
207	2.19*	2.18*	1.31*	0.78*	1.32*	—	—	—	3092.*	2912.*
208	2.40*	2.39*	1.44*	0.86*	1.45*	—	—	—	3392.*	3195.*
209	2.66	2.96*	1.79*	1.20*	1.71*	1520.	—	2.74*	3309.	2791.
210	2.93	3.26*	1.97*	1.32*	1.88*	1673.	—	3.04*	3643.	3072.
211	1.91*	1.29	1.13*	0.63*	1.16*	—	24.3*	1.11*	—	—
212	2.22*	1.50	1.31*	0.74*	1.35*	—	28.3*	1.29*	—	—
213	0.53*	0.54*	0.31*	0.16*	0.32*	—	10.0*	0.43*	—	—
214	2.12*	2.19*	1.23*	0.66*	1.29*	—	40.2*	1.73*	—	—
215	1.87*	1.74*	1.07*	0.56*	1.15*	—	44.4*	1.86*	—	—
216	2.03*	1.89*	1.16*	0.60*	1.24*	—	48.1*	2.02*	—	—
217	0.85*	0.81*	0.53*	0.37*	0.49*	—	—	—	1105.*	1053.*
218	3.70*	3.53*	2.33*	1.62*	2.16*	—	—	—	4831.*	4607.*
219	0.62*	0.62*	0.39*	0.27*	0.36*	—	—	—	808.	769.*
220	3.44*	3.47*	2.16*	1.48*	2.02*	—	—	—	4511.	4293.*
221	0.92*	0.97*	0.54*	0.31*	0.56*	—	19.4*	0.82*	—	—
222	2.24*	2.36*	1.32*	0.75*	1.36*	—	47.4*	1.99*	—	—
223	1.26*	1.33*	0.71*	0.36*	0.77*	—	22.9*	1.00*	—	—
224	1.98*	2.10*	1.13*	0.57*	1.22*	—	36.1*	1.58	—	—
225	—	0.66*	—	—	—	1551.	18.1*	0.89*	—	—
226	—	0.73*	—	—	—	1715.	20.0*	0.99*	—	—
227	1.78*	1.79*	1.02*	0.52*	1.10*	—	38.6*	1.64*	—	—
228	2.00*	2.00*	1.14*	0.58*	1.23*	—	43.2*	1.84*	—	—
229	0.71*	0.64*	0.42*	0.24*	0.41*	—	—	—	—	—
230	2.29*	2.07*	1.36*	0.78*	1.33*	—	—	—	—	—
231	1.89*	1.45*	1.09*	0.58*	1.15*	—	43.9*	1.84*	—	—
232	2.09*	1.61*	1.21*	0.64*	1.28*	—	48.5*	2.03*	—	—
233	—	—	—	—	—	—	—	—	—	—
234	—	—	—	—	—	—	—	—	—	—
235	—	—	—	—	—	—	—	—	—	—
236	—	—	—	—	—	—	—	—	—	—
								—		
237	2.13*	1.77*	1.30*	0.81*	1.28*	—	—	1.69*	—	—
238	2.60*	2.16*	1.59*	0.98*	1.56*	—	—	2.06*	—	—
239	1.90*	1.84*	1.12*	0.63*	1.15*	—	36.1*	1.38*	—	—
240	2.23*	2.16*	1.32*	0.74*	1.35*	—	42.3*	1.62*	—	—
241	1.62	1.62*	0.86*	0.37*	0.96*	—	27.8	1.25*	—	—
242	1.80	1.80*	0.96*	0.42*	1.07*	—	30.9	1.39*	—	—
243	2.82	3.21*	1.91*	1.30*	1.86*	2074.	—	3.26*	2746.	2636.
244	3.02	3.44*	2.05*	1.39*	1.99*	2219.	—	3.49	2938.	2820.
245	2.95	3.14*	2.03*	1.40*	1.89*	2533.	—	—	3237.*	2817.
246	3.22	3.43*	2.21*	1.52*	2.06*	2760.	—	—	3528.*	3070.

TABLE 3–2B (continued). Composition of Feeds—Energy Values

		IFN	DRY MATT %	TDN CTTL %	TDN SHEP %	DE CTTL Mcal /kg	DE SHEP Mcal /kg
	MAIZE, ZEA MAYS—Continued						
247	Distillers' solubles, dehydrated	5-02-844	92.9	81.6	79.3	3.35*	3.50*
248			100.0	87.9	85.4	3.60*	3.76*
249	Ears with husks, silage	4-02-839	43.4	31.2	34.4*	1.38*	1.52*
250			100.0	72.0	79.2*	3.17*	3.49*
251	Ears, ground (maize and cob meal)	4-02-849	86.5	72.7	71.4	3.20	3.15*
252			100.0	84.0	82.5	3.70	3.64*
253	Gluten, meal	5-02-900	91.3	76.0	79.4	3.11*	3.50*
254			100.0	83.2	87.0	3.41*	3.84*
255	Gluten with bran (maize gluten feed)	5-02-903	89.9	74.4	74.8	3.04*	3.30*
256			100.0	82.8	83.2	3.38*	3.67*
257	Grits byproduct (hominy feed)	4-02-887	90.2	84.8	83.5	3.73	3.68*
258			100.0	93.9	92.5	4.14	4.08*
259	Silage, well-earned	3-02-823	34.1	23.9	24.0*	1.05*	1.06*
260			100.0	70.0	70.5*	3.09*	3.11*
261	Silage, 30–50% dry matter	3-20-506	36.9	25.0	26.2*	1.07	1.15*
262			100.0	67.9	71.0*	2.90	3.13*
263	Silage, aerial part without ears without husks (stalklage)	3-02-836	30.7	17.3	16.3	0.73	0.72*
264	(stover)		100.0	56.4	53.3	2.38	2.35*
	MAIZE, DENT WHITE. ZEA MAYS INDENTATA						
265	Grits byproduct (hominy feed)	4-02-990	90.0	78.3	77.1*	3.45*	3.40*
266			100.0	87.0	85.7*	3.84*	3.78*
	MAIZE, DENT YELLOW. ZEA MAYS INDENTATA						
267	Grain	4-02-935	88.0	78.0	82.9	3.14*	3.59
268			100.0	88.7	94.2	3.57*	4.08
269	Grain, grade 2 69.5 kg/hl (54 lb/bushel)	4-02-931	87.3	78.4	82.3	3.46	3.63*
270			100.0	89.8	94.2	3.96	4.15*
271	Grain, grade 3 66.9 kg/hl (52 lb/bushel)	4-02-932	86.5	78.7	77.8*	3.47*	3.43*
272			100.0	91.0	89.9*	4.01*	3.96*
	MAIZE, DENT YELLOW, OPAQUE 2. ZEA MAYS INDENTATA						
273	Grain	4-28-253	91.0	74.7*	81.5*	3.29*	3.59*
274			100.0	82.1*	89.6*	3.62*	3.95*
	MAIZE, SWEET. ZEA MAYS SACCHARATA						
275	Process residue, silage	3-07-955	31.0	22.3	19.4*	0.98*	0.86*
276			100.0	72.0	62.6*	3.17*	2.76*
277	Process residue, wet (cannery residue)	2-02-975	76.8	53.8	51.2*	2.37*	2.26*
278			100.0	70.0	66.7*	3.09*	2.94*
	MANURE (SEE CATTLE; POULTRY)						
	MEAT (SEE ANIMAL)						
	MILK (SEE CATTLE)						
	MILLET. SETARIA SPP						
279	Grain	4-03-098	89.9	64.5	57.2	3.39	2.52*
280			100.0	71.7	63.6	3.77	2.80*
	MILLET, PEARL. PENNISETUM GLAUCUM						
281	Fresh	2-03-115	20.7	12.3*	12.9	0.54*	0.57*
282			100.0	59.4*	62.1	2.62*	2.74*
283	Silage	3-20-903	28.3	13.9	14.1	0.61*	0.62*
284			100.0	49.1	50.0	2.16*	2.20*
	MOLASSES (SEE BEET; SUGARCANE)						
	NAPIERGRASS. PENNISETUM PURPUREUM						
285	Fresh, late vegetative	2-03-158	20.2	11.5	10.9	0.51*	0.48*
286			100.0	57.0	53.6	2.51*	2.36*
287	Fresh, late bloom	2-03-162	23.0	13.1	11.3	0.58*	0.50*
288			100.0	57.0	49.1	2.51*	2.16*
	NEEDLE AND THREAD. STIPA COMATA						
289	Fresh, stem-cured	2-07-989	92.0	—	44.6	—	1.92
290			100.0	—	48.5	—	2.09

	ME CTTL Mcal /kg	ME SHEP Mcal /kg	NE-M CTTL Mcal /kg	NE-G CTTL Mcal /kg	NE-L CTTL Mcal /kg	ME-N CHKN kcal /kg	TDN HORS %	DE HORS Mcal /kg	DE SWIN kcal /kg	ME SWIN kcal /kg
247	2.96	3.11*	2.03*	1.40*	1.89*	2919.	—	—	3250.	3128.
248	3.19	3.35*	2.19*	1.50*	2.03*	3143.	—	—	3499.	3369.
249	1.20*	1.34*	0.74*	0.47*	0.71*	—	—	—	1526.*	1439.*
250	2.76*	3.08*	1.70*	1.08*	1.64*	—	—	—	3516.*	3317.*
251	2.77	2.79*	1.90*	1.31*	1.68*	2597.	—	2.85*	3109.	2779.
252	3.21	3.23*	2.20*	1.52*	1.94*	3001.	—	3.29*	3593.	3211.
253	2.73	3.13*	1.85*	1.25*	1.75*	3013.	—	—	3614.*	3166.
254	2.99	3.42*	2.03*	1.37*	1.92*	3300.	—	—	3958.*	3467.
255	2.67	2.93*	1.80*	1.21*	1.72*	1730.	—	—	3032.	2478.
256	2.96	3.25*	2.00*	1.35*	1.91*	1924.	—	—	3371.	2755.
257	2.85	3.31*	1.95*	1.34*	2.00*	2894.	—	—	3568.*	3381.
258	3.15	3.67*	2.16*	1.48*	2.22*	3208.	—	—	3955.*	3748.
259	0.91*	0.92*	0.56*	0.35*	0.54*	—	—	—	—	—
260	2.67*	2.69*	1.63*	1.03*	1.60*	—	—	—	—	—
261	0.91	1.00*	0.59*	0.36*	0.53*	—	—	—	—	—
262	2.48	2.71*	1.59*	0.99*	1.45*	—	—	—	—	—
263	0.60*	0.59*	0.34*	0.17*	0.34*	—	—	—	—	—
264	1.96*	1.92*	1.11*	0.55*	1.12*	—	—	—	—	—
265	3.08*	3.03*	1.93*	1.33*	1.81*	3008.	—	—	3577.*	3390.*
266	3.42*	3.37*	2.15*	1.47*	2.01*	3344.	—	—	3976.*	3769.*
267	2.78	3.23*	1.90*	1.31*	1.81*	3312.	—	3.38*	3443.	3235.
268	3.16	3.68*	2.16*	1.48*	2.05*	3765.	—	3.84*	3915.	3678.
269	3.10*	3.27*	1.95*	1.35*	1.84*	3425.	—	3.38	3492.	3359.
270	3.55*	3.75*	2.24*	1.55*	2.11*	3921.	—	3.87	3998.	3846.
271	3.12*	3.07*	1.96*	1.36*	1.83*	—	—	—	3419.*	3240.*
272	3.60*	3.55*	2.27*	1.57*	2.11*	—	—	—	3952.*	3745.*
273	2.92*	3.22*	1.82*	1.23*	1.72*	—	—	—	3615.*	3425.*
274	3.20*	3.54*	2.00*	1.35*	1.89*	—	—	—	3973.*	3765.*
275	0.85*	0.73*	0.53*	0.34*	0.51*	—	—	—	—	—
276	2.76*	2.34*	1.70*	1.08*	1.64*	—	—	—	—	—
277	2.05*	1.93*	1.25*	0.79*	1.23*	—	—	—	—	—
278	2.67*	2.52*	1.63*	1.03*	1.60*	—	—	—	—	—
279	3.02*	2.14*	1.90*	1.29*	1.79*	3178.	—	—	2900.	2698.
280	3.36*	2.38*	2.11*	1.44*	1.99*	3535.	—	—	3226.	3001.
281	0.45*	0.48*	0.27*	0.15*	0.28*	—	8.2*	0.35*	—	—
282	2.19*	2.32*	1.29*	0.72*	1.33*	—	39.7*	1.71*	—	—
283	0.49*	0.50*	0.26*	0.11*	0.31*	—	—	—	—	—
284	1.74*	1.78*	0.94*	0.39*	1.08*	—	—	—	—	—
285	0.42*	0.39*	0.24*	0.13*	0.26*	—	7.4*	0.32*	—	—
286	2.09*	1.94*	1.21*	0.64*	1.28*	—	36.6*	1.59*	—	—
287	0.48*	0.40*	0.28*	0.15*	0.29*	—	10.0*	0.42*	—	—
288	2.09*	1.74*	1.21*	0.64*	1.28*	—	43.5*	1.85*	—	—
289	—	1.67	—	—	—	—	—	—	—	—
290	—	1.81	—	—	—	—	—	—	—	—

TABLE 3–2B (continued).   Composition of Feeds—Energy Values

	IFN	DRY MATT %	TDN CTTL %	TDN SHEP %	DE CTTL Mcal /kg	DE SHEP Mcal /kg
OATS. AVENA SATIVA						
291 Cereal byproduct, less than 4% fiber (feeding oat meal) (oat	4-03-303	90.7	84.1	87.9	3.71*	3.88*
292 middlings)		100.0	92.8	97.0	4.09*	4.28*
293 Grain	4-03-309	89.2	68.9	68.5	3.01	3.02*
294		100.0	77.3	76.7	3.37	3.38*
295 Grain, Pacific coast	4-07-999	90.9	70.0	70.9	3.09*	3.13*
296		100.0	77.0	78.0	3.39*	3.44*
297 Groats	4-03-331	89.6	83.6	91.0	3.58*	4.01*
298		100.0	93.4	101.7	4.00*	4.48*
299 Hay, sun-cured	1-03-280	90.7	54.3	48.9	2.43*	2.16*
300		100.0	59.9	53.9	2.68*	2.38*
301 Hulls	1-03-281	92.4	31.8	32.0	1.40*	1.41*
302		100.0	34.4	34.6	1.52*	1.53*
303 Silage	3-03-298	30.5	17.6	18.4*	0.77*	0.81*
304		100.0	57.6	60.5*	2.54*	2.67*
305 Straw	1-03-283	92.2	47.9	43.7	2.11*	2.52
306		100.0	52.0	47.5	2.29*	2.74
ORANGE. CITRUS SINENSIS						
307 Pomace without fines, dehydrated (dried orange pulp)	4-01-254	88.1	66.0*	78.7	2.91*	3.47*
308		100.0	74.9*	89.2	3.30*	3.93*
ORCHARDGRASS. DACTYLIS GLOMERATA						
309 Fresh, early vegetative	2-03-439	22.6	15.1*	15.0	0.66*	0.66*
310		100.0	66.7*	66.5	2.94*	2.93*
311 Fresh, early bloom	2-03-442	23.5	15.5	14.3*	0.68*	0.63*
312		100.0	66.0	60.7*	2.91*	2.68*
313 Hay, sun-cured	1-03-438	89.6	51.5	51.6	2.27*	2.72
314		100.0	57.4	57.6	2.53*	3.03
OYSTER. CRASSOSTREA SPP-OSTREA SPP						
315 Shells, fine ground (oyster shell flour)	6-03-481	99.1	—	—	—	—
316		100.0	—	—	—	—
PANGLOAGRASS. DIGITARIA DECUMBENS						
317 Fresh	2-03-493	20.2	10.9	13.2	0.53	0.54
318		100.0	53.9	65.2	2.61	2.69
319 Hay, sun-cured	1-09-459	88.8	43.7	—	1.93*	—
320		100.0	49.2	—	2.17*	—
PAPER						
321 Waste, corrugated	1-28-257	90.0	—	—	—	—
322		100.0	—	—	—	—
PEA. PISUM SPP						
323 Haulm with pods, silage	3-03-596	24.5	13.7	14.0	0.60*	0.62*
324		100.0	56.0	57.1	2.47*	2.52*
325 Seeds	5-03-600	89.1	77.2	78.1*	3.40*	3.44*
326		100.0	86.7	87.6*	3.82*	3.86*
PEANUT. ARACHIS HYPOGAEA						
327 Hay sun-cured	1-03-619	90.7	44.7	50.6	1.97*	2.23*
328		100.0	49.3	55.8	2.17*	2.46*
329 Seeds without coats, meal, mechanical-extracted (peanut meal)	5-03-649	92.6	74.1	87.1	3.40	3.84*
330		100.0	80.0	94.1	3.67	4.15*
331 Seeds without coats, meal, solvent-extracted (peanut meal)	5-03-650	92.4	71.3*	74.7*	3.14	3.29*
332		100.0	77.1*	80.8*	3.40	3.56*
PEARL MILLET (SEE MILLET, PEARL)						
PINEAPPLE. ANANAS COMOSUS						
333 Process residue, dehydrated (pineapple bran)	4-03-722	87.1	64.9	62.5	2.86*	2.76*
334		100.0	74.6	71.8	3.29*	3.17*
POTATO. SOLANUM TUBEROSUM						
335 Process residue, dehydrated	4-03-775	90.0	70.2*	80.5	3.10*	3.55*
336		100.0	78.0*	89.5	3.44*	3.95*
337 Tubers, dehydrated	4-07-850	91.1	79.8	71.1	3.15*	3.14*
338		100.0	87.6	78.1	3.45*	3.44*
339 Tubers, fresh	4-03-787	21.9	17.4	18.2	0.77*	0.80*
340		100.0	79.2	83.0	3.49*	3.66*
341 Tubers, silage	4-03-768	24.7	21.0	20.3	0.92*	0.89*
342		100.0	85.0	82.2	3.75*	3.62*

	ME CTTL Mcal /kg	ME SHEP Mcal /kg	NE-M CTTL Mcal /kg	NE-G CTTL Mcal /kg	NE-L CTTL Mcal /kg	ME-N CHKN kcal /kg	TDN HORS %	DE HORS Mcal /kg	DE SWIN kcal /kg	ME SWIN kcal /kg
291	3.34*	3.51*	2.10*	1.47*	1.95*	3158.	—	—	3615.*	3426.
292	3.68*	3.87*	2.32*	1.62*	2.15*	3483.	—	—	3987.*	3779.
293	2.62	2.65*	1.77*	1.19*	1.55*	2543.	—	2.95*	2820.	2626.
294	2.94	2.97*	1.98*	1.33*	1.74*	2850.	—	3.30*	3160.	2943.
295	2.71*	2.75*	1.68*	1.11*	1.61*	2644.	—	2.91*	2797.*	2623.
296	2.98*	3.03*	1.85*	1.22*	1.77*	2909.	—	3.20*	3076.*	2886.
297	3.21	3.65*	2.24*	1.58*	1.94*	3251.	—	3.09*	3105.*	2928.
298	3.59	4.08*	2.50*	1.77*	2.17*	3630.	—	3.45*	3467.*	3269.
299	2.04	1.77*	1.26*	0.73*	1.22*	—	38.0*	1.75*	—	—
300	2.25	1.95*	1.39*	0.81*	1.35*	—	41.9*	1.92*	—	—
301	1.00*	1.01*	0.36*	-0.12*	0.67*	357.	22.6	1.59*	—	869.
302	1.08*	1.09*	0.39*	-0.13*	0.72*	386.	24.5	1.72*	—	940.
303	0.65*	0.68*	0.38*	0.20*	0.39*	—	—	—	—	—
304	2.12*	2.25*	1.23*	0.66*	1.29*	—	—	—	—	—
305	1.72	2.14*	0.94*	0.43*	1.06*	—	44.2	1.49*	—	—
306	1.86	2.32*	1.02*	0.47*	1.15*	—	47.9	1.62*	—	—
307	2.54*	3.11*	1.58*	1.02*	1.51*	—	—	—	3115.*	2939.*
308	2.89*	3.52*	1.79*	1.16*	1.72*	—	—	—	3534.*	3335.*
309	0.57*	0.57*	0.35*	0.21*	0.34*	—	11.1*	0.46*	—	—
310	2.52*	2.51*	1.53*	0.93*	1.51*	—	49.0*	2.05*	—	—
311	0.59*	0.53*	0.35*	0.22*	0.35*	—	9.3*	0.54*	—	—
312	2.49*	2.25*	1.51*	0.91*	1.50*	—	39.6*	2.29*	—	—
313	1.89*	2.34*	1.10*	0.59*	1.15*	—	40.0*	1.69	—	—
314	2.11*	2.61*	1.22*	0.66*	1.29*	—	44.6*	1.89	—	—
315	—	—	—	—	—	—	—	—	—	—
316	—	—	—	—	—	—	—	—	—	—
317	0.44*	0.46*	0.26*	0.14*	0.26*	—	8.5*	0.39*	—	—
318	2.18*	2.27*	1.28*	0.71*	1.26*	—	42.1*	1.95*	—	—
319	1.55*	—	0.83*	0.35*	0.96*	—	—	1.54*	—	—
320	1.74*	—	0.94*	0.39*	1.09*	—	—	1.74*	—	—
321	—	—	—	—	—	—	57.7*	2.34	—	—
322	—	—	—	—	—	—	64.1*	2.60	—	—
323	0.50*	0.51*	0.29*	0.15*	0.31*	—	—	—	—	—
324	2.04*	2.09*	1.18*	0.61*	1.25*	—	—	—	—	—
325	3.04*	3.07*	1.91*	1.31*	1.79*	2110.	—	3.07*	3261.	2844.*
326	3.41*	3.45*	2.14*	1.47*	2.00*	2368.	—	3.45*	3659.	3191.*
327	1.58*	1.84*	0.85*	0.36*	0.99*	—	32.1*	1.74*	—	—
328	1.74*	2.03*	0.94*	0.40*	1.09*	—	35.4*	1.91*	—	—
329	3.01*	3.46*	1.89*	1.28*	1.78*	2661.	—	—	4315.	4018.
330	3.25*	3.74*	2.04*	1.38*	1.92*	2873.	—	—	4659.	4339.
331	2.76*	2.91*	1.71*	1.13*	1.63*	2706.	—	3.00*	2854.	2785.
332	2.98*	3.15*	1.85*	1.22*	1.76*	2928.	—	3.25*	3088.	3013.
333	2.50*	2.39*	1.55*	1.00*	1.49*	—	—	—	2696.*	2530.*
334	2.87*	2.75*	1.78*	1.15*	1.71*	—	—	—	3097.*	2906.*
335	2.72*	3.18*	1.69*	1.12*	1.61*	—	—	—	3414.*	3230.*
336	3.03*	3.54*	1.88*	1.25*	1.79*	—	—	—	3794.*	3589.*
337	2.77	2.76*	1.88*	1.28*	1.85*	2822.	—	—	3447.*	3261.
338	3.04	3.03*	2.06*	1.40*	2.03*	3098.	—	—	3785.*	3581.
339	0.68*	0.71*	0.42*	0.28*	0.40*	—	—	—	820.*	775.*
340	3.08*	3.25*	1.92*	1.28*	1.82*	—	—	—	3736.*	3533.*
341	0.82*	0.79*	0.52*	0.35*	0.48*	— —	—	—	963.*	912.*
342	3.34*	3.21*	2.09*	1.42*	1.96*	—	—	—	3902.*	3695.*

TABLE 3–2B (continued).   Composition of Feeds—Energy Values

		IFN	DRY MATT %	TDN CTTL %	TDN SHEP %	DE CTTL Mcal /kg	DE SHEP Mcal /kg
	POULTRY						
343	Byproduct meal, rendered (viscera with feet with heads)	5-03-798	93.8	73.3	74.4	3.05*	3.28*
344			100.0	78.1	79.3	3.25*	3.50*
345	Feathers, meal, hydrolyzed	5-03-795	92.9	63.0	70.1	2.24*	4.14
346			100.0	67.8	75.5	2.41*	4.45
347	Manure, dehydrated	5-14-015	90.6	56.4*	43.5*	2.49*	1.92*
348			100.0	62.3*	48.1*	2.75*	2.12*
	PRAIRIE PLANTS, MIDWEST						
349	Hay, sun-cured, midbloom	1-07-956	94.1	47.7	50.0*	2.10*	2.20*
350			100.0	50.7	53.2*	2.24*	2.34*
351	Hay, sun-cured, late bloom	1-07-957	91.0	—	—	—	—
352			100.0	—	—	—	—
353	Hay, sun-cured, milk stage	1-03-185	89.6	47.7*	47.0*	2.10*	2.07*
354			100.0	53.2*	52.4*	2.35*	2.31*
	PRICKLY PEAR. OPUNTIA SPP						
355	Fresh	2-01-061	16.4	9.7*	9.5	0.43	0.42*
356			100.0	59.1*	57.8	2.60	2.55*
	RAPE. BRASSICA NAPUS						
357	Fresh	2-03-867	16.7	11.3*	13.2	0.50*	0.58*
358			100.0	67.9*	79.3	2.99*	3.49*
359	Seeds, meal, mechanical-extracted	5-03-870	92.0	72.3	68.8	3.19*	3.03*
360			100.0	78.6	74.8	3.47*	3.30*
361	Seeds, meal, solvent-extracted	5-03-871	91.2	62.9	68.2*	2.77*	3.01*
362			100.0	69.0	74.8*	3.04*	3.30*
	RAPE, SUMMER. BRASSICA NAPUS ANNUA						
363	Seeds, meal, boiled mechanical-extracted	5-08-136	94.0	69.1*	71.6*	3.05*	3.16*
364			100.0	73.5*	76.2*	3.24*	3.36*
365	Seeds, meal, prepressed solvent-extracted	5-08-135	91.6	68.9*	70.3*	3.04*	3.10*
366			100.0	75.2*	76.7*	3.32*	3.38*
	RAPE, TURNIP. BRASSICA CAMPESTRIS						
367	Seeds, meal, mechanical-extracted	5-07-871	93.7	71.8*	74.0*	3.17*	3.26*
368			100.0	76.7*	78.9*	3.38*	3.48*
369	Seeds, meal, solvent-extracted	5-07-870	90.9	63.8	68.7*	2.82	3.03*
370			100.0	70.2	75.6*	3.10	3.33*
	REDTOP. AGROSTIS ALBA						
371	Fresh, full bloom	2-03-891	26.3	16.3	16.2*	0.72*	0.71*
372			100.0	62.0	61.6*	2.73*	2.72*
373	Hay, sun-cured, midbloom	1-03-886	92.8	53.4*	53.3*	2.35*	2.35*
374			100.0	57.5*	57.5*	2.53*	2.53*
	RICE, ORYZA SATIVA						
375	Bran with germs (rice, bran)	4-03-928	90.5	60.6	67.4	2.41	2.97*
376			100.0	66.9	74.5	2.66	3.28*
377	Grain, ground	4-03-938	89.0	65.9*	69.2*	2.91*	3.05*
378			100.0	74.1*	77.8*	3.27*	3.43*
379	Groats, polished	4-03-942	88.6	74.4	80.8	3.51	3.56*
380			100.0	84.0	91.2	3.96	4.02*
381	Groats, polished broken	4-03-932	88.5	—	78.7	—	3.47*
382			100.0	—	88.9	—	3.92*
383	Hulls	1-08-075	91.9	11.5	9.9	0.76*	0.44*
384			100.0	12.5	10.8	0.82*	0.47*
385	Polishings	4-03-943	90.2	79.1	82.0	3.45	3.62*
386			100.0	87.7	90.9	3.82	4.01*
	RUSSIAN THISTLE, TUMBLING. SALSOLA KALI TENUIFOLIA						
387	Fresh, stem-cured	2-08-000	88.0	—	—	—	—
388			100.0	—	—	—	—
389	Hay, sun-cured	1-03-988	86.4	40.4	38.2	1.78*	1.68*
390			100.0	46.7	44.2	2.06*	1.95*

	ME CTTL Mcal /kg	ME SHEP Mcal /kg	NE-M CTTL Mcal /kg	NE-G CTTL Mcal /kg	NE-L CTTL Mcal /kg	ME-N CHKN kcal /kg	TDN HORS %	DE HORS Mcal /kg	DE SWIN kcal /kg	ME SWIN kcal /kg
343	2.66	2.89*	1.78*	1.18*	1.68*	2865.	—	—	3100.	2869.
344	2.83	3.08*	1.89*	1.25*	1.79*	3054.	—	—	3305.	3058.
345	1.84	3.76*	1.05*	0.53*	1.43*	2434.	—	—	2728.	2213.
346	1.98	4.05*	1.13*	0.57*	1.54*	2619.	—	—	2936.	2382.
347	2.11*	1.53*	1.25*	0.73*	1.27*	—	—	—	-355.*	-501.*
348	2.32*	1.69*	1.38*	0.80*	1.41*	—	—	—	-392.*	-553.*
349	1.70*	1.80*	0.93*	0.42*	1.06*	—	35.7*	1.55*	—	—
350	1.81*	1.92*	0.99*	0.44*	1.12*	—	37.9*	1.64*	—	—
351	—	—	—	—	—	—	—	—	—	—
352	—	—	—	—	—	—	—	—	—	—
353	1.72*	1.69*	0.97*	0.47*	1.06*	—	32.1*	1.40*	—	—
354	1.92*	1.88*	1.08*	0.53*	1.18*	—	35.8*	1.57*	—	—
355	0.36*	0.35*	0.21*	0.12*	0.21*	—	—	—	—	—
356	2.18*	2.12*	1.28*	0.71*	1.26*	—	—	—	—	—
357	0.43*	0.51*	0.26*	0.16*	0.26*	—	9.0*	0.37*	—	—
358	2.57*	3.08*	1.57*	0.97*	1.54*	—	53.9*	2.23*	—	—
359	2.80*	2.65*	1.75*	1.16*	1.66*	2002.	—	—	3016.*	2612.
360	3.05*	2.88*	1.90*	1.26*	1.81*	2177.	—	—	3279.*	2840.
361	2.39*	2.63*	1.46*	0.91*	1.43*	1754.	—	—	3007.	2745.
362	2.62*	2.88*	1.60*	1.00*	1.57*	1924.	—	—	3299.	3012.
363	2.65*	2.77*	1.64*	1.06*	1.58*	—	—	—	2717.*	2332.*
364	2.82*	2.94*	1.74*	1.12*	1.68*	—	—	—	2891.*	2481.*
365	2.66*	2.72*	1.65*	1.07*	1.58*	—	—	—	2724.*	2343.*
366	2.90*	2.97*	1.80*	1.17*	1.72*	—	—	—	2974.*	2558.*
367	2.78*	2.87*	1.72*	1.13*	1.65*	1570.	—	—	2874.*	2477.*
368	2.96*	3.06*	1.84*	1.21*	1.76*	1676.	—	—	3066.*	2643.*
369	2.39	2.65*	1.56*	1.00*	1.43*	2025.	—	—	3130.	2813.
370	2.63	2.92*	1.72*	1.10*	1.57*	2228.	—	—	3444.	3095.
371	0.61*	0.60*	0.36*	0.21*	0.37*	—	11.4*	0.48*	—	—
372	2.31*	2.29*	1.38*	0.80*	1.40*	—	43.2*	1.84*	—	—
373	1.96*	1.96*	1.14*	0.61*	1.20*	—	43.5*	1.83*	—	—
374	2.11*	2.11*	1.23*	0.66*	1.29*	—	46.9*	1.97*	—	—
375	2.17	2.59*	1.37*	0.84*	1.18*	2010.	—	2.62*	3103.	2967.
376	2.40	2.87*	1.52*	0.92*	1.30*	2221.	—	2.90*	3427.	3278.
377	2.53*	2.68*	1.57*	1.01*	1.51*	2668.	—	3.38*	3287.*	3107.
378	2.85*	3.02*	1.76*	1.14*	1.69*	2999.	—	3.80*	3695.*	3493.
379	3.15*	3.20*	1.98*	1.37*	1.87*	3085.	—	—	3741.	3656.
380	3.55*	3.61*	2.24*	1.55*	2.11*	3483.	—	—	4223.	4127.
381	—	3.11*	—	—	—	3092.	—	—	3719.	3314.
382	—	3.51*	—	—	—	3493.	—	—	4201.	3744.
383	0.35	0.03*	-0.57*	-1.04*	0.17*	79.	11.1*	1.15*	—	—
384	0.38	0.03*	-0.62*	-1.13*	0.19*	86.	12.1*	1.25*	—	—
385	3.08*	3.25*	1.93*	1.32*	1.82*	3044.	—	—	3712.	3427.
386	3.41*	3.60*	2.14*	1.47*	2.02*	3374.	—	—	4114.	3798.
387	—	—	—	—	—	—	—	—	—	—
388	—	—	—	—	—	—	—	—	—	—
389	1.41*	1.31*	0.73*	0.27*	0.89*	—	33.4*	1.44*	—	—
390	1.63*	1.52*	0.85*	0.31*	1.02*	—	38.6*	1.67*	—	—

TABLE 3–2B (continued). Composition of Feeds—Energy Values

	IFN	DRY MATT %	TDN CTTL %	TDN SHEP %	DE CTTL Mcal /kg	DE SHEP Mcal /kg
RYE. SECALE CEREALE						
391 Distiller's grains, dehydrated	5-04-023	91.9	49.1	59.0	2.16*	2.60*
392		100.0	53.4	64.2	2.35*	2.83*
393 Flour byproduct, less than 8.5% fiber (rye middlings)	4-04-031	89.5	77.6	71.4	3.42*	3.15*
394		100.0	86.7	79.8	3.82*	3.52*
395 Fresh, early vegetative	2-04-013	15.5	—	—	—	—
396		100.0	—	—	—	—
397 Grain	4-04-047	87.5	72.3	74.8	3.12	3.30*
398		100.0	82.6	85.4	3.57	3.77*
399 Mill run, less than 9.5% fiber	4-04-034	88.8	67.6*	66.2	2.98*	2.92*
400		100.0	76.2*	74.6	3.36*	3.29*
401 Silage	3-04-020	35.0	18.4	20.5*	0.81*	0.90*
402		100.0	52.7	58.5*	2.32*	2.58*
403 Straw	1-04-007	90.9	36.3	41.5	1.60*	1.83*
404		100.0	40.0	45.6	1.76*	2.01*
RYEGRASS, ITALIAN. LOLIUM MULTIFLORUM						
405 Fresh	2-04-073	22.6	14.0	13.7	0.62*	0.60*
406		100.0	62.0	60.5	2.73*	2.67*
RYEGRASS, PERENNIAL. LOLIUM PERENNE						
407 Fresh	2-04-086	26.6	16.9*	18.0	0.75*	0.79*
408		100.0	63.6*	67.7	2.81*	2.98*
409 Hay, sun-cured	1-04-077	86.6	48.2*	51.6	2.13*	2.27*
410		100.0	55.7*	59.6	2.45*	2.63*
SACATON, ALKALI. SPOROBOLUS AIROIDES						
411 Fresh, stem-cured	2-05-599	76.0	39.8*	26.6	1.76*	1.49
412		100.0	52.4*	35.0	2.31*	1.96
SAFFLOWER. CARTHAMUS TINCTORIUS						
413 Seeds	4-07-958	93.1	82.9	82.4	3.65*	2.36
414		100.0	89.0	88.5	3.92*	2.53
415 Seeds, meal, mechanical-extracted	5-04-109	91.6	52.2	56.7	2.31	2.50*
416		100.0	57.0	61.9	2.52	2.73*
417 Seeds, meal, solvent-extracted	5-04-110	91.7	52.2	51.4	2.21*	2.27*
418		100.0	56.9	56.1	2.41*	2.47*
419 Seeds without hulls, meal, solvent-extracted	5-07-959	91.0	69.2	64.6*	3.05*	2.85*
420		100.0	76.0	71.0*	3.35*	3.13*
SAGE, BLACK. SALVIA MELLIFERA						
421 Browse, fresh, stem-cured	2-05-564	65.0	—	31.1	—	1.38
422		100.0	—	49.4	—	2.12
SAGEBRUSH, BIG. ARTEMISIA TRIDENTATA						
423 Browse, fresh, stem-cured	2-07-992	65.0	—	27.5	—	1.46
424		100.0	—	42.3	—	2.25
SAGEBRUSH, BUD. ARTEMISIA SPINESCENS						
425 Browse, fresh, early vegetative	2-07-991	23.0	—	11.6	—	0.59
426		100.0	—	50.6	—	2.56
427 Browse, fresh, late vegetative	2-04-124	32.0	17.3*	18.9*	0.76*	0.83*
428		100.0	54.0*	59.0*	2.38*	2.60*
SAGEBRUSH, FRINGED. ARTEMISIA FRIGIDA						
429 Browse, fresh, midbloom	2-04-129	43.0	26.0*	26.7*	1.15*	1.18*
430		100.0	60.5*	62.0*	2.67*	2.74*
431 Browse, fresh, mature	2-04-130	60.0	29.9*	30.6*	1.32*	1.35*
432		100.0	49.8*	50.9*	2.19*	2.25*
SALTBUSH, NUTTAL. ATRIPLEX NUTTALLII						
433 Browse, fresh, stem-cured	2-07-993	55.0	—	20.0	—	0.82
434		100.0	—	36.3	—	1.49
SALTBUSH, SHADSCALE. A TRIPLEX CONFERTIFOLIA						
435 Browse, fresh, stem-cured	2-05-565	80.0	—	25.1	—	1.03
436		100.0	—	31.4	—	1.29
SALT GRASS. DISTICHLIS SPP						
437 Fresh	2-04-170	74.4	44.5*	45.3*	1.96*	2.00*
438		100.0	59.9*	60.9*	2.64*	2.68*
439 Hay, sun-cured	1-04-168	89.5	44.8*	45.2	1.98*	1.99*
440		100.0	50.1*	50.6	2.21*	2.23*
SALT GRASS, DESERT. DISTICHLIS STRICTA						
441 Fresh	2-04-171	29.0	17.6*	17.8*	0.77*	0.78*
442		100.0	60.6*	61.3*	2.67*	2.70*

	ME CTTL Mcal /kg	ME SHEP Mcal /kg	NE-M CTTL Mcal /kg	NE-G CTTL Mcal /kg	NE-L CTTL Mcal /kg	ME-N CHKN kcal /kg	TDN HORS %	DE HORS Mcal /kg	DE SWIN kcal /kg	ME SWIN kcal /kg
391	1.77*	2.21*	1.00*	0.49*	1.09*	—	—	—	3497.*	3057.*
392	1.93*	2.41*	1.09*	0.53*	1.19*	—	—	—	3807.*	3328.*
393	3.05*	2.78*	1.92*	1.31*	1.79*	—	—	—	3001.*	2825.*
394	3.41*	3.10*	2.14*	1.47*	2.00*	—	—	—	3353.*	3156.*
395	—	—	—	—	—	—	—	—	—	—
396	—	—	—	—	—	—	—	—	—	—
397	2.60	2.94*	1.75*	1.18*	1.63*	2652.	—	3.36*	3251.	2911.
398	2.97	3.35*	2.01*	1.35*	1.86*	3031.	—	3.84*	3716.	3327.
399	2.61*	2.55*	1.62*	1.06*	1.55*	—	—	—	2975.*	2801.*
400	2.94*	2.87*	1.83*	1.20*	1.75*	—	—	—	3351.*	3155.*
401	0.66*	0.75*	0.37*	0.18*	0.41*	—	—	—	—	—
402	1.89*	2.16*	1.06*	0.51*	1.17*	—	—	—	—	—
403	1.21*	1.44*	0.55*	0.07*	0.78*	—	31.9*	1.40*	—	—
404	1.33*	1.58*	0.60*	0.08*	0.86*	—	35.1*	1.54*	—	—
405	0.52*	0.51*	0.31*	0.18*	0.32*	—	9.1*	0.51*	—	—
406	2.31*	2.24*	1.38*	0.80*	1.40*	—	40.3*	2.20*	—	—
407	0.63*	0.68*	0.38*	0.22*	0.38*	—	10.0*	0.43*	—	—
408	2.38*	2.56*	1.43*	0.84*	1.44*	—	37.5*	1.63*	—	—
409	1.76*	1.91*	1.01*	0.52*	1.08*	—	35.9*	1.54*	—	—
410	2.03*	2.20*	1.16*	0.60*	1.24*	—	41.5*	1.77*	—	—
411	1.43*	1.25	0.70	0.38*	0.88*	—	24.2*	1.08*	—	—
412	1.88*	1.65	1.05*	0.50*	1.16*	—	31.8*	1.42*	—	—
413	3.27*	1.96*	2.06*	1.42*	1.92*	—	—	—	—	—
414	3.51*	2.11*	2.21*	1.52*	2.06*	—	—	—	—	—
415	1.92*	2.11*	1.11*	0.59*	1.11*	987.	—	—	—	—
416	2.09*	2.31*	1.21*	0.65*	1.21*	1077.	—	—	—	—
417	1.82	1.88*	1.04*	0.53*	1.17*	1186.	—	—	—	—
418	1.99	2.05*	1.14*	0.58*	1.27*	1294.	—	—	—	—
419	2.67*	2.47*	1.66*	1.08*	1.59*	1844.	—	—	2346.*	1995.*
420	2.93*	2.71*	1.82*	1.19*	1.74*	2026.	—	—	2578.*	2191.*
421	—	0.68	—	—	—	—	—	—	—	—
422	—	1.04	—	—	—	—	—	—	—	—
423	—	0.81	—	—	—	—	—	—	—	—
424	—	1.24	—	—	—	—	—	—	—	—
425	—	0.46	—	—	—	—	—	—	—	—
426	—	2.01	—	—	—	—	—	—	—	—
427	0.63*	0.70*	0.35*	0.18*	0.39*	—	12.5*	0.54*	—	—
428	1.96*	2.18*	1.11*	0.55*	1.20*	—	39.1*	1.69*	—	—
429	0.97*	0.99*	0.57*	0.32*	0.59*	—	19.3*	0.81*	—	—
430	2.25*	2.31*	1.33*	0.75*	1.36*	—	44.8*	1.89*	—	—
431	1.06*	1.09*	0.58*	0.25*	0.66*	—	12.9*	0.63*	—	—
432	1.77*	1.82*	0.96*	0.41*	1.10*	—	21.6*	1.04*	—	—
433	—	0.73	—	—	—	—	—	—	—	—
434	—	1.32	—	—	—	—	—	—	—	—
435	—	0.70	—	—	—	—	—	—	—	—
436	—	0.88	—	—	—	—	—	—	—	—
437	1.65*	1.68*	0.97*	0.54*	1.00*	—	32.4*	1.38*	—	—
438	2.22*	2.26*	1.31*	0.73*	1.35*	—	43.5*	1.85*	—	—
439	1.59*	1.61*	0.87*	0.38*	0.99*	—	32.5*	142*	—	—
440	1.78*	1.80*	0.97*	0.42*	1.11*	—	36.3*	1.58*	—	—
441	0.65*	0.66*	0.39*	0.22*	0.40*	—	12.9*	0.55*	—	—
442	2.25*	2.28*	1.33*	0.75*	1.36*	—	44.6*	1.89*	—	—

TABLE 3–2B (continued). Composition of Feeds—Energy Values

		IFN	DRY MATT %	TDN CTTL %	TDN SHEP %	DE CTTL Mcal /kg	DE SHEP Mcal /kg
	SCREENINGS (SEE BARLEY; CEREALS; WHEAT)						
	SEAWEED, KELP. LAMINARIALES (ORDER)—FUCALES (ORDER)						
443	Whole, meal	1-08-073	90.9	—	28.8	—	1.27*
444			100.0	—	31.7	—	1.40*
	SEDGE. CAREX SPP						
445	Hay, sun-cured	1-04-193	89.3	50.0*	46.7	2.21*	2.06*
446			100.0	56.0*	52.3	2.47*	2.31*
	SESAME. SESAMUM INDICUM						
447	Seeds, meal, mechanical-extracted	5-04-220	92.7	69.5	71.7	3.07*	3.16*
448			100.0	75.0	77.3	3.31*	3.41*
	SKIMMILK (SEE CATTLE)						
	SORGHUM. SORGHUM BICOLOR						
449	Aerial part without heads, sun-cured, mature	1-07-961	88.1	43.5	—	1.92*	—
450			100.0	49.4	—	2.18*	—
451	Distiller's grains, dehydrated	5-04-374	93.8	76.9	79.5	3.39*	3.51*
452			100.0	82.0	84.8	3.62*	3.74*
453	Grain	4-04-383	90.1	54.0	80.3	2.38*	3.54*
454			100.0	59.9	89.1	2.64*	3.93*
455	Grain, 9–12% protein	4-08-139	89.4	73.0	79.0*	2.82*	3.48*
456			100.0	81.7	88.4*	3.16*	3.90*
	SORGHUM, JOHNSONGRASS. SORGHUM HALEPENSE						
457	Hay, sun-cured	1-04-407	90.5	50.7	50.5	2.23*	2.23*
458			100.0	56.0	55.8	2.47*	2.46*
	SORGHUM, KAFIR. SORGHUM BICOLOR CAFFRORUM						
459	Grain	4-04-428	89.1	71.5*	75.5	3.15*	3.33*
460			100.0	80.2*	84.7	3.54*	3.73*
	SORGHUM, MILO. SORGHUM BICOLOR SUBGLABRESCENS						
461	Grain	4-04-444	88.5	75.0	77.7	3.11	3.43*
462			100.0	84.8	87.8	3.52	3.87*
463	Heads	4-04-446	89.9	69.9	75.9	3.08*	3.35*
464			100.0	77.8	84.4	3.43*	3.72*
	SORGHUM, SORGO. SORGHUM BICOLOR SACCHARATUM						
465	Silage	3-04-468	28.8	16.5	17.2	0.73*	0.76*
466			100.0	57.4	59.6	2.53*	2.63*
	SORGHUM, SUDANGRASS. SORGHUM BICOLOR SUDANENSE ≡						
467	Fresh, early vegetative	2-04-484	17.8	12.5	11.2*	0.55*	0.49*
468			100.0	70.0	62.7*	3.09*	2.77*
469	Fresh, midbloom	2-04-485	22.8	14.3	13.2*	0.63	0.58*
470			100.0	63.0	57.9*	2.79	2.55*
471	Hay, sun-cured	1-04-480	91.1	50.9	49.1*	2.24*	2.16*
472			100.0	55.8	53.8*	2.46*	2.37*
473	Silage	3-04-499	23.3	13.3	13.2	0.59*	0.58*
474			100.0	57.0	56.7	2.51*	2.50*
	SOYBEAN. GLYCINE MAX						
475	Hay, sun-cured	1-04-558	89.1	47.7	49.9	2.45	2.20*
476			100.0	53.5	56.0	2.75	2.47*
477	Seed coats (hulls)	1-04-560	90.3	69.1	51.1*	3.05*	2.25*
478			100.0	76.5	56.6*	3.37*	2.50*
479	Seeds	5-04-610	92.0	82.2	86.4	3.62*	3.81*
480			100.0	89.3	93.9	3.94*	4.14*
481	Seeds, heat-processed	5-04-597	92.6	92.3*	92.4*	4.07*	4.07*
482			100.0	99.7*	99.8*	4.40*	4.40*
483	Seeds, meal, mechanical-extracted	5-04-600	90.0	76.9	76.8	3.38	3.39*
484			100.0	85.4	85.4	3.76	3.76*
485	Seeds, meal, solvent-extracted	5-04-604	89.6	75.0	78.8	3.31*	3.47*
486			100.0	83.7	87.9	3.69*	3.88*
487	Seeds without hulls, meal, solvent-extracted	5-04-612	89.9	75.3*	77.9	3.32*	3.43*
488			100.0	83.7*	86.6	3.69*	3.82*
489	Silage	3-04-581	30.2	16.3	16.0	0.72*	0.70*
490			100.0	53.8	52.8	2.37*	2.33*
491	Straw	1-04-567	87.7	33.3	37.9	1.47*	1.67*
492			100.0	38.0	43.3	1.68*	1.91*

	ME CTTL Mcal /kg	ME SHEP Mcal /kg	NE-M CTTL Mcal /kg	NE-G CTTL Mcal /kg	NE-L CTTL Mcal /kg	ME-N CHKN kcal /kg	TDN HORS %	DE HORS Mcal /kg	DE SWIN kcal /kg	ME SWIN kcal /kg
443	—	0.87*	—	—	—	—	—	—	—	—
444	—	0.96*	—	—	—	—	—	—	—	—
445	1.83*	1.68*	1.05*	0.55*	1.12*	—	36.5*	1.56	—	—
446	2.04*	1.88*	1.18*	0.61*	1.25*	—	40.8*	1.75	—	—
447	2.68*	2.77*	1.66*	1.08*	1.59*	2174.	—	—	3428.	3029.
448	2.89*	2.99*	1.79*	1.16*	1.72*	2345.	—	—	3698.	3267.
449	1.54*	—	0.83*	0.35*	0.96*	—	—	—	—	—
450	1.75*	—	0.95*	0.40*	1.09*	—	—	—	—	—
451	3.00*	3.12*	1.88*	1.26*	1.77*	—	—	—	3385.*	2949.*
452	3.20*	3.33*	2.00*	1.35*	1.89*	—	—	—	3609.*	3144.*
453	2.00*	3.17*	1.18*	0.66*	1.21*	3411.	—	3.21*	3518.	3251.
454	2.22*	3.52*	1.31*	0.73*	1.35*	3787.	—	3.56*	3907.	3609.
455	2.45	3.11*	1.62*	1.06*	1.68*	3360.	—	—	3360.	3172.
456	2.74	3.48*	1.82*	1.19*	1.88*	3760.	—	—	3760.	3550.
457	1.85*	1.84.*	1.06*	0.55*	1.13*	—	34.9*	1.51*	—	—
458	2.04*	2.03*	1.18*	0.61*	1.25*	—	38.5*	1.66*	—	—
459	2.78*	2.96*	1.74*	1.16*	1.64*	3372.	—	—	3350.*	3169.
460	3.12*	3.32*	1.95*	1.30*	1.85*	3785.	—	—	3760.*	3557.
461	2.48	3.06*	1.65*	1.09*	1.62*	3226.	—	—	3347.	3247.
462	2.81	3.46*	1.87*	1.23*	1.83*	3645.	—	—	3782.	3669.
463	2.71*	2.98*	1.69*	1.11*	1.60*	—	—	—	3185.*	3005.*
464	3.01*	3.31*	1.87*	1.24*	1.79*	—	—	—	3543.*	3343.*
465	0.61*	0.64*	0.35*	0.19*	0.37*	—	—	—	—	—
466	2.10*	2.20*	1.22*	0.66*	1.29*	—	—	—	—	—
467	0.47*	0.42*	0.29*	0.18*	0.28*	—	7.7*	0.33*	—	—
468	2.67*	2.34*	1.63*	1.03*	1.60*	—	43.1*	1.83*	—	—
469	0.54*	0.48*	0.32*	0.19*	0.31*	—	8.3*	0.36*	—	—
470	2.36*	2.13*	1.42*	0.83*	1.38*	—	36.5*	1.59*	—	—
471	1.86*	1.77*	1.07*	0.55*	1.14*	—	41.3*	1.74*	—	—
472	2.04*	1.95*	1.17*	0.61*	1.25*	—	45.3*	1.91*	—	—
473	0.49*	0.48*	0.28*	0.15*	0.30*	—	—	—	—	—
474	2.09*	2.07*	1.21*	0.64*	1.28*	—	—	—	—	—
475	1.90	1.82*	1.14*	0.63*	1.21*	—	41.4*	1.74*	—	—
476	2.13	2.04*	1.27*	0.70*	1.36*	—	46.4*	1.95*	—	—
477	2.67*	1.87*	1.66*	1.09*	1.58*	663.	40.1*	1.70*	1945.	670.
478	2.96*	2.07*	1.84*	1.20*	1.75*	734.	44.4*	1.88*	2155.	742.
479	3.25*	3.43*	2.04*	1.41*	1.90*	3382.	—	—	4020.*	3540.
480	3.53*	3.73*	2.22*	1.53*	2.07*	3674.	—	—	4367.*	3846.
481	3.70*	3.70*	2.33*	1.65*	2.15*	—	—	3.52	4131.*	3642.
482	3.99*	3.99*	2.52*	1.78*	2.32*	—	—	3.80	4462.*	3933.
483	3.01*	3.02*	1.89*	1.29*	1.78*	2429.	—	—	3483.	2871.
484	3.34*	3.35*	2.10*	1.43*	1.98*	2699.	—	—	3870.	3190.
485	2.94*	3.11*	1.84*	1.25*	1.73*	2331.	—	3.15*	3494.	3006.
486	3.28*	3.47*	2.05*	1.39*	1.93*	2602.	—	3.52*	3900.	3355.
487	2.95*	3.06*	1.85*	1.25*	1.74*	2473.	—	3.36	3770.	3361.
488	3.28*	3.41*	2.05*	1.39*	1.93*	2750.	—	3.73	4192.	3737.
489	0.59*	0.57*	0.33*	0.16*	0.36*	—	—	—	—	—
490	1.95*	1.90*	1.10*	0.54*	1.20*	—	—	—	—	—
491	1.09*	1.29*	0.46*	0.00*	0.71*	—	28.5*	1.27*	—	—
492	1.24*	1.48*	0.53*	0.00*	0.81*	—	32.5*	1.45*	—	—

TABLE 3–2B (continued). Composition of Feeds—Energy Values

		IFN	DRY MATT %	TDN CTTL %	TDN SHEP %	DE CTTL Mcal /kg	DE SHEP Mcal /kg
	SQUIRREL TAIL. SITANION SPP						
493	Fresh, stem-cured	2-05-566	50.0	—	25.2	—	1.01
494			100.0	—	50.4	—	2.02
	SUDANGRASS (SEE SORGHUM)						
	SUGARCANE. SACCHARUM OFFICINARUM						
495	Molasses, dehydrated	4-04-695	94.4	66.2	69.2*	2.92*	3.05*
496			100.0	70.1	73.3*	3.09*	3.23*
497	Molasses, more than 46%invert sugars more than 79.5 degrees	4-04-696	74.3	61.1	58.7	2.76*	2.59*
498	brix		100.0	82.2	79.0	3.72*	3.48*
	SUNFLOWER, COMMON. HELIANTHUS ANNUUS						
499	Seeds without hulls, meal, mechanical-extracted	5-04-738	92.7	68.8	74.6	2.86	3.29
500			100.0	74.2	80.5	3.09	3.55
501	Seeds without hulls, meal, solvent-extracted	5-04-739	92.5	60.1	69.0*	2.65*	3.04*
502			100.0	65.0	74.6*	2.87*	3.29*
	STREETCLOVER, YELLOW. MELILOTUS OFFICINALIS						
503	Hay, sun-cured	1-04-754	88.5	50.5	47.5	2.23*	2.09*
504			100.0	57.0	53.6	2.51*	2.36*
	TANKAGE (SEE ANIMAL; ANIMAL-POULTRY)						
	TIMOTHY. PHLEUM PRATENSE						
505	Fresh, late vegetative	2-04-903	26.7	17.9	16.4	0.79*	0.72*
506			100.0	67.0	61.2	2.95*	2.70*
507	Fresh, midbloom	2-04-905	29.2	19.3	18.0	0.85*	0.79*
508			100.0	66.0	61.6	2.91*	2.72*
509	Hay, sun-cured, late vegetative	1-04-881	89.3	56.1	58.4	2.47*	2.58*
510			100.0	62.8	65.4	2.77*	2.89*
511	Hay, sun-cured, early bloom	1-04-882	89.1	52.7	50.4	2.32*	2.22*
512			100.0	59.1	56.5	2.61*	2.49*
513	Hay, sun-cured, midbloom	1-04-883	88.9	52.4	53.1	2.31*	2.34*
514			100.0	59.0	59.7	2.60*	2.63*
515	Hay, sun-cured, late bloom	1-04-885	88.3	50.7	48.9	2.24*	2.16*
516			100.0	57.4	55.4	2.53*	2.44*
517	Silage	3-04-922	33.5	19.9	20.0	0.88*	0.88
518			100.0	59.3	59.5	2.61*	2.63
	TOMATO. LYCOPERSICON ESCULENTUM						
519	Pomace, dehydrated	5-05-041	92.2	59.8	61.1	2.63*	2.69*
520			100.0	64.8	66.3	2.86*	2.92*
	TREFOIL, BIRDSFOOT. LOTUS CORNICULATUS						
521	Fresh	2-20-786	19.3	14.5	11.7	0.64*	0.52*
522			100.0	75.0	60.8	3.31*	2.68*
523	Hay, sun-cured	1-05-044	90.6	55.3	52.3	2.44*	2.01
524			100.0	61.0	57.8	2.69*	2.22
	TRITICALE. TRITICALE HEXAPLOIDE						
525	Grain	4-20-362	89.2	69.0*	75.3	3.04*	3.32*
526			100.0	77.4*	84.4	3.41*	3.72*
	TURNIP. BRASSICA RAPA RAPA						
527	Roots, fresh	4-05-067	9.2	7.7	7.8	0.34*	0.35*
528			100.0	84.0	85.7	3.70*	3.78*
	UREA						
529	45% nitrogen 281% protein equivalent	5-05-070	97.0	—	—	—	—
530			100.0	—	—	—	—
	VEGETABLE						
531	Oil	4-05-077	99.8	—	194.6	—	8.58*
532			100.0	—	195.0	—	8.60*
	VETCH. VICIA SPP						
533	Hay, sun-cured	1-05-106	88.7	55.0	54.4*	2.42*	2.40*
534			100.0	62.0	61.4*	2.73*	2.71*

	ME CTTL Mcal /kg	ME SHEP Mcal /kg	NE-M CTTL Mcal /kg	NE-G CTTL Mcal /kg	NE-L CTTL Mcal /kg	ME-N CHKN kcal /kg	TDN HORS %	DE HORS Mcal /kg	DE SWIN kcal /kg	ME SWIN kcal /kg
493	—	0.85	—	—	—	—	—	—	—	—
494	—	1.70	—	—	—	—	—	—	—	—
495	2.52*	2.66*	1.55*	0.97*	1.51*	2706.	—	3.21*	2661.*	2485.
496	2.67*	2.81*	1.64*	1.03*	1.60*	2866.	—	3.40*	2818.*	2632.
497	2.46	2.28*	1.70*	1.18*	1.41*	1902.	—	2.60*	2500.	2193.
498	3.31	3.07*	2.28*	1.58*	1.89*	2560.	—	3.50*	3364.	2950.
499	2.48*	2.91*	1.52*	0.95*	1.45*	2216.	—	—	3111.	2735.
500	2.67*	3.14*	1.64*	1.03*	1.57*	2391.	—	—	3357.	2951.
501	2.26*	2.66*	1.36*	0.82*	1.36*	2073.	—	2.59*	3043.*	2636.
502	2.44*	2.87*	1.47*	0.88*	1.47*	2242.	—	2.80*	3291.*	2851.
503	1.85*	1.71*	1.07*	0.57*	1.13*	—	42.8*	1.79*	—	—
504	2.09*	1.94*	1.21*	0.64*	1.28*	—	48.4*	2.03*	—	—
505	0.68*	0.61*	0.41*	0.25*	0.41*	—	10.5*	0.70*	—	—
506	2.53*	2.27*	1.54*	0.94*	1.52*	—	39.3*	2.37*	—	—
507	0.73*	0.67*	0.44*	0.27*	0.44*	—	11.6*	0.58*	—	—
508	2.49*	2.29*	1.51*	0.91*	1.50*	—	39.6*	2.00*	—	—
509	2.10*	2.20*	1.25*	0.73*	1.27*	—	56.4	2.29*	—	—
510	2.35*	2.46*	1.40*	0.82*	1.42*	—	63.2	2.57*	—	—
511	1.94*	1.84*	1.14*	0.63*	1.18*	—	38.8*	1.83*	—	—
512	2.18*	2.07*	1.28*	0.71*	1.33*	—	43.5*	2.06*	—	—
513	1.93*	1.96*	1.13*	0.63*	1.18*	—	50.6	1.77*	—	—
514	2.18*	2.21*	1.28*	0.71*	1.33*	—	56.9	1.99*	—	—
515	1.86*	1.78*	1.08*	0.58*	1.14*	—	37.0*	1.59*	—	—
516	2.11*	2.02*	1.22*	0.66*	1.29*	—	41.9*	1.80*	—	—
517	0.73*	0.74*	0.43*	0.24*	0.45*	—	—	—	—	—
518	2.19*	2.20*	1.28*	0.71*	1.33*	—	—	—	—	—
519	2.25*	2.31*	1.35*	0.81*	1.35*	1758.	—	—	2088.*	1754.*
520	2.44*	2.50*	1.47*	0.88*	1.47*	1907.	—	—	2265.*	1903.*
521	0.56*	0.44*	0.35*	0.22*	0.33*	—	10.1*	0.42*	—	—
522	2.89*	2.26*	1.79*	1.16*	1.72*	—	52.6.*	2.18*	—	—
523	2.05*	1.62*	1.22*	0.69*	1.25*	—	45.8*	1.99*	—	—
524	2.27*	1.79*	1.34*	0.77*	1.37*	—	50.5*	2.20*	—	—
525	2.67*	2.95*	1.66*	1.10*	1.58*	3140.	—	—	3213.	3139.
526	3.00*	3.31*	1.86*	1.23*	1.78*	3521.	—	—	3603.	3521.
527	0.30*	0.31*	0.19*	0.13*	0.18*	—	—	—	322.*	303.*
528	3.29*	3.37*	2.06*	1.40*	1.94*	—	—	—	3513.*	3314.*
529	—	—	—	—	—	—	—	2.45	—	—
530	—	—	—	—	—	—	—	2.52	—	—
531	—	8.22*	—	—	—	8041.	—	8.98*	8843.	8030.
532	—	8.23*	—	—	—	8058.	—	9.00*	8862.	8047.
533	2.05*	2.02*	1.22*	0.71*	1.24*	—	48.3*	2.00*	—	—
534	2.31*	2.28*	1.38*	0.80*	1.40*	—	54.5*	2.25*	—	—

TABLE 3–2B (continued). Composition of Feeds—Energy Values

		IFN	DRY MATT %	TDN CTTL %	TDN SHEP %	DE CTTL Mcal /kg	DE SHEP Mcal /kg
	WHEAT. TRITICUM AESTIVUM						
535	Bran	4-05-190	89.0	62.4	62.8	2.78	2.77*
536			100.0	70.1	70.5	3.12	3.11*
537	Flour byproduct, less than 4% fiber (wheat red dog)	4-05-203	88.3	71.6	82.0	3.24	3.62*
538			100.0	81.1	92.8	3.67	4.09*
539	Flour byproduct, less than 7% fiber (wheat shorts)	4-05-201	88.4	75.8	76.3	3.12*	3.36*
540			100.0	85.7	86.2	3.53*	3.80*
541	Flour byproduct, less than 9.5% fiber (wheat middlings)	4-05-205	88.9	75.1	72.6	2.94*	3.20*
542			100.0	84.4	81.7	3.31*	3.60*
543	Fresh, early vegetative	2-05-176	22.2	17.4	15.7*	0.77*	0.69*
544			100.0	78.2	70.6*	3.45*	3.11*
545	Germs, ground	5-05-218	88.4	84.0	82.1	3.70*	3.62*
546			100.0	95.0	92.9	4.19*	4.10*
547	Grain	4-05-211	89.0	78.6	79.1	3.40	3.49*
548			100.0	88.3	88.9	3.81	3.92*
549	Grain screenings	4-05-216	88.5	60.1	66.1	2.65*	2.91*
550			100.0	67.9	74.7	2.99*	3.29*
551	Hay, sun-cured	1-05-172	88.7	53.0	45.8	2.12*	2.02*
552			100.0	59.7	51.7	2.39*	2.28*
553	Mill run, less than 9.5% fiber	4-05-206	89.9	70.9	70.8	3.21	3.12*
554			100.0	78.8	78.8	3.58	3.47*
555	Silage, early vegetative	3-05-184	29.9	14.9*	20.6	0.66	0.91*
556			100.0	49.8*	69.0	2.20	3.04*
557	Straw	1-05-175	91.1	42.0	38.4	1.94	1.82
558			100.0	46.1	42.1	2.13	2.00
	WHEAT, DURUM. TRITICUM DURUM						
559	Grain	4-05-224	87.6	74.5	76.6*	3.28*	3.38*
560			100.0	85.0	87.4*	3.75*	3.85*
	WHEAT, HARD RED SPRING. TRITICUM AESTIVUM						
561	Grain	4-05-258	87.6	77.1	78.5	3.40*	3.46*
562			100.0	88.0	89.6	3.88*	3.95*
	WHEAT, HARD RED WINTER, TRITICUM AESTIVUM						
563	Grain	4-05-268	88.8	78.5	78.4	3.46*	3.46*
564			100.0	88.4	88.3	3.90*	3.89*
	WHEAT, SOFT RED WINTER, TRITICUM AESTIVUM						
565	Grain	4-05-294	88.4	78.7	78.0	3.47*	3.44*
566			100.0	89.0	88.2	3.92*	3.89*
	WHEAT, SOFT WHITE WINTER, TRITICUM AESTIVUM						
567	Grain	4-05-337	90.2	80.2	79.7*	3.54*	3.51*
568			100.0	88.9	88.3*	3.92*	3.90*
	WHEATGRASS, CRESTED. AGROPYRON DESERTORUM						
569	Fresh	2-05-429	45.0	28.4*	22.8	1.25*	1.00*
570			100.0	63.2*	50.6	2.78*	2.23*
571	Hay, sun-cured	1-05-418	91.8	50.9*	49.5	2.24*	2.18*
572			100.0	55.4*	53.9	2.44*	2.38*
	WHEY (SEE CATTLE)						
	YEAST, BREWER'S. SACCHAROMYCES CEREVISIAE						
573	Dehydrated	7-05-527	93.1	73.7	72.1	3.25*	3.18*
574			100.0	79.2	77.4	3.49*	3.41*
	YEAST, IRRADIATED. SACCHAROMYCES CEREVISIAE						
575	Dehydrated	7-05-529	93.9	—	71.7	—	3.16*
576			100.0	—	76.4	—	3.37*
	YEAST, TORULA. TORULOPSIS UTILIS						
577	Dehydrated	7-05-534	93.0	74.4	70.3	3.29	3.10*
578			100.0	80.0	75.6	3.53	3.33*

	ME CTTL Mcal /kg	ME SHEP Mcal /kg	NE-M CTTL Mcal /kg	NE-G CTTL Mcal /kg	NE-L CTTL Mcal /kg	ME-N CHKN kcal /kg	TDN HORS %	DE HORS Mcal /kg	DE SWIN kcal /kg	ME SWIN kcal /kg
535	2.24	2.40*	1.44*	0.90*	1.41*	1226.	44.1	2.94*	2648.	2372.
536	2.51	2.69*	1.62*	1.01*	1.58*	1377.	49.5	3.30*	2974.	2664.
537	2.87*	3.25*	1.80*	1.22*	1.70*	2593.	—	—	3153.	2880.
538	3.25*	3.68*	2.04*	1.38*	1.92*	2935.	—	—	3570.	3260.
539	2.75	3.00*	1.88*	1.28*	1.75*	2207.	—	—	3117.	2927.
540	3.11	3.39*	2.12*	1.45*	1.98*	2496.	—	—	3524.	3309.
541	2.57	2.83*	1.73*	1.15*	1.73*	2075.	—	3.04*	2932.	2712.
542	2.89	3.19*	1.94*	1.29*	1.95*	2333.	—	3.42*	3297.	3049.
543	0.67*	0.60*	0.42*	0.28*	0.40*	—	13.1*	0.64*	—	—
544	3.03*	2.69*	1.89*	1.25*	1.80*	—	58.9*	2.88*	—	—
545	3.34*	3.26*	2.11*	1.48*	1.95*	2696.	—	—	3815.*	3357.
546	3.78*	3.69*	2.38*	1.67*	2.21*	3051.	—	—	4316.*	3798.
547	2.82	3.12*	1.93*	1.33*	1.79*	3069.	—	—	3405.	3267.
548	3.17	3.51*	2.17*	1.49*	2.02*	3449.	—	—	3826.	3670.
549	2.28*	2.54*	1.39*	0.86*	1.37*	2800.	—	—	—	2334.
550	2.57*	2.88*	1.57*	0.97*	1.54*	3165.	—	—	—	2638.
551	1.74	1.64*	0.99*	0.49*	1.19*	—	39.8*	1.68*	—	—
552	1.96	1.85*	1.11*	0.56*	1.34*	—	44.8*	1.90*	—	—
553	2.52	2.75*	1.68*	1.11*	1.68*	1764.	—	—	3170.	2765.
554	2.80	3.06*	1.87*	1.23*	1.87*	1962.	—	—	3527.	3076.
555	0.53*	0.78*	0.29*	0.12*	0.30*	—	—	—	—	—
556	1.77*	2.62*	0.96*	0.41*	1.01*	—	—	—	—	—
557	1.56	1.43*	0.80*	0.31*	0.88*	—	36.8	1.48*	—	—
558	1.71	1.57*	0.88*	0.33*	0.96*	—	40.4	1.62*	—	—
559	2.92*	3.02*	1.83*	1.25*	1.72*	3200.	—	—	3093.*	2918.*
560	3.34*	3.44*	2.09*	1.42*	1.96*	3652.	—	—	3529.*	3330.*
561	3.04*	3.10*	1.91*	1.31*	1.78*	2701.	—	—	3098.	2952.
562	3.47*	3.54*	2.18*	1.50*	2.04*	3084.	—	—	3536.	3370.
563	3.10*	3.09*	1.95*	1.34*	1.82*	3214.	—	3.43*	3383.	3223.
564	3.49*	3.48*	2.19*	1.51*	2.05*	3620.	—	3.86*	3809.	3630.
565	3.11*	3.07*	1.95*	1.35*	1.82*	3093.	—	3.41*	3228.*	3049.*
566	3.51*	3.48*	2.21*	1.52*	2.06*	3499.	—	3.86*	3652.*	3450.*
567	3.17*	3.14*	1.99*	1.37*	1.86*	2858.	—	3.54*	3345.*	3162.*
568	3.51*	3.48*	2.21*	1.52*	2.06*	3167.	—	3.92*	3707.*	3505.*
569	1.06*	0.81*	0.64*	0.37*	0.64*	—	22.8*	0.95*	—	—
570	2.36*	1.80*	1.41*	0.83*	1.43*	—	50.5*	2.10*	—	—
571	1.85*	1.79*	1.06*	0.54*	1.14*	—	40.8*	1.73*	—	—
572	2.02*	1.95*	1.15*	0.59*	1.24*	—	44.4*	1.88*	—	—
573	2.86*	2.79*	1.79*	1.19*	1.69*	2071.	—	3.07*	—	2865.
574	3.08*	3.00*	1.92*	1.28*	1.82*	2225.	—	3.30*	—	3078.
575	—	2.77*	—	—	—	—	—	—	—	—
576	—	2.95*	—	—	—	—	—	—	—	—
577	2.90*	2.71*	1.81*	1.21*	1.71*	2141.	—	—	2836.	2416.
578	3.12*	2.92*	1.95*	1.30*	1.84*	2301.	—	—	3049.	2597.

TABLE 3–2C. Composition of Feeds—Mineral Values

	IFN	DRY MATT %	CA %	CL %	MG %	P %
ALFALFA. MEDICAGO SATIVA						
001 Fresh	2-00-196	26.0	0.40	0.12	0.09	0.07
002		100.0	1.52	0.46	0.34	0.28
003 Hay, sun-cured, late vegetative	1-00-054	89.7	1.34	0.31	0.19	0.30
004		100.0	1.50	0.34	0.21	0.33
005 Hay, sun-cured, early bloom	1-00-059	90.5	1.48	0.34	0.31	0.19
006		100.0	1.63	0.38	0.34	0.21
007 Hay, sun-cured, midbloom	1-00-063	91.0	1.24	0.34	0.32	0.22
008		100.0	1.37	0.38	0.35	0.24
009 Hay, sun-cured, full bloom	1-00-068	90.9	1.08	—	0.25	0.22
010		100.0	1.19	—	0.27	0.24
011 Hay, sun-cured, mature	1-00-071	91.2	1.07	—	0.23	0.19
012		100.0	1.18	—	0.26	0.21
013 Meal dehydrated, 15% protein	1-00-022	90.3	1.25	0.41	0.26	0.23
014		100.0	1.38	0.45	0.29	0.25
015 Meal dehydrated, 17% protein	1-00-023	91.8	1.38	0.47	0.29	0.23
016		100.0	1.51	0.52	0.32	0.25
017 Meal dehydrated, 20% protein	1-00-024	91.6	1.59	0.47	0.33	0.28
018		100.0	1.73	0.51	0.37	0.31
019 Silage, 30-50% dry matter	3-08-150	43.4	0.57	—	0.13	0.13
020		100.0	1.32	—	0.30	0.31
ALFALFA-BROME, SMOOTH. MEDICAGO SATIVA-BROMUS INERMIS						
021 Fresh	2-00-262	21.3	0.32	—	0.07	0.08
022		100.0	1.52	—	0.35	0.37
023 Silage, 30-50% dry matter	3-08-147	39.8	0.25	—	0.08	0.08
024		100.0	0.64	—	0.20	0.20
ALFALFA-GRASS. MEDICAGO SATIVA-GRASS						
025 Hay, sun-cured	1-08-331	91.4	1.34	—	0.28	0.22
026		100.0	1.47	—	0.30	0.25
ALFALFA-ORCHARDGRASS. MEDICAGO SATIVA-DACTYLIS GLOMERATA						
027 Silage, 30-50% dry matter	3-08-144	36.4	0.31	—	0.12	0.11
028		100.0	0.86	—	0.33	0.29
ANIMAL						
029 Blood, spray dehydrated	5-00-381	92.6	0.22	0.25	0.15	0.20
030		100.0	0.24	0.27	0.17	0.21
031 Bone, meal steamed	6-00-400	96.1	27.72	0.01	0.55	12.86
032		100.0	28.85	0.01	0.57	13.38
033 Meat, meal rendered	5-00-385	93.8	8.61	1.11	0.25	4.58
034		100.0	9.17	1.18	0.27	4.88
035 Meat with bone, meal rendered	5-00-388	93.3	10.00	0.75	1.02	4.96
036		100.0	10.71	0.80	1.09	5.32
037 Tankage, meal rendered	5-00-386	92.0	5.86	1.73	0.37	3.09
038		100.0	6.37	1.88	0.41	3.36
039 Tankage with bone, meal rendered	5-00-387	92.9	11.16	—	—	5.41
040		100.0	12.01	—	—	5.82
ANIMAL–POULTRY						
041 Fat	4-00-409	99.1	—	—	—	—
042		100.0	—	—	—	—
043 Tankage, meal rendered, 50% protein	5-00-410	94.2	11.32	—	—	5.32
044		100.0	12.01	—	—	5.65
APPLE. MALUS SPP						
045 Pomace, dehydrated	4-00-423	89.6	0.11	—	0.06	0.10
046		100.0	0.13	—	0.07	0.12
BAHIAGRASS. PASPALUM NOTATUM						
047 Fresh	2-00-464	28.7	0.13	—	0.07	0.09
048		100.0	0.44	—	0.25	0.30
049 Hay, sun-cured	1-00-462	90.0	0.45	—	0.17	0.20
050		100.0	0.50	—	0.19	0.22
BAKERY						
051 Waste, dehydrated (dried bakery product)	4-00-466	91.2	0.14	1.47	0.16	0.22
052		100.0	0.15	1.61	0.18	0.24

	K	NA	S	CO	CU	I	FE	MN	SE	ZN
	%	%	%	mg/kg	mg/kg	mg/kg	mg/kg	mg/kg	mg/kg	mg/kg
001	0.83	0.04	0.099	0.092	3.237	—	82.07	24.11	—	9.39
002	3.18	0.17	0.379	0.352	12.441	—	315.43	92.68	—	36.08
003	2.25	0.10	0.484	0.081	10.190	—	207.56	42.27	—	33.49
004	2.51	0.12	0.540	0.090	11.366	—	231.52	47.15	—	37.36
005	2.32	0.14	0.272	0.264	11.436	—	205.24	32.77	0.497	27.29
006	2.56	0.15	0.300	0.291	12.636	—	226.78	36.21	0.549	30.16
007	1.42	0.11	0.257	0.358	16.090	—	204.44	25.47	—	28.09
008	1.56	0.12	0.283	0.394	17.676	—	224.60	27.98	—	30.86
009	1.42	0.06	0.273	0.209	8.975	—	140.78	38.48	—	23.70
010	1.56	0.07	0.300	0.230	9.874	—	154.90	42.33	—	26.07
011	1.55	0.07	0.208	0.370	12.518	—	146.40	35.15	—	20.16
012	1.70	0.08	0.229	0.405	13.723	—	160.50	38.53	—	22.10
013	2.22	0.07	0.188	0.172	9.450	0.116	279.27	27.81	0.281	19.36
014	2.46	0.08	0.208	0.190	10.461	0.129	309.18	30.79	0.311	21.44
015	2.40	0.10	0.224	0.302	8.563	0.148	404.93	31.05	0.335	19.32
016	2.61	0.11	0.244	0.329	9.328	0.161	441.11	33.83	0.365	21.05
017	2.42	0.11	0.403	0.259	12.170	0.135	352.41	45.23	0.285	21.84
018	2.64	0.12	0.440	0.283	13.287	0.147	384.77	49.38	0.311	23.84
019	0.97	0.05	0.145	—	3.685	—	118.64	21.68	—	8.89
020	2.23	0.11	0.335	—	8.500	—	273.64	50.00	—	20.50
021	0.82	0.09	0.049	—	—	—	27.63	—	—	—
022	3.87	0.42	0.230	—	—	—	130.00	—	—	—
023	0.74	0.17	0.091	—	—	—	59.62	—	—	—
024	1.86	0.42	0.230	—	—	—	150.00	—	—	—
025	2.31	0.11	0.219	—	11.353	—	198.38	60.74	—	20.11
026	2.53	0.12	0.240	—	12.418	—	216.99	66.43	—	22.00
027	1.06	0.04	0.098	—	—	—	91.00	—	—	—
028	2.90	0.10	0.270	—	—	—	249.96	—	—	—
029	0.15	0.38	0.603	—	8.191	—	2772.21	6.42	—	—
030	0.16	0.42	0.651	—	8.845	—	2993.49	6.93	—	—
031	0.18	0.40	0.344	0.063	15.519	29.567	1030.75	37.28	—	365.81
032	0.19	0.42	0.358	0.065	16.149	30.767	1072.60	38.79	—	380.67
033	0.55	1.04	0.460	2.249	9.604	—	490.38	11.80	0.504	74.30
034	0.58	1.11	0.491	2.398	10.239	—	522.77	12.58	0.537	79.21
035	1.33	0.72	0.252	0.180	1.537	1.317	653.23	13.33	0.343	94.37
036	1.43	0.77	0.270	0.193	1.647	1.412	699.93	14.29	0.367	101.11
037	0.54	1.67	0.704	0.152	38.737	—	2100.90	19.14	—	—
038	0.59	1.81	0.765	0.165	42.095	—	2283.03	20.80	—	—
039	—	—	0.261	—	—	—	—	—	0.261	—
040	—	—	0.281	—	—	—	—	—	0.281	—
041	0.23	—	—	—	—	—	—	—	—	—
042	0.23	—	—	—	—	—	—	—	—	—
043	—	0.94	—	—	—	—	—	—	—	—
044	—	0.99	—	—	—	—	—	—	—	—
045	0.44	0.12	0.020	—	—	—	267.88	7.22	—	—
046	0.49	0.14	0.022	—	—	—	298.84	8.05	—	—
047	0.44	—	—	—	2.442	—	—	40.20	—	8.45
048	1.53	—	—	—	8.500	—	—	139.90	—	29.40
049	—	—	—	—	—	—	54.01	—	—	—
050	—	—	—	—	—	—	60.00	—	—	—
051	0.39	1.02	0.020	1.224	5.945	—	160.48	65.00	0.164	17.77
052	0.43	1.12	0.022	1.341	6.516	—	175.90	71.24	0.180	19.48

TABLE 3–2C. (continued). Composition of Feeds—Mineral Values

		IFN	DRY MATT %	CA %	CL %	MG %	P %
	BARLEY. HORDEUM VULGARE						
053	Grain	4-00-549	88.6	0.05	0.11	0.13	0.34
054			100.0	0.05	0.12	0.15	0.38
055	Grain, Pacific coast	4-07-939	88.6	0.05	0.15	0.12	0.34
056			100.0	0.05	0.17	0.14	0.38
057	Grain screenings	4-00-542	88.9	0.32	—	0.12	0.29
058			100.0	0.36	—	0.14	0.33
059	Hay, sun-cured	1-00-495	88.4	0.21	—	0.14	0.25
060			100.0	0.24	—	0.16	0.28
061	Straw	1-00-498	91.4	0.27	0.61	0.21	0.06
062			100.0	0.30	0.67	0.23	0.07
	BEAN, NAVY. PHASEOLUS VULGARIS						
063	Seeds	5-00-623	89.8	0.17	0.06	0.13	0.52
064			100.0	0.19	0.06	0.15	0.58
	BEET, MANGEL. BETA VULGARIS MACRORRHIZA						
065	Roots, fresh	4-00-637	11.0	0.02	0.15	0.02	0.02
066			100.0	0.18	1.41	0.20	0.22
	BEET, SUGAR. BETA VULGARIS ALTISSIMA						
067	Aerial part with crowns, fresh	2-00-649	16.5	0.17	0.09	0.18	0.04
068			100.0	1.01	0.56	1.07	0.22
069	Aerial part with crowns, silage	3-00-660	25.2	0.39	—	0.27	0.07
070			100.0	1.56	—	1.07	0.28
071	Molasses, more than 48% invert sugar more than 79.5 degrees	4-00-668	77.9	0.12	1.28	0.23	0.02
072	brix		100.0	0.15	1.64	0.29	0.03
073	Pulp, dehydrated	4-00-669	91.0	0.62	0.04	0.26	0.09
074			100.0	0.68	0.04	0.28	0.10
075	Pulp, wet	4-00-671	11.0	0.10	—	0.03	0.01
076			100.0	0.87	—	0.24	0.10
077	Pulp with molasses, dehydrated	4-00-672	91.8	0.56	—	0.15	0.09
078			100.0	0.61	—	0.16	0.10
	BERMUDAGRASS. CYNODON DACTYLON						
079	Fresh	2-00-712	28.9	0.14	—	0.07	0.08
080			100.0	0.49	—	0.23	0.29
081	Hay, sun-cured	1-00-703	91.2	0.43	—	0.16	0.16
082			100.0	0.47	—	0.17	0.17
	BERMUDAGRASS, COASTAL. CYNODON DACTYLON						
083	Fresh	2-00-719	30.3	0.15	—	—	0.08
084			100.0	0.49	—	—	0.27
085	Hay, sun-cured	1-00-716	91.6	0.39	—	0.18	0.19
086			100.0	0.43	—	0.19	0.21
	BLOOD (SEE ANIMAL)						
	BLUEGRASS, CANADA. POA COMPRESSA						
087	Fresh	2-00-764	31.4	0.12	—	0.05	0.12
088			100.0	0.39	—	0.16	0.39
089	Hay, sun-cured	1-00-762	91.7	0.28	—	0.30	0.24
090			100.0	0.30	—	0.33	0.26
	BLUEGRASS, KENTUCKY. POA PRATENSIS						
091	Fresh, early vegetative	2-00-777	30.8	0.15	—	0.05	0.14
092			100.0	0.50	—	0.18	0.44
093	Fresh, early bloom	2-00-779	35.0	0.16	—	0.04	0.14
094			100.0	0.46	—	0.11	0.39
095	Hay, sun-cured	1-00-776	88.9	0.28	0.55	0.35	0.16
096			100.0	0.31	0.62	0.39	0.18
	BLUESTEM. ANDROPOGON SPP						
097	Fresh, early vegetative	2-00-821	26.8	0.17	—	—	0.05
098			100.0	0.63	—	—	0.20
099	Fresh, mature	2-00-825	59.0	0.23	—	0.04	0.07
100			100.0	0.40	—	0.06	0.12
	BONES (SEE ANIMAL)						
	BREWERS GRAINS (SEE CEREALS)						

	K	NA	S	CO mg/kg	CU mg/kg	I mg/kg	FE mg/kg	MN mg/kg	SE mg/kg	ZN mg/kg
	%	%	%							
053	0.44	0.03	0.148	0.170	8.169	0.044	73.52	15.81	0.176	41.75
054	0.49	0.03	0.167	0.192	9.223	0.050	83.02	17.85	0.199	47.14
055	0.51	0.02	0.138	0.087	8.061	—	86.14	15.99	0.101	15.18
056	0.58	0.02	0.156	0.098	9.095	—	97.19	18.05	0.114	17.13
057	0.80	0.02	0.133	—	—	—	53.35	—	—	—
058	0.90	0.02	0.150	—	—	—	60.00	—	—	—
059	1.30	0.12	0.149	0.059	3.900	—	265.31	34.82	—	—
060	1.47	0.14	0.168	0.066	4.409	—	299.96	39.37	—	—
061	2.16	0.13	0.156	0.060	4.931	—	183.46	15.13	—	6.80
062	2.36	0.14	0.170	0.066	5.397	—	200.78	16.56	—	7.44
063	1.32	0.04	0.233	—	9.895	—	91.08	22.51	—	—
064	1.47	0.05	0.259	—	11.023	—	101.47	25.08	—	—
065	0.25	0.07	0.022	—	0.604	—	16.91	—	—	—
066	2.30	0.63	0.204	—	5.512	—	154.17	—	—	—
067	0.96	0.09	0.095	—	2.255	—	27.89	9.02	—	—
068	5.79	0.54	0.571	—	13.624	—	168.54	54.49	—	—
069	1.45	0.14	0.144	—	—	—	50.37	—	—	—
070	5.74	0.54	0.570	—	—	—	200.00	—	—	—
071	4.72	1.16	0.465	0.362	16.791	—	68.06	4.50	—	14.02
072	6.06	1.48	0.596	0.464	21.558	—	87.39	5.78	—	18.00
073	0.20	0.18	0.203	0.074	12.514	—	266.69	34.27	0.109	0.70
074	0.22	0.20	0.223	0.081	13.749	—	293.00	37.65	0.120	0.76
075	0.02	0.02	0.022	—	—	—	36.21	—	—	1.10
076	0.19	0.19	0.201	—	—	—	329.96	—	—	10.00
077	1.63	0.48	0.388	0.208	14.670	—	162.34	18.43	—	8.65
078	1.78	0.53	0.422	0.227	15.983	—	176.87	20.08	—	9.42
079	0.54	0.13	—	0.022	1.981	—	323.95	30.54	—	10.12
080	1.87	0.44	—	0.075	6.850	—	1119.96	105.60	—	35.00
081	1.40	0.07	0.190	0.110	24.291	0.105	264.31	99.37	—	53.00
082	1.53	0.08	0.209	0.121	26.649	0.115	289.96	109.02	—	58.14
083	—	—	—	—	—	—	—	—	—	—
084	—	—	—	—	—	—	—	—	—	—
085	1.73	0.22	0.190	—	2.442	—	191.39	37.55	—	15.57
086	1.89	0.24	0.208	—	2.667	—	208.99	41.00	—	17.00
087	0.64	0.04	0.053	—	—	—	94.04	24.81	—	—
088	2.04	0.14	0.170	—	—	—	299.96	79.13	—	—
089	1.73	0.10	0.119	—	—	—	275.13	84.93	—	—
090	1.88	0.11	0.130	—	—	—	299.96	92.59	—	—
091	0.70	0.04	0.052	—	—	—	92.46	—	—	—
092	2.27	0.14	0.170	—	—	—	299.96	—	—	—
093	0.70	0.05	0.060	—	—	—	105.03	—	—	—
094	2.01	0.14	0.170	—	—	—	299.96	—	—	—
095	1.85	0.10	0.119	—	9.767	—	248.68	216.82	0.032	68.75
096	2.08	0.11	0.134	—	10.981	—	279.60	243.78	0.036	77.30
097	0.46	—	—	—	12.587	—	239.97	28.49	—	—
098	1.72	—	—	—	46.967	—	895.42	106.29	—	—
099	0.30	—	—	—	9.496	—	383.44	21.71	—	—
100	0.51	—	—	—	16.097	—	649.98	36.81	—	—

TABLE 3–2C. (continued). Composition of Feeds—Mineral Values

		IFN	DRY MATT %	CA %	CL %	MG %	P %
	BROME. BROMUS SPP						
101	Fresh	2-00-900	36.0	—	—	—	0.10
102			100.0	—	—	—	0.28
103	Hay, sun-cured	1-00-890	90.8	0.32	—	0.09	0.15
104			100.0	0.36	—	0.10	0.16
	BROME, SMOOTH. BROMUS INERMIS						
105	Fresh	2-00-963	27.0	—	—	—	—
106			100.0	—	—	—	—
107	Hay, sun-cured	1-00-947	89.6	0.34	0.48	0.21	0.24
108			100.0	0.38	0.54	0.23	0.26
	BUCKWHEAT, COMMON. FAGOPYRUM SAGITTATUM						
109	Grain	4-00-994	88.4	0.08	0.04	0.16	0.31
110			100.0	0.10	0.05	0.18	0.36
	BUFFALOGRASS. BUCHLOE DACTYLOIDES						
111	Fresh	2-01-010	45.8	0.26	—	0.06	0.09
112			100.0	0.57	—	0.14	0.21
	BUTTERMILK (SEE CATTLE)						
	CANARYGRASS, REED. PHALARIS ARUNDINACEA						
113	Fresh	2-01-113	22.8	0.08	—	—	0.08
114			100.0	0.36	—	—	0.33
115	Hay, sun-cured	1-01-104	89.3	0.32	—	0.19	0.21
116			100.0	0.36	—	0.22	0.24
	CARROT. DAUCUS SPP						
117	Roots, fresh	4-01-145	11.5	0.05	0.06	0.02	0.04
118			100.0	0.40	0.50	0.20	0.35
	CASEIN						
119	Acid precipitated dehydrated	5-01-162	91.6	0.61	—	0.00	0.85
120			100.0	0.67	—	0.01	0.93
	CATTLE. BOS TAURUS						
121	Buttermilk, dehydrated	5-01-160	92.3	1.32	0.44	0.48	0.94
122			100.0	1.43	0.48	0.52	1.01
123	Manure, dehydrated	1-01-190	93.0	1.48	—	0.38	0.90
124			100.0	1.59	—	0.41	0.97
125	Milk, dehydrated	5-01-167	95.5	0.89	1.48	0.09	0.71
126			100.0	0.93	1.55	0.09	0.74
127	Milk, fresh	5-01-168	12.4	0.12	0.11	0.01	0.09
128			100.0	0.93	0.92	0.10	0.75
129	Skimmilk, dehydrated	5-01-175	94.1	1.28	0.90	0.12	1.02
130			100.0	1.36	0.96	0.13	1.09
131	Skimmilk, fresh	5-01-170	9.6	0.12	0.05	0.01	0.10
132			100.0	1.31	0.54	0.12	1.04
133	Whey, dehydrated	4-01-182	93.3	0.85	0.07	0.13	0.76
134			100.0	0.92	0.08	0.14	0.81
135	Whey, fresh	4-08-134	6.9	0.05	—	—	0.05
136			100.0	0.79	—	—	0.67
137	Whey low lactose, dehydrated (dried whey product)	4-01-186	93.7	1.50	1.03	0.22	1.11
138			100.0	1.60	1.10	0.23	1.18
	CEREALS						
139	Brewers grains, dehydrated	5-02-141	92.2	0.29	0.13	0.15	0.51
140			100.0	0.32	0.14	0.16	0.55
141	Grain screenings, uncleaned	4-02-153	92.1	0.37	—	0.22	0.41
142			100.0	0.40	—	0.24	0.45
	CITRUS. CITRUS SPP						
143	Pomace without fines, dehydrated (dried citrus pulp)	4-01-237	91.1	1.71	—	0.16	0.12
144			100.0	1.88	—	0.17	0.13
145	Pomace, silage (pulp)	4-01-234	20.8	0.42	—	0.03	0.03
146			100.0	2.04	—	0.16	0.15
147	Syrup (molasses)	4-01-241	66.9	1.18	0.07	0.14	0.09
148			100.0	1.76	0.11	0.21	0.14
	CLOVER, ALSIKE. TRIFOLIUM HYBRIDUM						
149	Fresh	2-01-316	22.5	0.31	0.17	0.07	0.06
150			100.0	1.36	0.77	0.32	0.29
151	Hay, sun-cured	1-01-313	87.7	1.14	0.68	0.39	0.22
152			100.0	1.30	0.78	0.45	0.25

	K %	NA %	S %	CO mg/kg	CU mg/kg	I mg/kg	FE mg/kg	MN mg/kg	SE mg/kg	ZN mg/kg
101	0.79	—	—	—	—	—	—	—	—	—
102	2.19	—	—	—	—	—	—	—	—	—
103	1.49	0.03	0.182	—	—	—	181.65	—	—	—
104	1.64	0.03	0.200	—	—	—	200.00	—	—	—
105	—	—	—	0.021	—	—	—	—	—	—
106	—	—	—	0.079	—	—	—	—	—	—
107	1.88	0.56	0.229	—	10.140	—	142.66	41.76	—	—
108	2.10	0.63	0.256	—	11.321	—	159.27	46.63	—	—
109	0.45	0.01	0.141	0.049	8.831	—	227.00	31.43	0.216	15.72
110	0.50	0.02	0.159	0.055	9.985	—	256.67	35.54	0.244	17.77
111	0.33	—	—	—	—	—	—	—	—	—
112	0.71	—	—	—	—	—	—	—	—	—
113	0.83	—	—	—	—	—	—	—	—	—
114	3.64	—	—	—	—	—	—	—	—	—
115	2.60	0.01	—	—	10.620	—	133.88	82.48	—	—
116	2.91	0.02	—	—	11.898	—	150.00	92.40	—	—
117	0.32	0.06	0.019	—	1.193	—	13.85	3.62	—	—
118	2.80	0.48	0.169	—	10.377	—	120.43	31.49	—	—
119	0.00	0.01	—	—	4.126	—	17.37	3.76	—	32.18
120	0.01	0.01	—	—	4.505	—	18.97	4.11	—	35.14
121	0.84	0.83	0.081	—	1.015	—	8.39	3.46	—	40.24
122	0.91	0.90	0.088	—	1.100	—	9.09	3.75	—	43.60
123	0.65	0.37	1.657	—	26.750	—	1454.26	177.01	—	165.51
124	0.70	0.40	1.782	—	28.773	—	1564.21	190.39	—	178.02
125	1.07	0.37	0.306	0.005	0.896	—	89.35	0.47	—	21.77
126	1.12	0.39	0.320	0.005	0.938	—	93.56	0.50	—	22.80
127	0.14	0.05	0.040	0.001	0.099	—	1.24	—	—	2.84
128	1.13	0.38	0.320	0.005	0.800	—	10.00	—	—	23.00
129	1.60	0.51	0.322	0.113	11.684	—	7.67	2.14	0.123	38.51
130	1.70	0.54	0.342	0.120	12.410	—	8.14	2.27	0.131	40.91
131	0.12	0.04	0.030	0.011	1.110	—	0.96	0.22	—	4.88
132	1.29	0.47	0.318	0.111	11.603	—	10.00	2.32	—	51.00
133	1.16	0.62	1.040	0.110	46.516	—	194.15	5.86	—	4.79
134	1.25	0.66	1.115	0.118	49.870	—	208.15	6.29	—	5.14
135	0.19	—	—	—	—	—	20.05	0.22	—	—
136	2.75	—	—	—	—	—	289.82	3.20	—	—
137	2.86	1.45	1.073	—	7.029	9.885	245.38	8.03	0.052	7.91
138	3.05	1.54	1.146	—	7.505	10.554	261.98	8.57	0.055	8.44
139	0.09	0.20	0.295	0.076	21.717	0.066	233.09	37.22	0.147	27.31
140	0.09	0.22	0.320	0.082	23.566	0.071	252.94	40.39	0.160	29.63
141	0.18	0.26	0.304	—	—	—	248.57	—	—	—
142	0.20	0.28	0.330	—	—	—	269.96	—	—	—
143	0.70	0.08	0.073	0.169	5.597	—	328.14	6.62	—	13.76
144	0.77	0.08	0.081	0.185	6.142	—	360.12	7.26	—	15.10
145	0.13	0.02	0.004	—	—	—	33.23	—	—	3.32
146	0.62	0.09	0.020	—	—	—	160.00	—	—	16.00
147	0.09	0.28	0.143	0.108	72.170	—	338.76	25.73	—	91.63
148	0.14	0.41	0.214	0.161	107.954	—	506.72	38.49	—	137.06
149	0.61	0.10	0.050	—	1.338	—	96.69	26.32	—	13.53
150	2.70	0.45	0.220	—	5.952	—	430.00	117.06	—	60.18
151	1.95	0.40	0.167	—	5.265	—	228.11	60.55	—	—
152	2.22	0.46	0.190	—	6.000	—	259.96	69.00	—	—

TABLE 3–2C. (continued). Composition of Feeds—Mineral Values

		IFN	DRY MATT %	CA %	CL %	MG %	P %
	CLOVER, CRIMSON. TRIFOLIUM INCARNATUM						
153	Fresh	2-01-336	17.5	0.24	0.11	0.05	0.05
154			100.0	1.38	0.61	0.29	0.29
155	Hay, sun-cured	1-01-328	87.8	1.23	0.55	0.25	0.19
156			100.0	1.40	0.63	0.28	0.22
	CLOVER, LADINO. TRIFOLIUM REPENS						
157	Fresh	2-01-383	17.7	0.22	—	0.09	0.07
158			100.0	1.27	—	0.48	0.42
159	Hay, sun-cured	1-01-378	89.1	1.30	0.27	0.42	0.30
160			100.0	1.45	0.30	0.47	0.33
	CLOVER, RED. TRIFOLIUM PRATENSE						
161	Fresh, early bloom	2-01-428	19.6	0.44	—	0.10	0.07
162			100.0	2.26	—	0.51	0.38
163	Fresh, full bloom	2-01-429	26.2	0.26	—	0.13	0.07
164			100.0	1.01	—	0.51	0.27
165	Hay, sun-cured	1-01-415	88.4	1.22	0.28	0.34	0.22
166			100.0	1.38	0.32	0.38	0.24
	COCONUT. COCOS NUCIFERA						
167	Kernels with coats, meal mechanical extracted (copra meal)	5-01-572	91.6	0.19	—	0.31	0.60
168			100.0	0.21	—	0.33	0.65
169	Kernels with coats, meal solvent extracted (copra meal)	5-01-573	91.1	0.17	0.03	0.31	0.65
170			100.0	0.19	0.03	0.34	0.71
	CORN (SEE MAIZE)						
	COTTON. GOSSYPIUM SPP						
171	Bolls, sun-cured	1-01-596	91.4	0.72	—	0.24	0.13
172			100.0	0.78	—	0.27	0.14
173	Hulls	1-01-599	90.4	0.13	0.02	0.13	0.08
174			100.0	0.15	0.02	0.14	0.09
175	Seeds, ground	5-01-608	92.2	0.14	—	0.32	0.67
176			100.0	0.15	—	0.35	0.73
177	Seeds, meal mechanical extracted	5-01-609	92.7	0.17	—	0.36	0.64
178			100.0	0.18	—	0.39	0.69
179	Seeds, meal mechanical extracted, 41% protein	5-01-617	92.6	0.19	0.05	0.53	1.08
180			100.0	0.20	0.06	0.57	1.17
181	Seeds, meal solvent extracted, 41% protein	5-01-621	91.0	0.17	0.05	0.54	1.11
182			100.0	0.18	0.05	0.59	1.22
183	Seeds without hulls, meal prepressed solvent extracted, 50% protein	5-07-874	93.0	0.18	0.05	0.46	1.16
184			100.0	0.19	0.05	0.50	1.24
	COWPEA, COMMON. VIGNA SINENSIS						
185	Hay, sun-cured	1-01-645	90.4	1.26	0.15	0.41	0.31
186			100.0	1.40	0.17	0.45	0.35
	DROPSEED, SAND. SPOROBOLUS CRYPTANDRUS						
187	Fresh, stem cured	2-05-596	88.0	0.40	—	0.06	0.07
188			100.0	0.45	—	0.06	0.08
	FAT (SEE ANIMAL–POULTRY)						
	FESCUE, MEADOW. FESTUCA ELATIOR						
189	Fresh	2-01-920	28.4	0.15	—	0.11	0.11
190			100.0	0.53	—	0.37	0.39
191	Hay, sun-cured	1-01-912	87.5	0.33	—	0.44	0.25
192			100.0	0.37	—	0.50	0.29
	FISH						
193	Solubles, condensed	5-01-969	50.4	0.17	3.00	0.03	0.56
194			100.0	0.33	5.95	0.06	1.11
195	Solubles, dehydrated	5-01-971	92.8	1.09	—	0.30	1.27
196			100.0	1.17	—	0.32	1.37
	FISH, ANCHOVETA. ENGRAULIS RINGEN						
197	Meal mechanical extracted	5-01-985	92.0	3.74	1.00	0.25	2.48
198			100.0	4.07	1.08	0.27	2.70
	FISH, HERRING. CLUPEA HARENGUS						
199	Meal mechanical extracted	5-02-000	91.8	2.19	0.99	0.14	1.67
200			100.0	2.39	1.08	0.16	1.82
	FISH, MENHADEN. BREVOORTIA TYRANNUS						
201	Meal mechanical extracted	5-02-009	91.7	5.01	0.55	0.15	2.87
202			100.0	5.46	0.60	0.16	3.14

132

	K	NA	S	CO	CU	I	FE	MN	SE	ZN
	%	%	%	mg/kg	mg/kg	mg/kg	mg/kg	mg/kg	mg/kg	mg/kg
153	0.54	0.07	0.050	—	—	—	—	43.14	—	—
154	3.10	0.40	0.284	—	—	—	—	245.80	—	—
155	2.10	0.34	0.246	—	—	0.058	614.74	183.27	—	—
156	2.40	0.39	0.280	—	—	0.066	699.96	208.67	—	—
157	0.33	0.02	0.021	—	—	—	63.80	12.66	—	—
158	1.87	0.12	0.121	—	—	—	361.45	71.70	—	—
159	2.17	0.12	0.188	0.144	8.385	0.267	418.66	109.73	—	15.15
160	2.44	0.13	0.211	0.161	9.407	0.300	469.68	123.10	—	17.00
161	0.49	0.04	0.033	—	—	—	58.79	—	—	—
162	2.49	0.20	0.170	—	—	—	299.96	—	—	—
163	0.51	0.05	0.045	—	—	—	78.59	—	—	—
164	1.96	0.20	0.170	—	—	—	299.96	—	—	—
165	1.60	0.16	0.145	0.138	18.768	0.217	210.82	95.15	—	32.47
166	1.81	0.18	0.164	0.156	21.234	0.245	238.51	107.65	—	36.73
167	1.67	0.04	0.335	0.127	16.690	—	679.64	70.11	—	48.53
168	1.82	0.04	0.366	0.138	18.228	—	742.24	76.57	—	53.00
169	1.41	0.04	—	—	—	—	—	54.47	—	—
170	1.55	0.04	—	—	—	—	—	59.78	—	—
171	2.42	—	—	—	—	—	—	—	—	—
172	2.64	—	—	—	—	—	—	—	—	—
173	0.79	0.02	0.076	0.018	11.989	—	118.65	107.76	0.081	19.84
174	0.88	0.02	0.084	0.019	13.264	—	131.27	119.22	0.090	21.95
175	1.10	0.03	0.239	—	50.279	—	138.72	8.79	—	—
176	1.20	0.03	0.259	—	54.549	—	150.51	9.54	—	—
177	1.25	0.03	0.241	—	—	—	139.12	—	—	—
178	1.35	0.03	0.260	—	—	—	150.00	—	—	—
179	1.34	0.04	0.394	0.626	18.540	—	170.63	22.30	0.130	61.88
180	1.44	0.05	0.426	0.676	20.012	—	184.18	24.07	0.140	66.80
181	1.38	0.04	0.263	0.483	19.496	—	187.71	20.66	0.373	60.74
182	1.52	0.05	0.289	0.531	21.424	—	206.27	22.70	0.410	66.75
183	1.45	0.05	0.521	0.042	14.511	—	111.63	23.05	—	73.82
184	1.56	0.06	0.560	0.045	15.598	—	120.00	24.78	—	79.35
185	2.04	0.24	0.316	0.063	—	—	542.19	438.37	—	—
186	2.26	0.27	0.350	0.070	—	—	599.98	485.10	—	—
187	0.28	0.01	—	0.503	13.464	0.598	424.23	41.36	—	36.78
188	0.32	0.01	—	0.571	15.300	0.680	482.08	47.00	—	41.80
189	0.67	—	—	0.038	1.137	—	—	—	—	—
190	2.34	—	—	0.135	4.000	—	—	—	—	—
191	1.61	—	—	0.118	—	—	—	21.44	—	—
192	1.84	—	—	0.135	—	—	—	24.50	—	—
193	1.66	2.50	0.125	0.069	47.381	1.110	325.29	12.58	—	43.14
194	3.29	4.97	0.247	0.136	93.961	2.202	645.08	24.95	—	85.56
195	—	0.37	—	—	—	—	—	50.44	—	76.65
196	—	0.40	—	—	—	—	—	54.35	—	82.59
197	0.72	0.88	0.715	0.173	9.082	3.137	214.74	10.99	1.355	105.03
198	0.78	0.95	0.778	0.188	9.875	3.410	233.48	11.95	1.473	114.20
199	1.08	0.59	0.499	0.054	5.603	0.524	114.34	4.80	1.953	124.72
200	1.18	0.65	0.543	0.059	6.101	0.571	124.50	5.22	2.127	135.80
201	0.71	0.41	0.529	0.153	10.342	1.092	544.58	37.02	2.148	144.32
202	0.77	0.44	0.577	0.167	11.283	1.191	594.15	40.38	2.344	157.45

TABLE 3–2C. (continued). Composition of Feeds—Mineral Values

		IFN	DRY MATT %	CA %	CL %	MG %	P %
	FISH, SARDINE. CLUPEA SPP-SARDINOPS SPP						
203	Meal mechanical extracted	5-02-015	92.9	4.60	0.41	0.10	2.68
204			100.0	4.95	0.44	0.11	2.88
	FISH, WHITE. GADIDAE (FAMILY)-LOPHIIDAE (FAMILY)-RAJIDAE (FAMILY)						
205	Meal mechanical extracted	5-02-025	91.2	6.64	—	0.18	3.97
206			100.0	7.27	—	0.20	4.35
	FLAX, COMMON. LINUM USITATISSIMUM						
207	Seed screenings	4-02-056	91.2	0.33	—	0.39	0.43
208			100.0	0.37	—	0.43	0.47
209	Seeds, meal mechanical extracted	5-02-045	90.8	0.41	0.04	0.58	0.87
210			100.0	0.45	0.04	0.63	0.96
	GALLETA. HILARIA JAMESII						
211	Fresh, stem cured	2-05-594	86.0	0.60	—	0.07	0.06
212			100.0	0.70	—	0.08	0.07
	GARBAGE, HOTEL AND RESTAURANT						
217	Boiled	4-07-865	22.9	0.10	—	0.01	0.06
218			100.0	0.44	—	0.06	0.28
	GRAMA. BOUTELOUA SPP						
221	Fresh, early vegetative	2-02-163	41.0	0.22	—	—	0.08
222			100.0	0.53	—	—	0.19
223	Fresh, mature	2-02-166	63.4	0.22	—	—	0.08
224			100.0	0.34	—	—	0.12
	GRAPE. VITIS SPP						
225	Pomace, dehydrated (marc)	1-02-208	90.4	0.52	0.01	0.09	0.15
226			100.0	0.58	0.01	0.10	0.17
	GRASS						
227	Hay, sun-cured, full bloom	1-02-244	89.3	0.51	—	0.16	0.21
228			100.0	0.57	—	0.18	0.24
	GRASS-LEGUME						
229	Silage	3-02-303	31.1	0.26	0.33	0.08	0.08
230			100.0	0.84	1.06	0.25	0.26
	HOMINY (SEE MAIZE)						
	LEGUME (SEE GRASS-LEGUME)						
	LESPEDEZA, CHINESE. LESPEDEZA CUNEATA						
231	Hay, sun-cured	1-02-607	90.4	0.93	—	0.20	0.22
232			100.0	1.03	—	0.22	0.25
	LINSEED (SEE FLAX)						
	LUCERNE (SEE ALFALFA)						
	MAIZE. ZEA MAYS						
237	Aerial part with ears, sun-cured, mature (fodder)	1-02-772	82.1	—	—	—	—
238			100.0	—	—	—	—
239	Aerial part without ears without husks, sun-cured (stover)	1-02-776	85.2	0.49	—	0.34	0.08
240	(straw)		100.0	0.57	—	0.40	0.10
241	Cobs, ground	1-02-782	89.8	0.11	—	0.06	0.04
242			100.0	0.12	—	0.07	0.04
243	Distillers grains, dehydrated	5-02-842	93.5	0.09	0.07	0.07	0.39
244			100.0	0.10	0.08	0.07	0.42
245	Distillers grains with solubles, dehydrated	5-02-843	91.8	0.16	0.17	0.18	0.69
246			100.0	0.17	0.18	0.20	0.75
247	Distillers solubles, dehydrated	5-02-844	92.9	0.30	0.26	0.60	1.30
248			100.0	0.32	0.28	0.65	1.40
249	Ears with husks, silage	4-02-839	43.4	0.04	—	0.05	0.12
250			100.0	0.10	—	0.12	0.29
251	Ears, ground (maize and cob meal)	4-02-849	86.5	0.06	0.04	0.12	0.24
252			100.0	0.07	0.05	0.14	0.27
253	Gluten, meal	5-02-900	91.3	0.15	0.06	0.06	0.46
254			100.0	0.16	0.07	0.06	0.51
255	Gluten with bran (maize gluten feed)	5-02-903	89.9	0.32	0.22	0.33	0.75
256			100.0	0.35	0.25	0.37	0.83
257	Grits by-product (hominy feed)	4-02-887	90.2	0.05	0.05	0.24	0.51
258			100.0	0.05	0.06	0.26	0.57
259	Silage, well eared	3-02-823	34.1	0.09	—	0.10	0.07
260			100.0	0.27	—	0.28	0.20

	K %	NA %	S %	CO mg/kg	CU mg/kg	I mg/kg	FE mg/kg	MN mg/kg	SE mg/kg	ZN mg/kg
203	0.32	0.18	0.306	0.182	20.134	—	297.79	23.12	1.768	—
204	0.35	0.19	0.329	0.196	21.669	—	320.48	24.88	1.902	—
205	0.53	0.62	—	3.365	4.092	—	250.93	10.04	1.616	68.43
206	0.58	0.68	—	3.688	4.485	—	275.03	11.00	1.771	75.00
207	0.77	—	0.228	—	—	—	91.16	—	—	—
208	0.84	—	0.250	—	—	—	100.00	—	—	—
209	1.22	0.11	0.371	0.414	26.358	0.066	176.55	38.44	0.807	32.97
210	1.34	0.12	0.408	0.455	29.015	0.073	194.35	42.32	0.889	36.30
211	0.41	0.01	0.086	0.591	16.340	—	439.73	67.65	—	19.49
212	0.48	0.01	0.100	0.687	19.000	—	511.32	78.67	—	22.67
217	—	—	—	—	5.030	—	97.40	5.03	—	—
218	—	—	—	—	22.000	—	425.96	22.00	—	—
221	—	—	—	—	2.255	—	—	18.16	—	—
222	—	—	—	—	5.500	—	—	44.30	—	—
223	0.22	—	—	0.115	8.114	—	823.98	30.05	—	—
224	0.35	—	—	0.181	12.798	—	1299.69	47.40	—	—
225	0.82	0.08	—	—	—	—	—	36.77	0.063	21.86
226	0.91	0.09	—	—	—	—	—	40.66	0.070	24.18
227	1.20	—	0.107	—	—	—	—	—	—	—
228	1.34	—	0.120	—	—	—	—	—	—	—
229	0.57	0.02	0.168	0.039	1.864	—	126.90	18.04	—	8.70
230	1.83	0.07	0.540	0.126	6.000	—	408.50	58.06	—	28.00
231	0.99	—	—	—	—	—	263.32	91.11	—	—
232	1.10	—	—	—	—	—	291.44	100.84	—	—
237	—	—	—	—	—	—	—	—	—	—
238	—	—	—	—	—	—	—	—	—	—
239	1.24	0.06	0.145	—	4.333	—	178.90	115.89	—	—
240	1.45	0.07	0.170	—	5.085	—	209.98	136.02	—	—
241	0.78	—	0.421	0.117	6.558	—	206.61	5.57	—	—
242	0.87	—	0.469	0.130	7.300	—	229.96	6.20	—	—
243	0.16	0.09	0.433	0.076	38.948	0.047	191.73	19.32	0.352	41.74
244	0.17	0.09	0.464	0.081	41.669	0.051	205.13	20.67	0.376	44.65
245	0.47	0.47	0.307	0.151	52.602	0.050	231.33	23.96	0.331	80.68
246	0.51	0.52	0.334	0.165	57.317	0.055	252.07	26.11	0.360	87.91
247	1.70	0.23	0.370	0.166	77.922	0.079	519.99	72.04	0.371	88.03
248	1.83	0.24	0.399	0.179	83.912	0.085	559.97	77.58	0.399	94.80
249	0.21	0.00	0.056	—	—	—	34.72	—	—	—
250	0.49	0.01	0.130	—	—	—	80.00	—	—	—
251	0.46	0.02	0.135	0.272	6.834	0.022	79.06	19.90	0.074	12.13
252	0.53	0.02	0.156	0.314	7.899	0.026	91.36	23.00	0.086	14.02
253	0.03	0.09	0.278	0.077	27.658	—	383.55	7.73	1.014	173.66
254	0.03	0.10	0.305	0.084	30.290	—	420.06	8.46	1.111	190.18
255	0.62	0.12	0.239	0.086	43.861	0.066	413.78	22.75	0.258	68.65
256	0.69	0.13	0.266	0.096	48.776	0.073	460.15	25.30	0.287	76.34
257	0.59	0.08	0.030	0.054	13.608	—	67.28	14.51	—	—
258	0.65	0.09	0.033	0.060	15.084	—	74.58	16.09	—	—
259	0.36	0.00	0.027	0.020	4.512	—	218.29	11.73	—	7.14
260	1.05	0.01	0.080	0.060	13.228	—	639.96	34.39	—	20.94

TABLE 3–2C. (continued). Composition of Feeds—Mineral Values

		IFN	DRY MATT %	CA %	CL %	MG %	P %
	MAIZE. ZEA MAYS—*Continued*						
261	Silage, 30–50% dry matter	3-20-506	36.9	0.08	—	0.07	0.09
262			100.0	0.22	—	0.18	0.24
263	Silage, aerial part without ears without husks (stalklage)	3-02-836	30.7	0.12	—	0.10	0.09
264	(stover)		100.0	0.38	—	0.31	0.31
	MAIZE, DENT WHITE. ZEA MAYS INDENTATA						
265	Grits by-product (hominy feed)	4-02-990	90.0	0.03	—	0.23	0.69
266			100.0	0.03	—	0.26	0.77
	MAIZE, DENT YELLOW. ZEA MAYS INDENTATA						
267	Grain	4-02-935	88.0	0.05	0.05	0.11	0.27
268			100.0	0.05	0.06	0.12	0.31
269	Grain, grade 2 69.5 kg/hl (54 lb/bushel)	4-02-931	87.3	0.02	0.04	0.11	0.29
270			100.0	0.02	0.05	0.13	0.33
271	Grain, grade 3 66.9 kg/hl (52 lb/bushel)	4-02-932	86.5	0.03	—	0.15	0.22
272			100.0	0.03	—	0.17	0.26
	MAIZE, DENT YELLOW, OPAQUE 2. ZEA MAYS INDENTATA						
273	Grain	4-28-253	91.0	0.03	—	0.12	0.33
274			100.0	0.04	—	0.13	0.36
	MAIZE, SWEET. ZEA MAYS SACCHARATA						
275	Process residue, silage	3-07-955	31.0	0.10	—	0.07	0.24
276			100.0	0.32	—	0.24	0.77
277	Process residue, wet (cannery residue)	2-02-975	76.8	0.25	—	0.18	0.54
278			100.0	0.32	—	0.24	0.70
	MANURE (SEE CATTLE; POULTRY)						
	MEAT (SEE ANIMAL)						
	MILK (SEE CATTLE)						
	MILLET. SETARIA SPP						
279	Grain	4-03-098	89.9	0.05	0.14	0.16	0.29
280			100.0	0.05	0.16	0.18	0.32
	MILLET, PEARL. PENNISETUM GLAUCUM						
281	Fresh	2-03-115	20.7	—	—	—	—
282			100.0	—	—	—	—
283	Silage	3-20-903	28.3	0.10	—	0.10	—
284			100.0	0.37	—	0.36	—
	MOLASSES (SEE BEET; SUGARCANE)						
	NAPIERGRASS. PENNISETUM PURPUREUM						
285	Fresh, late vegetative	2-03-158	20.2	0.12	—	0.05	0.08
286			100.0	0.60	—	0.26	0.41
287	Fresh, late bloom	2-03-162	23.0	0.08	—	0.06	0.07
288			100.0	0.35	—	0.26	0.30
	NEEDLEANDTHREAD. STIPA COMATA						
289	Fresh, stem cured	2-07-989	92.0	0.99	—	—	0.06
290			100.0	1.08	—	—	0.06
	OATS. AVENA SATIVA						
291	Cereal by-product, less than 4% fiber (feeding oat meal) (oat	4-03-303	90.7	0.07	0.05	0.12	0.44
292	middlings)		100.0	0.08	0.05	0.13	0.48
293	Grain	4-03-309	89.2	0.08	0.09	0.14	0.34
294			100.0	0.09	0.10	0.16	0.38
295	Grain, Pacific coast	4-07-999	90.9	0.10	0.12	0.17	0.31
296			100.0	0.11	0.13	0.19	0.34
297	Groats	4-03-331	89.6	0.08	0.09	0.11	0.42
298			100.0	0.09	0.10	0.13	0.47
299	Hay, sun-cured	1-03-280	90.7	0.29	0.47	0.26	0.23
300			100.0	0.32	0.52	0.29	0.25
301	Hulls	1-03-281	92.4	0.15	0.08	0.12	0.14
302			100.0	0.16	0.08	0.13	0.15
303	Silage	3-03-298	30.5	0.10	—	0.05	0.07
304			100.0	0.34	—	0.15	0.24
305	Straw	1-03-283	92.2	0.22	0.71	0.16	0.06
306			100.0	0.23	0.78	0.17	0.06
	ORANGE. CITRUS SINENSIS						
307	Pomace without fines, dehydrated (dried orange pulp)	4-01-254	88.1	0.62	—	—	0.10
308			100.0	0.71	—	—	0.11

	K	NA	S	CO	CU	I	FE	MN	SE	ZN
	%	%	%	mg/kg	mg/kg	mg/kg	mg/kg	mg/kg	mg/kg	mg/kg
261	0.39	0.00	0.037	0.395	2.470	—	81.89	22.66	0.049	9.28
262	1.07	0.01	0.102	1.071	6.700	—	222.16	61.47	0.133	25.19
263	0.47	0.01	0.034	—	—	—	55.19	—	—	—
264	1.54	0.03	0.110	—	—	—	180.00	—	—	—
265	0.64	0.13	0.027	0.020	13.251	—	71.04	13.91	—	—
266	0.71	0.14	0.030	0.022	14.730	—	78.97	15.46	—	—
267	0.32	0.01	0.111	0.126	3.670	—	30.96	5.47	0.120	19.02
268	0.37	0.01	0.126	0.143	4.172	—	35.20	6.22	0.137	21.63
269	0.31	0.02	0.123	0.028	3.751	—	25.37	5.31	—	13.64
270	0.36	0.02	0.140	0.032	4.295	—	29.04	6.08	—	15.61
271	0.33	0.01	0.121	—	—	—	21.63	5.53	—	—
272	0.38	0.01	0.140	—	—	—	25.00	6.39	—	—
273	—	—	—	—	—	—	53.14	7.82	—	67.79
274	—	—	—	—	—	—	58.40	8.60	—	74.50
275	0.36	0.01	0.034	—	—	—	62.00	—	—	—
276	1.15	0.03	0.110	—	—	—	200.00	—	—	—
277	0.88	0.02	0.100	—	5.376	—	153.60	—	—	—
278	1.15	0.03	0.130	—	7.000	—	200.00	—	—	—
279	0.43	0.04	0.128	0.044	21.751	—	57.44	29.67	—	13.88
280	0.48	0.04	0.142	0.049	24.193	—	63.89	33.00	—	15.43
281	—	—	—	—	—	—	—	—	—	—
282	—	—	—	—	—	—	—	—	—	—
283	0.98	0.11	—	—	—	—	—	—	—	—
284	3.47	0.38	—	—	—	—	—	—	—	—
285	0.27	0.00	0.020	—	—	—	—	—	—	—
286	1.31	0.01	0.100	—	—	—	—	—	—	—
287	0.30	0.00	0.023	—	—	—	—	—	—	—
288	1.31	0.01	0.100	—	—	—	—	—	—	—
289	—	—	—	—	—	—	—	—	—	—
290	—	—	—	—	—	—	—	—	—	—
291	0.51	0.09	0.196	0.045	5.197	—	381.64	43.83	—	139.46
292	0.56	0.10	0.216	0.050	5.732	—	420.94	48.35	—	153.82
293	0.40	0.05	0.208	0.056	5.952	0.112	65.45	36.14	0.215	35.03
294	0.45	0.06	0.233	0.062	6.670	0.125	73.35	40.51	0.241	39.25
295	0.38	0.06	0.200	—	—	—	72.72	38.00	0.076	—
296	0.42	0.07	0.220	—	—	—	80.00	41.80	0.083	—
297	0.36	0.03	0.198	—	6.020	0.108	71.00	30.77	0.452	33.58
298	0.40	0.04	0.221	—	6.723	0.120	79.29	34.37	0.505	37.50
299	1.35	0.17	0.212	0.066	4.376	—	368.66	89.56	—	40.82
300	1.49	0.18	0.233	0.073	4.824	—	406.43	98.74	—	45.00
301	0.55	0.06	0.093	—	6.597	—	128.00	25.35	0.397	26.89
302	0.59	0.07	0.101	—	7.138	—	138.51	27.43	0.430	29.10
303	0.81	0.06	0.057	0.024	1.391	—	85.82	27.71	0.026	7.30
304	2.66	0.18	0.188	0.080	4.564	—	281.60	90.94	0.084	23.94
305	2.33	0.39	0.206	—	9.468	—	151.21	28.99	—	5.47
306	2.53	0.42	0.223	—	10.271	—	164.04	31.45	—	5.94
307	—	—	—	—	—	—	—	—	—	—
308	—	—	—	—	—	—	—	—	—	—

TABLE 3–2C. (continued). Composition of Feeds—Mineral Values

		IFN	DRY MATT %	CA %	CL %	MG %	P %
	ORCHARDGRASS. DACTYLIS GLOMERATA						
309	Fresh, early vegetative	2-03-439	22.6	0.13	0.02	0.06	0.12
310			100.0	0.57	0.08	0.27	0.54
311	Fresh, early bloom	2-03-442	23.5	0.06	—	0.07	0.09
312			100.0	0.25	—	0.31	0.39
313	Hay, sun-cured	1-03-438	89.6	0.37	0.37	0.17	0.23
314			100.0	0.41	0.41	0.19	0.26
	OYSTER. CRASSOSTREA SPP-OSTREA SPP						
315	Shells, fine ground (oyster shell flour)	6-03-481	99.1	37.78	0.01	0.25	0.07
316			100.0	38.12	0.01	0.26	0.07
	PANGOLAGRASS. DIGITARIA DECUMBENS						
317	Fresh	2-03-493	20.2	0.08	—	0.04	0.04
318			100.0	0.38	—	0.18	0.22
	PEA. PISUM SPP						
323	Haulm with pods, silage	3-03-596	24.5	0.32	—	0.10	0.06
324			100.0	1.31	—	0.39	0.24
325	Seeds	5-03-600	89.1	0.12	0.05	0.12	0.41
326			100.0	0.14	0.06	0.14	0.46
	PEANUT. ARACHIS HYPOGAEA						
327	Hay, sun-cured	1-03-619	90.7	1.12	—	0.44	0.14
328			100.0	1.23	—	0.49	0.16
329	Seeds without coats, meal mechanical extracted (peanut meal)	5-03-649	92.6	0.20	0.03	0.27	0.56
330			100.0	0.22	0.03	0.29	0.60
331	Seeds without coats, meal solvent extracted (peanut meal)	5-03-650	92.4	0.36	0.03	0.27	0.61
332			100.0	0.39	0.03	0.30	0.66
	PEARLMILLET (SEE MILLET, PEARL)						
	PINEAPPLE. ANANAS COMOSUS						
333	Process residue, dehydrated (pineapple bran)	4-03-722	87.1	0.20	—	—	0.11
334			100.0	0.23	—	—	0.13
	POTATO. SOLANUM TUBEROSUM						
335	Process residue, dehydrated	4-03-775	90.0	0.10	0.21	0.06	0.19
336			100.0	0.11	0.24	0.07	0.22
337	Tubers, dehydrated	4-07-850	91.1	0.07	0.36	0.11	0.20
338			100.0	0.08	0.40	0.12	0.22
339	Tubers, fresh	4-03-787	21.9	0.01	0.06	0.03	0.05
340			100.0	0.04	0.28	0.14	0.24
341	Tubers, silage	4-03-768	24.7	0.01	—	—	0.06
342			100.0	0.04	—	—	0.23
	POULTRY						
343	By-product, meal rendered (viscera with feet with heads)	5-03-798	93.8	3.95	0.54	0.17	2.06
344			100.0	4.21	0.58	0.19	2.19
345	Feathers, meal hydrolyzed	5-03-795	92.9	0.35	0.28	0.17	0.51
346			100.0	0.38	0.30	0.19	0.55
347	Manure, dehydrated	5-14-015	90.6	5.70	0.87	0.54	1.94
348			100.0	6.29	0.96	0.59	2.14
	PRAIRIE PLANTS, MIDWEST						
349	Hay, sun-cured, midbloom	1-07-956	94.1	—	—	—	—
350			100.0	—	—	—	—
353	Hay, sun-cured, milk stage	1-03-185	89.6	0.47	—	0.35	0.07
354			100.0	0.52	—	0.39	0.08
	PRICKLYPEAR. OPUNTIA SPP						
355	Fresh	2-01-061	16.4	1.44	—	0.23	0.02
356			100.0	8.78	—	1.38	0.12
	RAPE. BRASSICA NAPUS						
357	Fresh	2-03-867	16.7	0.25	—	0.01	0.07
358			100.0	1.47	—	0.06	0.43
359	Seeds, meal mechanical extracted	5-03-870	92.0	0.66	—	0.50	1.04
360			100.0	0.72	—	0.54	1.14
361	Seeds, meal solvent extracted	5-03-871	91.2	0.60	0.10	0.43	0.96
362			100.0	0.66	0.11	0.48	1.06
	RAPE, SUMMER. BRASSICA NAPUS ANNUA						
363	Seeds, meal boiled mechanical extracted	5-08-136	94.0	0.71	—	—	1.00
364			100.0	0.76	—	—	1.06
365	Seeds, meal prepressed solvent extracted	5-08-135	91.6	0.67	—	0.64	1.05
366			100.0	0.73	—	0.70	1.15

138

	K	NA	S	CO	CU	I	FE	MN	SE	ZN
	%	%	%	mg/kg	mg/kg	mg/kg	mg/kg	mg/kg	mg/kg	mg/kg
309	0.73	—	—	—	1.582	—	—	7.10	—	—
310	3.21	—	—	—	7.000	—	—	31.40	—	—
311	0.80	0.01	0.061	—	7.784	—	184.76	24.49	—	—
312	3.38	0.04	0.260	—	33.069	—	784.96	104.06	—	—
313	2.69	0.01	0.212	0.340	12.982	—	134.01	163.40	0.041	29.86
314	3.00	0.02	0.236	0.379	14.481	—	149.48	182.27	0.046	33.31
315	0.10	0.21	—	—	14.826	—	2071.04	193.34	—	16.47
316	0.10	0.21	—	—	14.959	—	2089.63	195.08	—	16.62
317	0.29	—	—	—	—	—	—	—	—	—
318	1.42	—	—	—	—	—	—	—	—	—et
323	0.34	0.00	0.061	—	—	—	24.50	—	—	—
324	1.40	0.01	0.250	—	—	—	100.00	—	—	—
325	0.95	0.04	—	—	—	—	64.57	2.89	—	22.90
326	1.06	0.04	—	—	—	—	72.45	3.24	—	25.70
327	1.25	—	0.209	0.072	—	—	—	—	—	—
328	1.38	—	0.230	0.079	—	—	—	—	—	—
329	1.15	0.08	0.237	0.110	15.400	0.066	296.41	25.52	—	33.00
330	1.24	0.09	0.256	0.119	16.629	0.071	320.07	27.56	—	35.64
331	1.18	0.03	0.303	—	—	—	—	—	—	—
332	1.28	0.03	0.328	—	—	—	—	—	—	—
333	—	—	—	—	—	—	488.61	—	—	—
334	—	—	—	—	—	—	561.26	—	—	—
335	1.12	0.14	0.189	—	—	—	548.92	—	0.072	0.00
336	1.25	0.16	0.210	—	—	—	610.00	—	0.080	0.00
337	1.96	0.01	0.084	—	—	—	—	2.26	—	2.01
338	2.16	0.01	0.092	—	—	—	—	2.48	—	2.21
339	0.48	0.02	0.020	—	6.222	—	17.11	9.16	—	—
340	2.17	0.09	0.092	—	28.350	—	77.94	41.74	—	—
341	0.53	0.02	0.022	—	—	—	—	—	—	—
342	2.13	0.09	0.090	—	—	—	—	—	—	—
343	0.52	0.78	0.518	4.925	19.876	3.100	639.00	16.51	0.919	117.41
344	0.55	0.83	0.553	5.250	21.185	3.304	681.08	17.60	0.980	125.15
345	0.27	0.52	1.498	0.116	7.310	0.043	523.14	11.94	0.912	71.83
346	0.29	0.56	1.613	0.124	7.869	0.047	563.10	12.86	0.982	77.32
347	1.90	0.57	0.164	—	46.680	—	608.82	522.05	—	578.26
348	2.09	0.63	0.181	—	51.517	—	671.91	576.15	—	638.18
349	—	—	—	0.137	5.381	—	166.16	95.13	—	27.49
350	—	—	—	0.146	5.720	—	176.65	101.14	—	29.22
353	—	—	—	—	—	—	—	—	—	—
354	—	—	—	—	—	—	—	—	—	—
355	0.36	—	—	—	—	—	—	—	—	—
356	2.21	—	—	—	—	—	—	—	—	—
357	0.56	—	0.113	—	1.352	—	30.67	7.66	—	—
358	3.37	—	0.675	—	8.115	—	184.01	45.98	—	—
359	0.83	—	—	—	6.806	—	174.73	55.32	0.959	43.22
360	0.90	—	—	—	7.400	—	190.00	60.15	1.042	47.00
361	1.07	0.09	1.139	—	7.943	—	222.41	45.67	0.971	35.75
362	1.18	0.10	1.250	—	8.715	—	244.01	50.11	1.065	39.22
363	—	—	—	—	—	—	—	—	—	—
364	—	—	—	—	—	—	—	—	—	—
365	1.30	—	—	—	10.446	—	159.70	54.24	1.004	71.71
366	1.41	—	—	—	11.404	—	174.34	59.21	1.097	78.29

TABLE 3–2C. (continued). Composition of Feeds—Mineral Values

		IFN	DRY MATT %	CA %	CL %	MG %	P %
	RAPE, TURNIP. BRASSICA CAMPESTRIS						
367	Seeds, meal mechanical extracted	5-07-871	93.7	0.67	—	—	1.19
368			100.0	0.72	—	—	1.26
369	Seeds, meal solvent extracted	5-07-870	90.9	0.64	—	0.55	1.13
370			100.0	0.71	—	0.60	1.25
	REDTOP. AGROSTIS ALBA						
371	Fresh, full bloom	2-03-891	26.3	0.16	—	0.07	0.10
372			100.0	0.62	—	0.25	0.37
	RICE. ORYZA SATIVA						
375	Bran with germs (rice, bran)	4-03-928	90.5	0.09	0.07	0.88	1.57
376			100.0	0.10	0.08	0.97	1.73
377	Grain, ground	4-03-938	89.0	0.07	0.07	0.13	0.32
378			100.0	0.07	0.08	0.14	0.36
379	Groats, polished	4-03-942	88.6	0.02	0.04	0.09	0.11
380			100.0	0.03	0.04	0.10	0.13
381	Groats, polished broken	4-03-932	88.5	0.02	—	0.04	0.11
382			100.0	0.02	—	0.04	0.13
383	Hulls	1-08-075	91.9	0.11	0.07	0.34	0.07
384			100.0	0.12	0.08	0.37	0.07
385	Polishings	4-03-943	90.2	0.05	0.11	0.60	1.34
386			100.0	0.05	0.12	0.67	1.49
	RUSSIANTHISTLE, TUMBLING. SALSOLA KALI TENUIFOLIA						
387	Fresh, stem cured	2-08-000	88.0	—	—	—	0.09
388			100.0	—	—	—	0.10
389	Hay, sun-cured	1-03-988	86.4	1.31	—	0.65	0.22
390			100.0	1.51	—	0.75	0.25
	RYE. SECALE CEREALE						
391	Distillers grains, dehydrated	5-04-023	91.9	0.15	0.05	0.17	0.48
392			100.0	0.16	0.05	0.18	0.52
393	Flour by-product, less than 8.5% fiber (rye middlings)	4-04-031	89.5	0.06	—	—	0.63
394			100.0	0.07	—	—	0.70
395	Fresh, early vegetative	2-04-013	15.5	—	—	—	—
396			100.0	—	—	—	—
397	Grain	4-04-047	87.5	0.06	0.03	0.11	0.32
398			100.0	0.07	0.03	0.12	0.36
399	Mill run, less than 9.5% fiber	4-04-034	88.8	0.07	—	0.23	0.63
400			100.0	0.08	—	0.26	0.71
401	Silage	3-04-020	35.0	0.12	—	0.04	0.12
402			100.0	0.35	—	0.12	0.33
403	Straw	1-04-007	90.9	0.22	0.22	0.07	0.08
404			100.0	0.24	0.24	0.08	0.09
	RYEGRASS, ITALIAN. LOLIUM MULTIFLORUM						
405	Fresh	2-04-073	22.6	0.15	—	0.08	0.09
406			100.0	0.65	—	0.35	0.41
	RYEGRASS, PERENNIAL. LOLIUM PERENNE						
407	Fresh	2-04-086	26.6	0.12	—	—	0.05
408			100.0	0.45	—	—	0.21
409	Hay, sun-cured	1-04-077	86.6	—	—	—	0.24
410			100.0	—	—	—	0.27
	SACATON, ALKALI. SPOROBOLUS AIROIDES						
411	Fresh, stem cured	2-05-599	76.0	0.30	—	0.13	0.08
412			100.0	0.40	—	0.18	0.11
	SAFFLOWER. CARTHAMUS TINCTORIUS						
413	Seeds	4-07-958	93.1	0.24	—	0.34	0.57
414			100.0	0.26	—	0.36	0.61
415	Seeds, meal mechanical extracted	5-04-109	91.6	0.25	—	0.33	0.71
416			100.0	0.27	—	0.36	0.77
417	Seeds, meal solvent extracted	5-04-110	91.7	0.34	—	0.34	0.76
418			100.0	0.37	—	0.37	0.83
419	Seeds without hulls, meal solvent extracted	5-07-959	91.0	0.36	0.16	0.99	1.40
420			100.0	0.39	0.18	1.08	1.53
	SAGE, BLACK. SALVIA MELLIFERA						
421	Browse, fresh, stem cured	2-05-564	65.0	0.53	—	—	0.11
422			100.0	0.81	—	—	0.17

	K %	NA %	S %	CO mg/kg	CU mg/kg	I mg/kg	FE mg/kg	MN mg/kg	SE mg/kg	ZN mg/kg
367	—	—	—	—	—	—	—	—	—	—
368	—	—	—	—	—	—	—	—	—	—
369	1.11	0.00	0.000	—	6.947	—	93.10	48.01	0.943	65.35
370	1.22	0.00	0.000	—	7.643	—	102.43	52.82	1.037	71.90
371	0.62	0.01	0.042	—	—	—	52.60	—	—	—
372	2.35	0.05	0.160	—	—	—	200.00	—	—	—
375	1.71	0.03	0.181	1.381	10.997	—	207.39	358.47	—	43.81
376	1.89	0.03	0.200	1.526	12.148	—	229.09	395.98	—	48.39
377	0.44	0.06	0.044	—	—	—	—	17.99	—	14.99
378	0.49	0.07	0.049	—	—	—	—	20.22	—	16.85
379	0.23	0.01	0.080	0.846	5.396	—	17.39	29.57	—	9.70
380	0.26	0.02	0.090	0.955	6.092	—	19.63	33.39	—	10.95
381	0.10	0.00	0.040	—	2.851	—	10.71	8.66	0.273	8.59
382	0.11	0.00	0.045	—	3.220	—	12.09	9.79	0.308	9.70
383	0.60	0.02	0.074	—	3.108	—	91.28	294.08	0.138	21.95
384	0.65	0.02	0.080	—	3.383	—	99.37	320.11	0.150	23.89
385	1.25	0.04	0.171	—	7.977	—	81.00	126.74	—	63.18
386	1.39	0.05	0.189	—	8.841	—	89.77	140.47	—	70.03
387	—	—	—	—	—	—	—	—	—	—
388	—	—	—	—	—	—	—	—	—	—
389	5.92	—	—	—	—	—	—	—	—	—
390	6.85	—	—	—	—	—	—	—	—	—
391	0.07	0.17	0.436	—	—	—	—	18.36	—	—
392	0.08	0.18	0.475	—	—	—	—	19.99	—	—
393	0.63	—	—	—	—	—	—	43.95	—	—
394	0.70	—	—	—	—	—	—	49.10	—	—
395	—	0.01	—	—	—	—	—	—	—	—
396	—	0.07	—	—	—	—	—	—	—	—
397	0.45	0.02	0.147	—	7.553	—	62.83	72.01	—	28.47
398	0.51	0.03	0.168	—	8.633	—	71.80	82.30	—	32.54
399	0.82	—	0.039	—	—	—	—	—	—	—
400	0.92	—	0.044	—	—	—	—	—	—	—
401	1.06	—	0.046	—	3.149	—	39.18	17.49	—	8.05
402	3.02	—	0.130	—	9.000	—	112.00	50.00	—	23.00
403	0.88	0.12	0.099	—	3.638	—	—	6.03	—	—
404	0.97	0.13	0.109	—	4.003	—	—	6.63	—	—
405	0.45	0.00	0.023	—	—	—	225.79	—	—	—
406	2.00	0.01	0.100	—	—	—	1000.00	—	—	—
407	0.51	—	0.080	0.016	3.458	—	—	—	—	—
408	1.92	—	0.300	0.060	13.000	—	—	—	—	—
409	1.23	—	—	—	—	—	—	—	—	—
410	1.42	—	—	—	—	—	—	—	—	—
411	0.53	0.15	—	0.451	12.540	—	357.94	23.18	—	17.48
412	0.71	0.21	—	0.593	16.750	—	470.98	30.50	—	23.00
413	0.74	0.06	0.056	—	9.964	—	316.73	1.10	—	30.01
414	0.79	0.06	0.060	—	10.700	—	340.13	1.18	—	32.22
415	0.72	0.05	0.051	—	9.823	—	472.15	17.92	—	40.04
416	0.79	0.05	0.056	—	10.722	—	515.37	19.56	—	43.71
417	0.75	0.05	0.128	—	9.879	—	496.54	18.15	—	40.53
418	0.81	0.05	0.140	—	10.777	—	541.65	19.80	—	44.21
419	1.08	0.04	0.177	2.021	88.470	—	860.22	40.21	—	185.98
420	1.19	0.05	0.195	2.221	97.203	—	945.13	44.18	—	204.34
421	—	—	—	—	—	—	—	—	—	—
422	—	—	—	—	—	—	—	—	—	—

TABLE 3–2C. (continued). Composition of Feeds—Mineral Values

		IFN	DRY MATT %	CA %	CL %	MG %	P %
423	SAGEBRUSH, BIG. ARTEMISIA TRIDENTATA Browse, fresh, stem cured	2-07-992	65.0	0.32	—	—	0.09
424			100.0	0.50	—	—	0.14
	SAGEBRUSH, BUD. ARTEMISIA SPINESCENS						
425	Browse, fresh, early vegetative	2-07-991	23.0	0.22	—	—	0.08
426			100.0	0.97	—	—	0.33
427	Browse, fresh, late vegetative	2-04-124	32.0	0.19	—	0.16	0.13
428			100.0	0.60	—	0.49	0.42
	SALTBUSH, NUTTALL. ATRIPLEX NUTTALLII						
433	Browse, fresh, stem cured	2-07-993	55.0	1.22	—	—	0.06
434			100.0	2.21	—	—	0.12
	SALTBUSH, SHADSCALE. ATRIPLEX CONFERTIFOLIA						
435	Browse, fresh, stem cured	2-05-565	80.0	1.78	—	—	0.07
436			100.0	2.23	—	—	0.08
	SALTGRASS. DISTICHLIS SPP						
437	Fresh	2-04-170	74.4	0.16	—	—	0.07
438			100.0	0.21	—	—	0.09
	SALTGRASS, DESERT. DISTICHLIS STRICTA						
441	Fresh	2-04-171	29.0	0.07	0.16	0.03	0.04
442			100.0	0.23	0.55	0.10	0.13
	SCREENINGS (SEE BARLEY; CEREALS; WHEAT)						
	SEAWEED, KELP. LAMINARIALES (ORDER)-FUCALES (ORDER)						
443	Whole, meal	1-08-073	90.9	2.48	—	0.85	0.24
444			100.0	2.73	—	0.93	0.27
	SEDGE. CAREX SPP						
445	Hay, sun-cured	1-04-193	89.3	0.50	—	0.31	0.14
446			100.0	0.56	—	0.35	0.16
	SESAME. SESAMUM INDICUM						
447	Seeds, meal mechanical extracted	5-04-220	92.7	2.01	0.06	0.46	1.36
448			100.0	2.17	0.07	0.50	1.46
	SKIMMILK (SEE CATTLE)						
	SORGHUM. SORGHUM BICOLOR						
449	Aerial part without heads, sun-cured, mature	1-07-961	88.1	0.46	—	0.24	0.10
450			100.0	0.52	—	0.27	0.12
451	Distillers grains, dehydrated	5-04-374	93.8	0.15	—	0.18	0.69
452			100.0	0.16	—	0.19	0.74
453	Grain	4-04-383	90.1	0.04	0.07	0.15	0.32
454			100.0	0.04	0.08	0.17	0.36
455	Grain, 9-12% protein	4-08-139	89.4	0.03	—	0.15	0.27
456			100.0	0.04	—	0.17	0.30
	SORGHUM, JOHNSONGRASS. SORGHUM HALEPENSE						
457	Hay, sun-cured	1-04-407	90.5	0.80	—	0.31	0.27
458			100.0	0.89	—	0.35	0.30
	SORGHUM, KAFIR. SORGHUM BICOLOR CAFFRORUM						
459	Grain	4-04-428	89.1	0.03	0.10	0.15	0.31
460			100.0	0.04	0.11	0.17	0.35
	SORGHUM, MILO. SORGHUM BICOLOR SUBGLABRESCENS						
461	Grain	4-04-444	88.5	0.04	—	0.12	0.30
462			100.0	0.05	—	0.14	0.34
463	Heads	4-04-446	89.9	0.12	—	0.15	0.20
464			100.0	0.14	—	0.17	0.23
	SORGHUM, SORGO. SORGHUM BICOLOR SACCHARATUM						
465	Silage	3-04-468	28.8	0.10	0.02	0.08	0.06
466			100.0	0.35	0.06	0.27	0.21
	SORGHUM, SUDANGRASS. SORGHUM BICOLOR SUDANENSE						
467	Fresh, early vegetative	2-04-484	17.8	0.08	—	0.06	0.07
468			100.0	0.43	—	0.35	0.41
469	Fresh, midbloom	2-04-485	22.8	0.10	—	0.08	0.09
470			100.0	0.43	—	0.35	0.41
471	Hay, sun-cured	1-04-480	91.1	0.47	—	0.34	0.28
472			100.0	0.51	—	0.37	0.31
473	Silage	3-04-499	23.3	0.12	—	0.09	0.05
474			100.0	0.50	—	0.39	0.21

	K %	NA %	S %	CO mg/kg	CU mg/kg	I mg/kg	FE mg/kg	MN mg/kg	SE mg/kg	ZN mg/kg
423	—	—	0.060	—	—	—	—	—	—	—
424	—	—	0.093	—	—	—	—	—	—	—
425	—	—	—	—	—	—	—	—	—	—
426	—	—	—	—	—	—	—	—	—	—
427	—	—	—	—	—	—	—	—	—	—
428	—	—	—	—	—	—	—	—	—	—
433	—	—	—	—	—	—	—	—	—	—
434	—	—	—	—	—	—	—	—	—	—
435	—	—	—	—	—	—	—	—	—	—
436	—	—	—	—	—	—	—	—	—	—
437	0.18	—	—	—	—	—	141.33	115.14	—	—
438	0.24	—	—	—	—	—	190.00	154.78	—	—
441	0.19	0.05	—	—	—	—	—	—	—	—
442	0.64	0.16	—	—	—	—	—	—	—	—
443	1.29	4.18	—	—	6.273	—	891.18	1229.73	0.398	34.85
444	1.42	4.60	—	—	6.900	—	980.28	1352.68	0.438	38.34
445	0.58	—	0.094	—	2.977	—	57.16	112.09	—	17.42
446	0.65	—	0.105	—	3.333	—	64.00	125.50	—	19.50
447	1.27	0.05	0.352	—	—	—	92.70	47.76	—	99.67
448	1.37	0.05	0.380	—	—	—	100.00	51.51	—	107.51
449	0.72	0.06	—	—	—	—	206.92	—	—	—
450	0.82	0.06	—	—	—	—	235.00	—	—	—
451	0.36	0.05	0.169	—	—	—	46.89	—	—	—
452	0.38	0.05	0.180	—	—	—	50.00	—	—	—
453	0.37	0.01	0.135	0.274	5.411	—	57.17	12.33	0.411	26.77
454	0.41	0.01	0.150	0.304	6.008	—	63.48	13.70	0.457	29.73
455	0.34	0.02	0.140	0.066	9.737	0.022	31.39	15.49	—	13.72
456	0.38	0.02	0.157	0.074	10.897	0.025	35.13	17.34	—	15.36
457	1.22	0.01	0.091	—	—	—	534.05	—	—	—
458	1.35	0.01	0.100	—	—	—	590.00	—	—	—
459	0.34	0.05	0.160	0.387	6.950	—	63.25	15.83	0.796	13.53
460	0.38	0.06	0.179	0.434	7.800	—	70.99	17.76	0.894	15.19
461	0.31	0.04	0.107	0.885	4.087	0.061	48.34	15.93	0.200	16.96
462	0.35	0.04	0.121	1.000	4.618	0.069	54.62	18.00	0.226	19.16
463	0.50	—	0.120	—	—	—	—	—	—	—
464	0.56	—	0.133	—	—	—	—	—	—	—
465	0.32	0.04	0.029	—	8.979	—	57.22	17.60	—	—
466	1.12	0.15	0.100	—	31.137	—	198.42	61.03	—	—
467	0.38	0.00	0.020	—	—	—	35.60	—	—	—
468	2.14	0.01	0.110	—	—	—	200.00	—	—	—
469	0.49	0.00	0.025	—	—	—	45.55	—	—	—
470	2.14	0.01	0.110	—	—	—	200.00	—	—	—
471	1.90	0.01	0.055	0.116	28.641	—	149.47	69.51	—	34.63
472	2.08	0.02	0.060	0.127	31.425	—	164.00	76.26	—	38.00
473	0.52	0.00	0.023	0.063	6.241	—	39.20	14.76	—	9.78
474	2.21	0.01	0.100	0.270	26.798	—	168.33	63.38	—	42.00

TABLE 3-2C. (continued). Composition of Feeds—Mineral Values

		IFN	DRY MATT %	CA %	CL %	MG %	P %
	SOYBEAN. GLYCINE MAX						
475	Hay, sun-cured	1-04-558	89.1	1.13	0.13	0.72	0.22
476			100.0	1.26	0.15	0.80	0.24
477	Seed coats (hulls)	1-04-560	90.3	0.48	0.02	0.20	0.17
478			100.0	0.53	0.02	0.22	0.18
479	Seeds	5-04-610	92.0	0.34	0.01	0.37	0.63
480			100.0	0.37	0.01	0.40	0.68
481	Seeds, heat processed	5-04-597	92.6	0.26	—	0.22	0.61
482			100.0	0.28	—	0.23	0.66
483	Seeds, meal mechanical extracted	5-04-600	90.0	0.26	0.07	0.26	0.62
484			100.0	0.29	0.08	0.28	0.69
485	Seeds, meal solvent extracted	5-04-604	89.6	0.30	0.04	0.29	0.69
486			100.0	0.34	0.04	0.32	0.77
487	Seeds without hulls, meal solvent extracted	5-04-612	89.9	0.26	0.04	0.29	0.64
488			100.0	0.29	0.05	0.33	0.71
489	Silage	3-04-581	30.2	0.40	—	0.12	0.13
490			100.0	1.32	—	0.40	0.44
491	Straw	1-04-567	87.7	1.39	—	0.81	0.05
492			100.0	1.59	—	0.92	0.06
	SQUIRRELTAIL. SITANION SPP						
493	Fresh, stem cured	2-05-566	50.0	0.19	—	—	0.03
494			100.0	0.37	—	—	0.06
	SUDANGRASS—SEE SORGHUM						
	SUGARCANE. SACCHARUM OFFICINARUM						
495	Molasses, dehydrated	4-04-695	94.4	1.03	—	0.44	0.14
496			100.0	1.10	—	0.47	0.15
497	Molasses, more than 46% invert sugars more than 79.5 degrees	4-04-696	74.3	0.74	2.26	0.31	0.08
498	brix		100.0	1.00	3.04	0.42	0.10
	SUNFLOWER, COMMON. HELIANTHUS ANNUUS						
499	Seeds without hulls, meal mechanical extracted	5-04-738	92.7	0.40	0.17	0.67	1.07
500			100.0	0.43	0.18	0.72	1.15
501	Seeds without hulls, meal solvent extracted	5-04-739	92.5	0.42	0.15	0.65	0.94
502			100.0	0.45	0.17	0.70	1.02
	SWEETCLOVER, YELLOW. MELILOTUS OFFICINALIS						
503	Hay, sun-cured	1-04-754	88.5	1.44	0.33	0.39	0.24
504			100.0	1.63	0.37	0.44	0.27
	TANKANGE—SEE ANIMAL; ANIMAL–POULTRY						
	TIMOTHY. PHLEUM PRATENSE						
505	Fresh, late vegetative	2-04-903	26.7	0.11	—	0.04	0.07
506			100.0	0.40	—	0.16	0.26
507	Fresh, midbloom	2-04-905	29.2	0.11	0.19	0.04	0.09
508			100.0	0.38	0.63	0.14	0.30
509	Hay, sun-cured, late vegetative	1-04-881	89.3	0.41	—	0.10	0.36
510			100.0	0.45	—	0.11	0.40
511	Hay, sun-cured, early bloom	1-04-882	89.1	0.45	—	0.11	0.25
512			100.0	0.51	—	0.13	0.29
513	Hay, sun-cured, midbloom	1-04-883	88.9	0.32	—	0.12	0.20
514			100.0	0.36	—	0.13	0.23
515	Hay, sun-cured, late bloom	1-04-885	88.3	0.34	—	0.15	0.13
516			100.0	0.38	—	0.17	0.15
517	Silage	3-04-922	33.5	0.19	—	0.05	0.10
518			100.0	0.57	—	0.15	0.29
	TOMATO. LYCOPERSICON ESCULENTUM						
519	Pomace, dehydrated	5-05-041	92.2	0.39	—	0.18	0.56
520			100.0	0.42	—	0.20	0.60
	TREFOIL, BIRDSFOOT. LOTUS CORNICULATUS						
521	Fresh	2-20-786	19.3	0.33	—	0.08	0.05
522			100.0	1.74	—	0.40	0.26
523	Hay, sun-cured	1-05-044	90.6	1.54	—	0.46	0.21
524			100.0	1.70	—	0.51	0.23
	TRITICALE. TRITICALE HEXAPLOIDE						
525	Grain	4-20-362	89.2	0.05	—	0.23	0.33
526			100.0	0.05	—	0.26	0.37

	K	NA	S	CO	CU	I	FE	MN	SE	ZN
	%	%	%.	mg/kg	mg/kg	mg/kg	mg/kg	mg/kg	mg/kg	mg/kg
475	0.92	0.10	0.247	0.083	8.025	0.216	257.96	94.34	—	21.50
476	1.04	0.11	0.278	0.093	9.004	0.242	289.43	105.85	—	24.12
477	1.17	0.02	0.099	0.109	16.078	—	369.35	9.92	0.126	43.27
478	1.29	0.03	0.110	0.120	17.808	—	409.10	10.99	0.140	47.93
479	2.21	0.01	0.223	—	20.555	—	91.71	40.67	0.111	45.18
480	2.41	0.02	0.242	—	22.330	—	99.63	44.18	0.120	49.08
481	1.75	0.03	—	—	—	—	—	—	—	—
482	1.89	0.03	—	—	—	—	—	—	—	—
483	1.78	0.03	0.331	0.178	21.731	—	162.33	31.31	0.181	57.23
484	1.98	0.03	0.367	0.198	24.146	—	180.37	34.78	0.201	63.59
485	2.10	0.04	0.425	1.399	17.925	0.131	140.55	30.61	0.434	51.84
486	2.34	0.04	0.474	1.561	20.005	0.147	156.85	34.16	0.484	57.85
487	2.12	0.01	0.435	0.065	20.243	0.108	130.39	37.13	0.194	57.11
488	2.36	0.01	0.484	0.072	22.508	0.120	144.98	41.28	0.216	63.50
489	0.39	0.03	0.092	—	2.915	—	99.53	42.67	—	10.27
490	1.28	0.09	0.305	—	9.649	—	329.46	141.25	—	34.00
491	0.49	0.11	0.228	—	—	—	263.09	44.86	—	—
492	0.56	0.12	0.260	—	—	—	299.96	51.15	—	—
493	—	—	—	—	—	—	—	—	—	—
494	—	—	—	—	—	—	—	—	—	—
495	3.39	0.19	0.434	1.145	74.927	—	236.03	54.12	—	31.16
496	3.59	0.20	0.460	1.213	79.350	—	249.96	57.32	—	33.00
497	2.98	0.16	0.349	1.179	48.829	1.563	195.83	43.68	—	15.56
498	4.01	0.22	0.470	1.586	65.697	2.102	163.48	58.76	—	20.93
499	1.20	0.03	0.391	3.765	32.591	—	205.34	20.15	1.770	93.53
500	1.29	0.04	0.422	4.062	35.166	—	221.57	21.74	1.910	100.92
501	1.17	0.03	0.308	—	37.765	—	262.99	18.91	2.125	97.57
502	1.27	0.03	0.333	—	40.842	—	284.41	20.45	2.298	105.52
503	1.35	0.08	0.415	—	8.830	—	141.89	95.36	—	—
504	1.53	0.09	0.468	—	9.972	—	160.24	107.69	—	—
505	0.73	0.03	0.035	0.040	2.386	—	35.37	33.92	—	9.55
506	2.73	0.11	0.130	0.149	8.923	—	132.27	126.85	—	35.70
507	0.60	0.06	0.038	—	3.272	—	52.40	56.23	—	—
508	2.06	0.19	0.129	—	11.198	—	179.36	192.47	—	—
509	2.72	0.01	0.116	—	22.996	—	208.42	79.48	—	59.83
510	3.05	0.01	0.130	—	25.750	—	233.38	89.00	—	67.00
511	2.14	0.01	0.116	—	57.050	—	180.95	91.81	—	55.27
512	2.41	0.01	0.130	—	64.000	—	203.00	103.00	—	62.00
513	1.61	0.01	0.116	—	14.250	—	132.11	49.90	—	38.21
514	1.82	0.01	0.130	—	16.035	—	148.66	56.15	—	43.00
515	1.42	0.06	0.141	—	—	—	203.11	—	—	—
516	1.61	0.07	0.160	—	—	—	229.98	—	—	—
517	0.57	0.04	0.044	—	1.844	—	36.88	30.23	—	—
518	1.69	0.11	0.130	—	5.500	—	110.00	90.18	—	—
519	3.35	—	—	—	30.045	—	4239.14	47.17	—	—
520	3.63	—	—	—	32.600	—	4599.69	51.19	—	—
521	0.63	0.02	0.048	0.094	2.474	—	34.01	15.97	—	5.99
522	3.26	0.11	0.250	0.486	12.834	—	176.42	82.85	—	31.06
523	1.74	0.06	0.227	0.100	8.386	—	206.08	25.96	—	69.90
524	1.92	0.07	0.250	0.110	9.257	—	227.50	28.66	—	77.16
525	0.51	0.01	—	0.077	8.302	—	43.99	42.66	—	31.30
526	0.57	0.01	—	0.087	9.311	—	49.34	47.84	—	35.11

TABLE 3–2C. (continued). Composition of Feeds—Mineral Values

	IFN	DRY MATT %	CA %	CL %	MG %	P %
TURNIP. BRASSICA RAPA RAPA						
527 Roots, fresh	4-05-067	9.2	0.06	0.06	0.02	0.03
528		100.0	0.67	0.65	0.19	0.34
UREA						
529 45% nitrogen 281% protein equivalent	5-05-070	97.0	0.09	0.00	0.00	0.00
530		100.0	0.09	0.00	0.00	0.00
VETCH. VICIA SPP						
533 Hay, sun-cured	1-05-106	88.7	1.21	—	0.24	0.30
534		100.0	1.36	—	0.27	0.34
WHEAT. TRITICUM AESTIVUM						
535 Bran	4-05-190	89.0	0.13	0.05	0.56	1.13
536		100.0	0.14	0.06	0.63	1.27
537 Flour by-product, less than 4% fiber (wheat red dog)	4-05-203	88.3	0.06	0.11	0.19	0.53
538		100.0	0.07	0.13	0.21	0.60
539 Flour by-product, less than 7% fiber (wheat shorts)	4-05-201	88.4	0.08	0.05	0.27	0.90
540		100.0	0.10	0.06	0.31	1.01
541 Flour by-product, less than 9.5% fiber (wheat middlings)	4-05-205	88.9	0.13	0.04	0.34	0.89
542		100.0	0.14	0.05	0.38	1.00
543 Fresh, early vegetative	2-05-176	22.2	0.09	—	0.05	0.09
544		100.0	0.42	—	0.21	0.40
545 Germs, ground	5-05-218	88.4	0.06	0.06	0.25	0.95
546		100.0	0.06	0.07	0.28	1.07
547 Grain	4-05-211	89.0	0.05	0.07	0.13	0.35
548		100.0	0.05	0.08	0.15	0.39
549 Grain screenings	4-05-216	88.5	0.12	—	0.18	0.35
550		100.0	0.14	—	0.20	0.39
551 Hay, sun-cured	1-05-172	88.7	0.13	—	0.11	0.18
552		100.0	0.15	—	0.12	0.20
553 Mill run, less than 9.5% fiber	4-05-206	89.9	0.10	—	0.47	1.02
554		100.0	0.11	—	0.53	1.13
555 Silage, early vegetative	3-05-184	29.9	0.08	0.02	0.19	0.08
556		100.0	0.27	0.07	0.62	0.27
557 Straw	1-05-175	91.1	0.16	0.29	0.11	0.04
558		100.0	0.17	0.32	0.12	0.05
WHEAT, DURUM. TRITICUM DURUM						
559 Grain	4-05-224	87.6	0.10	—	0.16	0.35
560		100.0	0.11	—	0.18	0.41
WHEAT, HARD RED SPRING. TRITICUM AESTIVUM						
561 Grain	4-05-258	87.6	0.03	0.08	0.15	0.37
562		100.0	0.04	0.09	0.17	0.43
WHEAT, HARD RED WINTER. TRITICUM AESTIVUM						
563 Grain	4-05-268	88.8	0.04	0.05	0.13	0.38
564		100.0	0.05	0.06	0.14	0.42
WHEAT, SOFT RED WINTER. TRITICUM AESTIVUM						
565 Grain	4-05-294	88.4	0.05	0.07	0.10	0.36
566		100.0	0.06	0.08	0.11	0.40
WHEAT, SOFT WHITE WINTER. TRITICUM AESTIVUM						
567 Grain	4-05-337	90.2	—	—	0.10	0.30
568		100.0	—	—	0.11	0.33
WHEATGRASS, CRESTED. AGROPYRON DESERTORUM						
569 Fresh	2-05-429	45.0	0.20	—	0.05	0.07
570		100.0	0.43	—	0.12	0.16
571 Hay, sun-cured	1-05-418	91.8	0.24	—	—	0.14
572		100.0	0.26	—	—	0.15
WHEY (SEE CATTLE)						
YEAST, BREWERS. SACCHAROMYCES CEREVISIAE						
573 Dehydrated	7-05-527	93.1	0.14	0.07	0.24	1.36
574		100.0	0.15	0.07	0.26	1.47
YEAST, IRRADIATED. SACCHAROMYCES CEREVISIAE						
575 Dehydrated	7-05-529	93.9	0.07	—	—	1.42
576		100.0	0.07	—	—	1.51
YEAST, TORULA. TORULOPSIS UTILIS						
577 Dehydrated	7-05-534	93.0	0.55	0.02	0.14	1.61
578		100.0	0.59	0.02	0.15	1.73

	K	NA	S	CO mg/kg	CU mg/kg	I mg/kg	FE mg/kg	MN mg/kg	SE mg/kg	ZN mg/kg
	%	%	%							
527	0.25	0.01	0.039	—	1.954	—	9.96	3.91	—	2.65
528	2.74	0.13	0.430	—	21.333	—	108.76	42.67	—	28.97
529	0.00	0.02	0.000	—	6.792	—	175.58	—	—	6.79
530	0.00	0.02	0.000	—	7.000	—	180.96	—	—	7.00
533	1.88	0.46	0.133	0.315	8.780	0.436	434.63	53.93	—	—
534	2.12	0.52	0.150	0.355	9.898	0.492	490.00	60.80	—	—
535	1.22	0.05	0.212	0.074	12.610	0.065	144.80	119.34	0.507	97.71
536	1.37	0.06	0.238	0.084	14.165	0.073	162.65	134.05	0.569	109.76
537	0.52	0.01	0.222	0.117	6.278	—	49.65	52.17	0.368	65.07
538	0.58	0.01	0.252	0.132	7.107	—	56.20	59.06	0.416	73.66
539	0.93	0.02	0.191	0.105	11.512	—	73.48	114.18	0.475	102.49
540	1.05	0.03	0.215	0.118	13.016	—	83.07	129.10	0.537	115.87
541	0.98	0.02	0.167	0.102	15.894	0.109	89.78	114.16	0.736	97.08
542	1.10	0.02	0.187	0.114	17.870	0.122	100.94	128.35	0.828	109.14
543	0.78	0.04	0.048	—	—	—	22.20	—	—	—
544	3.50	0.18	0.215	—	—	—	100.00	—	—	—
545	0.94	0.02	0.275	0.120	9.199	—	58.53	133.07	0.541	119.40
546	1.06	0.03	0.311	0.135	10.407	—	66.21	150.54	0.612	135.08
547	0.41	0.06	0.176	0.353	5.794	0.089	52.68	41.89	0.265	30.93
548	0.46	0.06	0.197	0.396	6.510	0.100	59.19	47.07	0.298	34.76
549	0.80	0.05	0.195	1.249	4.769	—	316.31	17.17	0.531	34.80
550	0.91	0.05	0.220	1.411	5.390	—	357.52	19.40	0.600	39.33
551	0.88	0.19	0.191	—	—	—	177.49	—	—	—
552	0.99	0.21	0.215	—	—	—	200.00	—	—	—
553	1.20	—	0.173	0.208	18.478	—	94.73	104.06	0.629	—
554	1.33	—	0.192	0.232	20.559	—	105.41	115.78	0.700	—
555	0.42	0.02	0.072	0.014	4.126	—	56.81	38.57	—	—
556	1.39	0.07	0.240	0.045	13.798	—	190.00	129.00	—	—
557	1.28	0.13	0.170	0.041	3.296	—	143.33	37.26	—	5.88
558	1.40	0.14	0.186	0.045	3.616	—	157.27	40.88	—	6.46
559	0.44	—	—	0.079	6.817	—	40.89	30.65	—	19.28
560	0.50	—	—	0.090	7.779	—	46.67	34.98	—	22.00
561	0.36	0.02	0.149	0.122	6.161	—	56.40	36.72	0.228	43.95
562	0.41	0.02	0.170	0.140	7.033	—	64.38	41.92	0.261	50.17
563	0.43	0.02	0.130	0.145	4.915	—	36.34	34.65	0.252	32.98
564	0.48	0.02	0.147	0.163	5.535	—	40.93	39.02	0.284	37.14
565	0.41	0.01	0.106	0.103	7.030	—	28.98	33.39	0.041	42.14
566	0.46	0.01	0.120	0.116	7.953	—	32.79	37.77	0.047	47.67
567	0.39	0.02	0.119	0.135	7.038	—	36.09	36.09	0.045	27.07
568	0.43	0.02	0.131	0.150	7.800	—	40.00	40.00	0.050	30.00
569	0.78	0.05	—	0.062	1.303	—	29.68	17.11	—	10.99
570	1.72	0.12	—	0.137	2.892	—	65.90	38.00	—	24.42
571	—	—	—	0.219	—	—	—	—	—	—
572	—	—	—	0.238	—	—	—	—	—	—
573	1.68	0.07	0.438	0.505	38.444	0.357	83.20	6.74	0.911	38.97
574	1.81	0.08	0.470	0.543	41.311	0.384	89.40	7.24	0.978	41.88
575	2.14	—	—	—	—	—	—	—	—	—
576	2.28	—	—	—	—	—	—	—	—	—
577	1.91	0.01	0.482	0.030	12.088	2.501	105.96	9.66	0.050	99.48
578	2.06	0.01	0.518	0.032	12.992	2.688	113.88	10.38	0.054	106.92

147

TABLE 3–2D. Composition of Feeds—Vitamin Values

		IFN	DRY MATT %	CARO TENE mg/kg	VIT D2D3 IU/kg	VIT D3 ICU/kg	VIT E mg/kg
	ALFALFA. MEDICAGO SATIVA						
001	Fresh	2-00-196	26.0	—	0.0	—	—
002			100.0	—	0.2	—	—
003	Hay, sun-cured, late vegetative	1-00-054	89.7	—	—	—	—
004			100.0	—	—	—	—
005	Hay, sun-cured, early bloom	1-00-059	90.5	—	1.8	—	23.5
006			100.0	—	2.0	—	26.0
007	Hay, sun-cured, midbloom	1-00-063	91.0	—	1.4	—	—
008			100.0	—	1.5	—	—
009	Hay, sun-cured, full bloom	1-00-068	90.9	—	—	—	—
010			100.0	—	—	—	—
011	Hay, sun-cured, mature	1-00-071	91.2	—	1.3	—	—
012			100.0	—	1.4	—	—
013	Meal dehydrated, 15% protein	1-00-022	90.3	—	—	—	81.8
014			100.0	—	—	—	90.6
015	Meal dehydrated, 17% protein	1-00-023	91.8	—	—	—	105.9
016			100.0	—	—	—	115.3
017	Meal dehydrated, 20% protein	1-00-024	91.6	—	—	—	143.3
018			100.0	—	—	—	156.4
	ANIMAL						
029	Blood, spray dehydrated	5-00-381	92.6	—	—	—	—
030			100.0	—	—	—	—
031	Bone, meal steamed	6-00-400	96.1	—	—	—	—
032			100.0	—	—	—	—
033	Meat, meal rendered	5-00-385	93.8	—	—	—	0.9
034			100.0	—	—	—	1.0
035	Meat with bone, meal rendered	5-00-388	93.3	—	—	—	0.9
036			100.0	—	—	—	0.9
037	Tankage, meal rendered	5-00-386	92.0	—	—	—	—
038			100.0	—	—	—	—
039	Tankage with bone, meal rendered	5-00-387	92.9	—	—	—	0.8
040			100.0	—	—	—	0.9
	ANIMAL–POULTRY						
041	Fat	4-00-409	99.1	—	—	—	7.9
042			100.0	—	—	—	7.9
	BAKERY						
051	Waste, dehydrated (dried bakery product)	4-00-466	91.2	—	—	—	41.0
052			100.0	—	—	—	44.9
	BARLEY. HORDEUM VULGARE						
053	Grain	4-00-549	88.6	—	—	—	23.2
054			100.0	—	—	—	26.2
055	Grain, Pacific coast	4-07-939	88.6	—	—	—	26.2
056			100.0	—	—	—	29.6
057	Grain screenings	4-00-542	88.9	—	—	—	—
058			100.0	—	—	—	—
059	Hay, sun-cured	1-00-495	88.4	—	1.0	—	—
060			100.0	—	1.1	—	—
061	Straw	1-00-498	91.4	—	0.6	—	—
062			100.0	—	0.7	—	—
	BEAN, NAVY. PHASEOLUS VULGARIS						
063	Seeds	5-00-623	89.8	—	—	—	1.0
064			100.0	—	—	—	1.1
	BEET, MANGEL. BETA VULGARIS MACRORRHIZA						
065	Roots, fresh	4-00-637	11.0	—	—	—	—
066			100.0	—	—	—	—

148

	VIT K mg/kg	BIO TIN mg/kg	CHOL INE mg/kg	FOL ATE mg/kg	NIA CIN mg/kg	PANT OTHE mg/kg	RIBO FLVN mg/kg	THIA MINE mg/kg	VIT B6 mg/kg	VIT B12 ug/kg
001	—	0.126	374.4	0.64	15.4	8.9	4.6	1.65	1.66	
002	—	0.485	1439.0	2.47	59.1	34.3	17.5	6.35	6.38	
003	—	—	—	—	—	—	—	—	—	—
004	—	—	—	—	—	—	—	—	—	—
005	—	—	—	—	—	—	—	—	—	—
006	—	—	—	—	—	—	—	—	—	—
007	—	—	—	—	—	—	9.6	—	—	—
008	—	—	—	—	—	—	10.6	—	—	—
009	—	—	—	—	—	—	—	—	—	—
010	—	—	—	—	—	—	—	—	—	—
011	—	—	—	—	—	—	—	—	—	—
012	—	—	—	—	—	—	—	—	—	—
013	9.6	0.254	1571.1	1.56	41.5	20.7	10.6	2.99	6.27	—
014	10.6	0.281	1739.3	1.73	46.0	22.9	11.7	3.31	6.94	—
015	8.2	0.327	1349.6	4.37	37.0	29.8	12.9	3.39	7.19	—
016	9.0	0.357	1470.2	4.76	40.3	32.4	14.1	3.69	7.83	—
017	14.2	0.354	1417.2	2.96	48.0	35.5	15.2	5.36	8.72	—
018	15.5	0.386	1547.4	3.24	52.4	38.8	16.6	5.85	9.52	—
029	—	0.278	597.7	0.37	22.2	3.2	2.9	0.32	4.44	12.205
030	—	0.300	645.4	0.40	23.9	3.5	3.1	0.35	4.79	13.180
031	—	—	—	—	4.2	2.4	0.9	0.29	—	—
032	—	—	—	—	4.4	2.5	0.9	0.31	—	—
033	—	0.124	1979.3	0.39	56.0	6.0	5.2	0.22	4.23	75.126
034	—	0.132	2110.0	0.42	59.7	6.4	5.5	0.23	4.51	80.088
035	—	0.100	2049.0	0.37	51.3	5.5	4.7	0.16	5.86	18.415
036	—	0.107	2195.5	0.40	55.0	5.9	5.0	0.17	6.28	26.881
037	—	—	2199.8	1.54	37.6	3.2	2.2	0.33	—	89.330
038	—	—	2390.5	1.67	40.9	3.5	2.4	0.36	—	97.074
039	—	0.073	2066.8	0.57	58.5	4.8	5.0	0.21	—	04.389
040	—	0.079	2224.6	0.62	62.9	5.2	5.4	0.22	—	12.360
041	—	—	—	—	—	—	—	—	—	—
042	—	—	—	—	—	—	—	—	—	—
051	—	0.067	916.7	0.18	25.6	8.2	1.4	2.92	4.29	—
052	—	0.074	1004.8	0.20	28.0	9.0	1.5	3.20	4.70	—
053	0.2	0.154	1037.0	0.57	78.5	8.1	1.6	4.52	6.48	—
054	0.2	0.173	1170.9	0.64	88.6	9.1	1.8	5.11	7.32	—
055	—	0.149	976.3	0.50	46.7	7.1	1.5	4.19	2.89	—
056	—	0.169	1101.6	0.56	52.7	8.0	1.7	4.72	3.26	—
057	—	—	—	—	—	7.9	1.1	—	—	—
058	—	—	—	—	—	8.9	1.2	—	—	—
059	—	—	—	—	—	—	—	—	—	—
060	—	—	—	—	—	—	—	—	—	—
061	—	—	—	—	—	—	—	—	—	—
062	—	—	—	—	—	—	—	—	—	—
063	—	0.110	1020.0	1.30	24.2	2.4	1.7	6.35	0.30	—
064	—	0.122	1136.2	1.44	27.0	2.7	1.9	7.08	0.33	—
065	—	—	—	0.17	3.5	1.0	0.4	0.26	0.43	—
066	—	—	—	1.57	31.5	9.4	3.9	2.36	3.94	—

TABLE 3–2D. (continued). Composition of Feeds—Vitamin Values

		IFN	DRY MATT %	CARO TENE mg/kg	VIT D2D3 IU/kg	VIT D3 ICU/kg	VIT E mg/kg
	BEET, SUGAR. BETA VULGARIS ALTISSIMA						
067	Aerial part with crowns, fresh	2-00-649	16.5	—	—	—	—
068			100.0	—	—	—	—
071	Molasses, more than 48% invert sugar more than 79.5 degrees	4-00-668	77.9	—	—	—	4.0
072	brix		100.0	—	—	—	5.1
073	Pulp, dehydrated	4-00-669	91.0	—	0.6	—	—
074			100.0	—	0.6	—	—
077	Pulp with molasses, dehydrated	4-00-672	91.8	—	—	—	—
078			100.0	—	—	—	—
	BLOOD (SEE ANIMAL)						
	BLUEGRASS, KENTUCKY. POA PRATENSIS						
091	Fresh, early vegetative	2-00-777	30.8	—	—	—	47.8
092			100.0	—	—	—	155.0
095	Hay, sun-cured	1-00-776	88.9	—	—	—	—
096			100.0	—	—	—	—
	BONES (SEE ANIMAL)						
	BREWERS GRAINS (SEE CEREALS)						
	BROME. BROMUS SPP						
101	Fresh	2-00-900	36.0	—	—	—	—
102			100.0	—	—	—	—
	BROME, SMOOTH. BROMUS INERMIS						
105	Fresh	2-00-963	27.0	—	0.0	—	—
106			100.0	—	0.1	—	—
	BUCKWHEAT, COMMON. FAGOPYRUM SAGITTATUM						
109	Grain	4-00-994	88.4	—	—	—	—
110			100.0	—	—	—	—
	BUTTERMILK—SEE CATTLE						
	CANARYGRASS, REED. PHALARIS ARUNDINACEA						
115	Hay, sun-cured	1-01-104	89.3	—	—	—	—
116			100.0	—	—	—	—
	CARROT. DAUCUS SPP						
117	Roots, fresh	4-01-145	11.5	—	—	—	6.9
118			100.0	—	—	—	60.2
	CASEIN						
119	Acid precipitated dehydrated	5-01-162	91.6	—	—	—	—
120			100.0	—	—	—	—
	CATTLE. BOS TAURUS						
121	Buttermilk, dehydrated	5-01-160	92.3	—	—	—	6.3
122			100.0	—	—	—	6.8
123	Manure, dehydrated	1-01-190	93.0	—	—	—	—
124			100.0	—	—	—	—
125	Milk, dehydrated	5-01-167	95.5	—	0.3	—	—
126			100.0	—	0.4	—	—
127	Milk, fresh	5-01-168	12.4	—	—	—	—
128			100.0	—	—	—	—
129	Skimmilk, dehydrated	5-01-175	94.1	—	0.4	—	9.1
130			100.0	—	0.4	—	9.6
131	Skimmilk, fresh	5-01-170	9.6	—	—	—	—
132			100.0	—	—	—	—
133	Whey, dehydrated	4-01-182	93.3	—	—	—	0.2
134			100.0	—	—	—	0.2
135	Whey, fresh	4-08-134	6.9	—	—	—	—
136			100.0	—	—	—	—
137	Whey low lactose, dehydrated (dried whey product)	4-01-186	93.7	—	—	—	—
138			100.0	—	—	—	—
	CEREALS						
139	Brewers grains, dehydrated	5-02-141	92.2	—	—	—	26.7
140			100.0	—	—	—	29.0

	VIT K mg/kg	BIO TIN mg/kg	CHOL INE mg/kg	FOL ATE mg/kg	NIA CIN mg/kg	PANT OTHE mg/kg	RIBO FLVN mg/kg	THIA MINE mg/kg	VIT B6 mg/kg	VIT B12 ug/kg
067	—	—	—	—	—	—	1.1	—	—	—
068	—	—	—	—	—	—	6.6	—	—	—
071	—	—	827.5	—	41.0	4.5	2.3	—	—	—
072	—	—	1062.4	—	52.7	5.8	2.9	—	—	—
073	—	—	820.9	—	16.8	1.4	0.7	0.39	—	—
074	—	—	901.9	—	18.4	1.5	0.8	0.42	—	—
077	—	—	813.7	—	16.3	1.5	0.7	—	—	—
078	—	—	886.5	—	17.7	1.7	0.7	—	—	—
091	—	—	—	—	—	—	—	—	—	—
092	—	—	—	—	—	—	—	—	—	—
095	—	—	—	—	—	—	9.9	—	—	—
096	—	—	—	—	—	—	11.1	—	—	—
101	—	—	—	—	—	—	2.8	1.11	—	—
102	—	—	—	—	—	—	7.7	3.09	—	—
105	—	—	—	—	—	—	2.1	0.83	—	—
106	—	—	—	—	—	—	7.7	3.09	—	—
109	—	—	442.9	—	18.4	11.6	4.8	3.75	—	—
110	—	—	500.7	—	20.9	13.1	5.4	4.24	—	—
115	—	—	—	—	—	—	8.5	3.57	—	—
116	—	—	—	—	—	—	9.5	4.00	—	—
117	—	0.008	—	0.14	6.7	3.5	0.6	0.67	1.39	—
118	—	0.072	—	1.20	58.0	30.1	4.9	5.79	12.05	—
119	—	0.043	210.6	0.46	1.3	2.7	1.5	0.42	0.43	—
120	—	0.047	230.0	0.50	1.4	2.9	1.7	0.46	0.47	—
121	—	0.288	1745.6	0.39	8.6	37.0	30.6	3.40	2.47	19.561
122	—	0.312	1891.4	0.42	9.3	40.1	33.1	3.68	2.67	21.194
123	—	—	—	—	—	—	—	—	—	01.352
124	—	—	—	—	—	—	—	—	—	09.015
125	—	0.380	—	—	8.2	22.9	18.8	3.76	4.72	—
126	—	0.398	—	—	8.5	24.0	19.7	3.94	4.94	—
127	—	—	903.8	—	1.3	8.4	1.7	0.29	—	—
128	—	—	7310.9	—	10.1	68.0	13.8	2.36	—	—
129	—	0.329	1393.2	0.62	11.5	36.4	19.1	3.72	4.09	50.887
130	—	0.350	1479.8	0.66	12.2	38.6	20.3	3.95	4.35	54.051
131	—	—	—	—	1.1	3.5	2.0	0.44	—	—
132	—	—	—	—	11.5	36.9	20.8	4.64	—	—
133	—	0.353	1791.6	0.85	10.6	46.2	27.4	4.00	3.22	18.898
134	—	0.379	1920.8	0.91	11.4	49.6	29.4	4.29	3.45	20.260
135	—	—	—	—	0.9	5.3	1.4	0.30	—	—
136	—	—	—	—	13.6	76.7	20.3	4.35	—	—
137	—	0.505	4109.2	0.90	17.8	74.7	47.7	5.06	4.49	35.997
138	—	0.540	4387.3	0.96	19.0	79.8	50.9	5.40	4.79	38.433
139	—	0.444	1651.8	0.22	43.7	8.2	1.5	0.63	1.03	3.639
140	—	0.482	1792.5	0.24	47.4	8.9	1.6	0.68	1.11	3.949

TABLE 3–2D. (continued). Composition of Feeds—Vitamin Values

	IFN	DRY MATT %	CARO TENE mg/kg	VIT D2D3 IU/kg	VIT D3 ICU/kg	VIT E mg/kg
CITRUS. CITRUS SPP						
143 Pomace without fines, dehydrated (dried citrus pulp)	4-01-237	91.1	—	—	—	—
144		100.0	—	—	—	—
147 Syrup (molasses)	4-01-241	66.9	—	—	—	—
148		100.0	—	—	—	—
CLOVER, ALSIKE. TRIFOLIUM HYBRIDUM						
149 Fresh	2-01-316	22.5	—	—	—	—
150		100.0	—	—	—	—
151 Hay, sun-cured	1-01-313	87.7	—	—	—	—
152		100.0	—	—	—	—
CLOVER, LADINO. TRIFOLIUM REPENS						
157 Fresh	2-01-383	17.7	—	—	—	—
158		100.0	—	—	—	—
159 Hay, sun-cured	1-01-378	89.1	—	—	—	—
160		100.0	—	—	—	—
CLOVER, RED. TRIFOLIUM PRATENSE						
165 Hay, sun-cured	1-01-415	88.4	—	—	—	—
166		100.0	—	—	—	—
COCONUT. COCOS NUCIFERA						
167 Kernels with coats, meal mechanical extracted (copra meal)	5-01-572	91.6	—	—	—	—
168		100.0	—	—	—	—
169 Kernels with coats, meal solvent extracted (copra meal)	5-01-573	91.1	—	—	—	—
170		100.0	—	—	—	—
CORN (SEE MAIZE)						
COTTON. GOSSYPIUM SPP						
173 Hulls	1-01-599	90.4	—	—	—	—
174		100.0	—	—	—	—
177 Seeds, meal mechanical extracted	5-01-609	92.7	—	—	—	—
178		100.0	—	—	—	—
179 Seeds, meal mechanical extracted, 41% protein	5-01-617	92.6	—	—	—	32.4
180		100.0	—	—	—	34.9
181 Seeds, meal solvent extracted, 41% protein	5-01-621	91.0	—	—	—	14.6
182		100.0	—	—	—	16.1
183 Seeds without hulls, meal prepressed solvent extracted, 50%	5-07-874	93.0	—	—	—	11.3
184 protein		100.0	—	—	—	12.1
FAT (SEE ANIMAL–POULTRY)						
FESCUE, MEADOW. FESTUCA ELATIOR						
189 Fresh	2-01-920	28.4	—	—	—	46.9
190		100.0	—	—	—	165.1
191 Hay, sun-cured	1-01-912	87.5	—	—	—	118.6
192		100.0	—	—	—	135.6
FISH						
193 Solubles, condensed	5-01-969	50.4	—	—	—	—
194		100.0	—	—	—	—
195 Solubles, dehydrated	5-01-971	92.8	—	—	—	6.1
196		100.0	—	—	—	6.5
FISH, ANCHOVETA. ENGRAULIS RINGEN						
197 Meal mechanical extracted	5-01-985	92.0	—	—	—	3.7
198		100.0	—	—	—	4.0
FISH, HERRING. CLUPEA HARENGUS						
199 Meal mechanical extracted	5-02-000	91.8	—	—	—	22.0
200		100.0	—	—	—	24.0
FISH, MENHADEN. BREVOORTIA TYRANNUS						
201 Meal mechanical extracted	5-02-009	91.7	—	—	—	6.8
202		100.0	—	—	—	7.4
FISH, SARDINE. CLUPEA SPP-SARDINOPS SPP						
203 Meal mechanical extracted	5-02-015	92.9	—	—	—	—
204		100.0	—	—	—	—
FISH, WHITE. GADIDAE (FAMILY)-LOPHIIDAE (FAMILY)-RAJIDAE (FAMILY)						
205 Meal mechanical extracted	5-02-025	91.2	—	—	—	8.9
206		100.0	—	—	—	9.8

	VIT K mg/kg	BIO TIN mg/kg	CHOL INE mg/kg	FOL ATE mg/kg	NIA CIN mg/kg	PANT OTHE mg/kg	RIBO FLVN mg/kg	THIA MINE mg/kg	VIT B6 mg/kg	VIT B12 ug/kg
143	—	—	790.0	—	22.2	14.0	2.1	1.47	—	—
144	—	—	867.0	—	24.4	15.4	2.3	1.61	—	—
147	—	—	—	—	26.9	12.7	6.2	—	—	—
148	—	—	—	—	40.3	19.0	9.3	—	—	—
149	—	—	—	—	—	—	4.4	1.98	—	—
150	—	—	—	—	—	—	19.6	8.82	—	—
151	—	—	—	—	—	—	15.1	4.21	—	—
152	—	—	—	—	—	—	17.2	4.80	—	—
157	—	—	—	—	—	—	4.2	—	—	—
158	—	—	—	—	—	—	24.1	—	—	—
159	—	—	—	—	9.8	1.0	15.2	3.74	—	—
160	—	—	—	—	11.0	1.1	17.0	4.20	—	—
165	—	0.093	—	—	37.7	9.9	15.7	1.97	—	—
166	—	0.105	—	—	42.6	11.2	17.8	2.22	—	—
167	—	—	1046.9	1.08	27.3	6.1	3.3	0.82	—	—
168	—	—	1143.3	1.18	29.8	6.6	3.6	0.89	—	—
169	—	—	1089.4	0.30	23.8	6.5	3.5	—	4.36	—
170	—	—	1195.7	0.33	26.1	7.2	3.8	—	4.78	—
173	—	—	—	—	—	—	3.7	—	—	—
174	—	—	—	—	—	—	4.1	—	—	—
177	—	—	—	—	—	—	3.1	—	—	—
178	—	—	—	—	—	—	3.3	—	—	—
179	—	0.915	2755.5	2.45	35.2	10.2	5.2	7.07	5.01	—
180	—	0.988	2974.3	2.65	38.0	11.0	5.6	7.63	5.41	—
181	—	0.555	2782.9	2.55	40.9	13.7	4.7	7.32	5.41	—
182	—	0.610	3058.0	2.81	44.9	15.1	5.2	8.04	5.95	—
183	—	0.443	2961.6	0.93	44.8	14.3	4.9	8.20	6.29	—
184	—	0.477	3183.6	1.00	48.2	15.4	5.3	8.82	6.76	—
189	—	—	—	—	—	—	2.4	3.38	—	—
190	—	—	—	—	—	—	8.6	11.90	—	—
191	—	—	—	—	—	—	—	—	—	—
192	—	—	—	—	—	—	—	—	—	—
193	—	0.141	3314.9	0.22	175.9	35.6	12.7	5.53	12.20	06.382
194	—	0.279	6573.7	0.44	348.9	70.7	25.2	10.96	24.19	04.202
195	—	0.395	5525.3	0.57	255.8	50.4	13.5	7.39	19.71	85.930
196	—	0.425	5953.4	0.61	275.6	54.3	14.6	7.96	21.23	23.578
197	—	0.196	3700.4	0.16	80.5	10.0	7.3	0.52	4.71	14.478
198	—	0.213	4023.4	0.17	87.6	10.9	8.0	0.57	5.12	33.198
199	2.2	0.488	5262.4	0.34	85.2	17.3	9.7	0.38	4.65	28.794
200	2.3	0.532	5730.2	0.37	92.8	18.8	10.6	0.41	5.06	66.911
201	—	0.179	3114.5	0.15	54.6	8.6	4.8	0.57	3.81	22.128
202	—	0.196	3398.0	0.17	59.6	9.4	5.3	0.62	4.15	33.244
203	—	0.101	3268.6	—	74.8	10.9	5.4	0.32	—	37.430
204	—	0.109	3517.7	—	80.5	11.8	5.8	0.34	—	55.525
205	—	0.080	4305.2	0.35	59.4	9.9	9.1	1.68	5.32	84.514
206	—	0.087	4718.6	0.38	65.1	10.9	10.0	1.84	5.83	92.629

TABLE 3–2D. (continued). Composition of Feeds—Vitamin Values

		IFN	DRY MATT %	CARO TENE mg/kg	VIT D2D3 IU/kg	VIT D3 ICU/kg	VIT E mg/kg
	FLAX, COMMON. LINUM USITATISSIMUM						
207	Seed screenings	4-02-056	91.2	—	—	—	—
208			100.0	—	—	—	—
209	Seeds, meal mechanical extracted	5-02-045	90.8	—	—	—	7.8
210			100.0	—	—	—	8.6
	GRAPE. VITIS SPP						
225	Pomace, dehydrated (marc)	1-02-208	90.4	—	—	·	—
226			100.0	—	—	—	—
	GRASS–LEGUME						
229	Silage	3-02-303	31.1	—	0.1	—	—
230			100.0	—	0.3	—	—
	HOMINY (SEE MAIZE)						
	LEGUME (SEE GRASS–LEGUME)						
	LESPEDEZA, CHINESE. LESPEDEZA CUNEATA						
231	Hay, sun-cured	1-02-607	90.4	—	—	—	—
232			100.0	—	—	—	—
	LINSEED (SEE FLAX)						
	LUCERNE (SEE ALFALFA)						
	MAIZE. ZEA MAYS						
239	Aerial part without ears without husks, sun-cured (stover)	1-02-776	85.2	—	0.9	—	—
240	(straw)		100.0	—	1.1	—	—
241	Cobs, ground	1-02-782	89.8	—	—	—	—
242			100.0	—	—	—	—
243	Distillers grains, dehydrated	5-02-842	93.5	—	—	—	—
244			100.0	—	—	—	—
245	Distillers grains with solubles, dehydrated	5-02-843	91.8	—	0.6	—	39.8
246			100.0	—	0.6	—	43.4
247	Distillers solubles, dehydrated	5-02-844	92.9	—	—	—	45.9
248			100.0	—	—	—	49.4
251	Ears, ground (maize and cob meal)	4-02-849	86.5	—	—	—	17.5
252			100.0	—	—	—	20.2
253	Gluten, meal	5-02-900	91.3	—	—	—	29.2
254			100.0	—	—	—	32.0
255	Gluten with bran (maize gluten feed)	5-02-903	89.9	—	—	—	12.1
256			100.0	—	—	—	13.5
257	Grits by-product (hominy feed)	4-02-887	90.2	—	—	—	—
258			100.0	—	—	—	—
259	Silage, well eared	3-02-823	34.1	—	0.0	—	—
260			100.0	—	0.1	—	—
	MAIZE, DENT WHITE. ZEA MAYS INDENTATA						
265	Grits by-product (hominy feed)	4-02-990	90.0	—	—	—	—
266			100.0	—	—	—	—
	MAIZE, DENT YELLOW. ZEA MAYS INDENTATA						
267	Grain	4-02-935	88.0	—	—	—	20.9
268			100.0	—	—	—	23.8
269	Grain, grade 2 69.5 kg/hl (54 lb/bushel)	4-02-931	87.3	—	—	—	21.6
270			100.0	—	—	—	24.7
	MANURE (SEE CATTLE; POULTRY)						
	MEAT (SEE ANIMAL)						
	MILK (SEE CATTLE)						
	MILLET. SETARIA SPP						
279	Grain	4-03-098	89.9	—	—	—	—
280			100.0	—	—	—	—
	MOLASSES (SEE BEET; SUGARCANE)						

	VIT K mg/kg	BIO TIN mg/kg	CHOL INE mg/kg	FOL ATE mg/kg	NIA CIN mg/kg	PANT OTHE mg/kg	RIBO FLVN mg/kg	THIA MINE mg/kg	VIT B6 mg/kg	VIT B12 ug/kg
207	—	—	—	—	—	—	—	—	—	—
208	—	—	—	—	—	—	—	—	—	—
209	—	0.330	1782.1	2.84	37.5	14.4	3.2	4.20	5.50	—
210	—	0.363	1961.8	3.13	41.3	15.8	3.5	4.62	6.05	—
225	—	—	252.5	—	18.2	3.1	2.2	—	—	—
226	—	—	279.2	—	20.1	3.4	2.4	—	—	—
229	—	—	—	—	14.2	—	—	—	—	—
230	—	—	—	—	45.6	—	—	—	—	—
231	—	—	—	—	—	—	8.7	—	—	—
232	—	—	—	—	—	—	9.7	—	—	—
239	—	—	—	—	—	—	—	—	—	—
240	—	—	—	—	—	—	—	—	—	—
241	—	—	—	—	7.0	3.8	1.0	0.90	—	—
242	—	—	—	—	7.8	4.2	1.1	1.00	—	—
243	—	0.414	1113.1	1.00	38.3	11.3	5.0	1.77	4.22	0.253
244	—	0.443	1190.9	1.06	41.0	12.1	5.3	1.90	4.51	0.270
245	—	0.685	2581.6	0.91	73.4	13.8	8.5	3.01	4.74	1.513
246	—	0.747	2813.0	0.99	80.0	15.1	9.2	3.28	5.17	1.648
247	—	1.491	4750.9	1.34	123.6	23.3	15.1	6.76	9.41	4.187
248	—	1.606	5116.2	1.45	133.2	25.0	16.3	7.27	10.14	4.509
251	—	0.030	356.6	0.24	16.9	4.2	0.9	2.89	5.97	—
252	—	0.035	412.1	0.28	19.6	4.8	1.0	3.34	6.89	—
253	—	0.190	360.0	0.30	49.8	10.0	1.5	0.22	7.97	—
254	—	0.209	394.3	0.33	54.6	10.9	1.6	0.24	8.73	—
255	—	0.309	1587.9	0.27	70.5	13.6	2.2	1.99	13.93	—
256	—	0.343	1765.8	0.30	78.4	15.1	2.5	2.21	15.49	—
257	—	0.132	1154.6	0.31	46.9	8.2	2.1	8.05	10.95	—
258	—	0.146	1279.8	0.34	52.0	9.1	2.4	8.92	12.14	—
259	—	—	—	—	—	—	—	—	—	—
260	—	—	—	—	—	—	—	—	—	—
265	—	0.133	958.5	—	46.4	7.3	2.2	9.72	13.25	—
266	—	0.147	1065.6	—	51.6	8.1	2.5	10.80	14.73	—
267	0.2	0.068	495.9	0.31	22.5	5.1	1.1	3.73	6.16	—
268	0.2	0.077	563.8	0.35	25.6	5.8	1.2	4.24	7.01	—
269	—	0.059	568.1	0.35	24.5	3.9	1.3	3.47	6.87	—
270	—	0.067	650.4	0.40	28.0	4.5	1.5	3.97	7.86	—
279	—	—	746.9	0.22	49.0	8.8	1.8	6.58	—	—
280	—	—	830.8	0.25	54.5	9.8	2.0	7.32	—	—

TABLE 3–2D. (continued). Composition of Feeds—Vitamin Values

		IFN	DRY MATT %	CARO TENE mg/kg	VIT D2D3 IU/kg	VIT D3 ICU/kg	VIT E mg/kg
	OATS. AVENA SATIVA						
291	Cereal by-product, less than 4% fiber (feeding oat meal) (oat	4-03-303	90.7	—	—	—	23.7
292	middlings)		100.0	—	—	—	26.1
293	Grain	4-03-309	89.2	—	—	—	15.0
294			100.0	—	—	—	16.8
295	Grain, Pacific coast	4-07-999	90.9	—	—	—	20.2
296			100.0	—	—	—	22.2
297	Groats	4-03-331	89.6	—	—	—	14.8
298			100.0	—	—	—	16.5
299	Hay, sun-cured	1-03-280	90.7	—	1.4	—	—
300			100.0	—	1.5	—	—
301	Hulls	1-03-281	92.4	—	—	—	—
302			100.0	—	—	—	—
303	Silage	3-03-298	30.5	—	—	—	—
304			100.0	—	—	—	—
305	Straw	1-03-283	92.2	—	0.6	—	—
306			100.0	—	0.7	—	—
	ORCHARDGRASS. DACTYLIS GLOMERATA						
313	Hay, sun-cured	1-03-438	89.6	—	—	—	171.3
314			100.0	—	—	—	191.1
	PEA. PISUM SPP						
325	Seeds	5-03-600	89.1	—	—	—	3.0
326			100.0	—	—	—	3.3
	PEANUT. ARACHIS HYPOGAEA						
327	Hay, sun-cured	1-03-619	90.7	—	—	—	—
328			100.0	—	—	—	—
329	Seeds without coats, meal mechanical extracted (peanut meal)	5-03-649	92.6	—	—	—	2.4
330			100.0	—	—	—	2.6
331	Seeds without coats, meal solvent extracted (peanut meal)	5-03-650	92.4	—	—	—	2.9
332			100.0	—	—	—	3.2
	POTATO. SOLANUM TUBEROSUM						
337	Tubers, dehydrated	4-07-850	91.1	—	—	—	—
338			100.0	—	—	—	—
339	Tubers, fresh	4-03-787	21.9	—	—	—	—
340			100.0	—	—	—	—
	POULTRY						
343	By-product, meal rendered (viscera with feet with heads)	5-03-798	93.8	—	—	—	2.2
344			100.0	—	—	—	2.4
345	Feathers, meal hydrolyzed	5-03-795	92.9	—	—	—	—
346			100.0	—	—	—	—
347	Manure, dehydrated	5-14-015	90.6	—	—	—	—
348			100.0	—	—	—	—
	RAPE. BRASSICA NAPUS						
359	Seeds, meal mechanical extracted	5-03-870	92.0	—	—	—	18.8
360			100.0	—	—	—	20.4
361	Seeds, meal solvent extracted	5-03-871	91.2	—	—	—	—
362			100.0	—	—	—	—
	RAPE, SUMMER. BRASSICA NAPUS ANNUA						
363	Seeds, meal boiled mechanical extracted	5-08-136	94.0	—	—	—	—
364			100.0	—	—	—	—
365	Seeds, meal prepressed solvent extracted	5-08-135	91.6	—	—	—	—
366			100.0	—	—	—	—
	RAPE, TURNIP. BRASSICA CAMPESTRIS						
367	Seeds, meal mechanical extracted	5-07-871	93.7	—	—	—	—
368			100.0	—	—	—	—
369	Seeds, meal solvent extracted	5-07-870	90.9	—	—	—	—
370			100.0	—	—	—	—

	VIT K mg/kg	BIO TIN mg/kg	CHOL INE mg/kg	FOL ATE mg/kg	NIA CIN mg/kg	PANT OTHE mg/kg	RIBO FLVN mg/kg	THIA MINE mg/kg	VIT B6 mg/kg	VIT B12 ug/kg
291	—	0.219	1157.5	0.46	23.9	17.6	1.7	7.00	—	—
292	—	0.242	1276.7	0.51	26.3	19.4	1.9	7.72	—	—
293	—	0.265	967.6	0.39	14.0	9.7	1.4	6.13	2.61	—
294	—	0.297	1084.4	0.44	15.7	10.9	1.6	6.87	2.93	—
295	—	—	916.9	—	14.4	11.7	1.2	—	—	—
296	—	—	1008.7	—	15.8	12.8	1.3	—	—	—
297	—	—	1131.6	0.51	9.6	13.8	1.2	6.49	1.00	—
298	—	—	1263.7	0.56	10.7	15.4	1.3	7.25	1.12	—
299	—	—	—	—	—	—	—	—	—	—
300	—	—	—	—	—	—	—	—	—	—
301	—	—	260.0	0.96	9.2	3.1	1.5	0.61	2.19	—
302	—	—	281.3	1.04	10.0	3.4	1.7	0.66	2.37	—
303	—	—	302.0	—	—	—	—	—	—	—
304	—	—	991.0	—	—	—	—	—	—	—
305	—	—	—	—	—	—	—	—	—	—
306	—	—	—	—	—	—	—	—	—	—
313	—	—	—	—	—	—	6.1	2.57	—	—
314	—	—	—	—	—	—	6.8	2.87	—	—
325	—	0.178	545.8	0.22	30.5	27.7	1.8	4.61	1.96	—
326	—	0.200	612.4	0.25	34.3	31.1	2.0	5.17	2.20	—
327	—	—	—	—	—	—	8.8	—	—	—
328	—	—	—	—	—	—	9.7	—	—	—
329	—	0.330	1974.8	0.66	172.7	47.6	9.1	5.72	6.12	—
330	—	0.356	2132.5	0.71	186.4	51.4	9.8	6.18	6.61	—
331	—	—	1893.8	—	177.5	36.8	5.2	—	5.95	—
332	—	—	2048.9	—	192.0	39.8	5.7	—	6.43	—
337	—	0.101	2622.5	0.61	33.3	20.0	1.0	—	14.12	—
338	—	0.111	2879.2	0.66	36.5	22.0	1.1	—	15.50	—
339	—	—	—	0.17	17.3	—	0.5	0.86	1.81	—
340	—	—	—	0.77	78.6	—	2.1	3.92	8.26	—
343	—	0.089	6052.0	0.51	53.5	12.4	10.6	0.23	4.43	22.610
344	—	0.094	6450.5	0.54	57.1	13.2	11.3	0.24	4.72	24.104
345	—	0.044	894.0	0.22	21.1	8.9	2.0	0.11	4.39	80.356
346	—	0.047	962.3	0.23	22.7	9.6	2.2	0.12	4.72	86.495
347	—	—	—	—	19.3	—	11.8	—	—	21.205
348	—	—	—	—	21.3	—	13.0	—	—	23.402
359	—	—	6532.2	—	154.9	9.0	3.0	1.77	—	—
360	—	—	7102.9	—	168.5	9.8	3.3	1.92	—	—
361	—	—	6633.3	—	146.8	8.0	5.8	1.59	7.25	—
362	—	—	7277.3	—	161.1	8.8	6.4	1.74	7.95	—
363	—	—	70.0	—	167.0	9.9	4.2	1.90	—	—
364	—	—	74.5	—	177.7	10.5	4.5	2.02	—	—
365	—	0.904	6729.4	2.31	160.2	9.5	3.7	5.22	—	—
366	—	0.987	7346.5	2.52	174.9	10.4	4.1	5.70	—	—
367	—	—	6419.7	—	151.5	8.6	3.3	1.69	—	—
368	—	—	6850.1	—	161.7	9.1	3.5	1.81	—	—
369	—	0.818	6383.0	2.09	152.2	8.9	3.5	3.26	—	—
370	—	0.900	7022.7	2.30	167.4	9.8	3.9	3.59	—	—

TABLE 3–2D. (continued). Composition of Feeds—Vitamin Values

		IFN	DRY MATT %	CARO TENE mg/kg	VIT D2D3 IU/kg	VIT D3 ICU/kg	VIT E mg/kg
	RICE. ORYZA SATIVA						
375	Bran with germs (rice, bran)	4-03-928	90.5	—	—	—	85.3
376			100.0	—	—	—	94.2
377	Grain, ground	4-03-938	89.0	—	—	—	14.0
378			100.0	—	—	—	15.7
379	Groats, polished	4-03-942	88.6	—	—	—	3.5
380			100.0	—	—	—	4.0
381	Groats, polished broken	4-03-932	88.5	—	—	—	—
382			100.0	—	—	—	—
383	Hulls	1-08-075	91.9	—	—	—	7.5
384			100.0	—	—	—	8.1
385	Polishings	4-03-943	90.2	—	—	—	90.2
386			100.0	—	—	—	100.0
	RYE. SECALE CEREALE						
391	Distillers grains, dehydrated	5-04-023	91.9	—	—	—	—
392			100.0	—	—	—	—
393	Flour by-product, less than 8.5% fiber (rye middlings)	4-04-031	89.5	—	—	—	—
394			100.0	—	—	—	—
395	Fresh, early vegetative	2-04-013	15.5	—	—	—	—
396			100.0	—	—	—	—
397	Grain	4-04-047	87.5	—	—	—	14.5
398			100.0	—	—	—	16.6
	RYEGRASS, PERENNIAL. LOLIUM PERENNE						
407	Fresh	2-04-086	26.6	—	—	—	46.2
408			100.0	—	—	—	173.5
409	Hay, sun-cured	1-04-077	86.6	—	—	—	182.7
410			100.0	—	—	—	211.0
	SAFFLOWER. CARTHAMUS TINCTORIUS						
415	Seeds, meal mechanical extracted	5-04-109	91.6	—	—	—	0.9
416			100.0	—	—	—	1.0
417	Seeds, meal solvent extracted	5-04-110	91.7	—	—	—	0.8
418			100.0	—	—	—	0.9
419	Seeds without hulls, meal solvent extracted	5-07-959	91.0	—	—	—	0.7
420			100.0	—	—	—	0.8
	SCREENINGS—SEE BARLEY; CEREALS; WHEAT						
	SEAWEED, KELP. LAMINARIALES (ORDER)-FUCALES (ORDER)						
443	Whole, meal	1-08-073	90.9	—	—	—	—
444			100.0	—	—	—	—
	SESAME. SESAMUM INDICUM						
447	Seeds, meal mechanical extracted	5-04-220	92.7	—	—	—	—
448			100.0	—	—	—	—
	SKIMMILK (SEE CATTLE)						
	SORGHUM. SORGHUM BICOLOR						
451	Distillers grains, dehydrated	5-04-374	93.8	—	—	—	—
452			100.0	—	—	—	—
453	Grain	4-04-383	90.1	—	—	—	—
454			100.0	—	—	—	—
455	Grain, 9-12% protein	4-08-139	89.4	—	—	—	1.3
456			100.0	—	—	—	1.5
	SORGHUM, KAFIR. SORGHUM BICOLOR CAFFRORUM						
459	Grain	4-04-428	89.1	—	—	—	—
460			100.0	—	—	—	—
	SORGHUM, MILO. SORGHUM BICOLOR SUBGLABRESCENS						
461	Grain	4-04-444	88.5	—	0.0	—	12.2
462			100.0	—	0.0	—	13.8
463	Heads	4-04-446	89.9	—	—	—	—
464			100.0	—	—	—	—

	VIT K mg/kg	BIO TIN mg/kg	CHOL INE mg/kg	FOL ATE mg/kg	NIA CIN mg/kg	PANT OTHE mg/kg	RIBO FLVN mg/kg	THIA MINE mg/kg	VIT B6 mg/kg	VIT B12 ug/kg
375	—	0.422	1243.6	1.60	305.9	25.0	2.6	21.94	16.21	5.144
376	—	0.466	1373.7	1.77	337.9	27.6	2.8	24.23	17.90	5.682
377	—	—	926.5	0.25	40.0	7.1	0.7	—	—	—
378	—	—	1041.5	0.28	45.0	8.0	0.8	—	—	—
379	—	—	901.2	0.15	15.2	3.5	0.6	0.65	0.39	—
380	—	—	1017.4	0.17	17.2	3.9	0.6	0.74	0.45	—
381	—	—	877.0	—	22.6	3.3	0.4	1.39	—	—
382	—	—	990.7	—	25.5	3.7	0.5	1.57	—	—
383	—	—	—	—	28.1	7.9	0.5	2.21	0.07	—
384	—	—	—	—	30.6	8.6	0.6	2.41	0.08	—
385	—	0.616	1248.1	—	505.6	51.6	1.8	19.95	27.88	—
386	—	0.683	1383.3	—	560.4	57.2	2.0	22.11	30.90	—
391	—	—	—	—	16.8	5.2	3.3	1.31	—	—
392	—	—	—	—	18.3	5.7	3.6	1.43	—	—
393	—	—	—	—	16.9	23.0	2.4	3.29	—	—
394	—	—	—	—	18.9	25.7	2.7	3.67	—	—
395	—	—	—	—	—	—	—	—	—	—
396	—	—	—	—	—	—	—	—	—	—
397	—	0.053	419.1	0.58	14.1	7.2	1.8	4.51	2.97	—
398	—	0.061	479.0	0.66	16.1	8.3	2.0	5.16	3.40	—
407	—	—	—	—	—	—	—	—	—	—
408	—	—	—	—	—	—	—	—	—	—
409	—	—	—	—	—	—	—	—	—	—
410	—	—	—	—	—	—	—	—	—	—
415	—	1.394	1639.2	0.44	85.9	4.0	13.4	—	—	—
416	—	1.521	1789.2	0.48	93.8	4.4	14.6	—	—	—
417	—	1.426	814.8	0.45	61.9	26.2	2.0	—	—	—
418	—	1.555	888.8	0.49	67.5	28.5	2.2	—	—	—
419	—	1.709	3246.9	1.61	22.2	39.4	2.5	4.62	11.83	—
420	—	1.878	3567.5	1.77	24.4	43.3	2.7	5.08	13.00	—
443	—	—	273.8	—	22.9	7.0	2.5	—	—	—
444	—	—	301.2	—	25.2	7.7	2.7	—	—	—
447	—	—	1534.1	—	18.8	5.9	3.4	2.80	12.46	—
448	—	—	1654.8	—	20.3	6.4	3.6	3.02	13.44	—
451	—	0.308	805.0	—	—	—	—	—	8.36	—
452	—	0.328	858.3	—	—	—	—	—	8.91	—
453	—	0.261	692.5	0.22	46.6	10.2	1.2	4.52	5.40	—
454	—	0.290	768.9	0.24	51.8	11.3	1.4	5.02	6.00	—
455	—	0.288	765.7	0.22	48.2	12.8	1.3	4.43	4.62	—
456	—	0.322	856.9	0.25	54.0	14.4	1.5	4.95	5.17	—
459	—	0.236	438.9	0.20	37.9	12.0	1.2	3.82	6.68	—
460	—	0.265	492.7	0.22	42.6	13.4	1.4	4.28	7.50	—
461	0.2	0.253	647.4	0.21	36.0	11.2	1.1	4.34	4.92	—
462	0.2	0.286	731.5	0.23	40.7	12.7	1.3	4.90	5.56	—
463	—	—	—	—	—	—	2.0	—	—	—
464	—	—	—	—	—	—	2.2	—	—	—

TABLE 3–2D. (continued). Composition of Feeds—Vitamin Values

		IFN	DRY MATT %	CARO TENE mg/kg	VIT D2D3 IU/kg	VIT D3 ICU/kg	VIT E mg/kg
	SOYBEAN. GLYCINE MAX						
475	Hay, sun-cured	1-04-558	89.1	—	0.7	—	26.3
476			100.0	—	0.8	—	29.5
477	Seed coats (hulls)	1-04-560	90.3	—	—	—	6.6
478			100.0	—	—	—	7.3
479	Seeds	5-04-610	92.0	—	—	—	33.7
480			100.0	—	—	—	36.6
481	Seeds, heat processed	5-04-597	92.6	—	—	—	—
482			100.0	—	—	—	—
483	Seeds, meal mechanical extracted	5-04-600	90.0	—	—	—	6.5
484			100.0	—	—	—	7.3
485	Seeds, meal solvent extracted	5-04-604	89.6	—	—	—	2.4
486			100.0	—	—	—	2.7
487	Seeds without hulls, meal solvent extracted	5-04-612	89.9	—	—	—	3.3
488			100.0	—	—	—	3.7
	SUGARCANE. SACCHARUM OFFICINARUM						
495	Molasses, dehydrated	4-04-695	94.4	—		—	5.2
496			100.0	—		—	5.5
497	Molasses, more than 46% invert sugars more than 79.5 degrees	4-04-696	74.3	—	—	—	5.4
498	brix		100.0	—	—	—	7.3
	SUNFLOWER, COMMON. HELIANTHUS ANNUUS						
499	Seeds without hulls, meal mechanical extracted	5-04-738	92.7	—	—	—	—
500			100.0	—	—	—	—
501	Seeds without hulls, meal solvent extracted	5-04-739	92.5	—	—	—	11.1
502			100.0	—	—	—	12.0
	SWEETCLOVER, YELLOW. MELILOTUS OFFICINALIS						
503	Hay, sun-cured	1-04-754	88.5	—	1.7	—	—
504			100.0	—	1.9	—	—
	TANKANGE—SEE ANIMAL						
	TIMOTHY. PHLEUM PRATENSE						
511	Hay, sun-cured, early bloom	1-04-882	89.1	—	—	—	11.6
512			100.0	—	—	—	13.0
513	Hay, sun-cured, midbloom	1-04-883	88.9	—	1.8	—	—
514			100.0	—	2.0	—	—
	TOMATO. LYCOPERSICON ESCULENTUM						
519	Pomace, dehydrated	5-05-041	92.2	—	—	—	—
520			100.0	—	—	—	—
	TREFOIL, BIRDSFOOT. LOTUS CORNICULATUS						
523	Hay, sun-cured	1-05-044	90.6	—	1.4	—	—
524			100.0	—	1.5	—	—
	TRITICALE. TRITICALE HEXAPLOIDE						
525	Grain	4-20-362	89.2	—	—	—	—
526			100.0	—	—	—	—
	TURNIP. BRASSICA RAPA RAPA						
527	Roots, fresh	4-05-067	9.2	—	—	—	—
528			100.0	—	—	—	—

	VIT K mg/kg	BIO TIN mg/kg	CHOL INE mg/kg	FOL ATE mg/kg	NIA CIN mg/kg	PANT OTHE mg/kg	RIBO FLVN mg/kg	THIA MINE mg/kg	VIT B6 mg/kg	VIT B12 ug/kg
475	—	—	—	—	—	—	—	—	—	—
476	—	—	—	—	—	—	—	—	—	—
477	—	—	586.3	—	24.8	13.4	3.6	1.59	1.70	—
478	—	—	649.4	—	27.4	14.8	4.0	1.76	1.88	—
479	—	0.383	2931.1	—	22.5	16.0	2.9	11.26	11.05	—
480	—	0.416	3184.3	—	24.5	17.4	3.2	12.24	12.00	—
481	—	0.294	2489.7	3.62	22.6	16.1	2.7	6.11	—	—
482	—	0.318	2688.9	3.91	24.4	17.4	2.9	6.60	—	—
483	—	0.325	2624.1	6.39	30.7	14.3	3.4	3.89	7.22	—
484	—	0.362	2915.7	7.10	34.2	15.8	3.8	4.32	8.02	—
485	—	0.323	2619.6	0.55	27.7	16.3	2.9	5.98	6.00	—
486	—	0.361	2923.4	0.61	30.9	18.2	3.2	6.68	6.70	—
487	—	0.322	2746.5	0.74	21.5	14.8	2.9	3.10	4.92	—
488	—	0.359	3053.9	0.82	23.9	16.4	3.3	3.45	5.47	—
495	—	—	—	—	—	—	—	—	—	—
496	—	—	—	—	—	—	—	—	—	—
497	—	0.686	763.4	0.11	36.4	37.4	2.8	0.86	4.21	—
498	—	0.923	1027.1	0.15	49.0	50.3	3.8	1.16	5.67	—
499	—	—	2423.5	—	137.8	—	2.5	—	—	—
500	—	—	2615.0	—	148.7	—	2.6	—	—	—
501	—	—	3627.7	—	242.1	40.6	3.5	—	13.67	—
502	—	—	3923.3	—	261.9	43.9	3.8	—	14.78	—
503	—	—	—	—	—	—	—	—	—	—
504	—	—	—	—	—	—	—	—	—	—
511	—	—	—	—	—	—	—	—	—	—
512	—	—	—	—	—	—	—	—	—	—
513	—	—	—	—	—	—	—	—	—	—
514	—	—	—	—	—	—	—	—	—	—
519	—	—	—	—	—	—	6.1	11.34	—	—
520	—	—	—	—	—	—	6.7	12.30	—	—
523	—	—	—	—	—	—	14.6	6.16	—	—
524	—	—	—	—	—	—	16.1	6.80	—	—
525	—	0.058	458.7	0.60	14.7	6.8	2.3	8.25	4.21	—
526	—	0.065	514.4	0.67	16.5	7.6	2.6	9.25	4.73	—
527	—	—	92.4	0.26	6.7	1.7	0.4	0.87	—	—
528	—	—	1009.0	2.84	73.5	19.0	4.7	9.48	—	—

TABLE 3–2D. (continued). Composition of Feeds—Vitamin Values

		IFN	DRY MATT %	CARO TENE mg/kg	VIT D2D3 IU/kg	VIT D3 ICU/kg	VIT E mg/kg
	WHEAT. TRITICUM AESTIVUM						
535	Bran	4-05-190	89.0	—	—	—	14.3
536			100.0	—	—	—	16.0
537	Flour by-product, less than 4% fiber (wheat red dog)	4-05-203	88.3	—	—	—	37.4
538			100.0	—	—	—	42.4
539	Flour by-product, less than 7% fiber (wheat shorts)	4-05-201	88.4	—	—	—	36.0
540			100.0	—	—	—	40.7
541	Flour by-product, less than 9.5% fiber (wheat middlings)	4-05-205	88.9	—	—	—	23.9
542			100.0	—	—	—	26.9
543	Fresh, early vegetative	2-05-176	22.2	—	—	—	—
544			100.0	—	—	—	—
545	Germs, ground	5-05-218	88.4	—	—	—	141.2
546			100.0	—	—	—	159.7
547	Grain	4-05-211	89.0	0.2	—	—	15.8
548			100.0	0.2	—	—	17.8
549	Grain screenings	4-05-216	88.5	—	—	—	—
550			100.0	—	—	—	—
551	Hay, sun-cured	1-05-172	88.7	—	1.4	—	—
552			100.0	—	1.5	—	—
553	Mill run, less than 9.5% fiber	4-05-206	89.9	—	—	—	31.9
554			100.0	—	—	—	35.5
555	Silage, early vegetative	3-05-184	29.9	—	—	—	—
556			100.0	—	—	—	—
557	Straw	1-05-175	91.1	—	0.6	—	—
558			100.0	—	0.7	—	—
	WHEAT, DURUM. TRITICUM DURUM						
559	Grain	4-05-224	87.6	—	—	—	—
560			100.0	—	—	—	—
	WHEAT, HARD RED SPRING. TRITICUM AESTIVUM						
561	Grain	4-05-258	87.6	—	—	—	12.6
562			100.0	—	—	—	14.4
	WHEAT, HARD RED WINTER. TRITICUM AESTIVUM						
563	Grain	4-05-268	88.8	—	—	—	11.1
564			100.0	—	—	—	12.5
	WHEAT, SOFT RED WINTER. TRITICUM AESTIVUM						
565	Grain	4-05-294	88.4	—	—	—	15.6
566			100.0	—	—	—	17.7
	WHEAT, SOFT WHITE WINTER. TRITICUM AESTIVUM						
567	Grain	4-05-337	90.2	—	—	—	30.9
568			100.0	—	—	—	34.2
	WHEATGRASS, CRESTED. AGROPYRON DESERTORUM						
569	Fresh	2-05-429	45.0	—	—	—	—
570			100.0	—	—	—	—
	WHEY (SEE CATTLE)						
	YEAST, BREWERS. SACCHAROMYCES CEREVISIAE						
573	Dehydrated	7-05-527	93.1	—	—	—	2.1
574			100.0	—	—	—	2.3
	YEAST, IRRADIATED. SACCHAROMYCES CEREVISIAE						
575	Dehydrated	7-05-529	93.9	—	—	—	—
576			100.0	—	—	—	—
	YEAST, TORULA. TORULOPSIS UTILIS						
577	Dehydrated	7-05-534	93.0	—	—	—	—
578			100.0	—	—	—	—

	VIT K mg/kg	BIO TIN mg/kg	CHOL INE mg/kg	FOL ATE mg/kg	NIA CIN mg/kg	PANT OTHE mg/kg	RIBO FLVN mg/kg	THIA MINE mg/kg	VIT B6 mg/kg	VIT B12 ug/kg
535	—	0.378	1201.1	1.77	196.7	27.9	3.6	8.36	10.33	—
536	—	0.424	1349.2	1.98	221.0	31.4	4.0	9.39	11.61	—
537	—	0.108	1455.4	0.82	46.0	13.3	2.2	21.85	5.40	—
538	—	0.122	1647.6	0.93	52.1	15.0	2.5	24.73	6.12	—
539	—	—	1698.0	1.51	105.2	21.9	4.1	19.52	—	—
540	—	—	1919.8	1.71	119.0	24.8	4.6	22.07	—	—
541	—	0.242	1228.4	1.24	95.0	17.8	2.0	14.18	9.15	—
542	—	0.272	1381.1	1.39	106.8	20.0	2.3	15.94	10.29	—
543	—	—	—	—	12.6	4.7	6.1	—	—	—
544	—	—	—	—	56.9	21.2	27.6	—	—	—
545	—	0.216	3063.0	2.12	68.0	18.6	6.0	23.14	9.97	—
546	—	0.244	3465.3	2.40	77.0	21.0	6.8	26.18	11.28	—
547	—	0.099	895.4	0.43	58.7	11.5	1.2	4.30	3.74	0.876
548	—	0.111	1006.1	0.49	65.9	12.9	1.4	4.83	4.20	0.984
549	—	—	866.7	0.43	57.9	11.3	0.9	6.33	—	—
550	—	—	979.7	0.49	65.4	12.7	1.0	7.16	—	—
551	—	—	—	—	—	—	15.1	—	—	—
552	—	—	—	—	—	—	17.0	—	—	—
553	—	0.308	1004.7	1.08	115.6	13.7	2.1	15.25	11.09	—
554	—	0.343	1117.9	1.20	128.7	15.2	2.4	16.97	12.33	—
555	—	—	—	—	—	—	—	—	—	—
556	—	—	—	—	—	—	—	—	—	—
557	—	—	—	—	—	—	2.2	—	—	—
558	—	—	—	—	—	—	2.4	—	—	—
559	—	—	—	0.39	44.3	—	1.2	6.42	—	—
560	—	—	—	0.44	50.6	—	1.4	7.32	—	—
561	—	0.080	1009.7	0.43	57.2	9.5	1.4	4.22	5.16	—
562	—	0.092	1152.6	0.49	65.2	10.8	1.6	4.82	5.89	—
563	—	0.110	1006.3	0.38	53.0	10.1	1.3	4.52	3.02	—
564	—	0.124	1133.2	0.43	59.7	11.4	1.5	5.09	3.40	—
565	—	—	891.8	0.41	53.4	10.1	1.5	4.71	3.21	—
566	—	—	1008.9	0.46	60.4	11.4	1.7	5.33	3.63	—
567	—	—	—	—	62.1	11.1	—	—	4.77	—
568	—	—	—	—	68.8	12.3	—	—	5.29	—
569	—	—	972.9	—	—	—	3.9	1.49	—	—
570	—	—	2160.4	—	—	—	8.6	3.31	—	—
573	—	1.040	3847.0	9.69	443.3	81.5	33.6	85.21	36.66	1.064
574	—	1.117	4133.9	10.41	476.4	87.6	36.1	91.56	39.40	1.143
575	—	—	—	—	—	—	18.5	—	—	—
576	—	—	—	—	—	—	19.7	—	—	—
577	—	1.171	2962.7	25.66	509.1	104.4	50.0	6.66	34.48	4.002
578	—	1.259	3184.2	27.58	547.2	112.2	53.8	7.16	37.06	4.301

TABLE 3–2E. Composition of Feeds—Amino Acid Values

		IFN	DRY MATT %	ARG %	CYS %	GLY %	HIS %
	ALFALFA. MEDICAGO SATIVA						
001	Fresh	2-00-196	26.0	0.2	0.1	0.2	0.1
002			100.0	0.9	0.2	0.9	0.4
003	Hay, sun-cured, late vegetative	1-00-054	89.7	0.8	—	0.8	0.4
004			100.0	0.9	—	0.9	0.4
005	Hay, sun-cured, early bloom	1-00-059	90.5	0.7	—	0.7	0.3
006			100.0	0.8	—	0.8	0.4
009	Hay, sun-cured, full bloom	1-00-068	90.9	0.7	—	0.7	0.3
010			100.0	0.7	—	0.8	0.4
013	Meal dehydrated, 15% protein	1-00-022	90.3	0.6	0.2	0.7	0.3
014			100.0	0.6	0.2	0.8	0.3
015	Meal dehydrated, 17% protein	1-00-023	91.8	0.8	0.3	0.8	0.3
016			100.0	0.8	0.3	0.9	0.4
017	Meal dehydrated, 20% protein	1-00-024	91.6	0.9	0.3	1.0	0.4
018			100.0	1.0	0.3	1.1	0.4
	ANIMAL						
029	Blood, spray dehydrated	5-00-381	92.6	3.8	1.2	3.9	5.2
030			100.0	4.1	1.3	4.2	5.6
031	Bone, meal steamed	6-00-400	96.1	1.7	0.1	—	—
032			100.0	1.8	0.1	—	—
033	Meat, meal rendered	5-00-385	93.8	3.6	0.6	7.2	0.9
034			100.0	3.8	0.6	7.7	0.9
035	Meat with bone, meal rendered	5-00-388	93.3	3.5	0.6	6.6	1.0
036			100.0	3.8	0.6	7.0	1.1
037	Tankage, meal rendered	5-00-386	92.0	3.6	0.5	6.4	2.1
038			100.0	3.9	0.5	7.0	2.2
039	Tankage with bone, meal rendered	5-00-387	92.9	2.8	0.3	6.6	1.8
040			100.0	3.0	0.3	7.1	1.9
	BAKERY						
051	Waste, dehydrated (dried bakery product)	4-00-466	91.2	0.5	0.2	0.7	0.2
052			100.0	0.5	0.2	0.8	0.2
	BARLEY. HORDEUM VULGARE						
053	Grain	4-00-549	88.6	0.5	0.2	0.4	0.2
054			100.0	0.6	0.2	0.4	0.3
055	Grain, Pacific coast	4-07-939	88.6	0.4	0.2	0.3	0.2
056			100.0	0.5	0.2	0.3	0.2
057	Grain screenings	4-00-542	88.9	—	—	—	—
058			100.0	—	—	—	—
	BEAN, NAVY. PHASEOLUS VULGARIS						
063	Seeds	5-00-623	89.8	1.3	—	0.7	0.6
064			100.0	1.5	—	0.8	0.7
	BEET, MANGEL. BETA VULGARIS MACRORRHIZA						
065	Roots, fresh	4-00-637	11.0	—	—	—	—
066			100.0	—	—	—	—
	BEET, SUGAR. BETA VULGARIS ALTISSIMA						
073	Pulp, dehydrated	4-00-669	91.0	0.3	0.0	0.3	0.2
074			100.0	0.3	0.0	0.3	0.2
077	Pulp with molasses, dehydrated	4-00-672	91.8	0.3	—	—	—
078			100.0	0.3	—	—	—
	BLOOD (SEE ANIMAL)						
	BONES (SEE ANIMAL)						
	BREWERS GRAINS (SEE CEREALS)						
	BUCKWHEAT, COMMON. FAGOPYRUM SAGITTATUM						
109	Grain	4-00-994	88.4	1.0	0.2	0.7	0.3
110			100.0	1.1	0.2	0.7	0.3
	BUTTERMILK—SEE CATTLE						
	CASEIN						
119	Acid precipitated dehydrated	5-01-162	91.6	3.4	0.3	1.6	2.5
120			100.0	3.7	0.3	1.8	2.8

	ISO	LEU	LYS	MET	PHE	SER	THR	TRY	TYR	VAL
	%	%	%	%	%	%	%	%	%	%
001	0.2	0.4	0.2	0.1	0.3	0.2	0.2	—	0.2	0.3
002	1.0	1.5	1.0	0.3	1.0	0.9	0.9	—	0.6	1.1
003	0.8	1.4	1.0	0.3	0.8	0.8	—	—	0.6	1.0
004	0.9	1.5	1.1	0.3	0.9	0.9	—	—	0.6	1.1
005	0.6	1.1	0.8	0.2	0.7	0.7	0.7	—	0.5	0.9
006	0.7	1.2	0.9	0.2	0.8	0.7	0.8	—	0.5	1.0
009	0.6	1.1	0.8	0.2	0.7	0.6	0.6	—	0.5	0.8
010	0.7	1.2	0.9	0.2	0.8	0.7	0.6	—	0.5	0.9
013	0.6	1.0	0.6	0.2	0.6	0.6	0.6	0.4	0.4	0.7
014	0.7	1.1	0.7	0.2	0.7	0.7	0.6	0.4	0.5	0.8
015	0.8	1.3	0.8	0.3	0.8	0.7	0.7	0.3	0.5	0.9
016	0.9	1.4	0.9	0.3	0.9	0.8	0.8	0.4	0.6	1.0
017	0.9	1.4	0.9	0.3	0.9	0.9	0.8	0.4	0.6	1.1
018	1.0	1.6	1.0	0.3	1.0	1.0	0.9	0.4	0.7	1.2
029	1.0	10.8	7.6	1.1	6.0	4.4	3.8	1.0	2.5	6.9
030	1.0	11.7	8.2	1.2	6.4	4.7	4.1	1.1	2.7	7.5
031	—	—	0.9	0.2	—	—	—	0.1	—	—
032	—	—	0.9	0.2	—	—	—	0.1	—	—
033	1.6	3.1	3.0	0.7	1.7	2.3	1.7	0.3	1.1	2.4
034	1.7	3.3	3.2	0.7	1.9	2.5	1.8	0.4	1.2	2.6
035	1.6	3.1	2.8	0.7	1.8	2.0	1.6	0.3	1.1	2.4
036	1.7	3.3	3.0	0.7	1.9	2.2	1.8	0.3	1.1	2.5
037	1.8	5.1	3.9	0.7	2.6	2.8	2.3	0.7	1.4	3.8
038	2.0	5.5	4.2	0.8	2.8	3.1	2.5	0.7	1.5	4.2
039	1.9	5.3	3.3	0.7	2.3	—	2.2	0.6	—	3.4
040	2.0	5.7	3.6	0.7	2.5	—	2.3	0.7	—	3.7
051	0.4	0.8	0.3	0.2	0.4	0.6	0.4	0.1	0.3	0.5
052	0.5	0.8	0.3	0.2	0.5	0.7	0.5	0.1	0.4	0.5
053	0.4	0.7	0.4	0.2	0.6	0.4	0.4	0.2	0.3	0.5
054	0.5	0.8	0.4	0.2	0.6	0.5	0.4	0.2	0.4	0.6
055	0.4	0.6	0.3	0.1	0.5	0.3	0.3	0.1	0.3	0.5
056	0.5	0.7	0.3	0.2	0.5	0.4	0.3	0.1	0.3	0.5
057	—	—	—	—	—	—	0.4	—	—	—
058	—	—	—	—	—	—	0.4	—	—	—
063	0.8	1.5	1.6	0.5	1.1	1.2	0.9	0.3	0.7	0.9
064	0.9	1.7	1.8	0.6	1.2	1.4	1.0	0.3	0.8	1.0
065	—	—	—	—	—	—	—	—	—	—
066	—	—	—	—	—	—	—	—	—	—
073	0.3	0.6	0.5	0.0	0.3	0.4	0.4	0.1	0.4	0.4
074	0.3	0.6	0.6	0.0	0.3	0.4	0.4	0.1	0.4	0.4
077	—	—	0.6	0.0	—	—	0.2	0.1	—	—
078	—	—	0.7	0.0	—	—	0.3	0.1	—	—
109	0.4	0.7	0.6	0.2	0.5	0.5	0.4	0.2	0.2	0.5
110	0.4	0.7	0.7	0.2	0.5	0.6	0.5	0.2	0.3	0.6
119	5.0	8.5	6.9	2.7	4.5	5.0	3.8	1.2	4.6	6.1
120	5.5	9.2	7.6	2.9	4.9	5.5	4.2	1.3	5.0	6.7

TABLE 3–2E. (continued). Composition of Feeds—Amino Acid Values

		IFN	DRY MATT %	ARG %	CYS %	GLY %	HIS %
	CATTLE. BOS TAURUS						
121	Buttermilk, dehydrated	5-01-160	92.3	1.1	0.4	0.5	0.9
122			100.0	1.2	0.4	0.5	0.9
123	Manure, dehydrated	1-01-190	93.0	0.3	0.2	0.5	0.2
124			100.0	0.3	0.2	0.5	0.2
125	Milk, dehydrated	5-01-167	95.5	0.9	0.4	—	0.7
126			100.0	1.0	0.4	—	0.7
127	Milk, fresh	5-01-168	12.4	—	—	—	—
128			100.0	—	—	—	—
129	Skimmilk, dehydrated	5-01-175	94.1	1.2	0.4	0.3	0.9
130			100.0	1.2	0.5	0.3	0.9
131	Skimmilk, fresh	5-01-170	9.6	—	—	—	—
132			100.0	—	—	—	—
133	Whey, dehydrated	4-01-182	93.3	0.3	0.3	0.4	0.2
134			100.0	0.4	0.3	0.5	0.2
137	Whey low lactose, dehydrated (dried whey product)	4-01-186	93.7	0.6	0.4	0.7	0.3
138			100.0	0.6	0.5	0.8	0.3
	CEREALS						
139	Brewers grains, dehydrated	5-02-141	92.2	1.3	0.3	1.1	0.5
140			100.0	1.4	0.4	1.2	0.6
141	Grain screenings, uncleaned	4-02-153	92.1	0.7	—	0.6	0.3
142			100.0	0.7	—	0.7	0.3
	CITRUS. CITRUS SPP						
143	Pomace without fines, dehydrated (dried citrus pulp)	4-01-237	91.1	0.2	0.1	0.2	0.1
144			100.0	0.3	0.1	0.2	0.1
	COCONUT. COCOS NUCIFERA						
167	Kernels with coats, meal mechanical extracted (copra meal)	5-01-572	91.6	2.3	0.2	1.0	0.3
168			100.0	2.5	0.3	1.1	0.4
169	Kernels with coats, meal solvent extracted (copra meal)	5-01-573	91.1	2.6	0.4	0.9	0.4
170			100.0	2.8	0.4	1.0	0.5
	CORN—SEE MAIZE						
	COTTON. GOSSYPIUM SPP						
177	Seeds, meal mechanical extracted	5-01-609	92.7	3.8	0.7	1.6	1.0
178			100.0	4.1	0.7	1.7	1.0
179	Seeds, meal mechanical extracted, 41% protein	5-01-617	92.6	4.2	0.7	1.9	1.1
180			100.0	4.5	0.8	2.0	1.2
181	Seeds, meal solvent extracted, 41% protein	5-01-621	91.0	4.3	0.8	1.9	1.1
182			100.0	4.7	0.8	2.1	1.2
183	Seeds without hulls, meal prepressed solvent extracted, 50% protein	5-07-874	93.0	4.8	1.0	2.8	1.2
184			100.0	5.2	1.1	3.0	1.3
	COWPEA, COMMON. VIGNA SINENSIS						
185	Hay, sun-cured	1-01-645	90.4	1.1	—	—	0.5
186			100.0	1.2	—	—	0.5
	FISH						
193	Solubles, condensed	5-01-969	50.4	1.6	0.4	3.7	1.4
194			100.0	3.1	0.8	7.4	2.9
195	Solubles, dehydrated	5-01-971	92.8	3.1	0.6	5.8	2.1
196			100.0	3.3	0.7	6.2	2.3
	FISH, ANCHOVETA. ENGRAULIS RINGEN						
197	Meal mechanical extracted	5-01-985	92.0	3.8	0.6	3.7	1.6
198			100.0	4.1	0.7	4.0	1.7
	FISH, HERRING. CLUPEA HARENGUS						
199	Meal mechanical extracted	5-02-000	91.8	4.6	0.7	4.4	1.7
200			100.0	5.1	0.8	4.8	1.8
	FISH, MENHADEN. BREVOORTIA TYRANNUS						
201	Meal mechanical extracted	5-02-009	91.7	3.8	0.5	4.3	1.5
202			100.0	4.2	0.6	4.7	1.6
	FISH, SARDINE. CLUPEA SPP-SARDINOPS SPP						
203	Meal mechanical extracted	5-02-015	92.9	2.7	0.8	4.5	1.8
204			100.0	2.9	0.9	4.8	1.9
	FISH, WHITE. GADIDAE (FAMILY)-LOPHIIDAE (FAMILY)-RAJIDAE (FAMILY)						
205	Meal mechanical extracted	5-02-025	91.2	4.3	0.8	5.5	1.4
206			100.0	4.7	0.8	6.0	1.5

	ISO %	LEU %	LYS %	MET %	PHE %	SER %	THR %	TRY %	TYR %	VAL %
121	2.4	3.2	2.3	0.7	1.5	1.5	1.5	0.5	1.0	2.6
122	2.6	3.5	2.5	0.8	1.6	1.6	1.6	0.5	1.1	2.8
123	0.4	1.0	0.4	0.1	0.3	0.6	0.4	—	0.3	0.5
124	0.4	1.0	0.5	0.1	0.3	0.6	0.4	—	0.3	0.5
125	1.3	2.5	2.2	0.6	1.3	—	1.0	0.4	1.3	1.7
126	1.4	2.7	2.3	0.6	1.4	—	1.1	0.4	1.4	1.8
127	0.3	0.3	0.3	0.1	0.2	—	0.2	0.0	—	0.3
128	2.6	2.0	2.3	0.7	1.3	—	1.3	0.4	—	2.0
129	2.2	3.3	2.5	0.9	1.6	1.7	1.6	0.4	1.1	2.3
130	2.3	3.5	2.7	1.0	1.7	1.8	1.7	0.5	1.2	2.4
131	—	—	0.3	—	—	—	—	—	—	—
132	—	—	2.9	—	—	—	—	—	—	—
133	0.8	1.2	0.9	0.2	0.3	0.5	0.9	0.2	0.2	0.7
134	0.8	1.3	1.0	0.2	0.4	0.5	1.0	0.2	0.3	0.7
137	1.0	1.5	1.4	0.4	0.6	0.6	0.9	0.3	0.5	0.9
138	1.0	1.6	1.5	0.4	0.6	0.6	1.0	0.3	0.5	0.9
139	1.5	2.5	0.9	0.5	1.5	1.3	0.9	0.4	1.2	1.6
140	1.7	2.8	0.9	0.5	1.6	1.4	1.0	0.4	1.3	1.7
141	0.5	0.9	0.4	0.2	0.6	0.7	0.4	—	0.6	0.6
142	0.5	1.0	0.5	0.2	0.6	0.7	0.5	—	0.6	0.6
143	0.1	0.3	0.2	0.1	0.2	0.2	0.1	0.1	0.1	0.3
144	0.2	0.3	0.2	0.1	0.2	0.2	0.2	0.1	0.1	0.3
167	0.9	1.3	0.5	0.3	0.8	0.8	0.6	0.2	0.6	1.0
168	0.9	1.5	0.6	0.3	0.9	0.9	0.7	0.2	0.6	1.1
169	0.7	1.4	0.6	0.3	0.9	0.9	0.7	0.2	0.5	1.0
170	0.8	1.5	0.7	0.4	1.0	1.0	0.7	0.2	0.6	1.1
177	1.2	2.0	1.4	0.6	1.9	1.5	1.1	0.5	1.0	2.2
178	1.3	2.2	1.5	0.6	2.1	1.6	1.2	0.5	1.1	2.3
179	1.4	2.3	1.6	0.6	2.2	1.7	1.3	0.5	1.0	1.9
180	1.5	2.5	1.7	0.6	2.4	1.8	1.4	0.6	1.0	2.0
181	1.5	2.5	1.7	0.6	2.3	1.8	1.4	0.5	1.1	1.9
182	1.6	2.7	1.9	0.6	2.5	2.0	1.5	0.6	1.2	2.1
183	1.9	2.8	1.9	0.8	2.6	—	1.7	0.6	0.8	2.2
184	2.0	3.0	2.1	0.8	2.8	—	1.8	0.7	0.9	2.3
185	1.3	2.0	1.1	0.5	1.3	—	1.1	0.5	—	1.4
186	1.4	2.2	1.2	0.6	1.4	—	1.2	0.6	—	1.6
193	1.0	1.9	1.8	0.7	1.0	1.1	0.9	0.3	0.5	1.2
194	2.1	3.8	3.6	1.4	2.1	2.1	1.7	0.6	1.0	2.5
195	2.1	3.0	3.5	1.2	1.5	2.0	1.4	0.6	0.8	2.1
196	2.2	3.2	3.8	1.3	1.6	2.2	1.5	0.6	0.9	2.3
197	3.1	5.0	5.0	2.0	2.8	2.4	2.8	0.7	2.2	3.5
198	3.4	5.4	5.5	2.2	3.0	2.6	3.0	0.8	2.4	3.8
199	3.2	5.2	5.6	2.1	2.7	2.6	2.9	0.8	2.2	4.4
200	3.4	5.7	6.1	2.3	3.0	2.9	3.1	0.9	2.4	4.7
201	2.6	4.5	4.7	1.7	2.4	2.4	2.5	0.6	1.9	3.0
202	2.9	4.9	5.2	1.9	2.6	2.6	2.8	0.7	2.1	3.3
203	3.3	—	5.9	2.0	2.0	—	2.6	0.5	2.8	4.1
204	3.6	—	6.3	2.2	2.2	—	2.8	0.5	3.0	4.4
205	2.8	4.6	4.8	1.8	2.4	3.4	2.6	0.6	2.2	3.2
206	3.0	5.0	5.2	2.0	2.6	3.7	2.8	0.7	2.4	3.5

TABLE 3–2E. (continued). Composition of Feeds—Amino Acid Values

		IFN	DRY MATT %	ARG %	CYS %	GLY %	HIS %
	FLAX, COMMON. LINUM USITATISSIMUM						
209	Seeds, meal mechanical extracted	5-02-045	90.8	2.8	0.6	1.6	0.6
210			100.0	3.1	0.7	1.8	0.7
	GRAPE. VITIS SPP						
225	Pomace, dehydrated (marc)	1-02-208	90.4	0.7	0.2	0.9	0.3
226			100.0	0.7	0.2	1.0	0.3
	HOMINY (SEE MAIZE)						
	LINSEED (SEE FLAX)						
	LUCERNE (SEE ALFALFA)						
	MAIZE. ZEA MAYS						
243	Distillers grains, dehydrated	5-02-842	93.5	1.0	0.2	0.7	0.6
244			100.0	1.1	0.2	0.8	0.7
245	Distillers grains with solubles, dehydrated	5-02-843	91.8	1.0	0.3	0.6	0.6
246			100.0	1.1	0.3	0.6	0.7
247	Distillers solubles, dehydrated	5-02-844	92.9	1.0	0.4	1.1	0.7
248			100.0	1.1	0.5	1.2	0.7
251	Ears, ground (maize and cob meal)	4-02-849	86.5	0.4	0.1	0.3	0.2
252			100.0	0.4	0.1	0.4	0.2
253	Gluten, meal	5-02-900	91.3	1.4	0.7	1.5	1.0
254			100.0	1.5	0.7	1.7	1.1
255	Gluten with bran (maize gluten feed)	5-02-903	89.9	0.8	0.4	0.8	0.6
256			100.0	0.9	0.5	0.9	0.7
257	Grits by-product (hominy feed)	4-02-887	90.2	0.5	0.1	0.3	0.2
258			100.0	0.5	0.2	0.4	0.2
	MAIZE, DENT WHITE. ZEA MAYS INDENTATA						
265	Grits by-product (hominy feed)	4-02-990	90.0	0.4	0.1	0.3	0.2
266			100.0	0.5	0.1	0.3	0.2
	MAIZE, DENT YELLOW. ZEA MAYS INDENTATA						
267	Grain	4-02-935	88.0	0.4	0.1	0.4	0.3
268			100.0	0.5	0.1	0.4	0.3
269	Grain, grade 2 69.5 kg/hl (54 lb/bushel)	4-02-931	87.3	0.4	0.2	0.3	0.2
270			100.0	0.5	0.2	0.4	0.3
271	Grain, grade 3 66.9 kg/hl (52 lb/bushel)	4-02-932	86.5	0.3	0.1	0.2	0.2
272			100.0	0.4	0.1	0.3	0.2
	MAIZE, DENT YELLOW, OPAQUE 2. ZEA MAYS INDENTATA						
273	Grain	4-28-253	91.0	0.6	0.1	0.5	0.3
274			100.0	0.6	0.2	0.5	0.4
	MANURE—SEE CATTLE; POULTRY						
	MEAT—SEE ANIMAL						
	MILK—SEE CATTLE						
	MILLET. SETARIA SPP						
279	Grain	4-03-098	89.9	0.3	0.1	0.4	0.2
280			100.0	0.4	0.2	0.4	0.2
	OATS. AVENA SATIVA						
291	Cereal by-product, less than 4% fiber (feeding oat meal) (oat middlings)	4-03-303	90.7	0.8	0.3	0.6	0.3
292			100.0	0.9	0.3	0.7	0.3
293	Grain	4-03-309	89.2	0.7	0.2	0.5	0.2
294			100.0	0.8	0.2	0.6	0.2
295	Grain, Pacific coast	4-07-999	90.9	0.6	0.2	0.4	0.2
296			100.0	0.6	0.2	0.4	0.2
297	Groats	4-03-331	89.6	0.9	0.2	0.6	0.3
298			100.0	1.0	0.2	0.7	0.3
301	Hulls	1-03-281	92.4	0.2	0.1	0.2	0.1
302			100.0	0.2	0.1	0.2	0.1
	PEA. PISUM SPP						
325	Seeds	5-03-600	89.1	2.1	0.3	1.0	0.6
326			100.0	2.3	0.3	1.1	0.7
	PEANUT. ARACHIS HYPOGAEA						
329	Seeds without coats, meal mechanical extracted (peanut meal)	5-03-649	92.6	5.1	1.0	2.5	1.0
330			100.0	5.5	1.0	2.7	1.1
331	Seeds without coats, meal solvent extracted (peanut meal)	5-03-650	92.4	5.8	0.5	2.9	1.5
332			100.0	6.3	0.6	3.1	1.6

	ISO	LEU	LYS	MET	PHE	SER	THR	TRY	TYR	VAL
	%	%	%	%	%	%	%	%	%	%
209	1.7	1.9	1.2	0.6	1.4	1.9	1.1	0.5	1.0	1.6
210	1.9	2.1	1.3	0.6	1.5	2.1	1.3	0.6	1.1	1.8
225	0.5	1.6	0.5	0.2	0.5	—	0.4	0.1	0.2	1.1
226	0.6	1.8	0.6	0.2	0.6	—	0.4	0.1	0.2	1.2
243	1.0	3.0	0.8	0.4	1.0	1.0	0.6	0.2	0.8	1.2
244	1.1	3.2	0.8	0.4	1.1	1.1	0.6	0.2	0.9	1.3
245	1.3	2.3	0.7	0.5	1.5	1.2	0.9	0.2	0.7	1.5
246	1.4	2.5	0.8	0.5	1.6	1.3	1.0	0.2	0.8	1.6
247	1.3	2.4	0.9	0.6	1.5	1.2	1.0	0.2	0.9	1.5
248	1.4	2.6	1.0	0.6	1.6	1.3	1.1	0.3	0.9	1.7
251	0.3	0.9	0.2	0.1	0.4	—	0.3	0.1	0.3	0.3
252	0.4	1.0	0.2	0.2	0.4	—	0.3	0.1	0.4	0.4
253	2.2	7.4	0.8	1.0	2.8	1.7	1.4	0.2	1.0	2.2
254	2.5	8.1	0.9	1.1	3.1	1.9	1.6	0.2	1.1	2.4
255	0.9	2.2	0.6	0.4	0.8	0.8	0.8	0.2	0.7	1.1
256	1.0	2.4	0.7	0.4	0.9	0.9	0.9	0.2	0.8	1.2
257	0.4	0.8	0.4	0.2	0.3	—	0.4	0.1	0.5	0.5
258	0.4	0.9	0.4	0.2	0.4	—	0.4	0.1	0.6	0.5
265	0.3	0.8	0.4	0.1	0.3	—	0.3	0.1	0.4	0.4
266	0.4	0.9	0.4	0.1	0.4	—	0.4	0.1	0.4	0.5
267	0.4	1.2	0.3	0.2	0.5	0.5	0.4	0.1	0.3	0.5
268	0.4	1.3	0.3	0.2	0.5	0.6	0.4	0.1	0.3	0.5
269	0.3	1.0	0.3	0.2	0.4	0.4	0.3	0.1	0.3	0.4
270	0.4	1.1	0.3	0.2	0.5	0.5	0.4	0.1	0.4	0.4
271	0.2	1.0	0.2	0.1	0.3	0.4	0.3	0.1	0.2	0.3
272	0.3	1.1	0.3	0.2	0.4	0.4	0.3	0.1	0.3	0.4
273	0.3	0.9	0.4	0.1	0.4	0.4	0.4	—	0.3	0.5
274	0.3	0.9	0.5	0.1	0.5	0.5	0.4	—	0.3	0.5
279	0.5	1.2	0.3	0.3	0.6	—	0.4	0.1	—	0.6
280	0.5	1.3	0.3	0.3	0.6	—	0.4	0.2	—	0.6
291	0.6	1.1	0.5	0.2	0.7	0.7	0.5	0.2	0.7	0.7
292	0.6	1.2	0.6	0.2	0.8	0.8	0.5	0.2	0.8	0.8
293	0.5	0.9	0.4	0.2	0.6	0.5	0.4	0.2	0.4	0.6
294	0.5	1.0	0.4	0.2	0.6	0.6	0.4	0.2	0.5	0.7
295	0.4	0.7	0.3	0.1	0.4	0.4	0.3	0.1	0.7	0.5
296	0.4	0.8	0.4	0.1	0.5	0.4	0.3	0.1	0.8	0.5
297	0.5	1.0	0.6	0.3	0.7	0.6	0.4	0.2	0.5	0.7
298	0.6	1.1	0.6	0.3	0.8	0.6	0.5	0.2	0.5	0.8
301	0.2	0.3	0.2	0.1	0.2	0.3	0.2	0.1	0.1	0.2
302	0.2	0.3	0.2	0.1	0.2	0.3	0.2	0.1	0.2	0.2
325	1.0	1.7	1.7	0.2	1.1	1.1	0.9	0.2	0.7	1.1
326	1.1	1.9	1.9	0.3	1.3	1.2	1.0	0.2	0.8	1.3
329	1.8	3.1	1.7	0.5	2.4	1.4	1.3	0.5	1.6	2.3
330	1.9	3.4	1.8	0.5	2.6	1.6	1.4	0.5	1.7	2.5
331	1.8	3.3	1.4	0.4	2.1	3.1	1.4	0.5	—	2.2
332	2.0	3.5	1.6	0.5	2.3	3.4	1.5	0.5	—	2.3

TABLE 3-2E. (continued). Composition of Feeds—Amino Acid Values

		IFN	DRY MATT %	ARG %	CYS %	GLY %	HIS %
	POTATO. SOLANUM TUBEROSUM						
335	Process residue, dehydrated	4-03-775	90.0	0.2	0.1	0.2	0.1
336			100.0	0.2	0.1	0.2	0.1
337	Tubers, dehydrated	4-07-850	91.1	0.2	0.1	0.2	0.1
338			100.0	0.3	0.1	0.2	0.1
	POULTRY						
343	By-product, meal rendered (viscera with feet with heads)	5-03-798	93.8	4.0	0.8	6.1	1.1
344			100.0	4.3	0.9	6.5	1.2
345	Feathers, meal hydrolyzed	5-03-795	92.9	5.6	3.7	6.3	0.6
346			100.0	6.1	4.0	6.8	0.7
347	Manure, dehydrated	5-14-015	90.6	0.7	0.5	1.3	0.3
348			100.0	0.8	0.5	1.4	0.3
	RAPE. BRASSICA NAPUS						
359	Seeds, meal mechanical extracted	5-03-870	92.0	2.0	0.3	1.9	0.9
360			100.0	2.2	0.3	2.0	1.0
361	Seeds, meal solvent extracted	5-03-871	91.2	2.0	0.3	1.8	1.0
362			100.0	2.2	0.3	2.0	1.1
	RAPE, SUMMER. BRASSICA NAPUS ANNUA						
363	Seeds, meal boiled mechanical extracted	5-08-136	94.0	1.8	—	1.6	0.8
364			100.0	1.9	—	1.8	0.9
365	Seeds, meal prepressed solvent extracted	5-08-135	91.6	2.3	0.5	1.9	1.1
366			100.0	2.5	0.5	2.1	1.2
	RAPE, TURNIP. BRASSICA CAMPESTRIS						
367	Seeds, meal mechanical extracted	5-07-871	93.7	1.8	0.3	1.8	0.9
368			100.0	2.0	0.3	1.9	0.9
369	Seeds, meal solvent extracted	5-07-870	90.9	2.0	0.4	1.8	1.0
370			100.0	2.2	0.5	2.0	1.1
	RICE. ORYZA SATIVA						
375	Bran with germs (rice, bran)	4-03-928	90.5	0.9	0.2	0.8	0.3
376			100.0	1.0	0.2	0.8	0.4
377	Grain, ground	4-03-938	89.0	0.5	0.1	0.6	0.2
378			100.0	0.6	0.1	0.7	0.2
379	Groats, polished	4-03-942	88.6	0.5	0.1	0.4	0.2
380			100.0	0.6	0.1	0.5	0.2
381	Groats, polished broken	4-03-932	88.5	0.6	0.1	0.4	0.2
382			100.0	0.7	0.1	0.4	0.2
383	Hulls	1-08-075	91.9	0.1	0.0	0.1	0.0
384			100.0	0.1	0.0	0.1	0.0
385	Polishings	4-03-943	90.2	0.6	0.1	0.6	0.2
386			100.0	0.6	0.2	0.7	0.2
	RYE. SECALE CEREALE						
393	Flour by-product, less than 8.5% fiber (rye middlings)	4-04-031	89.5	0.8	0.3	0.8	0.3
394			100.0	0.9	0.3	0.8	0.4
397	Grain	4-04-047	87.5	0.5	0.2	0.4	0.2
398			100.0	0.6	0.2	0.5	0.3
	SAFFLOWER. CARTHAMUS TINCTORIUS						
415	Seeds, meal mechanical extracted	5-04-109	91.6	1.4	0.6	1.0	0.4
416			100.0	1.5	0.7	1.1	0.5
417	Seeds, meal solvent extracted	5-04-110	91.7	1.9	0.4	1.1	0.5
418			100.0	2.1	0.4	1.2	0.6
419	Seeds without hulls, meal solvent extracted	5-07-959	91.0	3.7	0.7	2.4	1.0
420			100.0	4.0	0.8	2.6	1.1
	SCREENINGS (SEE BARLEY; CEREALS; WHEAT)						
	SESAME. SESAMUM INDICUM						
447	Seeds, meal mechanical extracted	5-04-220	92.7	4.6	0.6	4.0	1.1
448			100.0	4.9	0.6	4.3	1.2
	SKIMMILK (SEE CATTLE)						
	SORGHUM. SORGHUM BICOLOR						
453	Grain	4-04-383	90.1	0.4	0.2	0.3	0.2
454			100.0	0.4	0.2	0.4	0.3
455	Grain, 9-12% protein	4-08-139	89.4	0.4	0.1	0.4	0.2
456			100.0	0.5	0.2	0.4	0.3

	ISO	LEU	LYS	MET	PHE	SER	THR	TRY	TYR	VAL
	%	%	%	%	%	%	%	%	%	%
335	0.2	0.3	0.3	0.1	0.2	0.2	0.2	0.1	0.1	0.4
336	0.2	0.4	0.3	0.1	0.2	0.3	0.2	0.1	0.2	0.4
337	0.2	0.5	0.4	0.1	0.3	0.2	0.3	0.1	0.2	0.4
338	0.3	0.5	0.4	0.1	0.4	0.3	0.4	0.2	0.3	0.4
343	2.3	4.1	3.1	1.1	2.1	2.8	2.1	0.5	1.8	2.9
344	2.5	4.3	3.3	1.2	2.2	3.0	2.2	0.5	2.0	3.1
345	3.5	6.6	1.8	0.6	3.7	9.4	3.7	0.5	2.4	5.7
346	3.8	7.1	2.0	0.6	4.0	10.1	4.0	0.6	2.6	6.1
347	0.5	0.9	0.6	0.3	0.5	0.7	0.6	0.5	0.5	0.7
348	0.6	1.0	0.6	0.3	0.6	0.8	0.6	0.6	0.5	0.8
359	1.4	2.4	1.7	0.7	1.4	1.5	1.5	0.5	0.8	1.8
360	1.5	2.6	1.8	0.7	1.5	1.6	1.7	0.5	0.9	2.0
361	1.3	2.5	2.0	0.7	1.4	1.6	1.6	0.4	0.8	1.8
362	1.5	2.7	2.2	0.8	1.5	1.7	1.7	0.5	0.9	2.0
363	1.3	2.3	1.5	0.7	1.3	1.4	1.4	0.3	0.8	1.7
364	1.4	2.4	1.6	0.7	1.4	1.5	1.5	0.4	0.8	1.8
365	1.5	2.7	2.2	0.7	1.5	1.7	1.7	0.5	0.8	1.9
366	1.6	2.9	2.4	0.8	1.7	1.8	1.9	0.5	0.9	2.1
367	1.4	2.3	1.6	0.6	1.3	1.4	1.4	0.4	0.8	1.7
368	1.5	2.4	1.7	0.7	1.4	1.5	1.5	0.4	0.9	1.8
369	1.4	2.5	2.1	0.7	1.4	1.6	1.6	0.4	0.8	1.8
370	1.5	2.7	2.3	0.8	1.5	1.7	1.7	0.5	0.9	2.0
375	0.5	0.9	0.6	0.3	0.6	0.7	0.5	0.1	0.5	0.7
376	0.5	1.0	0.6	0.3	0.6	0.8	0.5	0.1	0.6	0.8
377	0.3	0.5	0.2	0.1	0.3	0.5	0.2	0.1	0.6	0.4
378	0.3	0.6	0.3	0.2	0.3	0.6	0.3	0.1	0.7	0.4
379	0.3	0.5	0.2	0.2	0.3	0.3	0.2	0.1	0.2	0.4
380	0.3	0.5	0.3	0.2	0.3	0.3	0.3	0.1	0.3	0.5
381	0.4	0.7	0.3	0.2	0.5	0.4	0.3	0.1	0.4	0.5
382	0.4	0.8	0.3	0.2	0.5	0.5	0.3	0.1	0.5	0.6
383	0.1	0.2	0.1	0.0	0.1	0.1	0.1	0.0	0.0	0.1
384	0.1	0.2	0.1	0.0	0.1	0.1	0.1	0.0	0.0	0.1
385	0.4	0.7	0.5	0.2	0.4	0.5	0.3	0.1	0.5	0.7
386	0.4	0.8	0.6	0.2	0.5	0.5	0.4	0.1	0.5	0.8
393	0.5	0.9	0.7	0.2	0.5	0.6	0.5	—	0.2	0.8
394	0.6	1.1	0.7	0.2	0.6	0.7	0.5	—	0.2	0.9
397	0.5	0.7	0.4	0.2	0.5	0.5	0.3	0.1	0.2	0.6
398	0.5	0.8	0.5	0.2	0.6	0.6	0.4	0.1	0.3	0.6
415	0.6	1.1	0.6	0.3	0.9	0.7	0.6	0.3	0.4	0.9
416	0.7	1.2	0.7	0.4	1.0	0.8	0.6	0.3	0.4	1.0
417	0.3	1.2	0.7	0.3	1.0	—	0.5	0.3	—	1.0
418	0.3	1.3	0.8	0.4	1.1	—	0.6	0.3	—	1.1
419	1.6	2.5	1.3	0.7	1.8	—	1.3	0.6	1.0	2.2
420	1.8	2.7	1.4	0.7	1.9	—	1.4	0.6	1.1	2.4
447	2.0	3.2	1.3	1.4	2.1	2.9	1.6	0.7	1.9	2.3
448	2.1	3.5	1.4	1.5	2.3	3.2	1.7	0.8	2.0	2.5
453	0.4	1.5	0.3	0.1	0.6	0.5	0.4	0.1	0.4	0.5
454	0.5	1.6	0.3	0.1	0.6	0.5	0.4	0.1	0.4	0.6
455	0.4	1.4	0.2	0.1	0.5	0.5	0.3	0.1	0.4	0.5
456	0.5	1.6	0.3	0.2	0.6	0.5	0.4	0.1	0.5	0.6

TABLE 3–2E. (continued). Composition of Feeds—Amino Acid Values

		IFN	DRY MATT %	ARG %	CYS %	GLY %	HIS %
	SORGHUM, KAFIR. SORGHUM BICOLOR CAFFRORUM						
459	Grain	4-04-428	89.1	0.4	0.2	0.3	0.3
460			100.0	0.4	0.2	0.3	0.3
	SORGHUM, MILO. SORGHUM BICOLOR SUBGLABRESCENS						
461	Grain	4-04-444	88.5	0.4	0.1	0.3	0.2
462			100.0	0.4	0.2	0.4	0.3
	SOYBEAN. GLYCINE MAX						
477	Seed coats (hulls)	1-04-560	90.3	0.5	0.1	0.5	0.2
478			100.0	0.5	0.1	0.6	0.2
479	Seeds	5-04-610	92.0	2.8	0.7	1.4	0.9
480			100.0	3.0	0.7	1.5	1.0
481	Seeds, heat processed	5-04-597	92.6	2.5	0.3	4.3	0.9
482			100.0	2.7	0.4	4.6	0.9
483	Seeds, meal mechanical extracted	5-04-600	90.0	3.1	0.6	2.4	1.2
484			100.0	3.4	0.7	2.6	1.3
485	Seeds, meal solvent extracted	5-04-604	89.6	3.4	0.7	1.9	1.2
486			100.0	3.7	0.8	2.1	1.3
487	Seeds without hulls, meal solvent extracted	5-04-612	89.9	3.7	0.7	2.2	1.3
488			100.0	4.1	0.8	2.4	1.4
	SUNFLOWER, COMMON. HELIANTHUS ANNUUS						
499	Seeds without hulls, meal mechanical extracted	5-04-738	92.7	3.4	—	—	0.9
500			100.0	3.7	—	—	1.0
501	Seeds without hulls, meal solvent extracted	5-04-739	92.5	3.6	0.7	2.6	1.0
502			100.0	3.9	0.8	2.9	1.0
	TANKANGE (SEE ANIMAL; ANIMAL-POULTRY)						
	TIMOTHY. PHLEUM PRATENSE						
513	Hay, sun-cured, midbloom	1-04-883	88.9	0.3	—	—	—
514			100.0	0.4	—	—	—
	TOMATO. LYCOPERSICON ESCULENTUM						
519	Pomace, dehydrated	5-05-041	92.2	1.2	—	—	0.4
520			100.0	1.3	—	—	0.4
	TRITICALE. TRITICALE HEXAPLOIDE						
525	Grain	4-20-362	89.2	0.8	0.3	0.7	0.4
526			100.0	0.9	0.3	0.8	0.4
	WHEAT. TRITICUM AESTIVUM						
535	Bran	4-05-190	89.0	0.9	0.3	0.8	0.3
536			100.0	1.0	0.3	0.9	0.4
537	Flour by-product, less than 4% fiber (wheat red dog)	4-05-203	88.3	1.0	0.4	0.7	0.4
538			100.0	1.1	0.4	0.8	0.4
539	Flour by-product, less than 7% fiber (wheat shorts)	4-05-201	88.4	1.2	0.4	0.9	0.5
540			100.0	1.4	0.4	1.1	0.5
541	Flour by-product, less than 9.5% fiber (wheat middlings)	4-05-205	88.9	1.0	0.2	0.9	0.4
542			100.0	1.1	0.2	1.0	0.5
545	Germs, ground	5-05-218	88.4	1.8	0.5	1.4	0.6
546			100.0	2.1	0.5	1.6	0.7
547	Grain	4-05-211	89.0	0.6	0.2	0.6	0.3
548			100.0	0.7	0.2	0.7	0.3
549	Grain screenings	4-05-216	88.5	0.7	0.1	0.5	0.3
550			100.0	0.7	0.2	0.6	0.3
553	Mill run, less than 9.5% fiber	4-05-206	89.9	0.9	0.2	0.5	0.4
554			100.0	1.0	0.3	0.6	0.4
	WHEAT, DURUM. TRITICUM DURUM						
559	Grain	4-05-224	87.6	0.6	0.1	0.5	0.3
560			100.0	0.7	0.1	0.5	0.3
	WHEAT, HARD RED SPRING. TRITICUM AESTIVUM						
561	Grain	4-05-258	87.6	0.7	0.3	0.6	0.3
562			100.0	0.8	0.3	0.7	0.4
	WHEAT, HARD RED WINTER. TRITICUM AESTIVUM						
563	Grain	4-05-268	88.8	0.6	0.3	0.6	0.3
564			100.0	0.7	0.3	0.6	0.3
	WHEAT, SOFT RED WINTER. TRITICUM AESTIVUM						
565	Grain	4-05-294	88.4	0.6	0.4	0.5	0.3
566			100.0	0.7	0.4	0.6	0.4
	WHEY (SEE CATTLE)						

	ISO	LEU	LYS	MET	PHE	SER	THR	TRY	TYR	VAL
	%	%	%	%	%	%	%	%	%	%
459	0.6	1.6	0.3	0.2	0.6	—	0.4	0.1	—	0.6
460	0.6	1.8	0.3	0.2	0.7	—	0.5	0.2	—	0.7
461	0.4	1.3	0.2	0.2	0.5	0.5	0.3	0.1	0.3	0.5
462	0.5	1.4	0.3	0.2	0.6	0.5	0.4	0.1	0.4	0.6
477	0.3	0.5	0.5	0.1	0.3	0.6	0.3	0.1	0.3	0.3
478	0.3	0.6	0.5	0.1	0.4	0.7	0.4	0.1	0.3	0.3
479	1.5	2.6	2.1	0.3	1.6	1.7	1.3	0.6	1.1	1.5
480	1.6	2.8	2.3	0.3	1.7	1.9	1.4	0.7	1.2	1.6
481	1.6	2.6	2.2	0.5	1.7	1.8	1.4	0.5	1.3	1.6
482	1.7	2.8	2.4	0.5	1.9	2.0	1.5	0.6	1.4	1.7
483	2.7	3.7	2.8	0.7	2.2	2.4	1.8	0.6	1.5	2.2
484	3.0	4.1	3.2	0.7	2.5	2.7	2.0	0.7	1.7	2.4
485	2.0	3.5	2.8	0.6	2.2	2.4	1.8	0.6	1.6	2.1
486	2.3	3.9	3.2	0.6	2.5	2.7	2.0	0.7	1.7	2.3
487	2.2	3.6	3.1	0.7	2.4	2.5	1.9	0.7	1.7	2.4
488	2.4	4.0	3.4	0.8	2.7	2.8	2.1	0.8	1.9	2.6
499	1.8	2.5	1.6	0.9	1.8	—	1.4	0.5	1.0	2.0
500	1.9	2.7	1.7	1.0	1.9	—	1.5	0.5	1.1	2.2
501	2.0	2.8	1.7	0.8	2.1	1.9	1.6	0.6	0.7	2.2
502	2.1	3.0	1.8	0.9	2.3	2.0	1.8	0.6	0.7	2.4
513	—	—	—	—	—	—	—	—	—	—
514	—	—	—	—	—	—	—	—	—	—
519	0.7	1.7	1.6	0.1	0.9	—	0.7	0.2	0.9	1.0
520	0.8	1.8	1.7	0.1	1.0	—	0.8	0.2	1.0	1.1
525	0.6	1.1	0.5	0.2	0.7	0.7	0.5	0.2	0.5	0.8
526	0.6	1.2	0.6	0.3	0.8	0.8	0.6	0.2	0.5	0.9
535	0.5	0.9	0.6	0.2	0.5	0.7	0.4	0.2	0.4	0.7
536	0.6	1.0	0.6	0.2	0.6	0.8	0.5	0.3	0.4	0.8
537	0.6	1.1	0.6	0.2	0.7	0.8	0.5	0.2	0.5	0.7
538	0.7	1.2	0.7	0.2	0.7	0.9	0.6	0.2	0.5	0.8
539	0.5	1.1	0.8	0.3	0.7	0.8	0.6	0.2	0.5	0.8
540	0.6	1.2	0.9	0.3	0.8	0.9	0.7	0.3	0.5	0.9
541	0.7	1.1	0.7	0.2	0.7	0.8	0.6	0.2	0.4	0.8
542	0.7	1.2	0.8	0.2	0.8	0.9	0.6	0.2	0.5	0.9
545	0.9	1.5	1.5	0.4	0.9	1.1	0.9	0.3	0.7	1.2
546	1.1	1.7	1.7	0.5	1.0	1.3	1.1	0.3	0.8	1.3
547	0.5	0.9	0.4	0.2	0.6	0.6	0.4	0.1	0.4	0.6
548	0.5	1.0	0.4	0.2	0.7	0.7	0.5	0.2	0.4	0.7
549	0.4	0.8	0.4	0.3	0.5	0.4	0.3	0.1	0.3	0.5
550	0.5	0.9	0.5	0.3	0.6	0.5	0.4	0.1	0.3	0.6
553	0.7	1.2	0.6	0.3	—	—	0.5	0.2	0.5	0.8
554	0.8	1.3	0.6	0.4	—	—	0.6	0.2	0.6	0.9
559	0.5	0.8	0.3	0.1	0.5	0.4	0.4	0.3	0.3	0.5
560	0.5	0.9	0.4	0.2	0.6	0.5	0.4	0.3	0.3	0.6
561	0.5	1.0	0.4	0.2	0.7	0.7	0.4	0.1	0.4	0.6
562	0.6	1.1	0.5	0.2	0.8	0.8	0.5	0.2	0.5	0.7
563	0.5	0.9	0.4	0.2	0.6	0.6	0.4	0.2	0.4	0.6
564	0.6	1.0	0.4	0.2	0.7	0.7	0.4	0.2	0.5	0.7
565	0.5	0.9	0.4	0.2	0.6	0.6	0.4	0.3	0.4	0.6
566	0.5	1.0	0.4	0.2	0.7	0.7	0.4	0.3	0.4	0.7

TABLE 3–2E. (continued). Composition of Feeds—Amino Acid Values

		IFN	DRY MATT %	ARG %	CYS %	GLY %	HIS %
	YEAST, BREWERS. SACCHAROMYCES CEREVISIAE						
573	Dehydrated	7-05-527	93.1	2.3	0.5	1.8	1.1
574			100.0	2.4	0.6	1.9	1.2
	YEAST, IRRADIATED. SACCHAROMYCES CEREVISIAE						
575	Dehydrated	7-05-529	93.9	2.5	—	—	1.0
576			100.0	2.6	—	—	1.1
	YEAST, TORULA. TORULOPSIS UTILIS						
577	Dehydrated	7-05-534	93.0	2.5	0.6	2.6	1.3
578			100.0	2.7	0.6	2.8	1.4

TABLE 3–2F. Composition of Feeds—Mineral Supplements

		IFN	DRY MATT %	PRO TEIN %	CA %	CL %	MG %	P %
	AMMONIUM, PHOSPHATE,-DIBASIC, (NH4)2HPO4							
001		6-00-370	97.8	115.5	0.504	—	0.454	20.082
002			100.0	118.1	0.516	—	0.464	20.544
	AMMONIUM, PHOSPHATE,-MONOBASIC, (NH4)H2PO4							
003		6-09-338	97.7	71.0	0.380	—	0.460	24.411
004			100.0	72.7	0.389	—	0.471	24.994
	BONE							
005	Charcoal (bone black) (bone char)	6-00-402	90.0	4.2	27.095	—	0.531	12.728
006			100.0	4.7	30.111	—	0.590	14.144
007	Meal steamed	6-00-400	96.1	—	27.723	0.010	0.552	12.858
008			100.0	—	28.849	0.010	0.574	13.380
	CALCIUM, CARBONATE,-CaCO3							
009		6-01-069	99.6	—	37.881	0.034	0.350	0.020
010			100.0	—	38.052	0.034	0.352	0.020
	CALCIUM, PHOSPHATE,-DIBASIC, FROM DEFLUORINATED PHOSPHORIC ACID							
011		6-01-080	97.6	—	21.811	0.003	0.461	18.538
012			100.0	—	22.353	0.003	0.473	18.999
	CALCIUM, PHOSPHATE,-MONOBASIC, FROM DEFLUORINATED PHOSPHORIC ACID							
013		6-01-082	98.0	—	16.393	—	0.800	21.907
014			100.0	—	16.728	—	0.816	22.354
	CALCIUM, SULFATE,-ANHYDROUS, CaSO4							
015		6-01-087	79.1	—	23.274	0.002	—	—
016			100.0	—	29.429	0.003	—	—
	COBALT, CARBONATE,-CoCO3							
017		6-01-566	99.0	—	—	0.010	—	—
018			100.0	—	—	0.010	—	—
	COLLODIAL CLAY (SOFT ROCK PHOSPHATE)							
019		6-03-947	99.5	—	16.005	—	—	9.000
020			100.0	—	16.086	—	—	9.045
	CUPRIC, SULFATE,-PENTAHYDRATE, CuSO4.5H2O, CHEMICALLY PURE							
021		6-01-720	100.0	—	—	0.001	—	—
022			100.0	—	—	0.001	—	—
	CURACAO PHOSPHATE							
023	Ground	6-05-586	99.0	—	35.099	—	—	14.237
024			100.0	—	35.454	—	—	14.381
	FERROUS, SULFATE,-HEPTAHYDRATE, FeSO4.7H2O							
025		6-20-734	99.5	—	—	—	0.205	—
026			100.0	—	—	—	0.206	—
	LIMESTONE							
027	Dolomitic, ground (limestone, magnesium)	6-02-633	99.8	—	20.607	0.120	10.372	0.023
028			100.0	—	20.651	0.120	10.394	0.023
029	Ground	6-02-632	99.7	—	37.113	0.025	1.056	0.214
030			100.0	—	37.220	0.025	1.059	0.215
	MAGNESIUM, CARBONATE,-ANHYDROUS, MgCO3.MG(OH)2							
031		6-02-754	98.0	—	0.0200	—	30.810	—
032			100.0	—	0.0201	—	31.444	—

	ISO	LEU	LYS	MET	PHE	SER	THR	TRY	TYR	VAL
	%	%	%	%	%	%	%	%	%	%
573	2.0	2.9	3.0	0.7	1.6	—	2.1	0.5	1.5	2.2
574	2.2	3.1	3.2	0.7	1.7	—	2.2	0.5	1.6	2.4
575	2.9	3.6	3.7	1.0	2.8	—	2.4	0.7	—	3.1
576	3.1	3.8	3.9	1.1	2.9	—	2.6	0.8	—	3.3
577	2.7	3.4	3.7	0.8	2.7	2.8	2.7	0.5	1.9	2.9
578	2.9	3.7	3.9	0.8	2.9	3.0	2.9	0.6	2.1	3.1

	K	NA	S	CO	CU	I	FL	FE	MN	ZN
	%	%	%	%	%	%	%	%	%	%
001	—	0.040	2.4685	—	0.0081	—	0.1547	1.5122	0.0504	0.0302
002	—	0.041	2.5253	—	0.0083	—	0.1582	1.5470	0.0516	0.0309
003	0.138	0.078	0.8218	—	0.0086	—	0.1832	0.9902	0.0462	0.0639
004	0.141	0.080	0.8414	—	0.0088	—	0.1876	1.0138	0.0473	0.0655
005	0.144	—	—	—	—	—	—	—	—	—
006	0.160	—	—	—	—	—	—	—	—	—
007	0.182	0.403	0.3441	—	0.0016	0.0030	0.0643	0.1031	0.0037	0.0366
008	0.190	0.419	0.3581	—	0.0016	0.0031	0.0669	0.1073	0.0039	0.0381
009	0.063	0.082	0.0800	0.0003	0.0014	—	0.0000	0.0512	0.0135	0.0017
010	0.064	0.082	0.0804	0.0003	0.0014	—	0.0000	0.0514	0.0135	0.0017
011	0.086	1.377	0.6920	0.0008	0.0009	—	0.1040	0.8481	0.0255	0.0123
012	0.088	1.412	0.7092	0.0009	0.0009	—	0.1065	0.8692	0.0261	0.0126
013	0.400	0.060	0.8000	0.0005	0.0005	—	0.1400	1.0000	0.0200	0.0416
014	0.408	0.061	0.8164	0.0005	0.0005	—	0.1429	1.0204	0.0204	0.0425
015	—	—	18.6163	—	—	—	—	0.0020	—	—
016	—	—	23.5399	—	—	—	—	0.0025	—	—
017	—	0.250	0.0300	46.5000	0.0015	—	—	0.0200	0.0100	0.0015
018	—	0.253	0.0304	46.9697	0.0015	—	—	0.0202	0.0101	0.0015
019	—	—	—	—	—	—	1.2061	—	—	—
020	—	—	—	—	—	—	1.2121	—	—	—
021	—	—	12.8400	—	39.7400	—	—	0.0030	—	—
022	—	—	12.8400	—	39.7400	—	—	0.0030	—	—
023	—	—	—	—	—	—	—	—	—	—
024	—	—	—	—	—	—	—	—	—	—
025	—	—	10.9992	—	0.0100	—	—	20.8983	0.0001	0.0100
026	—	—	11.0553	—	0.0101	—	—	21.0050	0.0001	0.0101
027	0.265	0.379	0.0060	—	0.0020	—	—	0.0525	—	—
028	0.265	0.380	0.0060	—	0.0020	—	·	0.0526	—	—
029	0.112	0.055	0.0400	—	0.0011	—	—	0.1866	0.0123	0.0028
030	0.113	0.055	0.0401	—	0.0011	—	—	0.1871	0.0124	0.0028
031	—	—	—	—	—	0.0200	—	—	—	—
032	—	—	—	—	—	0.0201	—	—	—	—

TABLE 3–2F. (continued). Composition of Feeds—Mineral Supplements

	IFN	DRY MATT %	PRO TEIN %	CA %	CL %	MG %	P %
MAGNESIUM, OXIDE,-MgO							
033	6-02-756	98.3	—	1.660	—	55.188	—
034		100.0	—	1.689	—	56.147	—
MANGANOUS, OXIDE,-MnO, CHEMICALLY PURE							
035	6-03-056	99.0	—	—	—	—	—
036		100.0	—	—	—	—	—
PHOSPHATE							
037 Defluorinated	6-01-780	99.6	—	31.955	—	0.293	16.907
038		100.0	—	32.086	—	0.294	16.977
PHOSPHATE, ROCK							
039 Ground	6-03-945	100.0	—	35.000	—	—	13.000
040		100.0	—	35.000	—	—	13.000
POTASSIUM, BICARBONATE,-KHCO3, CHEMICALLY PURE							
041	6-29-493	99.0	—	—	—	—	—
042		100.0	—	—	—	—	—
POTASSIUM, IODIDE,-KI							
043	6-03-759	99.0	—	—	—	—	—
044		100.0	—	—	—	—	—
SALT							
045 Iodine added, 0.007% iodine	6-04-151	99.8	—	0.154	—	0.140	0.050
046		100.0	—	0.154	—	0.140	0.050
SALT,NaCl							
047	6-04-152	99.5	—	—	59.950	—	—
048		100.0	—	—	60.257	—	—
SODIUM, BICARBONATE,-NaHCO3							
049	6-04-272	99.7	—	0.010	—	—	—
050		100.0	—	0.010	—	—	—
SODIUM, PHOSPHATE,-MONOBASIC, MONOHYDRATE, NaH2PO4.H2O							
051	6-04-288	94.0	—	0.085	0.019	0.009	24.186
052		100.0	—	0.090	0.020	0.010	25.717
SODIUM, SELENITE,-Na2SeO3							
053	6-26-013	99.2	—	—	—	0.010	—
054		100.0	—	—	—	0.010	—
SODIUM, SULFATE,-DECAHYDRATE, NA2SO4.10H2O, CHEMICALLY PURE							
055	6-04-292	97.0	—	—	—	—	—
056		100.0	—	—	—	—	—
SODIUM, TRIPOLYPHOSPHATE,-Na5P3O10							
057	6-08-076	96.7	—	—	—	—	24.529
058		100.0	—	—	—	—	25.375
ZINC, OXIDE,-ZnO							
059	6-05-553	100.0	—	4.290	—	0.295	—
060		100.0	—	4.290	—	0.295	—
ZINC, SULFATE,-MONOHYDRATE, ZnSO4.H2O							
061	6-05-555	99.2	—	0.050	0.199	—	—
062		100.0	—	0.050	0.200	—	—

III. Characteristics of Common Concentrate Feedstuffs

A. *Carbonaceous Concentrates*

High-energy feeds; mostly feed grains and their byproducts. According to International Feed nomenclature, these products contain less than 20% protein and less than 18% fiber.

1. General nutritive characteristics
   a. High in energy (TDN or NE).
   b. Low in fiber.
   c. Low in protein (in relation to oil seeds and some mill feeds).
   d. Protein quality is variable and generally quite low.
   e. Mineral level
      (1) Fair in phosphorus (good when compared with forages).
      (2) Low in calcium.

	K	NA	S	CO	CU	I	FL	FE	MN	ZN
	%	%	%	%	%	%	%	%	%	%
033	—	—	0.0996	—	0.0005	—	0.0251	1.0477	0.0080	0.0009
034	—	—	0.1013	—	0.0005	—	0.0255	1.0660	0.0082	0.0009
035	—	—	—	—	—	—	—	—	76.670	—
036	—	—	—	—	—	—	—	—	77.450	—
037	0.100	2.067	0.1300	—	0.0040	—	0.1793	0.8392	0.0496	0.0090
038	0.100	2.075	0.1305	—	0.0041	—	0.1800	0.8426	0.0498	0.0090
039	—	—	—	—	—	—	3.7000	—	—	—
040	—	—	—	—	—	—	3.7000	—	—	—
041	38.65	—	—	—	—	—	—	—	—	—
042	39.05	—	—	—	—	—	—	—	—	—
043	21.00	0.100	—	—	68.170	—	—	—	—	—
044	21.02	0.100	—	—	68.238	—	—	—	—	—
045	0.007	40.525	—	—	0.0003	0.0070	—	0.0005	0.0001	0.0008
046	0.007	40.592	—	—	0.0003	0.0070	—	0.0006	0.0001	0.0008
047	—	38.811	—	—	—	—	—	—	—	—
048	—	39.011	—	—	—	—	—	—	—	—
049	—	28.094	—	—	—	—	—	—	—	—
050	—	28.188	—	—	—	—	—	—	—	—
051	0.094	21.461	—	—	0.0007	—	—	—	—	0.0005
052	0.100	22.820	—	—	0.0007	—	—	—	—	0.0005
053	—	26.799	—	—	0.0010	—	—	0.0301	—	—
054	—	27.003	—	—	0.0010	—	—	0.0303	—	—
055	—	13.840	9.6593	—	—	—	—	0.0010	—	—
056	—	14.268	9.9581	—	—	—	—	0.0010	—	—
057	—	30.184	—	—	—	—	0.0247	0.0039	—	—
058	—	31.225	—	—	—	—	0.0256	0.0040	—	—
059	—	—	1.0000	0.1500	0.0500	—	—	0.5500	0.0800	72.4968
060	—	—	1.0000	0.1500	0.0500	—	—	0.5500	0.0800	72.4968
061	—	—	17.6212	—	0.0055	—	—	0.0521	0.0169	35.9073
062	—	—	17.7558	—	0.0056	—	—	0.0525	0.0171	36.1815

f. Vitamin levels
   (1) Low in vitamin D and vitamin A activity (excluding yellow corn).
   (2) High in thiamine.
   (3) Low in riboflavin, B_{12} and pantothenic acid. ⎤
   (4) High in niacin, but niacin in grain is present in a bound   ⎬ Common additives to swine and poultry rations
       or unavailable form to the pig.
   (5) Fair in vitamin E. ⎦
2. Some of the more important carbonaceous concentrates
  a. Corn—most popular and most widely fed in Midwest
    (1) High in energy (80% TDN) and thiamine.
    (2) Fair in P and low in Ca.
    (3) High in niacin but in bound form for swine and poultry.
    (4) Low in crude protein (8% to 9%), lysine, tryptophan, vitamin D, riboflavin and pantothenic acid.

Corn (maize) ready for harvest.

    (5) Present activity in the area of breeding corn includes
        (a) Higher content of the amino acid lysine (high-lysine corn) (see Chapter 7, III-1).
        (b) Change in the type of carbohydrate (waxy corn).
        (c) Increase in the fat content (high-oil corn).
    (6) Only difference in value due to color is vitamin A potency (white vs yellow corn).
    (7) Grades of shelled corn. The value of corn per bushel can vary with water content and percentage of unsound kernels and foreign material. Therefore, grade permits purchase of a known amount of dry matter.
        No. 1—not more than 14.0% moisture
        No. 2—not more than 15.5% moisture
        No. 3—not more than 17.5% moisture
        No. 4—not more than 20.0% moisture
        No. 5—not more than 23.0% moisture
    (8) Fed in various forms
        (a) Air-dry (approximately 88% to 90% d.m.); weight—56 lb/bushel.
        (b) High moisture corn (20% to 34% moisture as stored)—satisfactory for farm animals. Beef claims only major response in improving feed efficiency (see discussion under feed processing in this chapter).
        (c) Whole ear (corn-and-cob meal).
  b. Sorghum grains—(milo, kafir, hybrids etc.)
    (1) Must be processed for maximum digestibility.
    (2) Somewhat lower in energy than corn (75% to 78% TDN); 95% to 98% the feeding value of corn for poultry and swine; 85% to 90% for cattle and sheep.
    (3) Higher but more variable in protein than corn (8% to 12%) and lower in carotene.
    (4) Low in Ca—fair in P.
    (5) Weight per bushel: threshed = 56 lb; in head = 80 lb. (However, it is sold on tonnage or cwt basis.)
    (6) Quite drouth resistant and grown in those areas inadequate in rainfall for corn production.
  c. Oats—(65% to 70% TDN: 12% CP)
    (1) Quite palatable—85% the feeding value of corn in most feeds. If hulled (oat groats), then equal to corn but generally too expensive for routine feeding.
    (2) Limit in beef finishing rations and for swine and poultry rations because of high fiber (11%) and low energy value.
    (3) Weight—32 lb/bushel; may vary from 27 lb/bu (light oats) to 36 lb/bu (heavy oats.)
    (4) General recommendations
        (a) Excellent for young animals and starting livestock on feed.
        (b) Finishing (cattle and sheep) rations—not more than 20% to 25% of grain.
        (c) Creep for calves—not more than 50–50 with corn.
        (d) Wintering cattle—feed as full replacement (lb for lb) with corn.
        (e) Swine growing and finishing—not more than 20% replacement for corn.

(f) Swine breeding herd—any level—use to regulate condition.

(g) Dairy rations—use to provide variety and bulk.

(h) Sheep and cattle breeding rations—equal to corn.

(i) Creep feeding lambs; 10% to 20% ration 1 to 4 weeks; remove after 4 weeks on creep to increase energy level.

(j) Excellent grain for horses to provide bulk.

(5) Widely grown, but most common in Midwestern and North Central states.

d. Barley—(70% to 75% TDN and 11% to 12% CP)

(1) 88% to 90% value of corn for cattle and sheep; 80% value of corn for swine.

(2) 48 lb/bushel.

(3) Limit in swine and poultry rations due to fiber content (5% to 6% fiber).

(4) Can be used as the only grain in "all concentrate" rations with cattle along with supplemental mineral fortification.

(5) Majority of barley is grown in North Central and Far West states.

e. Rye—(75% TDN and 12% CP)

(1) Least palatable of the grains.

(2) May be contaminated with ergot (black fungus which reduces palatability, causes abortion and reduces blood supply to the extremities resulting in necrosis of the hooves, lower limbs, tail and tips of the ears).

(3) Tends to cause digestive disturbances if ground too fine.

(4) Should not make up more than one-third of the ration.

f. Wheat—(80% TDN and 12% to 14% CP)

(1) 105% value of corn in limited amounts to swine and cattle (not over 50% of swine or beef ration).

(2) Good feed but packs in stomach.

(3) 60 lb/bushel.

(4) Widely grown in U.S., but in demand as a human food and generally too expensive for livestock diets.

g. Triticale—(78% TDN; 15% CP)

(1) Hybrid cereal derived from a cross of wheat and rye.

(2) Lower test weight (44 to 50 lb/bu) and yield than either wheat or rye.

(3) Contains higher quality protein than other cereals, but lysine is the most limiting amino acid.

(4) Unpalatable and may contain ergot similar to rye.

(5) Limit to no more than 50% of livestock ration.

h. Corn-and-cob meal—(73% TDN; 7% to 8% CP)

(1) Consists of whole ears of corn (cob and grain) ground to varying degrees of fineness.

(2) Partitioned by weight: 20% to 25% cob, 75% to 80% grain (assume 70 lb/bushel).

(3) Excellent feed for finishing ruminants (cattle and sheep) and horses; generally not fed to swine or poultry.

i. Dried beet pulp—(65% to 70% TDN; 8% to 10% CP)

(1) Residue from sugar beet processing.

(2) Contains 18% to 19% crude fiber, thus may be considered a roughage.

(3) Excellent for bulk and laxative in farrowing sow rations.

(4) Generally should not replace more than 15% to 20% of the grain in the ration.

j. Molasses—(55% to 75% TDN; 3% to 7% CP mostly NPN)

(1) Obtained as a byproduct of the manufacture of sugar from either sugar beets or, more commonly, sugarcane; molasses may also be produced from citrus and wood extracts.

(2) Commonly fed in the liquid form (70% to 80% dry matter); also available in dehydrated (dried) forms.

(3) Readily available source of energy and quite palatable (must contain more than 48% sugar after manufacturing process.

(4) Feed manufacturers often use molasses as a pellet binder, to reduce dustiness or both, and improve ration acceptability.

(5) Forms the basis for most liquid protein supplements containing urea.

(6) Should not be used at more than 10% of the replacement value of corn in livestock rations. Most commonly fed to ruminants or horses at 3% to 7% of the ration.

Corn (maize)

Grain sorghum (milo)

Oats

Barley

Hard red winter wheat

Rye

Feed grains used in livestock feeding.

(7) High levels of molasses impart a laxative effect due to high mineral salt content and may disrupt rumen microorganism activity; ration levels above 15% will also cause ration to become sticky and difficult to handle.

(8) The quality of molasses is expressed by the term *Brix*. Brix is determined by measuring the specific gravity of molasses, which then correlates to the level of sucrose. As sugar content increases, degrees Brix likewise increase.

k. Dried citrus pulp

(1) Residue of the citrus family (ground peel and occasional cull fruits) that have been dried, producing a coarse, flaky product.

(2) High in fiber (13%); low in crude protein (6% to 7%).

(3) Low in P (0.12%) but Ca content may be 2.0% or more because certain calcium compounds may be used as an aid in processing.

(4) Fed mainly to dairy cattle but may be fed to beef cattle.

(5) Generally does not make up more than 20% to 25% of the ration.

l. Animal fat

(1) Obtained from the tissues of mammals and/or poultry in the commercial processes of rendering or extracting.

(2) Usually treated with an antioxidant to prevent rancidity.

(3) Used in feed mixtures for the following reasons

(a) Increase the energy value.

(b) Decrease dustiness.

(c) Improve texture and palatability.

(d) Speed up pellet mill capacity and reduce machinery wear.

(4) Can be added at levels up to 5% in ruminant diets and up to 10% in nonruminant diets.

m. Dried bakery product

(1) Consists of stale bakery products (bread, cookies, cake, crackers etc.) and certain other bakery wastes that have been blended together, dried and ground into a meal.

(2) Similar to corn in nutrient composition except that it is usually much higher in fat (12% to 16%) and may contain a considerable amount of salt.

(3) Because of high salt content, should be limited to about 20% of the total diet of cattle or swine.

B. *Proteinaceous Concentrates*

1. Protein quality

a. Refers to kinds, amounts and ratio of amino acids in a feedstuff.

b. Variability in composition means that in nonruminants, supplementation of certain amino acids will be necessary or a mixture of feeds provided to ensure intake of all essential amino acids.

c. Nonprotein nitrogen source (as urea) may be fed to ruminants but not swine or poultry.

Remains of citrus fruit after the juice has been removed. This material is usually dehydrated and marketed as dried citrus pulp.

    d. While all amino acids are physiologically essential, some must be supplied in diet of nonruminants.

2. The essential amino acids and some sources
    a. Tryptophan
       (1) Corn and sorghum—very low.
       (2) Oats and barley—low; meat products—low.
       (3) Soybean meal, fish meal and milk products—high.
    b. Threonine, histidine, arginine, isoleucine, valine and phenylalanine—Grains fair source
    c. Lysine
       (1) Corn, barley, milo and oats—low (high lysine corn the exception).
       (2) Milk products—high.
       (3) Fish products—excellent.
       (4) Soybean meal and tankage—good source.
       (5) May be added from synthetic sources.
    d. Leucine
       Grain—good source.
    e. Methionine
       (1) Corn, milo, oats and barley—fair to low.
       (2) Fish meal products—excellent.
       (3) Soybean meal and packing house byproducts—fair source.
       (4) Sesame—good source.
       (5) May be added from synthetic sources.
    f. Glycine (required by the chick) and glutamic acid. Corn is a fairly good source.
    g. When feeding high-grain rations to swine or poultry, ration levels of four specific amino acids must be carefully checked. They are
       (1) Lysine.
       (2) Tryptophan.
       (3) Threonine.
       (4) Methionine.

3. Some protein supplements
    a. Sources of NPN
       (1) Urea—Not a protein supplement, but a source of nitrogen (42% to 45% N) for protein synthesis by rumen bacteria (1 lb of urea contains as much nitrogen as 2.62 to 2.81 lb of protein)
         (a) Works well in mixtures with plant proteins to lower protein cost. Thumb rule: 1 lb of urea and 6 lb of corn grain will replace 7 lb of soybean meal.
         (b) To avoid caking, some carrier is added so that 1 lb of a commercial urea is equivalent to either 2.62 or 2.81 lb of protein; we could say it has 262% ot 281% crude protein equivalent (CPE).
         (c) General rules for use of urea or other NPN in ruminant diets
            (1') No more than ⅓ of total nitrogen in ration.
            (2') No more than 1% of diet or 3% of concentrate mix should be urea.
            (3') No more than 10% to 15% of a typical protein supplement (meal or pellet form).
            (4') No more than 5% of a supplement to be used with low-grade roughages.
            (5') Refer to protein evaluation section in Chapter 8.
       (2) Biuret
         (a) Feed-grade biuret is defined as a mixture of the nitrogen compounds resulting from controlled pyrolysis of urea. The predominant component at this mixture is biuret or carbamylurea, $C_2H_5N_3O_2$.
         (b) The composition of feed-grade biuret as defined in the *Federal Register* (Vol. 34, No. 26, February 7, 1969, page 1826) is

Components	Percentage
Biuret	60 minimum
Urea	15 maximum
Cyanuric acid and triuret	21 maximum
Total N (equivalent to 218.75% crude protein)	35 minimum

(c) Feed-grade biuret has fewer feeding limitations and restrictions than urea; however, like urea, only supplies a source of nitrogen for protein synthesis by rumen bacteria.
(3) Diammonium phosphate
(a) White (to grey) crystalline product made by the treatment of phosphoric acid with ammonia.
(b) Contains at least 17% N (equivalent to 106% crude protein) and 20% to 24% P.
(c) Fed to ruminants as a source of P and N.
(d) May be used in monogastric diets as a source of P only.
(4) Monoammonium phosphate
(a) Product resulting from the neutralization of phosphoric acid with ammonia.
(b) Contains at least 9% N (equivalent to 56% crude protein) and 24% P.
(c) Fed to ruminants as a source of N and P.
(d) May be used in monogastric feeds as a source of P only.
(5) Ammonium sulfate
(a) Product resulting from the neutralization of sulfuric acid with ammonia.
(b) Contains at least 21% N (equivalent to 131% crude protein) and 24% sulfur.
(c) Fed to ruminants as a source of N and S.
b. Supplements of plant origin (oil meals)
(1) Manufacturing process for oil meals
(a) Expeller or hydraulic process (old or mechanical process)—seed is crushed, heated in steam or cooked, pressed and ground.
(b) Solvent process (new process)—seed in cracked, heated mildly, rolled, extracted with hexane; flakes are then toasted and ground.
(2) Effect of processing
(a) Hydraulic or expeller-processed meals contain more fat and less protein than solvent-extracted. (Solvent less than 1% fat—expeller 3% to 5% fat.)
(b) Cooking improves palatability, color and increases availability of some amino acids. However, overcooking can destroy some of the amino acids.
(3) Characteristics of the oil meals
(a) Soybean meal (SBM)
(1') Soybean meal is the most widely used oilseed meal in the United States. It is the product obtained by grinding the seeds that remain after removal of most of the oil; the protein content is generally standardized at 44% to 50% by dilution with soybean hulls.
(2') Nutritive characteristics
(a') Solvent-extracted—44% to 50% crude protein.
(b') Mechanically extracted—41% to 44% crude protein.
(c') Low in fiber—must not contain more than 7% crude fiber.
(d') TDN—71% to 80%.
(e') Low in Ca (0.30%); fair in P (0.65%).
(3') Must be heated for maximum feeding efficiency and safety; cooking destroys an antitrypsin factor that suppresses nonruminant growth and prevents action of the enzyme trypsin; cooking also destroys urease enzyme, which may break down urea present in cattle diets.
(4') Soybean meal is the most abundant as well as the most complete amino acid source used to supplement or balance the amino acid deficiencies and low amounts of protein in cereal feed grains and their byproducts. Properly processed SBM has become an almost universal yardstick by which other protein ingredients are measured (and priced) for use in scientific feed formulation.
(5') On-the-farm processed full-fat soybeans
(a') Nutritive characteristics: 37% to 38% crude protein, 17% to 18% fat, 84% to 92% TDN.
(b') Takes special equipment to uniformly heat beans to a constant temperature that will destroy the antitrypsin factor (harmful to poultry and swine) and the urease enzyme (undesirable in beef diet containing urea).
(c') Limited research has been conducted primarily with swine diets. Most studies indicate that whole cooked soybeans can be used in place of soybean

Soybeans ready for harvest.

meal very successfully in swine diets. Using whole cooked soybeans will likely improve feed efficiency (4% to 8%) due to higher fat content but will have little effect on daily gain. May influence swine carcass quality. Refer to further discussion in chapter on swine feeding (Chapter 7).

(b) Cottonseed meal (CSM)

   (1′) Cotton is grown primarily in southern United States; the meal results from the kernels being removed from the leathery hulls and crushed. Then, as much of the oil as possible is removed by one of the extraction processes.

   (2′) Nutritive characteristics

      (a′) 36% to 41% crude protein.

      (b′) TDN—61% to 70%.

      (c′) Low in Ca (0.17%).

      (d′) High in P (1.1%).

      (e′) No fiber limit—generally 10% to 14%.

      (f′) Protein quality is low, which limits it use in swine and poultry feed (limited lysine).

      (g′) Available as either a cake or meal.

   (3′) Danger in swine and poultry feeding

      (a′) Cottonseed meal contains gossypol (.03% to .2%), a toxic substance. Nonruminant tolerance depends upon the species and age and other components of the ration. Older animals are less sensitive than young. Poultry are less sensitive than swine.

      (b′) Broilers are not affected by a dietary level of 0.015% gossypol, and growing pigs tolerate dietary levels up to 0.01%; the addition of ferrous sulfate to the ration will help block the harmful effects of free gossypol; use a 1:1 ratio of iron to free gossypol.

      (c′) Symptoms of gossypol poisoning are similar to pneumonia, except fluid develops in abdominal cavity, which is typical of poisoning.

      (d′) Egg production is not affected by a dietary free gossypol levels up to 0.02%; however, levels above 0.005% will cause discoloration of eggs stored up to three months (greenish cast or color to the egg yolk). Certain cyclopropenoid fatty acids found in CSM are also undesirable because they lend a definitely pink tint to the egg white.

      (e′) Degossypolized meals are available; degossypolized CSM must not contain more than 0.04% free gossypol.

   (4′) Feeding recommendations

      (a′) CSM should not make up more than 25% to 30% of the protein supplement for swine or poultry diets because of low protein quality and poisoning potential.

      (b′) CSM serves as a satisfactory protein supplement in ruminant diets.

(c′) Whole cottonseed is frequently fed to lactating dairy cattle as an excellent supplement of protein (23% CP), energy (88% TDN; 20% fat) and fiber (17% to 24%).

(c) Linseed meal (LSM)

    (1′) Most flaxseed is grown in the north-central United States almost entirely as a cash crop for the production of linseed oil from the seed. The meal, used for livestock feed, is produced after one of the oil-extraction processes.

    (2′) Nutritive characteristics

        (a′) 34% to 38% crude protein; approximately 9% crude fiber.

        (b′) Fair in Ca (0.41%); high in P (0.87%); high in selenium (0.90%).

        (c′) Palatable and slightly laxative in nature.

        (d′) Contains poor quality of protein, being lower in its content of both lysine and tryptophan than SBM.

    (3′) Feeding characteristics

        (a′) Linseed meal may replace 25% to 33% of the SBM in swine and poultry diets.

        (b′) Linseed meal serves as a satisfactory protein supplement in ruminant and horse diets.

    (4′) Linseed meal contains a conditioning factor (mucin) that results in glossy haircoats or greater animal bloom.

(b) Sunflower meal (SNFM)

    (1′) Produced from what remains following the extraction of oil from dehulled sunflower seed.

    (2′) Nutritive characteristics

        (a′) Wide range in crude protein and fiber content depending on the oil extraction method and amount of hulls removed during processing.

Process	% CP	% Fiber
Whole pressed, solvent-extracted	32.2	23.7
Mechanically extracted, without hulls	40.5	12.7
Solvent-extracted, without hulls	45.2	11.7

        (b′) Fair in Ca (0.40%); high in phosphorus (1.0%).

        (c′) Rich source of the B-complex vitamins.

        (d′) Limiting in lysine and energy when compared to SBM.

        (e′) No nutritional toxins have been reported present in SNFM, however, some meals, if finely ground, may be quite dusty and result in reduced palatability or cause feed buildup around beaks of poultry.

    (3′) Feeding recommendations

        (a′) Poultry—30% to 50% of the SBM in layer or broiler rations can be replaced by SNFM.

        (b′) Swine—similar to use in poultry feeds; from 30% to 50% of the SBM in growing-finishing swine diets can be replaced by SNFM (preferably after pigs have reached 75 to 100 lb body weight).

        (c′) Ruminants—well received as a ruminant feed and should equal all other protein sources in supplementing range and feedlot cattle, dairy cows, sheep and other ruminants.

    (4′) Sunflower hulls are a useful course roughage for ruminants; they add bulk and readily absorb liquids such as molasses; hulls are reported to contain approximately 4% crude protein, 2% fat, 50% crude fiber and 2.5% ash.

(e) Safflower meal (SAFM)

    (1′) Safflower seed contains oil rich in unsaturated fatty acids; the meal is what remains after most of the hull and oil has been removed from the seed. Generally less palatable than other oilmeals.

    (2′) Nutritive characteristics

        (a′) Wide range in CP and fiber content, depending on the oil extraction method and whether decorticated (without hulls) or undecorticated (with hulls).

Process	% CP	% Fiber
Mechanically extracted, undecorticated	21.0	31.0
Solvent-extracted, undecorticated	23.0	30.0
Solvent-extracted, decorticated	42.0	15.0

      (b') Lysine is the first limiting amino acid; also limiting in sulfur amino acids.

      (c') Low in Ca (0.3%); fairly high in P (0.8%).

    (3') Feeding recommendations

      (a') Swine and broiler diets should not have decorticated safflower meal replacing more than about 30% of the SBM because of the lowered lysine and energy level; there are indications that hen and older turkey diets may contain high levels of decorticated safflower meal (replace up to 50% of the SBM).

      (b') With ruminants, safflower protein from decorticated meal appears equal to soybean meal.

  (f) Rapeseed meal (RSM) or canola meal

    (1') Rapeseed has been used as a protein feed for livestock in many parts of the world where cooler climatic conditions are such that other high protein oil seeds cannot grow. The meal is what remains after most of the hull and oil has been removed from the seed.

    (2') Nutritive characteristics

      (a') 35% to 40% crude protein; 13% to 14% crude fiber.

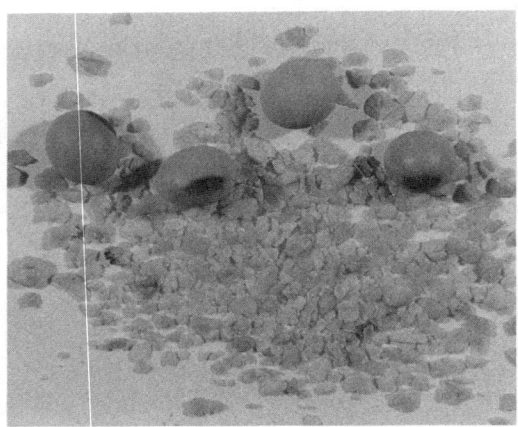

Soybean seed/Soybean meal

Flax seed/Linseed meal

Cottonseed/Cottonseed meal

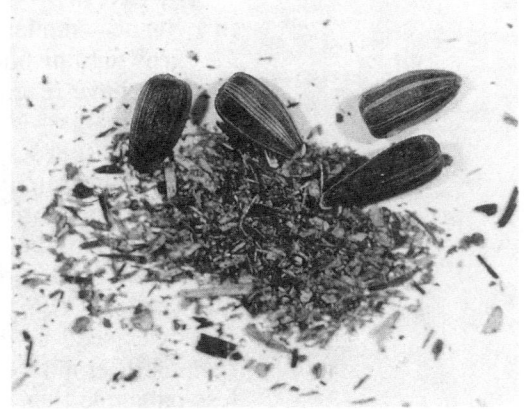

Sunflower seed/Sunflower meal

Examples of common plant protein supplements.

(b') Fairly high in Ca (0.60%); high in phosphorus (1.0%).
(c') Lower in palatability, energy and lysine than SBM.
(3') Danger in livestock feeding
(a') Whole, unprocessed rapeseed contains erucic acid and the enzyme myrosinase.
(b') Extraction of oil removes the erucic acid; in heat processing of rapeseed, it is important that temperature be raised quickly to 100°C and moisture content controlled between 6% and 10% to inactivate the myrosinase enzyme activity quickly in the rapeseed meal. This enzyme has the effect of changing certain chemicals in the rapeseed into toxic compounds that have thyrotoxic or goitrogenic activity (glucosinolates).
(c') Canola meal—genetically developed cultivar of rapeseed that contains a much lower erucic acid and glucosinolate content. In addition, during processing the seed is cooked, which inactivates the myrosinase enzyme and prevents the hydrolysis of the glucosinolates. To be called canola (rather than rape), oil must contain less than 5% erucic acid, while the meal must contain less than 3 mg/g of glucosinolates (Canola Council of Canada, 1983).
(4') Feeding recommendations
(a') Canola meal should be from a low erucic acid, low glucosinolate variety of rapeseed.
(b') For young swine and poultry, canola meal may be used up to 5% of the total diet. Older swine and poultry diets may contain up to 12% canola meal. For sows during gestation and lactation, it is recommended that canola meal be limited to 12% of the diet.
(c') Canola meal is not palatable to ruminants. As long as feed intake is not reduced, canola meal may be used at levels up to 10% of the ruminant diet.
(g) Peanut meal (PNM)
(1') Consists of fat-extracted peanut kernels ground to a meal with a certain amount of ground peanut hulls added.
(2') Nutritive characteristics
(a') Composition varies considerably depending on the quality of the nuts, the method of fat extraction used and the amount of hull included.
(b') 40% to 48% crude protein; approximately 6% to 13% crude fiber.
(c') Low in Ca (0.2%–0.3%); fair in P (0.60%).
(d') Poor amino acid balance; limiting in lysine and also methionine.
(3') Danger in livestock feeding
(a') In addition to the relatively poor amino acid balance, considerable danger still exists in PNM from certain aflatoxins produced from molds such as *Aspergillus flavus*.
(4') Feeding recommendations
(a') Swine and poultry—PNM can provide a fair quality protein and can replace part of SBM (5% to 10% of the diet or 30% to 50% of the SBM) when priced exceedingly cheap. One must expect, however, to obtain considerable growth retardation and increased feed requirements unless an unusually good sample of PNM has been obtained or unless the diet is supplemented with lysine and methionine.
(b') Ruminants—PNM free of molds and toxins appears equal to SBM in feeding value.
c. Supplements of animal and marine origin
(1) General characteristics
(a) Derived from meat or poultry packing or rendering plants, from surplus milk or milk byproducts, or from marine sources.
(b) Usually employed to improve the total protein of basal feeds, but in addition contribute a mixture of amino acids quite different from proteins of plant origin; plant proteins are usually seriously deficient in lysine. However, animal and fish proteins are relatively rich in this amino acid, though they are likely to be short of either the sulfur amino acids or tryptophan.

      (c) Although individual recommendations concerning product characteristics and animal requirements may be made, in general these products may be included at 5% to 10% of the diet as a replacement for plant proteins.

  (2) Tankage (meat meal tankage or digester tankage)

      (a) Byproduct of the meat packing industry—also available from dead animal rendering plants.

      (b) Consists of otherwise unusable animal tissue, including bones, which has either been cooked under steam pressure (wet-rendered) or cooked in steamjacketed kettles (dry-rendered), partially defatted, dried and ground.

      (c) Composition is quite variable due to

         (1') Blood meal content (often added to increase protein content).

         (2') Fat content (usually around 8%).

         (3') May have large amounts of connective tissue (reduces BV).

      (d) 55% to 60% crude protein (60% most common).

      (e) If more than 4.4% phosphorus, must be labeled "meat and bone tankage."

      (f) Contains blood meal, gut, tendon and connective tissue, which reduces BV of the protein.

      (g) Exclusive of hair, hoof, horn, manure, stomach contents and hide trimmings.

      (h) Used primarily in feeds for swine.

  (3) Meat scrap (meat meal)

      (a) Similar to tankage in origin except that it is usually cooked in steamjacketed kettles in its own fat and not under steam pressure.

      (b) Not as variable in protein quality as tankage because blood meal is normally not added, making meat scrap a more acceptable product than tankage.

      (c) Usually does not contain gut, tendon and connective tissue to the extent of tankage.

      (d) Contains 45% to 55% crude protein (50% most common).

      (e) If more than 4.4% phosphorus, must be labeled "meat and bone scrap" or "meat and bone meal."

      (f) Used primarily in rations for swine and poultry.

  (4) Blood meal

      (a) Blood meal is coagulated packing house blood that has been dried into a meal. Drying methods include drum, ring and flash.

      (b) High in crude protein (80%+) but may be low in digestibility and quality due to heat damage occuring in the drying process.

      (c) Unpalatable.

      (d) Blood meal contains highly undegradable (bypass) protein for ruminants (Table 1–2). For this reason, it is finding increased use in ruminant diets.

  (5) Fish protein sources

      (a) Fish meal—consists of whole fish or fish cuttings (byproducts) with or without the extraction of part of the oil, dried and ground into a meal.

         (1') Several types, depending on the type of fish used.

         (2') Crude protein ranges from 35% to 70% depending on type of product (whole fish or cuttings).

         (3') Excellent protein quality and source of B vitamins.

         (4') High in content of calcium and phosphorus.

      (b) Fish solubles—evaporated product of the aqueous portion obtained when oil is removed.

         (1') If dried, contains about 60% crude protein; if condensed (approximately 50% d.m.) contains about 30% crude protein.

         (2') Excellent protein quality and source of B vitamins.

         (3') Fish solubles may be marketed as such or mixed back into the fish meal.

      (c) Fish protein sources are used primarily in rations for swine and poultry.

  (6) Milk products

      (a) Whole milk normally constitutes the first diet of the young suckling mammal; composition ranges of whole milk of different animals is shown in Table 3–3.

TABLE 3–3. Milk Composition of Selected Mammals

	Cow	Ewe	Sow	Mare	Dog	Cat
Water, %	87–88	82–84	82–84	90–91	79–81	77–79
Fat, %	3.5–5.5	5.5–9.5	5.2–7.5	1.3–1.6	7.5–9.0	6.5–7.5
Protein, %	3.1–3.9	4.8–6.8	4.8–6.8	2.2–2.5	6.5–7.5	9.0–11.0
Lactose, %	4.5–5.1	3.8–4.8	4.9–5.3	5.8–6.1	3.5–4.0	4.0–5.0
Minerals, %	0.6–0.8	0.8–1.0	0.8–1.0	0.3–0.5	1.0–1.5	0.5–1.0

- (b) Whey is the product obtained as a fluid by separating the coagulum from milk, cream or skim milk in the manufacturing of cheese. Contains less than 10% dry matter and 1% crude protein.
- (c) Dried milk products are palatable and a source of excellent quality protein.
  - (1′) Dried skim milk—residue obtained by drying defatted milk. It contains less than 8% moisture and about 33% crude protein.
  - (2′) Dried buttermilk—residue obtained by drying buttermilk. It contains less than 8% moisture, 13% maximum ash, 5% minimum fat and 32% to 33% crude protein.
  - (3′) Dried whey—obtained by removing water from whey. It contains at least 11% crude protein and 61% lactose.
- (d) Dried milk products are used primarily in milk replacers or in starter diets for young pigs, ruminants or pets.
- (7) Poultry byproducts
  - (a) Poultry byproduct meal. Made from ground, dry-rendered or wet-rendered parts of the carcass—heads, feet, undeveloped eggs and intestines. Exclusive of feathers except in traces—must not contain more than 16% ash. Contains 55% to 65% crude protein.
  - (b) Hydrolyzed poultry feathers (Feathermeal). Pressure-treated, clean undecomposed feather from poultry. Contains 85% crude protein and 75% of the crude protein content should be guaranteed digestible. Primarily used in swine and poultry feeds in quantities limited to no more than 3% of the diet because it is low in the amino acids histidine, lysine, methionine and tryptophan. Ruminants can utilize feathermeal by benefiting from its high bypass protein.
- d. Miscellaneous protein sources
  - (1) Single-cell protein (SCP)
    - (a) Produced by yeast or bacteria that can use the carbon from petroleum byproducts as a source of energy. A culture is propagated by blending a nitrogen source, an appropriate mixture of minerals and petroleum hydrocarbons under controlled conditions.
    - (b) Nutrient composition varies: 50% to 85% CP, 5% to 8% EE, 7% to 9% CF, 6% to 8% ash, and is high in phosphorus (1.5% to 2.0%) but low in calcium (.1% to .2%).
    - (c) Feeding trials with swine indicate SCP to be highly digestible and a high quality protein source if supplemented with the amino acid methionine, which is generally limiting.
  - (2) Animal wastes
    - (a) Animal and poultry wastes consist of feces and urine (with or without bedding) fed primarily to ruminant animals in one of the following forms: unprocessed, solids fraction, ensiled or dehydrated.
    - (b) The nutrient composition, especially the CP, fiber, and energy content of animal waste varies, depending on the level of dry matter intake, roughage level and digestibility of rations fed different animals for different kinds of production. A summary of composition is presented in Table 3–4.
    - (c) Appreciable proportions (10% to 60%) of the protein in animal waste exist as NPN, which ruminants can utilize more efficiently than nonruminants.
    - (d) Limited research has shown that satisfactory performance can be obtained from feeding animal waste as a portion of the ruminant and nonruminant ration.

TABLE 3–4. Nutrient Composition of Animal Waste (dry matter basis)*

	Cage Layer Excreta, DPW	Broiler Litter	Cattle Manure Steer	Cattle Manure Cow	Swine
CP, %	28.0	31.3	20.3	12.7	22–27
CF, %	12.7	16.8	—†	37.5	—
EE, %	2.0	3.3	—	2.5	—
Ash, %	28.0	15.0	11.5	16.1	—
TDN, %	52.3	72.5	48.0	45.0	45.0

*Bhattacharya AN and JC Taylor, 1975, Recycling animal waste as a feedstuff: a review. *J Anim Sci,* 41:1438.
†No values reported.

    (1') Incorporation of 5% to 10% animal waste into poultry or swine diets does not appear to alter performance significantly.

    (2') Levels of 20% to 30% animal waste into ruminant diets have supported satisfactory gain when compared to plant proteins.

  (e) The US Food and Drug Administration has expressed concern over the potential for animal and human health hazards that may result from the use of these types of products in animal feeds. Accordingly, studies have been conducted and others are planned that have been designed to establish the safety of processed animal wastes to animal and man.

  (f) There have been no reports of infectious diseases attributed to the feeding of waste-containing rations; meat from steers produced on waste-containing rations has been just as acceptable as that of control animals; no excessive levels of harmful residues have been reported for the meat from steers finished on waste-containing rations.

 e. Mixed supplements

  (1) Any mixture of the above can be prepared.

  (2) Animal protein sources are generally used in swine and poultry rations. However, these emphasize SBM.

  (3) Ruminants are primarily vegetable proteins and urea.

C. *Feed Grain Byproducts*

 1. Corn byproducts

  a. Dry corn milling—process of preparing table corn meal, hominy and corn grits used as human food. Feed products resulting from this process include

    (1) Corn bran—outer coating of the kernel; 12% to 16% crude protein; 10% to 12% crude fiber.

    (2) Corn germ meal—ground corn germ from which most of the oil has been removed; 20% crude protein; 8% to 10% oil if not removed.

    (3) Hominy feed—a mixture of corn bran, corn germ and part of the starchy portion of the grain; must contain not less than 4% crude fat; contains 10% to 11% CP. Generally about equal to corn in feeding value.

  b. Wet corn milling—process of preparing starch, sugar, syrup and corn oil, all used as human food. Feed products resulting from this process include

    (1) Corn gluten meal—dried residue remaining after removal of larger part of starch, germ and bran; 40% to 60% crude protein.

    (2) Corn gluten feed—dried residue remaining after removal of the larger portion of the starch, gluten and germ; contains corn bran; 20% to 25% CP.

  c. Distiller's byproduct feeds—obtained by the processing of the residue remaining after removal of the alcohol and some water from a yeast-fermented mash. Corn is the predominant grain used in this industry with varying amounts of rye, rice and other grains. For product identification, the predominating grain in the mash mixture is used to complete the following product names:

    (1) . . . distiller's dried grains—made from the dried coarse grain fraction; contains 25% to 27% crude protein; 9% to 11% crude fiber.

(2) . . . distiller's dried solubles—made by condensing the thin stillage fraction and drying it; contains 25% to 27% CP; 4% CF.

(3) . . . distiller's dried grains with solubles—combination product containing 25% to 27% CP.

(4) Distiller's wet grains—generally fed as produced (wet) from the distillation of alcohol for "gasohol" production. The guaranteed analysis shall include the maximum moisture. Its water content limits its use to near the place of production.

(5) May be included as a part of ruminant or horse rations but seldom fed to poultry or swine because of high fiber content.

2. Sorghum grain byproducts

Sorghum grain (milo) is used to make starch, sugar and other products similar to those made by the wet milling processing of corn. The resulting byproduct feeds are milo gluten feed, milo gluten meal and milo germ meal, similar to the products from corn. It also is possible to use sorghum grain to make alcohol; sorghum distiller's grain results as the byproduct. By far the major portion of sorghum is used directly as livestock feed.

3. Wheat byproducts

The wheat flour milling byproducts have a long history in the feed industry. Many names were used for the wheat fractions resulting from various milling processes, but they now have been reduced to the following official names

a. Wheat bran—coarse outer coating of the kernel containing 15% to 17% CP and 9% to 11% CF.

(1) Included in swine farrowing rations or in diets of horses or cattle used for show because of its bulky and mild laxative properties.

(2) Generally limited to 10% to 15% of the ration (replacement for grain).

b. Wheat middlings—consists of the fine particles of wheat bran, wheat shorts, wheat germ, wheat flour and some of the offal from the "tail of the mill."

(1) Contain 16% to 18% CP and must contain not more than 9.5% CF.

(2) Most commonly fed in swine rations because the fine particle size of middlings makes them unpalatable to ruminants.

c. Wheat shorts—consists of the same components as wheat middlings but must contain not more than 7% CF.

d. Wheat germ meal—consists chiefly of wheat germ, together with some of the bran and middlings; contain 25% to 28% CP and not less than 7% crude fat.

e. Defatted wheat germ meal—obtained after removal of part of the oil from wheat germ and must contain not less than 30% CP.

4. Oat byproducts

Oats are processed primarily to separate the groat (kernel) from the hull. The groat is used for both food and feed products.

a. Oat groats—kernels produced from cleaned and dried oats with the hull removed; contain 16% to 17% CP and only about 3% CF.

(1) Approximately equal to corn in feeding value, but, too expensive for general livestock feeding.

(2) May be used in special diets such as early weaning rations for pigs.

b. Oat hulls—consists primarily of the outer covering of the oat; contain 5% to 6% CP and 26% to 28% CF.

c. Feeding oat meal—consists of broken rolled oat groats, oat groat chips and a small quantity of ground oat hulls; contain 15% to 17% CP and no more than 4% CF.

5. Barley byproducts

The chief byproduct feeds from barley are derived from the brewing industry.

a. Malter barley (barley malt)—made by sprouting good quality barley by means of moisture and a warm temperature. When the sprout has reached the proper length, the heat is increased and the barley malt is dried. Nutrient composition similar to barley grain.

b. Malt sprouts—obtained from the dried malted barley by the removal of the sprouts (rootlets), which may include some of the malt hulls or other parts of malt; contains 24% to 27% CP and about 14% CF.

Corn distillers dried grains

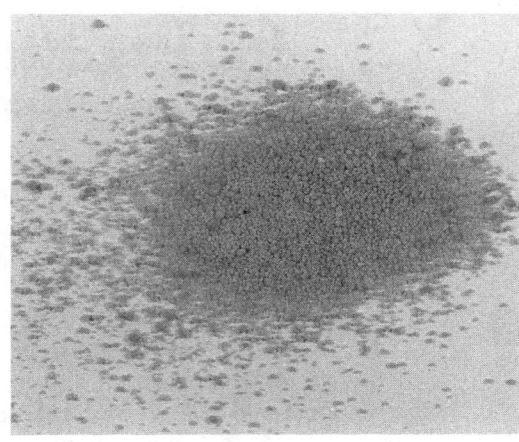

Corn gluten meal

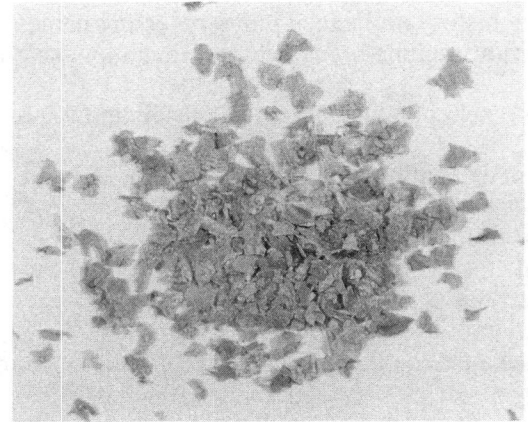

Wheat bran

Brewers dried grains

Examples of common grain by-product feeds.

      c. Brewer's dried grains—the dried extracted residue of barley malt alone or in mixture with other cereal grain or grain products resulting from the manufacturing of wort.
        (1) Contains 25% to 27% CP and 14% to 16% CF.
        (2) Commonly fed to dairy cattle up to about one-third of the grain mix.
        (3) Often included in horse rations but seldom fed to swine or poultry because of high fiber content.
    6. Rice bran (13% to 15% CP; 12% CF).
      a. Consists primarily of the seed coat and germ removed from rice grain in the manufacture of polished rice for human consumption.
      b. Comparable to wheat bran in feeding value, although lower in crude protein.
      c. Some fed to dairy cattle or used as a carrier in feed additive premixes.
    7. While many of the byproduct feeds are fairly high in CP content, it should be remembered that the protein quality is low as in the original grains and that these products are generally used to supply a part but not all of the protein and energy for livestock rations.

IV. Characteristics of Common Roughage Feedstuffs
    A. *General Characteristics*
      1. Feed materials low in energy and containing more than 18% CF.
      2. Includes pasture, hay, silage or soilage; the crop may be the same but preservation can influence the nutrient value to the animal.
      3. Higher in fiber than concentrates and usually lower in energy. Variable in protein content.
      4. Needed for bulk in ruminant rations.

5. Higher in calcium and trace mineral elements than most concentrates.
6. Legumes are higher in protein and B vitamins than some concentrates.
7. Better sources of fat-soluble vitamins than most concentrates.
8. Palatable to ruminants.
9. Limited or excluded in swine rations (not suited for finishing rations).
10. Limited in beef finishing rations and in some high energy lactating rations.
11. Required in lactating dairy cattle rations to help maintain normal fat level in milk.
12. More variable in nutritive content and acceptability than concentrates due to variation in stage of maturity and harvesting and storing procedures.

B. *Proteinaceous Roughages*

Contain more than 10% CP (d.m. basis); made up primarily of legumes and some immature grasses.
1. Legume forages
   a. Advantages of legume forage
      (1) Produces a high yield of palatable feed per acre.
      (2) Higher in protein content and quality than other forages (more than 10% CP, d.m. basis).
      (3) High in calcium (more than 0.9% Ca).
      (4) Excellent in vitamin A activity when harvested; sun-cured hay is rich in vitamin D.
      (5) Increases soil fertility.
      (6) Excellent in combination with grasses.
      (7) Legumes tend to have more lignin content than grasses, with the highest concentration in the stems; may also contain higher levels of certain estrogenlike compound.
   b. Alfalfa
      (1) Perennial plant varying in height from 18 in. to 3 ft.
      (2) Extensive root system, which results in alfalfa being quite drouth resistant.
      (3) Alfalfa requires a dry, hot climate with a deep, fairly productive, well-drained soil.
      (4) Grown extensively throughout the Midwest and western United States.
      (5) Probably no other forage crop produced in the United States has so many varied uses.
      (6) The hay crop is high in feeding value and is excellent for general feeding purposes
         (a) 15% to 25% CP.
         (b) More than 50% TDN.
         (c) High in calcium (1.3% to 1.5%); fair in phosphorus (0.24%).
      (7) About 80% of the crop is made into hay. It is also used for pasturage, soiling, silage and making meal; several cuttings (2 to 4) can be made per year.
         (a) Dehydrated alfalfa meal—dried product obtained after cutting and grinding the alfalfa plant.
            (1') Crop is cut and may or may not be wilted.
            (2') The freshly chopped alfalfa is dried by rapid passage (3 to 5 minutes) through a high temperature (about 1000° F) rotary drum drier.
            (3') Evaporation keeps the product cool for better nutrient preservation.
            (4') Pelleting and the addition of fat are used to reduce bulk and dustiness.
            (5') Contains 15% to 23% CP depending on the stage of maturity when harvested.
            (6') Some dehydrated alfalfa is stored in an atmosphere of inert gas (nitrogen) until ready for shipment to reduce carotene and other vitamin losses.
      (8) Alfalfa pasture is not very tolerant of continuous grazing, and there is often a considerable hazard of bloat for ruminants.
   c. Red clover
      (1) Perennial plant normally growing taller than alfalfa but does not have as extensive root development nor productive life.
      (2) Best suited to regions with abundant rainfall. An important crop in the northcentral and northeastern regions of the United States.
      (3) Used primarily for hay purposes and sown extensively with timothy.
      (4) Red clover hay has similar nutrient composition to alfalfa except for slightly lower CP content (12% to 22%)
   d. Alsike clover
      (1) Perennial plant similar to red clover but more leafy and has greater root development; under favorable conditions, the stems reach a length of 2.5 to 5 ft.

(2) Requires a cool climate with abundant moisture; very cold resistant; probably as resistant to dry weather as red clover; grown in similar regions to red clover.

(3) Produced largely where red clover yields are decreasing (although red clover normally outyields alsike).

(4) Especially suited for pastures, generally seeded with timothy or with timothy and red clover for hay.

(5) Compares favorably with red clover hay in feeding value.

e. White clover

(1) Long-lived, shallow-rooted perennial; plants spread rapidly, frequently form a dense turf and are seldom injured by mowing or grazing.

(2) Widely grown, occurring in all parts of the United States; thrives best under cool, moist climates.

(3) White clover is a pasture plant, usually seeded as a mixture with bluegrass.

(4) Highly palatable and nutritious (15% to 28% CP).

f. Ladino clover

(1) Ladino clover is a giant, rapid-growing, perennial type of common white clover, spreading by creeping fleshy stems that root at the nodes. The leaves, stems and flower heads grow from two to three times as large as those of common white clover.

(2) Ladino clover thrives in a temperate climate and favors moist fertile soils; does not grow well on extremely wet soils or on poor droughty soils.

(3) Has many uses, however; is primarily a grazing crop. Grows well in mixtures of orchard grass and tall fescue.

(4) Ladino clover does not withstand close or continuous grazing. It may also be winter killed or summer killed. It is difficult to mow and especially to cure. Hay yields are usually quite low.

(5) Similar feeding value, but has a higher livestock-carrying capacity compared to common white clover.

g. Sweet clover

(1) There are both biannual and annual forms; stems vary in height from 3 to 6 ft and bear numerous sweet-scented flowers.

(2) Sweet clover will grow almost anywhere, provided there is more than 17 in of annual rainfall and sufficient lime in the soil. It is very resistant to cold and drought.

(3) It may be used for pasture, hay, silage, soilage or for soil improvement.

(4) It is lower in CP (11% to 20%) than most other clovers and is too coarse and stemmy. Shatters its leaves too readily to make quality hay.

(5) Sweet clover leaves contain coumarin, which reduces palatability; if sweet clover is not properly cured before collecting as a hay feed, molds may develop and change the coumarin to dicoumerol-like compounds resulting in prolonged blood-clotting time in animals consuming large amounts of the moldy hay (called sweet clover poisoning).

h. Birdsfoot trefoil

(1) Long-lived perennial with a branching taplike root with a few or many stems growing from each crown.

(2) Grows under a wide variety of soil conditions, is moderately drought resistant and is winter hardy. It is better adapted to the northern than the southern half of the United States because it is susceptible to a summer disease known as rhizotonia, which kills stands during warm, humid summer weather.

(3) Satisfactory trefoil stands are hard to establish compared with alfalfa and other forages; however, once established, they provide excellent forage for pasture, hay or silage.

(4) Birdsfoot trefoil

(a) Produces well on soils less fertile than needed for alfalfa.

(b) Is not damaged by the alfalfa weevil.

(c) Is less likely to cause bloat in cattle and sheep.

(d) Compares favorably in feeding value with other common legumes (13% to 21% CP).

(e) Grows well in mixtures with orchard grass, Kentucky bluegrass, or timothy.

    i. Lespedeza
        (1) Summer annual that normally grows to a height of 4 to 6 in.; under favorable circumstances the crop readily reseeds itself, and for this reason a stand is maintained from year to year by natural seeding.
        (2) Lespedeza will grow on all kinds of soils as long as they are well drained near the surface. It thrives in hot weather and seldom begins growth in the spring until warm weather. Since it is sensitive to frost, it does not make early spring or late fall growth.
        (3) Lespedeza is used extensively for pasture, and in certain sections of southern United States it is important as a hay crop. If used for hay, it is relatively low yielding and good for only about one cutting per season. It might be of low quality because of shattering and a high content of weeds.
        (4) Feeding trials indicate that annual lespedeza hay approaches alfalfa and will normally contain 13% to 16% CP.
    j. Crown vetch
        (1) Long-lived productive perennial that spreads from fleshy rhizomes; stems are leafy, hollow and weak, forming a dense mat of growth.
        (2) Adapts well throughout the humid north on well-drained, eroded land.
        (3) Once established, crown vetch is excellent for erosion control and is widely used for road bank cover.
        (4) Because of its bloat-free characteristics and tolerance to alfalfa weevil, it is finding some use for permanent cattle pasture. Recovery growth is slow, and in some instances stands have been largely lost when cut for hay.
        (5) Contains β-nitropropionic acid, which appears to reduce its palatability compared to alfalfa when fed to cattle; this compound also appears to have a toxic effect when crown vetch is fed to monogastric animals (e.g., swine and poultry).
        (6) Contains 12% to 25% CP (d.m.) depending on stage of maturity (avg 18% CP).
    k. Soybean, cowpea and peanut forages
        (1) All three are annual legumes with somewhat similar nutrient composition (16%, 17% and 11% CP, respectively); however, nutrient composition and feeding value vary greatly depending on harvest procedure and conditions.
        (2) Soybean and cowpea must be freshly seeded for each cutting; peanut forage usually consists of remains of the peanut plant after the nuts, which grow underground, have been harvested.
        (3) Not used extensively; however, can be used as emergency hay crops.
  2. Nonlegume forages (primarily grasses)
    a. Most nonlegume grass forage will contain more than 10% CP (d.m. basis) if properly managed, e.g., adequate nitrogen fertilization and harvested in the immature stage. Improperly managed grass forages contain less than 10% CP.
    b. Nonlegume grass forages generally contain less than 0.9% calcium.
    c. Excellent in combination with legumes.
    d. See carbonaceous roughage section for a description of some of the grass forages.
C. *Carbonaceous Roughages*
Contain less than 10% CP (d.m. basis); made up of nonlegume forages and miscellaneous low quality roughages.
  1. Nonlegume forages—often contain less than 10% CP when grown under normal to poor management (e.g., no nitrogen fertilization) conditions. As mentioned in the previous section, they will contain more than 10% CP when properly fertilized and harvested. Thus, a wide range of CP composition is shown for each forage.
    a. Native perennial range or prairie grasses (warm-season grasses)
      These species have greater temperature and drouth tolerance and are considered more productive during summer months compared to the commonly grown cool-season species. They are very important as a source of forage in the Great Plains and western corn belt. The following species are some of the more important native warm-season grasses.
        (1) Big bluestem
            (a) Erect, robust, perennial bunchgrass that grows 3 to 6 ft tall and is often reddish-purple at maturity.

     (b) The leafy forage is highly palatable to all classes of livestock (especially after maturity), and it withstands moderately heavy grazing.

     (c) The hay is of good quality if mowed before the stemmy seed heads have formed.

     (d) Big bluestem is a very desirable grass where adapted because of high production of very palatable forage and hay.

     (e) Contains 4% to 11% CP (d.m.).

  (2) Indiangrass

     (a) Erect, robust perennial growing from 3 to 6 ft tall.

     (b) Normally grows in small patches or scattered bunches and is not tolerant of repeated close grazing.

     (c) It does not cure particularly well (for haymaking) and is generally considered only moderately palatable after maturity and fair forage for winter grazing.

     (d) Contains 5% to 11% CP (d.m.).

  (3) Switchgrass

     (a) Tall, coarse-stemmed, broad-leaved perennial that grows from 3 to 5 ft tall.

     (b) Spreads slowly by short rhizomes.

     (c) This species is easier to seed, seed cost is lower and is established more quickly than Indian grass, and side-oats grama.

     (d) In the immature stage, leaves and stems of switchgrass have a good forage value and are readily grazed by livestock. After heading, palatability and nutrient content decline quickly. Its value for standing winter feed is poor.

     (e) Contains 3% to 11% CP (d.m.).

  (4) Side-oats grama

     (a) Long-lived, erect perennial growing up to 3 ft tall and readily spreads by seed. and rhizomes; recognized by its long flower stalks with short, dangling purplish spikes.

     (b) Rarely forms pure stands, usually growing in association with big bluestem.

     (c) Grazed mostly in late summer and fall but remains moderately palatable into the winter; stems are unpalatable and are not normally grazed.

     (d) Side-oats grama is palatable to all classes of livestock, having about the same forage value as big bluestem.

     (e) Contains 7% to 13% CP (d.m.).

b. Cool-season perennial forage grasses

These grasses are mostly introduced perennial species that have been of major importance in forage-livestock pasture and hay programs in the humid northern regions of the United States. They are most productive during early spring, late spring to early summer and again during fall periods if moisture and temperature conditions are favorable. Although they are extremely valuable as a source of forage in the Midwest, all of these perennial cool-season grasses suffer some degree of dormancy during July and August due to high temperatures and drouthy conditions.

  (1) Kentucky bluegrass

     (a) Long-lived, perennial sodgrass with rhizomes, and grows 6 to 30 in. tall.

     (b) Often the earliest growing grass in the spring, goes into semidormancy in the summer, but is revived by late summer and fall rains. It is very sensitive to heat and summer drought.

     (c) Highly palatable and nutritious to all species of livestock. Few grasses are able to withstand continued heavy grazing as well as this grass.

     (d) Undesirable as a hay grass because of its low growth, low yield, and maturity before other grasses are ready to cut.

     (e) Main lawn grass in America.

     (f) Contains 12 to 17% CP (d.m.).

  (2) Smooth bromegrass

     (a) Erect, leafy, long-lived, drought-resistant perennial, 2 to 4 ft tall with many underground rootstocks; each plant may reach a diameter of a foot or more and commonly produces a dense sod.

     (b) Used for pasture, hay, silage and erosion control. Produces abundant herbage in the spring and late summer. Plants reach their best growth in the second and third year.

(c) When seeded on adapted sites, smooth brome is resistant to grazing, is cold tolerant, has high seedling vigor and will tolerate moderately salty soil.

(d) For all classes of livestock it is quite palatable, even after seedstock development.

(e) Often grown in combination with alfalfa, other legumes or both.

(f) Contains 4% to over 20% CP (d.m.) depending on maturity and fertilization.

(3) Orchardgrass

(a) Long-lived perennial bunchgrass that forms dense circular bunches that do not produce good sod. It commonly grows in clumps 2 to 4 ft tall.

(b) Shade tolerant, moderately heat and cold resistant and establishes a stand rapidly.

(c) Starts growth early in the spring, and new, immature growth is highly palatable to livestock. However, it grows and matures rapidly. As it matures, palatability and nutritive value decline. To keep orchardgrass green throughout the summer and to prevent large, unpalatable clumps from forming, it should be grazed or hayed while actively growing.

(d) Recovers from grazing or mowing more rapidly than smooth bromegrass and continues growth during midsummer when smooth bromegrass becomes somewhat dormant.

(e) Generally recognized as superior to smooth bromegrass as a deterrent to bloat when used in mixtures with alfalfa, because orchardgrass is more uniformly productive through the season rather than only in the early part of the summer.

(f) Contains 8% to 18% CP (d.m.).

(4) Timothy

(a) Perennial bunchgrass, 2 to 3½ ft tall, with a swollen or bulblike base but without rhizomes. The root system is relatively shallow and fibrous.

(b) Timothy is primarily a hay plant and does not stand heavy grazing. It is generally included, however, in mixtures for pastures. When grown in mixtures with clover or alfalfa, the first growth frequently is harvested for hay and the later aftergrowth pastured.

(c) Timothy is especially popular as a hay for horses and should be cut no later than the early bloom stage for maximum nutrient value.

(d) Contains 8% to 12% CP (d.m.).

(5) Reed Canarygrass

(a) Coarse, vigorous, long-lived perennial 2 to 6 ft tall, with leafy, short stems, tending to grow in dense bunches 2 or 3 ft in diameter and spreading underground by short, creeping rootstocks.

(b) Prefers a moist, cool site and thrives on land too wet for most other grasses. In addition, it is one of the most heat and drought tolerant of the cool-season grasses. Thus, under proper management this species also does well on upland sites.

(c) Used mostly as pasturage, but its use for hay and silage is increasing.

(d) For best quality pasture, excessive forage growth must be avoided. Keep it grazed to at least 12 to 18 in. during the flush growing period.

(e) Best quality hay is produced by mowing when the first heads begin to appear. Hay quality may be improved by early spring pasturing to delay maturity dates, thus reducing the coarseness of growth. Dense reed Canarygrass sod will support equipment where haying operations were formerly impossible.

(f) Has a long growing season and recovers quickly from grazing or mowing.

(g) In general, seeding a legume with reed Canarygrass has not been successful, due to the shading effect of the leafy, tall-growing grass.

(h) Contains 9% to 13% CP (d.m.).

(6) Tall fescue

(a) Tall fescue is a deeply rooted and strongly tuffed perennial bunchgrass with stems 3 to 4 ft tall. The numerous dark green basal leaves are broad and flat and glisten in the sunlight.

(b) It is adapted to moist, deep soils, tolerates moderately high soil salinity, is able to survive prolonged winter flooding, but is not tolerant of extended drought.

(c) Tall fescue is used for pasture and hay, but more widely for pasture—especially winter grazing.

(d) It is only fair in palatability for livestock and becomes coarse and low in quality if left ungrazed or lightly grazed and allowed to mature. Periodic close grazing (down to 2 to 4 in.) in a rotation program is recommended.

(e) When used for pasture, it is frequently sown with a legume such as ladino clover.

(f) In some areas, cattle grazing tall fescue have developed a lameness commonly called "fescue foot." It is characterized by loss of blood supply to the extremities similar to ergotism.

(g) Contains 10% to 15% CP (d.m.).

(h) "Summer slump" (summer toxicosis)—a time period during which growth or milk production of cattle grazing fescue pasture are depressed. Generally occurs as environmental temperature rises above 86° F and plant begins to develop head; these conditions tend to coincide with formation of a plant endophyte fungus and high plant content of the alkaloid perloline. These compounds reduce palatability and digestibility and cause an increase in body temperature and roughened hair coat of the cattle. Cattle prefer to stand in the shade or in water. Problems with "summer slump" may be lessened or eliminated by planting a variety of tall fescue, which is resistant to endophyte.

(7) Redtop

(a) Perennial grass, 2 to 3 ft tall, which spreads by shallow, vigorous rhizomes, 2 to 6 in. long and produces a coarse, open turf. Redtop has flat, sharp-pointed leaves and gets its name from the fact that the mature heads lend a reddish tinge to a meadow made up of this crop.

(b) Redtop probably has a wider range of adaptation to soil and climate than any other cultivated grass. It succeeds well over most of the United States, except in the drier regions and the extreme south.

(c) Used primarily as a hay crop although less palatable than many grasses. Redtop makes acceptable hay if cut in the early flowering stage, but it will quickly become stemmy and unpalatable if cutting is delayed.

(d) Contains 6% to 15% CP (d.m.).

(8) Perennial ryegrass

(a) Perennial ryegrass is short-lived, rapid-growing, leafy perennial that ordinarily attains a height of 1 to 2 ft. On poor soils the grass lives only 2 years, and frequently, when seeded in hay mixtures, perennial ryegrass will disappear after the first year as a result of shading by taller plants.

(b) Because it makes rapid germination and growth, it is frequently included in pasture and lawn mixtures. It serves excellently as a temporary covering, while the more valuable and permanent plants are being established. It is also used to sow on Bermuda pastures in the south for winter grazing.

(c) Contains 6% to 13% CP (d.m.).

c. Southern grasses

Climate in southern United States permits the growth of a wide variety of forage crops. Livestock production in this region has greatly increased, and much progress has been made in breeding and production of valuable crops as feed for these animals. Some of the grasses more commonly used in the southern states are herewith described.

(1) Common Bermudagrass

(a) Long-lived perennial spreading by runners, rootstocks or both, and by seeds. Runners may vary in length from a few inches to several feet. The stems of the plant are very leafy.

(b) Requires warm weather during the growing season and will bear intense heat without injury. It is not resistant to cold and will not stand shading well.

(c) The more common method of securing a stand is through the use of sprigs or pieces of sod, since Bermudagrass sets seed lightly.

(d) Most commonly used for pasture rather than for hay. When well fertilized and cut at proper maturity, Bermudagrass hay approaches coastal Bermudagrass hay in feeding value.

(e).Not as drought resistant or as high yielding as coastal Bermudagrass.

(f) Contains 6% to 15% CP (d.m.).

(2) Coastal Bermudagrass
   (a) Coastal Bermudagrass is a variety superior to common Bermudagrass that makes more growth, grows later in the fall, is more cold resistant and is more resistant to leaf diseases and root-knot nematode.
   (b) Coastal Bermudagrass is the most extensively produced hay crop over much of the Deep South.
   (c) When adequately fertilized and cut at the proper stage, it will make a high quality hay providing three or more cuttings per season. Its fine stems and low moisture content make it easier to cure than any other hay crop adapted to the South.
   (d) Coastal Bermudagrass is a hybrid and does not produce viable seed; thus, it must be reproduced vegetatively from sprigs.
   (e) Contains 7% to 18% CP (d.m.).
(3) Dallisgrass
   (a) A stout, upright-growing bunch grass reaching 2 to 4 ft in height. It has a deep, strong root system and, if properly fertilized and managed, remains productive for some years.
   (b) First grass to begin growth in the spring, makes continuous growth during warm weather, is not injured by moderate frosts and is the last grass to become dormant in the fall.
   (c) Used primarily for pasture, Dallisgrass and adapted legume combinations provide grazing over a longer period than any other combination.
   (d) Produces abundant seed, but the germination is often poor because of the fungus disease ergot which attacks and destroys the seed.
   (e) Contains 4% to 12% CP (d.m.).
(4) Bahiagrass
   (a) Deep-rooted perennial, with short, stout, woody, horizontal rhizomes, which form dense, tough sods. Culms range from 6 to 24 in. in height and occur in dense tuffs.
   (b) The crop is adapted in Florida and the lower coastal plain.
   (c) Bahiagrass is used primarily for pasture. It is medium in cold resistance and makes most of its growth during midsummer. It furnishes little grazing during the winter months.
   (d) If cut for hay, it has lower feeding value than coastal Bermudagrass and is used primarily as a wintering feed for the beef breeding herd.

d. Selected annual grasses
(1) Sorghums (Sudangrass; sorghum sudan hybrids; hybrid forage sorghums)
   (a) Annual, erect plants belonging to the sorghum group. Stems are solid and slender (less than 3/16 inch diameter) and average 3 to 5 ft in height when the crop is planted broadcast or drilled and 6 to 8 ft when grown in rows. The plants have numerous rather soft leaves; loose, open panicles; numerous tillers; rarely any branches and no rootstocks.
   (b) The production of numerous tillers is most apparent after the first cutting, and, as a result, the hay of the second cutting is of finer texture than of the first.
   (c) May be used for hay, soilage, silage and summer pasture. They are relished by all classes of livestock.
   (d) Sorghums grow most rapidly in the hotter summer months, when other kinds of pasture are scarce or lacking. Pasturing is best delayed until the plants have made a growth of about 18 in. or about 5 to 6 weeks after planting.
   (e) Sudangrass is best suited for pasture (grazing) and hay production, while hybrid forage sorghum and sorghum Sudans are better for silage production.
   (f) Primarily because of lower yields, summer annuals should not be harvested for silage or hay at the early vegetative stage.
   (g) Effect of maturity on yield and feeding value of summer sorghum forages. (Source: *Summer Annual Forages for Livestock Production in Kansas*. KSU Agr Expt Sta Bult 642.)

	Vegetative	Boot	Soft Dough
d.m., %	20–25	23–28	34–36
d.m. yield, T/acre	4.3–4.5	6.0–6.4	6.6–8.8
CP, %	13–18	10–13	5–9
ADF, %	32	34	38
IVDMD, %	66	63	56

(h) Sudangrass makes a hay of excellent quality; however, livestock tend to refuse the coarser stems. Yields are good. In the southern United States, two cuttings may be secured and sometimes as many as four. A hay conditioner may be used to facilitate drying.

(i) Sudangrass, like other sorghums, sometimes contains prussic acid or hydrocyanic acid, which may cause toxicity in animals grazing these plants. Silage and dry fodder may contain toxic amounts of prussic acid, but usually can be fed safely. The free acid volatilizes as the silage is being handled in feeding. Also, the poison decreases during the curing process of the fodder. The leaves and tillers are higher in prussic acid than other parts of the plant.

(2) Pearl millet

   (a) Tall, erect, annual grass that grows from 6 to 15 ft in height. Stems are rather thick, coarse, harsh and pithy.

   (b) Requires a rich soil and under favorable conditions produces enormous yields of green fodder, which can be cut several times in a season.

   (c) Can be used as a pasture crop or cut for hay. For best quality hay, pearl millet should be harvested when the heads begin to appear.

   (d) Fed primarily to cattle, as it is not satisfactory for lambs and should not be fed horses since it may cause joint swelling and lameness.

   (e) Pearl millet will generally be higher in CP and digestibility than the sorghum forages, and has the further advantage of freedom from prussic acid.

   (f) Contains 6% to 20% CP (d.m.) depending on stage of maturity.

(3) Small (cereal) grains (wheat, oats, rye, barley)

   (a) Most varieties commonly grown for grain are good varieties for forage (pasture, hay or silage). If made into hay, awnless varieties are more palatable and will cause less irritation and soreness to the mouths of livestock.

   (b) Use the same cereal crop seeding rates for silage production as for grain production. An increase of 30% to 100% above these rates is suggested for harvest of hay or early wilted silage. If they are to be fall grazed, seed about 2 weeks earlier than for grain.

   (c) Harvesting as pasture

      (1') Fall grazing of winter cereals should not be excessive; leave a 3-in. stubble height for winter protection.

      (2') Spring grazing of cereal grains is most common; if to be harvested for grain, grazing must not continue after "jointing," or when the developing head is 1 in. above the soil surface. Excessive or prolonged grazing will severely decrease grain yields and straw production. Grazing of small grains in the spring will also result in a 1- to 4-day delay in maturity.

      (3') Expected carrying capacity = 2 to 4 acres/cow.

   (d) Harvesting as hay or silage

      (1') Harvesting for hay or silage when about 20% of the stems reach the flower (bloom) stage of maturity appears to yield the maximum animal production per acre.

      (2') Silage should be wilted to 65% to 70% moisture if harvested before the dough stage of growth.

      (3') Forage for hay should be passed through a hay conditioner to encourage rapid curing.

      (4') Yields from all cereals may range from 2- to 3-ton dry matter per acre. Barley usually has the highest grain-to-forage ratio, followed in order by soft wheat, hard wheat, rye and oats.

(e) Nutrient content will vary greatly with stage of maturity and grain-to-forage ratio.

	Boot-early Flower	Milk	Dough
d.m., %	16–22	21–27	28–40
CP, %	10–22	8–16	5–15
TDN, %	62–72	50–60	50–60

(f) Feeding guidelines
  (1') Because of nutrient variability, a protein analysis of forage is valuable for most efficient use.
  (2') Where grain-to-forage ratio is high, the feed value of small grain forage may approach that of corn silage. When harvested at the flower or earlier stages, forage should be equal to grass hays of approximately the same protein content.
  (3') On occasion, small grain hays (particularly oats hay) have had enough nitrate content to cause death. Hays harvested at the boot to heading stage are apt to be higher in nitrate than those harvested at later stages. Nitrate poisoning seldom occurs from feeding small grain silages.
  (4') Low magnesium levels in some low quality small grain forages is a possibility and may cause grass tetany in grazing cattle. If these forages provide the main feed for long periods of time, may want to feed a high magnesium mineral supplement.

2. Miscellaneous carbonaceous roughages
  a. Corn silage
    (1) Most popular silage in the U.S. in areas where corn plant grows well.
    (2) Excellent succulent feed for most classes of livestock.
    (3) Moderate to high energy content (for forage material), but usually low in protein (7% to 9% CP on d.m. basis).
    (4) The grain in corn silage harvested at optimum maturity stage makes up about half the dry weight and two-thirds the nutritive value of the silage. Thus, when selecting a hybrid for silage, pick varieties that give high grain yields.
    (5) Comparison of energy value (based on NE system) of corn silage to other ensiled feeds for feedlot cattle (Figure 3–1).
    (6) Animals on a high corn silage ration do not consume as much dry matter or energy as they would from a high grain ration. This slows rate of gain and increases length of time necessary to reach a comparable slaughter grade. These disadvantages of high silage feeding must be balanced against the usual lower costs and high volume of production of silage.
  b. Sorghum silage
    (1) Includes grain sorghum (milo), forage sorghum, Sudangrass, and sorghum-Sudangrass hybrids.
    (2) Similar to corn in their adaptability for silage, preserve well when harvested at the right moisture content and make excellent feed.
    (3) Grain sorghum and the one-cut forage sorghums are best suited for silage.
    (4) Grain sorghums for silage should be selected on the basis of time needed for maturity and their ability to produce grain. Grain sorghum stalks are short, and the grain constitutes more than two-thirds of the feed energy produced.
    (5) Forage sorghums for silage should be selected for their ability to produce forage (dry matter) rather than for grain. Silage yields of forage sorghum are often equal or slightly higher than those of corn. However, their nutritive value is lower, and total feed energy produced per acre is usually less.
    (6) Harvesting grain sorghum too mature (beyond the dough stage) results in losss when fed to cattle. The grain is then so hard that much of it passes through the livestock undigested.
  c. Corn plant residue
    (1) Dry matter distribution of the corn plant when the grain is at 25% moisture (typical grain harvest stage) includes
      (a) Grain—50.4%.
      (b) Stalk—19.9%.

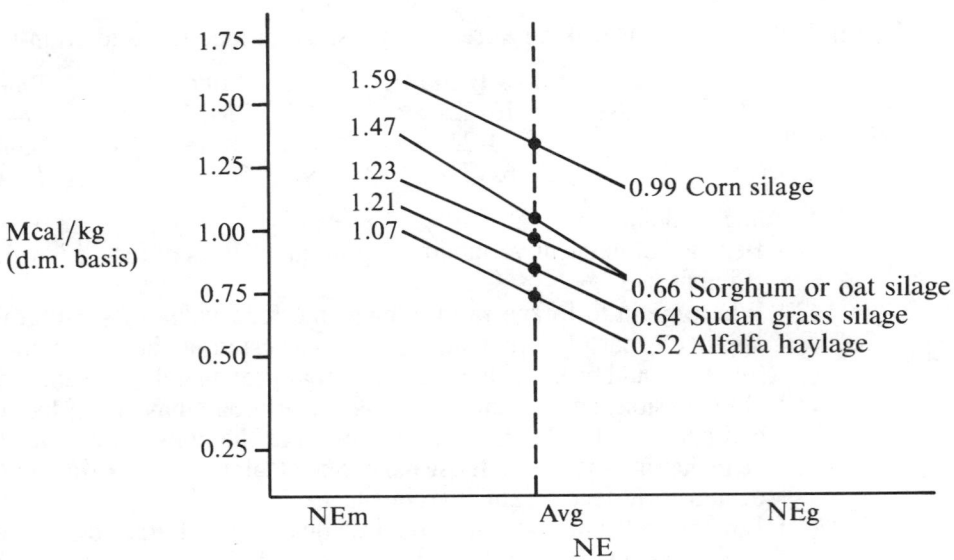

Relative energy value			
Corn silage	1.29/1.29	=	100
Sorghum silage	1.07/1.29	=	83
Oat silage	.95/1.29	=	74
Sudan grass silage	.92/1.29	=	71
Alfalfa haylage	.79/1.29	=	61

FIGURE 3–1.

(c) Cob—11.5%.

(d) Leaf—9.4%.

(e) Husk—8.8%.

(2) The residue products of corn harvesting are all low in feeding value; however, they can supply considerable amounts of dietary energy and, if properly supplemented, can serve as a useful roughage in rations for at least certain classes of ruminant animals.

(3) Cornstalk quality decreases after the grain reaches physiological maturity. Most of the decline is due to a reduction in cell solubles as the potentially digestible cell wall remains constant. When there is a lower grain yield, the stalks are of higher quality. However, in excellent grain-producing years, the feeding value of the stalk is lower. The rate of the cell-wall digestion may vary between varieties and cultural practices.

(4) A lamb digestion trial was performed at the University of Nebraska (*1980 Nebraska Beef Cattle Report*) to compare corn silage, corn stalks at physiological maturity, corn stalks at 35 days postmaturity (at high moisture grain harvest) and corn stalks at 63 days postmaturity (harvested after dry grain harvest). All feedstuffs were ensiled and allowed to ferment before being fed. Cornstalks harvested at the time of corn silage harvest were 91% as digestible by lambs as corn silage. Stalks harvested at the time of high-moisture grain harvest were 79% as digestible, while those harvested when dry were only 55% as digestible as corn silage. Thus, if stalks are being used, increased quality can be obtained with an earlier harvest date.

(5) Most corn grain is presently combined, leaving the plant residue in the field. The utilization of this residue (primarily cornstalks or stover) as a feedstuff is primarily by animal grazing or mechanical harvesting.

(6) In northern sections of the United States, grazing fields may become impossible due to snow cover. Thus, many cattle producers are using equipment designed to harvest, handle and feed residue products to cattle in drylot.

(7) Equipment is available for making the following two popular corn residue feeds for dry (nonensiled) storage

(a) Corn husklage

Husklage is the forage discharged from a combine; can be collected with a "straw buncher" pulled behind the combine. The husks, cobs and some grain carryover make an excellent feed, although volume is light because most of the stalk and leaves are left in the field. Using a "dump vehicle," husklage is usually left on headlands as the combine turns, parceled out for feeding by use of electric fence. Iowa State University research indicated that husklage yields nearly a ton of dry matter per acre, with about a 90% recovery rate. Contains 3.5% to 5% crude protein (d.m.)

(b) Corn stover (baled or stacked)

This system involves collection of corn plant residue from the ground after grain combining.

    (1') Advantages of the stover system compared to husklage

        (a') Stover harvest need not compete for operator time during grain harvest.

        (b') Most systems feature one-man harvesting operation.

        (c') Most of the stover systems provide enough compaction to permit no-cost outdoor storage.

    (2') Most farmers would prefer to combine corn when grain moisture is down around the 20% level. When the grain is at the 20% level, the stalks-leaves-husks, cobs are around 40% moisture, which is considered too high for safe "dry storage." After combining grain, delay stover harvest for a few days or a week. Drydown behind the combine is relatively fast. Long delays in stover harvest will reduce the percentage of roughage picked up and reduce feeding value.

    (3') An emerging pattern seems to be to harvest two rows of each four or six rows, leaving the remainder for field grazing (and erosion control). Some operators take the spreader off the combine, leave combine residue in windrows and harvest these center two rows for maximum stover yields.

    (4') Stover is commonly stored as large-round bales (ranging upward from 1500 lb); some stockmen prefer the big-package stacks (upwards from "one" ton each). Several companies are offering equipment along these lines.

    (5') Stover is usually too dry for storage in a silo, is not chopped fine enough for silage packing and, because of its light density, it takes as much as twice the silo capacity as regular whole-plant silage.

    (6') Crude protein content varies but is generally figured between 4% to 6% (d.m.).

(8) Corn stover silage (or stalklage)

(a) Refers to corn plant residue harvested at the time of or soon after grain harvest and stored in an anaerobic (silo) structure.

(b) To ensure proper packing and fermentation, stalklage must be finely chopped and the moisture content be 40% or greater.

Beef cows grazing cornstalks—the residue remaining in the field after harvesting the corn grain.

(c) Most research reports suggest ensiled corn stover is superior in nutrient content and digestibility compared to any other method of feeding the corn residue material. Corn stover silage has the potential of providing a major portion of the energy requirements of the gestating beef cow and has supported up to 80% the performance of growing-finishing cattle fed corn silage. Several developments must occur, however, before corn stover silage will be widely accepted. These include

    (1') Equipment for efficient simultaneous harvesting of grain and stover.

    (2') Genetic lines of corn that will result in optimum yields of grain and quality of stover.

    (3') Nutritional studies to determine

        (a') Proper processing of stover (chopping and storage).

        (b') Optimum stover-to-grain ratio and the required supplementation to maximize the feeding value of stover.

(9) Corn cobs offer a potential feed source in areas where ear corn is harvested. Although their crude protein content is low (2% to 3%), corn cobs are useful as an energy source for ruminants.

d. Sorghum stover

(1) Sorghum stover is the forage residue remaining in the field after grain sorghum (milo) harvest. Nearly 50% to 60% of the total plant dry matter of grain sorghum remains in the field, as compared with estimates of 45% to 50% of the corn plant remaining.

(2) Photosynthetic activity continues after seed maturation or grain harvest, and the stover plant accumulates water-soluble carbohydrates and CP until a killing frost. This unique characteristic allows the stover crop to maintain and perhaps improve its nutritional composition before utilization for winter grazing. Dry matter yields may increase from 10% to 13% between grain harvest and a killing frost. Generally, sorghum stover can remain in the field from 20 to 30 days after a killing frost without significant losses in dry matter.

(3) When grazed, only about 25% to 30% of the available residue is normally used by the animals because of selectivity and reduced availability by mud and snow. However, sorghum stubble is cut at a relatively high level, so there is more forage available and snow is less of a problem than is usual for other crops such as cornstalks.

(4) Studies have suggested that taller late maturing varieties resulted in more forage available for grazing when snow covered the field. Furthermore, these varieties produce more forage with more digestible leaves and resulted in better calf gains than early maturing varieties (*1985 Nebraska Beef Cattle Report*).

(5) When grazing sorghum stover, producers must be aware of possible prussic acid poisoning. It may be safely utilized by cattle after a killing frost if the cattle are kept from the stover for 3 days following the killing frost. The frost must also have completely killed any new shoots that may have appeared after grain harvest for cattle to safely graze the stover.

e. Soybean stover

(1) Refers to the stems (with few or no leaves) and pods available to be grazed or harvested from soybean plant residue left on the ground following combine harvest of soybeans.

(2) Soybean stover can be stacked or baled immediately after combining because the pods and stems will be about as dry as the soybeans.

(3) Contains 4% to 6% CP (d.m.).

(4) An average yield of soybean stover is one ton of dry matter per acre.

f. Oat, barley and wheat straws

(1) Straw is what remains after harvesting the grain from the mature oat, barley or wheat plant.

(2) Used more commonly for bedding than for feeding purposes because all three are very low in feeding value (2% to 5% CP, d.m.).

(3) Most commonly fed straw is oat. It is softer than wheat or barley straw.

g. Hulls

(1) Several products include:

    (a) Cottonseed hulls—byproduct of the oil extraction process; contains 4% to 5% CP (d.m.).

Corn plant residue gathered and compressed into stacks by hydraulically operated field machines.

    (b) Oat hulls—byproduct of the manufacture of oat groats and rolled oats from oat grain; contain 3% to 4% CP (d.m.).

    (c) Peanut hulls—byproduct of the manufacture of peanut butter, peanut oil and shelled peanuts; contain 5% to 8% CP (d.m.)

    (d) Rice hulls—byproduct from the manufacture of polished rice from rice grain; contains 3% to 4% CP (d.m.).

  (2) Hulls may provide limited energy value to a ruminant animal but must be mixed with more palatable feeds to induce their consumption.

  (3) All four are very low in nutritive value and are used primarily as a bulk or filler factor in the ration.

  (4) Cottonseeed hulls are generally considered the most palatable.

h. Soybean mill feed

  (1) Composed of soybean hulls and the offal from the tail of the mill that results from the manufacture of soy grits or flour.

  (2) Must contain not less than 13% CP and not more than 32% CF.

  (3) Used principally in livestock rations as a source of bulk or as a carrier agent in premix products.

i. Note: If ruminant animals are to make maximum use of the previous low grade roughages they must be supplemented with

  (1) Adequate protein.

  (2) Some readily available carbohydrate (starch from grain, sugar from molasses or both).

  (3) Additional minerals (major and trace minerals).

  (4) Vitamin A.

D. *Methods of Utilizing Forage Crops*

  1. Pasture

Pasture is the major feed in the United States for dairy and beef cattle, and for horses. Even swine and some poultry programs may use pasture. It is the most economical feed available for certain feeding programs. For these reasons, consideration is given here to some management practices that relate to the feeding value of pasture.

a. Essential qualities of pastures

  (1) It is generally desirable that pastures be made as enduring as possible.

  (2) A good pasture should start growth early in the season and continue to produce until late in the fall.

  (3) Plants should form a continuous, compact turf that will withstand much trampling by animals.

  (4) A variety of plants that will provide for growth under both moist and dry soil conditions is advantageous.

  (5) In seeding permanent pastures, a mixture of several kinds of grasses and legumes has many advantages.

(a) Legumes in the mixture help keep up the nitrogen supply of the soil.

(b) Mixtures usually result in a more uniform stand and greater yield because of the different soil conditions that may exist.

(c) Mixtures provide a more uniform production during the various seasons, because plants vary in their periods of lush growth.

(d) Mixtures of grasses and legumes provide a more balanced feed, because legumes contain more protein and minerals than grasses.

  (6) Further information regarding varieties and planting suggestions for grasses and legumes in permanent pastures can be obtained from university extension publications in your local area.

b. Stage of maturity effects nutritive composition of pasture (and hay cut from pasture). Increasing maturity may result in

(1) Reduced protein content.

(2) Reduced energy content (TDN or NE).

(3) Increased fiber content.

(4) Mineral content reduced.

(5) Lowered carotene content.

(6) Examples: (Values expressed on a dry matter basis)

	CP, %	TDN, %	NE_L, Mcal/kg	Ca, %
Grass, late veg.	12.2	67.0	1.52	.40
Grass, mature	7.8	57.0	1.29	.36
Legume, late veg.	22.6	63.0	1.42	1.50
Legume, mature	16.7	55.0	1.23	1.20

c. Effect of soil fertility on nutritive makeup

(1) Fertilizing for maximum yield usually gives maximum nutritive value and a more palatable pasture.

(2) Deficiency of Ca, P or trace minerals in soil can result in low values in the crop.

  (a) Low phosphorus soil; grass = .10% P (calcium not as greatly affected as P).

  (b) High phosphorus soil: grass = .20% to .40% P.

(3) Nitrogen content of soil

  (a) Yield suffers greatly from low soil nitrogen supply.

  (b) Nitrogen additions to the soil can increase nitrogen (or CP) content of the plant (primarily grasses).

d. Proper grazing can increase yield and nutritive value of the plant (relates back to stage of maturity).

(1) Avoid grazing any of the pasture while the land is wet in the spring and when the principal forage plants are just beginning growth.

(2) Limit the animals to the number it is believe the whole area will support in good grazing condition (carrying capacity). Proper stocking rate or carrying capacity is highly variable but the following are typical

  (a) Alfalfa-brome (or orchardgrass): 1 to 2 acres/animal unit equivalent.

  (b) Bluegrass: 3 to 5 acres/animal unit equivalent.

  (c) Tall range grasses: 5 to 6 acres/animal unit equivalent.

  (d) Short range grasses: 10 to 20 acres/animal unit equivalent.

Animal unit equivalents are defined as the number of animals assumed to consume approximately 12 to 16 lb of TDN daily. Examples

    1 mature beef or dairy cow

   1–2 yearling beef or dairy animal

   3–4 weaned calves (under 1 year of age)

    1 mature horse

   4–5 ewes, with lambs at side

 12–14 weaned lambs

   4–6 gestating sows

  20 growing pigs on ⅔ concentrate feed

(3) Keep animals from certain areas each year until after certain forage plants have seeded.

(4) Control and distribute the stock by fences, well-distributed watering places and salt troughs so as to minimize congregation in large herds.

(5) Watch the vegetation on the area as a whole to find out whether the best forage plants are increasing or decreasing, and increase or decrease the number of animals as may be necessary to bring the pasture, or each division of it, to its maximum production.

(6) Rotation grazing is sometimes recommended. Under this system the animals are allowed to graze on one area for a time and then move to another until the pasture on the one just grazed recovers.

  (a) Advantages

    (1') Provides a more nutritious and digestible forage because the plant is maintained at an ideal vegetative stage.

    (2') Improves stand persistence because the plant is allowed time to recover and build root reserves and leaves.

    (3') Reduces selective grazing by the animal and prevents overgrazing and under-grazing.

    (4') Increases carrying capacity. Greater amounts of plant nutrients are removed with reduced losses due to trampling.

  (b) Disadvantages

    (1') Higher input of management and capital than continuous grazing system.

    (2') Always a continuous decline in forage quality after animals turned into the grazing area.

e. Special problems of grazing livestock on pasture

  (1) Poisonous plants

    (a) Certain species of plants can be poisonous to animals that graze them.

    (b) Most apt to occur when pastures are in poor condition from overgrazing, from low soil fertility or other conditions of poor management.

  (2) Bloat

    (a) Bloat is a hazard for cattle and sheep grazing pastures that have a predominance of legumes, such as alfalfa or ladino clover.

    (b) The danger of bloat is reduced by seeding alfalfa or ladino in combination with a grass.

  (3) Allelochemical compounds

    (a) Tannins

      (1') Compounds found in some legumes (crown vetch, birdsfoot trefoil) with strong protein-binding properties.

      (2') Tend to reduce palatability because tannins have a bitter taste.

      (3') May reduce bloat because of ability to interact with soluble proteins.

    (b) Cyanogenic glycosides (prussic acid poisoning)

      (1') Poisoning can occur in all livestock grazing members of the sorghum family.

      (2') Concentration of the toxic acid will be greatest in second-growth sorghum following severe drouth, early frost or period of heavy trampling or physical damage.

    (c) Estrogenic flavonoid—white clover and alfalfa may contain estrogenic compounds at levels high enough to cause reproductive failure or poor reproductive performance.

    (d) Alkaloids

      (1') Certain alkaloid compounds, if ingested by a pregnant animal at a specific time during gestation, alter normal fetal development and result in a malformed fetus.

      (2') Other plant alkaloids have been implicated as vasoconstrictor compounds that may cause certain noninfectious diseases of livestock, such as "fescue foot" of cattle grazing tall fescue.

    (e) Coumarin

      (1') Compound found in sweetclover leaves, which reduces palatability.

      (2') Coumarin is converted to the hemorrhagic agent, dicoumarol, by microbial action during spoilage. This factor reduces the blood prothrombin level and thus prevents proper blood clotting (sweet clover poisoning).

Beef cows with calves at side utilizing rotational-managed pastures.

      (4) Nitrate poisoning
          (a) High accumulation of nitrates in forage may occur from nitrate fertilization or drought. Weeds and cultivated forages (corn, sorghum, oats) are often high in nitrate content.
          (b) Seldom encountered in native plants under range conditions; grasses (brome, orchard) and legumes store very little nitrate under normal growing conditions.
          (c) Causes abortion and even death in livestock.
      (5) Elemental deficiencies and imbalances
          (a) Generally associated with animal health problems.
          (b) One of the most widely recognized problems is hypomagnesemia or "grass tetany" in cattle grazing lush, new-growth forage.

2. Soilage (green chop)

Soilage refers to fresh forage that has been cut and chopped in the field and then fed directly to livestock in confinement. Plants used in this manner include the forage grasses, legumes, some sorghums, the corn plant and at times residues of food crops used for human consumption.

  a. Advantages
      (1) Produces maximum yield per acre; larger yield of nutrients in "hay stage" than if kept grazed.
      (2) Less loss of nutrients than with other harvesting procedures.
      (3) Less fencing than if pasture; crop not trampled by livestock (tall crops, e.g., Sudangrass).
      (4) Doesn't eliminate, but tends to reduce bloat problems with certain legume crops.
      (5) Used most commonly in large commercial dairy operations, although not usually better than good pasture for dairy or beef cattle.
  b. Disadvantages
      (1) Lack of uniform quality from day to day, depending on weather conditions and stage of maturity.
      (2) Weather can be a problem (muddy fields).
      (3) May be expensive due to labor and machinery unless handled in volume.
      (4) Could not have this crop year-round.

3. Hays

Hay is feed produced by dehydrating green forage to a moisture content of 15% or less.

  a. Steps in haymaking
      (1) Harvest the plant at the optimum stage of maturity that will provide a maximal yield of nutrients per unit of land without damage to the next crop. Most forages should be mowed just as soon after reaching an early bloom stage of maturity as circumstances will permit. Delay in harvesting will result in low quality hay.

(2) Proper curing refers to reducing the moisture content of green forages sufficiently to permit their safe storage without spoilage or serious nutrient loss. The maximum permissible moisture content for baling hay is around 18% to 22%. Machines such as crimpers crush the plant stems as they pass between rollers and thereby facilitate the drying process. This is called "conditioning" the crop and may reduce curing time as much as 50%.

(3) Raking the mowed forage into windrows should be completed before it is completely dry to avoid excessive shattering of leaves and overexposure to the sun. Turning the windrow, when necessary to facilitate drying, should be done when the dew is on, especially with hays subject to serious shattering.

(4) Collecting and storing the cured hay is most commonly in the form of bales or stacks. This should be carried out just as soon as the hay is sufficiently dry.

   (a) Square bales should be stored as soon as possible and in any event before they are rained on.

   (b) Round bales (large or small) will shed rain and may be left in the field for extended periods without serious damage to the hay.

   (c) Stacks should be fairly compressed and loaf-shaped to shed rainwater.

   (d) Grass hays normally suffer less damage than do legume hays from outside storage.

b. Common losses in haymaking

(1) Shattering leaves

   (a) Leaves contain two to three times as much protein as stems.

   (b) Leaves are also richer in carotene, B vitamins, minerals and energy.

   (c) Losses from leaf shattering may range up to 8% for grass hay and as high as 40% for legume hays; under ideal conditions one could expect as much as 5% and 20% leaf loss of grass and legume, respectively. With bad weather conditions, leaf losses in legume hay may range from 50% to 75%.

   (d) Hay crusher can be helpful in permitting limited drying time.

(2) Heat damage

   (a) Hay stored containing excess moisture (more than 25% to 30%) may tend to mold and contain bacterial growth and heat. Preventing excessive heating of hay is important to prevent spontaneous combustion and loss of nutrients.

      (1') Hay baled dry reaches a maximum temperature of about 84°F.

      (2') Above about 120°F, nutrient destruction or binding occurs. Proteins appear to be most vulnerable to heat damage.

      (3') Any time the temperature of stored hay reaches 160° to 165°F, there is a danger of spontaneous combustion.

(3) Fermentation or plant cell respiration

   (a) Converts sugars and starch to CO_2 and H_2O (this is lost).

   (b) Reduces energy value.

   (c) Destroys carotene.

   (d) Under good conditions accounts for 5% to 7% of loss in total dry matter.

   (e) Rapid drying is the key to low fermentation losses.

(4) Bleaching

   (a) Color loss is due to destruction of chlorophyll by sunlight.

   (b) Reduces carotene (related to greenness) or vitamin A activity.

(5) Leaching due to rainfall

c. Normal loss values from field to bunk. (Legume hay under good conditions)

(1) Field and barn cured hay

	Percentage Loss
d.m.	20–30
Protein	27–30
TDN	25–28
Carotene	90–97

(2) Dehydrated hay

	Percentage Loss
d.m.	16
Protein	18
TDN	13
Carotene	76

(3) Wilted silage

	Percentage Loss
d.m.	16
Protein	16
TDN	19
Carotene	81

d. Additives in haymaking
  (1) Preservatives
    (a) Chemical preservatives (propionic acid, acetic acid, sodium propionate or sodium diacetate) allow aerobic storage of hay at higher moisture content (25% to 30% moisture).
    (b) Chemical preservatives are applied onto the hay as it enters the bale chamber (about 1% of weight or 20 lb/ton). Complete and uniform coverage of the treated hay is essential for successful preservation.
    (c) Hay baled at a higher than normal moisture level has less leaf and respiration losses in the field, so it retains more of the original nutritional value. A preservative should reduce overall losses by inhibiting microbial activity (mold) during storage and by decreasing the time required between cutting and baling. However, some studies indicate that excess heating may still occur in the treated hay during storage.
  (2) Drying agents
    (a) Applied at the time of cutting, drying agents remove the waxy cutin layer from the stems of alfalfa and other legumes allowing a more rapid loss of plant water.
    (b) Drying agents are generally composed of sodium or potassium carbonate.
  (3) Anhydrous ammonia
    (a) Treating low-quality hay or crop residues with anhydrous ammonia will improve the forage energy availability, protein content and intake by livestock.
    (b) Baled or stacked forage is placed in a sealed environment. Anhydrous ammonia gas is allowed to slowly flow into the covered area over a period of 20 to 30 minutes until about 40 to 60 lb is released per ton of forage in the stack. The sealed environment should be maintained about 3 weeks to allow the gas to partially break down the fibrous cell walls of the forage.

Large round bales of hay (1200+ lb.) stored for winter dry lot feeding.

(c) Another treatment method would be to cover the forage with plastic and inject the bales with liquid anhydrous ammonia at the previous mentioned rate.

(d) Ammoniation works better on summer-harvested forages such as wheat straw or hay rather than fall-harvested residues such as cornstalks. This is because ammonia acts more effectively when temperatures are higher.

(e) Ammoniation of dry forages can increase CP 3% to 6% and improve digestibility as much as 10%.

(f) Ammoniated forage containing highly soluble carbohydrates or high-moisture residues, or treated with more than 60 lb ammonia per ton or both, may cause toxicity and erratic animal behavior. When soluble sugars from medium to high quality hays react with the ammonia in the presence of moisture and heat, toxic compounds called imidazoles apparently are produced. In general, no problems occur with dry or low quality ammoniated hay low in sugar content.

e. Federal grades of hay include: US No. 1, US No. 2, US No. 3 and US Sample. These grades, in general are based upon the percent of leafiness (in legumes), percent of green color, percent of foreign material, maturity or ripeness when cut, size and pliability of the stems and general condition of the hay.

f. The type of animal will largely determine the kind of hay used.

(1) Swine, poultry, sheep and horses need a good quality hay.

(2) Beef and dairy cattle can make use of poor quality hay except in highly productive rations.

4. Silages

a. Silage is the product of acid fermentation of green forage crops that have been compressed and stored under anaerobic conditions in a container called a silo.

b. Types of silos

(1) Upright or tower

(a) Cylindrical, vertical structures that normally have a roof.

(b) Semiairtight structures may be constructed of wood or concrete staves, poured concrete, tile block, brick or field stone masonry.

(c) Airtight structures may be constructed of protected metal with rubber-cemented joints or of fiberglass.

(2) Horizontal silos

(a) Trench

(1') Any simple, narrow beneath-ground excavation with the floor and walls varying from plain dirt to wood or concrete.

(2') They are normally narrower at the bottom than at the top to facilitate effective packing of silage material and the floor is sloped to permit drainage.

(3') When filled, the silage is covered in some manner (usually heavyweight polyethylene film) to reduce surface spoilage.

(b) Bunker

(1') Above-the-ground silo with retaining walls.

(2') Floor is usually concreted and walls may be either wood or concrete; ends are left open.

(3') Similar to trench silo in construction and use.

(3) Temporary silos may be constructed with snow fences, plastic films, wire fence and paper, baled hay or other materials that can provide a retaining wall; temporary storage is also available in commercially manufactured "plastic bags."

(4) Stack silos are formed by stacking forage directly on the ground or preferably on a simple concrete slab. Plastic film is usually used to cover these silos.

c. What occurs in the silo (Figure 3–2)

(1) Plant cells continue to respire

(a) Oxygen is consumed.

(b) Carbon dioxide is produced.

(c) Mold will not grow without oxygen.

(2) Temperature of silage increases.

(a) Desirable temperature of silage after respiration stops is 80° to 100° F.

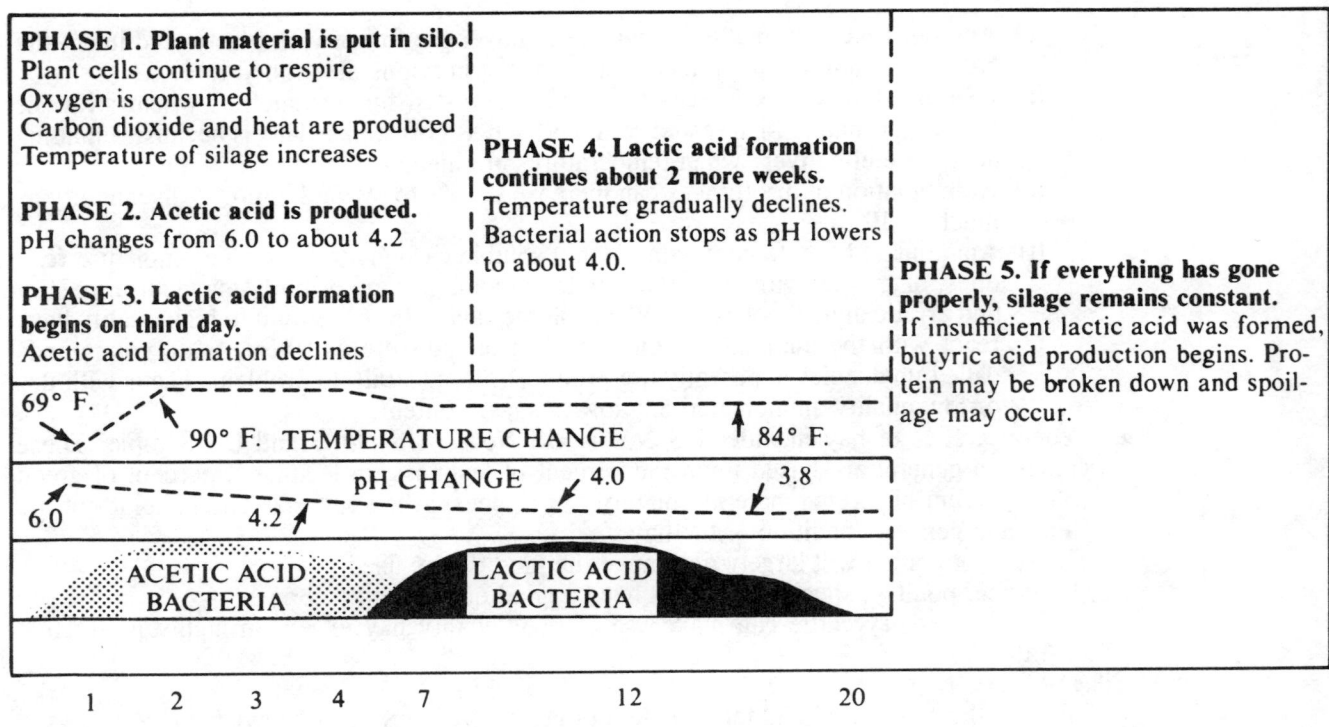

PHASE 1. Plant material is put in silo.
Plant cells continue to respire
Oxygen is consumed
Carbon dioxide and heat are produced
Temperature of silage increases

PHASE 2. Acetic acid is produced.
pH changes from 6.0 to about 4.2

PHASE 3. Lactic acid formation begins on third day.
Acetic acid formation declines

PHASE 4. Lactic acid formation continues about 2 more weeks.
Temperature gradually declines. Bacterial action stops as pH lowers to about 4.0.

PHASE 5. If everything has gone properly, silage remains constant. If insufficient lactic acid was formed, butyric acid production begins. Protein may be broken down and spoilage may occur.

69° F.   90° F. TEMPERATURE CHANGE   84° F.

pH CHANGE   4.0   3.8
6.0   4.2

ACETIC ACID BACTERIA   LACTIC ACID BACTERIA

1   2   3   4   7   12   20

AGE OF SILAGE (days)

FIGURE 3–2. A normal silage fermentation process (from Iowa State University Bulletin Pm-417).

    (b) Lower temperature—lactic acid-forming bacteria can't compete with butyric acid–forming bacteria.
    (c) Temperatures above 100° to 120°F result in sweet, tobacco-smelling, dark brown, "caramelized" silage being formed. Palatable, but much nutrient value has been lost (especially the availability of protein).
  (3) Fermentation
    (a) Anaerobic environment and warm temperature necessary.
    (b) Bacteria (normally present on all field crops) begin to multiply and attack plant sugars, forming the following acids
      (1′) Acetic acid—first acid produced; pH changes from 6.0 to about 4.2; formation declines about third day.
      (2′) Lactic acid—formation begins about the third day and continues for 2 weeks; most abundant acid formed in silage; bacterial action stops as pH lowers to about 4.0.
      (3′) Butyric acid—formed if insufficient lactic acid was formed; may also form as a result of excess moisture.
      (4′) Propionic acid—very little formed.
    (c) Other products of fermentation include some ethanol and fermentation gases—primarily CO_2, CH_4, CO, NO and NO_2 (toxic).
  (4) The lowered pH stops all bacterial growth, and silage remains in a fairly stable state in the silo providing it does not come in contact with oxygen.
    (a) Normal fermentation process lasts about 21 days.
    (b) Corn silage has been known to keep 12 years or more.
 d. Kinds of crops used for silage.
  Almost any crop can be ensiled, providing it contains an adequate level of moisture, sufficient amounts of readily fermentable carbohydrates and other nutrients and can be sufficiently packed.
  (1) Corn
    (a) Whole plant.

Corn being chopped for silage.

        (b) Corn stover silage.
        (c) Ear corn and shelled corn silage (high-moisture corn).
    (2) Sorghums
        (a) Grain sorghum (milo).
        (b) Forage sorghum.
    (3) Small grains—oat crop is best
    (4) Legumes or grasses
        (a) Silage—a forage crop carrying from 60% to 75% moisture is generally termed a silage.
        (b) Haylage—a forage intermediate between hay and silage in moisture content (40% to 60% moisture) is termed a haylage. Haylage material is drier and more difficult to pack than silage and is most successfully stored in an airtight or conventional silo.
e. Consideration for use of silage
    (1) Advantages
        (a) Feed more livestock per acre of land (less loss and more harvested TDN).
        (b) Forage harvesting, storing and feeding can be mechanized.
        (c) High quality, succulent feed and eaten practically without waste.
        (d) Earlier harvest; can be harvested over a range of moisture content and in almost any kind of weather.
    (2) Disadvantages
        (a) Once ensiled, you are committed to livestock feeding (no "off-farm" market).
        (b) Consumption of dry matter in the form of silage is nearly always lower than when the same crop is fed as hay or dry grain. This may result as a slower rate of gain and longer feeding period required to reach market weight.
        (c) Considerable costly equipment for the harvesting, storing and feeding of silage.
        (d) Must handle more water.
f. Prerequisites for making good quality silage
    (1) Proper moisture
        (a) 50% to 70% moisture range.
        (b) If too wet
            (1') Silage may sour.
            (2') Excess butyric acid formation.
            (3') Nutrient loss by seepage.
        (c) If too dry
            (1') Difficult to pack.
            (2') Mold, improper fermentation and heat may result.
        (d) Rules of thumb
            (1') 4 to 5 gallons of water added per ton of ensiled material will raise moisture level 1%.

(2') Mowed forage, lying in the swath, loses approximately 5% moisture per hour on a sunny day (more rapid if crimped).
(2) Proper stage of maturity
   (a) Important for two reasons
      (1') To have proper moisture content.
      (2') To get maximum TDN from crop.
   (b) Plants too young
      (1') Moisture high.
      (2') Kernel not completely filled.
      (3') Late maturing hybrid may be killed by frost.
   (c) Plant too old
      (1') Loss of dry matter in harvesting.
      (2') Need for additional water at ensiling time.
   (d) Desirable stage of maturity
      (1') Corn—grain is in the glazed to hard dent stage (50 to 60 days after silking); use the "black layer" test. When the grain reaches physiological maturity, several layers of cells near the tip of the kernel turn black forming the "black layer." This layer can be located by removing several kernels from the middle of the ear, split lengthwise or just cut off the tip and look for the "black layer" near the tip.
      (2') Grain or forage sorghum—grain is at the medium to hard-dough stage. If harvested at earlier stages, the sorghum usually has too high moisture content (75% to 85%) for proper storage. Harvesting when the grain is too mature (beyond the dough stage) results in feeding losses because the hard grain passes through livestock undigested.
      (3') Small (cereal) grains—oats can be harvested from late boot to late dough stage depending on the producers need for a high-protein silage or one yielding greater energy and total dry matter. When making silage from wheat or barley, harvest at least by the milk stage. Harvest rye by the late boot stage just before the heads show. Later harvest of these crops will produce dry fluffy silages, which are hard to keep and have low palatability.
      (4') Legumes and grasses—harvesting on time and at the right stage of maturity is important to quality and yield of legume or grass silage (Figure 3–3). Each day's delay decreases digestibility about 0.5%. Late harvesting of the first cutting often reduces yield of subsequent cuttings. With legume-grass mixtures where legume predominates, select a harvest stage suited to the legume.
         (a') Alfalfa—late bud to ¹⁄₁₀ bloom.
         (b') Clovers—early to ¼ bloom.
         (c') Birdsfoot trefoil—¼ bloom.
         (d') Most grasses—boot to early head.
         (e') Timothy—early head.
         (f') Smooth bromegrass—medium head.

Conventional tower (upright) concrete stave silos.

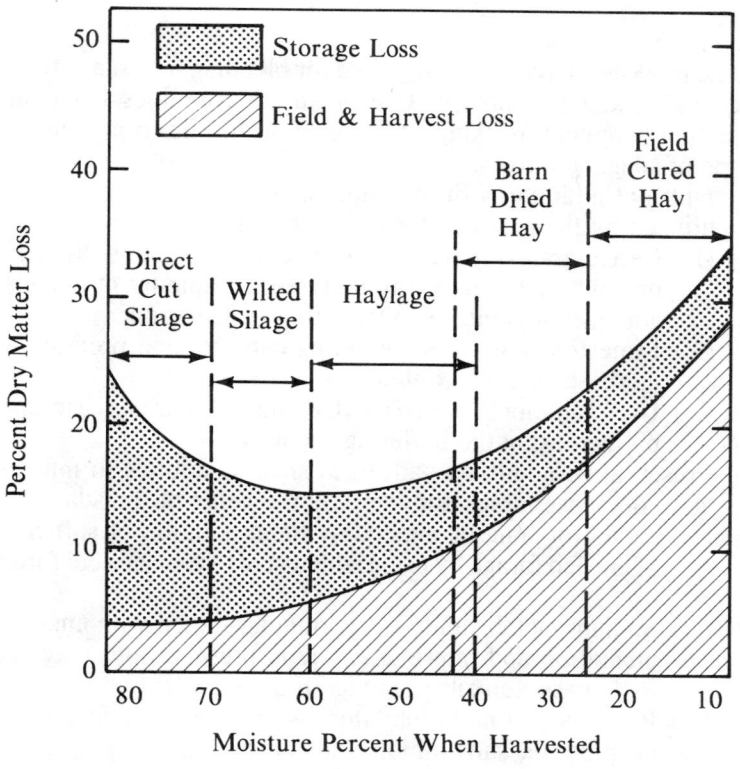

(from Michigan State University)

FIGURE 3–3. Estimated total field and harvest loss, and storage loss when legume-grass forages are harvested at varying moisture levels and by alternative harvesting methods.

(3) Proper packing
    (a) Length of cut—finer chop will pack firmer, be easier to handle and have less separation of coarse and fine material; pieces will usually vary from a fraction of an inch to more than an inch in length.
    (b) Even distribution of all material facilitates good packing.
    (c) Upright silos generally do not need tramping for a proper pack; packing in a bunker or trench silo should be obtained by use of a tractor as the silo is filled.
(4) Proper drainage—excess liquid should be drained from bottom of silo
(5) Exclude air
    (a) Seal cracks in wall.
    (b) Cover top.
g. Losses from ensiling
    (1) Some losses are incurred in the ensiling process regardless of the procedure used.
    (2) Field losses vary with the degree of field drying.
    (3) Losses in the silo can be grouped under three main headings
        (a) Gaseous or fermentation losses.
        (b) Surface spoilage.
        (c) Seepage.
    (4) Estimated average silage storage losses under good management (without additives)

Percent of Loss

Type of Silo	Average	Range
Gas tight	5	1–11
Concrete stave	6	2–12
Bunker or trench	15	10–25
Stack	20	12–25

(From Iowa State University Bulletin Pm-417)

h. Silage additives
  (1) Some materials have been suggested for blending into silage to improve its preservation, feed value and palatability. Use of an additive does not eliminate the necessity for effective chopping, packing and sealing of the silo to prevent the entrance of air
  (2) Type of silage additives
    (a) Nutrient additives—those materials that add nutrients to the silage which can be utilized by the animal after the fermentation process.
      (1') Cereal grains, molasses, citrus pulp, soybean hulls, straw, beet pulp and other dry materials—added to increase the amount of dry matter in the mass and to add carbohydrates available for fermentation.
      (2') Limestone—increase buffering capacity and promote greater acid production; increase calcium content.
      (3') Urea, ammonia—increase CP content and increase the extent of fermentation by increasing the buffering capacity.
    (b) Preservative additives—added in small quantities to influence the fermentation so as to decrease losses and increase the value of the silage
      (1') Acids (organic or mineral)—direct addition of sufficient acid for preservation of ensiled forage by using added acid to replace acid formation through fermentation.
      (2') Sodium metabisulfite—dry granular mixed with ensiled forage to decrease fermentation and improve carotene preservation; may reduce the production of toxic gasses in silage of high nitrate content.
      (3') Bacteria and mold inhibitors—certain antibiotics and mold inhibitors (sodium propionate, salt) inhibit undesirable microorganisms without preventing the action of bacteria that produce desirable acids.
      (4') Enzyme preparations—designed to increase the availability of carbohydrates in the plant material and promote the rapid growth of the naturally occurring lactic acid organisms.
      (5') Microorganism cultures—products containing cultures of *Lactobacillus* are added to provide an inoculum or to increase the numbers of these lactic acid-producing organisms and ensure rapid and desirable fermentation.
    (c) Combinations of the above types
  (3) General functions of the various silage additives
    (a) Adding dry matter and thus reducing moisture content.
    (b) Adding water to increase moisture content.
    (c) Altering rate, amount, and kind of acid production.
    (d) Acidifying the silage.
    (e) Inhibiting bacterial and mold growth.
    (f) Culturing the silage.
    (g) Increase nutrient content.

Large plastic bags for storage of ensiled forage.

216

i. Ensiling forage crops vs harvesting for hay
    (1) Advantages
        (a) Silage may be made during periods unfavorable for the field curing of hay.
        (b) A greater proportion of feed nutrients is saved, as losses due to rains, bleaching and shattering of leaves are reduced.
        (c) Early-season crops may be used to supplement short pastures.
        (d) Weed seeds are prevented from spreading, as most weed seeds are destroyed by the fermentation processes that occur in the silo.
        (e) Carotene content is better preserved.
        (f) Less storage space is required.
        (g) Fits into soil-conservation program better, as the necessity for growing soil-depleting crops is lessened.
        (h) Good "drought insurance" agent in shortage of green pastures.
    (2) Disadvantages
        (a) There are more tons of forage to be handled when the crop is harvested as silage than as hay.
        (b) If heavy manual labor is to be avoided in making silage, the outlay for machinery may be greater than when the crop is harvested as hay.
        (c) The use of preservatives is an added expense.
        (d) Grass silage exerts greater pressure on the silo than corn silage, and greater reinforcements may be required for upright silos.
    (3) Feeding value of haylage
        Approximately 1 lb of grain per 100 lb of steer weight must be added to a grass or legume silage or haylage ration in order to produce gain and finish equivalent to that obtained when steers are full-fed corn silage balanced for protein.

V. Characteristics of Common "Nutrient Additive" Feedstuffs
A. *Sources of Minerals*
    1. Major minerals
        a. The following sources of major minerals may be obtained individually or in some form of a mixture. When purchasing a mineral supplement, it is always advisable to check the feed tag for guaranteed level of elements contained in supplement.
        b. Some sources of calcium, phosphorus or both are shown in Table 3–5 together with their typical composition.
            (1) There appears to be little difference between calcium sources in the utilization or availability of calcium to the animal. Most sources are well utilized by different animal species.
            (2) Phosphorus sources differ greatly in availability, especially when fed to monogastric animals. Of the sources listed in Table 3–5, soft colloidal phosphate and Curacao phosphate are low in availability of P while raw rock phosphate is only fair and the others

TABLE 3–5. Common Sources of Calcium and/or Phosphorus

Mineral Source	Ca, %	P, %	Remarks
Limestone	33–38	0	
Oyster shell	37–39	0	
Soft colloidal phosphate	18–19	9–10	
Steamed bone meal	24–30	12–14	12–13% CP
Curacao phosphate	34–36	12–15	
Raw rock phosphate	29–35	12–15	2–4% F
Defluorinated rock phosphate	32–33	16–18	< 1 part F to 100 parts P
Dicalcium phosphate	23–28	18–22	
Diammonium phosphate	0	20–24	not < 17% N
Monosodium phosphate	0	22–26	16–19% Na
Phosphoric acid	0	22–31.6	

are good to excellent. Raw rock phosphate is high in fluorine content and should be defluorinated to prevent toxicity.

    (3) In plants, about ½ to ⅔ of the phosphorus is bound to phytic acid (called phytin P) and is poorly utilized by monogastric animals. The usual recommendation is to consider only ⅓ to ½ of plant phosphorus available to monogastric animals although ruminants utilize it very well.

  c. Salt (source of sodium and chlorine)

    (1) Common or plain salt.

    (2) Iodized salt—contains not less than 0.007% iodine.

    (3) Trace mineralized salt—defined as iodized salt to which have been added small quantities of other trace minerals. The mixture most generally contains compounds of cobalt, manganese, iron, zinc and copper, but additional elements may be included.

    (4) Any of the above salt sources may be purchased in either block or loose form and may be incorporated into the animal diet (usually at 0.2% to 0.5%), or fed free-choice, or both.

2. Trace minerals

  a. Differences exist in utilization or biological availability of some of the inorganic salts and oxides of different elements. Factors that should be considered in the biological value of an element include

    (1) Solubility in water or dilute acid.

    (2) Effectiveness in preventing or curing deficiency symptoms.

    (3) Tissue concentrating effect or comparative level at which toxicity is reached.

  b. Following are some of the trace minerals supplemented to certain livestock diets and the commonly used sources

    (1) Magnesium

      (a) Magnesium oxide—high availability.

      (b) Magnesium carbonate—high availability.

      (c) Magnesium sulfate—high availability.

    (2) Potassium

      (a) Potassium chloride—high availability.

      (b) Potassium sulfate—high availability.

      (c) Potassium carbonate—high availability.

    (3) Sulfur

      (a) Any element in the sulfate form will supply S; sodium sulfate (22.5% S) is commonly used in beef diets containing urea.

    (4) Iron

      (a) Ferrous sulfate—high availability.

      (b) Ferrous gluconate—high availability.

      (c) Ferrous fumarate—high availability.

      (d) Ferric ammonium citrate—high availability.

      (e) Ferric chloride—poor availability.

      (f) Ferrous carbonate—availability is quite variable depending on the source.

      (g) Ferric oxide—largely unavailable; often used as a coloring agent.

    (5) Copper

      (a) Copper sulfate—high availability; considered the best form for ruminants.

      (b) Copper carbonate—high availability.

      (c) Copper oxide—medium to poor availability.

      (d) Copper chloride—medium to poor availability.

    (6) Manganese

      (a) Manganese sulfate—high availability.

      (b) Manganous oxide—high availability.

      (c) Manganous carbonate—high availability.

      (d) Native manganese ores—poor availability.

      (e) Manganese dioxide—poor availability.

    (7) Zinc

      (a) Zinc oxide—high availability.

      (b) Zinc carbonate—high availability.

       (c) Zinc sulfate—high availability.

       (d) Zinc chloride—high availability.

       (e) Zinc sulfide (sphalerite)—poor availability.

    (8) Iodine

       (a) Calcium iodate and potassium iodide are high in availability but easily destroyed by oxidation when mixed in the livestock diet.

       (b) 3,5-diiodosalicylic acid (DIS)—highly available organic compound, resistant to oxidation but poorly retained within the body.

       (c) Ethylenediamine dihydriodide (EDDI)—stable and available source of organic iodine.

    (9) Cobalt

       (a) Cobalt carbonate—high availability.

       (b) Cobalt oxide—medium availability.

       (c) Cobalt chloride—medium availability.

       (d) Cobalt sulfate—medium availability.

   (10) Selenium

      Sodium selenite and sodium selenate are high in availability.

  c. Chelation of trace minerals to improve their stability and utilization is currently receiving interest. Chelates are compounds in which the mineral atom is bound to another molecule or substance, usually organic. If a mineral is chelated by a compound that will release it in ionic form at the intestinal wall, or which will be readily absorbed as the intact chelate, this chelate may greatly enhance absorption of the mineral. This enhanced absorption occurs because the chelate prevents conversion of the mineral to insoluble chemical compounds in the intestinal tract, or prevents its strong absorption on insoluble colloids in the intestine.

Limited research is available and further studies are needed to know how well chelated trace minerals will work out for all species.

B. *Sources of Vitamins*

  1. Vitamins may be purchased individually or as mixtures. The fat-soluble vitamins need a stabilizer or antioxidant in order to retain their potency. Most vitamin products on the market are stabilized. Some of the water-soluble vitamins are subject to destruction by heat or by continued warming or heating.

  2. Fat-soluble vitamins

    a. Vitamin A

     (1) Natural sources

       (a) Carotene (provitamin) found in green plants.

       (b) Vitamin A alcohols found in oils such as fish or shark liver oil and certain vegetable oils (carrot oil).

     (2) Chemically synthesized sources

    b. Vitamin D

     (1) Natural sources

       (a) Sunlight (sun-cured hays)—D_2.

       (b) Fish and fish liver oils—D_3.

     (2) Synthetic source

       (a) Synthetic D_3 is produced by the activation of a sterol fraction of animal origin (7-dehydrocholesterol) with ultraviolet light or other means. Labeled as "D-activated animal sterol (source of vitamin D_3)."

       (b) Synthetic D_2 is produced by activation of a sterol fraction of plant origin (ergosterol) with ultraviolet light or other means. Labeled as "D-activated plant sterol (source of vitamin D_2)."

       (c) Irradiated dried yeast (source of vitamin D_2) is the dried nonfermentative yeast, containing ergosterol, which has been subjected to ultraviolet rays.

    c. Vitamin E

     (1) Vitamin E is the term given to a group of closely related fat-soluble compounds called tocopherols. The D-α form has the highest potency.

  (2) Natural sources
    (a) Germs of grains—wheat germ meal or oil.
    (b) Green forages and well-cured hays—dehydrated alfalfa meal.
  (3) Synthetically produced sources
 d. Vitamin K
  (1) Natural sources—green forages; good hay, dehydrated alfalfa meal and most whole seeds.
  (2) Synthetically produced
    (a) Menadione sodium bisulfite complex (MSBC).
    (b) Menadione dimethylpyrimidinol bisulfite (MPB).
2. Water-soluble vitamins
 a. Riboflavin
  (1) Natural sources
    (a) Milk byproducts.
    (b) Liver meal.
    (c) Dried yeast.
    (d) Leafy legume hay.
    (e) Dried cereal grasses.
    (f) Distillers' and fermentation byproducts.
  (2) Microbiologically or chemically synthesized
    Dried or condensed fermentation solubles are derived from the liquid byproduct resulting from the action of the ferment on the basic medium of grain, molasses, whey, or other media, and must contain not less than 18 mg of riboflavin per lb (d.m.). Riboflavin concentrates are prepared by purifying the fermentation solubles and can be brought to any degree of concentration.
 b. Niacin
  (1) Natural source—some present in most feedstuffs, but that in cereal grains is largely unavailable, especially to swine.
  (2) Synthetically produced as the amide form of nicotinic acid, niacinamide or nicotinamide.
 c. Pantothenic acid
  (1) Natural source—contained in variable amounts in feedstuffs but is low in corn.
  (2) The synthetic product most commonly used is calcium pantothenate, which is composed of D and ℓ isomeric forms, of which the ℓ form is biologically inactive. Thus, guarantees of this vitamin must be made as D-pantothenic acid.
 d. Choline
  (1) Natural sources—it is present in variable amounts in most feeds although those of animal origin are usually considered to be the better sources.
  (2) Chemically synthesized sources include choline chloride (the most common compound) and choline xanthate.
 e. Vitamin B_{12}
  (1) Natural sources of this vitamin are primarily the animal protein materials—meat scraps, fish meal, fish solubles etc.
  (2) Commercial preparations of vitamin B_{12} are produced by fermentation processes. Although its structural composition is known, the vitamin has not been synthesized, but it is produced biologically by microorganisms.

VI. Methods of Feedstuff Preparation
 A. *Why Process Feedstuffs?*
  1. Increase efficiency of handling from an engineering standpoint.
  2. Increase efficiency of utilization.
   a. Palatability
   b. Digestibility
    (1) Increase surface area for greater bacterial or enzymatic activity or both.
    (2) Alter molecular structure to enhance digestion.

3. Alter the density of the feed or diet.
   a. Increase bulk of high concentrate diet (resulting in a "safer" feed because of reduced digestive system disorders).
   b. Reduce bulk of a high roughage diet (resulting in increased daily consumption by the animal).
4. Many processing techniques yield improvements in feed efficiency of 5% to 15% or more, and a few techniques may also yield some improvement in rate of gain, perhaps up to 5% to 10%. In many cases, feed/gain ratio (F/G) or feed efficiency is improved but not rate of gain. Roughly speaking, with today's feed prices, one needs to obtain about 1% improvement in feed efficiency for each $1 spent on processing cost per ton.
B. *Specific Processes*
   1. Grinding
      a. Grinding is a reduction in particle size by impact, shearing or attrition.
      b. Type of grinders
         (1) Hammer mill
            (a) Grinding is accomplished by a flywheel rotor with swing-type hammers in a chamber where a perforated screen determines the final size of the product. Feed material is beaten until it is fine enough to pass through the perforated screen.
            (b) Will grind hay, ear corn, small grain and mixtures.
            (c) Well suited for fine grinding.
            (d) Usually unable to produce a uniform coarse grind.
         (2) Burr mill
            (a) Grinding is accomplished by feed passing between two plates containing unmeshed burrs. Usually, one plate is revolving and the other stationary. Degree of fineness of grind is determined by distance between the two plates.
            (b) Produces a relatively uniform grind.
            (c) Not designed for roughage materials.
            (d) Maintenance cost are generally higher compared to hammer mill.
      c. Feed particles may separate after grinding; all the course material in the middle and the powder around the edges.
      d. Degree of fineness
         (1) Finer grind will be more subject to wind loss.
         (2) Finer grind will tend to "ball up" in animal's digestive system.
         (3) Finer grind may result in cattle going off feed (reduced palatability).
         (4) Finer grind does not "feed down" properly in a self-feeder.
         (5) Finer grind tends to have a faster rate of passage through digestive tract.
         (6) Finer grind may cause digestive disturbances.
            (a) Stomach ulcers in swine.
            (b) Ruminal parakeratosis in feedlot cattle.
      e. Grinding grain for swine—medium to fine grind for both corn and milo.
      f. Grinding grain for sheep—probably not necessary except for sorghum grains (milo), which could be coarsely ground.
      g. Fine grinding of grain for dairy cattle will result in lowered milk fat production; grains should be coarsely ground.
      h. Grinding grain for beef cattle
         (1) Because of a hard seed coat, it is essential to break milo grain before feeding.
            (a) Coarse grinding results in a poorer performance and efficiency than a finer grinding of milo.
            (b) Grind to fineness where palatability and consumption are not affected.
         (2) The grinding of corn depends upon the roughage/concentrate (R/C) ratio of the diet.
            (a) If corn is fed with dry roughage in the ration (more than 15% to 20% R), it should be coarsely ground.
            (b) If corn is fed with silage where palatability is not a problem, a medium-fine grind is preferred.
            (c) Processing corn loses its advantages when fed with a limited intake of roughage (less than 15% to 20% R); whole dry corn fed to cattle with limited roughage produces equal or superior gain compared to cattle fed processed corn with limited roughage.

i. Grinding hay for cattle and sheep
   (1) Definitions
       (a) Chopped = 2″ in length.
       (b) Ground hay = less than 1″
       (c) Finely ground hay = $\frac{1}{16}$–$\frac{1}{4}$″.
   (2) As hay is ground to a finer degree, there is a general decrease in digestibility of the nutrients (probably due to increased rate of passage through digestive tract).
   (3) Chopping hay may result in increased intake, gain and efficiency compared to long (unchopped) hay; ground hays are, as a rule, quite dusty and may not be consumed readily. Adding molasses, fat or water will usually improve intake.
   (4) Chopping a poor quality hay would be more advantageous than chopping a high quality hay (increased intake).
   (5) Chopping or grinding puts hay in a physical form for ready handling by some mechanical equipment and can be incorporated into complete ration mixtures.
   (6) Additional expense may be incurred by grinding and loss of dust may be appreciable from grinding in hammer mills.
   (7) Grinding their hay will cause a drop in the milk-fat level of dairy cows.

2. Dry rolling or cracking
   a. Rolling involves changing the shape, size or both of grain particles by compressing between rollers. This leaves the grain in the form of a flake.
   b. Crimping is accomplished in a manner similar to rolling, except that instead of rollers with smooth surfaces being used, rollers with corrugated surfaces are used. The end result is much the same.
   c. Dry rolling, cracking or coarse grinding are similar in physical nature and feeding value to the animal.

3. Pelleting
   a. Pelleting is accomplished by agglomerating ground feed material by compacting and forcing it through die openings. Some combination of heat, moisture and pressure are usually, but not always needed in the process. Pellets can be made in different diameters, lengths and hardnesses.
   b. General advantages to pelleting grain or hay
      (1) Reduced dustiness and increased palatability make for greater consumption.
      (2) Less loss of fine particle ingredients in transport.
      (3) Reduced selective eating and feed wastage.
      (4) Reduced storage space (especially for forages).
      (5) Increased utilization of fibrous fraction of feedstuffs.
      (6) Adaptable to bulk and mechanized feeding.
      (7) Partial gelatinization of starch in grain, increasing susceptibility to enzymatic action (or bacterial action) and better digestion.
   c. General disadvantages to pelleting grain or hay
      (1) Increased cost ($3 to $5+ per ton).
      (2) Pellet quality sometimes not ideal.
      (3) Added transportation if pelleting mills are not local.
      (4) Improper pelleting procedures may cause feed spoilage.
      (5) Cereal grains and hays require finer grinding before being pelleted.
      (6) Rations high in fat are difficult to pellet.
   d. Pelleted swine or poultry diets
      (1) Usually results in improvement in performance, partly because of increased feed consumption (compared to meal diet).
      (2) Generally improves rate of gain (0% to 5%) and feed efficiency (5% to 10%); efficiency may be improved partially because of less feed wastage with pellets.
      (3) Some studies suggest pelleting improves plant phosphorus (phytin) availability.
   e. Pelleted dairy diets—pelleting of the roughage portion, or especially the complete diet of the lactating dairy cow, generally is not recommended as it results in reduced rumen acetate production and lower milk-fat test.

f. Pelleted beef cattle diets—advantage is dependent upon the R/C ratio of the diet.
  (1) High roughage/low concentrate diet
    (a) Increased feed consumption.
    (b) Improved feed efficiency.
    (c) Should expect increased gain.
    (d) Pelleting of low quality roughage is more advantageous than of high quality roughage.
  (2) High concentrate/low roughage diet
    (a) Reduced feed consumption.
    (b) Similar feed efficiency.
    (c) Reduced gain.
g. Pelleted horse diets
  (1) Pelleted diets are especially useful as creep feeds or feeds for weanlings when there is a tendency for the young horse to separate out dietary components.
  (2) The grain mix or complete diet (grain plus hay) may be pelleted.
  (3) Pelleted diets composed of hay and grain intended as a complete feed should contain at least 60% coarsely ground hay. It is sometimes helpful to feed small amounts of grass hay with such complete pelleted diets to prevent wood chewing and mane and tail chewing when groups of horses are penned together.
h. Variations of pelleting
  (1) Granules or crumbles
    (a) Poultry diets are often in the granulated or crumbled form.
    (b) This is usually done by breaking or cutting pellets to the size wanted, and removing the fines for repelleting.
  (2) Cubed roughages
    (a) Cubing is a relatively new process in which dry hay is forced through dies producing a square product (about 1 in. in size) of varying length.
    (b) Grinding may not be required, but water is usually sprayed on the dry hay as it is cubed.
    (c) Alfalfa hay normally produces the best cubes.
    (d) Cubes lend themselves to convenience of mechanization and to feeding on the ground.
4. Heat treatment methods of grain processing
  a. Steam rolling
    (1) Short term exposure of air-dry grain to steam (1 to 8 minutes) followed by rolling.
    (2) Grain moisture content slightly increased.
    (3) Less starch alteration occurs than in steam flaking, thus, feeding value is not materially improved over dry rolling.
    (4) May improve physical form of some grains.
  b. Steam flaking
    (1) Long-term exposure of air-dry grain to steam (15 to 30 minutes), then rolling to produce a very thin flake containing 16% to 20% moisture and a density of 22 to 28 lb per/bushel.
    (2) Flaking gelatinizes some of the starch granules, rendering them more digestible.
  c. Pressure flaking
    (1) Air-dry grain is steamed under pressure for 1 to 2 minutes, cooled down to about 200°F and finally dried to 20% moisture before rolled to thin flakes.
    (2) More expensive process than steam flaking.
  d. Roasting
    (1) Air-dry grain is heated in a roaster to about 300°F, producing a puffed and slightly caramelized appearing product.
    (2) Partial starch gelatinization occurs during the heating.
    (3) Moisture content is decreased to about 5% or less, and the bulkiness of the grain is increased about 15%.
  e. Extruding—heat and pressure are applied to air-dry grain as it is pushed by a tapered spiral screw through an orifice, producing a ribbonlike product that breaks into thin flakes.
  f. Popping
    (1) Air-dry grain is exposed to very high temperatures for a short time—perhaps 700° to 800°F for 15 to 30 seconds.

          (2) Heat causes moisture in grain to steam, which gelatinizes and expands the starch granules.

          (3) Moisture is often reduced to 3%; thus, the product is usually rolled and moisture added before feeding.

   g. Exploding—Air-dry grain is placed in high tensile strength containers and subjected to steam under pressure at about 250 psi. After about 20 seconds the pressure is released and the grain escapes as expanded balls with hulls removed.

   h. Micronizing

          (1) Air-dry grain is heated to between 200° and 300°F by exposure to microwaves from an infrared generator.

          (2) The grain is then dropped into Knorling rollers which have spiral grooves that place diagonal and parallel pressure on the grain.

          (3) The grain does not pop, but the internal moisture in the grain volatilizes, causing expansion and gelatinization.

          (4) When the heated grain is rolled, an intact, flake like product is produced with a density of about 25 lb per/bushel.

5. High-moisture grain harvesting

   a. Grain (usually corn or milo) is harvested and stored in a high moisture state ranging from 20% to 35% moisture. The most typical moisture level is 25 to 30%.

   b. In the case of harvesting ear corn, the moisture of the ear will usually exceed that in the shelled corn by about 5%.

   c. The high-moisture grain is stored and fed either as a ground product in a trench, bunker, or upright silo or in the whole form in gas-tight silos.

   d. Fermentation does occur with high-moisture grain storage but at a much faster rate than forage silage because

          (1) High-moisture grains are higher in dry matter and lower in water content than forage silage, thus

              (a) Much faster fermentation rate because of available starches present in grains.

              (b) With less water, there is less dilution of the acids formed thus a much faster drop in pH.

          (2) High-moisture grains can generally be packed tighter allowing for greater elimination of oxygen.

   e. Nonnutritional factors with high-moisture grain system

          (1) Nonnutritional advantages

              (a) Early harvest.

              (b) Reduced shattering, lodging and hail losses.

              (c) Less bird and insect loss.

              (d) Dry matter yield increased, especially in corn, because of first three points.

              (e) Longer harvest season—can reconstitute grain that is harvested too dry.

              (f) No artificial drying required.

              (g) Storage cost lessened if bunker or trench silos are used.

              (h) Less damage from early frost.

          (2) Nonnutritional disadvantages

              (a) Moisture content makes it necessary to handle more weight.

              (b) Grain must be stored immediately.

              (c) Storage must be oxygen free.

              (d) Yield may be depressed in the case of milo; unable to shell wet milo out of the head.

              (e) The high-moisture grain must be merchandized through livestock.

   f. Nutritional factors with high-moisture grain fed to cattle

          (1) High-moisture corn stored whole can be satisfactorily fed in the whole form if low roughage levels are used.

              (a) Average daily gain—about the same as dry corn.

              (b) Feed efficiency of high moisture whole corn may be improved up to 5%.

          (2) High-moisture vs dry shelled corn (both rolled or ground).

              (a) Average daily gain—about the same.

              (b) Feed efficiency—an improvement of up to 8% is usually observed in cattle fed processed high-moisture vs processed dry corn.

(3) High-moisture vs dry ground ear corn
   (a) Average daily gain–about the same; some indicate about 3% increase in gain with high-moisture ground ear corn.
   (b) Feed efficiency—normally, 5% to 10% less feed is required per 100-lb gain in cattle fed high-moisture ground ear corn.
   (c) The increased efficiency may be resulting from the cobs being altered in the high-moisture silo.
(4) Whole high-moisture milo must be rolled or ground prior to feeding. Gain and efficiency relationships are similar to processed corn studies.
(5) Storage dry matter losses are generally greater in high-moisture grain (3% to 6% in an oxygen-limiting silo) than dry stored grain. Thus, greater storage d.m. losses with high-moisture grain may offset any advantages realized in improved feed efficiency.

  g. Nutritional factors with high-moisture grain fed to swine (refer also to Chapter 7, III-G).
    (1) High-moisture corn can be substituted for dry corn on a dry matter basis with little or no effect on gain and efficiency of swine if both are fed as a complete mixed diet. Likewise, there appears to be little difference in performance whether high-moisture corn is fed whole or processed.
    (2) Whole high-moisture milo must be rolled or ground prior to feeding swine and preferably be fed as a complete mixed diet rather than free-choice the supplement. Studies show little differences in performance between processed high-moisture or dry milo.

  6. Reconstituted high-moisture grain
    a. This process involves the addition of water to dry grain to raise the moisture up to 25% to 30% (higher for milo than corn) followed by storage in the whole form in an oxygen-limiting silo for at least 14 to 21 days before feeding.
    b. May improve efficiency of gain in cattle by approximately 2% to 3% (compared to dry grain), but will not equal harvested high-moisture grain.

  7. Acid-preserved high-moisture grain
    a. Organic acids, usually propionic or acetic-propionic mixtures, are thoroughly blended with high-moisture harvested grains (20% to 30% moisture).
    b. The added acid aids in preservation of the high-moisture grains when stored by nonsilo storage means.
    c. The quantity of acid required increases with moisture level of the grain and desired length of storage. Typical levels could range from 10 to 30 lb acid per ton of grain.

VII. Methods of Feed Mixing

A. *Buildings and Housekeeping*
  1 Mixing and storage facilities must be weather-tight, and rodents must be controlled.
  2. Separate room or enclosed area for storage of feed additives and premixes.
  3. Weighing and mixing area well lighted and ventilated.
  4. Floors should be smooth and solid to facilitate sweeping.
  5. Floor sweepings should either be added to immediate batch of feed being prepared or be destroyed before identity is lost; adding to a later batch of feed may contaminate that feed.

B. Weighing
  1. Scales must be of adequate sensitivity such that a difference of ± 2% of item's weight can be measured. An item weighing 5 lb must have scale reading in tenths (0.1) of a pound.
  2. Scales must be accurate; should be maintained and checked periodically with a set of standard weights.
  3. Scale weight pan or platform should be kept clean.
  4. The scale operator must be able to make calculations and conversions if needed—e.g., grams to pounds, from percent to grams per ton, and in many cases account for a "dilution factor."

C. *Types of Feed Mixers*
The many commercial brands and sizes of mixers available may be categorized into two types: batch feed mixers or continuous-flow feed mixers.
  1. Batch feed mixers
    a. Individual feed ingredients are ground (in the case of grain), weighed and mixed in batches.

b. Grinding and conveying equipment should be of high capacity, and batches should be large to reduce the number required.
c. Requires an operator to be present.
d. Types of batch mixers
   (1) Vertical mixers
       (a) Cylindrical, usually about 5 ft in diameter with an inverted cone-shaped bottom.
       (b) Contain an enclosed, high rpm vertical auger, (or twin augers), which lifts feed from the bottom of the cone to the top of the mixing chamber. The feed is then thrown by centrifugal force and spread over the top of the batch of material in the mixer.
       (c) These mixers have a relatively low initial cost and low power requirement; however, a long mixing time is required to obtain a completely mixed product.
   (2) Horizontal mixers
       (a) These mixers have a U-shaped trough mixing chamber and a horizontal shaft with paddle agitators to stir the contents of the mixing chamber.
       (b) This mixer can handle liquid molasses, silage and ground hay, which the vertical mixer cannot.
       (c) Original cost and power requirement are relatively high; however, these mixers do a thorough job of mixing in a short time.
   (3) Auger wagon mixers
       (a) These are a modification of the basic V-bottom tank-type auger wagons.
       (b) A horizontal tank auger provides the basic mixing action. You can usually arrange the vertical unloading auger so it will return the feed mixture to the top of the load.
       (c) Liquid molasses, silage and ground hay can be mixed in auger wagon mixers.
       (d) Relatively long mixing time is required to obtain a completely mixed product.
e. Batch mixers may be either stationary or portable (mobile). The stationary mixers are generally separate from a grinder, whereas the portable units are normally a grinder-mixer combination.
   (1) Stationary mixer and separate grinder.
       (a) Advantages
           (1') High accuracy in mixing and processing can be obtained.
           (2') Flexibility for mixing many different diets is present.
       (b) Disadvantages
           (1') Feed must be conveyed to place of use.
           (2') Storage for finished diets may be needed.
           (3') Labor is required while feed is being processed.
           (4') Electric service to processing point is required.
   (2) Portable grinder-mixer
       (a) Advantages
           (1') Feed can be ground, mixed and delivered anywhere on the farm. Can also take unit to another farm if you move.
           (2') Existing tractor power can be used; electric service is not needed.
           (3') Convenient to use as an auger wagon and is well adapted to fenceline feeding.
           (4') Feed is ground fresh each time you feed.
       (b) Disadvantages
           (1') Tractor is tied up providing power during mixing and delivery.
           (2') Labor is required while processing, and feed processing can't be switched to slack labor periods as with common stationary mills.
           (3') Complete unit is heavy (1 to 1½ tons empty), so solid footing is required in bad weather.
2. Continuous-flow mixers (volumetric blenders)
   a. Feed ingredients are simultaneously metered into a mill and mixed while the grain is ground. Automatic controls are built-in features of most continuous-flow mills.
   b. Advantages
      (1) Unattended operation resulting in low labor cost is possible because of high automation.
      (2) These mills require very little space and operate at a high level of reliability with low operating cost.
      (3) Processing can be low capacity compared to the large-capacity batch systems.

c. Disadvantages
  (1) These mills will not handle roughage.
  (2) Flexibility in rations may be limited slightly.
  (3) Storage for ingredients and finished diets may be needed at processing location; feed must be conveyed to place of use.
  (4) Electric service to processing point is needed.

D. *Sequence of Ingredient Addition to Batch Mixers*

The order in which ingredients are added to the batch mixer affects the time required to obtain a thorough mix. With monogastric corn-soybean meal diets, the mixer should be charged with 25% to 35% of the ground corn, followed by the vitamin, drug, mineral and protein sources and then the remaining ground corn. If liquid additions are made, their addition should be delayed until the critical ingredients (drugs, vitamins etc.) have mixed with the bulk ingredients for 60% to 70% of the minimum mixing period.

E. *Factors Affecting Mixing Efficiency*
  1. Type of mixer.
  2. Sequence of ingredient addition.
  3. Length of mixing time—in most cases adequate mixing is obtained in 15 to 20 minutes using a vertical mixer and 7 to 10 minutes using a horizontal mixer after the addition of the last ingredient. The longer times should be used when there are small additions of critical ingredients.
  4. Wide differences in ingredient particle size and density will slow the mixing process.
  5. Mixer cleanout—to prevent contamination of subsequent batches of feed, the mixer and auger system should be cleaned and flushed out with ground grain after mixing a medicated feed.

F. *Storage, Handling and Records*
  1. Store feed additives in tight, labeled containers.
  2. Keep inventory of feed additives.
  3. Prevent cross-contamination.
  4. Wash after using drugs.
  5. Label bags and bins of mixed feed.

VIII. Feed Storage

A. Feed ingredients must be properly stored after harvest, mixing or both until fed to livestock. Storage conditions determine ingredient damage, nutrient loss or influence on consuming livestock.

B. *Factors That Influence Feed in Storage*
  1. Moisture content of feed
    a. High moisture levels may initiate bacteria, mold growth or both, which can create heat.
      (1) Browning or heat damage protein in hay forages.
      (2) Spoilage and nutrient destruction in grains and forages.
      (3) Possible spontaneous combustion causing fire.
    b. Formation of certain fungi that may reduce palatability and be harmful or toxic to consuming livestock.
    c. Feeds high in moisture tend to cake or bridge in the storage structure.
    d. Desirable moisture levels of feeds stored under aerobic conditions
      (1) Whole grains—most grains should contain less than 13% moisture; shelled corn will tolerate up to about 15.5% moisture.
      (2) Ground or rolled feeds should be stored containing less than about 11% moisture.
      (3) Stacked or baled forage must be cured down to 18% to 20% moisture before storage.
    e. Methods to facilitate safe storage of high moisture-containing feeds
      (1) Artificial drying—generally requires use of expensive fossil fuels.
      (2) Field drying—crop remains in field longer before harvest; greater loss due to weather damage, predators, harvest loss and extended time of field occupancy.
      (3) Ensiling in oxygen-limiting structures or acid preservation (refer to previous discussion).
  2. Insect infestation
    a. Most commonly a problem in grains rather than forage crops.

Farm feed storage center including large bins for dried grains and oxygen-limiting silos for high-moisture grains.

    b. Damage may be in the form of grain destruction or contamination with insect eggs or excrement.

    c. The degree of insect damage is determined by length of storage time and temperature of the storage environment.

    d. Fumigation procedures may be necessary to reduce insect damage.

3. Rodent infestation

    a. Most commonly mice and rats.

    b. Although the amount of feed consumed by rodents constitutes only a limited economic loss to the producer, it is estimated that rodents contaminate with their feces and urine about ten times the amount of feed actually eaten.

    c. Contamination of feed by rodents reduces consumption and increases the potential for spread of livestock disease.

    d. Suggested program of rodent control

        (1) Store feed in rodent-proof containers or buildings.

        (2) When feeders or storage bins are not in use, empty and clean them.

        (3) Clean up spilled feed and eliminate shelter rodents can use for nesting.

        (4) Where rodents are present, the most successful control programs are those using a combination of methods including removal of feed and shelter, use of traps, and the proper use of rodenticides.

IX. Factors Affecting Feed (Dry Matter) Intake

A. *Availability of Free-Choice Water*

1. Inadequate water supply results in reduced dry matter intake.

2. Contaminated water, foul-tasting water or water containing a high level of total dissolved solids will result in reduced consumption of both water and feed.

B. *Palatability*

1. Summation of taste, odor, appearance, texture and temperature of a feedstuff determine its degree of acceptance.

2. Palatability factors may be affected by the chemical or physical nature of the feed, or both.

    a. Chemical factors

        (1) Taste is probably the major component of the palatability complex. The major taste responses are

            (a) Sweet taste (enhance palatability)—molasses; sucrose.

            (b) Salty taste (enhance palatability)—salt.

            (c) Bitter or acid (reduced palatability)—plant alkaloids, tannins, cyanogenic compounds (prussic acid).

(2) Odors are produced by volatile materials. The limited information available would suggest that odor may serve as an attractant, but may not highly influence total consumption.

   b. Physical factors

      (1) Various forages may contain physical defenses, such as prickliness or coarseness, which reduce palatability.

      (2) Silica is deposited in cell walls of many grass species, which contributes to physical roughness or harshness and may have negative effects on palatability.

      (3) Feed texture and particle size may influence consumption. Increased consumption is often observed in animals fed rolled or cracked grain compared to whole grain or when fed pelleted feed compared to meal form. Feed preparation procedures that reduce dustiness usually result in increased feed intake.

C. *Dietary Energy Level*

1. If feeds are sufficiently palatable, most animals will consume that quantity of feed needed to satisfy their energy requirements until the digestive tract can no longer accomodate the bulk in the diet.
2. An animal expending large amounts of energy will generally eat more feed than another animal not expending as much energy.
3. If a diet containing a low concentration of energy is fed to an animal, feed intake is high, whereas with a high energy containing diet, feed intake is reduced. In either case, the caloric intake by the animal may be about the same.

D. *Function, Age or Size of the Animal*

1. Young animals having rapid growth or mature lactating animals generally have greater energy requirements and will consume larger quantities of feed per unit of body size.
2. The larger the animal, the less feed it consumes per unit of body weight. Energy requirements of adult animals are generally related to body weight raised to the 0.75 power. This is known as metabolic body size (MBS = wt.$^{0.75}$). Metabolic body size is widely used in animal nutrition research to reflect the fact that energy need (and feed consumption) is related more nearly to lean body mass and basal metabolic rate than to total body weight.

E. *Environmental Temperature*

1. The metabolic rate of an animal is at a minimum when environmental temperatures are in a range such that no energy demand is made on the body temperature regulating mechanisms. This temperature range is known as the zone of thermoneutrality or comfort zone.
2. Hot ambient temperatures have a inhibitory effect on consumption, whereas cold temperatures usually stimulate appetite.
3. Specific dynamic action (SDA) refers to the extra heat produced by the animal as a result of feed ingestion and the metabolism of its nutrients. It is highest for protein and lowest for fat. Thus, under conditions of high temperatures the dietary protein may need to be reduced, and it may be desirable to increase dietary fat levels.
4. Heat increment (HI) is the sum of SDA and the heat of rumen fermentation. The heat increment is useful at temperatures below the comfort zone, and detrimental at high environmental temperatures.

F. *Health of the Animal*

1. Most infectious diseases and internal parasites will cause reduced feed consumption.
2. Certain metabolic problems (ketosis, bloat, diarrhea etc.) and stresses (crowding, handling, noise) will cause reduced feed consumption.
3. Deterioration and contamination of feed or water will cause reduced feed consumption.

G. *Expected Feed Intake*

Several factors may affect the amount of feed consumed by an animal. However, expected feed intake is important when diets are designed for a specific purpose. Thus, the following estimates of daily dry matter intake are made based on percentage of the animal body weight.

1. Poultry 5–6%
2. Swine 4–5%
3. Sheep 3–4%
4. Cattle 2–3%
5. Dogs/cats 2–3%
6. Horses 1.5–2.5%

Study Questions and Problems

Chapter 3

I. True or False Questions

1. Concentrate feeds are low in crude fiber (CF) and low in total digestible nutrients (TDN).
2. Relative to forages, grains tend to contain more calcium and less phosphorus.
3. Salt should never be provided for livestock in the loose or free-choice form.
4. Solvent extracted linseed meal (LSM) usually contains more than 40% crude protein (CP).
5. Defluorinated rock phosphate, steamed bone meal and ground limestone are all good sources of phosphorus for livestock.
6. Young plants are higher in TDN on a dry matter basis than at later stages of growth.
7. Gossypol is a toxic substance found in linseed meal.
8. When partitioned by weight, ground ear corn contains approximately 80% cob and 20% grain.
9. Triticale is a hybrid cereal grain derived from a cross of barley and rye.
10. One must know the origin or parent material of a feed when using the international feed vocabulary.
11. The early bloom maturity of forages is when $\frac{1}{10}$ to $\frac{2}{3}$ of the plants are in bloom.
12. Diammonium phosphate may be fed to ruminants as a source of phosphorus and nonprotein nitrogen.
13. Cottonseed meal (CSM) contains a conditioning factor that results in glossy hair coats or greater animal bloom.
14. Corn gluten meal and corn gluten feed both contain approximately 25% crude protein.
15. Red clover hay has similar nutrient composition to alfalfa except for significantly higher crude protein content.
16. Sweet clover poisoning can occur if animals overconsume large amounts of fresh sweetclover pasture.
17. Kentucky bluegrass is desirable both as a pasture as well as a hay grass.
18. Both legume and grass pastures require annual nitrogen fertilization for maximum forage production.
19. Grass hays normally suffer less damage than do legume hays from outside storage.
20. Enzyme preparations and mold inhibitors are examples of "preservative" silage additives.
21. Magnesium oxide is a highly available form of the mineral magnesium.
22. Irradiated dried yeast is an excellent source of vitamin D_3.
23. Corn grain ground too coarse may cause greater incidence of stomach ulcers in swine.
24. Vertical feed mixers have a U-shaped trough mixing chamber with high rpm mixing augers.
25. Stacked or baled forage must be cured down to at least 10% moisture before storage.

II. Multiple Choice (select the single best answer)

1. An example of a nonnutrient feed additive is
   a. Salt.
   b. Ground limestone.
   c. Water.
   d. Fish meal.
   e. Antibiotic.
2. Which of the following has the highest protein content?
   a. Barley grain.
   b. Milo grain.
   c. Cottenseed meal.
   d. Alfalfa hay.
   e. Oats grain.
3. Which of the following is not a facet of the International Feed Name?
   a. Origin or parent.
   b. Part fed to animals.
   c. Process or treatment.
   d. Stage of maturity.
   e. Cost per pound.

4. According to international feed nomenclature, products that contain 20% or more protein from animal or plant origin are classified as
   a. Energy feeds.
   b. Mineral supplements.
   c. Silages.
   d. Protein supplements.
   e. Vitamin supplements.
5. A stage of maturity in plants in which seeds are well formed but soft and immature is
   a. Late vegetative.
   b. Midbloom.
   c. Milk stage.
   d. Dough stage.
   e. Postripe.
6. Which B vitamin found in corn grain appears to be bound or unavailable to the pig?
   a. Riboflavin.
   b. Niacin.
   c. Pantothenic acid.
   d. Thiamin.
   e. Pyridoxine.
7. The addition of _?_ to the livestock ration will help block the harmful effects of free gossypol.
   a. Protein.
   b. Water.
   c. Copper oxide.
   d. Ferrous sulfate.
   e. Heat.
8. Which of the following feed grain byproducts is derived from wet corn milling?
   a. Corn distiller's dried solubles.
   b. Corn gluten feed.
   c. Corn germ meal.
   d. Hominy feed.
   e. Corn bran.
9. Which of the following feed grain byproducts is highest in content of crude protein?
   a. Corn gluten meal.
   b. Hominy feed.
   c. Corn distiller's dried solubles.
   d. Wheat germ meal.
   e. Corn gluten feed.
10. Which forage pasture may commonly cause bloat in ruminants?
    a. Bromegrass.
    b. Sudangrass.
    c. Birdsfoot trefoil.
    d. Alfalfa.
    e. Sweetclover.
11. Which of the following forages would be lowest yielding as a hay crop?
    a. Alfalfa.
    b. Red clover.
    c. White clover.
    d. Sweetclover.
    e. Birdsfoot trefoil.
12. Which of the following forages is a warm-season perennial grass?
    a. Smooth bromegrass.
    b. Switchgrass.
    c. Sudangrass.
    d. Timothy.
    e. Reed Canarygrass.

13. Which of the following grasses may contain undesirable level of prussic acid?
    a. Pearl millet.
    b. Perennial ryegrass.
    c. Coastal Bermudagrass.
    d. Sudangrass.
    e. Side-oats grama.
14. Which of the following conditions results from an elemental deficiency of the forage?
    a. Sweetclover poisoning.
    b. Fescue foot.
    c. Grass tetany.
    d. Bloat.
    e. Prussic acid poisoning.
15. Which forage would probably suffer the greatest loss from leaf shattering during haymaking?
    a. Timothy.
    b. Sweetclover.
    c. Bromegrass.
    d. Switchgrass.
    e. All about the same.
16. Which is the most abundant acid produced during silage fermentation?
    a. Acidic acid.
    b. Lactic acid.
    c. Butyric acid.
    d. Propionic acid.
    e. All about equal.
17. Forage materials ensiled too dry may result in?
    a. Nutrient loss by seepage.
    b. Mold and heat formation.
    c. Excessive butyric acid formation.
    d. Sour silage.
    e. All of the above.
18. Which type of silage structure would generally have the greatest ensiling loss?
    a. Gas-tight.
    b. Concrete stave.
    c. Bunker.
    d. Stack.
    e. All about the same.
19. Which phosphorus source contains the highest level of the toxic element fluorine?
    a. Dicalcium phosphate.
    b. Curacao phosphate.
    c. Raw rock phosphate.
    d. Defluorinated rock phosphate.
    e. Phosphoric acid.
20. An advantage of a continuous-flow mixer compared to a batch mixer might be
    a. Being able to handle roughage.
    b. Unattended operation resulting in lower labor cost.
    c. Portability of the mixing unit.
    d. Flexibility in ration ingredients.
    e. All of the above.

III. *Matching or Identification*
    A. Indicate the international nomenclature class of the following feeds

a. dry forage	e. protein supplement
b. fresh forage	f. mineral supplement
c. silage	g. vitamin supplement
d. energy feed	h. additive

1. Dehydrated alfalfa meal
2. Cottonseed meal
3. Red clover hay
4. Corn grain
5. Ground limestone
6. Corn silage
7. Orchardgrass pasture
8. Steamed bone meal
9. Soybean meal
10. Oat hulls
11. Vegetable oil

12. Wheat bran
13. Irradiated dried yeast
14. Ground oyster shell
15. Dried bakery products
16. Brewer's dried grains
17. Dried whey
18. Dicalcium phosphate
19. Urea
20. Sudangrass green chop
21. Alfalfa haylage
22. Salt

B. Indicate the approximate crude protein content (as-fed basis) of each of the following concentrate feeds

    a. Less than 8%
    b. 8–14%
    c. 15–23%
    d. 24–38%
    e. More than 38%

1. Fish meal
2. Wheat bran
3. Corn grain
4. Corn gluten meal
5. Molasses
6. Tankage
7. Linseed meal, solvent-extracted
8. Citrus pulp
9. Soybean meal, solvent-extracted
10. Wheat grain

11. Dried beet pulp
12. Animal fat
13. Barley grain
14. Brewer's dried grains
15. Sunflower meal, without hulls, solvent-extracted
16. Dried skim milk
17. Hominy feed
18. Corn distiller's grains
19. Oats
20. Wheat middlings

C. Indicate into which category the following forages would fall

    a. Legume, perennial
    b. Legume, annual
    c. Nonlegume, perennial, warm-season
    d. Nonlegume, perennial, cool-season
    e. Nonlegume, annual

1. Winter rye
2. White clover
3. Fescue
4. Millet
5. Alfalfa
6. Soybean
7. Reed Canarygrass
8. Switchgrass
9. Timothy
10. Birdsfoot trefoil
11. Ryegrass

12. Sudangrass
13. Crown vetch
14. Sweetclover
15. Kentucky bluegrass
16. Big bluestem
17. Smooth bromegrass
18. Lespedeza
19. Redtop
20. Side-oats grama
21. Orchardgrass

IV. *Short Answer*

1. What are the various processes involved in the oil extraction of soybeans?
2. What is reconstituted grain?
3. Name four common feedstuffs that contain more than 35% crude protein.
4. Define "concentrates" and "roughages" and give examples of each.
5. List some essentials of proper grazing management to secure high pasture yields.

6. Why should a mixture of grasses and legumes ordinarily be used for permanent pasture?
7. What are premixes?
8. Define the nutritional differences between grass and legume forages.
9. Define the four bloom stages of maturity in forage plants.
10. Discuss the various undesirable components found in certain oilseeds.

V. Problems

1. Calculate the following
   a. % Na and Cl in salt (NaCl)
   b. % Na in sodium bicarbonate ($NaHCO_3$)
   c. % P in monosodium phosphate ($NaH_2PO_4 \cdot H_2O$)
   d. % Se in sodium selenite (Na_2SeO_3)
   e. % N in biuret ($C_2H_5N_3O_2$)
2. Assume the following broiler formulation (as-fed basis)

Ingredient (IFN)	lb
Ground corn (4-02-931)	1255
Soybean meal (5-04-612)	500
Meat and bone meal (5-00-388)	80
Alfalfa meal, dehydrated (1-00-023)	40
Fat (4-00-409)	70
Vitamin-antibiotic mix	40
Mineral mix	15
	2000

   a. Calculate the percentage of crude protein.
   b. Calculate the metabolizable energy, value, kcal/kg.
   c. What is the protein/calorie ratio (g CP/1000 kcal ME) of the diet?
3. A cattle feeder is feeding the following diet to 800-lb finishing heifers (as-fed basis):

Ingredient (IFN)	%
Corn silage (3-20-506)	60
Corn grain (4-02-931)	35
Commercial supplement	5
	100

   a. If each heifer is consuming 2.5% of its body weight as total daily dry matter, how many dry matter pounds of each feed are consumed daily per heifer? Assume corn silage contains 40% d.m. and corn and supplement each contain 90% d.m.

   _____ lb corn silage
   _____ lb corn grain
   _____ lb supplement

   b. If the cattle feeder has 325 heifers in the feedlot and wants to feed them 60 more days before shipping to market, how many total as-fed pounds of each feedstuffs are needed?

   _____ lb corn silage
   _____ lb corn grain
   _____ lb supplement

   c. The cattle feeder has the following feed remaining in storage
   (1) Corn silage in a bunker silo measuring an average of 24 ft wide, 9 ft high and 50 ft long. Assume 35 lb per cubic foot. _____ lb

   (2) Corn grain in a circular bin measuring 20 ft in diameter and 15 ft high. Assume 56 lb per bushel. _____ lb

   (3) Supplement in a rectangular bin measuring 8 ft wide, 6 ft high and 10 ft long. Assume 56 lb per bushel. _____ lb

   Will the cattle feeder have enough feed in storage to finish the cattle? Indicate any shortage or excess.

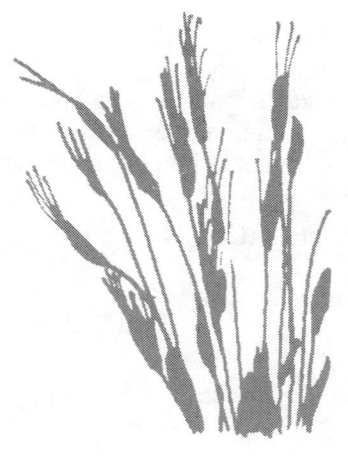

4

Procedures in Feed Formulation

Chapter Goals

- Examine feeding standard tables for various livestock.
- Describe and discuss mathematical solutions to animal diet formulation (algebra, Pearson square, substitution).
- Define the purpose of a premix and how these can be formulated to incorporate into animal diets.

I. Balancing Rations to Meet Daily Nutrient Requirements of Animals

A. *Materials Needed*

1. Nutrient requirement data from available sources
   a. National Research Council (NRC) requirement tables. These are available by species.
   b. University research and extension publications.
2. Feed analysis data from available sources
   a. National Research Council.
   b. University research and extension publications.
   c. Commercial publications (several firms publish feed analysis tables).

B. *Procedural Outline to Follow*

1. Consult any available texts or guides for developing rations and feeding programs for the animals involved.
2. Prepare a listing of the requirements of all nutrients to be considered.
3. Determine the feedstuffs available for the program and consult available material about the use of these feeds in the program planned. Remember that different feeds may have different feed value for different classes of animals.
4. Prepare a listing of the nutrient composition of the feeds to be used. You may need to consult several sources to cover all nutrients of interest.
5. Proceed to balance the ration, using guides for dry matter, crude or digestible protein and energy (TDN or other). Use the upper limit of the suggested dry matter intake as a general guide for estimated total dry feed allowance.
6. Check ration for other nutrients such as calcium, phosphorus and vitamins. You may need to add concentrated sources of vitamins and minerals to complete the requirement.
7. If the ration is complete ask the following questions
   a. Have all deficiencies been corrected?
   b. Are excesses present?
   c. Is the ration palatable and physically feasible to feed the animal?
   d. Does this appear to be the most economical combination of feeds?
   e. What is the cost of the ration per pound or ton, or what does it cost to feed this animal daily?
   f. What will be needed in addition? (free-choice salt, mineral etc.)

II. Simple Techniques in Ration Formulation

The techniques presented here will allow formulation of simple mixtures on the basis of a single nutrient (protein). These techniques can also be used with other procedures to accomplish more complex formulations of complete rations. Our approach shall be to first learn the techniques as applied to simple formulations and then apply variations that will allow their application to more complex formulations.

A. *Using Two Feed Sources*

Formulate 100 lb of a complete swine diet containing 16% crude protein (CP). The feeds to be used are corn (8.9% CP) and a commercial supplement containing 36% CP.

1. Algebraic equations—a system of two equations in two unknowns
   a. Mathematical procedure
      X = lb corn
      Y = lb supplement
      equation (1)   X +   Y = 100 lb diet
      equation (2) 0.089X + 0.360Y = 16.0 lb protein (16% of 100 lb)

      A third equation is developed to subtract from equation (2) in order to cancel either X or Y; equation (3) is developed by multiplying everything in equation (1) by a factor of 0.089, thus

$$
\begin{array}{rl}
\text{equation (2)} & 0.089X + 0.360Y = 16.0 \\
\text{(subtract) equation (3)} & -0.089X - 0.089Y = -8.9 \\
\hline
& 0 \qquad 0.271Y = 7.1
\end{array}
$$

$$Y = \frac{7.1}{0.271} = 26.2 \text{ (lb supplement)}$$

$$X = 100 - 26.2 = 73.8 \text{ (lb corn)}$$

   b. Check

      73.8 lb corn × 8.9% CP   = 6.57 lb CP
      26.2 lb supplement × 36.0% CP = 9.43 lb CP
      100.0 lb diet   16.00 lb CP

2. Pearson square—another technique to accomplish the same objective
   a. Place the percent protein desired in the combination of the two feeds in the center of a square and the percent protein content of each feed at the left corners.

      Corn   8.9%   20.0 parts corn
                              16%
      Supplement   36.0%   7.1 parts supplement
                                          27.1 total parts

   b. Subtract diagonally across the square, the smaller number from the larger without regard to sign and record the difference at the right corners.
   c. The parts of each feed can be expressed as a percent of the total, and these percentages can be applied to any quantity.

      $$\frac{20.0 \text{ parts corn}}{27.1 \text{ total parts}} (100) = 73.8\% \text{ corn and}$$

      $$\frac{7.1 \text{ part supplement}}{27.1 \text{ total parts}} (100) = 26.2\% \text{ supplement}$$

      73.8% × 100 lb = 73.8 lb corn
      26.2% × 100 lb = 26.2 lb supplement

   d. Check

      73.8 lb corn   × 8.9% CP = 6.57 lb CP
      26.2 lb supplement × 36.0% CP = 9.43 lb CP
      100.0 lb diet   16.00 lb CP

   e. Precautions about using the Pearson square
      (1) It can only be used for two feed materials; however, either or both of these can be mixtures.

(2) The number in the center of the square must be intermediate to the two numbers at the left corners. For example, any combination of a 8.9% protein corn and a 36% protein supplement would have to have a protein content between 8.9% and 36%. Always check this because the Pearson square will give an answer if the number in the center is not intermediate to the other two even though such an answer is incorrect. This precaution also applies to algebraic equations.

(3) The requirement must be expressed as a percent or proportion and can be used for any nutrient or expression of energy, e.g., percent protein, percent Ca, percent TDN, Mcal/lb, etc.

B. *Using Three or More Feed Sources*

Prepare 100 lb of diet containing 12% protein from a mixture of soybean meal (SBM) and tankage (3 parts SBM and 1 part tankage) with corn. Assume corn to contain 9.0% protein, SBM to contain 44% protein and tankage to contain 60% protein.

1. First, we must arrive at a weighted average protein percent for those ingredients that are most similar in protein content. In this case, the 3:1 mixture of SBM and tankage.

3 parts SBM	× 44% prot. =	1.32 parts protein
1 part tankage	× 60% prot. =	0.60 parts protein
4 parts mix		1.92 parts protein

$$\frac{1.92 \text{ parts protein}}{4 \text{ parts mix}} (100) = 48\% \text{ protein}$$

2. Now, the Pearson square can be used as before

Corn 9.0   36.0 $\frac{36}{39} (100) = 92.31\%$ corn

    12.0

Mix 48.0 $\frac{3.0}{39.0}$ $\frac{3}{39} (100) = 7.69\%$ mix

a. In 100 lb this means
92.31% × 100 lb = 92.31 lb corn
7.69% × 100 lb = 7.69 lb mix

b. The 7.69 lb mix must be divided into ¾ (75%) SBM and ¼ (25%) tankage, which complies with the initial proportions of each feed. Thus

c. Check

92.31 lb corn	× 0.09 =	8.31 lb protein
5.77 lb SBM	× 0.44 =	2.54 lb protein
1.92 lb tankage	× 0.60 =	1.15 lb protein
100.00 lb diet		12.00 lb protein

3. Algebraic equations could also be used to solve this problem

$$Y = \text{lb corn}$$
$$Y = \text{lb mix (3:1 mixture of SBM: tankage)}$$

1) $X + Y = 100.0$
2) $0.09X + 0.48Y = 12.0$
−3) $0.09X − 0.09Y = −9.0$

    $0 \quad\quad 0.39Y = 3.0$

$$Y = \frac{3.0}{0.39} = 7.69 \text{ lb mix}$$

$$X = 100 − 7.69 = 92.31 \text{ lb corn}$$

C. *Using Fixed Ingredients*

Prepare 1000 lb of diet from corn (8.9% CP), SBM (46% CP) and fixed ingredients totaling 10% of the diet (e.g., salt, limestone, dicalcium phosphate, trace mineral premix, vitamin premix etc.). The final diet should contain 14% CP. Assume no protein content in the fixed ingredients.

1. Use of Pearson square
   a. Find percent protein to use in center of square
      (1) The nonfixed portion (corn–SBM combination) is 900 lb (1000 lb × 90%) and will have to supply all the protein (1000 × 14% = 140 lb protein).
      (2) To do this by the Pearson square method, it is first necessary to calculate what percent protein will be needed in the corn-SBM combination to provide 140 lb of protein per 900 lb, as follows

$$\frac{140}{900}(100) = 15.56\% \text{ CP}$$

   b. This figure (15.56%) is then used in conjunction with the Pearson square as follows

   Corn 8.9   [15.56]   30.44 parts corn $\frac{30.44}{37.10}(100) = 82.05\%$

   SBM 46.0 $\frac{6.66}{37.10}$ parts SBM $\frac{6.66}{37.10}(100) = 17.95\%$

   900 lb × 82.05% = 738.45 lb corn
   900 lb × 17.95% = 161.55 lb SBM

   c. Check

738.45 lb corn	× 8.9%	= 65.72 lb protein
161.55 lb SBM	× 46.0%	= 74.31 lb protein
100.00 lb fixed	× 0	= 0
1000.00 lb ration		= 140.03 lb protein

2. Use of algebraic equations for the same problem

   $$X = \text{lb corn}$$
   $$Y = \text{lb SBM}$$

   1) $X + Y = 900.0$ lb corn-SBM
   2) $.089X + .460Y = 140.0$ lb CP
   −3) $.089X − .089Y = −80.1$
   —————————————————————————
       0     $.371Y = 59.9$

   $$Y = \frac{59.9}{.371} = 161.5 \text{ lb SBM}$$

   $$X = 900 − 161.5 = 738.5 \text{ lb corn}$$

3. If any of the fixed ingredients contain protein, the amount contributed to the diet is calculated and then subtracted from total quantity needed before formulation by either Pearson square or algebra. Assume the following example. Formulate 1 ton (2000 lb) of broiler diet to contain 20.0% crude protein using the following ingredients

Feedstuff	Amount, lb
Ground corn (9.0% CP)	?
SBM (44% CP)	?
Meat and bone meal (50% CP)	100.0
Fish meal (65% CP)	40.0
Alfalfa meal, dehydrated (17.5% CP)	40.0
Mineral premix (0% CP)	30.0
Vitamin premix (0% CP)	20.0
TOTAL	2000.0

   a. Determine the total amount of crude protein needed in the formulation.
      2000 lb diet × .20 = 400 lb CP needed

b. Determine the amount of ingredients fixed and the amount of crude protein they contribute.
Fixed ingredients

Meat and bone meal	100.0 lb	× .50	=	50.0 lb CP fixed
Fish meal	40.0 lb	× .65	=	26.0 lb CP fixed
Alfalfa meal, dehydrated	40.0 lb	× .175	=	7.0 lb CP fixed
Mineral premix	30.0 lb	× 0	=	0
Vitamin premix	20.0 lb	× 0	=	0
	230.0 lb feed			83.0 lb CP

Thus, 2000.0 lb − 230.0 lb fixed = 1770.0 lb nonfixed (corn-SBM).
400.0 lb − 83.0 lb fixed = 317.0 lb CP needed from nonfixed.

c. Solve by algebra.

$$X = \text{lb corn}$$
$$Y = \text{lb SBM}$$

1)	$X + Y = 1770.0$ lb corn-SBM	
2)	$.09X + .44Y = 317.0$ lb CP	
−3)	$.09X − .09Y = −159.3$	
	$0 \quad .35Y = 157.7$	

$$Y = \frac{157.7}{.35} = 450.6 \text{ lb SBM}$$
$$X = 1770.0 − 450.6 = 1319.4 \text{ lb corn}$$

d. Solve by Pearson square.
(1) The nonfixed portion (corn-SBM combination) is 1770.0 lb and will have to supply the remaining 317.0 lb CP not contributed by the fixed ingredients (400.0 − 83.0 = 317.0)
(2) To do this by the Pearson square, it is first necessary to calculate what percent CP will be needed in the corn-SBM combination to provide the 317.0 lb of protein per 1770.0 lb, as follows

$$\frac{317.0}{1770.0}(100) = 17.91\% \text{ CP}$$

(3) This figure (17.91% CP) is then used in conjunction with the Pearson square as follows

Corn 9.0   [17.91]   26.09 parts corn $\frac{26.09}{35.00}(100) = 74.54\%$

SBM 44.0 $\frac{8.91}{35.00}$ parts SBM $\frac{8.91}{35.00}(100) = 25.46\%$

1770.0 × 74.54% = 1319.4 lb corn
1770.0 × 25.46% = 450.6 lb SBM

(4) Check

1319.4 lb corn × 9.0% CP	=	118.75
450.6 lb SBM × 44.0% CP	=	198.26
230.0 lb fixed ingredients	=	83.00
2000.0 lb diet		400.01 lb CP

D. *Substitution Method*

A process of substituting amount of one ingredient for that amount of another or of substituting in a new ingredient.
1. Example of an original formulation

Ingredient	Amount, lb	% CP	CP, lb
Smooth brome hay	60.0	6.0	3.60
Ground corn	33.0	9.0	2.97
SBM	7.0	46.0	3.22
	100.0	Total	9.79

2. Assume you want to increase the crude protein content to 13% by substituting SBM for corn. Rather than using a trial-and-error approach, establish a one for one substitution

Add in 1 lb SBM	=	+0.46 lb CP
Remove 1 lb corn		−0.09 lb CP
Net change in protein	=	+0.37 lb CP

3. Since you want to increase from 9.79% to 13% CP, you will need 3.21 lb (13.0 − 9.79) additional protein in each 100-lb mixture.

4. Thus, if each one for one substitution increases CP by 0.37 lb, then

$$\frac{3.21}{0.37} = 8.68 \text{ lb SBM needed to substitute for 8.68 lb corn}$$

5. The revised formulation follows

Ingredient	Amount, lb	% CP	CP, lb
Smooth brome hay	60.00	6.0	3.60
Ground corn	24.32	9.0	2.19
SBM	15.68	46.0	7.21
	100.0	Total	13.00

6. Another possible substitution would be to replace some of the low protein containing brome hay (6% CP) with a high protein content alfalfa hay (16% CP). This substitution would cause a less drastic change in the energy value of the formulation.

a.
Add in 1 lb alfalfa hay	=	+0.16 lb CP
Remove 1 lb brome hay	=	−0.06 lb CP
Net change in protein	=	+0.10 lb CP

b. Thus, if each one for one substitution increases CP by 0.10 lb, then

$$\frac{3.21}{0.10} = 32.1 \text{ lb alfalfa hay to substitute for 32.1 brome hay}$$

c. This revised formulation follows

Ingredient	Amount, lb	% CP	CP, lb
Smooth brome hay	27.9	6.0	1.67
Alfalfa hay	32.1	16.0	5.14
Ground corn	33.0	9.0	2.97
SBM	7.0	46.0	3.22
	100.0	Total	13.00

E. *Simultaneous Algebraic Equations*

Useful in calculating what combination of two feeds (or two feed groups) will provide the required amount of each of two different nutrients.

1. Assume you want to determine the amounts of corn and SBM needed to meet the daily requirements of CP and metabolizable energy (ME) for a 27.3-kg (60-lb) pig.

2. Requirements and feed composition

	CP	ME
27.3 kg pig daily requirement—	0.272 kg	5390 kcal
Corn—	9.0%	3275 kcal/kg
SBM—	46.0%	2825 kcal/kg

3. Mathematical procedure

X = kg corn
Y = kg SBM
Protein equation (1) $0.09X + 0.46Y = 0.272 \text{ kg CP}$
Energy equation (2) $3275X + 2825Y = 5390 \text{ kcal ME}$

As discussed previously, a third equation is developed to subtract from equation (2) to cancel either X or Y; equation (3) is developed by multiplying everything in equation (1) by a factor of 36,388.89 (3275 ÷ 0.09), thus

equation (2)   3275X +   2825Y =   5390
(subtract) equation (3) − 3275X − 16,739Y = − 9898
0   − 13,914Y = − 4508

$$Y = \frac{-4,508}{-13,914} = 0.324 \text{ (kg SBM)}$$

Then solve for X

equation (1) 0.09X + 0.46Y = 0.272
0.09X + (0.46) (.324) = 0.272
0.09X + 0.149 = 0.272
0.09X = 0.123

$$X = \frac{0.123}{0.09} = 1.367 \text{ (kg corn)}$$

4. Check

1.367 kg corn × 9% CP   =   0.123 kg CP
0.324 kg SBM × 46% CP =   0.149 kg CP
0.272 kg CP
1.367 kg corn × 3275 kcal/kg = 4477
0.324 kg SBM × 2825 kcal/kg =   915
5392 kcal ME

5. The above amounts of corn and SBM could then be used as the basis for formulating a dietary mixture for 27.3-kg (60-lb) pigs

	Daily kg	% of Diet	lb per Ton
Corn	1.367	80.84	1616.8
SBM	0.324	19.16	383.2
Total	1.691	100.00	2000.0

6. To produce a properly balanced mixture, this diet should be supplemented with a vitamin-mineral supplement.

III. Formulating Vitamin Premixes

A. *Premixes are mixtures of microingredients and some type of carrier material.*
They are added to feeds at the time of mixing to ensure uniform distribution of these ingredients in mixed feeds.

B. *Commonly used carriers*
1. Soybean meal.
2. Ground grain.
3. Corn gluten meal.
4. Wheat middlings.
5. Several other mill feeds.

C. *A vitamin A and D premix could have these ingredients.*
1. Carrier.
2. Vitamin A concentrate.
3. Vitamin D concentrate.

D. *Development of a vitamin A premix for beef cattle*
Objective: Prepare 100 lb of a vitamin A premix to be used at a rate of 20 lb/ton of beef cattle supplement that is to contain 10,000 IU of vitamin A per pound. The vitamin A concentrate selected contains 2 million IU/gram. Soybean (SBM) will be used as a carrier.
1. Sufficient premix is being prepared to mix 5 tons of supplement; thus
5 × 2000 lb = 10,000 lb of finished supplement

2. The supplement is to contain 10,000 IU/lb then

   10,000 IU × 10,000 lb = 100,000,000 IU required in total
3. The amount of vitamin A concentrate required is

$$\frac{100,000,000 \text{ IU}}{2,000,000 \text{ IU/g}} = 50 \text{ g}$$

4. 50 g = 0.11 lb of vitamin A concentrate $\left(\dfrac{50}{454} = 0.11\right)$

5. The final premix formula is

Vitamin A concentrate (50 g)	0.11 lb
Carrier (SBM)	99.89 lb
Total	100.00 lb

6. The premix may now be packaged in 20-lb bags and one used for each ton of finished supplement prepared.

E. *Development of a B-vitamin–antibiotic premix for a swine diet*

   Objective: Prepare a premix to be added at a rate of 10 lb/ton of growing-finishing swine diet.

Nutrients or Medicants to Be Added	Concentration of Source	Level/lb of Diet to Be Supplied by Premix	Total Nutrient Required per Ton of Diet	Quantity of Source
Riboflavin	Pure	1.5 mg	3.0 g	3.0 g
Pantothenic acid	Pure	2.5 mg	5.0 g	5.0 g
Niacin	Pure	6.0 mg	12.0 g	12.0 g
B$_{12}$	20 mg/lb	0.005 mg	10.0 mg	227.0 g
Antibiotic	Pure	10.0 mg	20.0 g	20.0 g

Final premix formula:

Riboflavin	3.0 g	
Pantothenic acid	5.0 g	
Niacin	12.0 g	267 g or 0.59 lb
B$_{12}$ premix	227.0	
Antibiotic	20.0g	
Carrier (SBM)		9.41 lb
Total		10.00 lb

If it is desired to prepare this in large quantities, the above formula can be used by multiplying values by the factor desired.

IV. Formulating Trace Minerals Premixes

A. *In formulating trace mineral premixes, the compounds selected as sources of certain elements will contain other elements as well.*

   Atomic weights and formula weights of compounds will permit determination of the concentration of the element in the source material if the formula is known. Generally, the supplier furnishes a guaranteed analysis of the mineral content and this is then used in the formulation.

B. *Mineral compounds may also be variable with respect to water of hydration.*

   This must be known to arrive at the correct formula weight. For example, iron sulfate might be one of the following forms

	Molecular Wt	Percent Iron
$FeSO_4 \cdot H_2O$	169.92	32.87
$FeSO_4 \cdot 4H_2O$	223.96	24.94
$FeSO_4 \cdot 7H_2O$	278.01	20.09

(% calculations based on atomic wt of 55.85 for Fe)

C. *Development of a Trace Mineral Premix*

Objective: Prepare a trace mineral premix to be used at the rate of 10 lb/ton of beef cattle supplement. Ground limestone is to be used as a carrier.

Source	Atomic Wt	Formula Wt	% Element
$MnSO_4$	54.94 (Mn)	151.00	36.38 Mn
$CuSO_4$	63.54 (Cu)	159.60	39.81 Cu
$CoSO_4 \cdot H_2O$	58.93 (Co)	172.99	34.01 Co
$FeSO_4 \cdot 7H_2O$	55.85 (Fe)	278.01	20.10 Fe
$ZnSO_4$	65.37 (Zn)	161.43	40.49 Zn

Level to Be Added per lb of Supplement		Total Nutrient Needed per Ton (grams)	Quantity of Source (grams)
Mn	15.0 mg	30 g	82.46
Cu	5.0 mg	10 g	25.12
Co	0.5 mg	1 g	2.93
Fe	15.0 mg	30 g	149.25
Zn	5.0 mg	10 g	24.70
		Total	284.46 g
			(0.63 lb)

Final premix

Trace mineral sources	284.46 g or 0.63 lb
Ground limestone carrier	9.37 lb
Total	10.00 lb

This premix would be used to mix 1 ton of supplement. If it is desired to prepare enough premix for several tons of supplement, the above values can be increased as appropriate. Note that the use of ground limestone as the carrier will increase the calcium content of the ton of supplement approximately 0.15%.

V. Formulating a Complete Supplement

A. *A procedure outlined for formulating a complete ration (II, previously) can be used for this purpose.*

B. *Example Procedure*

1. Establish the feed to be supplemented with the formula to be prepared. Assume the animal in this case is a 1000 to 1100-lb beef cow (dry) being wintered on weathered bluestem grass.
2. Estimate intake of the feed to be supplemented. The following intake estimates for beef cows are usually made

Type of Roughage	Estimated Dry Matter Intake (% of Body Wt)
Low-grade (dry grass, straw etc.)	1.5
Average-grade (nonlegume hay)	2.0
High-grade (legume hays, green pasture etc.)	2.5

3. Determine nutrient deficiencies that must be met by determining the amount of nutrients supplied by estimated intake of the feed to be supplemented.

	CP lb	TDN lb	Ca lb	P lb	Vitamin A IU
Requirement, 1100-lb cow (dry	.97	8.4	.026	.026	20,000
Nutrients supplied by 16.5 lb of dry bluestem pasture	.74	10.2	.066	.018	0
Deficiency	.23	—	—	.008	20,000

243

4. Establish the specification for the supplement from the deficiencies above.

	Daily Deficiency	Amount Needed per lb of Supplement	Specifications for Supplement
Crude protein	.23 lb	.23 lb[1]	23%
Phosphorus	.008 lb	.008 lb	0.8%
Vitamin A	20,000 IU	20,000 IU	20,000 IU/lb

5. Using the above specifications in formulating the supplement

Example: Assume the supplement will contain 5.0% dehydrated alfalfa meal and 5.0% liquid molasses and that the following ingredients will be used to complete the formula.

Milo grain, ground
SBM (50% CP)
Dicalcium phosphate (27% calcium, 19% phosphorus)
Vitamin A concentrate (40,000 IU per g)

The "fixed portion" of the formula, and the amount of protein furnished by it, is

		Crude Protein	
Ingredients	% in Formula	% in Ingredient	% Furnished
Alfalfa meal	5.0	17.0	0.85 (17.0% × 5.0)
Molasses	5.0	4.3	0.21 (4.3% × 5.0)
	10.0		1.06

Thus:

23.0% − 1.06 = 21.94 protein to be furnished by 90.0% of the formula
(100.0% − 10.0%).

The combination of soybean meal and milo must contain 24.4% protein.
(21.94 ÷ 90.0%).

The "square" procedure can be used to calculate the proportions of SBM and milo needed

SBM—50% CP

Milo—9% CP

24.4

	Parts	%
	15.4	37.5
	25.6	62.5
	41.0	100.0

The square is set up by placing the percent protein of the two feeds at the two left corners and the percent protein needed in the mixture of the two feeds in the center. Solve by subtracting diagonally. The resulting figures show that 15.4 parts SBM with 25.6 parts milo will provide a mixture containing 24.4% protein. Convert proportions to percentages (25.6 ÷ 41.0 = 62.5%; 15.4 ÷ 41.0 = 37.5%) and multiply the answers by 90.0 to find percent SBM meal and percent milo needed in the formula:

62.5% × 90.0 = 56.3% milo
37.5% × 90.0 = 33.7% SBM

The tentative supplement, and its content of protein and phosphorus is

Ingredient	% in Formula	Crude Protein % in Ingredient	Crude Protein % Furnished	Phosphorus % in Ingredient	Phosphorus % Furnished
Alfalfa meal	5.0	17.0	0.85	0.26	.013
Molasses	5.0	4.3	0.21	0.08	.004
SBM	33.7	50.0	16.85	0.63	.212
Milo	56.3	9.0	5.07	0.28	.158
	100.0		22.98 or 23.0		.387

---

1. Missing nutrients to be supplied by 1 lb of mixed supplement daily.

The supplement furnishes 0.39% phosphorus, while 0.8% is needed. The difference, .8 − .39 = .41%, can be provided by dicalcium phosphate. To calculate the necessary amount of dicalcium phosphate, divide .41% (amount needed) by .19% (% phosphorus in dical), which equals 2.2%. Next, include the dicalcium phosphate in the fixed portion and recalculate the needed amounts of SBM and milo as before.

Ingredient	% in Formula	Crude Protein % in Ingredient	% Furnished
Alfalfa meal	5.0	17.0	0.85
Molasses	5.0	4.3	0.21
Dicalcium phosphate	2.2	0	0
	12.2		1.06

21.94 (i.e., 23.0 − 1.06) ÷ 87.8 (i.e., 100 − 12.2) = 25.0

		Parts	%
SBM 50		16.0	39.0
	25.0		
Milo 9		25.0	61.0
		41.0	100.0

39.0% × 87.8 = 34.2% SBM
61.0% × 87.8 = 53.5% milo

The final formula will be

Ingredient	% in Formula	CP % in Ingredient	CP % Furnished	Phosphorus % in Ingredient	Phosphorus % Furnished
Alfalfa meal	5.0	17.0	0.85	0.26	.013
Molasses	5.0	4.3	0.21	0.08	.004
Dicalcium phosphate	2.2	0	0	19.0	.418
SBM	34.2	50.0	17.1	0.63	.215
Milo	53.5	9.0	4.8	0.28	.150
	99.9		22.96		.800
	or		or		
	100.0		23.0		

The values for protein and phosphorus, which are close to the specifications, could be made more exact by minor ingredient adjustments if desired.

Finally, calculate the amount of vitamin A concentrate needed. Since 20,000 IU are needed per lb of supplement, a total of 2,000,000 IU are needed for 100 lb (20,000 × 100). To calculate the amount of vitamin A concentrate needed for 100 lb of supplement, divide 2,000,000 by 40,000 (potency of vitamin A concentrate per g). The answer is 50 grams. A total of 1000 g would be needed for a ton of the supplement (50 × 20). The formula can be converted to a ton basis for mixing purposes, and the estimated cost can be calculated, if desired, as illustrated by the following example

Ingredient	% in Formula	lb in Ton (lb)	Cost/lb Ingredient ($)	Cost Ingredient
Alfalfa meal	5.0	100.0	0.075	7.50
Molasses	5.0	100.0	0.040	4.00
Dicalcium phosphate	2.2	44.0	0.100	4.40
SBM	34.2	684.0	0.110	75.24
Milo	53.5	1070.0	0.050	53.50
Vitamin A		2.2*	1.000	2.20
				$146.84[†]

*1000 g ÷ 454 (g/lb) = 2.2 lb.
[†]Does not include cost of mixing and pelleting.

VI. Additional Examples of Diet Formulation Are Shown at the End of Chapters 7, 8 and 9

VII. Computer-Assisted Formulation

The acceptance of linear programming (LP) in the formulation of least-cost diets for livestock and poultry feeds has been extremely rapid and is becoming almost universal in its usage in the feed industry. The development of the microcomputer has placed the capabilities of least-cost programming within the range of small to medium-sized mills and individual producers who could not previously afford such a service.

A. *Linear Programming Defined*
  1. Linear programming is a mathematical technique for determining the optimum allocation of resources (different feedstuffs) to obtain a particular objective (meeting nutrient requirements, reducing costs, optimizing profits or exploring changes in nutrient specifications of the diet or in the nutrient content of feedstuffs).
  2. Basically, LP is a system by which a number of linear equations are solved on a simultaneous basis. One must establish a series of equations that describe in mathematical terms the conditions or requirements of the formula. These requirements must be measurable in numerical terms.

B. *Equipment and Programs Needed*
  1. Access to a large mainframe type computer or various microcomputers.
  2. Many excellent programs are available. The matrix capacity of the LP programs designed for the microcomputer is limited primarily by the memory available. It is recommended that at least 640K total memory be available to provide for adequate matrix size.

C. *Information Needed*
  In order to formulate a diet one must supply the following information to the computer.
  1. A list of feed ingredients available for use in the diet and their present cost. Often the prices of feed ingredients fluctuate enough from week to week to greatly influence the ingredients selected by a computer in a least-cost formulation. These marked ration ingredient changes may effect animal performance. Therefore, it should be pointed out that only ingredients known to be palatable and biologically available to the animal should be included.
  2. The nutrient content of each ingredient. There are many sources of information on average nutrient content of ingredients, and these may serve as a starting point for establishing matrix values. However, it must be emphasized that these values usually represent an average nutrient content; one should always strive to have values that accurately reflect the composition of ingredients actually available for use.
  3. The nutrient requirements of the animal in terms of minimum, maximum or exact quantities needed. There are several basic sources of information on nutrient requirements (NRC, university bulletins, feed company brochures etc.). When using any of these recommendations, know that they represent average ingredient values and feeding standards for average or better growing conditions. Each user must make proper adjustments for local conditions and quality of management.

  There is often a tendency for persons starting to use LP to attempt to include specifications for nearly every known nutrient. For most situations it is not necessary to include more than a dozen or so in the requirements table. This number will depend a great deal upon the quality and variety of ingredients available, and through experience the nutritionist will learn which nutrients must have minimum values expressed in the computer.

4. Physical, nonnutritive or nutritive usage limitations on certain ingredients. Some examples may include

a. Toxic properties in certain ingredients allow only limited usage in a diet.

b. Limitation of feeds containing undesirable properties that may reduce palatability or impart undesirable odors to the carcass, milk or egg.

c. Limitation on quantity available in inventory.

d. Adverse effects on physical texture and storage ability. Ingredients such as fats or molasses may be good sources of energy, but if fed at excessive levels may cause problems with bridging of feed in bins, ability of the feed to flow and other such problems.

e. Excessive levels of certain nutrients in the feed. In swine or poultry diets, ingredients such as dried bakery product containing large quantities of salt or dehydrated alfalfa meal's high fiber content may impair performance if fed at higher levels.

f. Variation in nutrient content of certain ingredients. Limitations are often placed on the amount of some ingredients because of a high degree of variability. Animal byproducts are an example of this limitation.

D. The role of the nutritionist has changed markedly as a result of the increased speed and capacity of the computer. Where once the nutritionist spent many hours daily in the routine of formulating or revising feed mixes, this task can now be accomplished in a fraction of the time. This should allow the nutritionist to concentrate upon other aspects of feed formulation and manufacturing such as ingredient evaluation, establishing nutrient requirements and quality control.

1. If you are mixing soybean meal (SBM) (44% CP) and ground corn (9% CP) together to make 2000 lb of a 16% CP diet, the amount of soybean meal required will be about
   a. 200 lb.
   b. 300 lb.
   c. 400 lb.
   d. 500 lb.
   e. 600 lb.

2. Assume you want to formulate 1 ton of a pig starter to contain 18% CP. How many lb of corn (9% CP) and soybean meal (44% CP) are needed in this mixture?
   a. 1440 lb corn; 560 lb SBM
   b. 1486 lb SBM; 514 lb corn
   c. 560 lb corn; 1440 lb SBM
   d. 1486 lb corn; 514 lb SBM
   e. 1391 lb corn; 609 lb SBM

3. Formulate 1000 kg of an 18% crude protein mix using ground milo (11% CP) and cottonseed meal (CSM) (41% CP).

4. Prepare 500 pounds of mineral mix to contain 27% Ca using ground limestone (36% Ca) and dicalcium phosphate (23% Ca, 18% P). What is the percent phosphorus in the final mix?

5. Formulate 1500 pounds of a 15% crude protein sow lactation diet using a mixture of 3 parts ground corn (9% CP) and 1 part oats (11% CP) in conjunction with soybean meal (46% CP).

6. Formulate 100 lb of mineral supplement to contain 7.0% phosphorus. Fix salt (0% P) at 40 lb, limestone (0% P) at 20 lb and have dicalcium phosphate (22% P) and rock phosphate (15% P) make up the remainder of the supplement.

Ingredient	lb	lb P
Salt	40	0
Limestone	20	0
Dical	——	——
Rock phosphate	——	——
	100	7.0

7. A Brown Swiss dairy cow requires 6 lb of crude protein and 28 Mcal NE_L daily. She is receiving 2.25 lb protein and 12 Mcal NE_L from intake of forages. What combination of corn grain and cottonseed meal are needed in a concentrate mix to meet her requirements?
   _____ lb corn (9% CP; 0.8 Mcal/lb)
   _____ lb CSM (40% CP; 0.7 Mcal/lb)
   _____ lb TOTAL

8. Formulate one ton (2000 lb) of a 15% protein swine diet containing milo (11% CP), soybean meal (48% CP) and the following fixed ingredients
   5% Alfalfa meal (20% CP)
   2% Dicalcium phosphate
   2% Trace mineralized salt
   1% Vitamin-antibiotic premix (50% CP)

   _____ lb milo
   _____ lb SBM
   _____ lb alfalfa meal
   _____ lb dicalcium phosphate
   _____ lb TM salt
   _____ lb vitamin-antibiotic premix
   2000 lb TOTAL

9. Prepare 1000 kg of a 16% crude protein swine feed from the following feedstuffs

   a. grain source: a mixture of three parts corn (8.9% CP) and one part oats (12.1% CP)

   b. protein balancer: a 32% CP supplement

   c. 10 kg of trace mineral premix

   d. 5 kg of vitamin premix

   e. 5 kg of salt

Assume that the premixes (items c and d) will add a total of 5.0 kg of protein to the 1000 kg formula.

List the final ingredients on a 1000-kg basis in the spaces provided below.

_____ kg corn

_____ kg oats

_____ kg supplement

_____ kg trace mineral premix

_____ kg vitamin premix

_____ kg salt

   1000   kg TOTAL

10. You have just begun working for the Super-Grow Feed Company. Your first assignment is to formulate a beef cattle supplement. Each pound of the supplement is to contain 30,000 IU vitamin A, 3000 IU vitamin D, 200 mg growth stimulant and 0.40 lb protein. Formulate 2000 lb of a supplement, and indicate the quantity of each of the following available ingredients

_____ lb soybean meal (48% CP)

_____ lb linseed meal (36% CP)

_____ lb vitamin A concentrate (50,000 IU/g)

_____ lb vitamin D concentrate (7000 IU/g)

_____ lb growth stimulant (50 g/lb)

   2000   lb TOTAL

Notes

SECTION III

COMMERCIAL FEEDS
AND ADDITIVES

5

Commercial Feeds: Laws and Regulations

Chapter Goals

- Describe the overall scope of the commercial feed industry.
- Highlight feed manufacturing laws, rules and regulations.
- Explain label format as to what has to be included on a label and how this format is interpreted.

I. General

The annual volume of primary feed manufactured in the United States amounts to an estimated 109+ million tons. Tonnage figures are estimated based on U.S. Department of Agriculture's studies of the formula feed industry.

Primary feed includes that which is processed and mixed with individual feed ingredients, sometimes with the addition of a premix at the rate of less than 100 lb/ton of finished feed.

In addition, 12+ million tons of secondary feed are mixed annually in the United States. Secondary feed is processed and mixed with one or more ingredients and a formula feed supplement. Supplements generally are used at the rate of 300 lb or more per ton of finished feed depending upon the protein content of the supplement and percentage of protein desired in the finished feed.

Farmers mix or feed free choice another 185+ million tons of feed concentrates on their own premises, bringing the total concentrates used annually to approximately 300+ million tons. In addition, livestock annually consume approximately 150 million tons of hay and 150 million tons of pasture.

By type of animal, here is approximately how the primary feed tonnage breaks down (1991):

- 17.1% Beef cattle and sheep
- 15.6% Dairy cows
- 26.3% Broilers
- 12.6% Starter/grower/layer/breeder
- 13.8% Hogs
- 7.8% Turkeys
- 6.8% Others (pet, horse, fish etc.)

Nearly 10,000 feed mills in the United States are registered with the Food and Drug Administration, according to a tabulation using FDA's list of Registered "Drug Establishments." These mills manufactured or mixed feed for sale, produced feed for integrated poultry operations or were part of large-scale livestock feeding operations. These mills register with the FDA when they manufacture or mix animal feeds that contain drugs. Not included in this total were a number of other mills that did not mix drugs in their formula feeds.

Nearly 2000 of all U.S. feed plants make all or most of their feed from the ground up using vitamin premixes. Most of the remaining mills are local units—including mobile mills—that use protein concentrates to mix feeds. However, some of these local mills may make primary feeds as well.

Of the 30,000 farm supply outlets in the United States, about 20,000 sell feed. These retail outlets may handle feed products in bags, or they may grind and mix bulk feed.

More than 60% of the primary feed manufactured is now sold in bulk; the remainder is marketed in bags. And approximately 66% of the formula feed is manufactured as complete feed, with the remaining 34% manufactured as supplements, concentrates, base mixes and premixes.

II. Types of Commercial Feeds and Their Ingredients
A. *Dry (Air-dry) Commercial feeds*

(See Table 5–1).
1. Usually prepared for a particular class of livestock.
2. Supplements (also called concentrates) are blends of vegetable, animal high protein feeds or both, and may include urea, if for ruminants.
3. Specialty products may include
   a. Antibiotics.
   b. Chemotherapeutic agents.
   c. Growth stimulants.
   d. Anthelmintics.
   e. Flavoring or coloring agents.
   f. Pellet-binding agents.
   g. Others.
4. Physical forms of dry commercial feeds
   a. Meal—mixture of feeds in which all of the ingredients are ground or otherwise reduced in particle size, have a particle size somewhat larger than flour or both.
   b. Cake—mass resulting from the pressing of seeds, meat or fish to remove oils, fats or other liquids.
   c. Pellets (cubes)—agglomerated feed formed by compacting and forcing through die openings by a mechanical process.
   d. Crumbles—pelleted feed reduced to granular form.
   e. Wafers—form of agglomerated feed based on fibrous ingredients in which the finished form usually has a diameter or cross-section measurement greater than its length.
   f. Range cubes (range wafers)—large pellets (¾ in.–1 in. in diameter and 1½ in.–2½ in. in length) designed to be fed on the ground.
   g. Blocks—agglomerated feed compressed into a solid mass cohesive enough to hold its form and weighing more than 2 lb. Generally marketed in 30-, 50-, 250-, and 500-lb units.
   h. Scratch (scratch grain, scratch feed)—whole, cracked or coarsely cut grain.

B. *Liquid Protein Supplements*
1. Defined as any product suitable for animal feeding that consists of a liquid vehicle plus one or more additives.
2. Generally utilize some type of molasses as the one vehicle or diluent but may also contain fermentation liquors, distiller's solubles or other nutritive liquids (propylene glycol, alcohol or even water).
3. The most common additive in liquid supplements is a source of NPN such as urea.

TABLE 5–1. Type of Feed

Ingredients	Complete Ruminant Feed	Complete Monogastric Feed	Supplement	Base-mix	Premix
Roughages	X				
Grains	X	X			
Plant proteins	X	X	X		
Animal proteins	X	X	X		
Grain byproducts	X	X	X	X	
Macrominerals	X	X	X	X	
Speciality products	X	X	X	X	X
Vitamins	X	X	X	X	X
Trace minerals	X	X	X	X	X

Grazing Cattle provided a liquid protein supplement in a manner whereby it is consumed by licking from a self-feeder.

4. Other additives include vitamins, minerals, antibiotics, drugs etc., depending on local feeding practices and requirements.
5. Extremely high in (or totally) NPN and are thus used only in ruminant diets.
6. May be blended into a complete diet for cattle or provided to such animals to consume by licking from a self-feeder.

III. Feed Laws and Regulations

The feed laws, rules and regulations for most of the different states are quite similar. This has largely resulted from the preparation and publication of the Uniform State Feed Bill by the Association of American Feed Control Officials. This bill was designed as a guideline to be followed by the different states in the writing of their respective feed laws, rules and regulations. The Uniform State Feed Bill as published in the 1987 *Official Publication of the AAFCO* reads as follows[1]:

1. Reprinted with the permission of the Association of American Control Officials.

UNIFORM STATE FEED BILL
EDITOR—JACK VAN STAVERN
Officially Adopted By
ASSOCIATION OF AMERICAN FEED CONTROL OFFICIALS
And Endorsed By
AMERICAN FEED MANUFACTURERS ASSOCIATION
NATIONAL FEED INGREDIENTS ASSOCIATION
PET FOOD INSTITUTE

Although this Bill and the Regulations have not been passed into law in all the states the subject matter covered herein does represent the official policy of this Association.

AN ACT

To regulate the manufacture and distribution of commercial feeds in the State of _____ BE IT ENACTED by the Legislature of the State of _____.

Section 1. Title

This Act shall be known as the "_____ Commercial Feed Law of 19_____".

Section 2. Enforcing Official

This Act shall be administered by the _____ of the State of _____, hereinafter referred to as the "_____".

Section 3. Definitions of Words and Terms

When used in this Act:

(a) The term "person" includes individual, partnership, corporation, and association.

(b) The term "distribute" means to offer for sale, sell, exchange, or barter, commercial feed; or to supply, furnish, or otherwise provide commercial feed to a contract feeder.

(c) The term "distributor" means any person who distributes.

(d) The term "commercial feed" means all materials except whole seeds unmixed or physically altered entire unmixed seeds, when not adulterated within the meaning of Sec. 7(a), which are distributed for use as feed or for mixing in feed: Provided, That the _____ by regulation may exempt from this definition, or from specific provisions of this Act, commodities such as hay, straw, stover, silage, cobs, husks, hulls, and individual chemical compounds or substance when such commodities, compounds or substances are not inter-mixed or mixed with other materials, and are not adulterated within the meaning of Section 7(a), of this Act.

(e) The term "feed ingredient" means each of the constituent materials making up a commercial feed.

(f) The term "mineral feed" means a commercial feed intended to supply primarily mineral elements or inorganic nutrients.

(g) The term "drug" means any article intended for use in the diagnosis, cure, mitigation, treatment, or prevention of disease in animals other than man and articles other than feed intended to affect the structure or any function of the animal body.

(h) The term "customer-formula feed" means commerical feed which consists of a mixture of commercial feeds and/or feed ingredients each batch of which is manufactured according to the specific instructions of the final purchaser.

(i) The term "manufacture" means to grind, mix or blend, or further process a commercial feed for distribution.

(j) The term "brand name" means any word, name, symbol, or device, or any combination thereof, identifying the commercial feed of a distributor or registrant and distinguishing it from that of others.

(k) The term "product name" means the name of the commercial feed which identifies it as to kind, class, or specific use.

(l) The term "label" means a display of written, printed, or graphic matter upon or affixed to the container in which a commercial feed is distributed, or on the invoice or delivery slip with which a commercial feed is distributed.

(m) The term "labeling" means all labels and other written, printed, or graphic matter (1) upon a commercial feed or any of its containers or wrapper or (2) accompanying such commercial feed.

(n) The term "ton" means a net weight of two thousand pounds avoirdupois.

(o) The term "per cent" or "percentages" means percentages by weights.

(p) The term "official sample" means a sample of feed taken by the _____ or his agent in accordance with the provisions of Section 11 (c), (e), or (f) of this Act.

(q) The term "contract feeder" means a person who as an independent contractor, feeds commercial feed to animals pursuant to a contract whereby such commercial feed is supplied, furnished, or otherwise provided to such person and whereby such person's remuneration is determined all or in part by feed consumption, mortality, profits, or amount or quality of product.

(r) The term "pet food" means any commercial feed prepared and distributed for consumption by pets.

(s) The term "pet" means any domesticated animal normally maintained in or near the household(s) of the owner(s) thereof.

(t) The term "specialty pet food" means any commercial feed prepared and distributed for consumption by specialty pets.

(u) The term "specialty pet" means any domesticated animal pet normally maintained in a cage or tank, such as, but not limited to, gerbils, hamsters, canaries, psittacine birds, mynahs, finches, tropical fish, goldfish, snakes and turtles.

Section 4. Registration

(a) No person shall manufacture a commercial feed in this State, unless he has filed with the _____ on forms provided by the _____ , his name, place of business and location of each manufacturing facility in this State.

(b) No person shall distribute in this State a commercial feed, except a customer-formula feed, which has not been registered pursuant to the provisions of this section. The applicant for registration shall be submitted in the manner prescribed by the _____. Upon approval by the _____ the registration shall be issued to the applicant. All registrations expire on the 31st day of December of each year. (Option: A registration shall continue in effect unless it is cancelled by the registrant or unless it is cancelled by the _____ pursuant to Subsection (c) of this section.)

(c) The _____ is empowered to refuse registration of any commercial feed not in compliance with the provisions of this Act and to cancel any registration subsequently found not to be in compliance with any provision of this Act: Provided, That no registration shall be refused or canceled unless the registrant shall have been given an opportunity to be heard before the _____ and to amend his application in order to comply with the requirements of this act.

Section 5. Labeling

A commercial feed shall be labeled as follows:

(a) In case of a commercial feed, except a customer-formula feed, it shall be accompanied by a label bearing the following information:

(1) The net weight (may be stated in metric units in addition to the required avoirdupois).

(2) The product name and the brand name, if any, under which the commercial feed is distributed.

(3) The guaranteed analysis stated in such terms as the _____ by regulation determines is required to advise the user of the composition of the feed or to support claims made in the labeling. In all cases the substances or elements must be determinable by laboratory methods such as the methods published by the Association of Official Analytical Chemists.

(4) The common or usual name of each ingredient used in the manufacture of the commercial feed: Provided, That the _____ by regulation may permit the use of a collective term for a group of ingredients which perform a similar function, or he may exempt such commercial feeds, or any group thereof, from this requirement of an ingredient statement if he finds that such statement is not required in the interest of consumers.

(5) The name and principal mailing address of the manufacturer or the person responsible for distributing the commercial feed.

(6) Adequate directions for use for all commercial feeds containing drugs and for such other feeds as the _____ may require by regulation as necessary for their safe and effective use.

(7) Such precautionary statements as the _____ by regulation determines are necessary for the safe and effective use of the commercial feed.

(b) In the case of a customer-formula feed, it shall be accompanied by a label, invoice, delivery slip, or other shipping document, bearing the following information:

(1) Name and address of the manufacturer.

(2) Name and address of the purchaser.

(3) Date of delivery.

(4) The product name and net weight (may be stated in metric units in addition to the required avoirdupois) of each commercial feed and each other ingredient used in the mixture.

(5) Adequate directions for use for all customer-formula feeds containing drugs and for such other feeds as the _____ may require by regulation as necessary for their safe and effective use.

(6) The directions for use and precautionary statements as required by Regulation 6 and 7.

(7) If a drug containing product is used:

I . The purpose of the medication (claim statement)

II. The established name of each active drug ingredient and the level of each drug used in the final mixture expressed in accordance with Regulation 4 (d).

Section 6. Misbranding

A commercial feed shall be deemed to be misbranded:

(a) If its labeling is false or misleading in any particular.

(b) If it is distributed under the name of another commercial feed.

(c) If is not labeled as required in Section 5 of this act.

(d) If it purports to be or is represented as a commercial feed, or if it purports to contain or is represented as containing a commercial feed ingredient, unless such commercial feed or feed ingredient conforms to the definition, if any, prescribed by regulation by the _____.

(e) If any word, statement, or other information required by or under authority of this Act to appear on the label or labeling is not prominently placed thereon with such conspicuousness (as compared with other words, statements, designs, or devices in the labeling) and in such terms as to render it likely to be read and understood by the ordinary individual under customary conditions of purchase and use.

Section 7. Adulteration

A commercial feed shall be deemed to be adulterated:

(a) (1) If it bears or contains any poisonous or deleterious substance which may render it injurious to health; but in case the substance is not an added substance, such commercial feed shall not be considered adulterated under this subsection if the quantity of such substance in such commercial feed does not ordinarily render in injurious to health; or

(2) If it bears or contains any added poisonous, added deleterious, or added nonnutritive substance which is unsafe within the meaning of Section 406 of the Federal Food, Drug, and Cosmetic Act (other than one which is (i) a pesticide chemical in or on a raw agricultural commodity: or (ii) a food additive); or

(3) If it is, or it bears or contains any food additive which is unsafe within the meaning of Section 409 of the Federal Food, Drug, and Cosmetic Act; or

(4) If it is a raw agricultural commodity and it bears or contains a pesticide chemical which is unsafe within the meaning of Section 408 (a) of the Federal Food, Drug, and Cosmetic Act; Provided, That where a pesticide chemical has been used in or on a raw agricultural commodity in conformity with an exemption granted or a tolerance prescribed under Section 408 of the Federal Food, Drug, and Cosmetic Act and such raw agricultural commodity has been subjected to processing such as canning, cooking, freezing, dehydrating, or milling, the residue of such pesticide chemical remaining in or on such processed feed shall not be deemed unsafe if such residue in or on the raw agricultural commodity has been removed to the extent possible in good manufacturing practice and the concentration of such residue in the processed feed is not greater than the tolerance prescribed for the raw agricultural commodity unless the feeding of such processed feed will result or is likely to result in a pesticide residue in the edible product of the animal, which is unsafe within the meaning of Section 408 (a) of the Federal Food, Drug, and Cosmetic Act.

(5) If it is, or it bears or contains any color additive which is unsafe within the meaning of Section 706 of the Federal Food, Drug and Cosmetic Act.

(6) If it is, or it bears or contains any new animal drug which is unsafe within the meaning of Section 512 of the Federal Food, Drug, & Cosmetic Act.

(b) If any valuable consituent has been in whole or in part omitted or abstracted therefrom or any less valuable substance substituted therefor.

(c) If its composition or quality falls below or differs from that which it is purported or is represented to possess by its labeling.

(d) If it contains a drug and the methods used in or the facilities or controls used for its manufacture, processing, or packaging do not conform to current good manufacturing practice regulations promulgated by the _____ to assure that the drug meets the requirement of this Act as to safety and has the identity and strength and meets the quality and purity characteristics which it purports or is represented to possess. In promulgating such regulations, the _____ shall adopt the current good manufacturing practice regulations Type A medicated Articles and Type B and Type C Medicated Feeds established under authority of the Federal Food, Drug and Cosmetic Act, unless he determines that they are not appropriate to the conditions which exist in this State.

(e) If it contains viable weed seeds in amounts exceeding the limits which the _____ shall establish by rule or regulation.

Section 8. Prohibited Acts

The following acts and the causing thereof within the State of _____ are hereby prohibited.

(a) The manufacture or distribution of any commercial feed that is adulterated or misbranded.

(b) The adulteration or misbranding of any commercial feed.

(c) The distribution of agricultural commodities such as whole seed, hay, straw, stover, silage, cobs, husks, and hulls, which are adulterated within the meaning of Section 7(a), of this Act.

(d) The removal or disposal of a commercial feed in violation of an order under Section 12 of this act.

(e) The failure or refusal to register in accordance with Section 4 of this Act.

(f) The violation of Section 13(f) of this Act.

(g) Failure to pay inspection fees and file reports as required by Section 9 of this Act.

Section 9. Inspection Fees and Reports

(a) An inspection fee at the rate of _____ cents per ton shall be paid on commercial feeds distributed in this State by the person whose name appears on the label as the manufacturer, guarantor or distributor, except that a person other than the manufacturer, guarantor or distributor may assume liability for the inspection fee, subject to the following:

(1) No fee shall be paid on a commercial feed if the payment has been made by a previous distributor.

(2) No fee shall be paid on customer-formula feeds if the inspection fee is paid on the commercial feeds which are used as ingredients therein.

(3) No fee shall be paid on commercial feeds which are used as ingredients for the manufacture of commercial feeds which are registered. If the fee has already been paid, credit shall be given for such payment.

(4) In the case of a commercial feed which is distributed in the State only in packages of ten pounds or less, an annual fee of _____ shall be paid in lieu of the inspection fee specified above.

(5) The minimum inspection fee shall be _____ per quarter.

(6) In the case of specialty pet food which is distributed in the state in packages of one pound or less, an annual fee of _____ shall be paid in lieu of an inspection fee.

(b) Each person who is liable for the payment of such fee shall:

(1) File, not later than the last day of January, April, July, and October of each year, a quarterly statement, setting forth the number of net tons of commercial feeds distributed in this State during the preceding calendar

quarter; and upon filing such statement shall pay the inspection fee at the rate stated in paragraph (a) of this Section. Inspection fees which are due and owing and have not been remitted to the _____ within 15 days following the date due shall have a penalty fee of _____ per cent (minimum _____) added to the amount due when payment is finally made. The assessment of this penalty fee shall not prevent the _____ from taking other actions as provided in this chapter.

   (2) Keep such records as may be necessary or required by the _____ to indicate accurately the tonnage of commercial feed distributed in this State, and the _____ shall have the right to examine such records to verify statements of tonnage. Failure to make an accurate statement of tonnage or to pay the inspection fee or comply as provided herein shall constitute sufficient cause for the cancellation of all registrations on file for the distributor.

(c) Fees collected shall constitute a fund for the payment of the costs of inspection, sampling, and analysis, and other expenses necessary for administration of this Act.

Section 10. Rules and Regulations

(a) The _____ is authorized to promulgate such rules and regulations for commercial feeds and pet foods as are specifically authorized in this Act and such other reasonable rules and regulations as may be necessary for the efficient enforcement of this Act. In the interest of uniformity the _____ shall by regulation adopt, unless he determines that they are inconsistent with the provisions of this Act or are not appropriate to conditions which exist in this state, the following:

   (1) The Official Definitions of Feed Ingredients and Official Feed Terms adopted by the Association of American Feed Control Officials and published in the Official Publication of that organization, and

   (2) Any regulation promulgated pursuant to the authority of the Federal Food, Drug, and Cosmetic Act (U.S.C. Sec. 301, **et seq.**): Provided, That the _____ would have the authority under this Act to promulgate such regulations.

(b) Before the issuance, amendment, or repeal of any rule or regulation authorized by this Act, the _____ shall publish the proposed regulation, amendment, or notice to repeal an existing regulation in a manner reasonably calculated to give interested parties, including all current registrants, adequate notice and shall afford all interested persons an opportunity to present their views thereon, orally or in writing, within a reasonable period of time. After consideration of all views presented by interested persons, the _____ shall take appropriate action to issue the proposed rule or regulation or to amend or repeal an existing rule or regulation. The provisions of this paragraph not withstanding, if the _____ pursuant to the authority of this Act, adopts the Official Definitions of Feed Ingredients or Official Feed Terms as adopted by the Association of American Feed Control Officials, or regulations promulgated pursuant to the authority of the Federal Food, Drug, and Cosmetic Act, any amendment or modification adopted by said Association or by the Secretary of the US Department of Health and Human Resources in the case of regulations promulgated pursuant to the Federal Food, Drug and Cosmetic Act, shall be adopted automatically under this Act without regard to the publication of the notice required by this paragraph (b), unless the _____, by order specifically determines that said amendment of modification shall not be adopted.

Section 11. Inspection, Sampling, and Analysis

(a) For the purpose of enforcement of this act, and in order to determine whether its provisions have been complied with, including whether or not any operations may be subject to such provisions, officers or employees duly designated by the _____ , upon presenting appropriate credentials, and a written notice to the owner, operator, or agent in charge, are authorized (1) to enter, during normal business hours, any factory, warehouse, or establishment within the State in which commercial feeds are manufactured, processed, packed, or held for distribution, or to enter any vehicle being used to transport or hold such feeds; and (2) to inspect at reasonable times and within reasonable limits and in a reasonable manner, such factory, warehouse, establishment or vehicle and all pertinent equipment, finished and unfinished materials, containers, and labeling therein. The inspection may include the verification of only such records, and production and control procedures as may be necessary to determine compliance with the Good Manufacturing Practice Regulations established under Section 7(d) of this Act.

(b) A separate notice shall be given for each such inspection, but a notice shall not be required for each entry made during the period covered by the inspection. Each such inspection shall be commenced and completed with reasonable promptness. Upon completion of the inspection, the person in charge of the facility or vehicle shall be so notified.

(c) If the officer or employee making such inspection of a factory, warehouse, or other establishment has obtained a sample in the course of the inspection, upon completion of the inspection and prior to leaving the premises he shall give to the owner, operator, or agent in charge a receipt describing the samples obtained.

(d) If the owner of any factory, warehouse, or establishment described in paragraph (a), or his agent, refuses to admit the _____ or his agent to inspect in accordance with paragraphs (a) and (b), the _____ is authorized to obtain from any State Court a warrant directing such owner or his agent to submit the premises described in such warrant to inspection.

(e) For the enforcement of this Act, the _____ or his duly designated agent is authorized to enter upon any public or private premises including any vehicle of transport during regular business hours to have access to, and to obtain samples, and to examine records relating to distribution of commercial feeds.

(f) Sampling and analysis shall be conducted in accordance with methods published by the Association of Official Analytical Chemists, or in accordance with other generally recognized methods.

(g) The results of all analyses of official samples shall be forwarded by the _____ to the person named on the label and to the purchaser. When the inspection and analysis of an official sample indicates a commercial feed has been adulterated or misbranded and upon request within 30 days following the receipt of the analysis of the _____ shall furnish to the registrant a portion of the sample concerned.

(h) The _____ , in determining for administrative purposes whether a commercial feed is deficient in any component, shall be guided by the official sample as defined in paragraph (p) of Section 3 and obtained and analyzed as provided for in paragraphs (c), (e), and (f) of Section 11 of this Act.

Section 12. Detained Commercial Feeds

(a) "Withdrawal from distribution" orders: When the _____ or his authorized agent has reasonable cause to believe any lot of commercial feed is being distributed in violation of any of the provisions of this Act or any of the prescribed regulations under This Act, he may issue and enforce a written or printed "withdrawal from distribution" order, warning the distributor not to dispose of the lot of commercial feed in any manner until written permission is given by the _____ or the Court. The _____ shall release the lot of commercial feed so withdrawn when said provisions and regulations have been complied with. If compliance is not obtained within 30 days, the _____ may begin, or upon request of the distributor or registrant shall begin, proceedings for condemnation.

(b) "Condemnation and Confiscation": Any lot of commercial feed not in compliance with said provisions and regulations shall be subject to seizure on complaint of _____ to a court of competent jurisdiction in the area in which said commercial feed is located. In the event the court finds the said commercial feed to be in violation of this Act and orders the condemnation of said commercial feed, it shall be disposed of in any manner consistent with the quality of the commercial feed and the laws of the State: Provided, That in no instance shall the disposition of said commercial feed be ordered by the court without first giving the claimant an opportunity to apply to the court for release of said commercial feed or for permission to process or re-label said commercial feed to bring it into compliance with this Act.

Section 13. Penalties

(a) Any person convicted of violating any of the provisions of this Act or who shall impede, hinder, or otherwise prevent, or attempt to prevent, said _____ or his duly authorized agent in performance of his duty in connection with the provisions of this Act, shall be adjudged guilty of a misdemeanor and shall be fined not less than _____ or more than _____ for the first violation, and not less than _____ or more than _____ for a subsequent violation.

(b) Nothing in this Act shall be construed as requiring the _____ or his representative to: (1) report for prosecution, or (2) institute seizure proceedings, or (3) issue a withdrawal from distribution order, as a result of minor violations of the Act, or when he believes the public interest will best be served by suitable notice of warning in writing.

(c) It shall be the duty of each _____ attorney to whom any violation is reported to cause appropriate proceedings to be instituted and prosecuted in a court of competent jurisdiction without delay. Before the _____ reports a violation for such prosecution, an opportunity shall be given the distributor to present his view to the _____.

(d) The _____ is hereby authorized to apply for and the court to grant a temporary or permanent injunction restraining any person from violating or continuing to violate any of the provisions of this act or any rule or regulation promulgated under the Act notwithstanding the existence of other remedies at law. Said injunction to be issued without bond.

(e) Any person adversely affected by an act, order, or ruling made pursuant to the provisions of this Act may within 45 days thereafter bring action in the (here name the particular Court in the county where the enforcement official has his office) for judicial review of such actions. The form of the proceeding shall be any which may be provided by statutes of this state to review decisions of administrative agencies, or in the absence or inadequacy thereof, any applicable form of legal action, including actions for declaratory judgments or writs of prohibitory or mandatory injunctions.

(f) Any person who uses to his own advantage, or reveals to other than the _____ , or officers of the _____ (appropriate departments of this State), or to the courts when relevant in any judicial proceeding, any information acquired under the authority of this Act, concerning any method, records, formulations, or processes which as a trade secret is entitled to protection, is guilty of a misdemeanor and shall on conviction thereof be fined not less than $_____ or imprisoned for not less than _____ year(s) or both: Provided, That this prohibition shall not be deemed as prohibiting the _____, or his duly authorized agent, from exchanging information of a regulatory nature with duly appointed officials of the United States Government, or of other States, who are similarly prohibited by law from revealing this information.

Section 14. Cooperation with other entities

The _____ may cooperate with and enter into agreements with governmental agencies of this State, other States, agencies of the Federal Government, and private associations in order to carry out the purpose and provisions of this act.

Section 15. Publication

The _____ shall publish at least annually, in such forms as he may deem proper, information concerning the sales of commercial feeds, together with such data on their production and use as he may consider advisable, and a report of the results of the analyses of official samples of commercial feeds sold within the State as compared with the analyses guaranteed in the registration and on the label; Provided, That the information concerning production and use of commercial feed shall not disclose the operations of any person.

Section 16. Constitutionality

If any clause, sentence, paragraph, or part of this Act shall for any reason be judged invalid by any court of competent jurisdiction, such judgment shall not affect, impair, or invalidate the remainder thereof but shall be confined in its operation to the clause, sentence, paragraph, or part thereof directly involved in the controversey in which such judgment shall have been rendered.

Section 17. Repeal

All laws and parts of laws in conflict with or inconsistent with the provisions of this Act are hereby repealed. (The specific statute and specific code sections to be repealed may have to be stated.)

Section 18. Effective Date

This Act shall take effect and be in force from and after the first day of

under the

UNIFORM STATE FEED BILL

Pursuant to due publication and public hearing required by the provisions of Chapter _____ of the Laws of this State, the _____ has adopted the following Rules and Regulations.

Regulation 1. Definition and Terms

(a) The names and definitions for commercial feeds shall be the Official Definition of Feed Ingredients adopted by the Association of American Feed Control Officials, except as the _____ designates otherwise in specific cases.

(b) The terms used in reference to commercial feeds shall be the Official Feed Terms adopted by the AAFCO, except as the _____ designates otherwise in specific cases.

(c) The following commodities are hereby declared exempt from the definition of commercial feed, under the provisions of Section 3(d) of the Act: Raw meat, hay, straw, stover, silages, cobs, husks, and hulls when unground and when not mixed or intermixed with other materials: Provided that these commodities are not adulterated within the meaning of Section 7(a), of the Act.

(d) Individual chemical compounds and substances are hereby declared exempt from the definition of Commercial Feed under the provisions of Section 3(d) of the Act. It has been determined that these products meet the following criteria:

   (1) There is an adopted AAFCO definition for the product.

   (2) The product is either GRAS or is not covered by a specific FDA Regulation.

   (3) The product is either a natural occurring product of relatively uniform chemical composition or is manufactured to meet the AAFCO defintion of the product.

   (4) The use of the product in the feed industry constitutes a minor portion of its total industrial use.

   (5) Small quantities of additives, which are intended to impart special desirable characteristics shall be permitted.

   (6) There is no need or problem of control of this product.

LIST OF EXEMPTED SUBSTANCES
Loose Salt

Regulation 2. Label Format

(a) Commercial feed, other than customer-formula feed, shall be labeled with the information prescribed in this regulation on the principal display panel of the product and in the following general format.

   (1) Net Weight may be stated in metric units in addition to the required avoirdupois units).

   (2) Product name and brand name if any.

   (3) If a drug is used:

      I. The word "medicated" shall appear directly following and below the product name in type size, no smaller than one-half the type size of the product name.

      II. The purpose of medication (claim statement).

III. An active drug ingredient statement listing the active drug ingredients by their established name and the amounts in accordance with Regulation 4(d).

IV. The required directions for use and precautionary statements or reference to their location if the detailed feeding directions and precautionary statements required by Regulations 6 and 7 appear elsewhere on the label.

(4) The guaranteed analysis of the feed as required under the provisions of Section 5(a)(3) of the Act include the following items, unless exempted in (IX) of this subsection, and in the order listed:

   I. Minimum percentage of crude protein.

   II. Maximum or minimum percentage of equivalent protein from non-protein nitrogen as required in Regulation 4(e).

   III. Minimum percentage of crude fat.

   IV. Maximum percentage of crude fiber.

   V. Minerals, to include, in the following order: (a) minimum and maximum percentages of calcium (Ca), (b) minimum percentages of phosphorus (P), (c) minimum and maximum percentages of salt (NaCl), and (d) other minerals.

   VI. Vitamins in such terms as specified in Regulation 4(c).

   VII. Total sugars as invert on dried molasses products or products being sold primarily for their sugar content.

   VIII. Viable lactic acid producing microorganisms for use in silages in terms specified in regulation 4(g).

   IX. Exemptions.

      a. Guarantees for minerals are not required when there are no specific label claims and when the commercial feed contains less than 6½% of Calcium, Phosphorus, Sodium and Chloride.

      b. Guarantees for vitamins are not required when the commercial feed is neither formulated for nor represented in any manner as a vitamin supplement.

      c. Guarantees for crude protein, crude fat, and crude fiber are not required when the commercial feed is intended for purposes other than to furnish these substances or they are of minor significance relating to the primary purpose of the product, such as drug premixes, mineral or vitamin supplements, and molasses.

      d. Guarantees for microorganisms are not required when the commercial feed is intended for a purpose other than to furnish these substances or they are of minor significance resisting to the primary purpose of the product, and no specific label claims are made.

(5) Feed ingredients, collective terms for the grouping of feed ingredients, or appropriate statements as provided under the provisions of Section 5(a)(4) of the Act.

   I. The name of each ingredient as defined in the Official Publication of the Association of American Feed Control Officials, common or usual name, or one approved by the _____.

   II. Collective terms for the grouping of feed ingredients as defined in the Official Definitions of Feed Ingredients published in the Official Publication of the Association of American Feed Control Officials in lieu of the individual ingredients; Provided that:

      a.   When a collective term for a group of ingredients is used on the label, individual ingredients within that group shall not be listed on the label.

      b.   The manufacturer shall provide the feed control official, upon request, with a list of individual ingredients, within a defined group, that are or have been used at manufacturing facilities distributing in or into the state.

III.   The registrant may affix the statement, "Ingredients as registered with the State" in lieu of the ingredient list on the label. The list of ingredients must be on file with the _____. This list shall be made available to the feed purchaser upon request.

(6)   Name and principal mailing address of the manufacturer or person responsible for distributing the feed. The principal mailing address shall include the street address, city, state, and zip code; however the street address may be omitted if it is shown in the current city directory or telephone directory.

(7)   The information required in Section 5(1)(1)-(5) of the Act must appear in its entirety on one side of the label or on one side of the container. The information required by Section 5(a)(6)-(7) of the Act shall be displayed in a prominent place on the label or container but not necessarily on the same side as the above information. When the information required by Section 5(1)(6)-(7) is placed on a different side of the label or container, it must be referenced on the front side with a statement such as "See back of label for directions for use." None of the information required by Section 5 of the Act shall be subordinated or obscured by other statements or designs.

(b)   Customer-formula feed shall be accompanied with the information prescribed in this regulation using labels, invoice, delivery ticket, or other shipping document bearing the following information.

(1)   The name and address of the manufacturer.

(2)   The name and address of the purchaser.

(3)   The date of sale or delivery.

(4)   The customer-formula feed name and brand name if any.

(5)   The product name and net weight (may be stated in metric units in addition to he required avoirdupois) of each registered commercial feed and each other ingredient used in the mixture.

(6)   The direction for use and precautionary statements as required by Regulations 6 and 7.

(7)   If a drug containing product is used:

   I.   The purpose of the medication (claim statement)

  II.   The established name of each active drug ingredient and the level of each drug used in the final mixture expressed in accordance with Regulation 4 (d).

Regulation 3. Brand and Product Names

(a)   The brand or product name must be appropriate for the intended use of the feed and must not be misleading. If the name indicates the feed is made for a specific use, the character of the feed must conform therewith. A mixture labeled "Dairy Feed," for example, must be suitable for that purpose.

(b)   Commercial, registered brand or trade names are not permitted in guarantees or ingredient listings and only in the product name of feeds produced by or for the firm holding the rights to such a name.

(c)   The name of a commerical feed shall not be derived from one or more ingredients of a mixture to the exclusion of other ingredients and shall not be one representing any components of a mixture unless all components are included in the name: Provided, That if any ingredient or combination of ingredients is intended to impart a distinctive characteristic to the product which is of significance to the purchaser, the name of that ingredient or combination of ingredients may be used as a part of the brand name or product name if the ingredients or combination of ingredients is quantitatively guaranteed in the guaranteed analysis, and the brand or product name is not otherwise false or misleading.

(d)   The word "protein" shall not be permitted in the product name of a feed that contains added non-protein nitrogen.

(e)   When the name carries a percentage value, it shall be understood to signify protein and/or equivalent protein content only, even though it may not explicitly modify the percentage with the word "protein": Provided, That other percentage values may be permitted if they are followed by the proper description and conform to good labeling practice. Digital numbers shall not be used in such a manner as to be misleading or confusing to the customer.

(f)   Single ingredient feeds shall have a product name in accordance with the designated definition of feed ingredients as recognized by the Association of American Feed Control Officials unless the _____ designates otherwise.

(g)   The word "vitamin", or a contraction thereof, or any word suggesting vitamin can be used only in the name of a feed which is represented to be a vitamin supplement, and which is labeled with the minimum content of each vitamin declared, as specified in Regulation 4(c).

(h)   The term "mineralized" shall not be used in the name of a feed except for "TRACE MINERALIZED SALT". When so used, the product must contain significant amounts of trace minerals which are recognized as essential for animal nutrition.

(i)   The term "meat" and "meat by-products" shall be qualified to designate the animal from which the meat and meat by-products is derived unless the meat and meat by-products are made from cattle, swine, sheep and goats.

Regulation 4. Expression of Guarantees

(a)   The guarantees for crude protein, equivalent protein from non-protein nitrogen, crude fat, crude fiber and mineral guarantees (when required) will be in terms of percentage.

(b)   Commercial feeds containing 6½% or more Calcium, Phosphorus, Sodium and Chloride shall include in the guaranteed analysis the minimum and maximum percentages of calcium (Ca), the minimum percentage of phosphorus (P), and if salt is added, the minimum and maximum percentage of salt (NaCl). Minerals, except salt (NaCl), shall be guaranteed in terms of percentage of the element. When calcium and/or salt guarantees are given in the guaranteed analysis such shall be stated and conform to the following.

(1)   When the minimum is 5.0% or less, the maximum shall not exceed the minimum by more than one percentage point.

(2)   When the minimum is above 5.0%, the maximum shall not exceed the minimum by more than 20% and in no case shall the maximum exceed the minimum by more than 5 percentage points.

(c) Guarantees for minimum vitamin content of commecial feeds shall be listed in the order specified and are stated in mg/lb unless otherwise specified:

    (1) Vitamin A, other than precursors of vitamin A, in International Units per pound.

    (2) Vitamin D-3 in products offered for poultry feeding, in International Chick Units per pound.

    (3) Vitamin D for other uses, International Units per pound.

    (4) Vitamin E, in International Units per pound.

    (5) Concentrated oils and feed additive premixes containing vitamins A, D and/or E may, at the option of the distributor be stated in units per gram instead of units per pound.

    (6) Vitamin B-12, in milligrams or micrograms per pound.

    (7) All other vitamin guarantees shall express the vitamin activity in milligrams per pound in terms of the following: menadione; riboflavin; d-pantothenic acid; thiamine; niacin; vitatmin B-6; folic acid, choline, biotin, inositol; p-amino benzoic acid; ascorbic acid; and carotene.

(d) Guarantees for drugs shall be stated in terms of percent by weight, except:

    (1) Antibiotics, present at less than 2,000 grams per ton (total) of commercial feed shall be stated in grams per ton of commercial feed.

    (2) Antibiotics present at 2,000 or more grams per ton (total) of commercial feed, shall be stated in grams per pound of commercial feed.

    (3) Labels for commercial feeds containing growth promotion and/or feed efficiency levels of antibiotics, which are to be fed continuously as the sole ration, are not required to make quantitative guarantees except as specifically noted in the Federal Food Additive Regulations for certain antibiotics, wherein, quantitative guarantees are required regardless of the level or purpose of the antibiotic.

    (4) The term "milligrams per pound" may be used for drugs or antibiotics in those cases where a dosage is given in "milligrams" in the feeding directions.

(e) Commercial feeds containing any added non-protein nitrogen shall be labeled as follows:

    (1) For ruminants

    a. Complete feeds, supplements, and concentrates containing added non-protein nitrogen and containing more than 5% protein from natural sources shall be guaranteed as follows:

        Crude Protein, minimum, _____%

        (This includes not more than _____ % equivalent protein from non-protein nitrogen).

    b. Mixed feed concentrates and supplements containing less than 5% protein from natural sources may be guaranteed as follows:

        Equivalent Crude Protein from Non-Protein Nitrogen, minimum, _____%

    c. Ingredient sources of non-protein nitrogen such as Urea, Di-Ammonium Phosphate, Ammonium Polyphosphate Solution, Ammoniated Rice Hulls, or other basic non-protein nitrogen ingredients defined by the Association of American Feed Control Officials shall be guaranteed as follows:

        Nitrogen, minimum, _____%

        Equivalent Crude Protein from Non-Protein Nitrogen, minimum, _____%

(2) For non-ruminants

    a. Complete feeds, supplements and concentrates containing crude protein from all forms of non-protein nitrogen, added as such, shall be labeled as follows:

        Crude protein, minimum _____%

        (This includes not more than _____ % equivalent crude protein which is not nutritionally available to (**species of animal for which feed is intended**).

    b. Premixes, concentrates or supplements intended for non-ruminants containing more than 1.25% equivalent crude protein from all forms of non-protein nitrogen, added as such, must contain adequate directions for use and a prominent statement: WARNING: This feed must be used only in accordance with directions furnished on the label.

(f) Mineral phosphatic materials for feeding purposes shall be labeled with the guarantee for minimum and maximum percentage of calcium (when present), the minimum percentage of phosphorus, and the maximum percentage of fluorine.

(g) Guarantees for microorganisms shall be stated in colony forming units per gram (CFU/g) when directions are for using the product in grams, or in colony forming units per pound (CFU/lb) when directions are for using the product in pounds. A parenthetical statement following the guarantee shall list each species in order of predominance.

Regulation 5. Ingredients

(a) The name of each ingredient or collective term for the grouping of ingredients, when required to be listed, shall be the name as defined in the Official Definitions of Feed Ingredients as published in the Official Publication of American Feed Control Officials, the common or usual name, or one approved by the _____.

(b) The name of each ingredient must be shown in letters or type of the same size.

(c) No reference to quality or grade of an ingredient shall appear in the ingredient statement of a feed.

(d) The term "dehydrated" may precede the name of any product that has been artificially dried.

(e) A single ingredient product defined by the Association of American Feed Control Officials is not required to have an ingredient statement.

(f) Tentative definitions for ingredients shall not be used until adopted as official, unless no official definition exists or the ingredient has a common accepted name that requires no definition, (i.e. sugar).

(g) When the word "iodized" is used in connection with a feed ingredient, the feed ingredient shall contain not less than 0.007% iodine, uniformly distributed.

Regulation 6. Directions for Use and Precautionary Statements

(a) Directions for use and precautionary statements on the labeling of all commercial feeds and customer-formula feeds containing additives (including drugs, special purpose additives, or non-nutritive additives) shall:

    (1) Be adequate to enable safe and effective use for the intended purposes by users with no special knowledge of the purpose and use of such articles; and,

    (2) Include, but not be limited to, all information described by all applicable regulations under the Federal Food, Drug and Cosmetic Act.

(b) Adequate directions for use and precautionary statements are required for feeds containing non-protein nitrogen as specified in Regulation 7.

(c) Adequate directions for use and precautionary statements necessary for safe and effective use are required on commercial feeds distributed to supply particular dietary needs or for supplementing or fortifying the usual diet or ration with any vitamin, mineral, or other dietary nutrient or compound.

Regulation 7. Non-Protein Nitrogen

(a) Urea and other non-protein nitrogen products defined in the Official Publication of the Association of American Feed Control Officials are acceptable ingredients only in commercial feeds for ruminant animals as a source of equivalent crude protein. If the commercial feed contains more than 8.75% of equivalent crude protein from all forms of non-protein nitrogen, added as such, or the equivalent crude protein from all forms of non-protein nitrogen, added as such, exceeds one-third of the total crude protein, the label shall bear adequate directions for the safe use of feeds and a precautionary statement: "CAUTION: USE AS DIRECTED." The directions for use and the caution statement shall be in type of such size so placed on the label that they will be read and understood by ordinary persons under customary conditions of purchase and use.

(b) Non-protein nitrogen defined in the Official Publication of the Association of American Feed Control Officials, when so indicated, are acceptable ingredients in commercial feeds distributed to non-ruminant animals as a source of nutrients other than equivalent crude protein. The maximum equivalent crude protein from non-protein nitrogen sources when used in non-ruminant rations shall not exceed 1.25% of the total daily ration.

(c) On labels such as those for medicated feeds which bear adequate feeding directions and/or warning statements, the presence of added non-protein nitrogen shall not require a duplication of the feeding directions or the precautionary statements as long as those statements include sufficient information to ensure the safe and effective use of this product due to the presence of non-protein nitrogen.

Regulation 8. Drug and Feed Additives

(a) Prior to approval of a registration application and/or approval of a label for commercial feed which contain additives (including drugs, other special purpose additives, or non-nutritive additives) the distributor may be required to submit evidence to prove the safety and efficacy of the commercial feed when used according to the directions furnished on the label.

(b) Satisfactory evidence of safety and efficacy of a commercial feed may be:

(1) When the commercial feed contains such additives, the use of which conforms to the requirements of the applicable regulation in the Code of Federal Regulations, Title 21, or which are "prior sanctioned" or "informal review sanctioned" or "generally recognized as safe" for such use, or

(2) When the commercial feed is itself a drug as defined in Section 3(g) of the Act and is generally recognized as safe and effective for the labeled use or is marketed subject to an application approved by the Food and Drug Administration under Title 21 U.S.C. 360(b).

(3) When one of the purposes for feeding a commercial feed is to impart immunity (that is to act through some immunological process) the constituents imparting immunity have been approved for the purpose through the Federal Virus, Serum and Toxins Act of 1913, administered by the A.P.H.I.S., U.S.D.A. The reason for the amendment is because of a problem encountered by Feed Control Officials in developing the basis for evaluation of such products that have a primary purpose to impart immunity. States in many cases are unable to make

correct judgments on the effectiveness on such products. The committee feels that this will definitely have more uniformity and effectiveness in handling product registrations.

Regulation 9. Adulterants

    (a)   For the purpose of Section 7(a)(1) of the Act, the terms "poisonous or deleterious substances" include but are not limited to the following:

        (1) Fluorine and any mineral or mineral mixture which is to be used directly for the feeding of domestic animals and in which the fluorine exceeds 0.20% for breeding and dairy cattle; 0.30% for slaughter cattle; 0.30% for sheep; 0.35% for lambs; 0.45% for swine; and 0.60% for poultry.

        (2) Fluorine bearing ingredients when used in such amounts that they raise the fluorine content of the total ration (exclusive of roughage) above the following amounts: 0.004% for breeding and dairy cattle; 0.009% for slaughter cattle; 0.006% for sheep; 0.01% for lambs; 0.015% for swine and 0.03% for poultry.

        (3) Fluorine bearing ingredients incorporated in any feed that is fed directly to cattle, sheep or goats consuming roughage (with or without) limited amounts of grain, that results in a daily fluorine intake in excess of 50 miligrams of Fluorine per 100 pounds of body weight.

        (4) Soybean meal, flakes or pellets or other vegetable meals, flakes or pellets which have been extracted with trichlorethylene or other chlorinated solvents.

        (5) Sulfur dioxide, Sulfurous acid, and salts of Sulfurous acid when used in or on feeds or feed ingredients which are considered or reported to be a significant source of vitamin B1 (Thiamine).

    (b)   All screenings or by-products of grains and seeds containing weed seeds, when used in commercial feed or sold as such to the ultimate consumer, shall be ground fine enough or otherwise treated to destroy the viability of such weed seeds so that the finished product contains no more than _____ viable prohibited weed seeds per pound and not more than _____ viable restricted weed seeds per pound

Regulation 10. Good Manufacturing Practices

    (a)   For the purposes of enforcement of Section 7(d) of the Act the _____ adopts the following as current good manufacturing practices:

        (1) The regulations prescribing good manufacturing practices for Type B and Type C medicated feeds as published in the Code of Federal Regulations, Title 21, Part 225, Sections 225.1-225.115.

        (2) The regulations prescribing good manufacturing practices for Type A Medicated Articles as published in the Code of Federal Regulations, Title 21, Part 226, Sections 226.1-226-115.

MODEL REGULATIONS FOR PROCESSED ANIMAL WASTE PRODUCTS AS ANIMAL FEED INGREDIENTS

The following Model Regulations have been developed by the Animal Waste Task Force after consideration of a number of state regulations on the same topic and after careful consideration of a number of regulatory options which might be open to a state control official. It represents the best judgment of the Task Force and is recommended, should any member state choose to adopt it for its use.

Any State Control Official proposing to adopt the following Model Regulations, or regulate Processed Animal Waste Products and Animal Feed Ingredients under his own state feed law and regulations, should read carefully all of the

Federal Register notice published by FDA on Recycled Animal Wastes (*F.R. 45*, No. 251, 86272-86276, Dec. 30, 1980), and the 1981 Recycled Animal Waste Committee Report.

Regulation 1. Legal Authority

Legal Authority (designated specifically by each state to meet legal requirements). Section 10, Uniform State Feed Bill).

Regulation 2. Definitions

Definitions (in addition to those listed in the current issue of the Official Publication of AAFCO).

Regulation 3. Registration

Registration Required (Section 4, Uniform State Feed Bill).

A. No person shall sell, offer or expose for sale, or distribute, in this state, any processed animal waste product intended, promoted, represented, advertised or distributed for use as a commercial feed as defined in Section II prior to registering same with _____ , as specified in Section 4 Uniform State Feed Bill.

B. Application for registration shall be made to the _____ on forms provided by the _____ and shall be accompanied by payment of the statutory registration fee as set forth in _____ .

C. Applications for registration shall be accompanied by the following:

1. A copy of the label or tag which the applicant proposes to use for the processed animal waste product.

2. A detailed description of the facilities, equipment and method of manufacture to be used in processing, manufacturing and testing of the processed animal waste product.

3. A sampling schedule, a full description of all tests made, and the results, thereby purporting to show the processed animal waste product meets the standards of _____ and these rules and regulations for registration.

Regulation 4. Registration Refused or Cancelled

(Section 4, Uniform Feed Bill)

A. General—Registration of a processed animal waste product shall be refused if:

1. Applicant or the processed animal waste product is determined to be in violation of any state or federal statute or state agency rule or regulation affecting or relating to the sale of commercial feeds.

2. The processed animal waste product contains any pathogenic organisms, drug residues, pesticide residues, harmful parasites, or other toxic or deleterious substances above levels permitted by (State regulations), Federal Feed, Drug and Cosmetic Act, Sections 406, 408, 409 and 706, or which could be harmful to animals, or which could result in residue in the tissue or by-products of animals above levels determined and promulgated in regulations by the _____ to be harmful.

3. The processed animal waste product does not meet the Quality Standards set forth in _____ Definitions, of this regulation.

4. The processed waste product is not labeled in compliance with law and agency rules and regulations, including Regulation 5 of these rules.

5. Applicant or registrant fails to perform the testing as specified in Regulation 6 of these rules, or to accurately maintain and display to the _____ or his designee, upon demand, the records required.

B. Registration may be refused pursuant to and in compliance with any statutory provisions authorizing the _____ to refuse registration.

C. Registration may be cancelled by the _____ if the product or registrant is found to be in violation of any provision of these regulations.

Regulation 5. Labeling Requirements.

A. The label, tag, or label invoice accompanying shipments of animal waste products shall contain all information as required by Regulation 2, Uniform Model Feed Bill and Regulations.

B. In addition, it shall include the following information, in the list of guarantees, in following order, in percentages:
1. maximum moisture, following fiber guarantee.
2. maximum ash, following moisture guarantees.

C. Special labeling or warnings required, as appropriate:
1. If the product contains drug residues, then the label shall contain the following statement in boldface type:

   "WARNING: THIS PRODUCT CONTAINS DRUG RESIDUES. DO NOT USE WITHIN 15 DAYS OF SLAUGHTER AND DO NOT USE 15 DAYS PRIOR TO OR DURING THE FOOD PRODUCTION PERIOD OF DAIRY ANIMALS AND LAYING HENS."

2. If the product contains high levels (25 ppm or greater) of copper, a maximum guarantee of copper and the following statement is required: "WARNING: CONTAINS HIGH LEVELS OF COPPER: DO NOT FEEED TO SHEEP."

3. If the product derives one-third (⅓) or more of the guaranteed total crude protein from non-protein nitrogen sources, the label shall provide adequate directions for safe use of the product and the precautionary statement: "CAUTION: USE ONLY AS DIRECTED."

Regulation 6. Testing Required

A. The purpose of the sampling and testing requirements of this section shall be to determine the presence of harmful materials or biological contaminants specified in (State regulation) and to assure compliance with the quality standards in _____ of these regulations.

B. Any person seeking or receiving registration of any processed animal waste product shall test, by representative sampling and assaying of such samples, and keep accurate records thereof, the processed animal waste product for which the registration is sought or received. The sample shall be of sufficient size so as to provide meaningful data, statistically reliable in carrying out the purpose of such sampling and analysis. For example, 10 one pound samples taken randomly from one day's production run or other identifiable lot, should be packaged in sealed air-tight bags for prompt shipment to the analytical laboratory.

C. The registrant, manufacturer, or producer of any such processed animal waste product ingredient shall conform to the following sample and analyses requirements:

1. Analyses specified by the _____ to meet the requirements of the quality standards of _____ of these regulations shall be conducted on three sequential production runs to establish that the feed ingredient is consistently within the limitations specified prior to registration and/or sale of the processed animal waste product.
   OPTIONAL In addition to quality standards, testing on the same production runs or lots should include potential hazardous substances such as the following:
   a. Drugs suspected or known to be used in the feed or as a therapeutic treatment of the animals.

b. Pesticides used on the animal, facilities, and wastes for pest control.

c. Pathogenic organisms at least to include Salmonella and E. Coli.

d. Heavy metals: arsenic, cadmium, copper, lead, mercury and selenium, at least.

e. Parasitic larva or ova.

f. Mycotoxins, such as aflatoxins.

2. Following the initial sequential testing, periodic analyses shall be conducted on production runs no less than one (1) each calendar quarter. Less frequent testing may be allowed where the analytical results show continued uniformity and a consistent margin of compliance. More frequent tests shall be required where the analytical results show a wide range, or show levels close to the established quality standards. Any processed animal waste product that does not meet the quality standards for the product shall be further processed until standards are met, shall be diverted to non-feed uses, or destroyed.

3. Sequential testing shall again be required when the periodic analyses required by paragraph (C)(2) of this section or other information available to the manufacturer of the ingredient indicates that:

a. The ingredients are not within the limitations established in ____ these regulations.

b. Changes are made in the manufacturing process.

c. New or expanded sources of the raw ingredients are used.

d. Changes occur in the drugs or pesticides used by the supplier(s) of the raw ingredient(s).

Regulation 7. Records Required

Any person seeking or receiving registration of any processed animal waste product shall keep for a period of two (2) years, accurate records of:

A. All sources of raw materials and date acquired, including information on drugs and pesticide usage.

B. All production output, including a code or other method to identify the date of production.

C. All sales and distribution, including the name and address of the purchaser or to whom distributed, date, quantity and production code.

D. Sampling and assay records of the testing required by Regulation 6 of this regulation.

OFFICIAL PET FOOD REGULATIONS
EDITOR—PAUL RAYNES

The following regulations prepared and approved by the Association of American Feed Control Officials and the Pet Food Institute are currently in official status, under the new Uniform Feed Bill approved by the Association.

Regulation PF1. Definitions and Terms

(a) **Principal Display Panel** means the part of a label that is most likely to be displayed, presented, shown or examined under normal and customary conditions of display for retail sale.

(b) **Ingredient Statements** means a collective and contiguous listing on the label of the ingredients of which the pet food is composed.

(c) **Immediate Container** means the unit, can, box, tin, bag, or other receptacle or covering in which a pet food is displayed for sale to retail purchasers, but does not include containers used as shipping containers.

Regulation PF2. Label Format and Labeling

(a) The statement of net content and product name must be shown on the principal display panel. All other required information may be placed elsewhere on the label but shall be sufficiently conspicuous as to render it easily read by the average purchaser under ordinary conditions of purchase and sale.

(b) The declaration of the net content shall be made in conformity with the United States "Fair Packaging and Labeling Act" and the regulations promulgated thereunder.

(c) The information which is required to appear in the "Guaranteed Analysis" shall be listed in the following Order:

Crude protein (Minimum Amount)
Crude fat (Minimum Amount)
Crude fiber (Maximum Amount)
Moisture (Maximum Amount)
Additional guarantees shall follow moisture.

(d) The label of a pet food shall specify the name and address of the manufacturer, packer, or distributor of the pet food. The statement of the place of business should include the street address, if any, of such place unless such street address is shown in a current city directory or telephone directory.

(e) If a person manufactures, packages, or distributes a pet food in a place other than his principal place of business, the label may state the principal place of business in lieu of the actual place where each package of such pet food was manufactured or packaged or is to be distributed, if such statement is not misleading in any particular.

(f) A vignette, graphic, or pictorial representation of a product on a pet food label shall not misrepresent the contents of the package.

(g) The use of the word "proven" in connection with label claims for a pet food is improper unless scientific or other empirical evidence establishing the claim represented as "proven" is available.

(h) No statement shall appear upon the label of a pet food which makes false or misleading comparisons between that pet food and any other pet food.

(i) Personal or commercial endorsements are permitted on pet food labels where said endorsements are factual and not otherwise misleading.

(j) When a pet food is enclosed in any outer container or wrapper which is intended for retail sale, all required label information must appear on such outside container or wrapper.

(k) The words "Dog Food", "Cat Food", or similar designations must appear conspicuously upon the principal display panels of the pet food labels.

(l) The label of a pet food shall not contain an unqualified representation or claim, directly or indirectly, that the pet food therein contained or a recommended feeding thereof, is or meets the requisites of a complete, perfect, scientific or balanced ration for dogs or cats unless such product or feeding:

(1) Contains ingredients in quantities sufficient to provide the estimated nutrient requirements for all stages of the life of a dog or cat, as the case may be, which have been established by a recognized authority on animal nutrition, such as the Committee on Animal Nutrition of the National Research Council of the National Academy of Sciences* or,

(2) Contains a combination of ingredients which when fed to a normal animal as the only source of nourishment will provide satisfactorily for fertility of females, gestation and lactation, normal growth from weaning to maturity without supplemental feeding, and will maintain the normal weight of an adult animal whether working or at rest and has had its capabilities in this regard demonstrated by adequate testing.

(m) Labels for products which are compounded for or which are suitable for only a limited purpose (i.e., a product designed for the feeding of puppies) may contain representations that said pet food product or recommended feeding thereof, is or meets the requisites of a complete, perfect, scientific or balanced ration for dogs or cats only:

(1) In conjunction with a statement of a limited purpose for which the product is intended or suitable (as, for example, in the statement 'a complete food for puppies'). Such representations and such required qualification therefor shall be juxtaposed on the same panel and in the same size, style and color print; and

(2) Such qualified representations may appear on pet food labels only if:

(a) The pet food contains ingredients in quantities sufficient to satisfy the estimated nutrient requirements established by a recognized authority on animal nutrition, such as the Committee on Animal Nutrition of the National Research Council of the National Academy of Sciences for such limited or qualified purpose; or

(b) The pet food product contains a combination of ingredients which when fed for such limited purpose will satisfy the nutrient requirements for such limited purpose and has had its capabilities in this regard demonstrated by adequate testing.

(n) Except as specified by Regulation PF 3(a), the name of any ingredient which appears on the label other than in the product name shall not be given undue emphasis so as to create the impression that such an ingredient is present in the product in a larger amount than is the fact, and if the names of more than one such ingredient are shown, they shall appear in the order of their respective predominance by weight in the product.

---

*To the extent that the product's ingredients provide nutrients in amounts which substantially deviate from those nutrient requirements estimated by such a recognized authority on animal nutrition, or in the event that no estimation has been made by a recognized authority on animal nutrition of the requirements of animals for one or more stages of said animals' lives, the product's represented capabilities in this regard must have been demonstrated by adequate testing.

(o) The label of a dog or cat food (other than one prominently identified as a snack or treat as part of the designation required upon the principal display panel under regulation PF2 (k) shall bear, on either the principal display panel or the information panel (as those terms are defined in 21 C.F.R. 501.1 and 501.2 respectively), in type of a size reasonably related to the largest type on the panel, a statement of the nutritional adequacy or purpose of the product. Such statement shall consist of one of the following:

(1) A claim that the pet food meets or exceeds the requirements of one or more of the recognized categories of nutritional adequacy: gestation, lactation, growth, maintenance, and complete for all life stages, as those categories are set forth in regulations PF2(l) and (m).

(2) A nutrition or dietary claim for purposes other than those listed in regulations PF2(l) and (m) if the claim is scientifically substantiated.

(3) The statement: "Use only as directed by your veterinarian", if it is a dietary animal food product intended for use by, or under the supervision or direction of a veterinarian.

(4) The statement: "this product is intended for intermittent or supplemental feeding only," if a product does not meet either the requirements of regulations PF2(l) and (m) or any other special nutritional or dietary need and so is suitable only for limited or intermittent or supplementary feeding.

Regulation PF3. Brand and Product Names

(a) No flavor designation shall be used on a pet food label unless the designated flavor is detectable by a recognized test method, or is one the presence of which provides a characteristic distinguishable by the pet. Any flavor designation on a pet food label must either conform to the name of its source as shown in the ingredient statement or the ingredient statement shall show the source of the flavor. The word flavor shall be printed in the same size type and with an equal degree of conspicuousness as the ingredient term(s) from which the flavor designation is derived.

Distributors of pet food employing such flavor designation or claims on the labels of the product distributed by them shall, upon request, supply verification of the designated or claimed flavor to the appropriate control official.

(b) The designation "100%" or "All" or words of similar connotation shall not be used in the brand or product name of a pet food if it contains more than one ingredient. However, for the purpose of this provisions, water sufficient for processing, required decharacterizing agents and trace amounts of preservatives and condiments shall not be considered ingredients.

(c) The term "meat" and "meat by-products" shall be qualified to designate the animal from which the meat and meat by-products are derived unless the meat and meat by-products are from cattle, swine, sheep and goats. For example, "horse-meat" and "horsemeat by-products."

(d) The name of the pet food shall not be derived from one or more ingredients of a mixture of a pet food product unless all components or ingredients are included in the name except as specified by Regulation PF3(a), (e), or (f); provided that the name of an ingredient or combination of ingredients may be used as a part of the product name if:

(1) The ingredient or combination of ingredients is present in sufficient quantity to impart a distinctive characteristic to the product or is present in amounts which have a material bearing upon the price of the product or upon acceptance of the product by the purchaser thereof: or

(2) It does not constitute a representation that the ingredient or combination of ingredients is present to the exclusion of other ingredients; or

(3) it is not otherwise false or misleading.

(e) When an ingredient or a combination of ingredients derived from animals, poultry, or fish constitutes 95% or more of the total weight of all ingredients of a pet food mixture, the name or names of such ingredient(s) may form a part of the product name of the pet food; provided that where more than one ingredient is part of such product name, then all such ingredient names shall be in the same size, style, and color print. For the purpose of this provisions, water sufficient for processing shall be excluded when calculating the percentage of the named ingredient(s). However, such named ingredient(s) shall constitute at least 70% of the total product.

(f) When an ingredient or a combination of ingredients derived from animals, poultry or fish constitutes at least 25% but less than 95% of the total weight of all ingredients of a pet food mixture, the name or names of such ingredient or ingredients may form a part of the product name of the pet food only if the product name also includes a primary descriptive term such as "meatballs" or "fishcakes" so that the product name describes the contents of the product in accordance with an established law, custom or usage or so that the product name is not misleading. All such ingredient names and the primary descriptive term shall be in the same size, style and color print. For the purpose of this provision, water sufficient for processing shall be excluded when calculating the percentage of the named ingredient(s). However, such named ingredient(s) shall constitute at least 10% of the total product.

(g) Contractions or coined names referring to ingredients shall not be used in the brand name of a pet food unless it is in compliance with Regulations PF 3(a), (d), (e), or (f).

Regulation PF4. Expression of Guarantees

(a) The sliding scale method of expressing a guaranteed analysis (for example, "protein 15-18%") is prohibited.

(b) Pursuant to Section 5(a) 3 of the Uniform State Feed Bill, the label of a pet food which is formulated as and represented to be a mineral additive supplement, shall include in the guaranteed analysis the maximum and minimum percentages of calcium, the minimum percentage of phosphorus and the maximum and minimum percentages of salt. The minimum content of all other essential nutrient elements recognized by NRC from sources declared in the ingredient statement shall be expressed as the element and in units of measurement established by a recognized authority of animal nutrition, such as the National Research Council.

(c) Pursuant to Section 5(a) (3) of the Uniform State Feed Bill, the label of pet food which is formulated as and represented to be a vitamin supplement, shall include a guarantee of the minimum content of each vitamin declared in the ingredient statement. Such guarantees shall be stated in units of measurements established by a recognized authority on animal nutrition such as the National Research Council.

(d) The vitamin potency of pet food products distributed in containers smaller than 1 lb. may be guaranteed in approved units per ounce.

(e) If the label of a pet food does not represent the pet food to be either a vitamin or a mineral supplement, but does include a table of comparison of a typical analysis of the vitamin, mineral, or nutrient content of the pet food with levels recommended by recognized animal nutrition authority, such comparison may be stated in the units of measurement used by the recognized authority on animal nutrition such as the National Research Council. The statement in a table of comparison of the vitamin, mineral, or nutrient content shall constitute a guarantee, but need not be repeated in the guaranteed analysis. Such table of comparison may appear on the label separate and apart from the guaranteed analysis.

Regulation PF5. Ingredients

(a) The maximum moisture in all pet foods shall be guaranteeed and shall not exceed 78.00% or the natural moisture content of the constitutent ingredients of the product, whichever is greater. Pet foods such as those consisting principally of stew, gravy, sauce, broth, juice or a milk replacer which are so labeled, may contain moisture in excess of 78.00%.

(b) Each ingredient of the pet food shall be listed in the ingredient statement, and names of all ingredients in the ingredient statement must be shown in letters or type of the same size. The failure to list the ingredients of a pet food in descending order by their predominance by weight in non-quantitative terms may be misleading. Any ingredient for which the Association of American Feed Control Officials has established a name and definition shall be identified by the name so established. Any ingredient for which no name and definition has been so established shall be identified by the common or usual name of the ingredient. Brand or trade names shall not be used in the ingredient statement.

(c) The term "dehydrated" may precede the name of any ingredient in the ingredient list that has been artifically dried.

(d) No reference to quality or grade of an ingredient shall appear in the ingredient statement of a pet food.

(e) A reference to the quality, nature, form, or other attribute of an ingredient shall not be made unless such designation is accurate and unless the ingredient imparts a distinctive characteristic to the pet food because it possesses that attribute.

Regulation PF6. Drugs and Pet Food Additives

(a) An artificial color may be used in a pet food only if it has been shown to be harmless to pets. The permanent or provisional listing of an artificial color in the United States Food and Drug Regulations as safe for use, together with the conditions, limitations, and tolerances, if any, incorporated therein, shall be deemed to be satisfactory evidence that the color is, when used pursuant to such regulations, harmless to pets.

(b) Prior to approval of a registration application and/or approval of a label for pet food, which contains additives, (including drugs, other special purpose additives, or non-nutritive additives) the distributor may be required to submit evidence to prove the safety and efficacy of the pet food, when used according to directions furnished on the label. Satisfactory evidence of the safety and efficacy of a pet food may be:

1. When the pet food contains such additives, the use of which conforms to the requirements of the applicable regulation in the Code of Federal Regulations, Title 21, or which are "prior sanctioned" or "Generally Recognized as Safe" for such use or

2. When the pet food itself is a drug as defined in Section 3 (g) of the Act and is generally recognized as safe and effective for label use or is marketed subject to an application approved by the Food and Drug Administration under Title 21, U.S.C. 360(b).

(c) The medicated labeling format recommended by Association of American Feed Control Officials shall be used to assure that adequate labeling is provided.

IV. Feed-Mixing Regulations

A. *Registration*

1. While plant registrations may vary according to state, most states require that manufacturers file with the state their name, place of business and location of each manufacturing facility in the state.
2. Practically all states require manufacturers to register their commercial feeds, except for "customer-formula feeds" manufactured according to the specific instructions of the final purchaser.
3. Registration forms are obtained from the Food and Drug Administration (FDA). The manufacturer should use Form 2656 for the first registration of the company or of an additional plant. Reregistration using Form 2656e is required annually.
4. Before mixing any drug into animal feed, further mixing medicated feed or even repackaging a medicated feed, the manufacturer should be aware of the FDA's approval requirements for such activities. A new medicated feed regulatory scheme was finalized by the FDA in 1986, which is called "Second Generation." The overall effect of this new program is to focus FDA requirements on those medicated feed manufacturers who use human-risk drug sources. A human-risk drug source is a drug premix that requires a Form FDA 1900 for its use. Under previous rules, most manufacturers who mixed medicated feeds were subject to the full requirements of establishment registration, good manufacturing practices (GMP) and FDA inspection regardless of human safety aspects of the drugs used. Under the new rules, medicated feed mixers who choose *not* to use human-risk drug sources are subject to a separate, relaxed set of GMPs and exempt from mandatory FDA inspections, establishment registration and drug assay requirements. Those feed mixers who *do* use any FDA 1900 medicated feed source will have their facility "regulated"; i.e., establishment registration, full GMPs and mandatory 2-year inspections.
5. A detailed second-generation listing of category and type of animal drugs along with warning statements, permitted drug levels, feeding directions and other regulatory requirements appears in the most recent edition of *Feed Additive Compendium,* published annually by the Miller Publishing Company, Minneapolis, Minnesota.

B. *Inspection Fees*

1. Most states support their feed control activities by assessing inspection fees on tonnage distributed. However, customer-formula feeds (or custom-mixed feeds) normally are not subject to inspection fees, providing that fees have been paid on the commercial feeds used to prepare the custom mix.
2. FDA is mandated to inspect all firms registered annually as "Drug Establishments" at least once each 2 years.
3. Unregistered mills will generally be subject to state and occasionally FDA inspection.

V. Feed Labeling

A. *Labeling is required for all commercial feed products.*

Labeling is the primary means of communication between the feed manufacturer and the purchasers of that feed. Labeling serves a number of critical needs

1. Identifies the product.
2. Informs the user of the nature of the product and its intended purpose.
3. Provides instruction on how to use the product.
4. Conveys any particular cautions pertaining to the product's use.

B. *Labeling requirements for nonmedicated feed are well established.*

The basics are found in the *Official Publication of the Association of American Feed Control Officials*. The format for medicated feed labels is also well established. It is simply an adaptation of the nonmedicated label to which certain drug-related information is added. It is suggested that the label include information in the following form (refer also to VI. Sample Medicated Label)

1. Net weight statement—expressed as the numerical quantity in pounds. If labeling is used for both bagged and bulk feed, the word "Bulk" should follow the quantity in pounds. If labeling is solely for bulk feed, use the words "Bulk Only." Labels for liquid feeds should include the

word "Liquid" following the quantity in pounds or use the words "Liquid Only." It is also a common practice for the bag to declare the net weight rather than the label. In this instance, the label must carry a statement such as "Net Weight Shown on Bag."

2. Company brand name or trade name (such as SuperGrow Feeds).
3. Product name—name of the feed that identifies it as to kind, class or specific use such as "Beef Start" (see sample label). It must be the same as shown on the Medicated Feed Application (Form FDA 1900) and/or the state registration. Code number may be used as a part of the product name. They should be at least three digits to prevent misinterpretation as protein level. The word "medicated" shall appear directly following and below the product name, in type size no smaller than half the type size of the product name.
4. Purpose statement.—the specific purpose of the medication, the species of animal for which intended and directions for use must be stated. It is important that the purpose statement be precisely as approved by FDA, but it is not necessary to list all permitted claims. The use directions should state whether the product is to be fed as the sole ration, continuously or for a specified period. It is permissible to include detailed feeding directions on the back of the label, provided there is a reference to them on the front, such as "See back of label for directions for use."
5. Warning or caution statement—such required statements must be prominent on the label. It is recommended that capital letters or bold-face type be used. Added emphasis may be achieved through the use of arrow symbols developed by the Animal Health Institute and American Feed Manufacturers Association. The statement must appear on the front side and can be repeated on the back, alone or in conjunction with feeding directions and other appropriate information.
6. Active drug ingredient(s)—listing of established names (not trade names) and amount of all drugs present in the feed, except amounts of certain low-level antibiotics for growth promotion and feed efficiency in feeds that are to be fed continuously as a sole ration. The amount of drug is usually stated as a percentage. Antibiotics at a level less than 2000 g of antibiotics per ton are to be expressed in grams per ton, and those at 2000 g or more are to be expressed in grams per pound. In those cases where a drug dosage is given as "milligrams per day" in the feeding directions, the term "milligrams per pound" may be used on the label in lieu of percentage declarations for drugs (including antibiotics).
7. Guaranteed analysis—protein, fat and fiber, plus any other required guarantees. Additional guarantees, over and above required guarantees, may be made if acceptable to control officials.
8. Feed ingredients—use the common name of each ingredient established by the Association of American Feed Control Officials or the corresponding "collective term" covering a group of feed ingredients. (Collective terms now permitted in every state except Florida, Hawaii and California. These states permit only partial use of the collective term concept.)
9. Company name and address—the name and principal mailing address of the manufacturer or person responsible for distributing the feed must be shown. The principal mailing address shall include the street address, city, state and zip code; however, the street address may be omitted if the firm is shown in the current city directory or telephone directory. The address of the "general office" is commonly shown. If not the manufacturer, use a statement such as "Manufactured For" or "Distributed By."
10. Detailed use directions—it may be desirable or necessary (because of potential product liability claims) to expand on any required directions for use. Instructions may include
    a. Purpose (if necessary to clarify or amplify feed use, including the weight of animal involved when necessary and degree of exposure to disease when applicable).
    b. When to feed (specific period necessary for effective use).
    c. How to feed, or mix for feeding (i.e. free-choice, sole ration, intermittent feeding, etc.)— mixing directions for premixes and concentrates or supplements, should give proper level of medication for purpose intended and include the final level of drug after mixing.
    d. Precautions (i.e., Not to be fed to bred heifers, etc.).
    e. Warning statement—a withdrawal period may be required for safety. For a combination medication involving two different withdrawal periods, use the longest period. Use statements exactly as approved by the FDA. These are shown in the Feed Additive Compendium.

f. Other feeds—it may be good practice to include reference to other feeds in the total program. The FDA desires that any such reference be specific as to the feed name, medication etc. Due to possible program alternatives and variation of conditions, it may not be practical to include this much information on the label. Reference to specific feeding literature may be more desirable.

VI. Sample Medicated Label

The drug listed was selected only for illustrative purposes, and it is not intended to imply that it is preferred over other drugs which may be used for the same purpose.

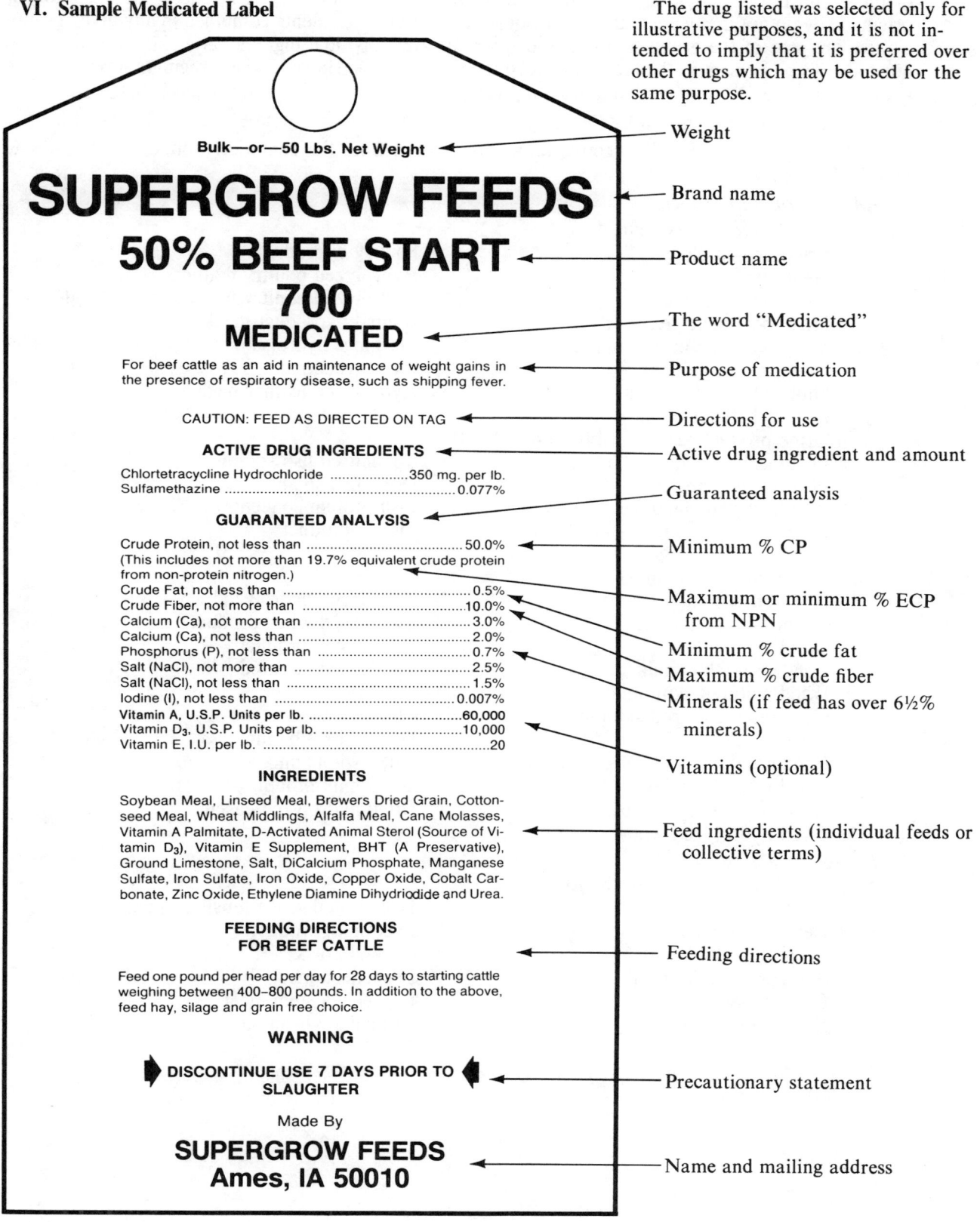

Bulk—or—50 Lbs. Net Weight ◄——— Weight

SUPERGROW FEEDS ◄——— Brand name

50% BEEF START ◄——— Product name
700
MEDICATED ◄——— The word "Medicated"

For beef cattle as an aid in maintenance of weight gains in the presence of respiratory disease, such as shipping fever. ◄——— Purpose of medication

CAUTION: FEED AS DIRECTED ON TAG ◄——— Directions for use

ACTIVE DRUG INGREDIENTS ◄——— Active drug ingredient and amount

Chlortetracycline Hydrochloride ....350 mg. per lb.
Sulfamethazine ....0.077% ◄——— Guaranteed analysis

GUARANTEED ANALYSIS

Crude Protein, not less than ....50.0% ◄——— Minimum % CP
(This includes not more than 19.7% equivalent crude protein from non-protein nitrogen.) ◄——— Maximum or minimum % ECP from NPN
Crude Fat, not less than ....0.5% ◄——— Minimum % crude fat
Crude Fiber, not more than ....10.0% ◄——— Maximum % crude fiber
Calcium (Ca), not more than 3.0%
Calcium (Ca), not less than 2.0%
Phosphorus (P), not less than ....0.7% ◄——— Minerals (if feed has over 6½% minerals)
Salt (NaCl), not more than 2.5%
Salt (NaCl), not less than 1.5%
Iodine (I), not less than 0.007%
Vitamin A, U.S.P. Units per lb. 60,000
Vitamin D₃, U.S.P. Units per lb. ....10,000 ◄——— Vitamins (optional)
Vitamin E, I.U. per lb. 20

INGREDIENTS

Soybean Meal, Linseed Meal, Brewers Dried Grain, Cottonseed Meal, Wheat Middlings, Alfalfa Meal, Cane Molasses, Vitamin A Palmitate, D-Activated Animal Sterol (Source of Vitamin D₃), Vitamin E Supplement, BHT (A Preservative), Ground Limestone, Salt, DiCalcium Phosphate, Manganese Sulfate, Iron Sulfate, Iron Oxide, Copper Oxide, Cobalt Carbonate, Zinc Oxide, Ethylene Diamine Dihydriodide and Urea. ◄——— Feed ingredients (individual feeds or collective terms)

FEEDING DIRECTIONS
FOR BEEF CATTLE ◄——— Feeding directions

Feed one pound per head per day for 28 days to starting cattle weighing between 400–800 pounds. In addition to the above, feed hay, silage and grain free choice.

WARNING

► DISCONTINUE USE 7 DAYS PRIOR TO ◄
SLAUGHTER ◄——— Precautionary statement

Made By
SUPERGROW FEEDS
Ames, IA 50010 ◄——— Name and mailing address

VII. Collective Terms

To prevent any possible need to reregister or submit new labels as ingredient composition of feeds changes, collective (or group) terms are utilized by many firms in declaring ingredients on labels. Such terms cover six groups of individual ingredients of the same general classification or origin. They do not imply equivalent nutritional value of the ingredients combined in any given group. To use any one of the group terms, one or more of the individual ingredients of the same general classification or origin. They do not imply equivalent nutritional value of the ingredients combined in any given group. To use any one of the group terms, one or more of the individual ingredients making up the group must always be present in the feed. An exception to this rule is Hawaii, where regulation requires the presence of at least two ingredients from a given group before the group term can be used.

Group terms are recognized by the FDA and are permitted (with some exceptions) in all states.

The following description of collective terms are derived from the definitions found in the AAFCO *Official Publication*.

A. *Animal Protein Products*
   1. Animal Products as defined

   Dried animal blood
   Meat meal
   Meat and bone meal
   Meat meal tankage
   Meat and bone meal tankage
   Dried meat solubles
   Poultry byproduct meal
   Poultry hatchery byproducts

   Poultry byproducts
   Hydrolyzed poultry feathers
   Hydrolyzed poultry byproduct aggregate
   Animal byproduct meal
   Fleshings hydrolsate
   Hydrolyzed hair
   Hydrolyzed leather meal

   2. Marine products as defined (except fish oil)

   Fish meal
   Fish residue meal
   Fish liver and glandular meal
   Crab meal
   Shrimp meal

   Condensed fish solubles
   Dried fish solubles
   Fish protein concentrate
   Fish byproduct

   3. Milk products as defined

   Dried buttermilk
   Condensed buttermilk
   Dried skim milk
   Condensed skim milk
   Dried cultured skim milk
   Condensed cultured skim milk
   Dried whey
   Condensed whey
   Dried whey solubles
   Condensed whey solubles
   Dried hydrolyzed whey

   Condensed hydrolyzed whey
   Condensed whey product
   Dried whey product
   Condensed cultured whey
   Casein
   Cheese rind
   Dried milk albumin
   Dried whole milk
   Dried milk protein
   Dried hydrolyzed casein

B. *Forage Products*
   May include one or more of the following

   Alfalfa leaf meal
   Dehydrated alfalfa meal
   Ground alfalfa hay
   Sun-cured alfalfa meal
   Coastal Bermudagrass hay
   Dehydrated corn plant

   Dehydrated silage (ensilage pellets)
   Flax plant product
   Ground grass
   Lespedeza meal
   Lespedeza stem meal
   Ground soybean hay

C. *Grain Products*

May include one or more of the following

Corn	Rice Ground brown
Grain sorghums	Ground paddy
Mixed feed oats	Ground rough
Barley	Broken or chipped Brewer's rice
Corn feed meal	Rye
Oats	Wheat

D. *Plant Protein Products*

May include one or more of the following

Algae meal	Soybean meal
Beans	Kibbled soybean meal
Coconut meal	Sunflower meal
Cottonseed meal	Dehulled sunflower meal
Cottonseed cake	Primary dried yeast
Cottonseed flakes	Heat-processed soybeans
Whole pressed cottonseed	Soy protein concentrate
Low gossypol cottonseed meal	Peas
Guar meal	Active dry yeast
Rapeseed meal	Dried yeast
Linseed meal	Brewer's yeast
Peanut meal	Grain distiller's dried yeast
Safflower meal	Molasses distillers dried yeast
Ground soybeans	Torula dried yeast
Soybean feed	Yeast culture

E. *Processed Grain Byproducts*

May include one or more of the following

Corn bran	Grain sorghum gluten meal
Peanut skins	Grain sorghum gluten feed
Rice bran	Corn gluten feed
Wheat bran	Distiller's dried solubles
Brewer's dried grains	Grain sorghum grits
Distiller's dried grains	Corn grits
Distiller's dried grains/solubles	Soy grits
Condensed fermented corn extractives with germ meal bran	Flour
Condensed distiller's solubles	Oat groats
Partially aspirated gelatinized sorghum grain flour	Hominy feed
Gelatinized sorghum grain flour	Malt sprouts
Corn flour	Buckwheat middlings
Soy grits	Rye middlings
Soy flour	Wheat middlings
Wheat feed flour	Wheat mill run
Grain sorghum germ cake	Pearl barley byproducts
Grain sorghum germ meal	Rice polishings
Corn germ meal (wet and dry milled)	Wheat shorts
Corn gluten meal	Aspirated grain fractions
Wheat germ meal	Wheat red dog
Defatted wheat germ meal	Feeding oat meal
	Grain sorghum mill feed

F. *Roughage Products*

May include one or more of the following

Corn cob fractions	Rye mill run
Barley hulls	Soybean hulls
Barley mill byproduct	Soybean mill feed
Malt hulls	Soybean mill run
Cottonseed hulls	Husks
Ground almond hulls	Dried citrus pulp
Buckwheat hulls	Dried citrus meal
Sunflower hulls	Citrus seed meal
Oat hulls	Corn plant pulp
Clipped oat byproduct	Dried apple pomace
Oat mill byproduct	Dried apple pectin pulp
Peanut hulls	Dried beet pulp
Rice hulls	Dried tomato pomace
Rice mill byproduct	Flax straw byproduct
Bagasse	Ground Straw

G. *Molasses Products*

May include one or more of the following

Beet molasses	Dried beet molasses product
Cane molasses	Condensed molasses fermentation solubles
Citrus molasses	Molasses yeast condensed solubles
Starch molasses	Molasses distiller's condensed solubles
Dried beet pulp with molasses	Molasses distiller's dried solubles

If individual ingredients are to be named on labels, the "and/or" listing of ingredients can be used for registration purposes. In lieu of listing of individual ingredients, the "and/or" approach lists alternate ingredients for registration purposes only. The six groups of alternate ingredients that may be used are

1. Corn, hominy feed, wheat, barley and grain sorghums.
2. Cottonseed meal, soybean meal, peanut meal, linseed meal and corn gluten meal.
3. Fish meal, meat and bone meal, tankage and poultry byproduct meal.
4. Beet molasses, corn sugar molasses, citrus molasses and cane molasses.
5. Wheat bran, wheat mill run, wheat middlings and rice bran.
6. Wheat shorts, wheat red dog, corn germ meal, corn gluten feed and grain sorghum gluten feed.

A few states do not permit this practice.

Three states do not require an ingredient listing. They are *Virginia, Missouri,* and *Oregon.*

There is a uniform registration form used by many of the states. It is described in the *Official Publication of the Association of American Feed Control Officials.* (Note: This annual publication is a most important reference and will prove useful in many ways.)

In three states (Alabama, Florida and Ohio) no specific form is required. Labels are simply submitted by letter for review and reference.

In Arizona, California, Ohio, Montana, New Jersey, Pennsylvania, Wisconsin and Tennessee, feed manufacturing plants must be registered. In all but Tennessee there is a plant registration fee.

In Alabama, Michigan, Iowa and Oregon the firm is registered for a fee. As in a number of states, Oregon has a combination system, also registering feeds by brand.

One state, Nevada, has no registration requirements. Nevada's first feed law, enacted in 1973, is confined to labeling requirements. These are the same as those found in the Uniform Feed Bill.

Study Questions and Problems

Chapter 5

I. True or False Questions

1. The term "feed ingredient," according to the Uniform State Feed Bill, refers to each of the constituent materials making up a commercial feed.
2. The label on a container of commercial feed must state the maximum percent of nitrogen-free extract (NFE) in the feed.
3. According to the Uniform State Feed Bill, urea may be used only in feeds for ruminants.
4. The term "brand name" means the name of the commercial feed, which identifies it as to kind, class or specific use.
5. The minimum percentage of crude fat must be shown on the commercial feed tag.
6. The label on a commercial feed shall indicate the class of livestock for which the feed has been prepared.
7. When the word "iodized" is used in connection with a feed ingredient, the feed ingredient shall contain not less than 0.007% iodine, uniformly distributed.
8. The label of a commercial pet food must list the maximum percent moisture.
9. The sliding scale method of expressing a guaranteed analysis (for example, "protein 15% to 18%") is allowed on labels of commercial pet foods.
10. Guarantees of the vitamin A content in commercial feeds should be stated on the label in milligrams per pound of feed.

II. Multiple Choice (select the single best answer)

1. A feed that consists of a mixture of commercial feeds, feed ingredients or both, each batch of which is manufactured according to the specific instructions of the final purchaser is called a
   a. Feed ingredient.
   b. Commercial feed.
   c. Customer-formula feed.
   d. Mineral feed.
   e. Medicated feed.
2. Any word, name, symbol or device, or any combination thereof, identifying the commercial feed of a distributor is called a
   a. Product name.
   b. Brand name.
   c. Mineral feed.
   d. Pet food.
   e. Customer-formula feed.
3. Which of the following nutrients found in a commercial poultry supplement can be expressed on a sliding scale?
   a. Crude protein
   b. Crude fat
   c. Crude fiber
   d. Phosphorus
   e. None of the above
4. The term referring to the name of a commercial feed that identifies it as to kind, class or specific use is called
   a. Label.
   b. Brand name.
   c. Registration.
   d. Product name.
   e. Customer-formula feed.
5. Which of the following items must be present on a feed tag?
   a. Minimum percentage of crude protein
   b. Maximum percentage of crude fat
   c. Maximum percentage of crude protein
   d. Minimum percentage of crude fiber
   e. Maximum percentage of phosphorus

6. Liquid protein supplements are extremely high in
   a. Calcium.
   b. Fiber.
   c. Fat.
   d. Nonprotein nitrogen (NPN).
   e. Feed additives.
7. The highest percent of primary feed in the United States is fed to
   a. Poultry.
   b. Dairy cows.
   c. Swine.
   d. Beef cattle.
   e. All about equal.

III. Matching

A. Match the appropriate word or value with the following questions related to a commercial swine finishing diet.

   a. Minimum 13%
   b. Pork Maker
   c. Minimum, 0.6%; maximum, 0.8%
   d. Maximum, 8%
   e. 50 lb

   f. 100 g/ton
   g. Minimum, 0.4%
   h. Super Grow
   i. 5 mg/lb
   j. 0%

1. % crude protein
2. Product name
3. Net weight of bag
4. % equivalent protein from NPN
5. Active drug level
6. % calcium
7. % crude fiber
8. % phosphorus
9. Brand name
10. Niacin

IV. Problems

1. A commercial beef cattle supplement label states a level of 46% crude protein with 33% equivalent protein from nonprotein nitrogen.
   a. How many total pounds of crude protein would be in 1500 lb of this supplement? _____
   b. How many pounds of crude protein equivalent from the nonprotein nitrogen source are in 1500 lb of this supplement? _____
   c. If urea (45% N) was used as the source on nonprotein nitrogen, how many pounds of urea are in 1500 lb of this supplement? _____

2. A commercial mineral supplement label indicates the following composition.

Ca (maximum)	15.0%
Ca (minimum)	13.0%
Fe	8.5%
Zn	6.0%
Mn	5.0%
Mg	3.1%
Cu	0.8%

Calculate the quantity of each ingredient present in 100 pounds of this mineral supplement.

Ingredient (Formula)		Pounds
Calcium carbonate	(37.9%)	_____
Ferrous sulfate	(20.9%)	_____
Zinc oxide	(72.5%)	_____
Manganous oxide	(76.7%)	_____
Magnesium oxide	(55.2%)	_____
Copper sulfate	(39.7%)	_____
		100.0 lb

3. Formulate a 50% crude protein beef supplement that contains 25% crude protein equivalent from urea. Use the following ingredients to total 2000 pounds.

  150.0 lb salt
  200.0 lb limestone
   50.0 lb vitamin premix
  _____ lb urea (281% CPE)
  _____ lb soybean meal (44% CP)
  _____ lb corn grain (9% CP)

2000.0 lb TOTAL

Notes

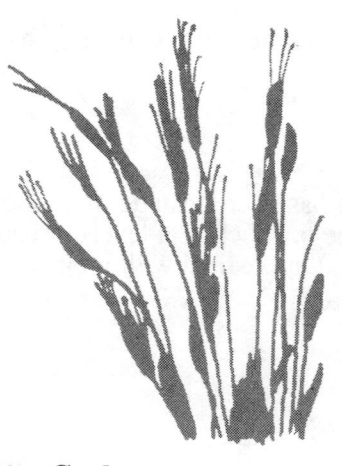

6
Feed Additives

Chapter Goals

- Describe the different types of feed additives used in livestock production
  —antibiotics
  —chemotherapeutics
  —anthelmintics
  —probiotics
- Identify which feed additives are used for particular purposes and with which animal.
- Identify implant products.

I. General Comments

Many commercially prepared additive products are included in animal diets. By our definition, we do not include those products that fall under the classification of feed nutrients (such as mineral or vitamin supplements).

The additives listed have either been approved for general use or have received considerable attention in recent research reports. This is only a partial list of additives and their claims. Additional detailed listings, descriptions and claims are available in the *Feed Additive Compendium*.[1] Also, withdrawal times are not presented for these additives because new research data may alter withdrawal times during the life of this textbook. In all cases the drug user should follow the instructions for proper use in the drug labeling, package inserts and on feed tags.

There may not be complete agreement on the claims made for all of these products, but they are listed for your information. Also, remember that we are concerned only with the products used as feed additives. It should be pointed out that many of these compounds are fed in combination.

Before mixing any additive into livestock feed, the producer or feed formulator should be aware of the Food and Drug Administration's (FDA) approval requirements for such activities. Since 1960, manufacturers have been required to demonstrate the safety of any proposed new additives before the FDA will certify their use. Tolerance levels are based on the quantity of drug shown to be safe in long-term studies. The margin of safety is fairly wide with the exception of those substances demonstrated to be carcinogenic. In 1958, Congress passed the Delaney Clause, which gave rise to the policy of "zero tolerance." That is, no substance can be used as a feed additive, even in minute amounts, if it has been in any way implicated as an inducer of cancer in either human or beast.

More than 1000 drug products are approved by the FDA for use in livestock and poultry feeds. They include products intended to protect animal health through the prevention or treatment of animal diseases, or to simply promote rapid growth, efficiency of feed utilization or both in animal production. Some of these drugs can leave potentially harmful residues in the animal tissue. For this reason, more than half of the drug products approved by FDA require withdrawal times from the diet before animal slaughter to protect consumers from drug residues. Thus, the feed formulator has the assignment of

---

1. *Feed Additive Compendium*, Miller Publishing Co., 2501 Wayzata Boulevard, Minneapolis, MN 55440.

choosing the right additive(s) to maximize rate and efficiency of production, while at the same time observing FDA regulations and protecting the consumer.

II. Growth Promotion and Feed Efficiency

A. *Antibiotics (Partial List)*

An antibiotic is a substance produced by a living organism that has bacteriostatic or bactericidal properties. In general, antibiotics are fed to reduce the incidence of subclinical levels of bacterial infections of the digestive and respiratory tracts, and thus are often credited with improving rate of gain and feed efficiency.
   1. Bacitracin methylene disalicylate—chicken, swine, turkey.
   2. Bacitracin, zinc—cattle, chicken, swine, turkey.
   3. Bambermycins—chicken, swine, turkey.
   4. Chlortetracycline—cattle, chicken, horse, sheep, swine, turkey.
   5. Erythromycin—chicken.
   6. Lasalocid—cattle, sheep.
   7. Lincomycin—chicken.
   8. Monensin—cattle.
   9. Oxytetracycline—cattle, chicken, sheep, swine, turkey.
   10. Penicillin—chicken, swine, turkey.
   11. Tylosin—chicken, swine.
   12. Virginiamycin—chicken, swine.

B. *Chemotherapeutic Agents*

A chemotherapeutic agent is an inorganic or organic compound that inhibits the growth of organisms but is not produced by a living organism. Again, we are primarily concerned with those credited with improving rate of gain and feed efficiency.
   1. Arsanilic acid—chicken, swine, turkey.
   2. Carbadox—swine.
   3. Ipronidazole—turkey.
   4. Roxarsone—chicken, turkey.

III. Medicinal Uses (Partial List)

A. *Coccidiostats*

A variety of drugs prevent and treat coccidiosis in chickens, and a smaller number are used with turkeys and other farm livestock. All manufacturer's directions and warnings should be carefully observed.
   1. Amprolium—cattle, chicken, turkey.
   2. Clopidol—chicken.
   3. Decoquinate—cattle, chicken.
   4. Halofuginone hydrombromide—chicken.
   5. Lasalocid—chicken, sheep.
   6. Monensin—chicken, cattle.
   7. Nicarbazin—chicken.
   8. Nitromide, sulfanitran, roxarsone—chicken.
   9. Robenidine hydrochloride—chicken.
   10. Salinomycin—chicken.
   11. Sulfadimethoxine and ormetoprim 5:3—chicken, turkey.
   12. Sulfaquinoxaline—chicken, turkey.
   13. Zoalene—chicken, turkey.

B. *Histostats*

Histomoniasis (blackhead) is a protozoan disease affecting turkeys and to a lesser extent, chickens. Turkeys of all ages are susceptible, but greatest mortality occurs in birds, younger than 12 weeks.
   1. Carbarsone—turkey.
   2. Dimetridazole—turkey.
   3. Nitarsone—chicken.

C. *Enteritis, Diarrhea, Dysentery*
1. Arsanilic acid—swine.
2. Bacitracin methylene disalicylate—chicken, swine, turkey.
3. Butynorate—turkey.
4. Carbadox—swine.
5. Chlortetracycline—cattle, chicken, swine, turkey.
6. Lincomycin—chicken, swine.
7. Nystatin—chicken, turkey.
8. Oxytetracycline—cattle, chicken, sheep, swine, turkey.
9. Penicillin—chicken, turkey.
10. Roxarsone—swine.
11. Tylosin—swine.
12. Virginiamycin—chicken, swine.

D. *Anthelmintic (Worming) Agents*
1. Coumaphos (Meldane, Baymix)—cattle, chicken.
2. Dichlorvos (Atgard)—swine.
3. Fenbendazole (Safe-Guard)—swine.
4. Hygromycin B (Hygromix)—chicken, swine.
5. Levamisole hydrochloride (Tramisol)—cattle, swine.
6. Morantel tartrate (Rumatel)—cattle.
7. Phenothiazine—cattle, chicken, horse, sheep, swine, turkey.
8. Piperazine—chicken, swine, turkey.
9. Pyrantel tartrate (Banminth)—swine.
10. Thiabendazole (TBZ)—cattle, sheep, swine.

E. *Miscellaneous Functions of Certain Feed Additives*
1. Liver abscesses in feedlot cattle
   a. Bacitracin methylene disalicylate.
   b. Chlortetracycline.
   c. Oxytetracycline.
   d. Tylosin.
2. Enterotoxemia (overeating disease) in sheep
   a. Chlortetracycline.
   b. Oxytetracycline.
3. Fly control in manure
   a. Methoprene-cattle.
   b. Phenothiazine (fly-control grade)—cattle.
   c. Rabon—cattle, horse, swine.
4. Bloat
   a. Oxytetracycline—dairy cattle.
   b. Poloxalene (Bloat Guard)—cattle.
5. Ketosis (acetonemia)
   a. Propylene glycol—dairy cattle.
6. Foot rot
   a. Chlortetracycline—cattle.
7. Stress
   a. Chlortetracycline—cattle, chicken, turkey.
   b. Erythromycin—chicken.
   c. Oxytetracycline—cattle, chicken, turkey.

IV. Other Feed Additives
A. *Buffers and Neutralizers*
Chemical compounds that lessen the decrease in pH caused by volatile fatty acid without causing any major increase in ruminal pH.
1. Sodium bicarbonate or sesquicarbonate.
2. Potassium bicarbonate.

3. Calcium carbonate.
4. Magnesium carbonate.
5. Magnesium oxide.
6. Sodium bentonite.

B. *Antioxidants*

Compounds that prevent oxidative rancidity of unsaturated fats. They may be used at a level of 0.25 lb per ton of feed.
1. Butylated hydroxytoluene (BHT).
2. Butylated hydroxyanisole (BHA).
3. Ethoxyquin.
4. Vitamin E.

C. *Chemical Preservatives (Partial List)*

Compounds used to retain nutritive factors and to prevent product deterioration (mold or bacteria inhibitors).
1. Ascorbic acid.
2. Calcium sorbate.
3. Citric acid.
4. Phosphoric acid.
5. Propionic acid.
6. Propyl gallate.
7. Propylene glycol.
8. Sodium metabisulfite.
9. Sodium nitrite.
10. Sodium propionate.
11. Sorbic acid.

D. *Pellet-Binding Agents*

1. Bentonite (calcium or sodium).
2. Attapulgite clay.
3. Ball clay.
4. Lignin sulfonate.
5. Hemicellulose extract.
6. Molasses.

E. *Probiotics*

1. Probiotics consist of specific microbial cultures, or ingredients or both that stimulate cultures capable of modifying the gastrointestinal environment to favor healthy tissue development.
2. Commercially available products may contain pure or combined cultures of
   a. *Lactobacillus*.
   b. *Streptococcus*.
   c. *Fungi (yeasts and molds)*.
   d. *Aspergillus*.
   e. *Bacillus*.
3. Reasons for using probiotics include
   a. Increase or balance the beneficial intestinal bacteria.
   b. Reduce toxic byproducts of digestion.
   c. Reduce intestinal pH.
   d. Improve appetite and digestion.
   e. Support rate of gain and feed efficiency.
   f. Alleviate the symptoms of stress.
4. Use of probiotics is most frequent in newborn foals, dairy replacement calves, growing/adult horses, lactating dairy cows, incoming feedlot cattle, sick pens and hospitals, newborn beef calves, weaned calves, newborn baby pigs and various pets. In general, young animals or animals in stress will be more likely to respond to microbial inoculation.

5. Probiotics are available as
   a. Feed additives (heat from pelleting or long storage may render certain products nonviable).
   b. Water dispensing.
   c. Bolus or gel form.

V. Hormonelike Products for Growth Stimulation

Hormones are substances formed in endocrine organs that activate specifically receptive organs when transported to them by the body fluids. Synthesized compounds that produce the same responses as those from naturally produced hormones can be used to promote animal growth when administered as either a feed additive or a subcutaneous implantation.

A. *Feed Additive*
   1. Melengestrol acetate (MGA)
      a. MGA is a synthesized steroid that is closely related structurally to progesterone.
      b. Feed continuously to feedlot heifers at 0.25 to 0.50 mg per head daily. Not effective in pregnant or spayed heifers or in steers.
      c. Claims include
         (1) Growth stimulatin (7% to 11%).
         (2) Improved feed efficiency (6% to 10%).
         (3) Suppression of estrus.
      d. Must be removed from the feed at least 48 hours before slaughter.
      e. Mode of action includes suppression of estrus as well as elevated endogenous estrogen levels that stimulate growth hormone release and growth promotion.

B. *Implants*

An implant is a substance implanted into the animal body and designed to release slowly, but constantly, the active chemical(s) for the purpose of growth promotion, improved feed efficiency or control of some physiological function.
   1. Ralgro—an implant pellet
      a. The active ingredient is zeranol which is made from zearalenone, a natural metabolite of the mold *Gibberella zeae*.
      b. Cleared for use in calves, growing cattle, feedlot steers and heifers, as well as feedlot lambs. Not cleared for breeding animals.
      c. Claimed to improve gain (8% to 12%) and improve feed efficiency (7% to 10%) in feedlot beef cattle and sheep.
      d. Cattle may be implanted anytime up to 65 days before slaughter. The duration of response is 70 to 110 days. Cattle should be reimplanted every 65 to 100 days for optimum results.
      e. Dosage level: 36 mg in cattle; 12 mg in sheep.
      f. Produces a potent anabolic property by increasing secretion of growth hormone. Enhances retention of nitrogen and promotes skeletal growth without an increased deposition of fat.
   2. Synovex—an implant pellet
      a. Synovex-S (for steers over 400 lb body weight)—each implant contains 20 mg of estradiol benzoate and 200 mg of progesterone.
      b. Synovex-H (for feedlot heifers over 400 lb body weight)—each implant contains 20 mg of estradiol benzoate and 200 mg of testosterone.
      c. Synovex-C (for steer or heifer calves 45 days to 400 lb body weight)—each implant contains 10 mg of estradiol benzoate and 100 mg of progesterone.
      d. Not for veal or replacement breeding animals.
      e. Claimed to improve gain (10% to 15%) and improve feed efficiency (7% to 12%).
      f. Reimplanting is recommended from 70 to 100 days after the previous implant.
      g. Mode of action includes stimulating muscle deposition at the cellular level and also increased secretion of growth hormone.
   3. Compudose—a drug-impregnated silicone rubber implant
      a. The active ingredient is a natural steroid, estradiol-17 B, which is consistently released over a 200-day period. Each implant contains 24 mg of estradiol.
      b. Approved for steers or heifers; beneficial to suckling calves, growing cattle on pasture, as well as feedlot cattle.

c. Claimed to increase average daily gain of suckling calves by 5%; 9% to 16% in growing cattle, and up to 17% in feedlot cattle. Also improves feed efficiency by 7% in feedlot cattle.

d. Molecules of estradiol are released into the circulatory system of the animal, thereby stimulating the pituitary gland to produce more growth hormone, which increases protein deposition. The estrogenic substances are a type of anabolic steroid that cause growth promotion.

4. Finaplix—an implant pellet

a. Finaplix-S (for growing and finishing feedlot steers)—each implant contains 140 mg of trenbolone acetate (TBA), a synthetic androgen.

b. Finaplix-H (for finishing feedlot heifers)—each implant contains 200 mg TBA; implant only during feedlot period approximately 63 days before slaughter.

c. Claimed to improve gain (5% to 10%) and improve feed efficiency (5%) in either feedlot steers or heifers compared to those with no implant.

d. Combination implanting with one containing estradiol results in additive gain and feed efficiency performance.

e. Trenbolone acetate appears to have a direct effect on stimulation of muscle growth and protein anabolism in the animal body. The duration of response is 60 to 80 days.

f. May reduce carcass grade (reduced marbling).

5. Revalor—an implant pellet

a. For growing and finishing feedlot steers; each implant contains 120 mg TBA and 24 mg estradiol.

b. Claimed to improve gain (15% to 25%) and improve feed efficiency (15%).

c. The duration of the response is approximately 100 days.

d. May reduce carcass grade (reduced marbling).

VI. Sample Uses in Livestock Diets

The following is a partial list of feed additives used for swine, beef cattle, sheep, poultry and horses. Users are advised to read the product label and adhere to recommendations of the manufacturer. This list does not carry drug withdrawal time information. You will find withdrawal time and other information important to the safe and effective use of these drugs in the drug labeling or the *Feed Additive Compendium*. The regulations governing the use of these drugs are subject to change by the FDA and may do so during the life of this textbook. Changes to these regulations appear in the *Federal Register* or the *Feed Additive Compendium*. The following information was taken from the 1992 *Feed Additive Compendium*.

A. *Feed Additives for Swine*

Drugs	Trade or Common Name	Claims	Use Level
Arsanilic acid	Pro-gen; Arsonic Powder	Promote growth and efficiency; swine dysentery control	45–90 g/ton cont.; 227–363 g/ton 5–6 days
Bacitracin methylene disalicylate	BMD; Solu-tracin; Fortracin	Promote growth and efficiency; swine dysentery control	10–30 g/ton; 250 g/ton
Bacitracin, zinc	Baciferm; Albac 50	Promote growth and efficiency.	20–40 g/ton
Bambermycins	Flavomycin	Increased wt gains and feed efficiency.	2 g/ton
Carbadox	Mecadox	Promote growth and efficiency; swine dysentery enteritis	10–25 g/ton; 50 g/ton
Chlortetracycline	Aureomycin; PfiChlor; CLTC; CTC	Promote growth and efficiency; various disease claims	10–50 g/ton; 50–400 g/ton

Drugs	Trade or Common Name	Claims	Use Level
Chlortetracycline, sulfa and penicillin	ASP-250 CSP-250	Promote growth and efficiency; various disease claims	C-100 g/ton S-100 g/ton P-50 g/ton
Dichlorvos	Atgard C	Wormer	0.0384% diet 2 days
Dichlorvos	Atgard C	Improved litter production; promote growth	0.0366–0.0550% diet last 30 days gestation; 1000 mg per head daily
Fenbendazole	Safe-Guard	Wormer	1.36 mg/lb body weight per day
Hygromycin B	Hygromix	Wormer	12 g/ton
Levamisole hydrochloride	Tramisol	Wormer	0.08% diet
Lincomycin	Lincomix	Control swine dysentery; treat swine dysentery	40 g/ton; 100 g/ton
Neomycin sulfate and oxytetracycline	Neo-terramycin; NEO/OXTC	Baby pig enteritis and dysentery	N-35–140 g/ton O–50–150 g/ton
Oxytetracycline	Terramycin; OXTC	Promote growth and efficiency; various disease claims	7.5–500 g/ton
Penicillin	Pro-pen	Promote growth and efficiency	10–50 g/ton
Penicillin and streptomycin	Strepcillin	Promote growth and efficiency; enteritis	1:5 ratio of P:S; 50–90 g/ton 90–270 g/ton for 14 days
Phenothiazine		Wormer	5–30 g/hd/day single dose
Piperazine	Wazine	Wormer	0.2–0.4% in feed
Pyrantel tartrate	Banminth	Wormer	96 g/ton continuous
Rabon	Rabon	Fly control in manure	0.05 g/cwt body weight/day
Roxarsone	Roxarsone; 3-Nitro	Promote growth and efficiency; dysentery	22.7–34.0 g/ton continuous; 181.6 g/ton 5–6 days
Thiabendazole	TBZ	Wormer	0.005–0.1% of diet continuous
Tiamulin	Denagard	Promote growth and efficiency; dysentery	10 g/ton; 35 g/ton
Tylosin	Tylan	Promote growth and efficiency; disease claims	10–100 g/ton; 40–100 g/ton
Tylosin and sulfamethazine	Tylan plus sulfa	Disease claims	100 g each/ton
Virginiamycin	Stafac	Promote growth and efficiency; swine dysentery	5–10 g/ton; 25–100 g/ton

B. *Feed Additives for Beef Cattle*

Drugs	Trade or Common Name	Claims	Use Level
Amprolium	Amprol	Prevent coccidiosis Treat coccidiosis	5 mg/kg body weight for 21 days 10 mg/kg body weight for 5 days
Bacitracin methylene disalicylate	BMD; Fortracin; Solu-tracin	Reduction of liver abscesses	70 mg/hd/day
Bacitracin, zinc	Baciferm; Albac 50	Increased wt gains and feed efficiency	35–70 mg/hd/day
Chlortetracycline	Aureomycin; PfiChlor; CLTC; CTC	Promote growth and feed efficiency; numerous disease claims; Reduction of liver abscesses	0.1 mg to 0.5 mg/lb body/weight day; 70–150 mg/hd/day
Chlortetracycline and sulfamethazine	AS-700	Shipping fever stress	350 mg each hd/day for 28 days
Coumaphos	Baymix; Meldane-2	Wormer	0.091 g/100 lb body weight/day for 6 days
Decoquinate	Deccox	Prevent coccidiosis	22.7 mg/100 lb body weight/day
Lasalocid	Bovatec	Improve feed efficiency; promote gain and improve feed efficiency	10–30 g/ton feed; 25–30 g/ton feed
Levamisole hydrochloride	Tramisol	Wormer	0.08–0.8% of diet
Melengestrol acetate	MGA	Promote growth and feed efficiency in heifers	0.25–0.50 mg/hd/day
Methoprene	Altosid CP-10	Fly control in manure	22.7–45.4 mg/100 lb body weight/month
Monensin sodium	Rumensin	Improved feed efficiency	5–30 g/ton complete feed continuous
Morantel Tartrate	Rumatel	Wormer	0.44 g/100 lb body weight
Oxytetracycline	Terramycin; OXTC	Promote growth and feed efficiency; numerous disease claims; reduction of liver abscesses	0.1 mg to 0.5 mg/lb body weight/day; 0.1 mg to 5.0 mg/lb body weight/day
Phenothiazine	Phenothiazine	Reduction of flies; wormer	0.25 g/100 lb body weight/day; 10–20 g/100 lb body weight single dose
Poloxalene	Bloat Guard	Bloat prevention	1.0–2.0 g/100 lb body weight/day
Rabon	Rabon	Fly control in manure	0.07 g/100 lb body weight per day

Drugs	Trade or Common Name	Claims	Use Level
Thiabendazole	TBZ	Wormer	3 or 5 g/100 lb body weight single dose
Tylosin	Tylan	Reduction of liver abscesses	60–90 mg/hd/day

C. *Feed Additives for Sheep*

Drugs	Trade or Common Name	Claims	Use Level
Chlortetracycline	Aureomycin; PfiChlor; CLTC; CTC	Promote growth and efficiency; enterotoxemia	20–50 g/ton; 20 g/ton
Lasalocid	Bovatec	Prevent coccidiosis	20–30 g/ton of feed
Oxytetracycline	Terramycin; OXTC	Increased wt gains and efficiency: diarrhea and dysentery; enterotoxemia	10–20 g/ton; 50–100 g/ton; 25 mg/hd/day
Phenothiazine	Phenothiazine	Control of certain internal parasites	1.0 g/hd/day continuous
Thiabendazole	TBZ	Wormer	2 g/100 lb body weight single dose

D. *Feed Additives for Lactating Dairy Cattle*

Drugs	Trade or Common Name	Claims	Use Level
Chlortetracycline	Aureomycin; PfiChlor; CLTC; CTC	Bacterial diarrhea; foot rot, respiratory infection	0.1 mg/lb body weight/day
Coumaphos	Baymix	Wormer	0.091 g/100 lb body weight daily for 6 days
Oxytetracycline	Terramycin; OXTC	Bacterial diarrhea; reduce bloat; increased milk production	75–100 mg/hd/day
Poloxalene	Bloat Guard	Bloat prevention	1–2 g/100 lb body weight per day
Propylene glycol	Sirlene	Prevent ketosis; treat ketosis	4–8 fl. oz./hd/d; 4–16 fl. oz./hd/d for 10 days
Rabon	Rabon	Fly control in manure	0.07 g/100 lb body weight daily

E. *Feed Additives for Horses*

Drugs	Trade or Common Name	Claims	Use Level
Chlortetracycline	Aureomycin	Promote growth and feed efficiency	85 mg/hd/day up to 1 year of age
Phenothiazine	Phenothiazine	Remove strongyles; control strongyles	2.5 g/100 lb body weight (max. 30 g) for 1 d; 2 g/hd/d for 21 days, repeat after 9 days
Rabon	Rabon	Fly control in manure	0.07 g/100 lb body weight daily

Several products are available for mixing with feed to reduce internal parasites. Consult your local veterinarian.

F. *Feed Additives for Chickens or Turkeys (Nutritional Uses)*

Drugs	Trade or Common Name	Claims	Use Level
Arsanilic acid	Pro-gen; Arsonic Powder	Bloom and feathering; egg production; feed efficiency; growth promotion; pigmentation	45–90 g/ton
Bacitracin methylene disalicylate	BMD; Fortracin; Solu-tracin	Egg production; feed efficiency; growth promotion	4–50 g/ton
Bacitracin, zinc	Baciferm; Albac 50	Feed efficiency; growth promotion	4–50 g/ton
Bambermycins	Flavomycin	Feed efficiency; growth promotion	1–2 g/ton
Chlortetracycline	Aureomycin; PfiChlor; CLTC; CTC	Early mortality; egg production; egg hatchability; feed efficiency; growth promotion	10–100 g/ton
Dimetridazole	Emtrymix	Feed efficiency; growth promotion	136–181 g/ton (turkeys only)
Erythromycin	Gallimycin	Egg production; feed efficiency; growth promotion	4.6–18.5 g/ton (chickens only)
Lincomysin	Lincomix	Feed efficiency; growth promotion	2–4 g/ton (chickens only)
Nitromide sulfanitran, Roxarsone	Unistat-3	Feed efficiency; growth promotion; pigmentation	N—0.025% S—0.030% R—0.005% (chickens only)
Oxytetracycline	Terramycin; OXTC	Egg production; egg hatchability; eggshell quality; feed efficiency; growth promotion; improve fertility; livability	5–100 g/ton
Penicillin	Pro-pen	Feed efficiency; growth promotion	2.4–50 g/ton
Roxarsone	Roxarsone; 3-Nitro	Feed efficiency; growth promotion; pigmentation	0.0025–0.005%
Tylosin	Tylan	Feed efficiency; growth promotion	4–50 g/ton (chickens only)
Virginiamycin	Stafac	Feed efficiency; growth promotion	5–15 g/ton (chickens only)

Many feed additives are available to mix into poultry diets for medicinal uses (blackhead, bluecomb, coccidiosis, chronic respiratory disease (CRD), stress, worms etc.). The author suggests the reader refer to the *Feed Additive Compendium* for detailed listing.

Study Questions and Problems
Chapter 6

I. True or False Questions

1. MGA and Ralgro are examples of feed additives used to promote gain in feedlot heifers.
2. Poloxalene is used to reduce problems with pasture bloat in cattle.
3. Chemotherapeutic compounds and antibiotics have essentially the same effect on microorganism growth.
4. Rumensin is a feed additive used in feedlot cattle diets to significantly improve rate of gain and improve feed efficiency.
5. Most of the feed additives approved for swine diets may also be fed to horses.
6. Tramisol and piperazine are examples of worming compounds used for swine.

II. Multiple Choice (select the single best answer)

1. A substance produced by a living organism that has bactericidal properties is called a(n)
   a. Antibiotic.
   b. Chemotherapeutic agent.
   c. Anthelmintic.
   d. Goitrogen.
   e. None of the above.
2. The incidence of liver abscesses in feedlot cattle may be reduced by the addition of
   a. Bacitracin.
   b. Chlortetracycline.
   c. Erythromycin.
   d. Poloxalene.
   e. All of the above.
3. The incidence of enterotoxemia in sheep may be reduced by the addition of
   a. Chlortetracycline.
   b. Ethylenediamine dihydriodide.
   c. Phenothiazine.
   d. Poloxalene.
   e. None of the above.
4. Which of the following is an anthelmintic agent for swine?
   a. Piperazine
   b. Tylosin
   c. Virginiamycin
   d. Oxytetracycline
   e. Erythromycin
5. Which of the following feed additives improves feed efficiency in feedlot cattle?
   a. Poloxalene
   b. Rabon
   c. Thiabendazole
   d. Rumensin
   e. All of the above
6. Which of the following feed additives claims increased milk production in dairy cows?
   a. Chlortetracycline
   b. Tylosin
   c. Oxytetracycline
   d. Rumensin
   e. Poloxalene

III. Matching or Identification

A. Identify the following feed additives

          a. Antibiotic
          b. Chemotherapeutic agent
          c. Antibiotic/chemotherapeutic mixture
          d. Anthelmintic
          e. Hormonelike product

1. Lasalocid
2. Bacitracin
3. Chlortetracycline
4. Arsanilic acid
5. ASP-250
6. Penicillin
7. Carbadox
8. Pyrantel tartrate
9. Tylosin
10. Dichlorvos
11. Oxytetracycline
12. Phenothiazine
13. Lincomycin
14. Monensin
15. Tramisol

IV. Short Answer

1. Describe the difference between antibiotics and chemotherapeutic agents.
2. Why do livestock producers use feed additives and implants?
3. What are buffers? Discuss factors that commonly lower rumen pH.
4. Discuss the role of implants and list examples of such products used in cattle.
5. Discuss the differences between antibiotics and probiotics in animal nutrition. Compare their modes of action.

Notes

Notes

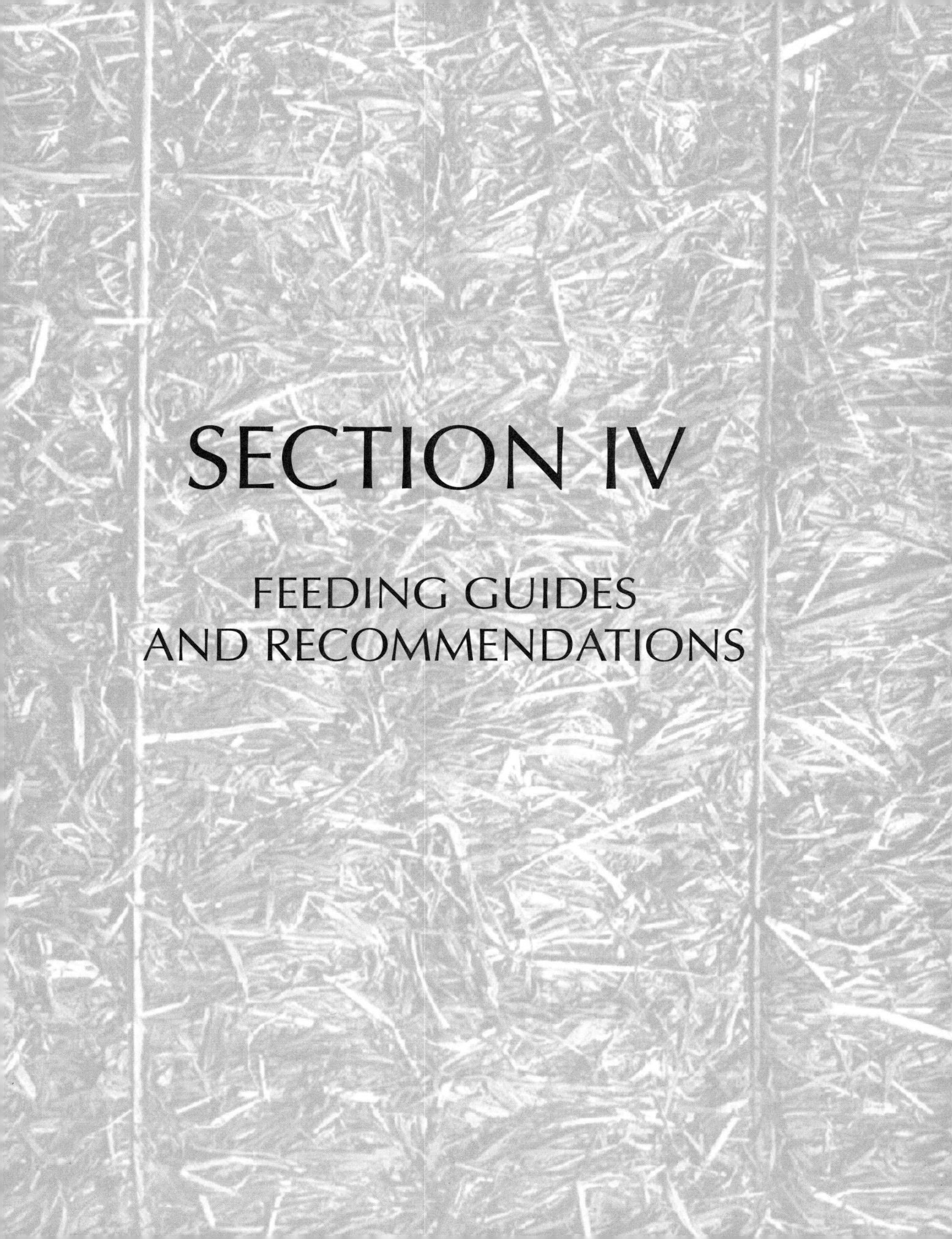

SECTION IV

FEEDING GUIDES
AND RECOMMENDATIONS

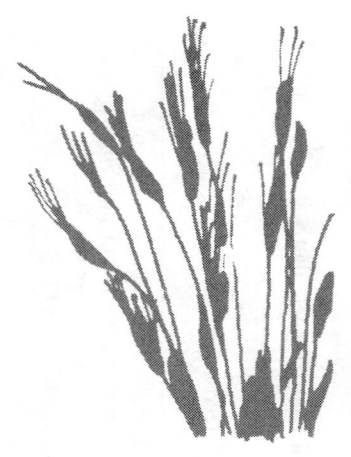

7

Swine Feeding Guides

Chapter Goals

- Outline and discuss the life-cycle swine feeding program.
- Identify specific nutrient needs or additives used within the life cycle.
- Explain feeding systems for the breeding herd and the growing-finishing herd.
- Formulate a swine diet.

The following material has been compiled from several texts and articles concerning swine management and nutrition. Table 7–1 (Life-Cycle Swine Feeding Program) was adapted from Iowa State University Publication PM–489. These requirements may exceed those in the NRC Swine Publication (Tables 7–2 through 7–5) in most cases only to provide additional safety factors.

I. The Breeding Herd

A. *The Boar*

1. Protein requirements: 50–125 lb b.w. (body weight)—18% crude protein (CP) ration; 125–200 lb b.w.—16% CP ration; mature boars—14% CP.
2. When a boar reaches about 180 to 200 lb, reduce ration energy level by increasing fiber content of the ration, or hand feed limited amounts. The boar is probably eating 6 to 8 lb of feed daily by this time. Energy level can be reduced by increasing the percentage alfalfa meal or hay in the ration, or by replacing half of the corn with oats.
3. During the nonbreeding season feed mature boars 4 to 6 lb of a 12% to 14% complete ration or 3 to 5 lb of grain + ¾ to 1.0 lb of 35% supplement daily.
4. About 10 to 14 days before breeding season, increase feed 50% (i.e., 4 up to 6 lb); then feed enough to hold in good condition without overconditioning.
5. Example rations are shown in Table 7–6.

B. *Sow Herd*

1. Keep in mind the stages of the reproductive cycle (Table 7–1) and the suggested protein levels. Sample gestation rations are shown in Table 7–6.
2. Pregestation
   In most swine operations, gilts to serve as replacements in the breeding herd are developed to 180 to 220 lb (5 to 6½ months) while in the market pens. These gilts are full-fed growing-finishing diets that allow maximum expression of the gilts' genetic potential for growth rate and carcass traits. Once gilts reach approximately 200 lb, they should be selected and separated from the market herd. Dietary energy intake should be restricted to 60% to 75% of *ad libitum* until breeding (not more than 6 lb/day). This will allow body growth and development (about 1 to 1½ lb/gain/day) but not excessive fat deposition.
3. Breeding
   a. Gilts should weigh 250 to 275 lb, be 7 to 8 months of age, and have undergone two to three estrus cycles by time of breeding. Gilts receiving a restricted energy intake during the prepubertal period will generally respond to "flushing."

TABLE 7-1. Life-Cycle Swine Feeding Program

Stage of Cycle	Length of Feeding Program	Season	Complete Ration		lb Corn or Grain/Day	lb Supplement/day (35–40% CP)
			% Protein	lb/Day		
Boars	From time purchased at 5–6 months of age	Summer	12–16	4–6	3–5	0.8–1.0
		Winter	12–16	5–7	4–6	
	Increase intake 1–2 lb during the heavy breeding season					
Gilts						
Pregestation	From time selected at 5–6 months until breeding at 7–9 months of age	Summer	12–16	4–6	3–5	0.8–1.0
		Winter	12–16	5–7	4–6	
Flushing and breeding	For 3 weeks before breeding (do not continue after breeding)		12–14	6–9	5–8	0.8–1.0
	Increase corn intake approximately 2 lb/day					
Gestation		Summer	12–14	4–5	3–4	0.8–1.0
		Winter	11–14	5–6	4–5	
	Increase intake 1–2 lb last 3 to 5 weeks of gestation if gilts appear to be too thin					
Lactation	Wean at 3–5 weeks after farrowing		13–16	10–14	Full feed complete diet	
Sows						
Breeding and gestation	Breed back at first heat period after weaning (flushing is not beneficial with sows)	Summer	12–14	3–4	2–3	0.8–1.0
		Winter	11–14	4–6	3–4	
	Increase intake 1–2 lb last 3 to 5 weeks of gestation if sows appear to be too thin					
Lactation	Wean 3–5 weeks after farrowing		13–16	11–15	Full feed complete diet	
Pigs						
Prestarter	Use only if weaned before 3 weeks and feed until pig weights 12 lb		20–24		Full feed complete diet	
Starter	Use as a creep feed and continue after weaning until pigs are 8 weeks of age or 40 lb		18–20		Full feed complete diet	
Grower-finisher	From 8 weeks of age or 40 lb until market weight		13–17	Full feed (may limit feed after 125 lb)	With free choice, corn consumption varies depending on weight of pig. Supplement intake should be approx. 0.75 lb regardless of the weight of the pig	

Source: I.S.U. Life Cycle Swine Nutrition, PM-489 (Rev.), 1988.

Flushing means increasing the energy intake prior to and during breeding. This can be accomplished by feeding 7 to 9 lb/hd/day starting 10 to 14 days before breeding. Flushing can provide

1) Improved health.
2) Increased ovulation rate.
3) Increased live embryos.

When using a flushing program, it is crucial to remember that feed intake of gilts must be reduced to 4 to 6 lb per head per day immediately after breeding. Continued feeding at high intake levels following breeding is associated with high embryo mortality. Fortifying the diet with an antibiotic 1 week before breeding and until bred has also increased the number of eggs ovulated. There is no advantage to flushing gilts that have been fed a high level of energy from market herd selection right up to time of breeding.

b. Sows—usually bred at first estrus after weaning (4–10 days postweaning). Under this type of breeding program, the sow has been receiving a high-energy lactation diet for about the last 3 to 5 weeks. Therefore, a producer simply reduces feed intake at weaning and feeds her 6 to 7 pounds of feed per day until mating. Some producers deprive the sow of any feed and/or water for 24 hours after weaning to help stimulate return to estrus.

4. Gestation

a. After breeding, energy intake should be restricted to prevent gilts or sows from becoming too fat during gestation. Excessive fat impedes reproduction. Energy intake is best controlled by hand-feeding a limited quantity of feed per head daily (i.e., 4 to 5 lb of a 12% to 14% protein mixture and providing 5500 to 6500 kcal ME per day). Restriction can also be attained through interval feeding (full feed every third day) or through feeding of a ration containing high-fiber ingredients such as alfalfa hay, dehydrated alfalfa meal or ground corn cobs. Other methods of restriction can be accomplished by feeding corn silage (10 to 12 lb daily) and limited intake of protein supplement (0.5 to 1.0 lb daily), or by the grazing of sows on good pasture and limiting the intake of protein supplement to 0.5 lb daily and grain to 2.0 lb daily. It is important to note that *only energy* is restricted during gestation. The sow's requirement for the other nutrients must be fulfilled in each of the above programs.

b. During the last third of gestation increase the daily feed intake of gilts to 6 to 7 lb of a 14% CP ration. For sows, a constant feed allowance throughout gestation appears to be as effective as other systems in which feed intake is varied.

c. Sows on lush pasture will need little grain after breeding until the tenth week of pregnancy (2 to 3 lb from there on), while gilts can be fed 2 to 3 lb of a 14% ration up to 10 weeks and 4 to 5 lb of the same ration until farrowing. Feed a mineralized salt mix free-choice.

d. Suggested mineralized salt mixture can be fed free-choice to breeding herd: 60 lb dicalcium phosphate, 20 lb trace mineralized salt, 20 lb ground limestone.

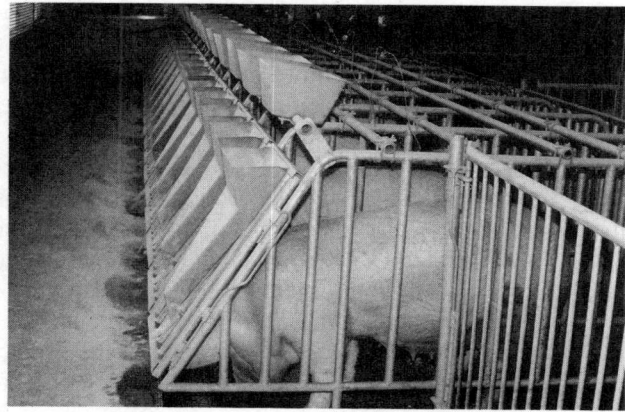

Individually feeding sows a measured quantity of feed daily offers the best control of nutrient intake during gestation.

e. Mature sows should be allowed a body weight gain of approximately 50 lb and gilts 75 to 100 lb from breeding to after they have farrowed. Excessive body weight gains are wasteful of feed and probably will reduce reproductive performance.

5. During farrowing (3 to 5 days before and after farrowing)
   a. A suggested program: 3 to 5 days before farrowing, increase bulk in sow's diet by substituting (for corn) 5% to 10% dried beet pulp or 10% to 15% wheat bran, or 10% to 20% of such feeds as oats or dehydrated alfalfa meal. This practice may help reduce problems with sow constipation.
      1) Feeding level of bulk ration—(14% CP) 6 to 8 lb.
      2) Continue this ration free choice for 3 to 5 days after farrowing. At this time return to regular ration-full feed.
   b. An alternate program: special rations not necessary. Just feed the regular ration; let the sow eat what she wants. (Many breeders use regular ration but limit intake, which again is not always necessary.) If a sow feels like eating after farrowing, there is little reason to restrict her feed intake.

6. Lactation (see suggested rations in Table 7–7)
   Increase feeding level within the first few days after parturition in order to meet the high nutrient and energy demands of milk production.
   a. Recommendations
      1) 13% to 16% CP diet, which is high in energy and low in fiber.
      2) Sow: 4 to 5 lb base plus 1 lb per nursing pig; dietary intake should provide 16 to 18 Mcal ME/day for maintaining sow weight, for postweaning estrous interval, and for litter performance to weaning.
      3) Antibiotic if needed or desired.
   b. Fat additions to the diet at farrowing
      1) Research trials have suggested from 6% to 20%+ added fat to diets fed for 0 to 10 days before farrowing and 0 to 21 days after farrowing. This fat has been substituted in the diet for corn and consequently raises the caloric density of the diet.
      2) Reported benefits of supplemental fat
         a) Increased energy density and yield of sow's colostrum and milk.
         b) Increased concentration of pig's liver glycogen or blood glucose levels at birth, or both.
         c) Improved pig survival (2% to 3%) in herds that normally have survival rates less than 80%.
         d) Heavier pig and litter weaning weights.
         e) Reduced sow weight loss during lactation.
      3) Considerations for use of supplemental fat
         a) Expensive if fed to entire herd.
         b) Feed only to sows that need it.
         c) Little effect on litter size or pig weight at birth.
         d) If a complete diet is fed, a minimum of 7.5% added fat is suggested and should be fed at least 5 days before farrowing and 14 to 21 days after farrowing.
      4) In many cases simply getting more feed into the sow is more economical than trying to increase caloric density of the feed by adding fat.
   c. The colostrum or first milk produced by the sow after farrowing contains important nutrients and antibodies. These antibodies give protection against disease until the young pig can synthesize its own antibodies at 3 to 4 weeks of age. These antibodies can only be absorbed for a short time after birth (<72 hours) so it is very important that each pig consume colostrum milk as soon as possible after birth.
   d. Anemia of suckling pigs occurs as a result of a deficiency of iron and, to some extent, copper. Sow's milk contains little iron. From birth until they begin to eat regularly, pigs raised inside have a deficient supply of available iron. Supplemental iron may be supplied to the pigs by
      1) Oral dosage—pigs should be given the first dose (50 to 100 mg) when they are 3 days old, and the dose repeated two or three times weekly for 3 weeks or until they are eating regularly.

High levels of feed intake during lactation are important to help reduce sow body weight loss, increase milk yield, increase pig growth rate and improve the sow's subsequent reproductive performance.

  2) Injection—iron dextran or other iron-carbohydrate complexes may be injected (150 mg) at 3 days of age. A second dose may be required if pigs do not start to eat at an early age.

 e. Length of nursing period

  Evidence supports early weaning of pigs (3 weeks), given a properly controlled climatic environment and a sound management program. Greater feed efficiency is accomplished, and it is not necessary for the sow to gain quite as much weight or condition during her dry period. Pigs will perform at least as well on a good prestarter or starter diet from 3 to 6 weeks of age as they will nursing their dam. Sows will normally come in heat within 1 week after pigs are weaned.

II. Feeding the Market Pig

 A. *Protein*

  Perhaps the most critical nutrient in swine rations. The key is balance of amino acids rather than percent protein. Diets similar in content of protein might result in different gains because of amino acid balance. Limiting amino acids in typical diets include lysine, tryptophan, threonine and methionine (up to 40% of methionine need can be met by cystine). The percentage of the different amino acids required varies with age (weight) of pig and the percent total protein fed (tables 7–2 and 7–3).

 B. *Types of Diets*

  1. Milk replacer diets—research at a number of experiment stations shows that baby pigs 3 to 5 days of age can be successfully weaned on liquid synthetic milk rations composed typically of dried skim milk, fat, vitamins, minerals and antibiotics. Although successful liquid sow-milk substitutes have been developed, they are generally not accepted by the farm operators because they are expensive, must be prepared daily and fed in a controlled environment. Another disadvantage is that all feeding equipment must be thoroughly cleaned and extreme sanitation maintained.

  2. Prestarters (20–24% CP) are complex and therefore often purchased from a commercial source. Generally fed to early weaned pigs (before 3 weeks of age) or to fortify the diet of a litter of pigs that are not receiving adequate amounts of milk from the sow. Commonly contain some added dried skim milk or dried whey, sugar (sucrose), fat and antibiotics and are fed dry and in the pellet or crumbled form. See example prestarter rations in Table 7–8.

  3. Starter (18% to 20% CP)—designed to be fed as a complete feed and fed from about the 3rd to 6th week of age. The pig starter ration may be used as a creep ration before weaning and fed after weaning until the pigs reach approximately 40 lb. Commonly contain 10% to 20% dried whey to improve protein quality and 1% to 3% stabilized fat to decrease dustiness and improve feed efficiency and palatability. At least half of the cereal grain should be corn and include an appropriate antibiotic. Sample pig starter rations are shown in Table 7–9.

4. Grower-finisher diets
   a. Grower diet (14% to 16% CP)—complete feed. Fed to pigs from about 40 to 120 pounds body weight. Commonly a corn-SBM formulation fortified with vitamins, minerals and an antibiotic. A mixture of grain and complete supplement is also sufficient. Sample pig grower rations are shown in Table 7–10.
   b. Finisher diet (13–14% CP)—complete feed; fed to pigs from about 120 lb to market weight. Antibiotic level is optional. Suggested formulations of swine finisher rations are shown in Table 7–10.

III. Feeding Systems

A. *Simplified versus Complex Ration*

Swine diets are becoming more simplified. The performance obtained with a simplified corn-SBM diet properly supplemented with minerals and vitamins will be essentially the same as the performance obtained with a more complex diet. Thus, because of ingredient cost, availability of supply and ease of preparation, a simplified corn-SBM diet has merit in an intensive swine operation where producers wish to grind and mix their own feed. On the other hand, more complex formulas may be used by commercial companies because of large-volume buying of ingredients and the use of computerized least-cost formulations.

B. *Complete Self-Fed Diets*

Complete diets are those in which all feedstuffs are combined into one feed mixture. Complete diets may be formulated by "building from the ground up," or they may be made by mixing balanced supplement with ground grain. The trend is toward the use of complete self-fed diets for growing and finishing pigs because they provide better control of nutrient intake than free-choice methods, result in faster gains than limit feeding and complement automation.

C. *Corn and Balanced Supplements Self-Fed Free-Choice*

A free-choice feeding is one in which the animal is free to eat two or more feeds at will. Although the free-choice system may offer the greatest simplicity for growing and finishing pigs, particularly for small producers, efficiency varies considerably both in cost and in time. Variation in the palatability of the grain or supplement may result in undereating or overeating of the supplement or grain. The free-choice system usually results in slower gains than a complete self-fed diet. Although some experimental trials have shown that the free-choice system may improve feed efficiency slightly, under field conditions there is probably little difference. The free-choice system demands greater supervision and, if it is not properly supervised, poor efficiency can easily occur.

D. *Limit Feeding*

Limit feeding implies feeding animals less than they would voluntarily consume. The primary purpose is to restrict energy intake sufficiently to maintain weight and growth but not to allow excess body fat deposition.
   1. Gilts and sows—replacement gilts should be started on a limit feeding program at 180 to 200 lb body weight. All gestating gilts and sows should be limit fed to prevent obesity and improve reproductive efficiency. Limit feeding the breeding herd may be accomplished by any one of the following methods
      a. Group feeding a measured quantity of feed to several sows. It is hoped that each sow will eat her proper amount of feed. However, often results in the "boss" sows getting too much and the "shy" sows getting too little.
      b. Individual feeding a measured quantity of feed to each sow. This requires sows to be fed in individual pens or crates or in tie stalls, tethered by a neck collar. This system offers the best control of intake.
      c. Feeding fibrous, bulky feed such as silage, haylage, hay, dehydrated alfalfa meal etc. is another way to lower the energy content of the ration. Such feeds must constitute one-third to half or more of the ration. This system requires more supervision to assure the level of fibrous feed is restricting energy intake and that feed costs are not excessive.
      d. Interval feeding is another method of limit feeding gestating gilts and sows. With this system the gilts or sows are allowed to eat every third day by turning them onto a self-feeder for 2 to 6 hours. This system may reduce the labor requirement compared to daily hand feeding. The daily feed consumption will average approximately 4 lb with gilts (12 lb consumed every

third day) and 5 lb with sows (15 lb every third day). Therefore, the restriction may be a little too severe for gilts in extremely cold weather and not strict enough for sows in the summer. The average intake can be regulated somewhat, either by limiting the time allowed at the self-feeder every third day or by feeding sows only twice a week during the summer months. Also, the gilts or sows can be hand fed a given amount every third day. Reproductive performance in several trials with interval feeding has been essentially the same as with the normal hand limit-feeding method.

2. Growing and finishing pigs—the use of limit feeding is questionable for growing and finishing pigs. The slower gains, increased labor or mechanization required, variability in performance that may occur and increased supervision needed probably more than offset the small beneficial effect produced on carcass fat-to-lean ratio. Thus, unless sufficient premium is paid for the modestly leaner carcasses, it cannot be justified. Research findings indicate little difference in feed efficiency between limit feeding of 5 lb/day from 125 lb to market and self-feeding when the pigs are fed in groups. Feeding on the floor (a type of limit feeding) may be justified in housing units with partially slotted floors or a flushing gutter system because it aids in manure disposal and maintaining pen cleanliness. Limit feeding of growing and finishing pigs may also be justified where the slower gain will result in hitting a substantially higher market if availability of facilities is not a problem. If limit feeding is used, it should not be used before 50 lb and preferably after 100 to 125 lb/b.w., and the feeding level should be maintained at approximately 85% to 95% of full feeding level.

E. *Liquid Feeding*

Liquid feeding is accomplished by blending water with a dry meal feed to form a slurry (about two to three parts water to 1 part feed) or paste (about one to two parts water to one part feed). Although there appears to be a slight advantage for liquid feeding when pigs are limit fed, there is no advantage for liquid feeding when pigs are full fed in groups. The rate of gain is essentially the same between liquid full-fed pigs and dry full-fed pigs. Liquid feeding has no effect on farm-to-market shrink, cooler shrink, carcass measurements or meat quality. Some studies suggest that newly weaned pigs may consume liquid feed more readily than a dry feed.

F. *Pelleting Complete Diets*

There has been considerable interest in the pelleting of complete diets for swine. Some of the advantages of pelleting are reduced feeding wastage, less storage space requirement, better control of nutrient intake, improved palatability and increased feed intake. The primary disadvantage is the additional cost of pelleting. The general opinion is that pelleting does not increase the nutritional value of feeds but simply allows the animal to consume more feed. On the other hand, there have been reports of increased feeding value for feeds containing large amounts of barley (improved fiber utilization). Some improvements have also been reported following pelleting of diets containing milo (starch gelatinization). Pelleting may also improve the availability of plant (phytin) phosphorus to the pig. Pelleting of a growing and finishing diet usually increases average daily gain by approximately 5% and improves feed efficiency by as much as 10%. Therefore, where a complete diet is already being purchased, pelleting the diet may be more economical than using a meal. However, the advantage of pelleting will not likely be large enough to offset the cost of hauling corn to town to use in the pelleted diet. Also, the cost of a pelleting machine probably cannot be justified for the volume normally handled by most producers.

G. *High-Moisture Corn*

High-moisture corn can be substituted for dry corn on a dry matter basis with little or no effect on performance. However, pigs fed high-moisture corn free-choice commonly over- or underconsume either the corn or supplement. Thus the pig does not properly balance its ration. Since complete ground and mixed diets using high-moisture corn cannot be stored in an aerobic self-feeder, a system that allows daily mixing of the high-moisture shelled corn and a pelleted supplement is probably the most feasible. If a meal supplement is used, the corn should be cracked or ground to prevent separation. There appears to be no nutritional advantage to grinding, rolling or cracking of high-moisture corn for growing and finishing pigs. Since the decision as to whether to use high-moisture corn is primarily an economic one rather than nutritional, the cost of high-moisture corn (including its storage and handling) must be balanced against the cost of dry corn (including drying, storage and handling).

The following comparison illustrates the pounds of high-moisture corn required to equal 100 lb of No. 2 dried corn (85% d.m.).

% Moisture in Corn	Pounds of High-Moisture Corn to equal 100 lb Dried Corn
15	100.00
16	101.19
17	102.41
18	103.66
19	104.94
20	106.25
21	107.59
22	108.97
23	110.39
24	111.84
25	113.33
26	114.86
27	116.44
28	118.06
29	119.72
30	121.43

H. *Whole Soybeans*

Numerous studies with pigs have shown that feeding raw soybeans to the growing pig will result in inhibited intestinal proteolysis, depressed feed intake and depressed pig growth. However, recent studies have suggested comparable reproductive performance when SBM was replaced with raw soybeans on an equal weight basis in diets fed to gestating or lactating sows. Thus, the adverse effect of the toxicants and inhibitors in the raw soybean appear to be mainly expressed in the growing-finishing stage of the swine life cycle.

Research has indicated that whole cooked soybeans can be used in place of SBM very successfully in swine diets. The use of whole cooked soybeans will probably have little effect on rate of gain but will likely improve feed efficiency by 5% to 10% due to the high fat content of the whole soybean, which results in a higher energy ration. But this improvement in feed efficiency may be offset by the fact that the whole soybeans are only 37% protein. Therefore, more whole cooked soybeans will be required to balance the diet for protein or amino acids in comparison with SBM. Also, since the use of whole cooked soybeans results in a higher energy diet, the protein level of the diet should be approximately 1% to 2% higher than that of a similar diet with SBM to maintain the same protein-to-energy ratio. Although the limited amount of work conducted on the effects of the use of whole cooked beans on carcass quality has not indicated any serious problems, more work is needed in this area. Since the use of whole cooked soybeans appears to be acceptable from the nutritional standpoint, the decision as to their use must be primarily based on their comparative costs with SBM and other protein sources.

I. *Feeding High-Lysine Corn (HLC)*

The lysine content of normal yellow corn averages about 0.20%. Plant breeders have developed certain strains of corn that may contain up to 0.55% lysine. Thus, the strain called HLC may contain over twice as much lysine and usually twice as much tryptophan as normal corn. The content of CP and the other amino acids may not vary between the two types of corn.

The use of HLC by the swine producer can save substantial amounts of SBM or other protein supplements. To make full use of its added lysine content, HLC must be analyzed to determine its lysine content. The diet should then be balanced relative to the pig's lysine requirement rather than protein. The following sample formulations show the amount of HLC and SBM needed to balance a ton (2000 lb) of complete feed for different weight pigs. Add 50 lb of a vitamin-mineral mix to each diet.

Actual Lysine Analysis of Corn, %

Percent Lysine in Final Ration		.20	.25	.30	.35	.40	.45
.75	Corn*	1540	1570	1600	1630	1660	1700
(35- to 75-lb pigs)	SBM†	410	380	350	320	290	250
.65	Corn	1610	1640	1675	1710	1745	1780
(75- to 125-lb pigs)	SBM	340	310	275	240	210	170
.55	Corn	1685	1720	1750	1790	1820	1860
(125- to 230-lb pigs)	SBM	265	230	200	160	130	90

*50 lb of minerals and vitamins are added to all rations.
†Assume SBM contains about 2.9% lysine.

IV. Miscellaneous

A. *Protein Supplements*

Supplements must be formulated to balance the deficiencies of grains such as corn, milo or barley whether fed as a mixed complete diet or on a free-choice basis. Any deviation from the principle will give poor results. All rations should be properly fortified with protein of high quality, essential mineral elements, vitamins and optional feed additives. Overfortification of supplements or complete diets is of no nutritional value, and often creates imbalances that are more harmful than creative. Balance and interrelationship of nutrients is as important as proper amounts. Excesses in any nutrient tend to inhibit the action and utilization of other nutrients. Hypernutrition is as destructive as hyponutrition.

B. *Crystalline (Synthetic) Amino Acids*

Swine actually require amino acids rather than protein. Considering availability and performance, SBM has become the most common natural source of supplemental amino acids to a grain-based swine diet. Soybean meal is an excellent source of lysine and tryptophan. Thus, simplified swine diets are composed of grain and SBM blended to a CP level to satisfy the amino acid requirements and then fortified with vitamins and minerals. When the cost of SBM increases, producers begin to search for various methods to lower feed costs. One such method may be to reduce SBM and then supplement the lowered protein diet with crystalline amino acids.

Numerous studies have demonstrated the feasibility of reducing protein by 2% from the generally recommended levels in corn- or milo-SBM diets, and supplementing lysine to meet the pig's requirement for this amino acid instead of feeding the higher protein level. Reducing the protein by 3% begins to limit a second amino acid, and that a 4% reduction causes a clear-cut deficiency of at least one other amino acid other than lysine. Current data suggest lysine to be the first limiting amino acid in a grain-SBM diet. The data also indicate the second limiting amino acid in such rations is tryptophan in corn-SBM diets and threonine in milo-SBM diets. These two amino acids, threonine and tryptophan, are quite likely to be third limiting in these respective diets. Methionine is considered fourth limiting in normal corn- or milo-SBM formulations. However, methionine may be second limiting in the growing swine diet using HLC as the grain source.

With two amino acids, lysine and methionine, commercially available and two others, tryptophan and threonine, available in reasonable quantities for further research, the supplementation of crystalline amino acids has received recent attention. Amino acids can be in two forms designated "D" and "L." The amino acids present in natural proteins are of the "L" form and are readily utilized by the pig and other animals. Thus, the most common feed-grade sources of crystalline amino acids available are

L-lysine HCL
L-tryptophan (or DL -)
L-threonine
DL-methionine

Individual crystalline amino acid supplementation can replace a part of the SBM supplementation in a swine diet by the following recommendations

1. Replace 5 parts SBM with
    4.85 parts corn
    +0.15 parts L-lysine

2. Replace 10 parts SBM with
    9.61 parts corn
    +0.30 parts L-lysine
    +0.03 parts L-tryptophan
    +0.06 parts L-threonine

3. Replace 15 parts SBM with
    14.35 parts corn
    +0.45 parts L-lysine
    +0.06 parts L-tryptophan
    +0.12 parts L-threonine
    +0.02 parts DL-methionine

The future availability and price will largely determine the practicality of crystalline amino acid supplementation as well as further research with and marketing of commercial quantities of tryptophan and threonine.

C. *Feed Additives*

Antibiotics and chemotherapeutics (with limits) are effective agents to stimulate the performance of swine during the growing and finishing phases of life but only have special uses during other periods. All rations fed to pigs from birth until market weight may be fortified with a bactericidal agent. Pigs up to 100 lb weight exhibit the greatest response to feed additives (about 10%+ faster gains on 5%+ less feed). There is no conclusive evidence that feeding antibiotics during gestation is of any special benefit. Limited research has indicated that high levels of antibiotics before and during breeding will improve conception rate and embryonic survival. Also, high levels of antibiotic may be helpful during outbreaks of certain types of pig scours. In addition to antibiotics and chemotherapeutic agents, copper sulfate (250 ppm) can be used as a feed additive to promote growth and improve feed efficiency. All compounds, alone or in combination with other antimicrobial compounds, should be used only at approved levels and for the specific purpose for which they are authorized. Follow label directions with care.

D. *Vitamins*

Riboflavin, niacin, pantothenic acid and vitamin B_{12} are the water-soluble vitamins most likely to be deficient in swine diets formulated with grains and plant protein. The primary fat-soluble vitamins of practical importance are vitamins A and D. Vitamins E and K may be of concern under some circumstances.

E. *Zinc*

Soybean meal contains phytic acid which, if fed at high levels, may tie up zinc in a swine ration and cause the clinical symptom of parakeratosis. High levels of calcium will also accentuate parakeratosis and increase the dietary zinc requirement. Since most swine rations are built around SBM as a supplement to grain, it is necessary to fortify all swine rations with an elemental source of zinc. For complete protection, complete swine rations should be supplemented with 50 to 100 parts per million (ppm) of zinc and protein supplements with 250 ppm of zinc. Research to date indicates that zinc is equally available from zinc oxide, zinc sulfate or zinc carbonate.

F. *Confinement Rearing*

A majority of growing-finishing swine today are raised in confinement (on concrete) from birth to market weight. For this stage of development it is more economical to raise hogs on producer-made rations than to rely on pasture to balance farm grains. There is a divided opinion on confining breeding stock to concrete the year around. Although this can be done, it requires a great deal more nutritional know-how to meet the sows' requirements. Therefore, under most farm conditions it is more practical to allow sows to glean part of their nutritional requirements from green pasture during the spring, summer and fall months. This does not eliminate the use of supplement and grain

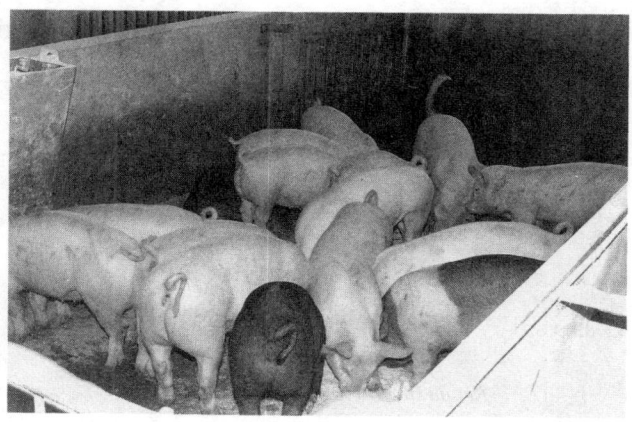

Modern confinement buildings provide a clean, climate-controlled environment designed for the comfort of pigs as well as the producer.

but does help prevent many nutritional and disease hazards inherent in complete confinement programs.

G. *Nutrient and Environment Interaction*

Daily feed intake and daily gain increase as floor space per pig increases. However, the effect of increasing floor space per pig on feed per gain ratio (efficiency) is variable and quite small. Thus, increased daily gain is primarily due to increased daily feed intake. The effect of floor space on pig performance becomes smaller as the pigs become larger.

Dietary energy digestibility is related directly to environmental temperature. This effect appears to be apart from other influences such as feed intake and digestive tract rate of passage. Energy used by the pig to drive maintenance and productive processes is called metabolizable energy (ME). The ME eventually becomes the energy of some product of the pig (meat, milk, fetus etc.) or else it is lost to the environment as heat. This heat originates in the maintenance processes themselves and in the heat increment (HI) of feeding. The HI is the heat produced during processing of the feed by the gastrointestinal tract and processing and use of absorbed nutrients by the body. The HI of feeding can be a burden to the pig in a warm environment or a help to maintain its body temperature in a cool environment. The HI of a swine diet will be lowered with added dietary fat, whereas the addition of dietary fiber results in greater HI. Thus, the value of intentional dietary alterations in relation to environmental temperature has been a recent area of research. Several studies have suggested the value of the addition of fat in the form of tallow or lard, which have low HI, to a swine diet at the rate of 5% is greater during warm weather than during cool. On the other hand, the value of adding fiber in the form of dehydrated alfalfa meal, which has a high HI, to a swine diet at the rate of 5% to 10% is greater during cool weather than warm.

The effect of environmental temperature on the efficient utilization of high fat or high fiber is important for at least two considerations

1. If high fiber containing feeds become economically available to the swine diet formulator, their use would be better justified in a colder environment.
2. Management decisions having to do with the economics of energy (feed energy and fossil fuel energy) are very important. Relative prices of feed and fuel affect decisions as to whether it would be more economical to keep environmental temperature in a swine building low and let the pigs compensate by eating more feed (or a higher fiber feed) or to hold the temperature at a higher point and thus spare some feed from being used as fuel.

V. Nutrient Requirements of Swine

The nutrient requirements for swine presented in tables 7–2 through 7–5 were adapted from *The Nutrient Requirements of Swine,* Publication ISBN 0-309-03779-4, Committee on Animal Nutrition, National Academy of Sciences—National Research Council, Washington, D.C., 1988. These values are expressed on an air-dry basis.

TABLE 7–2. Nutrient Requirements of Swine Allowed Feed Ad Libitum
(90 percent dry matter)

Intake and Performance Levels	Swine Liveweight (kg)				
	1–5	**5–10**	**10–20**	**20–50**	**50–110**
Expected weight gain (g/day)	200	250	450	700	820
Expected feed intake (g/day)	250	460	950	1,900	3,110
Expected efficiency (gain/feed)	0.800	0.543	0.474	0.368	0.264
Expected efficiency (feed/gain)	1.25	1.84	2.11	2.71	3.79
Digestible energy intake (kcal/day)	850	1,560	3,230	6,460	10,570
Metabolizable energy intake (kcal/day)	805	1,490	3,090	6,200	10,185
Energy concentration (kcal ME/kg diet)	3,220	3,240	3,250	3,260	3,275
Protein (%)	24	20	18	15	13

Requirement (% or amount/kg diet)*

Nutrient					
Indispensable amino acids (%)					
Arginine	0.60	0.50	0.40	0.25	0.10
Histidine	0.36	0.31	0.25	0.22	0.18
Isoleucine	0.76	0.65	0.53	0.46	0.38
Leucine	1.00	0.85	0.70	0.60	0.50
Lysine	1.40	1.15	0.95	0.75	0.60
Methionine + cystine	0.68	0.58	0.48	0.41	0.34
Phenylalanine + tyrosine	1.10	0.94	0.77	0.66	0.55
Threonine	0.80	0.68	0.56	0.48	0.40
Tryptophan	0.20	0.17	0.14	0.12	0.10
Valine	0.80	0.68	0.56	0.48	0.40
Linoleic acid (%)	0.1	0.1	0.1	0.1	0.1
Mineral elements					
Calcium (%)	0.90	0.80	0.70	0.60	0.50
Phosphorus, total (%)	0.70	0.65	0.60	0.50	0.40
Phosphorus, available (%)	0.55	0.40	0.32	0.23	0.15
Sodium (%)	0.10	0.10	0.10	0.10	0.10
Chlorine (%)	0.08	0.08	0.08	0.08	0.08
Magnesium (%)	0.04	0.04	0.04	0.04	0.04
Potassium (%)	0.30	0.28	0.26	0.23	0.17
Copper (mg)	6.0	6.0	5.0	4.0	3.0
Iodine (mg)	0.14	0.14	0.14	0.14	0.14
Iron (mg)	100	100	80	60	40
Manganese (mg)	4.0	4.0	3.0	2.0	2.0
Selenium (mg)	0.30	0.30	0.25	0.15	0.10
Zinc (mg)	100	100	80	60	50
Vitamins					
Vitamin A (IU)	2,200	2,200	1,750	1,300	1,300
Vitamin D (IU)	220	220	200	150	150
Vitamin E (IU)	16	16	11	11	11
Vitamin K (menadione) (mg)	0.5	0.5	0.5	0.5	0.5
Biotin (mg)	0.08	0.05	0.05	0.05	0.05
Choline (g)	0.6	0.5	0.4	0.3	0.3
Folacin (mg)	0.3	0.3	0.3	0.3	0.3
Niacin, available (mg)	20.0	15.0	12.5	10.0	7.0
Pantothenic acid (mg)	12.0	10.0	9.0	8.0	7.0
Riboflavin (mg)	4.0	3.5	3.0	2.5	2.0
Thiamin (mg)	1.5	1.0	1.0	1.0	1.0
Vitamin B_6 (mg)	2.0	1.5	1.5	1.0	1.0
Vitamin B_{12} (μg)	20.0	17.5	15.0	10.0	5.0

Note: The requirements listed are based upon the principles and assumptions described in the text of this publication. Knowledge of nutritional constraints and limiations is important for the proper use of this table.

*These requirements are based upon the following types of pigs and diets: 1- to 5-kg pigs, a diet that includes 25 to 75 percent milk products; 5- to 10-kg pigs, a corn-soybean meal diet that includes 5 to 25 percent milk products; 10- to 110-kg pigs, a corn-soybean meal diet. In the corn-soybean meal diets, the corn contains 8.5 percent protein; the soybean meal contains 44 percent.

From *Nutrient Requirements of Swine*, 9th ed rev, copyright © 1988 by the National Academy of Sciences, Washington D.C. Reprinted by permission.

TABLE 7–3. Daily Nutrient Intakes and Requirements of Swine Allowed Feed Ad Libitum

Intake and Performance Levels	Swine Liveweight (kg)				
	1–5	5–10	10–20	20–50	50–110
Expected weight gain (g/day)	200	250	450	700	820
Expected feed intake (g/day)	250	460	950	1,900	3,110
Expected efficiency (gain/feed)	0.800	0.543	0.474	0.368	0.264
Expected efficiency (feed/gain)	1.25	1.84	2.11	2.71	3.79
Digestible energy intake (kcal/day)	850	1,560	3,230	6,460	10,570
Metabolizable energy intake (kcal/day)	805	1,490	3,090	6,200	10,185
Energy concentration (kcal ME/kg diet)	3,220	3,240	3,250	3,260	3,275
Protein (g/day)	60	92	171	285	404
Requirement (amount/day)					
Nutrient					
Indispensable amino acids (g)					
Arginine	1.5	2.3	3.8	4.8	3.1
Histidine	0.9	1.4	2.4	4.2	5.6
Isoleucine	1.9	3.0	5.0	8.7	11.8
Leucine	2.5	3.9	6.6	11.4	15.6
Lysine	3.5	5.3	9.0	14.3	18.7
Methionine + cystine	1.7	2.7	4.6	7.8	10.6
Phenylalanine + tyrosine	2.8	4.3	7.3	12.5	17.1
Threonine	2.0	3.1	5.3	9.1	12.4
Tryptophan	0.5	0.8	1.3	2.3	3.1
Valine	2.0	3.1	5.3	9.1	12.4
Linoleic acid (g)	0.3	0.5	1.0	1.9	3.1
Mineral elements					
Calcium (g)	2.2	3.7	6.6	11.4	15.6
Phosphorus, total (g)	1.8	3.0	5.7	9.5	12.4
Phosphorus, available (g)	1.4	1.8	3.0	4.4	4.7
Sodium (g)	0.2	0.5	1.0	1.9	3.1
Chlorine (g)	0.2	0.4	0.8	1.5	2.5
Magnesium (g)	0.1	0.2	0.4	0.8	1.2
Potassium (g)	0.8	1.3	2.5	4.4	5.3
Copper (mg)	1.50	2.76	4.75	7.60	9.33
Iodine (mg)	0.04	0.06	0.13	0.27	0.44
Iron (mg)	25	46	76	114	124
Manganese (mg)	1.00	1.84	2.85	3.80	6.22
Selenium (mg)	0.08	0.14	0.24	0.28	0.31
Zinc (mg)	25	46	76	114	155
Vitamins					
Vitamin A (IU)	550	1,012	1,662	2,470	4,043
Vitamin D (IU)	55	101	190	285	466
Vitamin E (IU)	4	7	10	21	34
Vitamin K (menadione) (mg)	0.02	0.02	0.05	0.10	0.16
Biotin (mg)	0.02	0.02	0.05	0.10	0.16
Choline (g)	0.15	0.23	0.38	0.57	0.93
Folacin (mg)	0.08	0.14	0.28	0.57	0.93
Niacin, available (mg)	5.00	6.90	11.88	19.00	21.77
Pantothenic acid (mg)	3.00	4.60	8.55	15.20	21.77
Riboflavin (mg)	1.00	1.61	2.85	4.75	6.22
Thiamin (mg)	0.38	0.46	0.95	1.90	3.11
Vitamin B_6 (mg)	0.50	0.69	1.42	1.90	3.11
Vitamin B_{12} (μg)	5.00	8.05	14.25	19.00	15.55

From *Nutrient Requirements of Swine*, 9th ed rev, copyright © 1988 by the National Academy of Sciences, Washington D.C. Reprinted by permission.

TABLE 7–4. Nutrient Requirements of Breeding Swine: Percent of Amount per Kilogram of Diet*

Intake Levels	Bred Gilts, Sows, and Adult Boars	Lactating Gilts and Sows
Digestible energy (kcal/kg diet)	3,340	3,340
Metabolizable energy (kcal/kg diet)	3,210	3,210
Crude protein (%)	10	13

Requirement (% or amount/kg diet)*

Nutrient		
Indispensable amino acids (%)		
Arginine	0.00	0.40
Histidine	0.15	0.25
Isoleucine	0.30	0.39
Leucine	0.30	0.48
Lysine	0.43	0.60
Methionine + cystine	0.23	0.36
Phenylalanine + tyrosine	0.45	0.70
Threonine	0.30	0.43
Tryptophan	0.09	0.12
Valine	0.32	0.60
Linoleic acid (%)	0.1	0.1
Mineral elements		
Calcium (%)	0.75	0.75
Phosphorus, total (%)	0.60	0.60
Phosphorus, available (%)	0.35	0.35
Sodium (%)	0.15	0.20
Chlorine (%)	0.12	0.16
Magnesium (%)	0.04	0.04
Potassium (%)	0.20	0.20
Copper (mg)	5.00	5.00
Iodine (mg)	0.14	0.14
Iron (mg)	80.00	80.00
Manganese (mg)	10.00	10.00
Selenium (mg)	0.15	0.15
Zinc (mg)	50.00	50.00
Vitamins		
Vitamin A (IU)	4,000	2,000
Vitamin D (IU)	200	200
Vitamin E (IU)	22	22
Vitamin K (menadione) (mg)	0.50	0.50
Biotin (mg)	0.20	0.20
Choline (g)	1.25	1.00
Folacin (mg)	0.30	0.30
Niacin, available (mg)	10.00	10.00
Pantothenic acid (mg)	12.00	12.00
Riboflavin (mg)	3.75	3.75
Thiamin (mg)	1.00	1.00
Vitamin B_6 (mg)	1.00	1.00
Vitamin B_{12} (μg)	15.00	15.00

NOTE: The requirements listed are based upon the principles and assumptions described in the text of this publication. Knowledge of nutritional constraints and limitations is important for the proper use of this table.

*These requirements are based upon corn-soybean meal diets, feed intakes, and performance levels listed in Table 7–5. In the corn-soybean meal diets, the corn contains 8.5 percent protein; the soybean meal contains 44 percent.

TABLE 7–5a. Daily Nutrient Intakes and Requirements of Intermediate-Weight Breeding Animals

	Mean Gestation or Farrowing Weight (kg) of:	
	Bred Gilts, Sows, and Adult Boars	Lactating Gilts and Sows
Intake and Performance Levels	**162.5**	**165.0**
Daily feed intake (kg)	1.9	5.3
Digestible energy (Mcal/day)	6.3	17.7
Metabolizable energy (Mcal/day)	6.1	17.0
Crude protein (g/day)	228	689

Requirement (amount/day)

Nutrients		
Indispensable amino acids (g)		
Arginine	0.0	21.2
Histidine	2.8	13.2
Isoleucine	5.7	20.7
Leucine	5.7	25.4
Lysine	8.2	31.8
Methionine + cystine	4.4	19.1
Phenylalanine + tyrosine	8.6	37.1
Threonine	5.7	22.8
Tryptophan	1.7	6.4
Valine	6.1	31.8
Linoleic acid (g)	1.9	5.3
Mineral elements		
Calcium (g)	14.2	39.8
Phosphorus, total (g)	11.4	31.8
Phosphorus, available (g)	6.6	18.6
Sodium (g)	2.8	10.6
Chlorine (g)	2.3	8.5
Magnesium (g)	0.8	2.1
Potassium (g)	3.8	10.6
Copper (mg)	9.5	26.5
Iodine (mg)	0.3	0.7
Iron (mg)	152	424
Manganese (mg)	19	53
Selenium (mg)	0.3	0.8
Zinc (mg)	95	265
Vitamins		
Vitamin A (IU)	7,600	10,600
Vitamin D (IU)	380	1,060
Vitamin E (IU)	42	117
Vitamin K (menadione) (mg)	1.0	2.6
Biotin (mg)	0.4	1.1
Choline (g)	2.4	5.3
Folacin (mg)	0.6	1.6
Niacin, available (mg)	19.0	53.0
Pantothenic acid (mg)	22.8	63.6
Riboflavin (mg)	7.1	19.9
Thiamin (mg)	1.9	5.3
Vitamin B_6 (mg)	1.9	5.3
Vitamin B_{12} (μg)	28.5	79.5

TABLE 7–5b. Requirements for Several Nutrients of Breeding Herd Replacements Allowed Feed Ad Libitum

| | Weight (kg) of: | | | |
| | Developing Gilts | | Developing Boars | |
	20–50	50–110	20–50	50–110
Energy concentration (kcal ME/kg diet)	3,255	3,260	3,240	3,255
Crude protein (%)	16	15	18	16
Nutrient*				
Lysine (%)	0.80	0.70	0.90	0.75
Calcium (%)	0.65	0.55	0.70	0.60
Phosphorus, total (%)	0.55	0.45	0.60	0.50
Phosphrous, available (%)	0.28	0.20	0.33	0.25

*Sufficient data are not available to indicate that requirements for other nutrients are different from those in Table 7–2 for animals of these weights.

TABLE 7–5c. Daily Energy and Feed Requirements of Pregnant Gilts and Sows

| Intake and Performance Levels | Weight (kg) of Bred Gilts and Sows at Mating[a] | | |
	120	140	160
Mean gestation weight (kg)[b]	142.5	162.5	182.5
Energy required (Mcal DE/day)			
Maintenance[c]	4.53	5.00	5.47
Gestation weight gain[d]	1.29	1.29	1.29
Total	5.82	6.29	6.76
Feed required/day (kg)[e]	1.8	1.9	2.0

[a]Requirements are based on a 25-kg maternal weight gain plus 20-kg increase in weight due to the products of conception; the total weight gain is 45 kg.
[b]Mean gestation weight is weight at mating + (total weight gain/2).
[c]The animal's daily maintenance requirement is 110 kcal of DE/kg.$^{0.75}$
[d]The gestation weight gain is 1.10 Mcal of DE/day for maternal weight gain plus 0.19 Mcal of DE/day for conceptus gian.
[e]The feed required/day is based on a corn-soybean meal diet containing 3.34 Mcal of DE/kg.

TABLE 7–5d. Daily Energy and Feed Reguirements of Lactating Gilts and Sows

| Intake and Performance Levels | Weight (kg) of Lactating Gilts and Sows at Postfarrowing | | |
	145	165	185
Milk yield (kg)	5.0	6.25	7.5
Energy required (Mcal DE/day)			
Maintenance*	4.5	5.0	5.5
Milk production†	10.0	12.5	15.0
Total	14.5	17.5	20.5
Feed required/day (kg)‡	4.4	5.3	6.1

*The animal's daily maintenance required is 110 kcal of DE/kg$^{0.75}$
†Milk production requires 2.0 Mcal of DE/kg of milk
‡The feed required/day is based on a corn-soybean meal diet containing 3.34 Mcal of DE/kg.

VI. Steps Involved in Formulating a Swine Diet

A. *Specify Nutrient Requirements*

1. Assume we are to formulate 100 kg (as-fed basis) of a complete diet for pigs weighing an average of 30 kg (66 lb). The nutrient requirements for this weight pig are found in Table 7–2 and can be recorded on the swine ration worksheet found following this sample problem. A copy of this worksheet is also found in the appendix section at the back of the textbook.

B. *Select Feed Ingredients (as-fed composition in Table 3–2)*

1. Grain source—use equal parts of corn grain (IFN 4-02-931) and milo (sorghum grain, IFN 4-04-444).
2. Protein source—SBM, solvent-extracted (IFN 5-04-604).
3. Salt (plain)—fix at 0.5% of the diet.
4. Use the following commercially prepared vitamin and trace mineral premixes
   a. Swine vitamin premix

Guaranteed analysis	Amount per kg
Vitamin A	200,000 IU
Vitamin D	20,000 IU
Riboflavin	140 mg
Niacin	1,600 mg
D-calcium pantothenate	300 mg
Vitamin B_{12}	2 mg

   Direction for use
   For use in fortifying growing-finishing swine rations; use at the level of 1.0% of the complete diet.

   b. Swine trace mineral premix

Guaranteed analysis	% of premix
Iron (Fe)	6.000%
Zinc (Zn)	6.000%
Manganese (Mn)	0.200%
Magnesium (Mg)	0.340%
Copper (Cu)	0.400%
Iodine (I)	0.014%

   Direction for use
   Thoroughly mix into complete swine diets at the level of 0.10% of the diet.

5. Calcium and phosphorus sources
   a. Dicalcium phosphate (calcium phosphate, dibasic, IFN 6-01-080).
   b. Limestone (IFN 6-02-632).

C. *Estimate the parts (kg) of the total diet for fixed ingredients including Ca and P sources*

1. Salt   0.5 kg
2. Vitamin premix   1.0 kg
3. Trace mineral premix   0.1 kg
4. Dicalcium phosphate and Limestone   2.0 kg* (estimated)
5. The fixed ingredients sum to 3.6 kg. The remainder of the diet (nonfixed ingredients) will sum to 96.4 kg (100.0 − 3.6 = 96.4).

D. *Calculate the parts of CP from the fixed ingredients. Assume the fixed ingredients in our example contain no CP.*

E. *Balance for CP by Algebraic Solution or Pearson Square*

1. Amount of CP needed is 15 kg (15%).
2. CP supplied by fixed ingredients is none; thus the entire 15 kg CP must be supplied by the 96.4 kg of the grain and protein sources.
3. Algebraic solution
   a. X = kg grain mix; Y = kg SBM
   b. Grain mix is equal parts corn and milo with an average CP value of 9.4%.

$$\frac{8.8 + 10.0}{2} = 9.4\% \text{ avg CP}$$

c. Solution

$$
\begin{array}{rll}
X + & Y & = 96.40 \text{ kg} \\
0.094X + & 0.457Y & = 15.00 \text{ kg} \\
-0.094X - & 0.094Y & = -9.06 \text{ kg} \\
\hline
0 \quad\quad & 0.363Y & = 5.94
\end{array}
$$

Y = 5.94/0.363 = 16.4 kg SBM
X = 96.4 − 16.4 = 80.0 kg grain mix
(40.0 kg corn; 40.0 kg milo)

4. Pearson square solution
   a. 15 kg CP needed from 96.4 kg grain mix and SBM, thus

   15 ÷ 96.4(100) = 15.56% CP in nonfixed portion

   b. Solution
   Grain

   Mix   9.40   30.14 = 83.03% × 96.4 kg = 80.0 kg grain mix

                        15.56

   SBM   45.70 $\dfrac{6.16}{36.30}$ = 16.97% × 96.4 kg = 16.4 kg SBM

5. The quantities of corn, milo and SBM can be multiplied by their respective percent CP and these figures entered in the protein column of the worksheet.

F. *Determine Supplemental P Needs*

1. Calculate P from fixed and nonfixed ingredients by multiplying the quantity of these ingredients by their respective percent P

Corn supplies	0.116 kg
Milo supplies	0.120 kg
SBM supplies	0.113 kg
Total	0.349 kg

2. Subtract amount of P present in ingredients from amount required

   0.500 − 0.349 = 0.151 kg needed from dical
   thus

   $$\frac{0.151 \text{ kg P needed}}{0.1854 \ (\% \text{ P in dical})} = 0.81 \text{ kg dical needed}$$

G. *Determine Supplemental Ca Needs*

1. Calculate Ca from fixed and nonfixed ingredients by multiplying the quantity of these ingredients by their respective percent Ca

Corn supplies	0.008 kg
Milo supplies	0.016 kg
SBM supplies	0.049 kg
Dical supplies	0.177 kg
Total	0.250 kg

2. Subtract amount of Ca present in ingredients from amount required

   0.600 − 0.250 = 0.350 kg needed from limestone

   thus

   $$\frac{0.350 \text{ kg Ca needed}}{0.371 \ (\% \text{ Ca in lime})} = 0.94 \text{ kg limestone needed}$$

H. *Adjust the level of the energy source (±) to make the diet total 100 parts (kg).*

   Our estimated quantity of dical and limestone was 2.0 kg, but they only totaled 1.75 kg (0.81 + 0.94). Thus the total diet summed to 99.75 kg. This small amount of difference (0.25 kg) can be added to corn so the diet sums to 100. A small change in amount of grain source (<1.0) will not necessitate changing the CP, Ca and P values.

I. *Calculate Vitamin Level of Diet*

    1. Vitamins A and D—most swine diet formulators ignore any vitamin A or D value to the feedstuff ingredients and calculate these vitamins solely from the premix source. One kg of the vitamin premix in this example provides 200,000 IU of vitamin A and 20,000 IU of vitamin D.

    2. Riboflavin—the quantity of riboflavin supplied by each ingredient is added together and compared to the requirement.

    3. Niacin—grains are fairly high in niacin; however, research has shown that the niacin content of most plant products is relatively unavailable to swine. Therefore, the only source of niacin considered will be that from the vitamin premix.

    4. Pantothenic acid—the quantity of pantothenic acid supplied by each ingredient is added together and compared to the requirement.

    5. Vitamin B_{12}—the only natural sources of vitamin B_{12} are animal product feedstuffs. Thus, in this example, all of the vitamin B_{12} is being supplied by the vitamin premix.

J. *Calculate ME Level of Diet*

Metabolizable energy values for each ingredient (when available) are multiplied by the quantity of that ingredient and then the sum is compared to the requirement. This example problem is slightly low in ME but not enough to be of major concern.

K. *Calculate Amino Acid and Trace Mineral Level of Diet*

The reverse side of the swine ration worksheet accommodates calculation of three amino acids and seven trace minerals. Compute the amount of these nutrients supplied by each feed ingredient, and compare the sum to the requirements listed in Table 7–2.

L. *Calculate differences from needs of the nutrients and make adjustments as needed.*

VII. Suggested Rations for Various Stages of the Swine Life Cycle Shown in Tables 7–6 through 7–10

SWINE RATION WORKSHEET

Ration __30 kg pig__

Ingredient	kg.	Protein %	Calcium %	Total Phosphorus %	Vit. A units/kg	Vit. D units/kg	Riboflavin mg/kg	Niacin mg/kg	Pantothenic acid mg/kg	Vit. B₁₂ mcg/kg	Metabolizable energy kcal/kg
Nutritive Requirements		15.0	0.6	0.5	1,300	150	2.5	10.0	8.0	100	3,260
Amount Required		kg	kg	kg	IU	IU	mg	mg	mg	mcg	kcal
Corn	~~40.00~~ 40.25	3.52	.008	.116	—	—	52	—	157	—	135,200
Milo	40.00	4.00	.016	.120	—	—	44	—	448	—	129,880
SBM	10.40	1.49	.049	.113	—	—	47	—	267	—	49,298
Salt	0.50	—	—	—	—	—	—	—	—	—	—
Vitamin premix	1.00	—	—	—	200,000	20,000	140	1,600	300	2,000	—
TM premix	0.10	—	—	—	—	—	—	—	—	—	—
Dical PO₄	0.81	—	.177	.150	—	—	—	—	—	—	—
Limestone	0.94	—	.349	.002	—	—	—	—	—	—	—
TOTALS (kg, IU, mg, mcg)	100.00	15.01	.599	.501	200,000	20,00	283	1,600	1,172	2,000	314.378
Calculated Analysis: % or amt. per kg Feed	15.0	15.0	.6	.5	2,000	200	2.8	16.0	11.7	20.0	3,144
DIFFERENCE FROM NEEDS		—	—	—	+700	+50	+.3	+6.0	+3.7	+10.0	−116

Calculated Analysis

325

SWINE RATION WORKSHEET (CONTINUED)

Calculated Analysis

Ingredient	kg.	Lysine %	Threonine %	Tryptophan %	Cu	Fe	I	Mg	Mn	Se	Zn
Nutritive Requirements		0.75	0.48	0.12	4.0	60.0	0.14	400.0	2.0	0.15	60.0
Amount Required		kg	kg	kg	mg	mg	mg	mg	mg	mg	mg
								mg/kg			
Corn	40.25	.121	.121	.040	151	1,021	—	44,275	214	—	549
Milo	40.00	.080	.120	.040	163	1,934	2.44	48,000	631	8.00	678
SBM	16.40	.459	.275	.098	294	2,305	2.15	47,560	502	7.10	850
Salt	0.50	—	—	—	—	—	—	—	—	—	—
Vitamin premix	1.00	—	—	—	—	—	—	—	—	—	—
TM premix	0.10	—	—	—	400	6,000	14.00	340	200	—	6,000
Dical PO₄	0.81	—	—	—	7	6,870	—	5,734	207	—	100
Limestone	0.94	—	—	—	10	1,754	—	9,926	116	—	26
TOTALS (kg, IU, mg, mcg)	100.00	.660	.536	.178	1,025	19,884	18.59	153,835	1,876	15.10	8,205
Calculated Analysis: % or amt. per kg Feed		0.66	0.54	0.18	10.2	199.0	0.18	1,538	18.8	0.15	82.0
DIFFERENCE FROM NEEDS		-0.09	+0.06	+0.06	+6.2	+139.0	+0.04	+1,138	+16.8	—	+22.0

TABLE 7–6. Pregestation, Breeding and Gestation Rations for Boars, Sows or Gilts Being Fed 1.9 kg (4.2 lb) per Day[a]

Ingredient	IFN	1	2	3	4
Corn, gr.[b]	4–02–931	86.90	83.00	88.90	84.80
Soybean meal[c]	5–04–604	9.00	8.00	3.75	3.00
Alfalfa meal, dehydrated	1–00–023	—	5.00	—	5.00
Meat and bone meal	5–00–388	—	—	5.00	5.00
Limestone	6–02–632	0.95	0.75	0.35	0.20
Dicalcium phosphate	6–01–080	1.80	1.90	0.65	0.65
Salt, iodized	6–04–151	0.50	0.50	0.50	0.50
Trace mineral premix[d]		0.10	0.10	0.10	0.10
Vitamin premix[e]		0.75	0.75	0.75	0.75
Feed additives[f]		—	—	—	—
		100.00	100.00	100.00	100.00
Calculated analysis					
Crude protein, %		11.80	11.80	12.00	12.20
Calcium, %		0.80	0.81	0.80	0.81
Phosphorus, %		0.65	0.66	0.65	0.65
Lysine, %		0.51	0.51	0.53	0.53
Metabolizable energy, kcal/kg		3190	3095	3210	3120

[a]Formulations are on an as-fed basis. See Table 7–1 for recommended feeding levels for boars, gilts and sows when hand fed daily in drylot or confinement. These rations can be used for gilts on pasture during gestation since they require 3 to 4 lb of feed daily. These rations can also be used for interval-fed sows or gilts if the average daily intake is approximately 4 lb.

[b]Ground oats can replace corn up to 20% of the total ration. If more than 20% oats is used in the ration, the level of feeding should be increased because of the low energy content of oats. Ground milo, wheat or barley can replace the corn.

[c]If 48.5% CP soybean meal is used instead of 45.7%, use 7% less soybean meal and add the difference on to corn. If whole cooked soybeans are used instead of 45.7% CP soybean meal, use 30% more cooked beans and subtract the difference from corn.

[d]Trace mineral premix should supply the following per kg of ration: iron, 80 mg; zinc, 50 mg; manganese, 10 mg; copper, 5 mg; selenium, 0.1 mg.

[e]Vitamin premix should supply the following per kg of ration: vitamin A, 4000 IU; vitamin D, 200 IU; riboflavin, 3 mg; niacin, 10 mg; pantothenic acid, 12 mg; vitamin B_{12}, 0.015 mg.

[f]Feed additives are not generally recommended during gestation or for gilts during the developing period after selection unless specific disease problems exist. High levels of feed additives (100 to 300 g/ton) may be beneficial 2 to 3 weeks before breeding and 2 to 3 weeks before farrowing.

TABLE 7–7. Swine Lactation Rations[a]

Ingredient	IFN	1	2	3	4
Corn, gr.[b]	4–02–931	83.40	78.45	79.50	77.40
Soybean meal[c]	5–04–604	12.50	12.50	11.50	5.50
Beet pulp, dried	4–00–669	—	5.00	—	—
Wheat bran	4–05–190	—	—	5.00	5.00
Alfalfa meal, dehydrated	1–00–023	—	—	—	5.00
Meat and bone meal	5–00–388	—	—	—	5.00
Limestone	6–02–632	1.00	0.90	1.10	0.30
Dicalcium phosphate	6–01–080	1.45	1.50	1.25	0.15
Salt, iodized	6–04–151	0.50	0.50	0.50	0.50
Trace mineral premix[d]		0.15	0.15	0.15	0.15
Vitamin premix[e]		1.00	1.00	1.00	1.00
Feed additives[f]		—	—	—	—
		100.00	100.00	100.00	100.00
Calculated analysis					
Crude protein, %		13.00	13.00	13.00	13.50
Calcium, %		0.75	0.75	0.75	0.75
Phosphorus, %		0.60	0.60	0.60	0.60
Lysine, %		0.60	0.61	0.59	0.60
Metabolizable energy, kcal/kg		3180	3150	3135	3065

[a]Formulations are on an as-fed basis. These rations may be full fed from a few days after forrowing until the baby pigs are weaned.
[b]Ground oats can replace corn up to 15% of the total ration. Ground milo, wheat or barley can replace the corn.
[c]If 48.5% CP soybean meal is used instead of 45.7%, use 7% less soybean meal and add the difference on to corn. If whole cooked soybeans are used instead of 45.7% CP soybean meal, use 30% more cooked beans and subtract the difference from corn.
[d]Trace mineral premix should supply the following per kg of ration: iron, 80 mg; zinc, 50 mg; manganese, 10 mg; copper, 5 mg; selenium, 0.1 mg.
[e]Vitamin premix should supply the following per kg of ration: vitamin A, 2000 IU; vitamin D, 200 IU; riboflavin, 3 mg; niacin, 10 mg; pantothenic acid, 12 mg; vitamin B_{12}, 0.015 mg.
[f]A high level of feed additive (100 to 300 g/ton) may be beneficial 2 to 3 weeks before farrowing until 7 to 10 days after farrowing, but will be of little benefit thereafter.

TABLE 7–8. Swine Prestarter Rations[a]

Ingredient	IFN	1	2	3	4
Corn, gr.	4–02–931	39.00	37.10	19.25	29.50
Soybean meal, w/o hulls	5–04–612	35.50	—	—	35.00
Soybean meal	5–04–604	—	35.40	33.60	—
Oat grain rolled	4–03–309	—	—	19.25	10.00
Dried whey	4–01–182	20.00	20.00	20.00	20.00
Fish meal, Menhaden	5–02–009	—	2.50	3.00	—
Animal fat, stabilized[b]	4–00–409	2.00	2.00	2.00	2.00
Limestone	6–02–632	1.00	0.95	0.90	1.00
Dicalcium phosphate	6–01–080	1.10	0.65	0.60	1.10
Salt, iodized	6–04–151	0.25	0.25	0.25	0.25
Trace mineral premix[c]		0.15	0.15	0.15	0.15
Vitamin premix[d]		1.00	1.00	1.00	1.00
Feed additives (g/ton)[e]		100–300	100–300	100–300	100–300
		100.00	100.00	100.00	100.00
Calculated analysis					
Crude protein, %		23.50	23.60	23.80	23.60
Calcium, %		0.89	0.91	0.91	0.89
Phosphorus, %		0.70	0.70	0.70	0.70
Lysine, %		1.40	1.40	1.40	1.39
Metabolizable energy, kcal/kg		3280	3155	3020	3210

[a]Formulations are on an as-fed basis. The prestarter ration normally is fed in only limited amounts. It should be used for pigs weaned before 3 weeks of age until they reach approximately 12 lb. Then they can be switched to a starter ration. It is a good ration to feed orphan pigs when the sow dies, extreme disease outbreak occurs or the sow fails to produce milk.
[b]Plant seed oil may replace animal fat.
[c]Trace mineral premix should supply the following per kg of ration: iron, 150 mg; zinc, 100 mg; manganese, 4 mg; copper, 6 mg; selenium, 0.1 mg.
[d]Vitamin premix should supply the following per kg of ration: vitamin A, 2200 IU; vitamin D, 220 IU; riboflavin, 3 mg; niacin, 22 mg; pantothenic acid, 13 mg; vitamin B_{12}, 0.022 mg.
[e]The feed additive may be part of the vitamin premix, or if a separate premix, it should replace an equal amount of corn.

TABLE 7–9. Pig Starter Rations*

Ingredient	IFN	1	2	3	4	5
Corn, gr.	4–02–931	54.80	55.75	60.60	68.05	44.35
Soybean meal, without hulls	5–04–612	22.00	—	22.50	—	—
Soybean meal	5–04–604	—	26.00	—	28.25	—
Soybeans, whole cooked	5–04–597	—	—	—	—	37.50
Dried skim milk	5–01–175	—	—	2.50	—	—
Dried whey	4–01–182	20.00	15.00	10.00	—	15.00
Animal fat, stabilized	4–00–409	—	—	1.00	—	—
Limestone	6–02–632	0.75	0.80	0.75	0.90	0.75
Dicalcium phosphate	6–01–080	1.05	1.05	1.25	1.40	1.00
Salt, iodized	6–04–151	0.25	0.25	0.25	0.25	0.25
Trace mineral premix[†]		0.15	0.15	0.15	0.15	0.15
Vitamin premix[‡]		1.00	1.00	1.00	1.00	1.00
Feed additives (g/ton)[§]		100–300	100–300	100–300	100–300	100–300
		100.00	100.00	100.00	100.00	100.00
Calculated analysis						
Crude protein, %		18.3	18.7	18.6	18.9	19.6
Calcium, %		0.75	0.75	0.74	0.75	0.75
Phosphorus, %		0.65	0.65	0.65	0.65	0.65
Lysine, %		1.03	1.03	1.03	1.00	1.10
Metabolizable energy, kcal/kg		3200	3120	3270	3135	3320

*Formulations are on an as-fed basis. The pig starter ration should be used as a creep ration before weaning and fed after weaning until the pigs reach approximately 40 lb.
[†]Trace mineral premix should supply the following per kg of ration: iron, 110 mg; zinc, 90 mg; manganese, 3.5 mg; copper, 5.5 mg; selenium, 0.1 mg.
[‡]Vitamin premix should supply the following per kg of ration: vitamin A, 1975 IU; vitamin D, 210 IU; riboflavin, 3 mg; niacin, 20 mg; pantothenic acid, 12 mg; vitamin B_{12}, 0.019 mg.
[§]The feed additive may be part of the vitamin premix, or if a separate premix, it should replace an equal amount of corn.

TABLE 7–10. Swine Grower-Finisher Rations[a]

Ingredients	IFN	Pigs 44–110 lb (20–50 kd)			Pigs 110–242 lb (50–110 kg)		
		1	2	3	4	5	6
Corn, gr.[b,c]	4-02-931	83.15–79.65	75.75–70.50	83.65–80.40	85.25–83.35	79.35–77.35	86.75–83.75
Soybean meal[d]	5-04-604	14.00–17.50	—	10.00–13.25	12.10–14.00	—	7.00–10.00
Soybeans, whole cooked[e]	5-04-597	—	21.50–26.75	—	—	18.00–22.00	—
Meat and bone meal	5-00-388	—	—	5.00	—	—	5.00
Limestone	6-02-632	0.85	0.75	0.30	0.90	0.90	0.30
Dicalcium phosphate	6-01-080	1.15	1.15	0.20	0.90	0.90	0.10
Salt, iodized	6-04-151	0.25	0.25	0.25	0.25	0.25	0.25
Trace mineral premix[f]		0.10	0.10	0.10	0.10	0.10	0.10
Vitamin premix[g]		0.50	0.50	0.50	0.50	0.50	0.50
Feed additives (g/ton)[h]		0–100	0–100	0–100	0–100	0–100	0–100
		100.00	100.00	100.00	100.00	100.00	100.00
Calculated analysis							
Crude protein, %		13.7–15.0	14.5–16.0	14.4–15.6	13.0–13.7	13.6–14.7	13.3–14.4
Calcium, %		0.62–0.63	0.60–0.61	0.70–0.71	0.60–0.60	0.60–0.61	0.67–0.68
Phosphorus, %		0.55–0.57	0.56–0.58	0.60–0.61	0.50–0.51	0.50–0.52	0.57–0.58
Lysine, %		0.64–0.73	0.70–0.80	0.67–0.75	0.60–0.64	0.63–0.71	0.60–0.67
Metabolizable energy, kcal/kg		3210–3200	3330–3340	3220–3210	3230–3220	3320–3330	3240–3230

[a]Formulations are on an as-fed basis. Start with the higher level of soybean meal (lower level of corn) with lighter pigs in each group, and decrease the soybean meal (increase the corn) in 2.5 to 5.0% increments until you reach the lower level. If you prefer, one level of protein can be fed from 40 lb to 240 lb with similar results as with the varying levels. In this case use a level of soybean meal and corn which is approximately the high point of the listed range for growing pigs (for example, in ration No. 1 you might use 79.65% corn and 17.50% soybean meal). If barrows and gilts are separated, use the higher end of the range of soybean meal for the gilts and the lower end for the barrows.

[b]Ground milo, wheat or barley can replace the ground corn. Ground oats can replace corn up to 20% of the total ration.

[c]If the ration is to be pelleted, 1.25% to 2.50% of molasses or binder can replace a similar amount of corn.

[d]Three weight units (lb or kg) of L-lysine HCl and 97 weight units corn can be substituted for 100 weight units of soybean meal.

[e]Since the high fat content of whole cooked soybeans increases the energy content of the ration, the protein level in a ration utilizing whole cooked soybeans should be approximately 0.75% to 1.0% protein higher than a similar ration with soybean meal in order to maintain the same protein-to-energy ratio.

[f]Trace mineral premix should supply the following per kg of ration: iron, 50 mg; zinc, 50 mg; manganese, 2 mg; copper, 3 mg; selenium, 0.1 mg.

[g]Vitamin premix should supply the following per kg of ration: vitamin A, 1300 IU; vitamin D, 150 IU; riboflavin, 2.2 mg; niacin, 12 mg; pantothenic acid, 11 mg; vitamin B$_{12}$, 0.011 mg.

[h]The feed additive may be part of the vitamin premix, or if it is a separate premix, it should replace an equal amount of corn.

Study Questions and Problems

Chapter 7

I. True or False Questions

1. Baby pig prestarter diets are commonly mixed on the farm.
2. Finishing diets for swine usually contain 13% to 15% crude protein (CP).
3. Pigs self-fed free-choice diets usually gain more slowly than those fed complete self-fed diets.
4. Limit feeding sows during gestation refers to the use of bulky or fibrous feeds to lower the protein content of the ration.
5. Interval feeding of gestating gilts refers to a feeding system by which animals are given unlimited access to the separate components or groups of components of the diet.
6. Growing pigs fed pelleted complete diets generally are more efficient than those fed meal diets because they consume less daily feed.
7. Whole cooked soybeans can be substituted pound for pound for soybean meal (SBM) in a swine diet with little effect on the protein-to-energy ratio of the diet.
8. The use of high-lysine corn (HLC) by the swine producer can save substantial amounts of soybean meal or other protein supplements needed in the finishing swine diets.
9. Antibiotic and chemotherapeutic agents have their greatest benefits during the growing phase of the swine life cycle.
10. Soybean meal contains gossypol, which if fed at high levels may tie up zinc in a swine ration and cause the clinical symptom of parakeratosis.

II. Multiple Choice Questions (select the single best answer)

1. Of the following groups of B vitamins, which is most likely to be deficient in swine rations consisting of ordinary and average quality feeds?
   a. Riboflavin, thiamine, niacin, B_{12}
   b. Riboflavin, niacin, pantothenic acid, choline
   c. Riboflavin, niacin, pantothenic acid, B_{12}
   d. Riboflavin, niacin, B_{12}, choline
   e. Riboflavin, niacin, thiamine, pantothenic acid
2. An excessively high level of calcium in a swine diet could result in the deficiency of which trace mineral?
   a. Iodine
   b. Phosphorus
   c. Selenium
   d. Zinc
   e. All of the above
3. When flushing gilts, feed should be increased
   a. 1 week before and not more than 2 weeks after breeding.
   b. 1 to 2 weeks before and immediately after breeding.
   c. For one full heat cycle before breeding.
   d. Flushing is not recommended.
4. Pelleting diets for swine result in
   a. Improved feed efficiency.
   b. Increased wastage.
   c. Improved phosphorus availability.
   d. All of the above.
   e. Only answers a and c.
5. Of the folloiwing groups of trace minerals, which should be included in swine diets?
   a. Fe, Cu, Co, I, Zn, Mn
   b. Cu, I, Zn, Fe, S, Mg
   c. Fe, Cu, I, Mn, Zn, Mg
   d. Co, S, Mg, Se, I, Fe
   e. Fe, Cu, Ca, I, Zn, P

6. Supplemental fat would generally not be added in which of the following feeding programs?
   a. Corn-soybean meal diet, self fed
   b. Corn-whole cooked soybean diet, self fed
   c. High lysine corn-soybean meal diet, free-choice fed
   d. Corn-soybean meal diet, free-choice fed
   e. Fat is never added to a swine diet.

III. Matching or Identification

A. Indicate the approximate percent crude protein needed in the following diets

> a. 12% to 14%
> b. 14% to 16%
> c. 18% to 20%
> d. 20% to 24%

1. Gestating sows fed 2 kg feed per day
2. Grower pigs (20 to 50 kg)
3. Starter pigs (5 to 20 kg)
4. Finishing pigs (50 to 110 kg)
5. Pig prestarter (birth to 5 kg)
6. Lactating sows fed 4.5 to 6 kg feed per day

IV. Short Answer

1. What feeds would typically be found in swine milk replacer diets?
2. What B vitamins are most apt to be deficient in swine diets based on corn and soybean meal?
3. During what phase of the swine life cycle would pasture be best utilized?
4. List four ways to restrict energy intake of the pregnant gilt or sow.
5. List five feed additives that may commonly be included in growing-finishing swine diets.

V. Problems

1. 800 lb corn and 200 lb soybean meal will balance the protein needed in a given swine diet but will provide only 0.8 lb calcium and 1.6 lb available phosphorus. If 5.0 lb Ca and 4.0 lb P are needed, how much dicalcium phosphate (23% Ca, 18% P) and limestone (36% Ca, 0% P) would be needed in 1000 lb of diet?
2. A vitamin-antibiotic premix for swine is suggested for use in swine rations at the rate of 40 lb/ton. If the premix has the following potencies, determine the vitamin-antibiotic content supplied by the premix to the final ton of ration.

Premix Potencies	Ration Content
Vitamin A — 35,000 IU/lb premix	_____ IU/lb ration
Vitamin D — 4000 IU/lb premix	_____ IU/lb ration
Riboflavin — 20 mg/lb premix	_____ mg/lb ration
Pantothenic acid — 180 mg/lb premix	_____ mg/lb ration
Niacin — 250 mg/lb premix	_____ mg/lb ration
Vitamin B_{12} — 200 mcg/lb premix	_____ mg/lb ration
Antibiotic — 0.115%	_____ gm/ton ration

3. Formulate 100 kg of swine diet from milo (10% CP) and soybean meal (46% CP) and the following fixed ingredients

> 5.0% dehydrated alfalfa (20% CP)
> 2.0% dicalcium phosphate
> 1.5% limestone
> 0.5% TM salt
> 1.0% vitamin-antibiotic mix (50% CP)

The final formulation should contain 12% CP.

8

Beef Cattle Feeding Guides

Chapter Goals

- Outline and discuss a life-cycle beef cattle feeding program.
- Identify specific nutrient needs or additives used within the life cycle.
- Explain feeding systems for the beef breeding herd and for finishing market cattle.
- Introduce the merits of a system of protein evaluation for cattle, metabolizable protein.
- Formulate beef cattle diets.

The information presented should be considered only as a guide in formulating rations and in making general feeding recommendations for beef cattle.

The material is presented as guides for the *Breeding Herd* and for *Finishing Cattle*. Additional and more detailed information may be outlined by consulting publications distributed from various land grant universities and other colleges. Nutrient requirements, feed allowances and other pertinent information on beef cattle is provided in tables 8–1 through 8–14.

BEEF BREEDING HERD

I. The Cow Herd

The one most important consideration in a profitable cow and calf operation is a high percentage calf crop. A live calf from each cow each year should be the goal. For 280 to 285 days of a year, the cow is pregnant. Therefore, to have a calf every 365 days, she must recover from calving, come into heat and conceive (settle) within an 80- to 85-day period. Older cows (3 years and older) recover from calving stress and come into heat faster than first-calf heifers. So, breed yearling heifers 20 days earlier than old cows. This gives first-calf heifers more time to recover from calving stress, come into heat and conceive, consequently keeping them on schedule with the older cows when the heifers drop their second calves.

Nutrition influences breeding. Cows gaining just before and during the breeding season will show a shorter period between calving and the first heat period and tend to have high conception rates. Experiments and practical observations reveal that the period during which calf crop percentage is affected most by nutrition extends from 30 days before calving until 70 days after calving (until after rebreeding), a period of about 100 days. The nutritional needs for the cow during the more critical production period are approximately equal to her needs for the remainder of the year.

Most cow and calf producers will choose either a spring (March through April) or fall (September through October) calving program, thus avoiding the severely cold weather of many winters and the extreme heat of summer. Each individual manager has to size up the advantages and disadvantages of either program. In general, producers favor the spring calving season. Their main reasons are minimum housing needs and use of less harvested feed.

Supplemental minerals provided to cattle on pasture.

Nutrient requirements of breeding cattle are shown in Table 8–4. The level of energy and nutrients varies depending on the frame size and condition of the cow, stage of production and environmental conditions. Rations should supply approximately the following quantities on a dry matter basis:

Ration Component	Pregnant Cows	Lactating Cows
Protein (% of ration)	7–9	9.5–12.5
Protein (lb/day)	1.1–1.7	1.8–2.9
TDN (% of ration)	49–64	55–68
TDN (lb/day)	8–13	10.5–16
ME (Mcal/lb of ration)	.8–1.1	0.9–1.3
ME (Mcal/day)	14–21	18–28
Ca & P (% of ration)	0.2–0.3	0.25–0.5
Ca & P (g/day)	15–25	25–40

A. *Summer*
1. Pasture will supply most of the nutrients needed.
2. Mineral supplementation will be necessary (free-choice).
   a. A current recommendation is a mixture of: 1 part steamed-bone meal or 1 part dicalcium phosphate with 1 part plain salt.
   b. Bone meal and dical can both be included although either of these Ca and P sources mixed with salt (to improve palatability) are satisfactory.
   c. Salt may also be fed free-choice. This can be block or granular salt. Vitamin A can be included if pasture grass is dry and mature.
   d. Spring calving cows should calve early enough so that calves will be 1 to 2 months old by turnout to pasture. Breeding season will normally start within 4 to 6 weeks after turnout.
   e. Milk production by beef cows will average about 10 to 25 lb/day over a 175- to 200-day lactation (depending on breed of cattle and energy level of diet). Maximum milk production of beef cows occurs during the first 2 months after calving, then declines.
   f. Cows nursing calves should receive approximately 50% more feed than dry cows of comparable weight and condition.
3. If pasture is short or inadequate, supplemental energy will be needed. Cows cannot milk well and heifers will not grow rapidly without adequate energy.
   a. Carrying capacity can be increased by feeding silage, hay or both. When supplemental feed is necessary for the mature stock, you can follow these general guides: 15 lb of corn silage per head per day will substitute for about one third of the pasture acreage normally needed; 30 lb/day will make up for two thirds of the usual pasture acreage. Five to ten lb of good-quality hay per head daily will give the same results.
   b. Hay feeding on lush legume pasture may help control bloat.
   c. Concentrates can be used most efficiently by creep feeding calves.
   d. Pasture management is important. Consider renovation, fertilization, species of pasture plants and rotation grazing.

4. Temporary or supplemental pasture crops include Sudangrass, Sudan-sorghum crossed (i.e., Sudax), oats, rye and wheat. Best yield per acre can be obtained by rotation grazing.
5. Fall calving cows will lose weight rapidly during the winter lactation. The cows should be permitted sufficient time on a reasonably good pasture or feeding program to recover after the calf is weaned before calving time again.

B. *Winter*

1. When grazing forage is not available, cows can be fed hay, crop residues or silage. Average quality forages (7% to 8 + % CP) are adequate as the main or sole feed. Poor quality roughages may need supplementation with energy or protein sources.
2. The daily roughage is usually provided in either one or two feedings, with any supplement being fed at a single feeding. At least half of the ration is best fed in the evening during cold weather.
3. Suggested rations for wintering drylot beef cows are shown in Table 8–6 and Table 8–7.
    a. Spring calving cows in average fall condition should be wintered at a level so that no more than about 15% of fall weight will be lost through calving and until grass is ample in the spring. Cows normally lose about 100 to 130 lb at calving.
    b. Dry cows wintered on dry grass pastures should be fed 1½ to 2 lb of 40% to 44% crude protein supplement to meet protein needs. If additional energy is needed, provide 1 to 2 lb of grain along with the protein supplement.
    c. In the case of fall calving cows on dry grass pasture, protein supplement and energy levels higher than those listed above may be needed.
    d. If legume hay is used as a protein supplement, about 3 lb is equivalent to 1 lb of 40% to 44% supplement on a protein basis.
    e. If dry cows are wintered on harvested feeds, the following guides appear appropriate:
        (1) Total dry feed allowance
            (a) Cows in fleshy condition 1.50 lb/100 lb b.w.
            (b) Cows in average condition 1.75 lb/100 lb b.w.
            (c) Cows in thin condition 2.0–2.5 lb/100 lb b.w.
            Note: If silage is used, take moisture content into consideration in above recommendations. Often it takes about 3 lb of as-fed silage to provide the dry equivalent of 1 lb of hay or other dry roughage
        (2) The level of supplement required will depend on the roughage used. From 1½ to 2 lb of 40% to 44% supplement will usually be adequate even when low quality roughages are used.
        (3) Adjustments to "typical" daily intakes
            (a) Body condition—cows in good condition are more tolerant to the stresses of winter and require less maintenance energy per unit of weight than do cows in poor condition. Energy requirements for cows in different body conditions may be adjusted by calculating a ratio between the body weight of a cow (kg) to the height at the hooks (cm). The average weight: height ratio for a cow in good condition is 4.0:1. The adjustment in metabolizable energy (ME) required by mature cows in different condition can be determined using the following formula

            $$(4.0 - \text{weight/height}) \times 1.716 = \text{Mcal/day adjustment}$$

            Changes in winter height and heart girth circumference along with weight change can also be used to separate growth from change in body condition of bred heifers.
            (b) Cold stress—the lower "critical temperature" for beef cows (assuming an average dry winter hair-coat thickness) is about 20°F. Rule of thumb: increase the amount of feed 1% for each degree of coldness below critical temperature.
    f. Urea can be used as the only source of supplemental protein when cows are wintered on corn silage. When urea is used to supplement other types of low-protein roughage best results are usually obtained in a 20% to 25% protein supplement that contains some grain and molasses as readily digestible carbohydrate to aid in the utilization of the urea.

g. The following scheme represents the life-cycle weight-change feeding approach for the replacement female in a spring calving program to calve first at 2 years of age.
   (1) 1st winter   Gain 200 to 300 lb
   7 to 15 months of age
   (open heifer)
   (2) 2nd winter   Gain enough to offset calving
   (bred heifer)   loss which is about 125 to 150 lb
   Fall wt—750 to 850 lb
   Precalving wt—900 to 1000 lb
   (3) 3rd winter   Lose no more than 5% to 10% of
   (bred heifer)   fall weight including calving loss
   Fall wt—900 to 1000 lb
   Precalving wt—900 to 1050 lb
   Spring wt—850 to 950 lb
   (4) 4th and subsequent winters   Lose no more than 15% to 20%
   (bred cow)   of fall weight including calving loss

4. Additional forage materials such as hay, silage or a combination of the two may be used to supplement poor quality pasture. However, two common errors in winter management are furnishing cows high quality feeds that are too expensive and giving them too much of the more costly feeds. Alfalfa hay or haylage, corn silage and grain are too palatable to give the cows unlimited access. The cows will eat more than they need and more than you can afford in terms of low wintering costs.
   a. Can use corn cobs to the extent of about 65% of the total dry matter as an extender for corn silage (cobs should be ground to about ⅜ in.)
   b. Good results have been obtained using standing or stacked sorghums in a winter feeding program. Lowers maintenance costs while maintaining excellent reproductive performance.

5. If low grade roughages (poor hay, cobs, straw, stalks etc.) are fed over extended periods of time the ration must be supplemented with the following various nutrients.
   a. Protein—1–1½ lb of a 35% to 44% crude protein complete supplement daily should be adequate. (Depends on protein in rest of ration.) This supplement can be a vegetable protein such as soybean meal or may contain a nonprotein nitrogen source as urea. Be especially careful in feeding high-urea supplement to cows where it is bulk fed.
   b. Calcium, phosphorus, and trace minerals should be provided free choice in suitable salt and mineral mixtures.
   c. Readily digestible carbohydrates normally are supplied most economically by grain. 1 to 2 lb of grain or an equivalent amount of grain in corn silage should be adequate to ensure good utilization of the low-grade roughages. Molasses may be used if cost permits.

Round bale forage properly fed in a steel rack reducing loss by wastage and trampling.

338

d. Vitamin A—supplemental Vitamin A should be provided at 20,000 IU daily in the supplement when low-grade roughages or long-stored hays are used as a major source of energy in wintering rations. There are mineral mixes on the market that contain a stabilized form of vitamin A.

II. The Calf Crop

A. The basic ration for calves will include the dam's milk plus access to pasture or forage that the cows are fed.

B. *Creep Feeding*

1. Creep feeding is usually some means of supplying additional energy to the calves in such a manner that only the calves can get to it.
2. Advantages of creep feeding
   a. Heavier weight calves at weaning.
   b. Improves the condition and uniformity of the calves at weaning.
   c. Less weight loss by cows.
3. Disadvantages of creep feeding
   a. Requires extra labor, equipment, feed and management.
   b. Higher feed cost.
   c. Replacement heifer calves may become too fat, resulting in lower lifetime milk production.
   d. May alter the maternal ranking of cows by weaning weight, as creep feed consumption has its greatest affect on calves from poor milking cows.
4. Creep feeds for calves can be varied greatly as to ingredient makeup. Basically, creep rations will contain up to 80% to 90% grain, often 3% to 5% molasses and up to about 10% oil meal or commercial supplement. Sample creep rations are found in Table 8–8.
5. Creep fed calves will weigh 30 to 50 extra lb by weaning time. Each of those extra pounds will require 8 to 10 lb of creep feed. For example, it may take over 500 lb of grain per calf to get an average of 50 more lb at weaning.
6. Creep feeding can be expected to give the greatest response when pasture is short or quality is poor. Generally, calves consume about 3 to 4 lb creep feed per head daily.
7. When antibiotic is added to creep rations, the concentration should be sufficient to provide 25 to 75 mg/head/day. Refer to *Feed Additive Compendium* for specific recommendations.
8. Implanted growth stimulants can increase weight gains and therefore weaning weights. Implanting both steer and heifer calves at 60 to 90 days (150 to 200 lb) will boost weaning weights 20 to 40 lb in addition to that of creep feeding. Implanting breeding heifers is not recommended. However, experimental results have shown no effect on the breeding performance of heifers when implanted at recommended levels at 3 months of age.

C. *Weaning*

1. Weaning beef calves at 6 to 8 months of age is common. Weaning at this age fits well with calfhood vaccinations and performance testing for replacement heifers or for feeder calf sales. Three cases call for earlier weaning
   a. Calves from first-calf heifers—reduces stress and increases the possibility of earlier heat and conception.
   b. Fall-dropped calves—reduces expensive harvested feed being fed the cow.
   c. Bull calves, because they begin riding other animals at 7 to 8 months of age. Remove them from the herd earlier and put them on feed.
2. To wean, separate the calves from the cows and do not allow the cows and calves to come back together once they have been separated. The cow dries off most effectively when there is a definite halt in nursing. The calves will suffer less stress and shrink if they are eating a creep feed when weaning occurs. Weaned calves may enter the feedlot directly after weaning, or they may be raised on roughage for a year or more before entering the feedlot.

D. *Preconditioned Calves*

Refers to a schedule of practices preparing the feeder calf for feedlot adaptation before it leaves the production site. The steps used in preconditioning may vary somewhat in different areas of the country; however, the following is a suggested program.

1. Mandatory regulations
    a. Wean calves and start on feed (grain) no less than 30 days before the sale or shipment date. Adjust to feed bunks and water troughs and start on a ration similar to that which they will get in the feedlot.
    b. Castrate, dehorn and treat for grubs no less than 3 weeks before the sale date.
    c. Vaccinate calves for infectious bovine rhinotracheitis (IBR or red nose), para-influenza (PI$_3$), *Pasteurella* spp and *Clostridial* spp no less than 3 weeks before the sale date.
2. Optional regulations
    a. Treat the calves for worms before the sale date.
    b. Vaccinate calves with leptospirosis before the sale date.
    c. Vaccinate calves with bovine virus diarrhea (BVD) and *Haemophilus somnus* (bovine respiratory disease) no less than 3 weeks prior to the sale date.
    d. Owned by the seller for at least 60 days prior to the sale date.

Preconditioned calves should be in better health, more efficient and therefore more beneficial to the feedlot operator. In most cases at least a $3 to $5 per cwt. premium will be returned to the producer to recover their costs.

E. Stocker cattle—weaned calves (steers or heifers) which are forage-fed for a period of time before being sold as cattle to enter the feedlot. Stocker cattle are generally thin or carrying little finish.
   1. May be bought in the fall to be wintered on high-roughage diets in drylot and sold in the spring for further pasture grazing or enter a feedlot finishing program.
   2. May be bought in the fall and grazed on small winter grain forage (usually wheat or winter oats) or grass (fescue) pasture. They may then enter the feedlot as 600- to 800-lb feeder cattle.
   3. May be bought in the fall to graze stock fields (corn or sorghum), where available, and then moved into the drylot for the winter and fed corn or sorghum silage supplemented with a legume hay or protein supplement; then finished on a high-energy ration in the drylot for slaughter in the summer or fall.

F. Backgrounded cattle—weaned calves placed in the drylot or pasture (or both) with more emphasis on growing than described for the stocker calf. Backgrounded cattle are fed grain in addition to roughage to reach around an 800-lb weight and then moved onto a high-energy finishing ration. The animals should be in good health, bunk broke and ready to go on a full feed of grain.

III. Replacement Heifers

A. *Advisability of creep-feeding replacement heifers is debatable.*
   Creep feeding allows better appraisal of growth potential but cost more and increase in weight gain is not marketed.

B. *Developing Replacement Females*
   1. Replacement females should be developed with levels of feed, pasture or both that permits them to attain about 55% to 65% of their anticipated mature weight at 15 months of age for breeding to calve first at 2 years of age.
   2. This usually dictates a post weaning rate of gain from 7 to 15 months of age, between 0.75 and 1.25 lb daily.
   3. Low levels of gain may delay puberty and reduce reproductive efficiency.
   4. High levels of gain may damage mammary development.

C. *Postweaning Management*
   1. Weaning ration—free choice hay and 3 to 4 lb grain/day with free choice minerals—calf usually weighs 425 to 500+ lb (depending on breed).
   2. Winter ration for spring-born heifers under one year of age: Feed to gain about 1.0 to 1.25 lb/day to reach 600 to 700 lb by May 1. Will need from 5.3 to 7.2 lb TDN and 1.1 to 1.3 lb CP plus free choice source of Ca, P and salt and perhaps trace minerals. If roughage is not hay, feed from 7000 to 10,000 IU of vitamin A daily.
   3. Possible adjustment that must be made to the wintering ration include
       a. Variation in environmental temperature—under cold conditions, heifers need more feed energy to maintain body heat. Rule of thumb: for every one-degree drop in temperature (F) below freezing, the nutrient requirement for energy increases by 1%. If a heifer is out in

340

the wind without a windbreak and the temperature is cold, the wind-chill index may be a better guide than the temperature reading alone. For example, if the temperature (or wind-chill index) is minus 10°F, then the energy (TDN) requirements for a 500-pound heifer would be 42% higher than listed in the requirement table.

    b. Feed wastage—Cattle often waste 5% to 15% of their food. During the spring thaw it is not uncommon for cattle fed on the ground to tramp into the mud 40% to 50% of the feed offered them.

4. Pasture (plus minerals) is adequate for yearling heifers to gain 1 to 1½ lb/day from May until breeding time in mid-July (75 days). Breed weighing about 650 to 750 lb at 15 months of age.

5. Bred heifers can be fed and managed like the cow herd, and should not be fed with steers. They will usually gain about 1 to 1¼ lb/day calving at about 900 to 1100 lb (depending on breed and frame size) without being overconditioned.

6. After calving, feed first-calf heifers liberally so they can breed back on schedule.

7. Suggested rations for first and second winter replacement heifers are shown in Tables 8–9 and 8–10, respectively.

IV. The Bull

A. *Young Bulls*

Young bulls should be creep-fed and then full-fed a high energy ration (70% to 85% concentrate) from weaning to 12 to 14 months of age. Such animals should be ready for use when 15 to 18 months old.

This will necessitate a daily dry feed allowance equal to about 2½% of their body weight and a gain rate of at least 2½ lb daily. From about 15 months to 3 years old, bulls should make a daily gain of 1¾ to 2¼ lb and receive a dry feed allowance equal to 1¾% to 2¼% of their body weight, with the proportion of roughage increased after the first year.

B. *Mature Breeding Bulls*

Can usually maintain condition on the same kind of pasture and wintering management provided the cow herd.

C. *Winter Feeding*

1. One-half lb of grain/100 lb body weight with protein supplement and free choice roughage. Adjust grain level to attain desired condition.

2. If roughage is corn silage, adjust level of grain and protein supplement as needed.

3. Free-choice minerals—salt, Ca, P (and trace minerals if poor quality forage).

4. 20,000 units vitamin A daily if any question about level in feed.

D. *Summer Feeding*

1. Free-choice minerals and pasture as in the case of cow herd.

2. Supplement energy feed only if required to maintain satisfactory condition—especially in the case of young bulls.

FINISHING RATIONS FOR MARKET CATTLE

I. General Information

A. *Finishing of Cattle*

Refers to the time period within the life cycle in which there is an increase in the total muscle (red meat) and fat mass of the body. The ultimate purpose of the finishing process is to produce beef that is desirable to the food consumer. Most of the finished cattle come to market between 1 to 2 years of age and weighing over 1000 lb. Some may go into the finishing feedlot at 200 to 300 lb at a few months of age (calves), some may weight 600 to 800 lb (yearlings), while others may be 900 lb or more and be older than 1 year of age. More steers than heifers are fed, because more of them are available. A portion of the heifers is held back for replacement purposes. Thus, nutrient requirements differ among the different kinds of cattle. Complete (mixed) rations for finishing cattle should meet the following general nutrient specifications (dry matter basis).

1. Crude protein—9% to 14%; requirements are greatly affected by age, size of animal and growth rate. Thus, young cattle require appreciably higher dietary protein (as a percent of the diet) if fed the same type of ration as older cattle.
2. Energy—high concentrate diets (TDN = 65% to 85% or NEg = .4 to .65 Mcal/lb) increase rate of gain, improve carcass characteristics, and generally cost less per unit of usable energy than diets high in roughages. However, cattle fed high concentrate diets are more prone to develop rumen acidosis, founder and liver abscesses. It requires more attention in management and feed processing to avoid some of the problems mentioned.
3. Calcium—0.3% to 0.6%; supplemental calcium is needed when cattle are fed heavily on concentrates and limited quantities of roughage. Calcium-phosphorus ratios of 2:1 have been shown to be beneficial in reducing urinary calculi in steers. When calculi problems are encountered, even higher levels of calcium may be advisable.
4. Phosphorus—0.2% to 0.4%; grains and high-protein supplements are fairly high in phosphorus, so rations high in such ingredients require little or no phosphorus supplementation. Calcium-phosphorus ratios of 2:1 or wider are beneficial in reducing urinary calculi in steers.

B. *Thumb Rules for Estimating Feed Intake*
1. Total dry feed intake will commonly be from 2% to 3 + % of body wt (400-lb cattle and heavier). In the case of cattle that have been through a period of restricted gain, feed intake may average higher.
2. A full feed of grain is commonly considered to be approximately 2 lb of grain per 100 lb of body weight, plus supplement and forages. 1½ lb shelled corn is enough with high corn silage rations but 2 lb of ear corn/100 lb b.w. should be fed—especially on more typical rations.
3. A full feed of corn silage during the early phase of the feeding period is usually equivalent to at least 3 to 5 lb of silage (as-fed basis) per 100 lb of body weight along with limited grain and supplement. (Use more for a young calf. Example: 400-lb calf on wintering rations gets 4 to 5 lb grain and 20 lb corn silage.) If no grain is fed in addition to silage, intake may be as high as 6% to 7% of body weight (as-fed basis).
4. Ratio of roughage to concentrate in finishing ration varies. In high concentrate rations, the ratio may be around 15:85 or higher.
   a. The consumption level of corn silage, hay or both will vary with animal size. The decision to use either corn silage, hay or both, or corn grain is dependent on the cost of corn grain. When corn price is in area of $2.00 to $2.50/bu or less, it is generally more favorable than corn silage in economy of gain. Cost of ensiling corn and in feeding is a factor. Current research indicates that even though gains are slower, rations including high levels of corn silage can be profitable.
   b. Faster gains can be expected on higher energy rations, and this may be more important than feed costs per pound of gain if labor costs are higher or you want to reach a higher grade earlier to meet expected marketing advantage.
   c. Feed intake is regulated to a great degree by energy content of the ration. As a result, if the energy content of a typical finishing ration is increased by 10%, for example, intake may be expected to drop by about 10%. Thus, the concentration of protein and other nutrients should be adjusted proportionately.

C. *Calculating Feed needs of Finishing Cattle*

Feed consumption and daily gains of beef cattle vary with management practices, age, condition of cattle, palatability of feeds and weather conditions.

A general rule of thumb for daily feed consumption of finishing cattle is 2% to 3% (dry matter basis) of their body weight. Thus a 600-lb steer on full feed would be expected to consume 12 to 18 lb of dry feed, while a 1000-lb steer would eat 20 to 30 lb daily.

A rough estimate of the total feed required during a feeding period may be obtained by multiplying the average weight by 2½ to 3% and then multiplying this figure by the estimated length of the feeding period.

Example: The following example illustrates the calculation of total feed required per head for a 120-day feeding period for 640-lb steers expected to gain an average of 3 lb/day.

Weight of steers: 640 lb
Planned length of feeding period: 120 days
Daily gain expected: 3 lb

$\dfrac{\text{Starting weight} + \text{ending weight}}{2} \times 3\% \times \text{days on feed}$

$\dfrac{640 + 1,000}{2} = 820$

$820 \times 3\% = 24.6$ lb dry feed per head daily
24.6×120 days $= 2952$ total lb of dry feed per head
The final figure includes the total ration, concentrates, roughages and supplements.

If the daily dry ration includes 1½ lb of protein supplement, 4 lb of hay and the remainder of dry matter intake supplied by ground yellow corn, the following amounts of each would be required for the 120-day feeding period.
1. $1½ \times 120 = 180$ lb protein supplement
2. $4 \times 120 = 480$ lb hay
3. $(24.6 - 5.5 = 19.1$ lb corn$)$ $19.1 \times 120 = 2292$ lb corn or 41 bushels
4. Total feed = 2952 lb

D. *Sample Rations*
1. Wintering yearling steers and heifers weighing 600 to 700 lb for an expected gain of 1.5 lb/day
a. 25 to 30 lb corn silage and 1 lb of a 32% supplement (or ¾ lb SBM).
b. 25 to 30 lb corn silage and 3 lb mixed hay.
c. In either case feed 3 to 4 lb corn grain and minerals and salt free-choice or mixed into the diet.
2. Wintering 300 to 600 lb calves for an expected gain of 1 to 1½ lb/day. 20 to 25 lb corn silage, 3 to 4 lb ground ear corn and 1 lb of a 32% supp. (or ¾ lb SBM) plus minerals and salt
3. Finishing 800 to 1000-lb yearling steers and heifers to gain 2.5 to 3 lb/day
a. 4 lb good hay, 1 lb supplement and 16 to 20 lb shelled corn.
b. Full feed corn silage and 10 to 12 lb corn grain plus ¼ lb 44% supplement.
c. Full feed ground ear corn, 1½ lb of 44% supplement and 12 to 15 lb corn silage (optional).
d. In all of the above rations provide adequate minerals and salt.
4. More detailed suggested formulations for feedlot cattle at different body weights and expected rates of gain are shown in Tables 8–11 through 8–14.

E. *Nutritional Management of Finishing Cattle*
1. Starting cattle on feed
a. The primary goal of the cattle feeder during the first few days after cattle arrival must be minimizing disease and death loss.
b. There is not a uniform way of getting recently purchased cattle on a full feed of grain. Usually cattle are confined to lots or small pasture for health examination, etc., for a few days. Care should be exercised in getting cattle on full feed, but this is not a reason to delay feeding a high performance ration. One concept is to start cattle on a high roughage (e.g., hay) complete ration fed free-choice and gradually replace roughage with grain. High levels of vitamin A (30,000 to 50,000 + IU/day) and antibiotic may be desirable the first 10 days to 3 weeks. While highly stressed calves may require 2 to 3 weeks before all are healthy and eating to capacity, the majority of yearling cattle will be satisfactorily eating and receiving ration by the fifth or sixth day.
c. Guides for starting "green" cattle on feed. These are only guides—individuals situations may vary considerably.
(1) Hand or mechanized feeding on a daily allowance basis
(a) Observe and correct any health problems as early as possible.
(b) Offer hay or other medium quality roughage on a free choice basis and protein supplement if needed as soon as the cattle are unloaded. If legume or legume-grass hay is used, additional protein may not be required. Grass hay is often favored as an initial roughage over legume hay as the legume hay, particularly alfalfa, is apt to cause diarrhea in unadapted animals.

343

Crossbred feeder cattle consuming a diet containing silage and corn grain from a fenceline feed bunk.

(c) Within the first 2 to 4 days, or as soon as the cattle have taken a fill of roughage, start grain at a rate of ½ lb/100 lb of body weight along with free choice roughage. Protein supplement should be started as soon as possible at or slightly below the level that is to be used throughout the feeding period.

(d) If silage is to be used as a major ingredient in the ration, it can be offered as the cattle will take it after a fill of dry roughage. The amount of grain in the silage will influence the amount of grain that should be provided initially.

(e) If high-level antibiotic (stress or conditioner) supplements are used, these normally should be discontinued after 10 to 20 days.

(f) Bring to a full feed of grain by increasing at a rate of 1 lb daily until the cattle are consuming a total quantity of grain equal to 1% to 1⅓% of body weight. Further increases may be made as the cattle will take the additional grain but normally an increase of about ½ lb/head daily is satisfactory. As the grain is increased, it will be essential to reduce the roughage allowance accordingly. Generally, at least 2 to 3 weeks should be taken to get cattle up to a maximum intake of concentrates when they have not had any grain previously.

(2) Self-feeding a mixed ration

(a) Observe and correct any health problems as early as possible.

(b) Fill cattle with hay or other roughage of medium quality during a 2 to 4-day rest period.

(c) Start cattle on a complete, self-fed ration properly formulated to meet requirements and alter ratio of roughage to concentrate in the ration as follows

Ratio of Roughage to Concentrate	No. of Days at Each Ratio
50:50	3–4 days
40:60	4–5 days
30:70	4–5 days
20:80	4–5 days
15:85	4–5 days
10:90	4–5 days

(d) The cattle feeder may elect to continue feeding one of the suggested R/C ratios for the remainder of the feeding period.

2. Energy in growing and finishing rations

a. Once cattle have been adapted to the finishing ration, the goal of the feeding program is to maintain maximum feed intake without causing digestive upsets.

b. Research results indicate that cattle (or rumen bacteria) do not efficiently utilize the forage and grain components of rations at certain R/C ratios. This is because of negative associative effects of rations at 60% to 75% concentrate. Associative effects are the effects that one feed can have on the utilization of another in the same ration. Therefore, it is suggested that the feeding program be one of the following types.

(1) Two-phase feeding period
  (a) Phase 1—ration contains 50% to 60% concentrate and is fed to cattle from about 450 to 800 lb body wt; this exemplifies the feeding of a diet consisting primarily of corn silage and supplement.
  (b) Phase 2—feed cattle from 800 lb body wt until market a diet containing over 75% concentrate; this would require increasing the grain and reducing intake of silage and forage.
(2) Single phase feeding period—after safely adjusted to grain, new cattle be increased to 75% or more concentrate and continued on this program until market. When determining the percent concentrate in a ration, remember that grain and supplement are 100% concentrate, corn silage is 50% concentrate and hay is 0% concentrate (dry matter basis).

3. Protein in growing and finishing rations
  a. Protein needs of feedlot cattle may be expressed as either crude protein or metabolizable protein. The equation used to calculate the requirements for each system is shown below.
    (1) Crude protein requirement (CP); taken from NRC, 1984.

    $$= [(.0334 \text{ DMI}) + (2.75(\text{wt}, \text{Kg}^{0.5})) + (.2(\text{wt}, \text{Kg}^{0.6})) + (268 - (29.4)(\text{NE}_g)(\text{ADG})] \div .594$$

    CP = Crude protein requirement in g/day
    DMI = Dry matter intake in g/day
    Wt = Weight of the animal in kg
    NE_g = Net energy for gain in Mcal/kg
    ADG = Average daily gain in kg

    (2) Metabolizable protein requirement (MP); taken from Burroughs, et al., 1974.

    $$= 0.0526 ((24 \text{ wt}^{0.734})) + ((3527 - \text{wt}) \text{ ADG})$$

    MP = Metabolize protein requirement in g/day
    Wt = Cattle weight in lb
    ADG = Average daily gain in lb

  b. The greatest difference between the MP requirement and the 1984 NRC's CP requirement for feedlot cattle is in the early phases of a calf feeding program. The MP system suggests higher levels of supplemental protein at lighter cattle body weights. Both systems are similar for heavier weight cattle. Both systems indicate that when expressed as percent crude protein in the diet levels fed early in the feeding period (500 lb b.w.) are 13.0% to 14.0% CP while late in the finishing period (more than 900 lb b.w.) both systems require only slightly above 8.0% CP. This is a wider range in protein level than is normally fed in the feedlot industry.
  c. Protein sources are usually priced higher than feed grains, but experience will quickly teach that protein deficiency is much more expensive than a slight protein surplus. When protein is deficient, energy is not well utilized and performance suffers.
  d. The protein supplements for feeder cattle also serve other purposes as carriers for minerals, vitamins, and feed additives. The normal 1 to 2 lb/head/day of supplement is a convenient vehicle for delivering small amounts of these ingredients and ensures that they are distributed throughout the ration.
  e. Supplemental protein for feedlot cattle may be of natural or nonprotein nitrogen (NPN) origin. The principal NPN source in use today in livestock feeding is urea. The appropriate time of the life cycle and quantity of urea to include in feedlot cattle diets is best defined by the system of protein evaluation for cattle and sheep development by Burroughs, et al., presented at the end of this chapter.
  f. In general, NPN is best utilized when the following conditions are present.
    (1) The ration contains a high percentage of readily fermentable carbohydrates.
    (2) The soluble nitrogen levels of the ration are fairly low.
    (3) The cattle being fed the ration have a relatively low protein requirement.
  g. The previous conditions prescribe urea to be *least efficiently utilized* if fed to
    (1) Young growing calves with a high protein requirement relative to their consumption.
    (2) Calves that have been under stress and in transit and that produce less microbial protein.
    (3) Beef cows grazing an all range forage ration (low in readily available carbonhydrates).
    (4) Cattle consuming rations containing feeds with a high level of soluble nitrogen (e.g., legumes).

h. Urea is most efficiently utilized by feedlot cattle consuming relatively high levels of dry grain. Between 0.10 and 0.15 lb/head/day seems to be a common urea intake range for many rations. The urea fermentation potential (UFP) values for feedstuffs described by Burroughs and associates at the end of this chapter help establish the quantity of urea to effectively supplement with different feedstuffs.

II. Miscellaneous Finishing Ration Ingredients

A. *Minerals*

   1. Calcium

      a. Generally adequate when forage (especially legumes) make up over 25% of the ration.

      b. A deficiency of calcium is most likely to occur when high concentrate rations, such as finishing rations, are fed. Heavy corn silage rations may also be borderline for calcium.

      c. Feedlot steers may experience problems with urinary calculi unless a Ca:P ratio of 2:1 or higher is contained in the diet.

   2. Phosphorus

      a. Deficiency unlikely in a high concentrate feedlot diet but likely to occur in cattle wintering on all-roughage rations or on mature, cured grass.

      b. Phosphorus deficiency is usually associated with use of feeds grown in phosphorus-deficient soils.

   3. Salt (NaCl)

      a. The salt or sodium requirement of full-fed beef cattle is met by including 0.25% salt in the total ration.

      b. Salt may be satisfactorily fed free-choice rather than as part of a mixed ration. Cattle that have free access to granulated, flake or loose salt will consume approximately twice as much as they will when salt is furnished in the form of compressed blocks. Weathering loss is also at least twice as great when in the loose form.

   4. Potassium

      a. Requirements for potassium are reported to be 0.6% to 0.8% of ration dry matter. Cattle diets high in forage are generally adequate in potassium.

      b. Finishing cattle on high concentrate diets (or all concentrate) all likely to require supplemental potassium as grains often contain less than 0.5% of potassium.

      c. Potassium requirements of stressed calves may be twice that of normal cattle. Therefore, a supplement fortified with potassium is recommended for calves that have been stressed.

   5. Sulfur

      a. Practical feedlot rations containing roughage and natural protein sources are considered adequate in sulfur content.

      b. When urea (or other NPN) supplies a considerable portion of the N in the ration, sulfur, either in organic or inorganic form, has been shown to be beneficial as a mineral additive.

      c. A sulfur requirement of 0.1% of ration dry matter is suggested, or supplement sulfur maintaining a 10:1 urea nitrogen:sulfur ratio.

   6. Trace minerals

      a. Trace mineral deficiency is principally a geographical problem where soils may be low in or devoid of certain elements.

      b. Cattle fed high-concentrate rations with little or not roughage may have need for supplemental trace minerals.

B. *Vitamins*

   1. Cattle have physiological requirements for most vitamins required by other mammals. The ability by microorganisms in the rumen to synthesize B vitamins and vitamin K develops rapidly when solid feed is introduced into the diet.

   2. Practical B-vitamin deficiency is limited to situations where an antagonist is present or ruminal synthesis is limited by lack of precursors or other problems.

   3. Vitamin D is synthesized by animals exposed to direct sunlight and is found in large amounts in sun-cured forages. Supplemental levels may be given at approximately 10% of the supplemental vitamin A level.

4. High quality forages contain large amounts of vitamin A precursors and vitamin E.
   a. Rations containing 5 lb or more good quality alfalfa will generally provide enough vitamin A (from carotene).
   b. Actual amount of vitamin A needed as a supplement will vary with level in forages and grain, feeding period and stress conditions.
   c. In general, finishing rations are supplemented with 20,000 to 30,000 IU of vitamin A daily during a heavy grain feeding period.
   d. Conditioner (stress) levels of vitamin A are commonly 30,000 to 50,000+ IU/day. This would include conditioning recently purchased cattle for the first 2 to 3 weeks after purchase or when going on finishing rations. This level is more than required, but it is intended to help build up body reserves in case they are low.

C. *Nonnutritive Feed Additives*
   1. Consider previous coverage of this topic in Chapter 6. Refer also to the *Feed Additive Compendium* published annually by Miller Publishing Co., Minneapolis, Minnesota.
   2. Antibiotics are added at low levels to diets to improve rate of gain and feed efficiency. Several antibiotics reduce the incidence of liver abscesses in cattle fed high-grain diets. Antibiotics are also approved at higher levels for beef cattle for the first 14 to 28 days after transport to help control shipping fever.
   3. Melengestrol acetate (MGA) is a feed additive marketed for feedlot heifers and acts to suppress estrus, increase gains and improve feed efficiency.
   4. Ionophores approved by the FDA for use in finishing rations include Rumensin (monensin) and Bovatec (lasalocid). Feeding ionophores results in reduced feed intake and a tendency to increased rate of gain. The combination of these two factors has a significant effect on improving feed efficiency. An additional benefit from feeding these additives is that each seems to reduce the risk of acidosis and bloat on high concentrate finishing rations. They also have coccidiostatic action, and the incidence of coccidiosis in feedlots using ionophores is now quite low.
   5. Several products generally classified as buffers have been added to high-concentrate finishing diets to control low rumen pH found when high-grain diets are fed. These products include sodium bicarbonate, sodium sesquicarbonate, calcium carbonate, bentonite and magnesium oxide.

D. *Implants for Finishing Cattle*
   Refer to discussion in Chapter 6.
   1. As of this writing, implants cleared by FDA for use in finishing cattle included
      a. Ralgro.
      b. Synovex (S, H, C).
      c. Compudose.
      d. Finaplix (S, H).
      e. Revalor (S).
   2. These implants are active for approximately 60 to 100 days after insertion into the ear. Cattle that will be fed for periods much longer than this should be reimplanted. Compudose is a longer-lasting implant that is active for at least 200 days after insertion.
   3. Implanting steers on finishing diets has increased gains by 8% to 12% and improved feed efficiency by 5% to 8%. Similarly, heifer gains have been increased by 6% to 10% and feed efficiency by 4% to 7%.
   4. Some implants carry withdrawal periods before slaughter.

III. Bloat Prevention

A. *Pasture Bloat*
   1. Usually occurs in ruminants grazing lush legume pasture (alfalfa, red clover, Ladino) or fed green chopped legumes in the feedlot.
   2. Ways to reduce the occurance of pasture bloat
      a. Plant and manage pastures for no more than 50% legume forage.
      b. Fill cattle on dry roughage or grass pasture before turning onto legume pasture.
      c. Do not turn cattle onto pasture wet with dew or rain.

d. Feed an antifoaming agent such as Poloxalene.
   (1) Prevents pasture bloat for 12 hours if consumed in proper amounts.
   (2) Can be fed as a top dressing on feed, in a grain mixture fed free choice, or in a molasses-salt block.

B. *Feedlot Bloat*
   1. Occurs less frequently than pasture bloat and usually in only a few of the cattle in the lot.
   2. Poloxalene generally is not effective.
   3. Feeding ionophores may reduce the incidence of feedlot bloat.
   4. The following management changes may prove effective in reducing the frequency and severity of feedlot bloat.
      a. Coarse chop hay and mix with the grain.
      b. Increase the hay to 15% to 20% of the ration dry matter.
      c. Feed coarsely rolled or whole corn grain.
      d. Minimize separation of ration ingredients.

NUTRIENT REQUIREMENTS OF BEEF CATTLE

The nutrient requirements for beef cattle presented in tables 8–1 through 8–5 were adapted from the *Nutrient Requirements of Beef Cattle,* Publication ISBN 0-309-03447-7, Committee on Animal Nutrition, National Academy of Sciences—National Research Council, Washington, D.C., 1984. These values are expressed on a dry-matter basis.

TABLE 8–1. Tables of Daily Requirements: Energy, Protein, Calcium and Phosphorus

TABLE A Net Energy Requirements of Growing and Finishing Beef Cattle (Mcal/day)[a]

Body Weight, kg:	150	200	250	300	350	400	450	500	550	600
NE$_m$ Required:	3.30	4.10	4.84	5.55	6.24	6.89	7.52	8.14	8.75	9.33
Daily gain, kg	NE$_g$ Required									
Medium-frame steer calves										
0.2	0.41	0.50	0.60	0.69	0.77	0.85	0.93	1.01	1.08	
0.4	0.87	1.08	1.28	1.47	1.65	1.82	1.99	2.16	2.32	
0.6	1.36	1.69	2.00	2.29	2.57	2.84	3.11	3.36	3.61	
0.8	1.87	2.32	2.74	3.14	3.53	3.90	4.26	4.61	4.95	
1.0	2.39	2.96	3.50	4.02	4.51	4.98	5.44	5.89	6.23	
1.2	2.91	3.62	4.28	4.90	5.50	6.69	6.65	7.19	7.73	
Large-frame steers, compensating medium-frame yearling steers and medium-frame bulls										
0.2	0.36	0.45	0.53	0.61	0.68	0.75	0.82	0.89	0.96	1.02
0.4	0.77	0.96	1.13	1.30	1.46	1.61	1.76	1.91	2.05	2.19
0.6	1.21	1.50	1.77	2.03	2.28	2.52	2.75	2.98	3.20	3.41
0.8	1.65	2.06	2.43	2.78	3.12	3.45	3.77	4.08	4.38	4.68
1.0	2.11	2.62	3.10	3.55	3.99	4.41	4.81	5.21	5.60	5.98
1.2	2.58	3.20	3.78	4.34	4.87	5.38	5.88	6.37	6.84	7.30
1.4	3.06	3.79	4.48	5.14	5.77	6.38	6.97	7.54	8.10	8.64
1.6	3.53	4.39	5.19	5.95	6.68	7.38	8.07	8.73	9.38	10.01
Large-frame bull calves and compensating large-frame yearling steers										
0.2	0.32	0.40	0.47	0.54	0.60	0.67	0.73	0.79	0.85	0.91
0.4	0.69	0.85	1.01	1.15	1.29	1.43	1.56	1.69	1.82	1.94
0.6	1.07	1.33	1.57	1.80	2.02	2.23	2.44	2.64	2.83	3.02
0.8	1.47	1.82	2.15	2.47	2.77	3.06	3.34	3.62	3.88	4.15
1.0	1.87	2.32	2.75	3.15	3.54	3.91	4.27	4.62	4.96	5.30
1.2	2.29	2.84	3.36	3.85	4.32	4.77	5.21	5.64	6.06	6.47
1.4	2.71	3.36	3.97	4.56	5.11	5.65	6.18	6.68	7.18	7.66
1.6	3.14	3.89	4.60	5.28	5.92	6.55	7.15	7.74	8.31	8.87
1.8	3.56	4.43	5.23	6.00	6.74	7.45	8.13	8.80	9.46	10.10
Medium-frame heifer calves										
0.2	0.49	0.60	0.71	0.82	0.92	1.01	1.11	1.20	1.29	
0.4	1.05	1.31	1.55	1.77	1.99	2.20	2.40	2.60	2.79	
0.6	1.66	2.06	2.44	2.79	3.13	3.46	3.78	4.10	4.40	
0.8	2.29	2.84	3.36	3.85	4.32	4.78	5.22	5.65	6.07	
1.0	2.94	3.65	4.31	4.94	5.55	6.14	6.70	7.25	7.79	
Large-frame heifer calves and compensating medium-frame yearling heifers										
0.2	0.43	0.53	0.63	0.72	0.81	0.90	0.98	1.06	1.14	1.21
0.4	0.93	1.16	1.37	1.57	1.76	1.95	2.13	2.31	2.47	2.64
0.6	1.47	1.83	2.16	2.47	2.78	3.07	3.35	3.63	3.90	4.16
0.8	2.03	2.62	2.98	3.41	3.83	4.24	4.63	5.01	5.38	5.74
1.0	2.61	3.23	3.82	4.38	4.92	5.44	5.94	6.43	6.91	7.37
1.2	3.19	3.97	4.69	5.37	5.03	6.67	7.28	7.88	8.47	9.03

[a]Shrunk liveweight basis

TABLE B   Protein Requirements of Growing and Finishing Cattle (g/day)[a]

Body weight, kg:	150	200	250	300	350	400	450	500	550	600
Medium-frame steer calves										
Daily gain, kg										
0.2	343	399	450	499	545	590	633	675	715	
0.4	428	482	532	580	625	668	710	751	790	
0.6	503	554	601	646	688	728	767	805	842	
0.8	575	621	664	704	743	780	815	849	883	
1.0	642	682	720	755	789	821	852	882	911	
1.2	702	735	766	794	822	848	873	897	921	
Large-frame steer calves and compensating medium-frame yearling steers										
0.2	361	421	476	529	579	627	673	719	762	805
0.4	441	499	552	603	651	697	742	785	827	867
0.6	522	576	628	676	722	766	809	850	890	930
0.8	598	650	698	743	786	828	867	906	944	980
1.0	671	718	762	804	843	881	918	953	988	1021
1.2	740	782	822	859	895	929	961	993	1023	1053
1.4	806	842	877	908	938	967	995	1022	1048	1073
1.6	863	892	919	943	967	989	1011	1031	1052	1071
Medium-frame bulls										
0.2	345	401	454	503	550	595	638	680	721	761
0.4	430	485	536	584	629	673	716	757	797	835
0.6	509	561	609	655	698	740	780	819	856	893
0.8	583	632	677	719	759	798	835	871	906	940
1.0	655	698	739	777	813	849	881	914	945	976
1.2	722	760	795	828	860	890	919	947	974	1001
1.4	782	813	841	868	893	917	941	963	985	1006
Large-frame bull calves and compensating large-frame yearling steers										
0.2	355	414	468	519	568	615	661	705	747	789
0.4	438	494	547	597	644	689	733	776	817	857
0.6	519	574	624	672	718	761	803	844	884	923
0.8	597	649	697	741	795	826	866	905	942	979
1.0	673	721	765	807	847	885	922	958	994	1027
1.2	745	789	830	868	904	939	973	1005	1037	1067
1.4	815	854	890	924	956	986	1016	1045	1072	1099
1.6	880	912	943	971	998	1024	1048	1072	1095	1117
1.8	922	942	962	980	997	1013	1028	1043	1057	1071
Medium-frame heifer calves										
0.2	323	374	421	465	508	549	588	626	662	
0.4	409	459	505	549	591	630	669	706	742	
0.6	477	522	563	602	638	674	708	741	773	
0.8	537	574	608	640	670	700	728	755	781	
1.0	562	583	603	621	638	654	670	685	700	
Large-frame heifer calves and compensating medium-frame yearling heifers										
0.2	342	397	449	497	543	588	631	672	712	751
0.4	426	480	530	577	622	665	707	747	787	825
0.6	500	549	596	639	681	721	759	796	832	867
0.8	568	613	654	693	730	765	799	833	865	896
1.0	630	668	703	735	767	797	826	854	881	907
1.2	680	708	734	758	781	803	824	844	864	883

[a]Shrunk liveweight basis

TABLE 8–1 (continued). Tables of Daily Requirements: Energy, Protein, Calcium and Phosphorus

TABLE C Calcium and Phosphorus Requirements of Growing and Finishing Cattle (g/day)[a]

Body Weight, kg	Mineral	150	200	250	300	350	400	450	500	550	600
Medium-frame steer calves											
Daily gain, kg											
0.2	Ca	11	12	13	14	15	16	17	19	20	
	P	7	9	10	12	13	15	16	18	19	
0.4	Ca	16	17	17	18	19	19	20	21	22	
	P	9	10	12	13	14	16	17	18	20	
0.6	Ca	21	21	21	22	22	22	22	23	23	
	P	11	12	13	14	15	17	18	19	20	
0.8	Ca	27	26	25	25	25	25	24	24	24	
	P	12	13	14	15	16	17	19	20	21	
1.0	Ca	32	31	29	29	28	27	26	26	25	
	P	14	15	16	16	17	18	19	20	21	
1.2	Ca	37	35	33	32	31	29	28	27	26	
	P	16	16	17	17	18	19	20	21	21	
1.4	Ca	42	39	37	35	33	32	30	29	27	
	P	17	18	18	19	19	20	20	21	22	
Large-frame steer calves, compensating medium-frame yearling steers, and medium-frame bulls											
0.2	Ca	11	12	13	14	16	17	18	19	20	22
	P	7	9	10	12	13	15	16	18	20	21
0.4	Ca	17	17	18	19	19	20	21	22	23	24
	P	9	10	12	13	15	16	17	19	20	22
0.6	Ca	22	22	23	23	23	24	24	24	25	25
	P	11	12	13	15	16	17	18	20	21	22
0.8	Ca	28	27	27	27	27	27	27	27	27	27
	P	13	14	15	16	17	18	19	20	22	23
1.0	Ca	33	32	31	31	30	30	29	29	29	28
	P	14	15	16	17	18	19	20	21	22	23
1.2	Ca	38	37	36	35	34	33	32	31	30	30
	P	16	17	18	18	19	20	21	22	23	24
1.4	Ca	44	42	40	38	37	36	34	33	32	31
	P	18	18	19	20	20	21	22	22	23	24
1.6	Ca	49	47	44	42	40	38	37	35	34	32
	P	20	20	20	21	21	22	22	23	24	24
Large-frame bull calves and compensating large-frame yearling steers											
0.2	Ca	11	12	13	15	16	17	18	20	21	22
	P	7	9	10	12	13	15	17	18	20	21
0.4	Ca	17	18	19	19	20	21	22	23	24	25
	P	9	11	12	13	15	16	18	19	21	22
0.6	Ca	23	23	23	24	24	25	25	26	27	27
	P	11	12	14	15	16	18	19	20	22	23
0.8	Ca	28	28	28	28	28	29	29	29	29	30
	P	13	14	15	16	18	19	20	21	22	24
1.0	Ca	34	34	33	33	32	32	32	32	32	32
	P	15	16	17	18	19	20	21	22	23	24
1.2	Ca	40	39	38	37	36	36	35	35	34	34
	P	17	17	18	19	20	21	22	23	24	25
1.4	Ca	45	44	42	41	40	39	38	37	36	36
	P	18	19	20	20	21	22	23	24	25	26
1.6	Ca	51	49	47	45	44	42	41	40	39	38
	P	20	21	21	22	23	23	24	25	25	26
1.8	Ca	56	54	51	49	47	45	44	42	41	39
	P	22	22	22	23	23	24	25	25	26	26

TABLE 8–1 (continued). Tables of Daily Requirements: Energy, Protein, Calcium and Phosphorus

TABLE C   Calcium and Phosphorus Requirements of Growing and Finishing Cattle (g/day)[a]—*Continued*

Body Weight, kg	Mineral	150	200	250	300	350	400	450	500	550	600
Medium-frame heifer calves											
0.2	Ca	10	11	12	13	14	16	17	18	19	
	P	7	9	10	11	13	14	16	17	19	
0.4	Ca	15	16	16	16	17	17	18	19	19	
	P	9	10	11	12	14	15	16	18	19	
0.6	Ca	20	20	19	19	19	19	19	19	19	
	P	10	11	12	13	14	16	17	18	19	
0.8	Ca	25	23	23	22	21	20	20	19	19	
	P	12	12	13	14	15	16	17	18	19	
1.0	Ca	29	27	26	24	23	22	20	19	19	
	P	13	14	14	15	16	16	17	18	19	
Large-frame heifer calves and compensating medium-frame yearling heifers											
0.2	Ca	11	12	13	14	15	16	17	18	20	21
	P	7	9	10	12	13	15	16	18	19	21
0.4	Ca	16	16	17	17	18	19	19	20	21	22
	P	9	10	11	13	14	15	17	18	20	21
0.6	Ca	21	21	21	21	21	21	21	21	22	22
	P	10	12	13	14	15	16	17	19	20	21
0.8	Ca	26	25	24	24	23	23	23	22	22	22
	P	12	13	14	15	16	17	18	19	20	21
1.0	Ca	31	29	28	27	26	25	24	23	23	22
	P	14	14	15	16	17	18	18	19	20	21
1.2	Ca	35	33	31	30	28	27	25	24	23	22
	P	15	16	16	17	17	18	19	20	20	21

[a]Shrunk liveweight basis

TABLE 8–2
Mineral Requirements and Maximum Tolerable Levels for Beef Cattle

Mineral	Requirement — Suggested Value	Range[a]	Maximum Tolerable Level[b]
Calcium, %	—	See Tables 3 and 7.	2
Cobalt, ppm	0.10	0.07 to 0.11	5
Copper, ppm	8	4 to 10	115
Iodine, ppm	0.5	0.20 to 2.0	50
Iron, ppm	50	50 to 100	1000
Magnesium, %	0.10	0.05 to 0.25	0.40
Manganese, ppm	40	20 to 50	1000
Molybdenum, ppm	—	—	6
Phosphorus, %	—	See Tables 3 and 7.	1
Potassium, %	0.65	0.5 to 0.7	3
Selenium, ppm	0.20	0.05 to 0.30	2
Sodium, %	0.08	0.06 to 0.10	10[c]
Chlorine, %	—	—	—
Sulfur, %	0.10	0.08 to 0.15	0.40
Zinc, ppm	30	20 to 40	500

[a]The listing of a range in which requirements are likely to be met recognizes that requirements for most minerals are affected by a variety of dietary and animal (body weight, sex, rate of gain) factors. Thus, it may be better to evaluate rations based on a range of mineral requirements and for content of interfering substances than to meet a specific dietary value.

[b]From NRC (1980).

[c]10% sodium chloride.

TABLE 8–3
Maximum Tolerable Levels of Certain Toxic Elements[a]

Element	Maximum Tolerable Level, ppm
Aluminum	1,000
Arsenic	50 (100 for organic forms)
Bromine	200
Cadmium	00.5
Fluorine	20 to 100
Lead	30
Mercury	2
Strontium	2,000

[a]NRC (1980), Table 4, Mineral Requirements and Maximum Tolerable Levels for Beef Cattle.

TABLE 8-4. Nutrient Requirements of Breeding Cattle (metric)

| | | | Energy | | | | | | | | Total Protein | | Calcium | | Phosphorus | | Vitamin A[d] |
| | | | Daily | | | | In Diet DM | | | | | | | | | | |
Weight[a] (kg)	Daily Gain[b] (kg)	Daily DM[c] (kg)	ME (Mcal)	TDN (kg)	NE_m (Mcal)	NE_g (Mcal)	ME (Mcal/kg)	TDN (%)	NE_m (Mcal/kg)	NE_g (Mcal/kg)	Daily (g)	In Diet DM (%)	Daily (g)	In Diet DM (%)	Daily (g)	In Diet DM (%)	Daily (1000's IU)
Pregnant yearling heifers—Last third of pregnancy																	
325	0.4	7.1	14.2	3.9	8.04	NA^e	2.00	55.2	1.15	NA^e	591	8.4	19	0.27	14	0.20	20
325	0.6	7.3	15.7	4.3	8.04	0.77	2.15	59.3	1.29	0.72	649	8.9	23	0.32	15	0.21	20
325	0.8	7.3	17.2	4.8	8.04	1.67	2.35	64.9	1.47	0.88	697	9.5	27	0.37	16	0.22	20
350	0.4	7.5	14.8	4.1	8.38	NA	1.99	55.0	1.14	NA	616	8.3	20	0.27	15	0.21	21
350	0.6	7.7	16.5	4.6	8.38	0.81	2.14	59.1	1.28	0.71	674	8.8	24	0.32	16	0.21	22
350	0.8	7.8	18.1	5.0	8.38	1.76	2.34	64.6	1.46	0.88	720	9.3	27	0.35	17	0.22	22
375	0.4	7.8	15.5	4.3	8.71	NA	1.98	54.7	1.13	NA	641	8.2	21	0.27	15	0.19	22
375	0.6	8.1	17.2	4.8	8.71	0.86	2.13	58.8	1.27	0.70	697	8.6	25	0.31	17	0.21	23
375	0.8	8.2	19.0	5.2	8.71	1.86	2.32	64.1	1.45	0.86	743	9.1	27	0.33	18	0.22	23
400	0.4	8.2	16.1	4.5	9.04	NA	1.97	54.4	1.12	NA	664	8.1	22	0.27	16	0.20	23
400	0.6	8.5	18.0	5.0	9.04	0.90	2.12	58.6	1.26	0.69	721	8.5	25	0.30	18	0.21	24
400	0.8	8.6	19.8	5.5	9.04	1.95	2.31	63.8	1.44	0.85	764	8.9	28	0.33	18	0.20	24
425	0.4	8.6	16.8	4.6	9.36	NA	1.96	54.1	1.11	NA	687	8.0	23	0.27	17	0.20	24
425	0.6	8.9	18.7	5.2	9.36	0.94	2.11	58.3	1.25	0.69	743	8.4	26	0.30	18	0.20	25
425	0.8	9.0	20.7	5.7	9.36	2.04	2.30	63.5	1.43	0.84	786	8.8	28	0.31	19	0.21	25
450	0.4	8.9	17.3	4.8	9.67	NA	1.95	53.9	1.10	NA	710	8.0	23	0.26	18	0.20	25
450	0.6	9.2	19.4	5.4	9.67	0.98	2.10	58.0	1.25	0.68	765	8.3	26	0.29	19	0.21	26
450	0.8	9.4	21.5	5.9	9.67	2.13	2.29	63.3	1.42	0.84	807	8.6	28	0.30	20	0.21	26
Dry pregnant mature cows—Middle third of pregnancy																	
350	0.0	6.8	11.9	3.3	6.23	NA	1.76	48.6	0.92	NA	478	7.1	12	0.16	12	0.18	19
400	0.0	7.5	13.1	3.6	6.89	NA	1.76	48.6	0.92	NA	525	7.0	13	0.17	13	0.17	21
450	0.0	8.2	14.3	4.0	7.52	NA	1.76	48.6	0.92	NA	570	7.0	15	0.17	15	0.18	23
500	0.0	8.8	15.5	4.3	8.14	NA	1.76	48.6	0.92	NA	614	7.0	17	0.19	17	0.19	25
550	0.0	9.5	16.7	4.6	8.75	NA	1.76	48.6	0.92	NA	657	6.9	18	0.19	18	0.19	27
600	0.0	10.1	17.8	4.9	9.33	NA	1.76	48.6	0.92	NA	698	6.9	20	0.20	20	0.20	28
650	0.0	10.7	18.9	5.2	9.91	NA	1.76	48.6	0.92	NA	739	6.9	22	0.21	22	0.21	30
Dry pregnant mature cows—Last third of pregnancy																	
350	0.4	7.4	14.7	4.1	8.38	NA	1.98	54.7	1.13	NA	609	8.2	20	0.27	15	0.20	21
400	0.4	8.2	16.0	4.4	9.04	NA	1.96	54.1	1.11	NA	657	8.0	22	0.27	16	0.20	23
450	0.4	8.9	17.2	4.8	9.67	NA	1.94	53.6	1.10	NA	703	7.9	23	0.26	18	0.21	24
500	0.4	9.5	18.3	5.1	10.29	NA	1.92	53.1	1.08	NA	746	7.8	25	0.26	20	0.21	27
550	0.4	10.2	19.5	5.4	10.90	NA	1.91	52.8	1.07	NA	790	7.8	26	0.25	21	0.21	29
600	0.4	10.8	20.6	5.7	11.48	NA	1.90	52.5	1.06	NA	832	7.7	28	0.26	23	0.21	30
650	0.4	11.5	21.7	6.0	12.06	NA	1.89	52.2	1.05	NA	872	7.6	30	0.26	25	0.22	32
Two-year-old heifers nursing calves—First 3–4 months postpartum—5.0 kg milk/day																	
300	0.2	6.9	16.6	4.6	9.30^f	0.72	2.41	66.6	1.53	0.93	814^g	11.8	26	0.38	17	0.25	27
325	0.2	7.3	17.4	4.8	9.64^f	0.77	2.37	65.5	1.49	0.90	841^g	11.5	27	0.37	18	0.25	28
350	0.2	7.8	18.1	5.0	9.98^f	0.81	2.34	64.6	1.46	0.88	866^g	11.2	27	0.35	19	0.24	30
375	0.2	8.2	18.9	5.2	10.31^f	0.86	2.31	63.8	1.44	0.85	892^g	10.9	28	0.34	19	0.23	32
400	0.2	8.6	19.7	5.4	10.64^f	0.90	2.29	63.3	1.42	0.84	916^g	10.7	28	0.33	20	0.23	34
425	0.2	9.0	20.4	5.6	10.96^f	0.94	2.27	62.7	1.40	0.82	939^g	10.5	29	0.32	21	0.23	35
450	0.2	9.4	21.1	5.8	11.27^f	0.98	2.25	62.2	1.38	0.80	963^g	10.3	29	0.31	22	0.23	37

TABLE 8–4 (continued). Nutrient Requirements of Breeding Cattle (metric)

Weight[a] (kg)	Daily Gain[b] (kg)	Daily DM[c] (kg)	Energy Daily ME (Mcal)	Energy Daily TDN (kg)	Energy Daily NEm (Mcal)	Energy Daily NEg (Mcal)	In Diet DM ME (Mcal/kg)	In Diet DM TDN (%)	In Diet DM NEm (Mcal/kg)	In Diet DM NEg (Mcal/kg)	Total Protein Daily (g)	Total Protein In Diet DM (%)	Calcium Daily (g)	Calcium In Diet DM (%)	Phosphorus Daily (g)	Phosphorus In Diet DM (%)	Vitamin A[d] Daily (1000's IU)
Cows nursing calves—Average milking ability—First 3–4 months postpartum—5.0 kg milk/day																	
350	0.0	7.7	16.6	4.6	9.98[f]	NA	2.15	59.4	1.29	NA	814[g]	10.6	23	0.30	18	0.23	30
400	0.0	8.5	17.9	4.9	10.64[f]	NA	2.11	58.3	1.25	NA	864[g]	10.2	25	0.29	19	0.22	33
450	0.0	9.2	19.1	5.3	11.27[f]	NA	2.08	57.5	1.23	NA	911[g]	9.9	26	0.28	21	0.23	36
500	0.0	9.9	20.3	5.6	11.89[f]	NA	2.05	56.6	1.20	NA	957[g]	9.7	28	0.28	22	0.22	39
550	0.0	10.6	21.5	5.9	12.50[f]	NA	2.03	56.1	1.18	NA	1001[g]	9.5	29	0.27	24	0.23	41
600	0.0	11.2	22.6	6.2	13.08[f]	NA	2.01	55.5	1.16	NA	1044[g]	9.3	31	0.28	26	0.23	44
650	0.0	11.9	23.9	6.6	13.66[f]	NA	2.00	55.3	1.15	NA	1086[g]	9.1	33	0.28	27	0.23	46
Cows nursing calves—Superior milking ability—First 3–4 months postpartum—10.0 kg milk/day																	
350	0.0	6.2	18.5	5.1	13.73[f]	NA	3.00	82.9	2.03	NA	1009[g]	16.4	36	0.58	24	0.39	24
400	0.0	7.6	21.4	5.9	14.39[f]	NA	2.80	77.4	1.86	NA	1099[g]	14.4	37	0.49	25	0.33	30
450	0.0	9.1	23.2	6.4	15.02[f]	NA	2.56	70.7	1.66	NA	1186[g]	13.1	39	0.43	26	0.29	35
500	0.0	10.0	24.6	6.8	15.64[f]	NA	2.45	67.7	1.56	NA	1246[g]	12.4	40	0.40	28	0.28	39
550	0.0	10.9	25.8	7.1	16.25[f]	NA	2.38	65.8	1.50	NA	1299[g]	12.0	42	0.39	30	0.27	42
600	0.0	11.6	27.0	7.5	16.83[f]	NA	2.32	64.1	1.45	NA	1348[g]	11.6	43	0.37	31	0.27	45
650	0.0	12.4	28.2	7.8	17.41[f]	NA	2.28	63.0	1.41	NA	1394[g]	11.3	45	0.36	33	0.26	48
Bulls, maintenance and regaining body condition																	
For growth and development use requirements for bulls in Tables 1, 2, and 3.																	
<650																	
650	0.4	12.3	24.3	6.7	9.91	2.06	1.98	54.8	1.13	0.57	904	7.4	25	0.20	23	0.19	48
650	0.6	12.6	26.7	7.4	9.91	3.21	2.11	58.4	1.25	0.69	957	7.6	27	0.21	24	0.19	49
650	0.8	12.8	28.7	7.9	9.91	4.40	2.24	62.0	1.37	0.79	998	7.8	29	0.23	25	0.20	50
700	0.4	13.0	25.7	7.1	10.48	2.18	1.98	54.8	1.13	0.57	942	7.3	26	0.20	25	0.20	51
700	0.6	13.4	28.2	7.8	10.48	3.40	2.11	58.4	1.25	0.69	994	7.4	29	0.22	26	0.20	52
700	0.8	13.5	30.3	8.4	10.48	4.66	2.24	62.0	1.37	0.79	1032	7.6	30	0.22	26	0.19	53
800	0.0	12.9	22.6	6.3	11.58	NA	1.75	48.4	0.91	NA	882	6.8	27	0.21	27	0.21	50
800	0.2	13.7	25.5	7.1	11.58	1.12	1.86	51.5	1.02	0.47	956	7.0	27	0.20	27	0.20	53
900	0.0	14.1	24.7	6.8	12.65	NA	1.75	48.4	0.91	NA	958	6.8	30	0.21	30	0.21	55
900	0.2	15.0	27.9	7.7	12.65	1.23	1.86	51.5	1.02	0.47	1031	6.9	31	0.21	31	0.21	58
1000	0.0	15.3	26.8	7.4	13.69	NA	1.75	48.4	0.91	NA	1032	6.8	33	0.22	33	0.22	60

[a] Average weight for a feeding period.

[b] Approximately 0.4 ± 0.1 kg of weight gain/day over the last third of pregnancy is accounted for by the products of conception. Daily 2.15 Mcal of NE_m and 55 g of protein are provided for this requirement for a calf with a birth weight of 36 kg.

[c] Dry matter consumption should vary depending on the energy concentration of the diet and environmental conditions. These intakes are based on the energy concentration shown in the table and assuming a thermoneutral environment without snow or mud conditions. If the energy concentrations of the diet to be fed exceed the tabular value, limit feeding may be required.

[d] Vitamin A requirements per kilogram of diet are 2800 IU for pregnant heifers and cows and 3900 IU for lactating cows and breeding bulls.

[e] Not applicable.

[f] Includes .75 Mcal NE_m/kg of milk produced.

[g] Includes 33.5 g protein/kg of milk produced.

TABLE 8–5. Nutrient Requirements for Growing and Finishing Cattle (Nutrient Concentration in Diet Dry Matter, Avoirdupois System)*, †, ‡

Weight (lb)	Daily Gain (lb)	Dry Matter Intake (lb)	Protein Intake (lb)	Protein (%)	ME (Mcal/lb)	NE_m (Mcal/lb)	NE_g (Mcal/lb)	TDN (%)	Ca (%)	P (%)
Medium-frame steer calves										
300	0.5	7.8	0.75	9.6	0.89	0.50	0.25	54.0	0.31	0.20
	1.0	8.4	0.95	11.4	0.96	0.57	0.31	58.5	0.45	0.24
	1.5	8.7	1.14	13.2	1.04	0.64	0.38	63.0	0.58	0.28
	2.0	8.9	1.32	14.8	1.11	0.70	0.44	67.5	0.72	0.32
	2.5	8.9	1.48	16.7	1.21	0.79	0.51	73.5	0.87	0.37
	3.0	8.0	1.60	19.9	1.39	0.95	0.64	85.0	1.13	0.47
400	0.5	9.7	0.87	8.9	0.89	0.50	0.25	54.0	0.27	0.18
	1.0	10.4	1.06	10.3	0.96	0.57	0.31	58.5	0.38	0.21
	1.5	10.8	1.24	11.5	1.04	0.64	0.38	63.0	0.47	0.25
	2.0	11.0	1.41	12.7	1.11	0.70	0.44	67.5	0.56	0.26
	2.5	11.0	1.56	14.2	1.21	0.79	0.51	73.5	0.68	0.30
	3.0	10.0	1.65	16.6	1.39	0.95	0.64	85.0	0.86	0.37
500	0.5	11.5	0.98	8.5	0.89	0.50	0.25	54.0	0.25	0.17
	1.0	12.3	1.16	9.5	0.96	0.57	0.31	58.5	0.32	0.20
	1.5	12.8	1.33	10.5	1.04	0.64	0.38	63.0	0.40	0.22
	2.0	13.1	1.49	11.4	1.11	0.70	0.44	67.5	0.47	0.24
	2.5	13.0	1.63	12.5	1.21	0.79	0.51	73.5	0.56	0.27
	3.0	11.8	1.69	14.4	1.39	0.95	0.64	85.0	0.69	0.32
600	0.5	13.2	1.08	8.2	0.89	0.50	0.25	54.0	0.23	0.18
	1.0	14.1	1.26	9.0	0.96	0.57	0.31	58.5	0.28	0.19
	1.5	14.7	1.42	9.8	1.04	0.64	0.38	63.0	0.35	0.21
	2.0	15.0	1.57	10.5	1.11	0.70	0.44	67.5	0.40	0.22
	2.5	14.9	1.69	11.4	1.21	0.79	0.51	73.5	0.46	0.24
	3.0	13.5	1.73	12.9	1.39	0.95	0.64	85.0	0.57	0.29
700	0.5	14.8	1.18	7.9	0.89	0.50	0.25	54.0	0.22	0.18
	1.0	15.8	1.35	8.6	0.96	0.57	0.31	58.5	0.27	0.18
	1.5	16.5	1.50	9.2	1.04	0.64	0.38	63.0	0.31	0.20
	2.0	16.8	1.65	9.8	1.11	0.70	0.44	67.5	0.34	0.21
	2.5	16.7	1.75	10.5	1.21	0.79	0.51	73.5	0.40	0.22
	3.0	15.2	1.77	11.7	1.39	0.95	0.64	85.0	0.49	0.26
800	0.5	16.4	1.27	7.7	0.89	0.50	0.25	54.0	0.22	0.17
	1.0	17.5	1.44	8.3	0.96	0.57	0.31	58.5	0.24	0.19
	1.5	18.2	1.58	8.8	1.04	0.64	0.38	63.0	0.28	0.19
	2.0	18.6	1.72	9.2	1.11	0.70	0.44	67.5	0.31	0.20
	2.5	18.5	1.81	9.8	1.21	0.79	0.51	73.5	0.35	0.21
	3.0	16.8	1.81	10.8	1.39	0.95	0.64	85.0	0.42	0.25
900	0.5	17.9	1.36	7.6	0.89	0.50	0.25	54.0	0.21	0.18
	1.0	19.1	1.52	8.0	0.96	0.57	0.31	58.5	0.23	0.18
	1.5	19.9	1.66	8.4	1.04	0.64	0.38	63.0	0.25	0.19
	2.0	20.3	1.79	8.8	1.11	0.70	0.44	67.5	0.28	0.20
	2.5	20.2	1.87	9.3	1.21	0.79	0.51	73.5	0.31	0.20
	3.0	18.3	1.85	10.1	1.39	0.95	0.64	85.0	0.37	0.23
1000	0.5	19.3	1.45	7.5	0.89	0.50	0.25	54.0	0.21	0.18
	1.0	20.7	1.60	7.8	0.96	0.57	0.31	58.5	0.21	0.18
	1.5	21.5	1.74	8.1	1.04	0.64	0.38	63.0	0.24	0.18
	2.0	22.0	1.85	8.4	1.11	0.70	0.44	67.5	0.25	0.19
	2.5	21.9	1.92	8.8	1.21	0.79	0.51	73.5	0.27	0.19
	3.0	19.8	1.88	9.5	1.39	0.95	0.64	85.0	0.32	0.22

TABLE 8–5 (continued). Nutrient Requirements for Growing and Finishing Cattle (Nutrient Concentration in Diet Dry Matter, Avoirdupois System)*, †, ‡

Weight (lb)	Daily Gain (lb)	Dry Matter Intake (lb)	Protein Intake (lb)	Protein (%)	ME (Mcal/lb)	NE_m (Mcal/lb)	NE_g (Mcal/lb)	TDN (%)	Ca (%)	P (%)
Large-frame steer calves and compensating medium-frame yearling steers										
300	0.5	8.2	0.77	9.5	0.86	0.48	0.23	52.5	0.30	0.19
	1.0	8.7	0.99	11.3	0.92	0.54	0.28	56.0	0.46	0.23
	1.5	9.1	1.19	12.9	0.98	0.59	0.33	59.5	0.58	0.27
	2.0	9.4	1.37	14.6	1.04	0.64	0.38	63.5	0.70	0.30
	2.5	9.6	1.55	16.3	1.11	0.70	0.44	67.5	0.85	0.34
	3.0	9.6	1.73	18.0	1.18	0.77	0.49	72.0	0.99	0.39
	3.5	9.3	1.88	20.3	1.29	0.86	0.57	78.5	1.16	0.45
400	0.5	10.1	0.89	8.9	0.86	0.48	0.23	52.5	0.26	0.17
	1.0	10.8	1.10	10.2	0.92	0.54	0.28	56.0	0.37	0.20
	1.5	11.3	1.30	11.4	0.98	0.59	0.33	59.5	0.47	0.23
	2.0	11.7	1.47	12.7	1.04	0.64	0.38	63.5	0.57	0.26
	2.5	11.9	1.64	13.9	1.11	0.70	0.44	67.5	0.65	0.30
	3.0	11.9	1.81	15.2	1.18	0.77	0.49	72.0	0.76	0.33
	3.5	11.5	1.94	16.9	1.29	0.86	0.57	78.5	0.90	0.36
500	0.5	12.0	1.0	8.5	0.86	0.48	0.23	52.5	0.24	0.17
	1.0	12.8	1.21	9.5	0.92	0.54	0.28	56.0	0.33	0.19
	1.5	13.4	1.40	10.4	0.98	0.59	0.33	59.5	0.39	0.21
	2.0	13.8	1.57	11.4	1.04	0.64	0.38	63.5	0.46	0.24
	2.5	14.0	1.73	12.4	1.11	0.70	0.44	67.5	0.55	0.25
	3.0	14.0	1.88	13.4	1.18	0.77	0.49	72.0	0.63	0.28
	3.5	13.6	2.00	14.7	1.29	0.86	0.57	78.5	0.73	0.32
600	0.5	13.8	1.11	8.2	0.86	0.48	0.23	52.5	0.22	0.18
	1.0	14.6	1.31	9.0	0.92	0.54	0.28	56.0	0.29	0.18
	1.5	15.3	1.50	9.7	0.98	0.59	0.33	59.5	0.35	0.20
	2.0	15.8	1.66	10.5	1.04	0.64	0.38	63.5	0.40	0.22
	2.5	16.1	1.81	11.3	1.11	0.70	0.44	67.5	0.47	0.23
	3.0	16.1	1.95	12.1	1.18	0.77	0.49	72.0	0.52	0.26
	3.5	15.6	2.05	13.2	1.29	0.86	0.57	78.5	0.61	0.28
700	0.5	15.4	1.21	7.9	0.86	0.48	0.23	52.5	0.21	0.17
	1.0	16.4	1.41	8.6	0.92	0.54	0.28	56.0	0.27	0.19
	1.5	17.2	1.59	9.2	0.98	0.59	0.33	59.5	0.31	0.19
	2.0	17.8	1.74	9.8	1.04	0.64	0.38	63.5	0.36	0.21
	2.5	18.0	1.88	10.5	1.11	0.70	0.44	67.5	0.40	0.22
	3.0	18.0	2.01	11.1	1.18	0.77	0.49	72.0	0.45	0.23
	3.5	17.5	2.10	12.0	1.29	0.86	0.57	78.5	0.52	0.26
800	0.5	17.1	1.31	7.7	0.86	0.48	0.23	52.5	0.21	0.18
	1.0	18.2	1.51	8.3	0.92	0.54	0.28	56.0	0.24	0.18
	1.5	19.0	1.68	8.8	0.98	0.59	0.33	59.5	0.28	0.19
	2.0	19.6	1.82	9.3	1.04	0.64	0.38	63.5	0.32	0.20
	2.5	19.9	1.96	9.8	1.11	0.70	0.44	67.5	0.35	0.21
	3.0	19.9	2.07	10.4	1.18	0.77	0.49	72.0	0.40	0.22
	3.5	19.3	2.15	11.1	1.29	0.86	0.57	78.5	0.45	0.24
900	0.5	18.6	1.40	7.6	0.86	0.48	0.23	52.5	0.20	0.18
	1.0	19.8	1.60	8.0	0.92	0.54	0.28	56.0	0.23	0.18
	1.5	20.8	1.77	8.5	0.98	0.59	0.33	59.5	0.27	0.18
	2.0	21.4	1.90	8.9	1.04	0.64	0.38	63.5	0.29	0.20
	2.5	21.8	2.03	9.3	1.11	0.70	0.44	67.5	0.31	0.20
	3.0	21.7	2.13	9.8	1.18	0.77	0.49	72.0	0.36	0.21
	3.5	21.1	2.19	10.4	1.29	0.86	0.57	78.5	0.40	0.23
1000	0.5	20.2	1.49	7.5	0.86	0.48	0.23	52.5	0.20	0.17
	1.0	21.5	1.69	7.8	0.92	0.54	0.28	56.0	0.23	0.17
	1.5	22.5	1.85	8.2	0.98	0.59	0.33	59.5	0.25	0.18
	2.0	23.2	1.98	8.6	1.04	0.64	0.38	63.5	0.27	0.18
	2.5	23.6	2.09	8.9	1.11	0.70	0.44	67.5	0.29	0.19
	3.0	23.6	2.19	9.3	1.18	0.77	0.49	72.0	0.32	0.20
	3.5	22.8	2.24	9.8	1.29	0.86	0.57	78.5	0.35	0.21

TABLE 8–5 (continued). Nutrient Requirements for Growing and Finishing Cattle
(Nutrient Concentration in Diet Dry Matter, Avoirdupois System)*, †, ‡

Weight (lb)	Daily Gain (lb)	Dry Matter Intake (lb)	Protein Intake (lb)	Protein (%)	ME (Mcal/lb)	NE$_m$ (Mcal/lb)	NE$_g$ (Mcal/lb)	TDN (%)	Ca (%)	P (%)
1100	0.5	21.7	1.58	7.4	0.86	0.48	0.23	52.5	0.19	0.18
	1.0	23.1	1.77	7.7	0.92	0.54	0.28	56.0	0.21	0.18
	1.5	24.1	1.93	8.0	0.98	0.59	0.33	59.5	0.23	0.18
	2.0	24.9	2.05	8.3	1.04	0.64	0.38	63.5	0.25	0.18
	2.5	25.3	2.16	8.5	1.11	0.70	0.44	67.5	0.26	0.18
	3.0	25.3	2.25	8.9	1.18	0.77	0.49	72.0	0.29	0.19
	3.5	24.5	2.28	9.3	1.29	0.86	0.57	78.5	0.32	0.21
Medium-frame bulls										
300	0.5	7.8	0.76	9.7	0.88	0.49	0.24	53.5	0.31	0.20
	1.0	8.3	0.96	11.6	0.94	0.56	0.30	57.5	0.48	0.24
	1.5	8.6	1.15	13.4	1.01	0.62	0.35	61.5	0.62	0.28
	2.0	8.8	1.34	15.2	1.08	0.68	0.41	65.5	0.75	0.33
	2.5	8.9	1.52	17.0	1.15	0.74	0.47	70.0	0.92	0.37
	3.0	8.7	1.68	19.3	1.26	0.84	0.54	76.5	1.09	0.43
400	0.5	9.6	0.87	9.0	0.88	0.49	0.24	53.5	0.28	0.18
	1.0	10.3	1.07	10.4	0.94	0.56	0.30	57.5	0.39	0.21
	1.5	10.7	1.26	11.8	1.01	0.62	0.35	61.5	0.49	0.25
	2.0	11.0	1.44	13.1	1.08	0.68	0.41	65.5	0.60	0.28
	2.5	11.1	1.60	14.4	1.15	0.74	0.47	70.0	0.70	0.32
	3.0	10.8	1.74	16.1	1.26	0.84	0.54	76.5	0.84	0.37
500	0.5	11.4	0.98	8.6	0.88	0.49	0.24	53.5	0.25	0.17
	1.0	12.1	1.17	9.7	0.94	0.56	0.30	57.5	0.35	0.20
	1.5	12.7	1.35	10.7	1.01	0.62	0.35	61.5	0.42	0.23
	2.0	13.0	1.52	11.7	1.08	0.68	0.41	65.5	0.49	0.25
	2.5	13.1	1.68	12.8	1.15	0.74	0.47	70.0	0.59	0.27
	3.0	12.8	1.81	14.1	1.26	0.84	0.54	76.5	0.69	0.31
600	0.5	13.1	1.08	8.3	0.88	0.49	0.24	53.5	0.24	0.19
	1.0	13.9	1.27	9.2	0.94	0.56	0.30	57.5	0.30	0.19
	1.5	14.5	1.44	10.0	1.01	0.62	0.35	61.5	0.36	0.21
	2.0	14.9	1.61	10.8	1.08	0.68	0.41	65.5	0.43	0.24
	2.5	15.0	1.75	11.6	1.15	0.74	0.47	70.0	0.50	0.25
	3.0	14.7	1.86	12.7	1.26	0.84	0.54	76.5	0.57	0.29
700	0.5	14.7	1.18	8.0	0.88	0.49	0.24	53.5	0.23	0.18
	1.0	15.6	1.37	8.8	0.94	0.56	0.30	57.5	0.28	0.20
	1.5	16.3	1.53	9.4	1.01	0.62	0.35	61.5	0.32	0.20
	2.0	16.7	1.69	10.1	1.08	0.68	0.41	65.5	0.38	0.22
	2.5	16.8	1.82	10.8	1.15	0.74	0.47	70.0	0.43	0.24
	3.0	16.5	1.92	11.7	1.26	0.84	0.54	76.5	0.49	0.25
800	0.5	16.2	1.27	7.8	0.88	0.49	0.24	53.5	0.22	0.19
	1.0	17.3	1.45	8.4	0.94	0.56	0.30	57.5	0.25	0.19
	1.5	18.0	1.61	9.0	1.01	0.62	0.35	61.5	0.29	0.20
	2.0	18.5	1.76	9.5	1.08	0.68	0.41	65.5	0.33	0.21
	2.5	18.6	1.89	10.1	1.15	0.74	0.47	70.0	0.38	0.23
	3.0	18.2	1.97	10.8	1.26	0.84	0.54	76.5	0.44	0.24
900	0.5	17.7	1.36	7.7	0.88	0.49	0.24	53.5	0.21	0.19
	1.0	18.9	1.54	8.2	0.94	0.56	0.30	57.5	0.25	0.19
	1.5	19.7	1.69	8.6	1.01	0.62	0.35	61.5	0.28	0.19
	2.0	20.2	1.83	9.1	1.08	0.68	0.41	65.5	0.31	0.21
	2.5	20.3	1.95	9.6	1.15	0.74	0.47	70.0	0.34	0.22
	3.0	19.9	2.02	10.2	1.26	0.84	0.54	76.5	0.39	0.23

TABLE 8–5 (continued). Nutrient Requirements for Growing and Finishing Cattle (Nutrient Concentration in Diet Dry Matter, Avoirdupois System)*, †, ‡

Weight (lb)	Daily Gain (lb)	Dry Matter Intake (lb)	Protein Intake (lb)	Protein (%)	ME (Mcal/lb)	NE$_m$ (Mcal/lb)	NE$_g$ (Mcal/lb)	TDN (%)	Ca (%)	P (%)
1000	0.5	19.2	1.45	7.5	0.88	0.49	0.24	53.5	0.21	0.18
	1.0	20.4	1.62	8.0	0.94	0.56	0.30	57.5	0.24	0.18
	1.5	21.3	1.77	8.4	1.01	0.62	0.35	61.5	0.26	0.19
	2.0	21.8	1.90	8.7	1.08	0.68	0.41	65.5	0.28	0.19
	2.5	22.0	2.01	9.1	1.15	0.74	0.47	70.0	0.31	0.20
	3.0	21.5	2.07	9.6	1.26	0.84	0.54	76.5	0.35	0.22
1100	0.5	20.6	1.54	7.4	0.88	0.49	0.24	53.5	0.20	0.19
	1.0	21.9	1.70	7.8	0.94	0.56	0.30	57.5	0.22	0.19
	1.5	22.9	1.85	8.1	1.01	0.62	0.35	61.5	0.24	0.19
	2.0	23.4	1.97	8.4	1.08	0.68	0.41	65.5	0.26	0.19
	2.5	23.6	2.07	8.7	1.15	0.74	0.47	70.0	0.28	0.20
	3.0	23.1	2.11	9.2	1.26	0.84	0.54	76.5	0.32	0.21

Large-frame bull calves and compensating large-frame yearling steers

Weight (lb)	Daily Gain (lb)	Dry Matter Intake (lb)	Protein Intake (lb)	Protein (%)	ME (Mcal/lb)	NE$_m$ (Mcal/lb)	NE$_g$ (Mcal/lb)	TDN (%)	Ca (%)	P (%)
300	0.5	7.9	0.77	9.7	0.86	0.48	0.23	52.5	0.31	0.20
	1.0	8.4	0.98	11.7	0.92	0.54	0.28	56.0	0.47	0.24
	1.5	8.8	1.18	13.5	0.98	0.59	0.33	59.5	0.63	0.28
	2.0	9.0	1.38	15.1	1.03	0.63	0.37	62.5	0.76	0.32
	2.5	9.2	1.56	17.0	1.09	0.69	0.42	66.5	0.91	0.36
	3.0	9.2	1.74	18.8	1.16	0.75	0.47	70.5	1.08	0.43
	3.5	9.1	1.91	20.9	1.24	0.82	0.53	75.5	1.24	0.48
	4.0	8.2	2.04	24.7	1.41	0.96	0.66	86.0	1.53	0.59
400	0.5	9.8	0.89	9.0	0.86	0.48	0.23	52.5	0.27	0.18
	1.0	10.4	1.09	10.5	0.92	0.54	0.28	56.0	0.40	0.21
	1.5	10.9	1.29	11.9	0.98	0.59	0.33	59.5	0.51	0.24
	2.0	11.2	1.48	13.1	1.03	0.63	0.37	62.5	0.61	0.28
	2.5	11.4	1.65	14.5	1.09	0.69	0.42	66.5	0.72	0.31
	3.0	11.5	1.82	15.9	1.16	0.75	0.47	70.5	0.82	0.35
	3.5	11.3	1.98	17.5	1.24	0.82	0.53	75.5	0.96	0.39
	4.0	10.2	2.08	20.3	1.41	0.96	0.66	86.0	1.19	0.48
500	0.5	11.6	1.00	8.6	0.86	0.48	0.23	52.5	0.25	0.19
	1.0	12.3	1.20	9.8	0.92	0.54	0.28	56.0	0.36	0.21
	1.5	12.9	1.39	10.9	0.98	0.59	0.33	59.5	0.43	0.22
	2.0	13.2	1.58	11.8	1.03	0.63	0.37	62.5	0.52	0.25
	2.5	13.5	1.74	12.9	1.09	0.69	0.42	66.5	0.59	0.28
	3.0	13.6	1.90	14.0	1.16	0.75	0.47	70.5	0.68	0.31
	3.5	13.4	2.05	15.3	1.24	0.82	0.53	75.5	0.77	0.35
	4.0	12.0	2.13	17.5	1.41	0.96	0.66	86.0	0.97	0.40
600	0.5	13.3	1.10	8.3	0.86	0.48	0.23	52.5	0.23	0.18
	1.0	14.1	1.30	9.2	0.92	0.54	0.28	56.0	0.31	0.20
	1.5	14.8	1.48	10.1	0.98	0.59	0.33	59.5	0.37	0.21
	2.0	15.2	1.67	10.9	1.03	0.63	0.37	62.5	0.44	0.23
	2.5	15.5	1.82	11.8	1.09	0.69	0.42	66.5	0.51	0.26
	3.0	15.5	1.97	12.7	1.16	0.75	0.47	70.5	0.58	0.27
	3.5	15.3	2.11	13.7	1.24	0.82	0.53	75.5	0.66	0.30
	4.0	13.8	2.16	15.6	1.41	0.96	0.66	86.0	0.81	0.37
700	0.5	14.9	1.20	8.0	0.86	0.48	0.23	52.5	0.22	0.18
	1.0	15.9	1.40	8.8	0.92	0.54	0.28	56.0	0.29	0.19
	1.5	16.6	1.57	9.6	0.98	0.59	0.33	59.5	0.35	0.21
	2.0	17.0	1.75	10.2	1.03	0.63	0.37	62.5	0.39	0.22
	2.5	17.4	1.90	11.0	1.09	0.69	0.42	66.5	0.44	0.24
	3.0	17.5	2.04	11.7	1.16	0.75	0.47	70.5	0.50	0.25
	3.5	17.2	2.16	12.5	1.24	0.82	0.53	75.5	0.56	0.28
	4.0	15.5	2.20	14.1	1.41	0.96	0.66	86.0	0.70	0.33

Weight (lb)	Daily Gain (lb)	Dry Matter Intake (lb)	Protein Intake (lb)	Protein (%)	ME (Mcal/lb)	NE$_m$ (Mcal/lb)	NE$_g$ (Mcal/lb)	TDN (%)	Ca (%)	P (%)
800	0.5	16.5	1.30	7.9	0.86	0.48	0.23	52.5	0.21	0.19
	1.0	17.5	1.49	8.5	0.92	0.54	0.28	56.0	0.26	0.19
	1.5	18.3	1.66	9.1	0.98	0.59	0.33	59.5	0.31	0.20
	2.0	18.8	1.84	9.7	1.03	0.63	0.37	62.5	0.35	0.21
	2.5	19.2	1.97	10.3	1.09	0.69	0.42	66.5	0.40	0.23
	3.0	19.3	2.11	10.9	1.16	0.75	0.47	70.5	0.45	0.24
	3.5	19.0	2.22	11.6	1.24	0.82	0.53	75.5	0.50	0.26
	4.0	17.1	2.24	13.0	1.41	0.96	0.66	86.0	0.61	0.31
900	0.5	18.0	1.39	7.7	0.86	0.48	0.23	52.5	0.22	0.18
	1.0	19.2	1.58	8.3	0.92	0.54	0.28	56.0	0.25	0.18
	1.5	20.0	1.74	8.8	0.98	0.59	0.33	59.5	0.29	0.20
	2.0	20.6	1.92	9.2	1.03	0.63	0.37	62.5	0.32	0.20
	2.5	21.0	2.04	9.8	1.09	0.69	0.42	66.5	0.36	0.21
	3.0	21.1	2.17	10.3	1.16	0.75	0.47	70.5	0.40	0.23
	3.5	20.8	2.27	10.9	1.24	0.82	0.53	75.5	0.45	0.24
	4.0	18.7	2.27	12.1	1.41	0.96	0.66	86.0	0.53	0.28
1000	0.5	19.5	1.48	7.6	0.86	0.48	0.23	52.5	0.21	0.18
	1.0	20.7	1.66	8.1	0.92	0.54	0.28	56.0	0.25	0.19
	1.5	21.7	1.83	8.5	0.98	0.59	0.33	59.5	0.27	0.19
	2.0	22.3	1.99	8.9	1.03	0.63	0.37	62.5	0.30	0.20
	2.5	22.7	2.11	9.3	1.09	0.69	0.42	66.5	0.33	0.20
	3.0	22.8	2.23	9.7	1.16	0.75	0.47	70.5	0.36	0.21
	3.5	22.5	2.32	10.3	1.24	0.82	0.53	75.5	0.40	0.24
	4.0	20.2	2.30	11.3	1.41	0.96	0.66	86.0	0.48	0.27
1100	0.5	20.9	1.57	7.5	0.86	0.48	0.23	52.5	0.21	0.19
	1.0	22.3	1.75	7.9	0.92	0.54	0.28	56.0	0.23	0.19
	1.5	23.3	1.91	8.3	0.98	0.59	0.33	59.5	0.26	0.19
	2.0	23.9	2.07	8.6	1.03	0.63	0.37	62.5	0.28	0.19
	2.5	24.2	2.18	9.0	1.09	0.69	0.42	66.5	0.30	0.20
	3.0	24.5	2.29	9.3	1.16	0.75	0.47	70.5	0.32	0.21
	3.5	24.1	2.37	9.8	1.24	0.82	0.53	75.5	0.36	0.22
	4.0	21.7	2.33	10.7	1.41	0.96	0.66	86.0	0.43	0.25
Medium-frame heifer calves										
300	0.5	7.5	0.73	9.6	0.92	0.54	0.28	56.0	0.29	0.21
	1.0	8.0	0.91	11.4	1.02	0.63	0.36	62.0	0.44	0.22
	1.5	8.2	1.08	13.1	1.13	0.72	0.44	68.5	0.59	0.27
	2.0	8.0	1.22	15.1	1.26	0.84	0.55	77.0	0.74	0.33
400	0.5	9.3	0.84	8.9	0.92	0.54	0.28	56.0	0.26	0.19
	1.0	9.9	1.01	10.2	1.02	0.63	0.36	62.0	0.36	0.20
	1.5	10.2	1.17	11.4	1.13	0.72	0.44	68.5	0.45	0.24
	2.0	10.0	1.29	12.9	1.26	0.84	0.55	77.0	0.57	0.29
500	0.5	11.0	0.94	8.5	0.92	0.54	0.28	56.0	0.24	0.18
	1.0	11.8	1.11	9.4	1.02	0.63	0.36	62.0	0.30	0.21
	1.5	12.1	1.25	10.3	1.13	0.72	0.44	68.5	0.38	0.22
	2.0	11.8	1.35	11.4	1.26	0.84	0.55	77.0	0.45	0.24
600	0.5	12.6	1.04	8.1	0.92	0.54	0.28	56.0	0.23	0.18
	1.0	13.5	1.19	8.8	1.02	0.63	0.36	62.0	0.28	0.20
	1.5	13.8	1.32	9.5	1.13	0.72	0.44	68.5	0.32	0.21
	2.0	13.5	1.41	10.4	1.26	0.84	0.55	77.0	0.38	0.23
700	0.5	14.1	1.13	7.9	0.92	0.54	0.28	56.0	0.22	0.19
	1.0	15.1	1.28	8.4	1.02	0.63	0.36	62.0	0.25	0.19
	1.5	15.5	1.39	9.0	1.13	0.72	0.44	68.5	0.28	0.20
	2.0	15.2	1.46	9.6	1.26	0.84	0.55	77.0	0.32	0.22

TABLE 8–5 (continued). Nutrient Requirements for Growing and Finishing Cattle (Nutrient Concentration in Diet Dry Matter, Avoirdupois System)[*, †, ‡]

Weight (lb)	Daily Gain (lb)	Dry Matter Intake (lb)	Protein Intake (lb)	Protein (%)	ME (Mcal/lb)	NE_m (Mcal/lb)	NE_g (Mcal/lb)	TDN (%)	Ca (%)	P (%)
800	0.5	15.6	1.22	7.7	0.92	0.54	0.28	56.0	0.21	0.18
	1.0	16.7	1.36	8.1	1.02	0.63	0.36	62.0	0.22	0.18
	1.5	17.2	1.46	8.5	1.13	0.72	0.44	68.5	0.24	0.19
	2.0	16.8	1.51	9.0	1.26	0.84	0.55	77.0	0.28	0.20
900	0.5	17.1	1.31	7.5	0.92	0.54	0.28	56.0	0.21	0.18
	1.0	18.3	1.44	7.8	1.02	0.63	0.36	62.0	0.22	0.18
	1.5	18.8	1.53	8.1	1.13	0.72	0.44	68.5	0.22	0.19
	2.0	18.3	1.56	8.5	1.26	0.84	0.55	77.0	0.25	0.19
1000	0.5	18.5	1.39	7.4	0.92	0.54	0.28	56.0	0.20	0.19
	1.0	19.8	1.51	7.6	1.02	0.63	0.36	62.0	0.20	0.18
	1.5	20.3	1.59	7.8	1.13	0.72	0.44	68.5	0.21	0.18
	2.0	19.8	1.61	8.1	1.26	0.84	0.55	77.0	0.22	0.19

Large-frame heifer calves and compensating medium-frame yearling heifers

Weight (lb)	Daily Gain (lb)	Dry Matter Intake (lb)	Protein Intake (lb)	Protein (%)	ME (Mcal/lb)	NE_m (Mcal/lb)	NE_g (Mcal/lb)	TDN (%)	Ca (%)	P (%)
300	0.5	7.8	0.76	9.5	0.89	0.50	0.25	54.0	0.31	0.20
	1.0	8.4	0.95	11.3	0.98	0.58	0.32	59.0	0.45	0.24
	1.5	8.8	1.13	13.0	1.05	0.65	0.39	64.0	0.58	0.25
	2.0	8.9	1.30	14.6	1.14	0.74	0.46	69.5	0.69	0.30
	2.5	8.7	1.45	16.7	1.26	0.84	0.55	77.0	0.86	0.35
400	0.5	9.7	0.87	8.9	0.89	0.50	0.25	54.0	0.27	0.18
	1.0	10.5	1.06	10.1	0.98	0.58	0.32	59.0	0.36	0.21
	1.5	10.9	1.23	11.3	1.05	0.65	0.39	64.0	0.45	0.22
	2.0	11.1	1.38	12.6	1.14	0.74	0.46	69.5	0.54	0.26
	2.5	10.8	1.51	14.1	1.26	0.84	0.55	77.0	0.65	0.31
500	0.5	11.5	0.98	8.4	0.89	0.50	0.25	54.0	0.23	0.17
	1.0	12.4	1.16	9.4	0.98	0.58	0.32	59.0	0.30	0.20
	1.5	12.9	1.32	10.3	1.05	0.65	0.39	64.0	0.38	0.20
	2.0	13.1	1.46	11.2	1.14	0.74	0.46	69.5	0.44	0.24
	2.5	12.8	1.57	12.4	1.26	0.84	0.55	77.0	0.53	0.26
600	0.5	13.2	1.08	8.1	0.89	0.50	0.25	54.0	0.22	0.18
	1.0	14.1	1.25	8.9	0.98	0.58	0.32	59.0	0.28	0.19
	1.5	14.8	1.41	9.6	1.05	0.65	0.39	64.0	0.33	0.19
	2.0	15.0	1.54	10.3	1.14	0.74	0.46	69.5	0.38	0.22
	2.5	14.6	1.63	11.2	1.26	0.84	0.55	77.0	0.44	0.24
700	0.5	14.8	1.18	7.9	0.89	0.50	0.25	54.0	0.21	0.18
	1.0	15.9	1.34	8.5	0.98	0.58	0.32	59.0	0.25	0.18
	1.5	16.6	1.49	9.0	1.05	0.65	0.39	64.0	0.29	0.19
	2.0	16.8	1.61	9.6	1.14	0.74	0.46	69.5	0.33	0.20
	2.5	16.4	1.68	10.3	1.26	0.84	0.55	77.0	0.38	0.22
800	0.5	16.4	1.27	7.7	0.89	0.50	0.25	54.0	0.20	0.17
	1.0	17.6	1.43	8.2	0.98	0.58	0.32	59.0	0.24	0.18
	1.5	18.3	1.57	8.6	1.05	0.65	0.39	64.0	0.25	0.18
	2.0	18.6	1.67	9.0	1.14	0.74	0.46	69.5	0.28	0.19
	2.5	18.1	1.74	9.6	1.26	0.84	0.55	77.0	0.33	0.21
900	0.5	17.8	1.36	7.5	0.89	0.50	0.25	54.0	0.20	0.18
	1.0	19.2	1.52	7.9	0.98	0.58	0.32	59.0	0.22	0.18
	1.5	20.0	1.64	8.2	1.05	0.65	0.39	64.0	0.23	0.18
	2.0	20.3	1.74	8.6	1.14	0.74	0.46	69.5	0.26	0.18
	2.5	19.8	1.78	9.0	1.26	0.84	0.55	77.0	0.29	0.20

TABLE 8–5 (continued). Nutrient Requirements for Growing and Finishing Cattle (Nutrient Concentration in Diet Dry Matter, Avoirdupois System)[*, †, ‡]

Weight (lb)	Daily Gain (lb)	Dry Matter Intake (lb)	Protein Intake (lb)	Protein (%)	ME (Mcal/lb)	NE$_m$ (Mcal/lb)	NE$_g$ (Mcal/lb)	TDN (%)	Ca (%)	P (%)
1000	0.5	19.3	1.45	7.4	0.89	0.50	0.25	54.0	0.19	0.18
	1.0	20.8	1.60	7.7	0.98	0.58	0.32	59.0	0.21	0.18
	1.5	21.7	1.71	8.0	1.05	0.65	0.39	64.0	0.21	0.18
	2.0	22.0	1.80	8.2	1.14	0.74	0.46	69.5	0.23	0.18
	2.5	21.5	1.83	8.6	1.26	0.84	0.55	77.0	0.25	0.18
1100	0.5	20.8	1.54	7.3	0.89	0.50	0.25	54.0	0.19	0.18
	1.0	22.3	1.68	7.5	0.98	0.58	0.32	59.0	0.20	0.18
	1.5	23.3	1.78	7.7	1.05	0.65	0.39	64.0	0.20	0.18
	2.0	23.6	1.86	7.9	1.14	0.74	0.46	69.5	0.21	0.18
	2.5	23.1	1.88	8.2	1.26	0.84	0.55	77.0	0.22	0.18

*Shrunk liveweight basis.

†Vitamin A requirements are 1000 IU per pound of diet.

‡This table gives reasonable examples of nutrient concentrations that should be suitable to formulate diets for specific management goals. It does not imply that diets with other nutrient concentrations when consumed in sufficient amounts would be inadequate to meet nutrient requirements.

TABLE 8–6. Rations for Wintering Drylot Beef Cows—Nonlactating[1]

Ingredient	IFN	1	2	3	4	5	6
Corn stover[2]	1–02–776	61	—	—	—	80	—
Corncobs	1–02–782	—	42	—	—	—	34
Legume—grass hay[3]	1–08–331	39	36	24	40	—	—
Corn silage	3–02–823	—	—	76	—	—	60
Sorghum stover	1–07–961	—	—	—	60	—	—
Shelled corn	4–02–931	—	22	—	—	15	—
Protein supplement[4]		—	—	—	—	5	6
		100	100	100	100	100	100

Calculated analysis:[5]

	1	2	3	4	5	6
1) As-fed basis:						
Crude protein, %	8.9	8.3	5.6	8.1	7.3	4.6
TDN, %	51.5	54.2	30.6	46.5	55.8	32.8
NE$_m$, Mcal/kg	1.06	1.14	0.66	0.94	1.25	0.71
NE$_g$, Mcal/kg	0.56	0.62	0.38	0.32	0.75	0.39
Calcium, %	0.82	0.53	0.39	0.81	0.52	0.24
Phosphorus, %	0.13	0.16	0.11	0.15	0.20	0.16
Dry matter, %	87.6	89.8	47.8	89.4	85.7	56.4
2) Dry matter basis:						
Crude protein, %	10.2	9.3	11.7	9.1	8.5	8.1
TDN, %	58.8	60.4	63.9	52.0	65.1	58.1
NE$_m$, Mcal/kg	1.21	1.14	1.37	1.05	1.46	1.25
NE$_g$, Mcal/kg	0.64	0.62	0.79	0.36	0.88	0.69
Calcium, %	0.94	0.59	0.82	0.91	0.61	0.43
Phosphorus, %	0.15	0.18	0.22	0.17	0.23	0.28

1. Formulations are on an as-fed basis. Cows should consume approximately 1.75 to 2.0% of their body weight as daily dry matter. Daily feeding allowances should be made for waste in feeding, mixing, etc., and for inclement weather. A good vitamin- mineral mix should be fed free-choice with all rations.

2. Corn stover silage will substitute for dry stover if corrected for moisture content.

3. A typical legume-grass mixture may be alfalfa-brome or clover-timothy. Haylage will substitute for hay on an equal basis if corrected for moisture content.

4. Protein supplement should contain approximately 30 to 35% crude protein (up to 1/3 of the protein may be derived from non-protein nitrogen), 2.5 to 3.0% calcium and 1.5 to 2.0% phosphorus.

5. There is some variation in energy intake (TDN or net energy) between these rations. This was the result of judgement evaluation of some feedstuffs relative to palatability and resulting possible intake problems.

TABLE 8–7. Rations for Wintering the Beef Cow While Lactating and Rebreeding[1,2]

Ingredients	IFN	1	2	3	4
Corn silage	3–02–823	96	—	—	—
Corn stover	1–02–776	—	—	70	—
Alfalfa haylage	3–08–150	—	—	—	82
Legume-grass hay[3]	1–08–331	—	79	—	—
Shelled corn	4–02–931	—	21	20	18
Protein supplement[4]		4	—	10	—
		100	100	100	100

Calculated Analysis:

1) As-fed basis
Crude protein, %		4.0	13.3	8.8	8.1
TDN, %		25.3	57.4	57.6	33.0
NE_m, Mcal/kg		0.59	1.17	1.31	0.73
NE_g, Mcal/kg		0.37	0.65	0.80	0.42
Calcium, %		0.19	1.06	0.60	0.47
Phorphorus, %		0.14	0.23	0.29	0.16
Dry matter, %		36.3	90.5	86.1	51.3

2) Dry matter basis
Crude protein, %		11.0	14.7	10.2	15.9
TDN, %		69.8	63.4	66.9	64.3
NE_m, Mcal/kg		1.62	1.29	1.52	1.42
NE_g, Mcal/kg		1.02	0.71	0.93	0.83
Calcium, %		0.51	1.17	0.69	0.92
Phosphorus, %		0.38	0.26	0.34	0.31

1. Formulations are on an as-fed basis. The average 1000 lb. (454 kg) mature lactating cow of average milking ability should consume about 2.25% to 2.5% of their body weight as daily dry matter. Large, heavy-milking cows will need to be fed heavier than this.

2. A good vitamin-mineral mix should be fed free-choice with all rations.

3. A typical legume-grass hay mixture may be alfalfa-brome or clover-timothy.

4. Protein supplement should contain approximately 30 to 35% crude protein (up to 1/3 of the protein may be derived from non-protein nitrogen), 2.5 to 3.0% calcium and 1.5 to 2.0% phosphorus.

TABLE 8–8. Creep Rations for Beef Cattle[1]

Ingredient	IFN	1	2	3	4	5	6	7
Corn, cracked	4–02–931	85	—	65	—	90	—	35
Oats	4–03–309	—	100	35	70	—	—	38
Barley, rolled	4–00–549	—	—	—	30	—		
Protein supplement[2]		10	—	—	—	10	—	—
Alfalfa, dehy., pelleted	1–00–023	—	—	—	—	—	60	—
Soybean meal	5–04–604	—	—	—	—	—	—	15
Molasses, liquid	4–04–696	5	—	—	—	—	—	10
Dicalcium phosphate	6–01–080	—	—	—	—	—	—	1
Trace-mineralized salt		—	—	—	—	—	1	1
		100	100	100	100	100	100	100
Calculated analysis, %								
1) As-fed basis:								
Crude protein		10.9	11.8	9.9	11.7	11.2	14.9	14.8
TDN		75.7	68.9	75.1	70.3	76.6	62.1	71.0
NE_m, Mcal/kg		1.87	1.77	1.89	1.76	1.89	1.44	1.80
NE_g, Mcal/kg		1.29	1.19	1.29	1.18	1.30	0.89	1.23
Calcium		0.30	0.08	0.04	0.07	0.27	0.85	0.37
Phosphorus		0.43	0.34	0.31	0.34	0.44	0.27	0.53
Dry matter		86.9	89.2	88.0	89.0	87.6	90.6	87.3
2) Dry matter basis:								
Crude protein		12.6	13.2	11.2	13.2	12.8	16.5	17.0
TDN		87.1	77.2	85.4	79.0	87.4	68.5	81.3
NE_m, Mcal/kg		2.16	1.98	2.15	1.97	2.15	1.59	2.06
NE_g, Mcal/kg		1.49	1.33	1.47	1.33	1.49	0.98	1.41
Calcium		0.35	0.09	0.05	0.08	0.31	0.94	0.43
Phosphorus		0.49	0.38	0.35	0.38	0.50	0.30	0.60

1. Formulations are on an as-fed basis. Creep feeding is usually a means of supplying energy. Milk is rich in protein and minerals but relatively low in energy. Therefore, a simple creep ration like Rations 2, 3 or 4 will suffice if the calf is receiving normal milk and pasture. If pasture and milk are in short supply, the calf needs a more complex creep ration like Rations 1 or 5 to meet vitamin and mineral needs. Rations 6 and 7 are used for early-weaned calves. Vitamins A, D, and E should be added to these rations in quantities of 5000 IU/kg, 2500 IU/kg and 100 IU/kg, respectively.

2. All natural protein supplement containing approximately 30 to 35% crude protein, 2.5 to 3.0% calcium and 1.5 to 2.0% phosphorus.

TABLE 8–9. Rations for 500-lb (227-kg) First Winter Replacement Heifers[1]

Ingredient	IFN	1	2	3	4	5	6
Alfalfa hay[2]	1–00–063	—	84	—	—	17	—
Grass hay[2,3]	1–03–438	—	—	76	—	—	33
Corn, shelled[4]	4–02–931	—	16	24	—	—	15
Corn silage	3–02–823	96	—	—	—	83	—
Sorghum silage	3–04–468	—	—	—	96	—	50
Protein supplement[5]		4	—	—	4	—	2
		100	100	100	100	100	100

Calculated analysis:

1) As-fed basis

		1	2	3	4	5	6
Crude protein, %		4.0	15.7	10.1	2.9	5.2	6.3
TDN, %		25.3	56.3	58.0	17.8	28.7	38.0
NE_m, Mcal/kg		0.59	1.29	1.30	0.44	0.66	0.88
NE_g, Mcal/kg		0.37	0.76	0.77	0.21	0.40	0.50
Calcium, %		0.19	1.04	0.29	0.19	0.29	0.22
Phosphorus, %		0.14	0.23	0.24	0.12	0.10	0.18
Dry matter, %		36.3	90.4	89.0	29.5	43.8	58.0

2) Dry-matter basis

		1	2	3	4	5	6
Crude protein, %		11.0	17.3	11.3	9.8	11.9	10.9
TDN, %		69.8	62.3	65.1	60.3	65.5	65.5
NE_m, Mcal/kg		1.62	1.43	1.46	1.49	1.52	1.52
NE_g, Mcal/kg		1.02	0.84	0.87	0.71	0.92	0.86
Calcium, %		0.51	1.16	0.32	0.64	0.65	0.38
Phosphorus, %		0.38	0.26	0.27	0.41	0.22	0.31

1. Formulations are on an as-fed basis. Heifers should consume approximately 2.5% of their body weight as daily dry matter. Daily feeding allowances should be made for waste in mixing and feeding and for inclement weather. A good vitamin-mineral mix should be fed free-choice with all rations.

2. Haylage will substitute for hay on an equal basis but must be corrected for moisture content.

3. Grass hay may be bromegrass (1–00–944), orchardgrass (1–03–438) or canarygrass (1–01–104).

4. Oats or ground ear corn will substitute at rates of 1.2 and 1.0 pounds (or kg), respectively, for each weight unit of shelled corn.

5. An all natural protein supplement containing approximately 30 to 35% crude protein, 2.5 to 3.0% calcium and 1.5 to 2.0% phosphorus is recommended for heifers of this weight.

TABLE 8–10. Rations for 900-lb (409-kg) Second Winter Pregnant Replacement Heifers[1]

Ingredient	IFN	1	2	3	4	5	6	7
Alfalfa, hay[2]	1–00–063	—	—	86	—	—	9	—
Grass hay[2,3]	1–03–438	—	—	—	77	11	—	—
Corn, shelled[4]	4–02–931	—	—	14	23	—	—	25
Corn silage	3–02–823	98	—	—	—	89	—	—
Corn stover	1–02–776	—	—	—	—	—	—	70
Sorghum silage	3–04–468	—	98	—	—	—	91	—
Protein supplement[5]		2	2	—	—	—	—	5
		100	100	100	100	100	100	100

Calculated analysis:

1) As-fed basis

		1	2	3	4	5	6	7
Crude protein, %		3.4	2.3	15.8	10.1	3.6	2.9	7.6
TDN,%		24.6	16.9	55.8	57.7	26.9	19.2	58.5
NE_m, Mcal/kg		0.58	0.42	1.28	1.30	0.62	0.45	1.34
NE_g, Mcal/kg		0.36	0.19	0.75	0.76	0.38	0.21	0.82
Calcium, %		0.14	0.14	1.07	0.29	0.12	0.19	0.47
Phosphorus,%		0.10	0.08	0.23	0.24	0.09	0.07	0.22
Dry matter, %		35.2	28.3	90.5	89.1	40.2	32.7	86.0

2) Dry matter basis

		1	2	3	4	5	6	7
Crude protein, %		9.6	8.1	17.5	11.3	9.1	8.9	8.9
TDN, %		69.9	59.7	61.6	64.8	67.0	58.7	68.1
NE_m, Mcal/kg		1.63	1.48	1.41	1.45	1.54	1.38	1.56
NE_g, Mcal/kg		1.02	0.67	0.83	0.86	0.94	0.64	0.96
Calcium, %		0.39	0.50	1.18	0.33	0.30	0.58	0.55
Phosphorus, %		0.29	0.28	0.25	0.27	0.22	0.21	0.25

1. Formulations are on an as-fed basis. Heifers should consume approximately 2.5% of their body weight as daily dry matter. Daily feeding allowances should be made for waste in mixing and feeding and for inclement weather. A good vitamin-mineral mix should be fed free-choice with all rations.

2. Haylage will substitute for hay on an equal basis but must be corrected for moisture content.

3. Grass hay may be bromegrass (1–00–944), orchardgrass (1–03–438) or canarygrass (1–01–104).

4. Oats or ground ear corn will substitute at rates of 1.2 and 1.0 pounds (or kg), respectively, for each weight unit of shelled corn.

5. Protein supplement should contain approximately 30 to 35% crude protein (up to 1/3 of the protein may be derived from non-protein nitrogen), 2.5 to 3.0% calcium and 1.5 to 2.0% phosphorus.

TABLE 8–11. Rations for Weaning 440-lb (200-kg) Calves That Are to Go into a Backgrounding Program with Gains of 1.3 to 1.7 lb (0.6 to 0.8 kg) Daily[1]

Ingredient	IFN	1	2	3	4	5	6
Corn	4–02–931	35	—	—	—	22	—
Ground ear corn	4–02–849	—	—	25	—	—	—
Corn silage	3–02–823	—	94	—	—	—	70
Corn stover, gr.	1–02–776	—	—	—	—	63	20
Oats	4–03–309	—	—	—	35	—	—
Alfalfa hay[2]	1–00–063	65	—	75	65	—	—
Protein supplement[3]		—	6	—	—	15	10
		100	100	100	100	100	100

Calculated analysis:

1) As-fed basis:

		1	2	3	4	5	6
Crude protein, %		14.1	4.6	14.7	15.2	10.2	6.3
TDN,%		61.3	26.1	57.2	58.0	58.6	33.0
NE_m, Mcal/kg		1.44	0.61	1.35	1.38	1.33	0.75
NE_g, Mcal/kg		0.90	0.38	0.82	0.84	0.83	0.46
Calcium, %		0.81	0.23	0.94	0.83	0.69	0.41
Phosphorus, %		0.24	0.17	0.23	0.26	0.38	0.24
Dry matter, %		89.7	37.4	89.9	90.4	86.4	49.9

2) Dry matter basis:

		1	2	3	4	5	6
Crude protein		15.7	12.2	16.4	16.8	11.8	12.6
TDN, %		68.3	69.6	63.7	64.2	67.8	66.1
NE_m, Mcal/kg		1.61	1.62	1.50	1.53	1.54	1.50
NE_g, Mcal/kg		1.00	1.02	0.91	0.93	0.96	0.92
Calcium, %		0.91	0.63	1.05	0.92	0.80	0.82
Phosphorus, %		0.27	0.46	0.25	0.29	0.44	0.48

1. Formulations are on an as-fed basis. Calves should consume approximately 2.5 to 3.0% of their body weight as daily dry matter. Small calves may not consume adequate amounts of rations 2 and 6 in cold weather. Add 2 to 4 lb (0.9 to 1.8 kg) of grain during cold weather.

2. A mixed grass-legume hay may be used rather than alfalfa hay.

3. All natural protein supplement containing approximately 30 to 35% crude protein, 2.5 to 3.0% calcium and 1.5 to 2.0% phosphorus.

TABLE 8–12. Rations for Weaning 440-lb (220-kg) Calves Gaining 1.8 to 2.2 lb (0.8 to 1.0 kg) Daily[1]

Ingredient	IFN	1	2	3	4	5	6
Corn	4–02–931	50	10	19	—	—	40
Ground ear corn	4–02–849	—	—	—	55	—	—
Corn silage	3–02–823	—	84	—	—	—	—
Corn stover, gr.	1–02–776	—	—	—	—	—	50
Oats	4–03–309	—	—	54	—	75	—
Alfalfa hay[2]	1–00–063	45	—	27	40	22	—
Protein supplement[3]		5	6	—	5	3	10
		100	100	100	100	100	100

Calculated analysis:

1) As-fed basis

		1	2	3	4	5	6
Crude protein, %		13.7	5.2	12.6	12.7	13.6	9.5
TDN,%		65.7	31.5	66.2	63.8	64.9	63.0
NE_m, Mcal/kg		1.57	0.74	1.64	1.58	1.62	1.47
NE_g, Mcal/kg		1.01	0.48	1.07	1.02	1.06	0.94
Calcium,%		0.69	0.23	0.38	0.65	0.41	0.50
Phosphorus,%		0.33	0.19	0.30	0.31	0.36	0.33
Dry matter,%		89.1	42.8	89.3	88.5	89.6	86.5

2) Dry matter basis:

		1	2	3	4	5	6
Crude protein,%		15.3	12.1	14.1	14.3	15.1	11.0
TDN,%		73.7	73.7	74.1	72.1	72.5	72.8
NE_m, Mcal/kg		1.76	1.74	1.84	1.78	1.81	1.70
NE_g, Mcal/kg		1.14	1.13	1.20	1.16	1.18	1.09
Calcium, %		0.78	0.53	0.43	0.74	0.46	0.58
Phosphorus,%		0.37	0.45	0.33	0.35	0.40	0.38

1. Formulations are on an as-fed basis. Calves should consume approximately 2.5 to 3.0% of their body weight as daily dry matter. The hay and corn stover rations (ration 1 and 6) which contain more grain should be worked onto gradually.

2. A mixed grass-legume hay may be used rather than alfalfa hay.

3. All natural protein supplement containing approximately 30 to 35% crude protein, 2.5 to 3.0% calcium and 1.5 to 2.0% phosphorus.

TABLE 8-13. Rations for 660-lb (300-kg) Feedlot Calves and Yearlings[1]

Expected daily gain =		2.0 to 2.5 lb (0.9 to 1.1 kg)				2.50 to 3.0 lb (1.1 to 1.4 kg)				2.75 to 3.25 lb (1.25 to 1.5 kg)		
Ingredient	IFN	1	2	3	4	5	6	7	8	9	10	11
Corn, cracked	4-02-931	10	—	20	53	26	—	40	70	50	—	80
Ground ear corn	4-02-849	—	60	—	—	—	75	—	—	—	85	—
Corn silage	3-02-823	86	—	65	—	70	—	48	—	45	—	—
Legume-grass hay[2]	1-08-331	—	36	11	44	—	21	8	27	—	9	16
Protein supplement[3]		4	4	4	3	4	4	4	3	5	6	4
		100	100	100	100	100	100	100	100	100	100	100

Calculated analysis:

1) As-fed basis

		1	2	3	4	5	6	7	8	9	10	11
Crude protein,%		4.6	11.2	6.5	12.0	5.5	10.2	7.3	11.0	7.3	9.9	10.7
TDN,%		30.8	64.7	39.3	66.1	39.5	67.8	49.4	70.7	53.0	70.1	73.4
NE_m, Mcal/kg		0.73	1.54	0.91	1.50	0.95	1.68	1.18	1.66	1.29	1.78	1.77
NE_g, Mcal/kg		0.47	0.99	0.58	0.94	0.63	1.11	0.78	1.10	0.88	1.21	1.19
Calcium,%		0.18	0.62	0.31	0.68	0.17	0.43	0.26	0.45	0.18	0.32	0.33
Phosphorus,%		0.16	0.29	0.20	0.30	0.19	0.30	0.24	0.31	0.26	0.33	0.34
Dry matter,%		41.7	88.4	53.3	89.2	50.2	87.7	62.2	88.5	63.5	87.2	88.1

2) Dry matter basis

		1	2	3	4	5	6	7	8	9	10	11
Crude protein,%		11.0	12.7	12.1	13.5	11.1	11.6	11.8	12.5	11.5	11.3	12.1
TDN,%		73.9	73.1	73.8	74.2	78.8	77.3	79.4	79.9	83.4	80.4	83.4
NE_m, Mcal/kg		1.75	1.74	1.71	1.68	1.90	1.92	1.89	1.88	2.04	2.04	2.01
NE_g, Mcal/kg		1.13	1.12	1.09	1.06	1.26	1.27	1.25	1.24	1.38	1.39	1.35
Calcium,%		0.43	0.70	0.58	0.76	0.34	0.49	0.42	0.51	0.28	0.37	0.38
Phosphorus,%		0.38	0.33	0.37	0.34	0.39	0.34	0.38	0.36	0.42	0.38	0.38

1. Formulations are on an as-fed basis. Calves or yearlings should consume approximately 2.25 to 2.75% of their body weight as daily dry matter. Yearling stocker cattle will generally consume greater quantities of daily dry matter (resulting in greater daily gain) than grain backgrounded calves. Salt may be fed free-choice with all rations. Ground limestone (or other source of calcium) may be fed free-choice with rations 9, 10, and 11.

2. A typical legume-grass hay mixture may be alfalfa-brome or clover- timothy. Haylage will substitute for hay on an equal basis if corrected for moisture content.

3. Protein supplement should contain approximately 30 to 35% crude protein (up to 1/3 of the protein may be derived from non-protein nitrogen), 2.75 to 3.25% calcium and 1.5 to 2.0% phosphorus.

TABLE 8–14. Rations for 880 lb (400 kg) Feedlot Cattle[1]

Expected daily gain =		2.25–2.75 lb (1–1.25 kg)			2.75–3.25 lb(1.25–1.5kg)		
Ingredient	IFN	1	2	3	4	5	6
Corn	4–02–931	22	67	—	60	70	—
Ground ear corn	4–02–849	—	—	72	—	—	90
Corn silage	3–02–823	75	—	—	36	20	—
Legume-grass hay[2]	1–08–331	—	30	25	—	7	5
Protein supplement[3]		3	3	3	4	3	5
		100	100	100	100	100	100

Calculated analysis:

1) As-fed basis

Crude protein,%		5.0	11.2	10.2	7.6	8.7	9.4
TDN, %		37.0	69.9	67.1	58.0	65.1	71.0
NE_m, Mcal/kg		0.89	1.63	1.65	1.42	1.58	1.82
NE_g, Mcal/kg		0.59	1.07	1.08	0.97	1.07	1.25
Calcium,%		0.15	0.49	0.45	0.14	0.20	0.28
Phosphorus,%N		0.17	0.31	0.28	0.27	0.28	0.36
Dry matter,%		47.5	88.6	87.8	68.3	77.0	86.9

2) Dry matter basis

Crude protein,%		10.6	12.6	11.6	11.1	11.3	10.8
TDN,%		77.9	78.9	76.4	85.0	84.5	81.7
NE_m, Mcal/kg		1.87	1.84	1.88	2.09	2.06	2.10
NE_g, Mcal/kg		1.23	1.21	1.23	1.42	1.39	1.43
Calcium, %		0.31	0.55	0.52	0.21	0.26	0.28
Phosphorus,%		0.36	0.35	0.32	0.39	0.37	0.36

1. Formulations are on an as-fed basis. Cattle should consume approximately 2.0 to 2.5% of their body weight as daily dry matter. Salt may be fed free-choice with all rations. Ground limestone (or other source of calcium) may be fed free-choice with rations 4, 5 and 6.

2. A typical legume-grass hay mixture may be alfalfa-brome or clover-timothy. Haylage will substitute for hay on an equal basis if corrected for moisture content.

3. Protein supplement should contain approximately 30 to 35% crude protein (up to 1/2 of the protein may be derived from non-protein nitrogen), 2.75 to 3.25% calcium and 1.5 to 2.0% phosphorus.

SYSTEM OF PROTEIN EVALUATION FOR CATTLE AND SHEEP

Reevaluation of protein and amino acid requirements for cattle and sheep has become desirable because of two primary considerations: the widespread use of urea in feeding practice, and the worldwide growth in technical knowledge involving rumen fermentation. This knowledge provides a basis for understanding why, under certain feeding regimens, urea is very valuable, less valuable under other feeding regimens and in some instances useless in partly satisfying body amino acid requirements. Defining these feeding regimens or ration formulas is sufficiently complex to require new measurements for feedstuffs as well as for body requirements.

The measurements proposed are metabolizable protein (MP) and metabolizable amino acid (MAA) requirements of ruminants and the MP, MAA and urea fermentation potential (UFP) of feedstuffs.

The following explanation of this system was taken from Iowa State University A.S. Leaflet R190 prepared for Cattle Feeders Day at Iowa State University, July 1974.

METABOLIZABLE PROTEIN (AMINO ACID) FEEDING STANDARD FOR CATTLE AND SHEEP FED RATIONS CONTAINING EITHER α-AMINO OR NONPROTEIN NITROGEN

The metabolizable protein feeding standard began some years ago[1,2] as a system of evaluating protein nutrition of feedlot cattle. The need for this system arose primarily because of the expanded use of urea in feedlots and the failure of conventional feeding standards to consistently predict accurate urea utilization in feedlot rations.

A similar need was recognized for other ruminants, and the system has recently been proposed for use with lactating cows and feedlot lambs.[3,4] A growing interest in applying the system to a larger number of feedstuffs commonly used in cattle and sheep rations has prompted the preparation of this report.

The measurements of metabolizable protein (MP) and amino acids (MAA) for feedstuffs and body requirements of animals, plus the urea fermentation potential (UFP) of feedstuffs, have the same meaning in this report as in earlier reports. Some earlier values have been revised due to new research results and more simplified methods in computing values. These methods are also presented below.

Metabolizable Protein and Metabolizable Amino Acid Requirements and Feed Values Established

Metabolizable protein and MAA values are the quantity of protein digested or amino acid(s) absorbed in the postruminal portion of the ruminant digestive tract. This MP system deals with body tissue requirements for amino acids (AA) and their fulfillment from absorbable AA in digested ration ingredients escaping rumen destruction and digested proteins arising from rumen microbial synthesis. The digestion for both takes place in the true stomach and small intestine of ruminants. The MP system therefore parallels the measurement of digestible protein (DP) in nonruminants. The system avoids the assumption that absorbable AA arise from protein undergoing ruminal destruction without resynthesis into microbial protein within the rumen.

The first step in the establishment of MP and MAA requirements of animals (tables 1, 2 and 3) used a factorial system involving the net protein and AA deposited in milk, wool and body tissues, plus those needed daily for net body protein maintenance. This system (apparent net protein) is a modification of the net protein system developed by HH Mitchell.[5] The net body maintenance needs for protein (grams per animal daily) for animals of different sizes were established by using the formula developed by DB Smuts[6]:

$$\text{maintenance}_g = (0.0125)(70.4 w_{kg}^{0.734})$$

The second step in determining requirements was to convert total net protein needs into total metabolizable needs on the basis of losses of MP in metabolism. Estimated by growth and lactation trials with purified[7-9] and natural diets[10,11] fed to cattle and sheep, the values are

1. 60% in forming sheep body tissue growth, wool and maintenance.
2. 53% in forming cattle tissue growth and maintenance.
3. None in forming fecal protein of body origin in cattle and sheep.
4. 5% in forming cattle milk protein.

The MAA requirements for body growth and maintenance were assigned according to the AA composition of whole mammal body tissues. This AA composition data were adapted from body composition values and from feed composition values for meat scraps.[12-14] The AA requirements for wool growth were formulated from the AA composition of wool.[15] AA requirements for milk protein formation were assessed from the AA composition of milk[13] protein (Table 4).

Satisfying each of the more critical MAA requirements of cattle and sheep listed in tables 1, 2 and 3 often requires 20% more MP than shown. Therefore, final consideration should be given to filling each AA requirement besides satisfying the MP values given in tables 1, 2 or 3.

The first step in establishing MP values for feedstuffs was estimating the amounts of undegraded protein reaching the abomasum, as presented in tables 5 and 6. The second step was estimating the amount of microbial protein reaching the abomasum. This estimate is based on 10.44% of the TDN in a given feedstuff having adequate rumen ammonia.[16] The formula used in computing MP values in Table 6 is

$$MP_{(g/kg)} = (P_1 \times .90) + ([P_2 - 15.0] \times 0.80)$$

where P_1 and P_2 are the estimated grams of undegraded and microbial protein, respectively, available for productive purposes multiplied by true digestibility conversion values. The value of 15.0 represents feed used in digestion to form fecal metabolic protein when 1 kg of feed dry matter is consumed. Thus, the MP value of 47.6 for alfalfa containing 19.3% protein as listed in Table 6 was computed.

$$MP = (9.6 \times .90) + ([63.7 - 15.0] \times 0.80) = 47.6$$

The conversion values (Table 5) of degraded protein into microbial protein were based upon 100% when roughage protein and concentrate protein did not exceed 7.1% and 13.0% of DM, respectively. When feedstuffs had protein levels in excess of these 7.1% or 13.0% values, the excess nitrogen from degraded protein was considered totally useless unless the feedstuff was mixed with feedstuffs having positive UFP values. The excess protein degraded to NPN would then have a value similar to that of urea NPN as discussed later.

The MP values developed for feedstuffs were transformed into AA values by using AA[13,17] composition values for the undegraded protein (Table 4) and AA composition values for rumen bacteria and protozoa.[18] Equal protoplasmic mass of the two types of microorganisms was assumed. Metabolizable protein and MAA values for approximately 75 feeds are presented in Table 6.

Urea Fermentation Potential of Feeds Established

The UFP of feeds quantitatively evaluates the amount of useful urea in a cattle or sheep ration. A positive UFP value becomes an integral part of the MP or MAA system of evaluating cattle nutrition when urea is part of the ration. A positive UFP value of a feed or ration is the estimated grams of urea, per kilogram of dry matter consumed, that can be transformed into microbial protein in the forepart of the digestive tract of ruminants. All nitrogen composing the UFP was considered transformed into microbial protein in this report. Establishing UFP values involves the amount of fermentable energy present in a feed as reflected by its TDN content and the amount of ammonia formed from feed protein degraded in the rumen.

The formula used in establishing UFP values is

$$UFP = (1.044 \; TDN - B)/2.8$$

The UFP values are expressed as grams of urea (44.8% nitrogen) per kilogram of dry matter consumed. The value 1.044 represents the estimated potential net grams of rumen microbial protein (10.44% of TDN) from consumption of 10 g of TDN.[16] The letter B represents the estimated grams of protein in 1 kg of feed consumed that were degraded in the rumen and whose ammonia contributed to the total rumen pool. The 2.8 value in the formula transforms protein to urea nitrogen equivalence.

The UFP values obtained for each of the 75 feeds referred to earlier are also presented in Table 6. A negative UFP value indicates the estimated excess performed protein in 1 kg of feed degraded by rumen fermentation and incapable of being resynthesized into microbial protein with the energy present in the feed, expressed as grams of urea equivalent. This nitrogen would become useful in fermentation only to the extent in which combined with feeds having equal or greater positive UFP values.

Discussion

The present MP feeding standard for ruminants is believed superior to conventional standards. The traditional measurements of total protein[19] (N × 6.25) or DP[12] in systems evaluating protein nutrition in cattle and sheep as employed by the U.S. National Research Council[20-22] and the British Agricultural Research Council would be reasonably sound measurement tools if only small quantities of NPN, as in normal feedstuffs, were incorporated into ruminant diets. When larger and more variable quantities of NPN (urea) are fed, these measurement tools become inadequate. They do not reflect when protein equivalent from NPN is fully not as useful as α-amino protein (AAP) equivalent.

This situation generally occurs in younger developing calves and lambs, as well as in cows at higher lactation levels. Their ration protein needs (protein percentage of diet) are highest in relation to ration energy consumption. That NPN protein equivalence is not always as useful as true protein (expressed as AAP) becomes most evident in the performance of young calves,[7] lambs[8] and high-lactating cows[9] on diets devoid of AAP, but containing adequate-to-excessive levels of NPN. No greater than 65% to 70% of total expected performance results, based on 100% performance when adequate AAP is fed. Thus, if one were to establish a feeding standard using total nitrogen × 6.25 or DP based on the lowest NPN level promoting maximum cattle or sheep performance, the standard would underestimate animal performance by about one-third when AAP rations are fed. By contrast, if one established a feeding standard using N × 6.25 or DP based on the lowest AAP level regularly promoting maximum cattle or sheep performance, the standard would overestimate animal performance by about one-third when only NPN rations are fed.

Although the MP system predicts the type of ration with which supplemental urea will be totally useless, the system also predicts the types of rations with which urea will be useful, as well as the maximum beneficial quantity. Table 6 reveals that 20 major feedstuffs have fermentation characteristics (positive UFP values) whereby urea supplementation will prove beneficial provided there is a body need for supplemental nitrogen. Half of these feedstuffs are forages or roughages with less than 6.5% protein. The other ten feedstuffs are primarily concentrates with less than 10% to 12% protein. Feedstuffs from the corn plant dominate all others in their ability to be benefited by urea supplementation.

Looking to the future, perhaps another major benefit of the MP system over conventional systems will be the prediction of singular amino acid deficiencies in specific ruminant diets and their corrections by supplementation. The system presently is sufficiently precise to predict only groups of amino acids that are most limiting. In the case of diets composed chiefly of corn grain, corn silage and urea, the system predicts that for lightweight feedlot cattle and lambs, the most limiting AAs are arginine, lysine and the sulfur-containing AAs. For high-lactating animals, the system predicts that histidine, lysine and the sulfur-containing AAs are in shortest supply when a corn grain, corn silage and urea ration is fed.

The formula using TDN values in computing UFP values for feedstuffs in this report is believed more accurate than the formula[2] used earlier containing concentrate equivalent values, due to the greater precision of established TDN values. However, they are quite similar when considering all UFP values developed by the two formulae. The same is true for the similarity between MP and MAA values presented in this report and in the earlier report, in which methods employed were slightly different.

Summary

An apparent net protein system suitable as a feeding standard is described which evaluates amino acid nutrition of cattle and sheep consuming diets containing both nonprotein nitrogen (urea) and natural protein. The system includes body requirement standards and feedstuff values making use of two unique measurements: (1) the quantity of protein (amino acids) absorbed postruminally, designated metabolizable protein (MP) of amino acids (MMA), and (2) the urea fermentation potential (UFP) of each feedstuff. The UFP is a measurement of the urea transformed into rumen microbial protein when fed with a specific quantity of a given feedstuff to cattle or sheep. Values for approximately 75 cattle and sheep feeds are presented in tables, and methods are described for extending these values to other feedstuffs. The system in its present development is discussed as being superior to current feeding standards in common use for ruminant species.

References

1. Burroughs W, AH Trenkle and RL Vetter, 1971, Some new concepts of protein nutrition of feedlot cattle. *Vet Med Small Anim Clin* 66:238 and 598.
2. Burroughs W, AH Trenkle and RL Vetter, 1972, *Proposed new system of evaluating protein nutrition of feedlot cattle* (Metabolizable protein and urea fermentation potential [UFP] of feeds), Iowa State Univ Coop Ext Serv Leaflet EC–7771.
3. Burroughs W, AH Trenkle and RL Vetter, 1974, A system of protein evaluation for cattle and sheep involving metabolizable protein (aminoacids) and urea fermentation potential of feedstuffs. *Vet Med Small Anim Clin*, (in press).
4. Burroughs W, NL Jacobson and DK Nelson, 1974, Production efficiency in the high-producing cows: Evaluation of protein nutrition by metabolizable protein and urea fermentation potential. *J Dairy Sci*, (in press).
5. Mitchell HH, 1929, The minimum protein requirements of cattle. National Research Council Bull, 67, Washington, D.C.
6. Smuts DB, 1935, The relation between the basal metabolism and the endogenous nitrogen metabolism, with particular reference to the maintenance requirement of protein, *J Nutr*, 9:403.
7. Oltjen RR, LL Slyter and RL Wilson, 1972, Urea levels, protein and diethylstilbestrol for growing steers fed purified diets. *J Nutr*, 104:479.
8. Price WD, JA Brown, EE Menvielle and WH Smith, 1972, Effect of high levels of urea in purified diets for lambs: Growth and metabolism, *J Anim Sci*, 35:848.
9. Virtanen AI, 1966, Milk production of cows on protein-free feed, *Science*, 153:1603.
10. Burroughs W, AH Trenkle and RL Vetter, 1973, *Demonstration of the variable feeding value of urea as predicted by the new metabolizable protein system of evaluating cattle feeds and rations in satisfying tissue amino acid requirements*, Iowa State Univ Coop Ext Serv Leaflet R 173.
11. Johnson BC, TS Hamilton, WB Robinson and JC Garey, 1944, On the mechanism of nonprotein-nitrogen utilization by ruminants. *J Anim Sci*, 3:287.
12. Block RJ and KW Weiss, 1956, *The Amino Acid Handbook*. Springfield, Ill: Charles C. Thomas.
13. National Research Council, 1964, *Joint United States-Canadian Tables of Feed Composition*, National Research Council Pub 1232, Washington, D.C.
14. Thomas EE, 1974, *Amino acid composition of loin muscle protein as a predictor of whole-body amino acid composition of cattle protein*, M.S. thesis, Iowa State University, Ames, Iowa.
15. Broad A, JM Gillespie and PJ Reis, 1970, The influence of sulfur-containing amino acids on the biosynthesis of high-sulfur wool proteins. *Aust J Biol Sci*, 23:149.
16. Pitzen DF, 1974, *Quantitative microbial protein synthesis in the bovine rumen*, Ph.D. thesis, Iowa State University, Ames, Iowa.
17. Morrison FB, 1956, Feeds and Feeding. The Morrison Publishing Co., Ithaca, N.Y.
18. Meyer RM, EE Bartley, CW Deyoe and VF Colenbrander, 1967, Feed processing. I. Ration effects on rumen microbial protein synthesis and amino acid composition. *J Dairy Sci*, 50:1327.
19. Agricultural Research Council, 1965, The nutrient requirements of farm livestock. No. 1. Ruminants, Agricultural Research Council, London.
20. National Research Council, 1968, *Nutrient requirements of domesticated animals, no. 5, Nutrient requirements of sheep*, National Research Council, Washington, D.C.
21. National Research Council, 1970, *Nutrient requirements of domestic animals, no. 4, Nutrient requirements of beef cattle*, National Research Council, Washington, D.C.
22. National Research Council, 1971, *Nutrient requirements of domestic animals, no. 3, Nutrient requirements of dairy cattle*, National Research Council, Washington, D.C.
23. Prepared by W Burroughs, AH Trenkle and RL Vetter, professors of animal science, with the assistance of MR Geasler, DK Nelson and TW Wickersham, extension livestock specialists, and DR Mertens, assistant professor of dairy science.
24. Fowler MA, 1968, *Development of a feedlot-slaughter technique for net energy determinations of finishing rations in beef cattle with and without stilbestrol*. M.S. thesis. Iowa State University, Ames, Iowa.
25. Haecker TL, 1920, *Investigation in beef production*. Minn Agr Exp. Sta. Bull. 193.

26. Moulten CR, PF Trowbridge and LD Haigh, 1922, *Studies in animal nutrition*. Mo Agr Exp Sta Bull 55.
27. Jacobson NL, 1970, The mammary gland and lactation, *In* MJ Swenson (ed.), *Duke Physiology of Domestic Animals*. Cornell University Press, Ithaca, N.Y., and London.
28. Mitchell HH, 1962, *Comparative Nutrition of Man and Domestic Animals*, vol. 1, Academic Press, New York, N.Y.
29. Mitchell HH, WG Kammlade and TS Hamilton, 1926, *A technical study of the maintenance and fattening of sheep and their utilization of alfalfa hay*. Ill Agr Exp Sta Bull, 283.
30. Mitchell HH, WG Kammlade and TS Hamilton, 1928, *A technical study of the maintenance and fattening of lambs*. Ill Agr Exp Sta Bull, 314.
31. Mitchell HH, WG Kammlade and TS Hamilton, 1928, *Relative energy value of alfalfa, clover and timothy hay for the maintenance of sheep*. Ill Agr Exp Sta Bull, 317.

A.S. R190 TABLE 1. Metabolizable Protein and Selected Metabolizable Amino Acid Requirements for Feedlot Cattle

1	2	3	4	5	6	7	8	9	10	11
	Daily	Net Protein				Metabolizable				
Body Wt.	Live Wt. Gain[a]	Gain[b]	Maint.[c]	Total[d]	Prot.[e]	Arg.[f]	Cys.[f]	His.[f]	Lys.[f]	Met.[f]
kg	kg	g	g	g	g	g	g	g	g	g
Finishing Steers and Heifers										
150	0.70	102	35	136	348	20	4	6	21	6
	0.80	116	35	151	385	22	5	7	23	7
	0.90	131	35	165	423	24	5	7	25	7
	1.00	145	35	180	460	26	6	8	27	8
	1.10	160	35	194	497	28	6	8	29	9
200	0.80	112	43	155	396	22	5	7	23	7
	0.90	126	43	169	432	24	6	7	25	8
	1.00	140	43	183	467	26	6	8	28	8
	1.10	154	43	197	504	28	7	9	30	9
	1.20	168	43	211	539	30	7	9	32	9
300	0.90	117	60	175	447	25	6	8	26	8
	1.00	130	60	188	480	27	6	8	28	8
	1.10	143	60	201	514	29	7	9	30	9
	1.20	156	60	214	547	31	7	9	32	10
	1.30	169	60	227	580	32	8	10	34	10
400	1.10	132	72	204	521	29	7	9	31	9
	1.20	144	72	216	551	31	7	9	33	10
	1.30	156	72	228	582	33	8	10	34	10
	1.40	168	72	240	612	34	8	10	36	11
	1.50	180	72	252	643	36	8	11	38	11
500	1.00	110	84	194	496	28	6	8	29	9
	1.10	121	84	205	525	29	7	9	31	9
	1.20	132	84	216	552	31	7	9	33	10
	1.30	143	84	227	581	33	8	10	34	10
	1.40	154	84	238	609	34	8	10	36	11

[a]Adapted from National Research Council (21).

[b]Column 2 × 150, 140, 130, 120, and 110 for 150, 200, 300, 400, and 500 kg cattle, respectively, Fowler (24), Haecker (25), Moulton (26), and Thomas (14).

[c]Maintenance protein$_g$ = (0.0125) (70.4 × body weight $_{kg}^{0.734}$).

[d]Column 3 + column 4.

[e](Column 5 × 2.13) × 1.2 which assumes a loss of 53% in metabolism of each absorbed amino acid and when calculating diets with only MP, an average loss 20% higher in insuring adequacy of the more essential amino acids of the ruminant diet.

[f](Column 6 ÷ 1.2) × (% amino acid present in beef protein, A.S. R 190 table 4/100). Abbreviations are: Arg. = Arginine; Cys. = Cystine; His. = Histidine; Lys. = Lysine; Met. = Methionine. They also apply to A.S. R190 tables 2 and 3.

374

A.S. R190 TABLE 2. Metabolizable Protein and Selected Metabolizable Amino Acid Requirements for Lactating Cows

1	2	3	4	5	6	7	8	9	10
			Amounts Per Animal Per Day						
Body Wt.	Dry Feed[a]	Net Protein[b]	Metabolizable						
			Protein[c]	Arg.[d]	Cys.[d]	His.[d]	Lys.[d]	Met.[d]	TDN
kg	kg	g	g	g	g	g	g	g	kg

Maintenance of Mature Lactating Cows

350	5.0	66	168	9.5	2.1	2.8	9.9	2.9	2.8
400	5.5	72	184	10.4	2.3	3.0	10.9	3.2	3.1
450	6.0	78	199	11.3	2.5	3.3	11.8	3.5	3.4
500	6.5	85	217	12.3	2.7	3.6	12.9	3.8	3.7
550	7.0	92	235	13.3	2.9	4.0	13.9	4.1	4.0
600	7.5	98	251	14.2	3.1	4.2	14.8	4.4	4.2
650	8.0	104	265	15.0	3.3	4.4	15.7	4.6	4.5
700	8.5	110	281	15.9	3.5	4.6	16.6	4.9	4.8
750	9.0	116	296	16.8	3.7	4.8	17.5	5.2	5.0
800	9.5	121	308	17.5	3.9	5.0	18.2	5.4	5.3

Maintenance and Pregnancy (Last 2 Months of Gestation)

350	6.4	116	296	16.8	3.7	5.0	17.5	5.2	3.6
400	7.2	128	326	18.5	4.1	5.4	19.3	5.7	4.0
450	7.9	140	358	20.3	4.5	6.0	21.2	6.3	4.4
500	8.6	153	391	22.2	4.9	6.6	23.1	6.8	4.8
550	9.3	166	424	24.0	5.3	7.0	25.1	7.4	5.2
600	10.0	178	455	25.8	5.7	7.6	26.9	8.0	5.6
650	10.6	190	485	27.5	6.1	8.0	28.7	8.5	6.0
700	11.3	202	516	29.2	6.4	8.6	30.5	9.0	6.3
750	12.0	214	546	30.9	6.8	9.1	32.3	9.6	6.7
800	12.6	225	575	32.6	7.2	9.6	34.0	10.0	7.1

Milk Production (Nutrients Required Per Kg of Milk)
% Fat

2.5		28	36	1.1	0.4	0.8	2.7	0.8	
3.0		30	38	1.2	0.4	0.9	2.8	0.8	
3.5		32	41	1.2	0.4	0.9	3.0	0.9	
4.0		34	43	1.3	0.5	1.0	3.2	0.9	
4.5		36	46	1.4	0.5	1.1	3.4	1.0	
5.0		38	48	1.4	0.5	1.1	3.6	1.0	
5.5		39	49	1.5	0.5	1.1	3.6	1.0	
6.0		40	50	1.5	0.5	1.2	3.7	1.1	

[a]Adapted from National Research Council (1971) Nutrient Requirements of Dairy Cattle (22).

[b]Maintenance protein$_g$ = (0.0125) (70.4 × body weight$_{kg}^{0.734}$). Milk protein estimated from (13) and from page 1367 of Duke's 1970 edition: Physiology of Domestic Animals (27). Growth of bovine fetus and membranes estimated from page 556, Vol. 1, of H.H. Mitchell: Comparative Nutrition (28).

[c](Column 3 × 2.13) × 1.2 which assumes a loss of 53% in metabolism of each absorbed amino acid and when calculating diets with only MP, an average loss 20% higher in ensuring adequacy of the more essential amino acids of the ruminant diet. For milk production, (column 3 × 1.05) × 1.2 which assumes 5% loss in metabolism.

[d](Column 4 ÷ 1.2) × (% amino acid present in body or milk protein/100).

A.S. R190 TABLE 3. Metabolizable Protein and Selected Metabolizable Amino Acid Requirements for Feedlot Lambs

1	2	3	4	5	6	7	8	9	10	11	12	13	14
	Daily gain[a]		Prot. content[b]		Net prot.				Metabolizable				
Body Wt.	Body	Wool	Body	Wool	Body[c]	Wool[d]	Maint.[e]	Prot.[f]	Arg.[g]	Cys.[g]	His.[g]	Lys.[g]	Met.[g]
kg	g	g	%	%	g	g	g	g	g	g	g	g	g
25	103	7	14	70	14.4	4.9	10.1	88	2.0	0.8	0.6	1.8	0.7
	140	10	14	70	19.6	7.0	10.1	110	2.5	1.1	0.7	2.1	0.9
	178	12	14	70	24.9	8.4	10.1	130	3.0	1.3	0.8	2.5	1.1
30	122	8	13	67	15.9	5.4	11.7	98	2.2	0.9	0.7	2.0	0.8
	159	11	13	67	20.7	7.4	11.7	119	2.7	1.2	0.8	2.3	1.0
	197	13	13	67	25.6	8.7	11.7	138	3.1	1.5	0.9	2.6	1.2
35	150	10	12	64	18.0	6.4	13.1	113	2.5	1.1	0.7	2.2	0.9
	187	13	12	64	22.4	8.3	13.1	130	3.0	1.3	0.8	2.5	1.1
	225	15	12	64	27.0	9.6	13.1	149	3.3	1.6	0.9	2.9	1.3
40	150	10	11	61	16.5	6.1	14.4	112	2.5	1.1	0.7	2.2	0.9
	187	13	11	61	20.6	7.9	14.4	128	3.0	1.3	0.8	2.5	1.1
	225	15	11	61	24.7	9.1	14.4	144	3.3	1.5	0.8	2.9	1.3
45	131	9	10	58	13.1	5.2	15.8	102	2.3	0.8	0.6	2.0	0.8
	168	12	10	58	16.8	7.0	15.8	119	2.7	1.2	0.7	2.3	1.0
	206	14	10	58	20.6	8.1	15.8	133	3.1	1.3	0.8	2.6	1.1

Amounts Per Animal Per Day

[a]Adapted from National Research Council (20).

[b]Adapted from Mitchell et al. (29–31).

[c]Column 2 × column 4/100.

[d]Column 3 × column 5/100.

[e]Maintenance protein$_g$ = $(0.0125)(70.4 \times$ body weight$^{0.734}_{kg})$.

[f](Columns 6, 7, 8 × 2.5) × 1.2 which assumes a loss 60% loss in metabolism of each absorbed amino acid and when calculating diets with only MP, an average loss 20% higher in insuring adequacy of the more essential amino acids of the ruminant diet.

[g](Column 6 and 8 ÷ 2.5) × (2.5 × % amino acid present in beef protein/100) + (column 7 ÷ 1.2) × (2.5 × % amino acid in wool protein/100).

A.S. R190 TABLE 4. Amino Acid Composition of Proteins Expressed as a Percentage of Dry Matter

Protein	Arg.	Cys.	His.	Iso.	Leu.	Lys.	Met.	Al.	The.	Try.	Val.
Alfalfa, aerial part, dehy	3.2	2.0	2.2	5.4	6.5	3.8	1.1	4.9	3.8	1.1	4.3
Animal blood, dehy	4.0	1.8	4.7	1.1	12.4	6.7	1.3	6.5	4.8	1.3	8.6
Animal meat scraps	7.1	1.3	3.5	3.5	9.5	6.9	1.5	5.0	4.5	0.6	7.8
Animal meat and bone meal	7.2	1.2	1.8	3.4	6.2	6.7	1.3	3.6	3.6	0.6	4.8
Bakery refuse	4.1	1.4	0.9	(3.3)	7.3	2.8	1.3	(4.5)	5.5	0.8	(3.9)
Barley grain	(2.9)	1.6	2.3	4.6	6.9	4.0	1.5	5.4	3.1	1.4	5.4
Barley malt	2.9	1.4	2.2	4.4	5.1	3.6	1.5	4.4	2.9	1.5	5.1
Bean, kidney, leaves	5.7	0.7	2.7	5.4	7.4	4.7	0.7	4.7	4.7	1.4	6.1
Bean, lima, leaves	4.6	(2.0)	1.2	3.5	6.9	3.5	1.3	7.5	4.0	1.8	6.2
Beet, leaves	4.3	(1.9)	1.2	4.3	6.7	5.5	1.6	3.9	3.9	1.2	5.1
Beet pulp	3.4	0.1	2.3	3.4	6.9	7.4	0.1	3.4	4.6	1.0	4.6
Beet soluble prd.	(3.4)	(0.1)	2.3	2.1	3.1	(7.4)	(0.1)	0.5	(4.6)	1.1	1.6
Bermuda grass, coastal	(5.5)	0.5	(2.6)	(8.3)	(12.2)	4.8	1.2	(5.8)	(9.3)	1.3	(7.9)
Bread, bakery	2.7	1.1	0.9	3.3	7.3	1.8	0.8	4.5	5.6	0.7	3.9
Broccoli leaves	7.3	(1.9)	2.4	4.9	9.7	6.9	2.8	9.0	5.2	2.1	6.9
Buckwheat grain	9.8	(1.9)	2.5	3.3	5.0	5.7	1.9	4.1	4.1	1.7	5.0
Carrot leaves	5.5	(1.9)	1.2	6.1	9.2	5.5	2.4	8.5	4.2	1.8	7.3
Cattle whey	2.2	2.7	1.2	6.0	8.3	6.7	1.3	2.3	8.6	1.1	4.6
Cattle whey soluble	5.8	(1.3)	0.6	1.7	1.2	7.2	0.6	0.6	2.9	0.6	1.7
Cattle buttermilk	3.3	1.3	2.5	6.8	9.8	6.9	2.2	4.5	4.5	1.5	7.5
Cattle skim milk	3.4	1.3	2.7	6.8	9.8	7.6	2.7	4.4	4.1	1.2	6.5
Clover, ladino	5.4	2.0	2.4	5.9	10.3	5.9	1.5	5.9	6.4	2.4	6.4
Coconut meal	9.4	(1.5)	1.3	4.3	6.0	2.1	1.3	3.4	2.6	0.9	4.3
Corn, ears	5.0	1.7	(2.0)	(4.7)	(10.4)	2.5	1.7	(4.5)	(3.5)	0.9	(4.3)
Corn grain	4.0	1.0	2.0	4.7	10.4	2.5	1.4	4.5	3.5	1.0	4.2
Corn steep water solubles	4.4	(1.3)	3.7	3.6	9.3	4.2	2.2	3.6	4.4	0.7	6.5
Corn distillers grains w sol.	3.8	1.3	2.6	6.3	8.2	2.8	2.0	6.3	3.7	0.7	6.0
Corn distillers solubles	3.6	1.6	2.4	5.1	7.2	3.1	1.9	5.1	3.4	0.8	5.1
Corn gulten meal	3.4	1.7	2.6	5.9	10.2	1.9	2.6	7.5	3.6	0.5	5.7
Corn hominy feed	4.5	0.9	1.8	3.6	7.2	3.6	0.9	2.7	3.6	0.9	4.5
Cottonseed meal, mech-extd. grnd.	10.0	1.7	2.4	2.9	5.1	3.6	1.2	4.9	2.9	1.2	4.2
Cow pea browse	6.0	(2.0)	2.7	7.1	10.9	6.0	2.7	7.1	6.0	2.7	7.6
Fish solubles	4.1	1.2	4.1	2.7	4.2	4.7	1.9	2.0	1.9	1.0	3.0
Fish meal, anchovy	5.8	0.9	2.4	4.8	7.6	7.8	3.0	4.3	4.3	1.2	5.3
Fish meal, herring	5.9	1.1	2.3	4.5	7.2	7.8	2.9	3.9	4.0	1.1	5.4
Fish meal, menhadden	6.8	1.0	2.6	5.0	8.0	8.5	3.1	4.4	4.4	1.2	5.7
Flax, linseed meal	8.8	1.8	1.9	5.2	5.5	3.6	1.4	4.1	3.3	1.4	4.6
Grains, brewers	5.2	1.1	2.4	5.8	9.9	3.9	1.8	5.6	4.0	1.3	6.5
Grains, distillers	3.7	(1.4)	1.9	5.9	5.1	2.9	1.5	4.0	2.9	0.7	4.4
Grains, distillers solubles	4.6	1.4	2.5	4.9	9.8	4.2	1.8	4.9	3.9	0.7	5.6
Grass, nonspecified	6.8	(1.3)	3.1	9.4	13.5	7.3	2.1	8.9	6.8	2.1	10.4
Kale leaves	6.2	(1.9)	1.9	3.8	7.7	3.8	1.0	5.3	4.3	1.4	5.8
Lespedeza, aerial part	3.6	(2.0)	2.9	13.0	8.7	7.2	0.7	(6.2)	4.3	1.4	5.8
Millet, pruso, grain	2.9	2.2	1.6	3.9	9.4	2.0	2.5	4.7	3.1	1.5	4.7

377

A.S. R190 TABLE 4 (continued). Amino Acid Composition of Proteins Expressed as a Percentage of Dry Matter

Protein	Arg.	Cys.	His.	Iso.	Leu.	Lys.	Met.	Al.	The.	Try.	Val.
Oatgrass, tall Italian	6.1	(2.1)	2.4	7.3	11.0	6.1	2.4	6.1	7.3	2.4	9.8
Oats, aerial part	(5.5)	(2.0)	(2.6)	12.0	16.6	13.0	1.8	(5.8)	14.8	1.8	11.1
Oats rolled grain	6.2	1.6	1.9	3.1	6.8	3.7	1.2	4.4	3.1	1.2	4.4
Oats grain	4.9	1.3	1.8	(3.1)	(6.8)	2.9	1.5	(4.4)	(3.1)	1.1	(4.4)
Orchard grass leaves	21.1	2.0	3.3	(8.3)	12.0	8.3	1.6	3.3	(9.3)	2.5	5.8
Peanut meal, solv-extd.	12.4	1.5	2.5	4.2	7.8	4.9	0.9	5.7	3.2	1.1	5.9
Pea vine leaves	5.6	(2.0)	1.9	5.2	9.4	5.6	0.9	7.0	5.2	1.9	6.6
Rape seed meal, solv-extd.	5.5	0.7	2.6	3.7	6.7	5.3	2.0	3.7	4.1	1.4	5.0
Rice bran w germ	3.9	0.8	1.5	3.1	4.6	3.9	(2.2)	3.1	3.1	0.8	4.6
Rice grain w hull	7.1	1.2	1.1	3.9	6.7	3.5	2.2	3.9	2.8	1.4	5.6
Rye, aerial part	(5.5)	(2.1)	2.6	8.2	12.6	6.8	1.4	5.3	11.1	1.0	6.8
Rye grain	5.0	1.2	2.3	4.7	6.2	4.1	1.6	5.5	3.1	1.1	5.5
Rye distillers grain w sol.	3.7	1.2	2.6	5.5	7.7	3.7	1.5	4.8	4.0	1.1	5.9
Rye distillers solubles	2.9	(1.2)	2.0	5.1	5.1	1.7	1.4	4.8	3.2	0.6	5.4
Ryegrass, annual, aerial part	(5.5)	(2.1)	(2.6)	9.9	14.2	9.3	1.8	7.4	12.9	1.8	8.0
Safflower meal, mech-extd.	6.7	2.4	(2.6)	(6.7)	(8.6)	3.0	1.7	5.0	(4.0)	1.3	(5.2)
Sorghum gluten meal	2.9	(1.7)	1.9	5.5	17.8	1.9	1.4	6.3	3.4	1.0	6.0
Sorghum grain, grain var.	4.3	1.7	3.0	6.0	16.0	3.2	1.4	5.0	3.0	1.1	6.0
Sorghum grain, kafir	3.3	(1.7)	2.5	5.0	14.9	2.5	1.7	5.8	4.1	1.7	5.8
Sorghum grain, milo	3.2	1.6	2.4	4.8	13.1	2.4	0.8	4.1	2.4	0.8	4.8
Soybean flour, solv-extd.	6.1	1.2	1.4	(5.2)	(8.2)	8.3	1.8	3.6	(4.2)	2.0	0.6
Soybean meal, mech-extd.	6.9	1.5	2.6	6.7	8.6	8.6	1.7	5.0	4.1	1.4	5.2
Soybean meal, solv-extd.	7.4	1.5	2.7	5.2	8.2	6.7	1.2	5.3	4.2	1.4	5.2
Soybean meal, wo hulls sol-extd.	7.1	1.6	(2.7)	(5.2)	(8.2)	6.3	1.5	(5.3)	(4.2)	1.3	(5.2)
Spelt straw pulp	3.8	(2.1)	1.5	3.0	5.3	2.3	1.5	3.8	3.0	0.8	3.8
Spinach leaves	3.7	1.9	1.2	3.1	5.9	4.0	1.8	4.0	3.4	0.9	4.3
Sunflower meal, extd.	11.5	2.1	2.2	4.2	6.6	5.2	4.9	4.0	2.6	1.8	(5.2)
Tomato pulp	5.4	(0.8)	1.8	3.2	7.7	7.3	0.5	4.1	3.2	0.9	4.6
Turnip leaves	4.3	(1.9)	1.4	(4.4)	3.6	2.9	2.1	5.0	3.9	1.1	4.6
Wheat bran, dry milled	6.3	1.9	1.9	3.8	5.7	3.8	0.6	3.2	2.5	1.9	4.4
Wheat middlings	5.6	1.7	2.4	4.0	6.8	3.9	1.2	4.0	3.3	1.2	4.6
Wheat grain	4.6	1.9	2.1	3.3	6.4	2.8	1.2	4.5	2.8	(1.5)	3.9
Wheat distillers grain w sol	3.4	(1.9)	2.5	5.9	6.2	2.5	1.6	5.9	3.1	1.2	5.9
Wheat distillers solubles	3.2	(1.9)	2.6	5.1	4.8	2.3	1.3	5.5	3.2	1.6	4.8
Wheat gluten	3.6	2.1	2.0	4.1	6.7	1.9	1.5	5.1	2.6	0.9	4.7
Yeast, brewers	4.6	1.0	2.5	4.6	6.7	7.0	1.5	3.7	4.8	1.1	5.1
Yeast, torula	5.5	1.3	3.0	6.1	7.4	8.0	1.7	6.3	5.5	1.1	6.1
Seven grasses, avg aerial part	5.5	2.1	2.6	8.3	12.2	7.2	1.8	5.8	9.3	1.7	8.0
Seven legumes, avg aerial part	4.9	2.0	2.3	6.5	8.6	5.2	1.3	6.2	4.9	1.8	6.2
Rumen microbes	6.3	0.9	2.2	7.3	9.4	10.0	2.5	6.8	6.4	1.3	7.2
Whole body, mammal	6.8	1.5	2.0	3.7	7.0	7.1	2.1	3.7	3.9	0.8	5.2
Wool, sheep	9.7	11.2	0.8	4.4	8.2	2.8	0.6	3.2	6.5	1.5	5.1

[a]Values in parentheses are estimated values. Words abbreviated in table headings are: Arg. = Arginine; Cys. = Cystine; The. = Threonine; Try. = Tryptophane; and Val. = Valine.

A.S. R190 TABLE 5. Estimation of Rumen Undegraded Feed Protein and Degraded Protein Reformed into Microbial Protein

	Percent Protein Undegraded	Degraded Protein Reformed Microbially
	%	%
1. Concentrate feedstuffs with 13% or less protein in dry matter (DM)	25-50	90-100
2. Concentrates with more than 13% protein in DM	15-30	Less 90
3. Roughages and forages with less than 7% protein in DM	25	100
4. Roughages and forages with 7 to 13% protein in DM	15	50-95
5. Roughages and forages with 13 to 19% protein in DM	5	30-45
6. Roughages and forages with 19% or more protein in DM	0	Less 30
7. Vegetable oil meals and meat meals	25	10-20

A.S. R190 TABLE 6. Metabolizable Protein and Selected Amino Acids Available for Body Purposes Plus Urea Fermentation Potential of Feedstuffs[a]

Feed Description	TDN	Abom. Prot.			Metabolizable						Rumen		UFP	MP of UFP
		Prot.	Und.	Mcb.	Prot.	Arg.	Cys.	His.	Lys.	Met.	Prot.	Degraded		
	%	%	g	g	g	g	g	g	g	g	%	g	g	g
Alfalfa, aerial part	61	19.3	9.6	63.7	47.6	2.7	0.5	1.0	4.2	1.1	95	183.4	−42.8	
Alfalfa, aerial part	57	16.3	8.2	59.5	43.0	2.5	0.5	0.9	3.8	1.0	95	154.8	−34.0	
Animal meat scraps	76	57.1	142.8	79.3	180.1	12.4	2.1	6.7	14.0	3.2	75	428.2	−124.6	
Animal meat + bone meal	72	53.8	134.5	75.2	169.2	11.7	1.9	3.2	12.9	2.8	75	403.5	−117.3	
Barley grain	83	13.0	39.0	86.6	92.4	4.6	2.3	2.1	7.1	2.0	70	91.0	−1.6	
Barley, aerial part	57	8.9	13.4	59.5	47.7	2.9	0.6	1.1	4.4	1.1	85	75.6	−5.8	
Barley straw	41	4.1	10.2	30.8	21.8	1.3	0.3	0.5	1.9	0.5	75	30.8	+4.3	9.6
Beet, aerial part + crown	54	12.7	25.4	56.4	56.0	3.2	0.8	1.0	4.7	1.2	80	101.6	−16.1	
Beet molasses	89	8.7	4.4	82.6	58.0	3.6	0.5	1.3	5.7	1.4	95	82.6	+3.7	8.2
Beet pulp	72	10.0	20.0	75.2	66.2	3.6	0.4	1.5	6.2	1.2	80	80.0	−1.7	
Bermuda grass, aerial part	44	9.5	14.2	45.9	37.5	2.3	0.3	0.9	3.1	0.8	85	80.8	−12.4	
Bluegrass, aerial part	72	17.3	8.6	72.3	53.6	3.3	0.6	1.2	5.1	1.3	95	164.4	−31.9	
Bluegrass, aerial part	69	14.8	7.4	72.0	52.3	3.2	0.6	1.2	5.0	1.3	95	140.6	−24.5	
Bluestem grass, aerial part	64	11.0	16.5	66.8	26.3	3.4	0.7	1.3	6.2	1.3	85	93.5	−9.5	
Bluestem grass, aerial part	62	4.5	11.2	33.8	25.1	1.5	0.4	0.6	2.2	0.5	75	33.8	+11.0	24.5
Bromegrass, aerial part	68	20.3	10.2	71.0	53.0	3.3	0.6	1.2	5.1	1.3	95	192.8	−43.5	
Bromegrass, aerial part	65	6.4	16.0	48.0	40.8	2.5	0.5	1.0	3.7	0.9	75	48.0	+7.1	15.8
Buffalograss, aerial part	59	9.2	13.8	61.6	49.7	3.0	0.6	1.1	4.6	1.2	85	78.2	−5.9	
Canarygrass, Reed, aerial part	66	13.2	13.2	68.9	55.0	3.4	0.6	1.3	5.2	1.3	90	118.8	−17.8	
Canarygrass, Reed, aerial part	56	8.8	13.2	58.5	46.7	2.8	0.6	1.1	4.3	1.1	85	74.8	−5.8	
Cattle milk, whole	130	25.8	38.7	135.7	131.4	7.3	1.3	3.1	12.3	3.3	85	219.3	−29.8	
Cattle milk, skimmed	93	28.5	42.8	97.1	104.2	5.5	1.1	2.5	9.5	2.7	85	242.2	−51.8	
Citrus pulp	88	7.1	17.8	53.2	46.6	2.6	0.4	1.0	3.5	1.0	75	53.2	+13.8	
Citrus molasses	77	10.9	11.0	80.4	62.2	3.7	0.6	1.4	5.5	1.4	90	98.1	−6.3	
Clover, alsike aerial part	60	14.7	14.7	62.6	51.3	3.0	0.6	1.1	4.5	1.1	90	132.3	−24.9	
Clover, crimson aerial part	60	16.9	16.9	62.6	53.3	3.1	0.6	1.2	4.6	1.1	90	152.1	−32.0	
Clover, ladino aerial part	61	23.0	11.5	63.7	49.3	3.0	0.6	1.1	4.5	1.1	95	218.5	−55.3	
Clover, red aerial part	64	17.3	8.6	66.8	49.2	3.0	0.5	1.1	4.6	1.1	95	164.4	−34.8	
Clover, red aerial part	59	14.9	7.4	61.6	43.9	2.7	0.5	1.0	4.1	1.0	95	141.6	−28.6	
Coconut meal, solv extd.	74	23.1	57.8	77.3	101.9	8.0	1.2	1.8	6.1	1.9	75	173.2	−34.3	
Corn, aerial part ensiled	70	8.1	25.9	55.1	55.4	3.0	0.5	1.2	3.8	1.2	68	55.1	+6.4	14.2
Corn cobs	47	2.8	7.0	21.0	11.1	0.6	0.1	0.2	0.6	0.2	75	21.0	+10.0	22.2
Corn ears	89	9.3	32.6	60.4	65.8	3.5	0.6	1.4	4.4	1.3	65	60.4	+11.6	25.8
Corn dis. grain w sol.	88	29.8	74.5	91.9	128.6	6.4	1.5	3.1	8.0	2.9	75	223.5	−47.0	
Corn gluten w bran	82	28.1	70.2	85.6	119.6	5.7	1.6	2.9	6.9	3.0	75	210.8	−44.7	
Corn grain, dent, yellow	91	10.0	38.0	62.0	71.8	3.7	0.7	1.5	4.6	1.4	62	62.0	+11.8	26.3
Corn stover, aerial part wo ears	59	5.9	14.8	44.2	38.7	2.1	0.4	0.8	2.9	0.8	75	44.2	+6.2	13.8
Corn, sweet, cannery res.	70	8.8	17.6	70.4	60.2	3.4	0.6	1.3	4.8	1.3	80	70.4	+1.0	2.2
Cottonseed hulls	41	4.3	10.8	32.2	23.5	1.9	0.3	0.6	1.7	0.5	75	32.2	+3.8	8.5
Cottonseed meal, solv-extd.	75	44.8	112.0	78.3	151.4	13.3	2.2	3.6	8.6	2.5	75	336.0	−92.0	
Cow peas, aerial part	63	18.4	9.2	65.8	48.9	3.1	0.5	1.1	4.6	1.2	95	174.8	−38.9	
Fescue, meadow, aerial part	62	10.5	15.8	64.7	54.0	3.2	0.6	1.2	4.7	1.1	85	89.2	−8.7	
Fish, herring meal	76	76.7	191.8	79.3	224.1	13.3	2.3	5.1	18.6	6.3	75	575.2	−177.1	
Fish, menhadden	75	66.6	166.5	78.3	200.5	13.3	2.2	5.0	17.8	6.0	75	499.5	−150.4	
Flax seed screenings	64	17.4	52.2	66.8	88.3	6.7	1.2	1.8	5.8	1.7	70	121.8	−19.6	
Flax, linseed meal	81	38.8	97.0	84.6	143.0	11.2	2.1	2.9	8.7	4.6	75	291.0	−73.7	
Grains, brewers	66	28.1	70.2	68.9	106.3	6.0	1.1	2.5	6.7	2.2	75	210.8	−50.7	

A.S. R190 TABLE 6 (continued). Metabolizable Protein and Selected Amino Acids Available for Body Purposes Plus Urea Fermentation Potential of Feedstuffs[a]

Feed Description	TDN	Prot.	Abom. Prot. Und.	Abom. Prot. Mcb.	Metabolizable Prot.	Arg.	Cys.	His.	Lys.	Met.	Rumen Prot.	Rumen Degraded	UFP	MP of UFP
	%	%	g	g	g	g	g	g	g	g	%	g	g	g
Lespedeza, aerial part	63	17.8	8.9	65.8	48.6	2.8	0.5	1.1	4.6	1.1	95	169.1	−36.9	
Lespedeza, aerial part	55	13.4	6.7	57.4	40.0	2.4	0.4	0.9	3.8	0.9	95	127.3	−25.0	
Napiergrass, aerial part	52	7.8	11.7	54.3	42.0	2.6	0.5	1.0	3.9	0.9	85	66.3	−4.3	
Prairie grass, aerial part	50	8.1	12.1	52.2	40.6	2.5	0.5	0.9	3.8	0.9	85	68.9	−6.0	
Prairie grass, aerial part	54	4.6	11.5	34.5	26.0	1.6	0.4	0.6	2.3	0.6	75	34.5	+7.8	17.4
Oats, aerial part	61	9.2	13.8	63.7	51.4	3.1	0.6	1.2	4.8	1.2	85	78.2	−5.2	
Oats, straw	52	4.4	11.0	33.0	24.3	1.5	0.3	0.6	2.2	0.5	75	33.0	+7.6	16.9
Oats, grain	76	13.2	39.6	79.3	87.1	5.0	0.9	1.8	6.0	1.8	70	92.4	−4.7	
Oatgrass, tall, aerial part	57	8.2	12.3	59.5	46.7	2.9	0.6	1.1	5.0	1.1	85	69.7	−3.6	
Orchard grass, aerial part	66	13.1	13.1	68.9	54.9	3.4	0.6	1.3	5.3	1.3	90	117.9	−17.5	
Orchard grass, aerial part	57	9.7	14.6	59.5	48.7	3.0	0.6	1.2	4.6	1.1	85	82.4	−8.2	
Peanut meal, solv-extd.	77	51.5	128.8	80.4	168.2	17.7	2.2	4.1	10.8	2.3	75	386.2	−109.2	
Rape seed meal, solv-extd.	69	43.6	109.0	72.0	143.7	8.2	1.0	3.5	9.8	3.1	75	327.0	−91.1	
Redtop grass, aerial part	62	8.1	12.2	64.7	50.7	3.1	0.6	1.2	4.8	1.2	85	68.8	−1.4	
Rice grain, polished	84	8.2	24.6	57.4	56.1	3.0	0.5	1.1	4.2	1.4	70	57.4	+10.8	24.0
Rye straw	31	3.0	7.5	22.5	12.8	0.8	0.2	0.3	1.1	0.3	75	22.5	+3.5	7.8
Rye grain	85	13.4	40.2	88.7	95.1	5.5	1.0	2.2	7.4	2.1	70	93.8	−1.8	
Rye, distillers grain	46	24.1	60.2	48.0	80.6	3.7	0.9	2.0	4.6	1.4	75	180.8	−47.4	
Ryegrass, Italian, aerial part	62	16.3	8.2	64.7	47.2	2.9	0.5	1.0	4.5	1.1	95	154.8	−32.2	
Safflower meal, solv-exted.	55	23.3	58.2	57.4	86.3	5.6	1.6	1.7	5.0	1.7	75	174.8	−41.9	
Sorghum, grain var, grain	83	12.5	60.0	65.0	94.0	4.8	1.3	2.5	5.7	1.8	52	65.0	+7.7	17.1
Sorghum distillers grain	82	33.2	83.0	85.6	131.2	6.5	1.8	3.9	8.4	2.9	75	249.0	−58.4	
Sorghum milo grain	80	12.4	59.5	64.5	93.2	4.6	1.3	2.8	5.9	2.1	52	64.5	+6.8	15.1
Sudangrass, aerial part	59	12.7	12.7	61.6	48.7	3.0	0.6	1.1	4.5	1.1	90	114.3	−18.8	
Sudangrass, aerial part	63	8.7	13.0	65.8	52.3	3.2	0.6	1.2	4.9	1.2	85	74.0	−2.9	
Soybean plant, aerial part	52	16.3	8.2	54.3	38.8	2.4	0.4	0.8	3.5	0.9	95	154.8	−35.9	
Soybean hulls	45	13.7	6.8	47.0	31.7	1.9	0.4	0.7	2.9	0.7	95	130.2	−29.7	
Soybean straw	38	5.5	13.8	39.7	32.2	1.8	0.4	0.7	2.6	0.6	75	41.2	−0.6	
Soybean meal, solv-extd.	81	51.5	128.8	84.6	171.6	12.1	2.2	4.4	13.3	2.8	75	386.2	−107.7	
Sugarcane molasses	68	10.7	16.0	71.0	59.2	3.4	0.5	1.3	4.8	1.3	85	91.0	−7.2	
Sugarcane molasses	91	4.3	4.3	38.7	22.8	1.4	0.2	0.5	2.0	0.5	90	38.7	+20.1	44.7
Sunflower meal, solv-extd.	65	50.3	125.8	67.9	155.5	15.7	2.8	5.2	10.1	6.6	75	377.2	−110.5	
Timothy, aerial part	62	12.3	18.4	64.7	56.3	3.4	0.7	1.3	5.2	1.3	85	104.6	−14.2	
Timothy, aerial part	59	8.7	13.0	61.6	49.0	3.0	0.6	1.1	4.6	1.1	85	74.0	−4.4	
Trefoil, Birdsft., aerial part	75	28.0	14.0	78.3	63.2	3.8	0.7	1.4	5.7	1.4	95	266.0	−67.0	
Trefoil, Birdstf., aerial part	61	15.6	7.8	63.7	46.0	2.8	0.5	1.0	4.3	1.1	95	148.2	−30.2	
Urea, limited to + UFP values	0	280.0	−	−	2225.0	140.2	20.0	49.0	225.5	55.6	−	−	−	
Wheat bran	70	18.0	54.0	73.1	95.1	6.0	1.3	1.9	6.5	1.5	70	126.0	−18.9	
Wheat grain	88	14.3	42.9	91.9	100.1	5.6	1.3	2.1	7.2	2.0	70	100.1	−2.9	
Wheat grain screenings	77	16.9	50.7	80.4	98.0	5.4	1.3	2.1	6.5	1.9	70	118.3	−13.5	
Wheat mill run	81	17.0	51.0	84.6	101.6	5.6	1.4	2.2	7.2	1.9	70	119.0	−12.3	
Wheatgrass, aerial part	61	5.3	13.2	39.8	31.7	1.9	0.4	0.8	2.8	0.7	75	39.8	+8.5	18.9
Yeast, brewers	78	47.9	119.8	81.4	160.9	8.3	1.6	3.9	12.8	2.9	75	359.2	−99.2	

[a]Amounts in percent or grams per kilogram of dry matter. Words abbreviated in table headings are: TDN = Total Digestible Nutrients; Prot. = Protein; Abom. = Abomasal; Und. = Undegraded; Mcb. = Microbial; Arg. = Arginine; Cys. = Cystine; His. = Histidine; Lys. = Lysine; Met. = Methionine; UFP = Urea Fermentation Potential; and MP = Metabolizable Protein.

USE OF METABOLIZABLE PROTEIN AND UREA FERMENTATION VALUES IN FORMULATING RATIONS

The only effective way to understand the concepts previously presented is to apply the principles to some typical rations. Following is a series of example rations for feedlot cattle with a brief discussion about each ration and the computations made in determining the proper protein supplementation on that ration. In the following examples the rations are not recommended feeding programs but are used only as examples for calculation purposes.

A sample worksheet is provided with the appendix section at the end of this textbook.

Example 1

Assume a 200-kg large-frame heifer calf consumes 2.7% b.w. as daily dry matter (5.4 kg d.m.). The ration is 10% alfalfa hay, 20% gr. corn cobs and 70% gr. corn (dry matter basis). This means a daily intake of 0.54 kg alfalfa hay, 1.08 kg gr. corn cobs and 3.78 kg gr. corn.

The net energy system as described in Chapter 2 is used to calculate energy intake and predict an expected daily gain of 1.2 kg (refer to tables 3–2 and 8–1 for NE feed values and NE requirements, respectively). This animal needs 539.0 g of MP to realize this rate of gain (see A.S. R190 Table 1).

The MP and UFP values from the example ration can now be calculated. The alfalfa hay provides 23.2 g of MP (alfalfa hay, 57% TDN, contains 43.0 g MP/kg from A.S. R190 Table 6; therefore, 0.54 kg × 43.0 = 23.2). Gr. corn cobs provide 12.0 g MP

1.08 kg cobs × 11.1 g/kg MP = 12.0

and gr. corn supplies 271.4 g MP

3.78 kg corn × 71.8 g/kg MP = 271.4

As for the UFP values, alfalfa hay has a −34 g/kg UFP value (A.S. R190 Table 6). This multiplied by 0.54 kg of hay gives a −18.4 g of urea as a UFP value from the alfalfa hay. The gr. corn cobs and gr. corn have positive UFP values of +10.8 and +44.6, respectively. These were determined by multiplying the UFP value of each feed by the kg of that feedstuff the animal consumed daily. Thus,

1.08 kg gr. corn cobs × 10.0 = 10.8 and 3.78 kg gr corn × 11.8 = 44.6.

Step 6 of the worksheet indicates that if there is a negative UFP from the energy feeds (meaning there is excess ammonia in the ammonia pool of the rumen), it should be considered as a source of MP from satisfaction of the positive UFP value of the energy feeds. This ration contained −18.4 UFP, which is multiplied by 2.225 (the MP value of urea calculated as g of MP/g urea) to yield 40.9 g MP from the negative UFP feed. Thus, the total MP supplied by the ration is

23.2 + 12.0 + 271.4 + 40.9 = 347.5 g.

Since it took 18.4 g of positive UFP to balance out the −18.4 UFP from hay, the net positive UFP remaining is +37.0 g.

The MP needed from a supplemental source (calculated in Step 7) is 191.5 g

539 − 347.5 = 191.5.

If this was supplied from urea (containing 2.225 g MP/g urea as found in A.S. R190 Table 6), then

191.5 ÷ 2.225 = 86.1 g of urea.

However, the net positive UFP remaining was only 37.0 g. Thus, the total grams of urea needed to supply the MP exceed the net positive UFP remaining and urea cannot be fed to achieve predicted daily gain. This means that this ration does not have enough energy to ferment the 86.1 g of needed urea, and the entire supplemental MP source must come from a natural protein such as soybean meal. A.S. 190 Table 6 indicates that soybean meal contains 171.6 g/kg MP and a −107.7 g/kg UFP value. The 37.0 g of net positive UFP from the energy feeds can be satisfied by the negative UFP value of the soybean meal. Therefore

37.0 × 2.225 = 82.3 g of MP

from the net positive UFP. Now, the original MP needed minus the MP from satisfying the positive UFP leaves 109.2 g MP to come from the soybean meal

$(191.5 - 82.3 = 109.2).$

By dividing 109.2 g MP needed from natural protein sources by 171.6 (MP values of soybean meal), we find that 0.64 kg of soybean meal is needed to provide the supplemental MP needs

$(109.2 \div 171.6 = 0.64).$

Example 2

Assume a 300-kg medium-frame yearling heifer consumes 2.5% b.w. as daily dry matter (7.5 kg d.m.). If the ration is 50% shelled corn and 50% corn silage (dry matter basis), this means a daily intake of 3.75 kg shelled corn and 3.75 kg corn silage. The net energy system as described in Chapter 2 is again used to calculate energy intake and predict an expected daily gain of 1.3 kg (refer to tables 3–2 and 8–1 for NE feed values and NE requirements, respectively). This animal needs 584 g of MP to realize this rate of gain (see A.S. R190 Table 1).

The ration fed in this case is a combination of shelled corn and corn silage. The shelled corn provides 269.2 g of MP (corn contains 71.8 g/kg MP from A.S. R190 Table 6, therefore, 3.75 × 71.8 = 269.2) and a positive UFP value of 44.2 g (corn has a UFP value of +11.8 g/kg from A.S. R190 Table 6, therefore, 3.75 kg × 11.8 = 44.2). The corn silage provides 207.8 g MP

$(3.75 \text{ kg} \times 55.4 \text{ g/kg})$

and a positive UFP value of 24.0 g

$(3.75 \text{ kg} \times 6.4).$

Since there was no negative UFP to satisfy any of the positive UFP from these feeds, the total MP provided was 477 g and there were 68.2 g of net positive UFP remaining. Thus, this ration has the energy capable of fermenting 68.2 g of urea into microbial protein.

Urea contains 2.225 g of MP/g (from A.S. R190 Table 6), therefore, the supplemental MP needs of 103 g

$(580 - 477 = 103)$

can be met by 46.3 of urea

$(107 \div 2.225 = 46.3).$

This amount of urea can be satisfactorily fed since it satisfies the MP requirement of the animal and does not exceed the net positive UFP value of the ration.

Example 3

Assume a 400-kg feedlot large-frame steer consumes 2.0% b.w. as daily dry matter (8.0 kg d.m.). In this ration, assume 75% shelled corn and 25% corn silage (dry matter basis) or a total daily intake of 6.0 kg shelled corn and 2.0 kg corn silage.

The net energy system as described in Chapter 2 is again used to calculate energy intake and predict an expected daily gain of approximately 1.4 kg (refer to tables 3–2 and 8–1 for NE feed values and NE requirements, respectively). This animal will require 612.0 g MP as found in A.S. R190 Table 1 for 400-kg steers.

The shelled corn provides 430.8 g MP

$(6.0 \times 71.8 = 430.8)$

and the corn silage provides 110.8 g MP

$(2.0 \times 55.4 = 110.8).$

There is no negative UFP value to contribute additional MP, therefore the total MP from the energy sources is 541.6 g

$(430.8 + 110.8 = 541.6)$

and there were 83.6 g of positive UFP remaining. Thus, the ration has the energy capable of fermenting 83.6 g of urea into microbial protein.

Urea contains 2.225 g of MP/g (from A.S. R190 Table 6), therefore, the supplemental MP needs of 70.4 g

$$(612.0 - 541.6 = 70.4)$$

can be met by feeding 31.6 g urea

$$(70.4 \div 2.225 = 31.6).$$

This amount of urea can be satisfactorily fed since it satisfies the MP requirement of the animal and does not exceed the net positive UFP value of the ration.

In each of these examples, the amount of energy feeds were established and if a supplemental protein source was needed, it was simply added to the existing ration. This may have some influence on feed or energy intake but generally not enough to warrant reformulation of the entire ration. It should be noted that additional emphasis must be placed on fortification of the rations with minerals and vitamins.

EXAMPLE 1

METABOLIZABLE PROTEIN FORMULATION WORKSHEET FOR CATTLE AND SHEEP
(When Individual Amino Acid Requirements Are Not Computed)

1. Kind of Animal _Large-frame heifer calf_   Body Weight ___200___ kg
   Daily Dry Matter Consumption: ___2.7___   % of Body Weight = ___5.4___ kg

2. Daily Ration Ingredients Available in Satisfying Energy Needs:

IFN	Ration Ingredient	Amount, %	D.M. kg	NE_m Mcal	NE_g Mcal	NE_L Mcal
(1-00-063)	Alfalfa hay	10.0	0.54	0.69	0.38	
(1-02-782)	Gr. corn cobs	20.0	1.08	1.03	0.45	
(4-02-931)	Gr. corn	70.0	3.78	8.46	5.85	
	TOTAL	100.0	(a) 5.40	(b) 10.18	(c) 6.68	(d)
	Mcal/kg =			(e) 1.88	(f) 1.24	(h)
				(b ÷ a)	(c ÷ a)	(d ÷ a)

3. Daily Net Energy Requirement from NRC Tables:

   NE_m required, Mcal (i) ___4.10___ (from tables)
   Total kg. ration D.M. (a) ___5.40___ (a)
   Kg. ration to supply NE_m (j) ___2.18___ (i ÷ e)
   Kg. ration left for NE_g or NE_L (k) ___3.22___ (a − j)
   NE_g or NE_L available, Mcal (m) ___3.99___ (k x f or h)
   Expected daily gain or milk prod., kg ___1.20___ (m compared to tables)

4. MP Required for Expected Daily Gain or Milk Production (n) ___539___ g. (from tables)

5. MP and UFP Supplied by Above Daily Ration:

Ration Ingredient	Amount, D.M., kg	MP, g	UFP, g
Alfalfa hay	0.54	23.2	−18.4
Gr. corn cobs	1.08	12.0	+10.8
Gr. corn	3.78	271.4	+44.6

6. MP (from satisfaction of positive UFP value) to come from negative UFP feeds [total negative UFP (not exceeding the positive UFP)] ___18.4___ x 2.225 = ___40.9___ g from neg. UFP feeds

   Total MP = (o) ___347.5___
   Net positive UFP remaining (p) ___+37.0___

7. Supplemental MP Needs:
   a. From urea: MP needed (n − o) = (q) ___191.5___ g ÷ 2.225 = ___86.1___ g urea*
      *Note: If total g of urea needed exceed the net positive UFP remaining (p), urea *cannot* be fed in order to achieve predicted gain or milk production. Thus, feed natural protein.
   b. From natural protein source: ___SBM___ containing ___171.6___ g /kg MP and ___−107.7___ UFP
      If there is a net positive UFP remaining (p), then this must be multiplied by 2.225 to predict the amount of MP which will be realized from the negative UFP value of the natural protein source to be added. Thus:
      (1) Net pos. UFP remaining (p) ___37.0___ x 2.225 = (r) ___82.3___ g MP from pos. UFP
      (2) MP needed (q) ___191.5___ g − (r) ___82.3___ = (s) ___109.2___ g MP from natural protein source
      (3) (s) ___109.2___ ÷ ___171.6___ (MP value of natural protein) = ___0.64___ kg. natural protein

8. Final Supplemented Ration:

Ingredient	kg	%
Alfalfa hay	0.54	8.94
Gr. corn cobs	1.08	17.88
Gr. corn	3.78	62.58
SBM	0.64	10.60
TOTAL	6.04	100%

EXAMPLE 2

METABOLIZABLE PROTEIN FORMULATION WORKSHEET FOR CATTLE AND SHEEP
(When Individual Amino Acid Requirements Are Not Computed)

1. Kind of Animal _Medium-frame yearling heifer_ Body Weight _300_ kg
   Daily Dry Matter Consumption: _2.5_ % of Body Weight = _7.5_ kg

2. Daily Ration Ingredients Available in Satisfying Energy Needs:

IFN	Ration Ingredient	Amount, %	D.M. kg	NE_m Mcal	NE_g Mcal	NE_L Mcal
(4-02-931)	Shelled corn	50.0	3.75	8.40	5.81	
(3-02-823)	Corn silage	50.0	3.75	6.11	3.86	
	TOTAL	100.0	(a) 7.50	(b) 14.51	(c) 9.67	(d)
			Mcal/kg =	(e) 1.93	(f) 1.29	(h)
				(b ÷ a)	(c ÷ a)	(d ÷ a)

3. Daily Net Energy Requirement from NRC Tables:

   NE_m required, Mcal (i) _5.55_ (from tables)
   Total kg. ration D.M. (a) _7.50_ (a)
   Kg. ration to supply NE_m (j) _2.88_ (i ÷ e)
   Kg. ration left for NE_g or NE_L (k) _4.62_ (a − j)
   NE_g or NE_L available, Mcal (m) _5.96_ (k x f or h)
   Expected daily gain or milk prod., kg _1.30_ (m compared to tables)

4. MP Required for Expected Daily Gain or Milk Production (n) _580_ g. (from tables)

5. MP and UFP Supplied by Above Daily Ration:

Ration Ingredient	Amount, D.M., kg	MP, g	UFP, g
Shelled corn	3.75	269.2	+ 44.2
Corn silage	3.75	207.8	+ 24.0

6. MP (from satisfaction of positive UFP value) to come from negative UFP feeds [total negative UFP (not exceeding the positive UFP)] _0_ x 2.225 = _0_ g from neg. UFP feeds

   Total MP = (o) _477.0_
   Net positive UFP remaining (p) _+68.2_

7. Supplemental MP Needs:
   a. From urea: MP needed (n − o) = (q) _103.0_ g ÷ 2.225 = _46.3_ g urea*
      *Note: If total g of urea needed exceed the net positive UFP remaining (p), urea *cannot* be fed in order to achieve predicted gain or milk production. Thus, feed natural protein.

   b. From natural protein source: _—_ containing _—_ g /kg MP and _—_ UFP
      If there is a net positive UFP remaining (p), then this must be multiplied by 2.225 to predict the amount of MP which will be realized from the negative UFP value of the natural protein source to be added. Thus:
      (1) Net pos. UFP remaining (p) _—_ x 2.225 = (r) _—_ g MP from pos. UFP
      (2) MP needed (q) _—_ g − (r) _—_ = (s) _—_ g MP from natural protein source
      (3) (s) _—_ ÷ _—_ (MP value of natural protein) = _—_ kg. natural protein

8. Final Supplemented Ration:

Ingredient	kg	%
Shelled corn	3.750	49.70
Corn silage	3.750	49.70
Urea	0.046	0.61
TOTAL	7.546	100%

EXAMPLE 3

METABOLIZABLE PROTEIN FORMULATION WORKSHEET FOR CATTLE AND SHEEP
(When Individual Amino Acid Requirements Are Not Computed)

1. Kind of Animal _Large-frame steer_   Body Weight _400_ kg
   Daily Dry Matter Consumption: _2.0_   % of Body Weight = _8.0_ kg

2. Daily Ration Ingredients Available in Satisfying Energy Needs:

IFN	Ration Ingredient	Amount, %	D.M. kg	NE_m Mcal	NE_g Mcal	NE_L Mcal
(4-02-931)	Shelled corn	75.0	6.0	13.44	9.30	
(3-02-823)	Corn silage	25.0	2.0	3.26	2.06	
	TOTAL	100.0	(a) 8.0	(b) 16.70	(c) 11.36	(d)
			Mcal/kg =	(e) 2.08	(f) 1.42	(h)
				(b ÷ a)	(c ÷ a)	(d ÷ a)

3. Daily Net Energy Requirement from NRC Tables:

   NE_m required, Mcal   (i) _6.89_ (from tables)

   Total kg. ration D.M.   (a) _8.00_ (a)

   Kg. ration to supply NE_m (j) _3.31_ (i ÷ e)

   Kg. ration left for NE_g or NE_L (k) _4.69_ (a − j)

   NE_g or NE_L available, Mcal   (m) _6.65_ (k x f or h)

   Expected daily gain or milk prod., kg   _1.4_ (m compared to tables)

4. MP Required for Expected Daily Gain or Milk Production (n) _612_ g. (from tables)

5. MP and UFP Supplied by Above Daily Ration:

Ration Ingredient	Amount, D.M., kg	MP, g	UFP, g
Shelled corn	6.0	430.8	+ 70.8
Corn silage	2.0	110.8	+ 12.8

6. MP (from satisfaction of positive UFP value) to come from negative UFP feeds [total negative UFP (not exceeding the positive UFP)] _0_ x 2.225 = _0_ g from neg. UFP feeds

        Total MP = (o) _541.6_
        Net positive UFP remaining (p) _+83.6_

7. Supplemental MP Needs:
   a. From urea: MP needed (n − o) = (q) _70.4_ g ÷ 2.225 = _31.6_ g urea*
      *Note: If total g of urea needed exceed the net positive UFP remaining (p), urea *cannot* be fed in order to achieve predicted gain or milk production. Thus, feed natural protein.

   b. From natural protein source: _—_ containing _—_ g /kg MP and _—_ UFP
      If there is a net positive UFP remaining (p), then this must be multiplied by 2.225 to predict the amount of MP which will be realized from the negative UFP value of the natural protein source to be added. Thus:
      (1) Net pos. UFP remaining (p) _—_ x 2.225 = (r) _—_ g MP from pos. UFP
      (2) MP needed (q) _—_ g − (r) _—_ = (s) _—_ g MP from natural protein source
      (3) (s) _—_ ÷ _—_ (MP value of natural protein) = _—_ kg. natural protein

8. Final Supplemented Ration:

Ingredient	kg	%
Shelled corn	6.000	74.70
Corn silage	2.000	24.90
Urea	0.032	0.40
TOTAL	8.032	100%

Study Questions and Problems

Chapter 8

I. True or False Questions

1. Milk production by beef cows will average about 10 to 25 lb/head daily over approximately a 200-day lactation period.
2. Creep feeding beef calves can be expected to give the greatest response when pasture is short or quality is poor.
3. Preconditioned calves have been accustomed to grain consumption.
4. Under extremely cold environmental conditions, beef heifers need more dietary protein in order to maintain body heat.
5. High levels of dietary calcium may cause a greater incidence of urinary calculi in feedlot steers.
6. A full feed of grain for feedlot cattle is commonly considered to be approximately 2 lb of grain per 100 lb of body weight, plus supplement and forages.
7. Conditioner levels of vitamin A fed to recently purchased beef cattle are commonly 30,000 to 50,000 IUs per head daily.
8. Feeding certain antibiotics to finishing steers on high-concentrate diets can reduce the incidence of abscessed livers.
9. Bred heifers after the age of 15 months can be fed and managed like the cow herd, and should not be fed with steers.
10. Beef cows on mixed grass-hay in the winter time are less likely to need limestone in their mineral mix than are steers on a corn grain-corn silage ration.

II. Multiple Choice (select the single best answer)

1. Feedlot beef cattle consuming only corn silage may consume as much as ___?___ % of their body weight daily (as-fed basis).
   a. 2% to 3%
   b. 3% to 5%
   c. 6% to 7%
   d. 7% to 9%
   e. More than any of the above
2. Metabolizable protein is synonymous with
   a. Crude protein in the pig.
   b. Net protein.
   c. Digestible protein in the pig.
   d. Metabolizable energy.
   e. Answers 3 and 4.
3. The UFP value of a beef diet predicts
   a. The amount of urea in the feed.
   b. The amount of urea that can be utilized by available energy.
   c. The amount of urea in the metabolizable protein.
   d. The amount of urea recycled through the saliva from a given feed.
   e. None of the above.
4. Which intake of a high grain diet would be most appropriate for an 800 lb steer (dry matter as percent of body weight)?
   a. 1½ to 2
   b. 2 to 2½
   c. 2½ to 3
   d. 3 to 3½
   e. More than any of the above
5. When would you recommend creep feeding calves?
   a. For replacement heifers
   b. For steers going immediately into the feedlot on a grain diet
   c. For steers to be fed a high roughage growing ration before finishing
   d. For steers to be used to glean stock fields before entering the feedlot
   e. Only answers 3 and 4

6. Which of the following roughages for growing feeder calves generally needs to be supplemented with a protein supplement?
   a. Alfalfa hay and legume silage
   b. Alfalfa hay, corn silage and sorghum silage
   c. Legume silage and corn silage
   d. Corn and sorghum silage
   e. All of the above
7. Trace mineral supplementation of beef cattle rations would probably be needed most with
   a. High-grain finishing rations.
   b. High-hay growing rations.
   c. Summer pasture.
   d. Alfalfa hay-corn silage wintering rations.
   e. All of the above.
8. Gestating beef cows should consume approximately __?__ lb of TDN per day.
   a. 2 to 6
   b. 6 to 10
   c. 10 to 15
   d. 15 to 20
   e. Over 20
9. Approximately how many lb of creep feed will be required to get an average of 50 more lb of body weight at weaning?
   a. 50 to 100 lb
   b. 100 to 300 lb
   c. 300 to 400 lb
   d. 400 to 500 lb
   e. Over 500 lb
10. Developing replacement bulls should be fed a diet to produce __?__ lb of daily gain from weaning to 14 months of age.
   a. 3/4 to 1 lb
   b. 1 to 2 lb
   c. 2 to 3 lb
   d. 3 to 4 lb
   e. Less than any of the above

III. *Matching or Identification*

Match the appropriate answer to the following statements

a. 0	f. 6 to 8
b. 0.2 to 0.3	g. 9 to 14
c. 0.75 to 1.25	h. 10 to 25
d. 2	i. 30 to 50
e. 2 to 3	j. none of the above

1. Full feed of grain, % of body weight
2. Rate of daily gain for replacement heifers from weaning to 15 months of age, lb
3. Percent Ca and P in pregnant cow diet
4. Age at which young calves may be first implanted, months
5. Average age at which beef calves are weaned, months
6. Average milk production by beef cows, lb/day
7. Maximum daily consumption of urea by creep-fed calves, lb/day
8. Percent protein in feedlot cattle diets
9. Percent protein in pregnant beef cow diets
10. Conditioner levels of vitamin A fed newly purchased feeder calves, 1000s IU

IV. Short Answer Questions

1. Give free-choice mineral recommendations for
   a. cattle on liberal grain feeding.
   b. cattle primarily on forage.
2. What nutrient deficiencies need to be recognized for dry pregnant cows to utilize corn stalks or husklage?
3. What are the advantages and disadvantages of creep feeding calves?
4. Describe a preconditioning program.
5. Outline a procedure for getting cattle on a full feed of grain.
6. Why is the amount of ration energy important when urea is fed to cattle?
7. Define "metabolizable protein" for ruminants.
8. What determines whether a feed has a positive or negative urea fermentation potential (UFP)?

V. Problems

1. A farmer is feeding a concentrate mix to 900 lb steers that has a NE_m value of 0.80 Mcal/lb and NE_g value of 0.50 Mcal/lb. How much should be fed daily if the maintenance requirement is 7.06 Mcal/day and the requirement for 2.2 lb daily gain is 5.40 Mcal?
2. The metabolizable protein (MP) requirement is 400 g, and the MP of the ration is 380 g. The ration contains feed having a -40 g urea fermentation potential (UFP) and $+10$ g UFP. Should an additional source of protein be fed to meet the MP requirement? If so, what should be fed?
3. A cattle feeder needs a 45% crude protein supplement to properly balance a ration for his feeder calves. He has available all the feeds listed below and has asked you to prepare this supplement. Use *all* of the protein feeds to formulate your mixture. Note the fixed ingredients and have urea supply 20% crude protein equivalent in the mixture.

Feedstuffs available

Hominy feed	10% CP
Brewer's dried grains	23% CP
Linseed meal	32% CP
Corn gluten meal	38% CP
Urea	281% CPE

Composition of the supplement

Amount, lb		Protein, lb
4.0	lb salt	0.0
6.0	lb limestone	0.0
3.0	lb vitamin-TM premix	0.0
	lb urea	
	lb hominy feed	
	lb brewer's dried grains	
	lb linseed meal	
	lb corn gluten meal	
100.0	lb TOTAL	lb.

Notes

Notes

9

Dairy Cattle Feeding Guides

Chapter Goals

- Outline and discuss a life-cycle dairy cattle feeding program.
- Identify specific nutrient needs or additives used within the life cycle.
- Identify nutrient-related diseases or disorders within the dairy cattle life cycle.
- Formulate a lactating dairy cow diet.

Modern dairy nutrition encompasses a wide field of activities involving a constantly increasing application of many sciences. Only the briefest reference can be made here to its numerous phases.

The material is presented as guides for *Feeding for Milk Production* and *Feeding for Dairy Calves*. Additional and more detailed information may be obtained from publications of various land grant universities and other colleges.

FEEDING FOR MILK PRODUCTION

I. General Comments

A. The dairy industry is presently successfully utilizing a considerable number of widely different systems for handling milk-producing cows. The type of system used is partially dependent upon the geographic area and feedstuff supply. The traditional pasture system is continuing in areas of sparse human population, while drylot systems with minimum roughage and higher quantities of less bulky feeds such as concentrates are being used surrounding some of the larger cities.

Feeding, more than any other single factor, determines the productivity of lactating dairy cows. Approximately 75% of the differences in milk production between cows is determined by environmental factors, with feed making up the largest portion. Feed represents about 50% of the total costs. Therefore, a good feeding program is necessary for profitable milk production.

B. Nutrient requirements for lactation are considerable; they are often several times the maintenance requirements. In reviewing feedstuff characteristics in Chapter 3, one should remember that

1. There is more variation in protein content than in energy content among the various forages and concentrates.
2. In balancing rations for dairy cows, we are primarily concerned with energy ($NE_{lactation}$ or TDN) and either crude or digestible protein needs. Major mineral needs may be taken care of by adding 1% to 2% dicalcium phosphate or steamed bone meal plus 1% salt to the concentrate mix. These minerals should also be offered free-choice.

II. Forage Consumption

A. While the rumen capacity of a dairy cow is considerable, she cannot eat enough forage to meet her nutrient needs during lactation.

1. Estimated daily intake for forages is based on *body weight* and on *forage quality*. As a guide for estimating consumption of hay (dry matter basis) fed free choice the following thumb rules are suggested

Forage Quality	Daily Intake (% of b.w.)
Excellent	3.0
Good	2.5
Average	2.0
Fair	1.5
Poor	1.0

2. If cows are allowed to consume all the forage they will, they may not have sufficient capacity to consume enough grain to meet the energy needs of high milk production. Consequently, sometimes a maximum forage consumption level is established. In some states the Dairy Herd Improvement Association (DHIA) uses 1.75% to 2.0% of body weight as such a maximum.
3. Silage consumption (as-fed basis) can be estimated at 3 lb for each 1 lb of hay intake expected.
4. Pasture intake will usually be higher than silage at the same dry matter percentages.

B. There is some evidence to indicate that forage intake can be stimulated by feeding several times daily and by providing a variety of forages. There appears to be a difference in forage appetite between cows that is not accounted for by body size.

III. Concentrates for Dairy Cattle
A. *Concentrate Mixture*

The concentrate mixture will include grains, mill feeds, protein supplements and minerals.
1. The kind of mixture fed will vary with the kind of forage fed (a high-protein mix will be needed with a low-protein forage); the availability and cost of ingredients must also be considered.
2. The level or amount of concentrate mix fed will depend on
   a. The amount of forage consumed.
   b. The amount of milk produced.
   c. The composition (fat %) of the milk produced.
   d. Limit % of concentrates to a maximum of about 60% regardless of comparative cost of grains and roughages. Rations with more than about 60% concentrates generally result in changes in proportion of ruminal volatile fatty acids which in turn result in changes in milk composition (reduced fat content).

B. Intake is affected by palatability and the time available to consume concentrates in the barn or milking parlor.
C. Some suggested concentrate mixtures are given in Table 9–1.
D. Concentrate feeding levels are variable and while rules for feeding are available, they tend to overfeed the low producer and underfeed the high producer.

IV. Steps in Balancing a Ration for Lactating Dairy Cows (see Classroom Exercise at end of chapter)
A. *Establish Needs*

Maintenance *plus* milk production, and reproduction needs.
B. *Determine Feeds Available and their Composition*
1. Forage
   a. Kind—(legume vs grass) will determine kind of concentrate mix needed.
   b. Quality—will determine consumption level.
2. Concentrate mix
   a. Kind—protein level needed is determined by kind of forage fed (protein level of forage).
   b. Amount—determined by amount of energy needed (beyond that supplied by forages). The concentrate mix will usually contain 1% to 2% dicalcium phosphate or steamed bone meal and 1% to 2% salt; minerals and salt will also be fed free-choice. Lower levels will be adequate for cattle on forages containing a high proportion of legumes. The mineral addition will lower the energy and protein content in the concentrate mixture, which can be compensated for by feeding more.

C. *Establish Feeding Levels*
1. Forage—Use rules of thumb according to body weight and forage quality.
2. Concentrate mix—to supply additional energy needs (NE_L or TDN).

TABLE 9–1. Suggested Concentrate Mixtures for Lactating Dairy Cows Fed Different Quality Forages[1]

Ingredients	IFN	Feed with High-Protein Feed Forages		Feed with Medium-Protein Forages			Feed with Low Protein Forages	
		1	2	3	4	5	6	7
Corn, gr.	4–02–931	—	70	—	—	—	50	—
Ground ear corn	4–02–849	92	—	85	74	78	—	61
Oats, gr. or rolled	4–03–309	—	28	—	—	—	—	—
Wheat bran	4–05–190	—	—	—	—	—	23	—
Molasses, liquid	4–04–696	—	—	—	—	—	—	6
Urea (281% CPE)[2]		—	—	1	—	—	—	—
Soybean meal[3]	5–04–604	6	—	12	—	20	24	30
Soybeans, cracked	5–04–610	—	—	—	24	—	—	—
Dicalcium phosphate[4]	6–01–080	1	1	1	1	1	1	1
Limestone	6–02–632	—	—	—	—	—	1	1
TM salt & vitamin premix		1	1	1	1	1	1	1
		100	100	100	100	100	100	100

Calculated analysis:
1) *As-fed basis*

		1	2	3	4	5	6	7
Crude protein, %		9.9	9.5	14.9	15.2	15.2	18.9	18.7
TDN, %		71.4	74.2	70.8	73.5	71.7	71.6	70.5
NE_L, Mcal/kg		1.65	1.72	1.63	1.70	1.65	1.66	1.63
Calcium, %		0.29	0.25	0.30	0.34	0.32	0.70	0.76
Phosphorus, %		0.45	0.48	0.47	0.51	0.51	0.76	0.55
Dry matter, %		86.9	88.1	87.3	88.1	87.4	88.6	87.1

2) *Dry matter basis*

		1	2	3	4	5	6	7
Crude protein, %		11.4	10.8	17.1	17.2	17.4	21.3	21.4
TDN, %		82.2	84.2	81.1	83.4	82.0	80.8	80.9
NE_L, Mcal/kg		1.90	1.95	1.87	1.93	1.89	1.87	1.87
Calcium, %		0.33	0.28	0.34	0.38	0.37	0.79	0.87
Phosphorus, %		0.52	0.54	0.54	0.58	0.58	0.86	0.63

1. Formulations are on an as-fed basis.
2. Urea may be included up to 1% of the concentrate mix to supply protein.
3. Other high-protein feeds or commercial supplements can be substituted for soybean meal on a protein basis.
4. Other high Ca-P mineral mixes as steamed bone meal or commercial mixtures can replace dicalcium phosphate.

D. *Balance*
  1. Subtract nutrients supplied by forage from total needs to determine balance needed from concentrates.
  2. Using a concentrate mix appropriate to the forage (high protein concentrate mix for lower-protein forage etc.) supply the additional energy needs.
  3. Check to be sure both protein and energy needs are met.

V. Feeding Guidelines

  A. Figure 9–1 illustrates the level of milk production, milkfat test, dry matter intake and change in body weight measured in Minnesota research herds during the entire lactation. Based on Figure 9–1, four feeding periods can be defined.
   1. Early lactation—70 days after calving (increasing milk production).
   2. Peak dry matter intake—70 to 140 days after calving (declining milk production).
   3. Mid to late lactation—140 to 305 days after calving (declining milk production).
   4. Dry period—45 to 60 days before the next lactation.

High quality forage being consumed by the lactating cow herd.

B. Nutrient requirements and feeding programs vary with the phase of lactation and gestation illustrated in Figure 9–1.
  1. Early lactation—70 days postpartum
    a. Milk production increases rapidly, peaking at 6 to 8 weeks after calving.
    b. Increase grain intake 1 to 2 lb per day after calving to meet energy requirement. Use grains high in energy content.
    c. Avoid excessive levels of grain (more than 65% of the total dry matter) and maintain 17% to 19% acid detergent fiber level in diet to reduce rumen disorders.

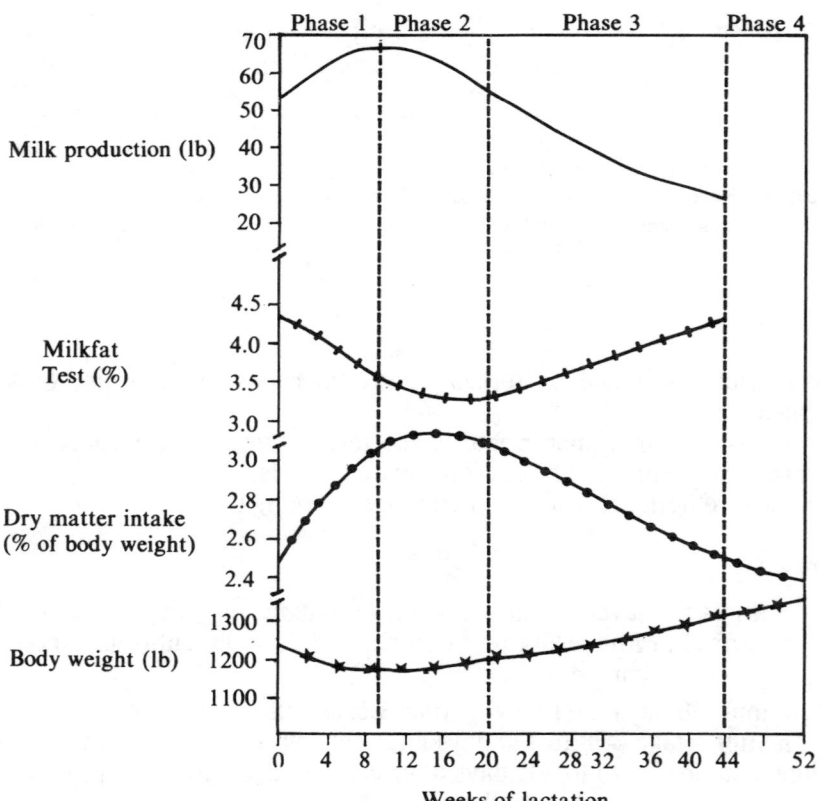

Figure 9–1. Various feeding periods with milk production, dry matter intake, and body weight change curves.

d. Extra dietary protein permits more efficient use of body fat for milk production because cows are usually losing body weight. A low rumen degradable protein source (bypass protein) is recommended for high producing cows in early lactation.

e. Limit urea to 0.2 to 0.4 lb/day. Some research indicates urea is utilized less efficiently when total ration protein level is high.

f. Increasing the energy density of the ration may help cows more nearly meet energy requirement. This can be achieved by feeding the cow 1 to 1.5 lb of added fat per day; which amounts to about 6% added fat to the concentrate (grain) mix.

2. Peak dry matter intake—second 10 weeks postpartum

a. Maximum dry matter intake is reached at 12 to 14 weeks and cows should be kept at peak milk production as long as possible.

b. Grain intake may reach 2½% of cow's body weight; forage intake (dry matter) should be at least 1% to 1¼% of cow's body weight to maintain rumen function and milk-fat test.

c. Feed forages and grain several times daily.

d. Current research suggests that high-producing cows (i.e., >70 lb 4% FCM) should be fed only natural protein and not urea).

3. Mid- to late lactation—days 140 to 305 postpartum

a. Milk production is declining; the cow is pregnant.

b. Match grain intake to milk production; avoid wasteful grain feeding to low producers.

c. Feed extra nutrients, if needed, to replace any body tissue lost in early lactation. Cows are more efficient in replacing body tissue while lactating than during the dry period. Avoid overconditioning cows.

d. Use recommended levels of urea (0.4 to 0.5 lb/cow/day) if it is needed.

4. Dry period—45 to 60 days before the next lactation

a. Quality of forage not as critical during the dry period. Cows need sufficient feed to support both the unborn calf and to meet body reserves not replaced in period 3.

b. Dry matter intake is approximately 2% of body weight. Dry cows may be fed up to 4 to 6 lb of grain per day (0.5% of body weight) depending on the condition of the cow. Cows in good condition should be fed little or no grain. Soybean meal may be fed to balance protein needs when dry cows are fed only corn silage. At about 2 weeks before calving increase grain feeding so cows are consuming 12 to 16 lb grain/day at calving (1% of body weight). This helps cows get accustomed to high grain intake needed after calving and will reduce the occurrence of ketosis, a metabolic disease.

c. Feeding high levels of corn silage or grain during the dry period may cause excess fat deposits in the liver region (fat cow syndrome) or cause a higher incidence of displaced abomasum. Hay feeding during the dry period may reduce these problems.

d. Problems with cows developing milk fever at calving time can be reduced by feeding a low calcium ration (<0.20% Ca) about 2 weeks before parturition and then returning to a normal diet at calving.

e. Cows should be brought into peak milk production as soon as possible after calving. This can best be done by feeding slightly more grain than recommended until there is no increase in production and then adjusting the amount of grain fed to the milk production. This practice of increasing the grain offered to the dairy cow beginning about 2 weeks before calving and continuing until milk flow starts to drop is referred to as lead or challenge feeding.

f. Cows should have 4 to 8 weeks dry period depending on age and physical condition.

g. Many cows will lose up to 200 lb of weight during early lactation so need to be in good condition at calving.

VI. Things to Remember in Feeding For Milk Production

A. Feed young growing cows enough to allow for growth as well as for maintenance and milk production. It is best to group cows by production level as a means of challenge feeding.

B. Additional nutrient allowances are usually made during late pregnancy.

C. More cows are underfed on energy than on protein.

D. Most lactation rations will contain 13% to 17% crude protein and 60% to 70% TDN (.6 to .8 Mcal/lb NE_L).

E. The concentrate mix contains
  1. Energy feed(s).
  2. Protein feed(s).
  3. 0.5% to 1% trace mineralized salt.
  4. 1% to 2% calcium-phosphorus supplement.
  5. No more than 6% molasses.
  6. No more than 1% urea.
  7. Up to 8% added fat.
F. Finely ground or pelleted forages or grains should not be fed alone to lactating cattle—lowers milk-fat test.
G. It is best to feed some hay with silage.
H. Cows in full production will consume 3 to 5 lb of water for each 1 lb of milk produced—including water in feed. Have water available at all times and warm in winter.
I. Give due consideration to the relationship of feeds to milk flavor. (For example, cows must be removed from wheat pasture several hours before milking to prevent an off-flavor problem.)
J. *Added Fat*
  1. High-producing cows during the first 12 to 16 weeks of lactation will benefit most from added fat. Cows under heat stress may also benefit from added dietary fat.
  2. Lactating cows can be fed 1 to 1.5 lb of added fat per day in order to increase the energy density of the diet and help meet energy needs.
     a. Supplemental fat can be blended into the concentrate mix at levels up to 8% (or up to 4% to the total ration). Higher levels of added fat may reduce feed intake, reduce fiber digestibility and cause digestive upsets. Unsaturated fatty acids are generally a greater problem than are saturated fatty acids.
     b. Whole or processed oilseeds, such as cottonseed, soybean or sunflower seeds may be fed as a source of added fat, even though they contain polyunsaturated fat, because they are slowly digested and the oil is gradually released into the rumen. This allows for saturation of the fatty acids and less chance of reduced fiber digestibility or milk fat depression. These oilseeds also provide some protein and fiber, which can be considered when formulating diets.
        (1) Feed 5 to 7 lb oilseeds per head daily.
        (2) May be fed whole or rolled.
        (3) Heat-treated soybeans may have greater protein bypass properties than unheated soybeans.
  3. When feeding added fat, increase the dietary calcium to 0.9%+, the magnesium to 0.3%, the acid detergent fiber to 20% and increase the protein content by 1% or 2%.
K. *Protein*

The National Research Council provides estimates (Table 9–6) of the rumen degradable and nondegradable (bypass) protein requirements, in which the ingested protein is split into the rumen undegraded intake protein (UIP) and (rumen) degraded intake protein (DIP) categories. The advantage of using UIP and DIP values is that it may be possible to formulate a ration with less total crude protein for a given milk yield, providing an economic benefit to the producer. The success of this system will depend on refinement of estimates of feed protein degradability in the rumen (refer to Table 1–2 in Chapter 1).

The relative proportion of ruminally degradable protein in the typical dairy cow diet is approximately 60%. The remainder of the ingested protein (40%) is undegraded (bypass) protein. High-producing cows (in general cows producing more than 5 lb of milk/100 lb of body weight) will likely benefit from diets formulated to contain greater than normal proportions of bypass protein.

FEEDING FOR DAIRY CALVES

I. From Birth to 4 Months of Age

A. *Common Feeds for Calves*

Include colostrum, whole milk replacers and calf starters along with hay or pasture.

1. Colostrum feeding is important. Usually the calf is left with its dam less than 24 hours and then placed on one of several milk feeding programs, which are described later. Since the calf will not be able to consume all the colostrum produced by the cow, some dairy producers will store the extra colostrum in a cooler or freezer or allow it to sour (ferment) in containers held at barn temperature. After the calf has had fresh colostrum, switch to the stored product and feed a quart of stored colostrum mixed with 1 quart of warm water twice daily. Three quarts may be used for larger calves (8% to 10% of body weight).
2. Whole milk—an excellent feed on which calves do quite well. However, whole milk is quite expensive especially in areas where a good milk market exists.
3. Milk replacers are high milk byproduct feeds sold as a powder and reconstituted with water for feeding (1 part dry replacer to 6 to 9 parts water or according to manufacturer's recommendation).
   a. Generally, milk replacers are too complex for home mixing and should be purchased.
   b. Typical composition includes dried skim milk or whey or both with 10% to 30% animal fat (lard) for energy and additives including vitamins, trace minerals and antibiotic. Protein content is usually 20% to 24%.
4. Calf starters are high-energy, high-protein (16% to 20%), low-fiber grain mixes fed to young calves.
   a. Emphasize corn and SBM with oats added for bulk and palatability.
   b. Sources of Ca, P, trace minerals and salt are usually added.
   c. The chief value of low levels of antibiotic (10 mg/lb starter) is to improve appetite. Therapeutic levels (100 to 500 mg/day) are used to combat scours.
   d. Can be mixed at home or purchased. Grains should be rolled or coarsely ground.
   e. Suggested calf starter rations are shown in Table 9–2.

B. *Milk Feeding Programs*

These are two general types.

1. Liberal milk system
   a. Veal calves—calves fed for veal given maximum levels of milk or milk replacer (all they will consume without developing diarrhea) for 6 to 8 weeks. Many milk replacers for veal feeding contain high levels of fat to increase energy intake. Also, they are quite expensive.
   b. Herd replacements
      (1) An expensive system where milk is sold.
      (2) Calves do quite well on liberal milk feeding.
      (3) Feeding rate is usually 8% to 10% of body weight (or an equivalent amount of milk replacer) until 3 to 4 months of age.
      (4) Hay, grain and salt are also fed.
2. Limited milk system
   a. Conventional
      (1) Calf is fed milk, milk replacer or stored colostrum at 8% to 10% of body weight until it is consuming calf starter at the rate of 2 to 3 lb/day at which time milk feeding can be decreased rapidly.
      (2) Calf is off milk entirely by 4 to 7 weeks of age.
      (3) Hay feeding can be started at 1 week of age or delayed until 1 month of age to encourage early starter consumption if desired.
      (4) This system is probably most economical under midwest conditions of grain supply and milk prices.

TABLE 9–2. Suggested Calf Starter Rations[1,2]

Ingredients	IFN	1	2	3	4	5	6
Corn, rolled	4–02–931	50	39	54	50	34	28
Oats, rolled	4–03–309	35	—	12	26	34	30
Barley, rolled	4–00–549	—	39	—	—	—	—
Beet pulp	4–00–669	—	—	—	—	—	20
Corn cobs, gr.	1–02–782	—	—	—	—	14	—
Wheat bran	4–05–190	—	10	11	—	—	—
Soybean meal	5–04–604	13	10	8	17	16	15
Linseed meal	5–02–045	—	—	8	—	—	—
Molasses, liquid	4–04–696	—	—	5	5	—	5
Dicalcium phosphate	6–01–080	1	1	1	1	1	1
TM salt & vitamin premix[3]		1	1	1	1	1	1
		100	100	100	100	100	100

Calculated analysis:
1) *As-fed basis*

		1	2	3	4	5	6
Crude protein, %		14.5	14.0	14.5	15.4	14.7	14.8
TDN, %		73.1	73.0	72.5	72.9	68.2	70.5
NE_m, Mcal/kg		1.83	1.76	1.80	1.83	1.68	1.75
NE_g, Mcal/kg		1.25	1.19	1.22	1.25	1.11	1.19
Calcium, %		0.29	0.29	0.35	0.34	0.32	0.45
Phosphorus, %		0.54	0.61	0.64	0.54	0.52	0.49
Dry matter, %		88.5	88.4	87.8	87.8	88.9	88.5

2) *Dry matter basis*

		1	2	3	4	5	6
Crude protein,%		16.4	15.8	16.5	17.5	16.5	16.7
TDN, %		82.6	82.6	82.5	83.0	76.7	79.7
NE_m, Mcal/kg		2.07	1.99	2.05	2.08	1.89	1.98
NE_g, Mcal/kg		1.41	1.35	1.39	1.42	1.25	1.34
Calcium, %		0.33	0.33	0.40	0.39	0.36	0.51
Phosphorus, %		0.61	0.69	0.73	0.61	0.58	0.55

1. Formulations are on an as-fed basis. Rations 1, 2, 3, and 4 recommended for calves weaned after 4 weeks of age and receiving forage. Rations 5 and 6 recommended for calves weaned after 4 weeks and not receiving forage.

2. Calf starter should be fed from about 3 days of age until 12 weeks of age. Intake should be limited to about 3 to 4 lb (1.4 to 1.8 kg) per calf daily.

3. Vitamin premix should supply the following per pound (or kg) of ration: vitamin A, 2000 IU (900); vitamin D, 500 IU (225).

b. Early weaning
   (1) Differs from conventional in that calves are entirely off milk by 1 month of age.
   (2) Requires good management and early adjustment to starter feeding.
   (3) Calves do not appear as thrifty at 1 month of age but usually are no different from others at 3 to 4 months.
   (4) Suggested milk feeding program follows.

Age, Days	Amount of Milk* Fed Daily[†]
0–3	4–6 lb[‡] colostrum or nurse dam
4–24	5–7 lb
25–31	3–4 lb

*Whole milk or an equivalent amount of replacer, or stored colostrum.
[†]Divided into two equal feedings.
[‡]Low end of range for small-breed calves and high end for large-breed calves.

   (5) When the amount of milk is limited by early weaning, the calf is encouraged to eat dry feed at an early age. The calf should be consuming dry feed at the rate of 1.5% of body weight, in addition to milk, at time of weaning.

400

Dairy calves being reared in individual isolated huts.
This procedure reduces disease problems associated
with group confinement housing.

C. Calf scours may develop in small calves. Early detection of sickness and prompt corrective action is important to prevent scours. When a calf has only a mild case of scours (not off-feed, not depressed and no fever), feeding an oral electrolyte solution usually is beneficial. A good procedure to follow is to
   1. Remove or drastically reduce the amount of milk or milk replacer offered.
   2. Feed only water containing an "electrolyte" for three to six feedings, depending on how soon feces become firm. Oral electrolyte solutions can be purchased commercially or can be made by combining the following ingredients
       4 teaspoons of table salt
       3 teaspoons of baking soda
       ½ cup of "light" corn syrup
       1 gallon of water
   3. Frequent feeding of smaller volumes is advantageous. A 100-lb calf should consume about 5 qt (10% of body weight) daily.
D. Hay or silage for the young calf
   1. While the calf may begin nibbling on good quality hay as early as 5 to 10 days of age, it probably will not consume appreciable quantities before 8 to 10 weeks of age. If forages are inconvenient because of the housing and management system, it may be desirable to incorporate a forage factor (fiber) into the starter ration (20% to 25%). Adequate fiber is essential for proper health of the rumen papillae and calves will crave roughage.
   2. Silages should be limited before 3 months of age because of the high moisture content, which can limit intake and growth.

II. **Heifers from 4 Months to 12 Months of Age**
   A. Rumen capacity at this age is not sufficient to allow the animal to meet energy needs from forages alone. Therefore, some grain feeding is necessary until 1 year of age.
       1. Summer—pasture, hay and grain mix (3 to 7 lb daily depending on body size and forage quality).
       2. Winter—hay, silage and grain mix (3 to 7 lb daily depending on body size and forage quality).
   B. The same forage and, consequently, the same grain mix as used for the milking herd can be used for heifers.
       1. The protein content of the grain mix should vary inversely as the protein content of the forage varies.
       2. A free-choice mineral mix is recommended. The mix should include Ca, P, salt and if poor forage, trace minerals should also be included.

3. Suggested grain mixes for the growing calf are shown in Table 9–3. These should be limited to no more than 5 to 7 pounds daily along with free-choice forage consumption.

C. If necessary, limit grain to keep calves from becoming overly fat. Excessive fat can result in breeding problems and also deposit fatty tissue in the udder. Overconditioned heifers produce less in later life than those reared on a more moderate level of nutrition.

III. Heifers from 12 Months of Age to Calving

A. By this age heifers have sufficient rumen capacity to meet their nutrient needs from good quality forages. Feed grain mix only if forages are poor in quality or limited in amount. Heifers should gain 1.5 to 1.8 lb per day.

1. Summer—pasture and hay; grain mix only if needed (2 to 8 lb daily depending on body size and forage intake).
2. Winter—hay and silage; grain mix only if needed (2 to 8 lb daily depending on body size and forage intake).
3. Free-choice minerals (Ca, P, salt and if poor forage, trace minerals).

B. For breeding at 15 months, heifers should weigh 550 to 800 lb depending on breed (550 for Jerseys—800 for Holstein and Brown Swiss). In order to accomplish this, the heifer has to gain about 1.75 lb/day from birth.

C. Avoid overconditioning
1. Fat heifers have impaired reproductive efficiency.
2. Due to fatty deposits in the udder, fat heifers become low-producing cows.

TABLE 9–3. Suggested Grain Mixes in Grower Rations of 440-lb (200-kg) Dairy Calves (4–12 months of age)[1, 2]

Ingredients	IFN	1	2	3	4
Corn, cracked	4–02–931	78	—	—	50
Oats, rolled	4–03–309	20	35	—	27
Barley, rolled	4–00–530	—	50	—	—
Gr. ear corn	4–02–849	—	—	76	—
Molasses, liquid	4–04–696	—	5	5	—
Soybean meal	5–04–604	—	8	17	20
Limestone	6–02–632	—	—	—	1
Dicalcium phosphate	6–01–080	1	1	1	1
Trace mineral salt		1	1	1	1
		100	100	100	100

Calculated analysis:
1) *As-fed basis*

		1	2	3	4
Crude protein, %		9.2	13.8	13.9	16.7
TDN, %		74.9	70.0	71.1	72.8
NE_m, Mcal/kg		1.87	1.71	1.84	1.82
NE_g, Mcal/kg		1.29	1.16	1.27	1.25
Calcium, %		0.25	0.33	0.35	0.68
Phosphorus, %		0.48	0.56	0.49	0.56
Dry matter, %		87.9	88.4	86.7	88.6

2) *Dry matter basis*

		1	2	3	4
Crude protein, %		10.5	15.6	16.0	18.8
TDN, %		85.2	79.2	82.0	82.2
NE_m, Mcal/kg		2.13	1.93	2.12	2.05
NE_g, Mcal/kg		1.47	1.31	1.46	1.41
Calcium, %		0.28	0.37	0.40	0.77
Phosphorus, %		0.55	0.63	0.56	0.63

1. Formulations are on an as-fed basis. Ration 1 is recommended to be fed with legume hay (14–17% crude protein). Rations 2 and 3 should be fed with a legume-grass mixed hay (10–13% crude protein). Ration 4 is recommended to be fed with a grass hay (6–9% crude protein).
2. Dairy calves should consume daily: 2.0 to 2.5% of their body weight as dry matter forage and 0.5 to 1.0% as dry matter grain mix.

D. About 2 months before calving, start feeding grain and increase gradually to accustome heifers to high grain intake that will be required for lactation after calving.

IV. Feeding Dairy Bulls

A. Bull calves raised for breeding purposes should be fed and handled much the same as heifers. Bulls will grow faster than heifers and should receive more feed than heifers of the same age. Because of the widespread use of artificial insemination in the dairy industry, only a few dairy bull calves are raised for breeding purposes.

B. Older bulls should be kept in thrifty, vigorous condition, but they should not be permitted to become too fat. Mature bulls can be maintained on forage with about 0.5 lb of grain per 100 lb of body weight, if needed. The grain ration can be the same as fed to the lactating cows.

V. Feeding Dairy Beef

A. Dairy calves not developed as replacement heifers or bulls are generally fed and marketed as beef. Two different finishing programs and market weights for dairy beef are most common.

1. High-energy diet; light market weights. These cattle are full fed a high-grain diet from about 300 lb to market weights of 800 to 1000 lb.

2. High-roughage; heavy market weights. These cattle are grown on roughages (corn or sorghum stalks; wheat or other excess pasture) to 600 to 800 lb, then fed a high-grain diet during a finishing period in the feedlot. They are generally marketed at weights of 1150 to 1400 lb.

STEPS INVOLVED IN FORMULATING A LACTATING DAIRY RATION

I. Problem

Balance the ration of a 650-kg mature cow in early lactation (nonpregnant) producing 32 kg of 3.5% (fat) milk daily. Assume the cow will consume alfalfa hay and corn silage as a forage mix plus an amount of concentrate mixture needed to meet any remaining energy, protein or mineral needs. Work the entire problem on a dry matter basis using the dairy ration worksheet following this example problem. A sample worksheet is also found in the appendix section at the back of the textbook.

II. Establish Dry Matter Nutrient Requirements (NRC requirements in Table 9–5)

Record on the worksheet (worksheet A_{1-4}) the requirements for maintenance and for milk production. Remember the NRC values given for milk production are stated as per kg. Thus, each NRC nutrient requirment value must be multiplied by the number of kg of milk being produced.

III. Determine Feeds Available and Their Nutrient Composition (dry matter basis) from Table 3–2

A. *Forage ration*

Use equal dry matter parts of alfalfa hay (IFN 1–00–063) and corn silage (IFN 3–02–823)

B. *Concentrate Mixture*

1. Ground corn (IFN 4–02–931).
2. Soybean meal (IFN 5–04–604).
3. Fix 1% salt and 1.5% dicalcium phosphate (calcium phosphate, dibasic, IFN 6–01–080) into the concentrate mixture.

IV. Establish Feeding Levels

A. *Total Daily Dry Matter*

1. Maximum dry matter intake must be considered when formulating lactating cattle rations. There is considerable variation between cows depending on cow size, level of milk production, and quality of feed ingredients, particularly forages. A dry matter intake range of 2% to 4% of body weight may be observed within a herd. Guidelines for estimating maximum dry matter intake for cows weighing between 400 to 800 kg, and producing from 10 to 45 kg of milk daily, are shown in Table 9–7.

Dairy heifers consuming good pasture require no grain or other forage.

2. The data in Table 9–7 estimate maximum dry matter intake based upon body weight and amount of 4% fat-corrected-milk (FCM) produced daily. The cow in this example weighed 650 kg and was producing 32 kg of 3.5% milk. Converting this to 4% fat-corrected-milk (FCM) is as follows.

   $$4\% \text{ FCM} = (0.4 \times \text{kg milk}) + (15 \times \text{kg fat})$$
   $$= (0.4 \times 32) + (15 \times 1.12)$$
   $$= 12.8 + 16.8$$
   $$= 29.6 \text{ (or 30) kg}$$

3. Table 9–7 indicates a 650 kg cow producing 30 kg FCM could consume a maximum of 3.1% of b.w. as daily d.m. Thus: 650 kg b.w. $\times$ .031 = 20.15 kg d.m./day.

B. *Daily Forage Intake*

1. The quantity of forage intake will vary with the amount of NE_L required. The cow in this example required 32.38 Mcal NE_L/day (from worksheet A_4) and can consume a total of 20.15 kg dm/day. Thus, the total ration must contain 1.607 Mcal/kg NE_L (32.38 Mcal $\div$ 20.15 kg).

2. The NE_L values for alfalfa hay and corn silage in the example are 1.27 and 1.60 Mcal/kg, respectively. If about half of the forage d.m. comes from alfalfa hay, and half form corn silage, the average NE_L of the forage mix will be 1.435 Mcal/kg [(1.27 + 1.60) $\div$ 2 = 1.453].

3. Corn grain and soybean meal have NE_L values of 2.11 and 1.93 Mcal/kg, respectively. If about 75% of the concentrate mix d.m. consists of corn grain and about 25% is soybean meal, the average NE_L of the concentrate mix will be 2.065 Mcal/kg [(.75(2.11) + (.25)(1.93) = 2.065].

4. Letting F = forage mix and 1-F = concentrate mix, the proportion of forage that can be included in the ration and still meet the recommended NE_L requirement in the total ration d.m. is calculated as follows.

   $$1.435F + 2.065 (1\text{-}F) = 1.607 \ (NE_L)$$
   $$1.435F + 2.065 - 2.065F = 1.607$$
   $$-0.630F = -0.458$$
   $$F = 0.727$$
   $$= 72.7\% \text{ forage in ration d.m.}$$

5. From this, the maximum forage mix intake can be calculated as follows: 20.15 kg dm $\times$ 72.7% forage = 14.65 kg forage mix. Since alfalfa hay and corn silage are to be fed in equal parts, the cow should be given 7.32 kg of each forage (14.65 $\div$ 2 = 7.32).

6. Enter these values and complete worksheet B_{1-4}.

V. Formulation of Concentrate Mixture

A. Determine the additional nutrients needed from concentrate mixture (worksheet C).

	Crude protein (kg)	NE_L (Mcal)
Daily requirement	3.116	32.380
From forage mix	1.962	21.380
Still need from concentrate	1.154	11.372

B. Establish the amounts of each ingredient in 100.0 kg of concentrate mix.

1. It is given that 1.0% and 1.5% of the concentrate mix be salt and dicalcium phosphate, respectively. Therefore, any mixture of ground corn (GC) and soybean meal (SBM) will constitute 97.5% of the final concentrate mix.

2. It was previously established (IV.B.3) that a combination of GC and SBM will contain an average of 2.065 Mcal/kg NE_L. If these two feeds comprise 97.5% of the final concentrate mix, then the final mix will contain about 2.0 Mcal/kg NE_L (2.065 × 97.5% = approximately 2.0).

3. Since 11.372 Mcal NE_L are needed from a concentrate mix containing approximately 2.0 Mcal/kg NE_L, this will require 5.69 kg concentrate mix (11.37 ÷ 2.0 = 5.69). Enter this value on worksheet D.

4. Next, calculate the quantity of GC and SBM. If 1.154 kg CP are needed from the 5.69 kg of concentration mix, then the final mix should contain 20.28% CP (1.154 ÷ 5.69 (100) = 20.28). Enter this value on worksheet E.

5. Algebraic solution

$$X = kg\ GC;\quad Y = kg\ SBM$$

$$
\begin{aligned}
1)\quad & X + Y = 97.5\ kg\ mix \\
2)\quad & .101X + .510Y = 20.28\ kg\ CP \\
3)\quad & \underline{-.101X - .101Y = -9.85} \\
& 0\qquad\ .409Y = 10.43
\end{aligned}
$$

$$Y = 10.43 \div .409 = 25.50\ kg\ SBM$$
$$X = 97.5 - 25.5 = 72.0\ kg\ GC$$

6. Pearson square solution
   a. 20.28 kg CP needed from 97.5 kg concentrate mix, thus 20.28 ÷ 97.5 (100) = 20.8% CP in the GC—SBM fraction
   b. Solution

   GC 10.1 $\qquad\boxed{20.8}\qquad$ 30.2 = 73.84% × 97.5 kg = 72.0 kg GC

   SBM 51.0 $\qquad\qquad$ $\dfrac{10.7}{40.9}$ = 26.16% × 97.5 kg = 25.5 kg SBM

7. The quantities of GC, SBM, dical and salt can be multiplied by their nutrient values and these figures entered in the columns of the concentrate mixture worksheet (worksheet F).

VI. Overall Composition of the Daily Ration

A. The daily amount of each ingredient is entered into the final portion of the worksheet (worksheet G) along with its nutrient contribution. The 5.69 kg of concentrate mixture is fractioned into the four feedstuffs by multiplying 5.69 by the percent of each ingredient determined previously.

B. When comparing the nutrients provided by the ration to the requirements (worksheet A_4), the formulation is adequate to meet the animal's needs.

NUTRIENT REQUIREMENTS OF DAIRY CATTLE

The nutrient requirements for dairy cattle presented in Tables 9–4, 9–5, and 9–6 were adapted from the *Nutrient Requirements of Dairy Cattle,* Publication ISBN 0–309–03826–X, Committee on Animal Nutrition, National Academy of Sciences—National Research Council, Washington, D.C., 1988. These values are expressed on a dry-matter basis.

DAIRY RATION WORKSHEET

	(a) Crude Protein kg	(b) Net Energy for Lactation Mcal	(c) Calcium gm	(d) Phosphorus gm
A. *Daily Requirements for Dairy Animal* (dry-matter basis)				
1. For maintenance, cow weighing _650_ kg	0.428	10.30	26.0	19.0
2. For maintenance + pregnancy (last 2 months), cow weighing _____ kg	0	0	0	0
3. For milk, _32_ kg per day testing _3.5_ % fat	2.688	22.08	95.0	58.5
4. Total Daily Requirements	3.116	32.38	121.0	77.5
B. *Forage Ration* (dry-matter basis)				
1. _7.32_ kg of _Alfalfa hay_	1.369	9.30	100.3	17.6
2. _7.32_ kg of _Corn silage_	0.593	11.71	19.8	14.6
3. _____ kg of _____	=	=	=	=
4. Total Nutrients from Forage	1.962	21.01	120.1	32.2
C. *Remaining,* to be supplied from concentrate mix ("Total Daily Requirements" less "Total Nutrients from Forage") ($A_4 - B_4$)	1.154	11.37	0.9	45.3

D. *Amount of Concentrate Mix Required*
A good dairy concentrate will contain approximately 1.8–2.0 Mcal/kg NE_L on a dry matter basis.

_11.37_ Mcal NE_L needed ÷ _2.0_ Mcal/kg NE_L =

_5.69_ kg

E. *Percent Crude Protein Needed in Concentrate Mix*
Divide Crude Protein to be supplied from concentrates ($C_{(a)}$) by concentrates required (**D**) times 100 = % Crude Protein needed in concentrate mix. ($C_{(a)}$) ÷ **D** × 100 = %

_1.154_ kg CP needed ÷ _5.69_ kg mix × 100 =

_20.28_ % CP

F. Concentrate Mixture

		Dry-Matter Basis			
Ingredient	Amount kg	Crude Protein kg	NE_L Mcal	Ca gm	P gm
Gr. corn	72.0	7.27	151.92	14.4	237.6
Soybean meal	.25.5	13.00	49.21	86.7	196.4
Dical PO_4	1.50	0	0	335.3	285.0
Salt	1.00	0	0	0	0
Totals (kg, Mcal, g)	100.00	20.27	201.13	436.4	719.0
Calculated Analysis: % or amt per kg		20.27 %	2.01 Mcal/kg	0.44%	0.72%

G. Overall Composition of Daily Ration

		Dry-Matter Basis			
Ingredient	Amount kg	Crude Protein kg	NE_L Mcal	Ca gm	P gm
Alfalfa hay	7.320	1.369	9.30	100.3	17.6
Corn silage	7.320	0.593	11.71	19.8	14.6
Gr. corn	4.097	0.414	8.64	0.8	13.5
Soybean meal	1.451	0.740	2.80	4.9	11.2
Dical PO_4	0.085	0	0	19.0	16.1
Salt	0.057	0	0	0	0
Totals (kg, Mcal, g)	20.330	3.116	32.45	144.8	73.0
Calculated Analysis: % or amt per kg		15.33 %	1.60 Mcal/kg	0.71%	0.36 %
Difference from needs (A_4)		OK	OK	+23.8g	−4.5g

TABLE 9-4. Daily Nutrient Requirements of Growing Dairy Cattle and Mature Bulls

Live Weight (kg)	Gain (g)	Dry Matter Intake[a] (kg)	NEM (Mcal)	NEG (Mcal)	ME (Mcal)	DE (Mcal)	TDN (kg)	UIP (g)	DIP (g)	CP (g)	Ca (g)	P (g)	A (1,000 IU)	D
			Energy					Protein			Minerals		Vitamins	
Growing Large-Breed Calves Fed Only Milk or Milk Replacer														
40	200	0.48	1.37	0.41	2.54	2.73	0.62	—	—	105	7	4	1.70	0.26
45	300	0.54	1.49	0.56	2.86	3.07	0.70	—	—	120	8	5	1.94	0.30
Growing Large-Breed Calves Fed Milk Plus Starter Mix														
50	500	1.30	1.62	0.72	5.90	6.42	1.46	—	—	290	9	6	2.10	0.33
75	800	1.98	2.19	1.30	8.98	9.78	2.22	—	—	435	16	8	3.20	0.50
Growing Small-Breed Calves Fed Only Milk or Milk Replacer														
25	200	0.38	0.96	0.37	2.01	2.16	0.49	—	—	84	6	4	1.10	0.16
30	300	0.51	1.10	0.52	2.70	2.90	0.66	—	—	112	7	4	1.30	0.20
Growing Small-Breed Calves Fed Milk Plus Starter Mix														
50	500	1.43	1.62	0.72	6.49	7.06	1.60	—	—	315	10	6	2.10	0.33
75	600	1.76	2.19	0.96	7.98	8.69	1.97	—	—	387	14	8	3.20	0.50
Growing Veal Calves Fed Only Milk or Milk Replacer														
40	200	0.45	1.37	0.55	1.89	2.07	0.47	—	—	100	7	4	1.70	0.26
50	400	0.57	1.62	0.57	2.39	2.63	0.59	—	—	125	9	5	2.10	0.33
60	540	0.80	1.85	0.81	2.84	3.17	0.71	—	—	176	13	8	2.60	0.40
75	900	1.36	2.19	1.47	4.82	5.39	1.21	—	—	300	16	9	3.20	0.50
100	1,250	2.00	2.72	2.26	6.22	7.06	1.58	—	—	440	20	11	4.20	0.66
125	1,250	2.38	3.21	2.44	7.40	8.40	1.88	—	—	524	22	13	5.30	0.82
150	1,100	2.72	3.69	2.29	8.46	9.60	2.15	—	—	598	24	15	6.40	0.99
Large-Breed Growing Females														
100	600	2.63	2.72	1.22	7.03	8.13	1.84	317	57	421	17	9	4.24	0.66
100	700	2.82	2.72	1.44	7.54	8.72	1.98	346	75	452	18	9	4.24	0.66
100	800	3.02	2.72	1.66	8.06	9.32	2.11	374	92	483	18	10	4.24	0.66
150	600	3.51	3.69	1.45	9.14	10.61	2.41	283	150	562	19	11	6.36	0.99
150	700	3.75	3.69	1.71	9.76	11.33	2.57	307	173	600	19	12	6.36	0.99
150	800	3.99	3.69	1.97	10.39	12.07	2.74	331	196	639	20	12	6.36	0.99
200	600	4.39	4.57	1.65	11.14	12.99	2.95	254	239	631	20	14	8.48	1.32
200	700	4.68	4.57	1.95	11.87	13.84	3.14	274	267	686	21	14	8.48	1.32
200	800	4.97	4.57	2.25	12.62	14.71	3.34	294	295	741	22	15	8.48	1.32
250	600	5.31	5.41	1.84	13.10	15.33	3.48	229	326	637	22	16	10.60	1.65
250	700	5.65	5.41	2.18	13.94	16.32	3.70	246	359	678	23	17	10.60	1.65
250	800	5.99	5.41	2.51	14.79	17.32	3.93	263	393	726	24	17	10.60	1.65
300	600	6.26	6.20	2.02	15.05	17.69	4.01	209	413	752	23	17	12.72	1.98
300	700	6.66	6.20	2.39	16.00	18.81	4.27	223	452	799	24	18	12.72	1.98
300	800	7.06	6.20	2.77	16.97	19.95	4.52	236	490	848	25	19	12.72	1.98
350	600	7.29	6.96	2.20	17.01	20.09	4.56	193	501	874	24	18	14.84	2.31
350	700	7.75	6.96	2.60	18.09	21.36	4.84	204	545	930	25	19	14.84	2.31
350	800	8.21	6.96	3.01	19.18	22.64	5.14	214	590	985	26	20	14.84	2.31
400	600	8.39	7.69	2.37	19.03	22.58	5.12	182	592	1,007	25	19	16.96	2.64
400	700	8.92	7.69	2.80	20.23	24.00	5.44	190	641	1,070	26	20	16.96	2.64
400	800	9.46	7.69	3.24	21.44	25.44	5.77	198	692	1,135	26	21	16.96	2.64
450	600	9.59	8.40	2.53	21.12	25.18	5.71	176	686	1,151	28	19	19.08	2.97
450	700	10.20	8.40	2.99	22.46	26.78	6.07	182	742	1,224	28	20	19.08	2.97
450	800	10.82	8.40	3.46	23.81	28.40	6.44	187	799	1,298	29	21	19.08	2.97
500	600	10.93	9.09	2.69	23.32	27.96	6.34	175	785	1,311	28	20	21.20	3.30
500	700	11.63	9.09	3.18	24.81	29.74	6.75	179	848	1,395	28	20	21.20	3.30
500	800	12.33	9.09	3.68	26.32	31.55	7.16	182	913	1,480	29	21	21.20	3.30
550	600	12.42	9.77	2.84	25.67	30.95	7.02	180	891	1,490	28	20	23.32	3.63
550	700	13.22	9.77	3.37	27.33	32.95	7.47	183	963	1,587	28	20	23.32	3.63
550	800	14.04	9.77	3.90	29.02	34.99	7.94	185	1,035	1,685	29	21	23.32	3.63
600	600	14.11	10.43	3.00	28.23	34.24	7.77	193	1,007	1,694	28	20	25.44	3.96
600	700	15.05	10.43	3.55	30.09	36.50	8.28	194	1,088	1,805	28	21	25.44	3.96
600	800	15.99	10.43	4.11	31.98	38.79	8.80	195	1,170	1,919	29	21	25.44	3.96

Continues

From _Nutrient Requirements of Dairy Cattle_ Sixth Revised Edition. Copyright © 1989 by the National Academy of Sciences, Washington D.C. Reprinted by permission.

TABLE 9–4 (continued). Daily Nutrient Requirements of Growing Dairy Cattle and Mature Bulls

Live Weight (kg)	Gain (g)	Dry Matter Intake[a] (kg)	Energy					Protein			Minerals		Vitamins	
			NEM (Mcal)	NEG (Mcal)	ME (Mcal)	DE (Mcal)	TDN (kg)	UIP (g)	DIP (g)	CP (g)	Ca (g)	P (g)	A (1,000 IU)	D
								_Small-Breed Growing Females_						
100	400	2.41	2.72	0.91	6.34	7.35	1.67	249	38	386	15	8	4.24	0.66
100	500	2.64	2.72	1.16	6.92	8.03	1.82	275	59	422	16	8	4.24	0.66
100	600	2.86	2.72	1.40	7.51	8.71	1.98	300	80	458	17	9	4.24	0.66
150	400	3.31	3.69	1.09	8.39	9.78	2.22	222	129	512	17	10	6.36	0.99
150	500	3.60	3.69	1.39	9.12	10.63	2.41	243	156	567	18	11	6.36	0.99
150	600	3.89	3.69	1.69	9.86	11.50	2.61	263	185	622	19	11	6.36	0.99
200	400	4.24	4.57	1.26	10.38	12.16	2.76	201	217	513	19	13	8.48	1.32
200	500	4.60	4.57	1.60	11.25	13.19	2.99	217	251	562	20	13	8.48	1.32
200	600	4.96	4.57	1.95	12.14	14.23	3.23	232	286	611	20	14	8.48	1.32
250	400	5.24	5.41	1.41	12.36	14.57	3.30	185	305	629	21	15	10.60	1.65
250	500	5.68	5.41	1.80	13.38	15.78	3.58	197	346	681	21	16	10.60	1.65
250	600	6.12	5.41	2.20	14.43	17.01	3.86	209	389	735	22	16	10.60	1.65
300	400	6.34	6.20	1.56	14.38	17.06	3.87	176	395	761	22	16	12.72	1.98
300	500	6.87	6.20	1.99	15.57	18.48	4.19	184	445	824	23	17	12.72	1.98
300	600	7.40	6.20	2.43	16.79	19.92	4.52	192	495	888	23	17	12.72	1.98
350	400	7.57	6.96	1.71	16.50	19.71	4.47	173	490	909	23	17	14.84	2.31
350	500	8.20	6.96	2.18	17.87	21.35	4.84	178	548	985	23	18	14.84	2.31
350	600	8.85	6.96	2.66	19.28	23.03	5.22	183	608	1,062	24	18	14.84	2.31
400	400	8.98	7.69	1.84	18.77	22.58	5.12	177	592	1,078	24	18	16.96	2.64
400	500	9.74	7.69	2.35	20.36	24.50	5.56	181	661	1,169	24	19	16.96	2.64
400	600	10.52	7.69	2.87	21.98	26.45	6.00	183	730	1,263	25	19	16.96	2.64
450	400	10.64	8.40	1.98	21.27	25.80	5.85	191	706	1,276	27	18	19.08	2.97
450	500	11.56	8.40	2.52	23.12	28.04	6.36	193	786	1,387	28	19	19.08	2.97
450	600	12.50	8.40	3.08	25.01	30.33	6.88	194	867	1,500	28	19	19.08	2.97
								_Large-Breed Growing Males_						
100	800	2.80	2.72	1.42	7.48	8.66	1.96	401	65	448	18	10	4.24	0.66
100	900	2.97	2.72	1.60	7.92	9.16	2.08	433	79	475	19	10	4.24	0.66
100	1,000	3.13	2.72	1.79	8.36	9.67	2.19	465	93	501	20	11	4.24	0.66
150	800	3.60	3.69	1.64	9.52	11.03	2.50	364	155	576	20	12	6.36	0.99
150	900	3.80	3.69	1.85	10.03	11.63	2.64	393	172	607	21	13	6.36	0.99
150	1,000	3.99	3.69	2.07	10.55	12.22	2.77	422	190	639	22	13	6.36	0.99
200	800	4.43	4.57	1.84	11.48	13.34	3.03	333	241	709	22	15	8.48	1.32
200	900	4.66	4.57	2.08	12.06	14.02	3.18	359	262	745	23	15	8.48	1.32
200	1,000	4.89	4.57	2.33	12.66	14.71	3.34	385	284	782	24	16	8.48	1.32
250	800	5.27	5.41	2.03	13.37	15.58	3.53	305	325	778	24	17	10.60	1.65
250	900	5.53	5.41	2.30	14.03	16.35	3.71	329	350	837	25	18	10.60	1.65
250	1,000	5.80	5.41	2.57	14.70	17.13	3.89	352	375	897	26	18	10.60	1.65
300	800	6.13	6.20	2.21	15.22	17.80	4.04	281	408	771	25	19	12.72	1.98
300	900	6.43	6.20	2.51	15.96	18.66	4.23	302	436	827	25	19	12.72	1.98
300	1,000	6.73	6.20	2.80	16.70	19.53	4.43	323	464	884	26	20	12.72	1.98
350	800	7.02	6.96	2.38	17.06	20.02	4.54	261	490	843	26	20	14.84	2.31
350	900	7.36	6.96	2.70	17.88	20.98	4.76	280	522	883	26	20	14.84	2.31
350	1,000	7.70	6.96	3.02	18.70	21.94	4.98	298	554	924	27	21	14.84	2.31
400	800	7.96	7.69	2.55	18.91	22.27	5.05	244	572	955	26	21	16.96	2.64
400	900	8.34	7.69	2.89	19.80	23.32	5.29	260	608	1,001	27	21	16.96	2.64
400	1,000	8.72	7.69	3.24	20.71	24.39	5.53	277	644	1,046	28	22	16.96	2.64
450	800	8.95	8.40	2.71	20.78	24.56	5.57	230	656	1,074	29	21	19.08	2.97
450	900	9.37	8.40	3.08	21.76	25.72	5.83	245	696	1,125	29	22	19.08	2.97
450	1,000	9.80	8.40	3.44	22.75	26.89	6.10	259	736	1,176	29	23	19.08	2.97
500	800	10.00	9.09	2.87	22.69	26.92	6.11	220	742	1,201	29	21	21.20	3.30
500	900	10.48	9.09	3.25	23.76	28.19	6.39	233	786	1,257	29	22	21.20	3.30
500	1,000	10.95	9.09	3.64	24.84	29.47	6.68	246	830	1,314	29	23	21.20	3.30
550	800	11.14	9.77	3.02	24.66	29.38	6.66	213	831	1,336	29	21	23.32	3.63
550	900	11.66	9.77	3.43	25.82	30.76	6.98	225	879	1,399	29	22	23.32	3.63
550	1,000	12.19	9.77	3.84	27.00	32.16	7.29	236	927	1,463	30	23	23.32	3.63

TABLE 9–4 (continued). Daily Nutrient Requirements of Growing Dairy Cattle and Mature Bulls

Live Weight (kg)	Gain (g)	Dry Matter Intake[a] (kg)	Energy NEM (Mcal)	NEG (Mcal)	ME (Mcal)	DE (Mcal)	TDN (kg)	Protein UIP (g)	DIP (g)	CP (g)	Minerals Ca (g)	P (g)	Vitamins A (1,000 IU)	D (1,000 IU)
600	800	12.36	10.43	3.17	26.71	31.95	7.25	211	923	1,483	29	21	25.44	3.96
600	900	12.95	10.43	3.60	27.97	33.47	7.59	221	976	1,554	29	22	25.44	3.96
600	1,000	13.54	10.43	4.03	29.25	34.99	7.94	231	1,029	1,624	30	23	25.44	3.96
650	800	13.69	11.07	3.32	28.86	34.67	7.86	212	1,020	1,643	29	21	27.56	4.29
650	900	14.35	11.07	3.77	30.24	36.33	8.24	222	1,078	1,722	29	22	27.56	4.29
650	1,000	15.01	11.07	4.22	31.63	38.00	8.62	230	1,137	1,801	30	23	27.56	4.29
700	800	15.16	11.70	3.46	31.14	37.59	8.52	219	1,124	1,820	29	22	29.68	4.62
700	900	15.90	11.70	3.93	32.64	39.40	8.94	227	1,187	1,907	29	22	29.68	4.62
700	1,000	16.63	11.70	4.40	34.16	41.23	9.35	235	1,252	1,996	30	23	29.68	4.62
750	800	16.79	12.33	3.60	33.59	40.73	9.24	232	1,235	2,015	29	22	31.80	4.95
750	900	17.62	12.33	4.09	35.23	42.73	9.69	239	1,305	2,114	29	23	31.80	4.95
750	1,000	18.45	12.33	4.58	36.89	44.74	10.15	246	1,376	2,213	30	23	31.80	4.95
800	800	17.56	12.94	3.74	35.12	42.59	9.66	216	1,303	2,107	29	22	33.92	5.28
800	900	18.41	12.94	4.25	36.83	44.67	10.13	221	1,377	2,210	29	23	33.92	5.28
800	1,000	19.28	12.94	4.76	38.55	46.76	10.61	227	1,451	2,313	30	23	33.92	5.28

Small-Breed Growing Males

Live Weight (kg)	Gain (g)	Dry Matter Intake[a] (kg)	Energy NEM (Mcal)	NEG (Mcal)	ME (Mcal)	DE (Mcal)	TDN (kg)	Protein UIP (g)	DIP (g)	CP (g)	Minerals Ca (g)	P (g)	Vitamins A (1,000 IU)	D (1,000 IU)
100	500	2.45	2.72	1.02	6.54	7.56	1.72	287	41	392	16	8	4.24	0.66
100	600	2.64	2.72	1.23	7.04	8.15	1.85	316	58	422	17	9	4.24	0.66
100	700	2.83	2.72	1.45	7.55	8.74	1.98	345	75	453	18	9	4.24	0.66
150	500	3.28	3.69	1.20	8.55	9.92	2.25	257	129	525	18	11	6.36	0.99
150	600	3.52	3.69	1.46	9.16	10.64	2.41	282	151	563	19	11	6.36	0.99
150	700	3.76	3.69	1.71	9.78	11.36	2.58	306	174	601	19	12	6.36	0.99
200	500	4.12	4.57	1.37	10.45	12.18	2.76	232	213	573	20	13	8.48	1.32
200	600	4.40	4.57	1.66	11.17	13.02	2.95	252	241	629	20	14	8.48	1.32
200	700	4.69	4.57	1.96	11.90	13.87	3.15	273	268	684	21	14	8.48	1.32
250	500	4.99	5.41	1.53	12.31	14.41	3.27	210	296	598	22	16	10.60	1.65
250	600	5.32	5.41	1.86	13.14	15.38	3.49	228	328	638	22	16	10.60	1.65
250	700	5.66	5.41	2.19	13.97	16.35	3.71	245	361	679	23	17	10.60	1.65
300	500	5.89	6.20	1.68	14.15	16.64	3.77	193	378	707	23	17	12.72	1.98
300	600	6.28	6.20	2.04	15.09	17.74	4.02	207	415	754	23	17	12.72	1.98
300	700	6.68	6.20	2.41	16.04	18.85	4.28	221	453	801	24	18	12.72	1.98
350	500	6.86	6.96	1.82	16.01	18.91	4.29	180	461	823	23	18	14.84	2.31
350	600	7.31	6.96	2.22	17.06	20.15	4.57	191	503	877	24	18	14.84	2.31
350	700	7.76	6.96	2.62	18.13	21.41	4.86	203	547	932	25	19	14.84	2.31
400	500	7.90	7.69	1.96	17.91	21.25	4.82	171	545	947	24	19	16.96	2.64
400	600	8.41	7.69	2.39	19.08	22.64	5.14	180	594	1,010	25	19	16.96	2.64
400	700	8.94	7.69	2.82	20.27	24.06	5.46	189	644	1,073	26	20	16.96	2.64
450	500	9.03	8.40	2.10	19.87	23.70	5.37	166	634	1,083	28	19	19.08	2.97
450	600	9.62	8.40	2.55	21.18	25.26	5.73	174	689	1,155	28	19	19.08	2.97
450	700	10.23	8.40	3.01	22.51	26.84	6.09	180	744	1,227	28	20	19.08	2.97
500	500	10.28	9.09	2.23	21.93	26.29	5.96	167	726	1,233	28	19	21.20	3.30
500	600	10.96	9.09	2.71	23.39	28.04	6.36	173	788	1,315	28	20	21.20	3.30
500	700	11.65	9.09	3.20	24.87	29.81	6.76	177	851	1,398	28	20	21.20	3.30
550	500	11.67	9.77	2.36	24.12	29.08	6.60	174	825	1,400	28	19	23.32	3.63
550	600	12.46	9.77	2.87	25.75	31.05	7.04	178	895	1,495	28	20	23.32	3.63
550	700	13.26	9.77	3.39	27.40	33.03	7.49	181	966	1,591	28	20	23.32	3.63
600	500	13.25	10.43	2.48	26.50	32.14	7.29	187	933	1,590	28	19	25.44	3.96
600	600	14.16	10.43	3.02	28.32	34.35	7.79	190	1,012	1,699	28	20	25.44	3.96
600	700	15.08	10.43	3.57	30.17	36.59	8.30	192	1,091	1,810	28	21	25.44	3.96

Maintenance of Mature Breeding Bulls

Live Weight (kg)	Gain (g)	Dry Matter Intake[a] (kg)	Energy NEM (Mcal)	NEG (Mcal)	ME (Mcal)	DE (Mcal)	TDN (kg)	Protein UIP (g)	DIP (g)	CP (g)	Minerals Ca (g)	P (g)	Vitamins A (1,000 IU)	D (1,000 IU)
500	—	7.89	9.09	—	15.79	19.15	4.34	161	472	789	20	12	21.20	3.30
600	—	9.05	10.43	—	18.10	21.95	4.98	155	573	905	24	15	25.44	3.96
700	—	10.16	11.70	—	20.32	24.64	5.59	148	670	1,016	28	18	29.68	4.62
800	—	11.23	12.94	—	22.46	27.24	6.18	142	764	1,123	32	20	33.92	5.28
900	—	12.27	14.13	—	24.53	29.76	6.75	135	854	1,227	36	22	38.16	5.94
1,000	—	13.28	15.29	—	26.55	32.20	7.30	129	943	1,328	41	25	42.40	6.60

Continues

TABLE 9–4 (continued). Daily Nutrient Requirements of Growing Dairy Cattle and Mature Bulls

Live Weight (kg)	Gain (g)	Dry Matter Intake[a] (kg)	Energy NEM (Mcal)	NEG (Mcal)	ME (Mcal)	DE (Mcal)	TDN (kg)	Protein UIP (g)	DIP (g)	CP (g)	Minerals Ca (g)	P (g)	Vitamins A (1,000 IU)	D (1,000 IU)
1,100	—	14.26	16.43	—	28.52	34.59	7.85	122	1,029	1,426	45	28	46.64	7.26
1,200	—	15.22	17.53	—	30.44	36.92	8.37	115	1,113	1,522	49	30	50.88	7.92
1,300	—	16.16	18.62	—	32.32	39.21	8.89	108	1,196	1,616	53	32	55.12	8.58
1,400	—	17.09	19.68	—	34.17	41.45	9.40	102	1,277	1,709	57	35	59.36	9.24

NOTE: The following abbreviations were used: NEM, net energy for maintenance; NEG, net energy for gain; ME, metabolizable energy; DE, digestible energy; TDN, total digestible nutrients; UIP, undegraded intake protein; DIP, degraded intake protein; CP, crude protein.

[a]The data for DMI are not requirements per se, unlike the requirements for net energy maintenance, net energy gain, and absorbed protein. They are not intended to be estimates of voluntary intake but are consistent with the specified dietary energy concentrations. The use of diets with decreased energy concentrations will increase dry matter intake needs; metabolizable energy, digestible energy, and total digestible nutrient needs; and crude protein needs. The use of diets with increased energy concentrations will have opposite effects on these needs.

TABLE 9–5. Daily Nutrient Requirements of Lactating and Pregnant Cows

Live Weight (kg)	Energy NEL (Mcal)	ME (Mcal)	DE (Mcal)	TDN (kg)	Total Crude Protein (g)	Minerals Ca (g)	P (g)	Vitamins A (1,000 IU)	D (1,000 IU)
Maintenance of Mature Lactating Cows[a]									
400	7.16	12.01	13.80	3.13	318	16	11	30	12
450	7.82	13.12	15.08	3.42	341	18	13	34	14
500	8.46	14.20	16.32	3.70	364	20	14	38	15
550	9.09	15.25	17.53	3.97	386	22	16	42	17
600	9.70	16.28	18.71	4.24	406	24	17	46	18
650	10.30	17.29	19.86	4.51	428	26	19	49	20
700	10.89	18.28	21.00	4.76	449	28	20	53	21
750	11.47	19.25	22.12	5.02	468	30	21	57	23
800	12.03	20.20	23.21	5.26	486	32	23	61	24
Maintenance Plus Last 2 Months of Gestation of Mature Dry Cows[b]									
400	9.30	15.26	18.23	4.15	890	26	16	30	12
450	10.16	16.66	19.91	4.53	973	30	18	34	14
500	11.00	18.04	21.55	4.90	1,053	33	20	38	15
550	11.81	19.37	23.14	5.27	1,131	36	22	42	17
600	12.61	20.68	24.71	5.62	1,207	39	24	46	18
650	13.39	21.96	26.23	5.97	1,281	43	26	49	20
700	14.15	23.21	27.73	6.31	1,355	46	28	53	21
750	14.90	24.44	29.21	6.65	1,427	49	30	57	23
800	15.64	25.66	30.65	6.98	1,497	53	32	61	24
Milk Production—Nutrients/kg of Milk of Different Fat Percentages									
(Fat %)									
3.0	0.64	1.07	1.23	0.280	78	2.73	1.68	—	—
3.5	0.69	1.15	1.33	0.301	84	2.97	1.83	—	—
4.0	0.74	1.24	1.42	0.322	90	3.21	1.98	—	—
4.5	0.78	1.32	1.51	0.343	96	3.45	2.13	—	—
5.0	0.83	1.40	1.61	0.364	101	3.69	2.28	—	—
5.5	0.88	1.48	1.70	0.385	107	3.93	2.43	—	—
Live Weight Change During Lactation—Nutrients/kg of Weight Change[c]									
Weight loss	−4.92	−8.25	−9.55	−2.17	−320	—	—	—	—
Weight gain	5.12	8.55	9.96	2.26	320	—	—	—	—

NOTE: The following abbreviations were used: NEL, net energy for lactation; ME, metabolizable energy; DE, digestible energy; TDN, total digestible nutrients.

[a]To allow for growth of young lactating cows, increase the maintenance allowances for all nutrients except vitamins A and D by 20 percent during the first lactation and 10 percent during the second lactation.

[b]Values for calcium assume that the cow is in calcium balance at the beginning of the last 2 months of gestation. If the cow is not in balance, then the calcium requirement can be increased from 25 to 33 percent.

[c]No allowance is made for mobilized calcium and phosphorus associated with live weight loss or with live weight gain. The maximum daily nitrogen available from weight loss is assumed to be 30 g or 234 g of crude protein.

From *Nutrient Requirements of Dairy Cattle* Sixth Revised Edition. Copyright © 1989 by the National Academy of Sciences, Washington D.C. Reprinted by permission.

TABLE 9–6. Recommended Nutrient Content of Diets for Dairy Cattle

Lactating Cow Diets — Milk Yield (kg/d) key:

Cow Wt (kg)	Fat (%)	Wt Gain (kg/d)	MY 1	MY 2	MY 3	MY 4	MY 5
400	5.0	0.220	7	13	20	26	33
500	4.5	0.275	8	17	25	33	41
600	4.0	0.330	10	20	30	40	50
700	3.5	0.385	12	24	36	48	60
800	3.5	0.440	13	27	40	53	67

Nutrient	Lactating 1	Lactating 2	Lactating 3	Lactating 4	Lactating 5	Early Lactation (wks 0–3)	Dry, Pregnant Cows	Calf Milk Replacer	Calf Starter Mix	Growing 3–6 Mos [a]	Growing 6–12 Mos [a]	Growing >12 Mos [a]	Mature Bulls	Maximum Tolerable Levels [b,c]
Energy														
NEL, Mcal/kg	1.42	1.52	1.62	1.72	1.72	1.67	1.25	—	—	—	—	—	—	—
NEM, Mcal/kg	—	—	—	—	—	—	—	2.40	1.90	1.70	1.58	1.40	1.15	—
NEG, Mcal/kg	—	—	—	—	—	—	—	1.55	1.20	1.08	0.98	0.82	—	—
ME, Mcal/kg	2.35	2.53	2.71	2.89	2.89	2.80	2.04	3.78	3.11	2.60	2.47	2.27	2.00	—
DE, Mcal/kg	2.77	2.95	3.13	3.31	3.31	3.22	2.47	4.19	3.53	3.02	2.89	2.69	2.43	—
TDN, % of DM	63	67	71	75	75	73	56	95	80	69	66	61	55	—
Protein equivalent														
Crude protein, %	12	15	16	17	18	19	12	22	18	16	12	12	10	—
UIP, %	4.4	5.2	5.7	5.9	6.2	7.0	—	—	—	8.2	4.4	2.1	—	—
DIP, %	7.8	8.7	9.6	10.3	10.4	9.7	—	—	—	4.6	6.4	7.2	—	—
Fiber content (min.) [d]														
Crude fiber, %	17	17	17	15	15	17	22	—	—	13	15	15	15	—
Acid detergent fiber, %	21	21	21	19	19	21	27	—	—	16	19	19	19	—
Neutral detergent fiber, %	28	28	28	25	25	28	35	—	—	23	25	25	25	—
Ether extract (min.), %	3	3	3	3	3	3	3	10	3	3	3	3	3	—
Minerals														
Calcium, %	0.43	0.51	0.58	0.64	0.66	0.77	0.39 [e]	0.70	0.60	0.52	0.41	0.29	0.30	2.00
Phosphorus, %	0.28	0.33	0.37	0.41	0.41	0.48	0.24	0.60	0.40	0.31	0.30	0.23	0.19	1.00
Magnesium, % [f]	0.20	0.20	0.20	0.25	0.25	0.25	0.16	0.07	0.10	0.16	0.16	0.16	0.16	0.50
Potassium, % [g]	0.90	0.90	0.90	1.00	1.00	1.00	0.65	0.65	0.65	0.65	0.65	0.65	0.65	3.00
Sodium, %	0.18	0.18	0.18	0.18	0.18	0.18	0.10	0.10	0.10	0.10	0.10	0.10	0.10	—
Chlorine, %	0.25	0.25	0.25	0.25	0.25	0.25	0.20	0.20	0.20	0.20	0.20	0.20	0.20	—
Sulfur, %	0.20	0.20	0.20	0.20	0.20	0.25	0.16	0.29	0.20	0.16	0.16	0.16	0.16	0.40
Iron, ppm	50	50	50	50	50	50	50	100	50	50	50	50	50	1,000
Cobalt, ppm	0.10	0.10	0.10	0.10	0.10	0.10	0.10	0.10	0.10	0.10	0.10	0.10	0.10	10.00
Copper, ppm [h]	10	10	10	10	10	10	10	10	10	10	10	10	10	100
Manganese, ppm	40	40	40	40	40	40	40	40	40	40	40	40	40	1,000
Zinc, ppm	40	40	40	40	40	40	40	40	40	40	40	40	40	500
Iodine, ppm [i]	0.60	0.60	0.60	0.60	0.60	0.60	0.25	0.25	0.25	0.25	0.25	0.25	0.25	50.00 [j]
Selenium, ppm	0.30	0.30	0.30	0.30	0.30	0.30	0.30	0.30	0.30	0.30	0.30	0.30	0.30	2.00
Vitamins [k]														
A, IU/kg	3,200	3,200	3,200	3,200	3,200	4,000	4,000	3,800	2,200	2,200	2,200	2,200	3,200	66,000
D, IU/kg	1,000	1,000	1,000	1,000	1,000	1,000	1,200	600	300	300	300	300	300	10,000
E, IU/kg	15	15	15	15	15	15	15	40	25	25	25	25	15	2,000

NOTE: The values presented in this table are intended as guidelines for the use of professionals in diet formulation. Because of the many factors affecting such values, they are not intended and should not be used as a legal or regulatory base.

[a] The approximate weight for growing heifers and bulls at 3–6 mos is 150 kg; at 6–12 mos, it is 250 kg; and at more than 12 mos it is 400 kg. The approximate average daily gain is 700 g/day.

[b] The maximum safe levels for many of the mineral elements are not well defined and may be substantially affected by specific feeding conditions. Additional information is available in *Mineral Tolerance of Domestic Animals* (NRC, 1980).

[c] Vitamin tolerances are discussed in detail in *Vitamin Tolerance of Animals* (NRC, 1987b).

[d] It is recommended that 75 percent of the NDF in lactating cow diets be provided as forage. If this recommendation is not followed, a depression in milk fat may occur.

[e] The value for calcium assumes the cow is in calcium balance at the beginning of the dry period. If the cow is not in balance, then the dietary calcium requirement should be increased by 25 to 33 percent.

[f] Under conditions conducive to grass tetany ___, magnesium should be increased to 0.25 or 0.30 percent.

[g] Under conditions of heat stress, potassium should be increased to 1.2 percent

[h] The cow's copper requirement is influenced by molybdenum and sulfur in the diet

[i] If the diet contains as much as 25 percent strongly goitrogenic feed on a dry basis, the iodine provided should be increased two times or more.

[j] Although cattle can tolerate this level of iodine, lower levels may be desirable to reduce the iodine content of milk.

[k] The following minimum quantities of B-complex vitamins are suggested per unit of milk replacer: niacin, 2.6 ppm; pantothenic acid, 13 ppm; riboflavin, 6.5 ppm; pyridoxine, 6.5 ppm; folic acid, 0.5 ppm; biotin, 0.1 ppm; vitamin B_{12}, 0.07 ppm; thiamin, 6.5 ppm; and choline, 0.26 percent. It appears that adequate amounts of these vitamins are furnished when calves have functional rumens (usually at 6 weeks of age) by a combination of rumen synthesis and natural feedstuffs.

From *Nutrient Requirements of Dairy Cattle* Sixth Revised Edition. Copyright © 1989 by the National Academy of Sciences, Washington D.C.

TABLE 9–7. Maximum Dry Matter Intake Guidelines[a]

Body Wt (kg)	400	500	600	700	800
FCM (4% milk)			% of Body Wt		
10	2.5	2.3	2.2	2.1	2.0
15	2.8	2.5	2.4	2.3	2.2
20	3.1	2.8	2.7	2.6	2.4
25	3.4	3.1	3.0	2.8	2.6
30	3.7	3.4	3.2	3.0	2.8
35	4.0	3.6	3.4	3.2	3.0
40	—	3.8	3.6	3.4	3.2
45	—	4.0	3.8	3.6	3.4

[a]Derived from:

(1) Chandler, P.T., and C.A. Brown. 1975. Adjusting nutrient concentrations according to season of year and make up of group or herd. Unpublished report
at Dairy Feed Conference Board, University of Delaware.

(2) Smith, N.E. 1971. Feed efficiency in intensive milk production. Proc. 10th Annual Dairy Cattle Day, p. 40. Dep. Anim. Sci., Univ. Calif., Davis.

(3) Swanson, E.W., S.A. Hinton, and J.T. Miles. 1967. Full lactation response on restricted vs. *ad libitum* roughage diets with liberal concentrate feeding. J. Dairy Sci. 50:1147–1152.

(4) Trimberger, G.W., H.G. Gray, W.L. Johnson, M.J. Wright, D. Van Vleck, and C.R. Henderson. 1963. Forage appetite in dairy cattle. Proc. Cornell Nutr. Conf. Feed Manuf., pp. 33–43.

(5) Trimberger, G.W., H.F. Tyrrell, D.A. Morrow, J.T. Reid, M.J. Wright, W.F. Shipe, W.G. Merrill, J.K. Loosli, C.E. Coppock, L.A. Moore, and C.H. Gordon. 1972. Effects of liberal concentrate feeding on health, reproductive efficiency, economy of milk production, and other related responses of the dairy cow. N.Y. Food Life Serv. Bull. No. 8.

Study Questions and Problems

Chapter 9

I. True or False Questions

1. The energy requirement for each pound of milk produced from dairy cattle, increases with fat percentage in the milk.
2. Most thumb rules for feeding grain to dairy cows tend to overfeed low producers and underfeed high producers.
3. Feeding legume hays during the dry (nonlactating) period is a good nutritional method of reducing the incidence of milk fever in dairy cattle.
4. Dairy heifers should be fed at least 6 lb of grain daily from the time they are 12 months of age to freshening to ensure adequate growth.
5. Lactating dairy cows reach peak milk production about 3 to 4 weeks after calving.
6. Dairy cows in good condition should be fed little or no grain during the dry period.
7. The occurrence of ketosis in lactating dairy cows can be reduced if the cow is consuming approximately 1% of her body weight as grain by calving time.
8. Fat cow syndrome occurs most frequently in those dairy cows fed excessive grain during early lactation.
9. A 100-lb dairy calf should consume about 5 qt (10% of body weight) of milk or milk replacer daily.
10. The rumen capacity of a 9-month-old dairy heifer is not sufficient to allow the animal to meet energy needs from forage alone.

II. Multiple Choice (select the single best answer)

1. By common definition, a milk replacer for dairy calves is
   a. A high-energy, high-protein, low-fiber grain mix.
   b. High milk-byproduct feed, sold as a powder and reconstituted with water.
   c. A medicated feed additive to be mixed with grains and fed to calves instead of milk.
   d. All of the above.
   e. None of the above.
2. A dairy cow producing 80 lb of milk per day requires about how many pounds of water daily?
   a. 20
   b. 40
   c. 80
   d. 160
   e. 320
3. Dairy heifers reared as replacement stock should gain approximately ___?___ pounds per day from birth to freshening.
   a. 0.5–1.0
   b. 1.0–1.5
   c. 1.5–2.0
   d. 2.0–2.5
   e. 2.5–3.0
4. What level of crude protein would you recommend in the concentrate mix fed to a lactating cow if corn silage (8% crude protein) were used as the source of forage (dry matter basis)?
   a. 6–9
   b. 10–13
   c. 14–17
   d. 18–21
   e. 21–24
5. Fiber in the dairy diet promotes the production of which rumen volatile fatty acid?
   a. Acetic
   b. Propionic
   c. Butyric
   d. Lactic
   e. All of the above

6. A dairy farmer told you his cow has suddenly dropped to zero milk production. He is not clear about other symptoms but knows that it has been 3 weeks since the cow has calved and he is feeding a 1-lb can of oats to her twice a day plus hay. Which of the following disorders would you expect?
   a. Milk fever
   b. Ketosis
   c. Fat cow syndrome
   d. Displaced abomasum
   e. Bloat
7. Challenge or lead feeding in dairy cows is
   a. Leading your cows to the feed bunk.
   b. Feeding a lot of grain during the dry period.
   c. Increasing grain allotment as long as milk production increases.
   d. Feeding more grain during the end of lactation to lead into the next lactation.
   e. Reduce dietary crude protein to challenge bacterial protein synthesis in the rumen.
8. A specific dairy cow needs 20 lb of TDN and receives 13 lb of TDN from forages. How many pounds of a 65% TDN concentrate mixture must be fed to meet her remaining TDN needs?
   a. 4.6
   b. 7.0
   c. 10.8
   d. 11.4
   e. None of the above

III. Matching or Identification

Identify the following situations.

a. Ketosis	f. Scours
b. Milk fever	g. Peak in milk production
c. Displaced abomasum	h. Colostrum
d. Fat cow syndrome	i. Dry period
e. Bloat	j. Rumen acidosis

1. May occur when cows graze fresh alfalfa pasture.
2. May occur in calves given high levels of milk replacer.
3. Problem can be reduced by feeding a low calcium diet about 2 weeks before calving.
4. Excess fat in the liver and other internal organ region.
5. Excess is stored by cooling or fermenting and consumed by the young calf.
6. May occur in lactating cows consuming inadequate energy.
7. Time during which cows in good condition should be fed little or no grain.
8. Hay feeding during the dry period may reduce this digestive tract disturbance.
9. Occurs about 6 to 8 weeks after calving.
10. May occur in cows fed high grain diets during early lactation.

IV. Short Answer

1. Why is fiber important in lactating dairy diets?
2. What is the cause of milk fever and the method of prevention?
3. What can be done through dairy feeding management to prevent or minimize fat cow syndrome?
4. What is the cause of ketosis? State a method of prevention and one of treatment.
5. Define "lead" or "challenge" feeding of dairy cows.

V. Problems

1. Formulate 1 ton of calf starter diet to contain 14% crude protein (as-fed basis). Use some combination of corn (9% CP) and oats (12% CP) as the grain source together with soybean meal (46% CP) as the protein fraction. Assume a mineral and vitamin premix are fixed at 3% of the diet (0% CP).

2. A 1400-lb dairy cow will consume 42 lb of dry matter from alfalfa hay cut at first bloom, but only 28 lb dry matter from alfalfa hay cut at full bloom.
   a. If intake is not limiting for grain, how much corn would have to be fed with the 28 lb of mature hay dry matter to provide the same amount of energy as supplied by the 42 lb dry matter from the first-flower alfalfa?

	% DM	% TDN (DM basis)	$/lb
Alfalfa hay, first flower	87	63.2	.0325
Alfalfa hay, full bloom	88	55.6	.0275
Corn grain	89	91.0	.0460

   b. Which would be more economical feeding the 1400-lb cow the first flower hay or the combination of full bloom hay and corn to feed the same amount of energy? Show cost.
3. A 1210-lb dairy cow is producing an average of 37 lb of milk with 4.5% fat content. Formulate a typical daily ration for this cow using average quality alfalfa-brome hay, ground ear corn and soybean meal. Assume a mineral and vitamin mixture will be fed free-choice.
   a. Information needed (dry matter basis)

	CP	NE_L
Daily requirements for maintenance and milk production	5.13 lb	25.2 Mcal
Feedstuff composition		
Alfalfa-brome hay	15.0%	0.55 Mcal/lb
Ground ear corn	9.5%	1.00 Mcal/lb
Soybean meal	47.0%	0.96 Mcal/lb

   b. Formulated daily ration (dry matter basis)

Ingredient	Amount, lb	CP, lb	NE_L, Mcal
Alfalfa-brome hay	_____	_____	_____
Ground ear corn	_____	_____	_____
Soybean meal	_____	_____	_____
TOTAL	_____	_____	_____

Notes

Notes

10
Sheep Feeding Guides

Chapter Goals

- Outline and discuss a life-cycle sheep feeding program.
- Identify specific nutrient needs or additives used within the life cycle.
- Identify nutrient-related diseases or disorders within the sheep life cycle.
- Formulate a feedlot sheep diet.

Feed represents the largest single production cost in all types of sheep operations. Diets must be formulated and fed that support optimum production, are efficient, economical to feed and minimize nutritionally related problems.

This section outlines guides for *The Breeding Flock* and *Feeding of Lambs*. Additional and more complete information may be obtained from various publications distributed by land grant universities and other colleges. Nutrient requirements, feed allowances and other pertinent information for sheep and lambs are given in tables 10–1 through 10–8.

I. The Breeding Flock

Ewes are the backbone of the sheep enterprise. They raise the lambs and produce the wool. These two cash crops are greatly influenced by feeding management.

Breeding and selection programs should not be overlooked, but feeding and management practices have more immediate impact upon production. Often the income of spring lamb production is more than that realized by other meat animal programs for capital invested. The return is quite high per animal unit.

An animal unit is generally considered as a mature cow and her calf until weaning. For sheep, an animal unit is 4 to 6 ewes and their lambs.

A recommended flock size is 100 or more ewes. The minimum size is 35 ewes (1 ram unit).

A. *Choosing a Lambing System*

One of the first management decisions is whether to follow an early or late lambing system. These are points to consider.

1. Early lambing (January through February)
   a. Lamb prices are highest in May and June when most early lambs can be marketed.
   b. More labor is available to give needed extra care and attention during lambing.
   c. Parasite problems in early lambs are generally less serious.
   d. More ewes can be carried in the flock for a given acreage of pasture.
   e. Lambing facilities should be adequate but need not be fancy or expensive.
   f. Good care and management is required during flushing, breeding and lambing periods.
2. Late lambing (March through April)
   a. Roughages provide most feed needed for both ewes and lambs.
   b. Lambing facilities need not be as good as for early lambs.
   c. Less care and management is required before and during the breeding season for good conception.
   d. Lambs can be marketed from pasture or cornfields with a minimum of concentrate feeding.

    e. Lamb prices are substantially lower in the fall and early winter when late lambs are marketed.

    f. High quality pastures are necessary.

    g. Parasite problems in late lambs are serious. Prevention and control are needed.

B. *A Summary of Feed Requirements*

A ewe and her lamb(s) need about 4 bushels of grain and 800 lb of hay every year. In addition, a ewe needs about 5 months of good pasture grazing and 2 months of crop residue or winter pasture grazing. (On a hay basis this would mean about 800 lb more hay or a total of 1600 lb of hay a year if pasture is not availabl). If the hay is not good quality alfalfa, protein supplement is necessary. In this case, it would take 20 to 25 lb of 44% protein supplement per ewe, per year.

One ram unit (35 ewes) requires per year: 130 to 150 bushels of corn or its equivalent; 14 to 15 ton of good quality alfalfa hay and, if the hay is poor quality or grass hay, 700 to 800 lb of 44% protein supplement.

Grains need not be ground for most sheep. If you have a few broken-mouthed ewes, grinding may be essential, but for good-mouth ewes, it is not necessary.

C. *Feeding for Maintenance*

The mature (3 to 8 years) ewe needs to be fed only well enough to maintain her normal physiological weight from the time her lambs are weaned until she is about 15 weeks into her next pregnancy. This assumes she has been fed such that she has not lost much weight during the previous lactation.

Pasture and other field feeds, when available, are adequate for maintaining ewes and would normally be used because they are the cheapest source of nutrients. When harvested feed must be fed, a variety of feeds or combination can be used. Hay, haylage, corn or sorghum silage all could be used as the basic feed. If grain is cheap relative to these feeds, it can be substituted for part of the roughage.

Based on energy needs (TDN), the following amounts of these feed (as-fed basis) should meet the daily maintenance requirement of 140-pound ewes

Legume hay	3.5 lb
Grass hay	4.0 lb
Haylage	6.0 lb
Corn silage	7.5 lb
Sorghum (forage) silage	8.0 lb
Legume hay 1.75 lb + corn	1.0 lb
Mixed hay 2.0 lb + corn	1.0 lb

Protein supplementation ⅓ lb) would be necessary for the corn and sorghum silage. A small amount (¼ lb) of grain may need to be fed with the haylage to increase the energy intake to a maintenance level.

D. *Feeding and Care at Breeding Time (Flushing)*

Take ewes off legume pasture approximately 2 weeks before breeding season. Some research indicates that the estrogens in some types of legumes keep the ewes from "settling."

At about the same time, attempt to condition the ewe's reproductive system so that all breeding can be done within a 3- to 4-week period. One reason for conditioning the reproductive system is to try to lamb ewes close together. Another is that this practice helps ovulate more eggs and may result in more twin births.

This process of conditioning ewes is called flushing. Flushing can be done several ways, depending on the condition of the ewe and the program.

Condition	**Ways to Flush**
1. Ewes on dry pasture, no green chop or grain	(a) Turn on Sudan or lush grass pasture 2 weeks before breeding season
	(b) Feed 0.5 to 1.0 lb of grain (corn, oats, milo) 2 weeks before and 1 week after breeding season starts
2. Ewes on good pasture (other than legume)	(a) Change pasture. Example: If on brome, change to Sudan and vice versa, 2 weeks before breeding
	(b) Feed 0.5–1.0 pound grain daily for 2 weeks before and one week after breeding season starts

E. *Feeding after Breeding (2 Periods)*

The gestation period of sheep is 147 to 150 days (5 months).

1. First 3½ months of pregnancy—After ewes are bred, there is very little nutritional worry for the next 3½ months. Try to keep ewes about the same weight and in good health. Pasture or hay is satisfactory as described for maintenance. If pasture is not available, several suggested ewe gestation rations fed as complete feeds are shown in Table 10–9.

2. Last 1½ months of pregnancy—when ewes reach this stage of pregnancy, it is the nutritionally critical period of the year's cycle. Poor care in this period means lost dollars the rest of the year. The producer loses in

    a. Lambing paralysis or pregnancy disease (possible loss of ewe and loss of lamb).
    b. Weak lambs.
    c. Drop in milk production.
    d. Low wool clip.
    e. Light wool clip.

    Energy requirements increase rapidly during this period. The need for protein, minerals and perhaps vitamins also increases. These requirements can be easily met by making gradual increases in the amount of feed fed.

    The following rations should be adequate for average size ewes.

    Wintering and spring lambing ewes:
    4–4½ lb hay + ½–¾ lb shelled corn
    2–2½ lb hay + 2 lb shelled corn
    7–8 lb corn silage + 1 lb shelled corn + ¼–½ lb supplement
    6–7 lb haylage + 1 lb shelled corn
    (Note: Hay should be at least average quality legume or legume-grass.)

    Fall lambing ewes:
    Pasture + ½–¾ lb grain/ewe daily. If pasture is not available, feed as for winter lambing ewes. Amount fed may be reduced slightly.

    The pounds of feed given are average amounts for the 40 to 50 days of late gestation. This means that early in this period the amount fed would be less and late in the period the amount would be greater. If ewes are larger, the amount fed would need to be somewhat greater; if small, somewhat less. If the ewes enter this period thin, then the amount fed would need to be increased.

    Trace mineral salt could be included in a free-choice mineral mixture made of 2 parts dicalcium phosphate, 2 parts plain salt and 1 part of trace mineralized salt. There is some possibility of a copper buildup and toxicity if trace mineralized salt is used as the only salt and if ewes are self-fed highly palatable mineral mixtures that contain copper in the sulfate form.

F. *Range Ewe Nutrition*[1]

1. Present and previous stocking rate greatly affects range forage production and composition. Overgrazed ranges are not only unproductive but many of the more palatable and nutritious range plants disappear and are replaced by less desirable and often poisonous plants.

2. Range forages
    Three classes of forages are found in range areas. All three may be present on some ranges while forage on other ranges will be predominantly of one type.

    a. Grass plants
       Mature grasses are lowest in protein and phosphorus but are highest in energy-yielding cellulose. Protein and TDN content decreases while lignin and cellulose increase with maturity. The grazing season is extended when both cool weather and warm weather grasses are utilized. Cool weather grasses provide grazing in the spring, mature during the summer and become less productive. The warm weather grasses grow during the summer (when moisture conditions allow) and mature in late summer or fall.

---

1. Summarized from *The Sheepman's Production Handbook* published for the Sheep Industry Development Program, Inc., Denver, CO.

Pasture provides adequate nutrition for ewes during early gestation.

   b. Browse

Browse is a broad-leafed woody plant, a shrub, a bush or a tree of small stature. These plants are relatively high in protein and low in cellulose. Protein content of browse decreases only slightly with maturity while the increase of lignin and cellulose is moderate. The phosphorus content of browse also has a slight tendency to increase with maturity.

   c. Forbs

Forbs are broad-leafed, herbaceous plants, commonly referred to as weeds. They are intermediate between grasses and browse in most respects. Protein content declines only slightly with maturity but phosphorus content declines more rapidly.

   d. Volume and quality of forage found on the range are affected by season of the year, maturity of the plant, present and previous stocking rate and moisture condition.

During the growing season, most ranges, properly stocked, will provide the necessary nutrient requirements of sheep during any phase of production. Inadequate or excessive rainfall may alter this situation.

As the plants mature, need for supplementation will increase according to type of vegetation found on the range. Since in most range operations the ewe is dry or in early gestation during this period of the year, a slight loss of weight and condition can be tolerated and supplementation will only be required when range condition is extremely poor.

3. Range supplements

   a. The amount and kind of supplement necessary in range sheep operations is extremely variable between ranges and between years. The successful operator must be able to accurately assess the condition of the range, the condition of the sheep and be able to determine what supplements and at what levels can be supplied at the least cost to assure optimum production. Supplementation usually increases production. However, it is commonly not economical to supplement range sheep for maximum production because the increased costs are usually not commensurate with output in saleable animal products. However, sheep must be maintained in reasonable condition throughout the year.

Range sheep production is entirely dependent upon weather and moisture conditions. Under drought conditions, sheep may require supplementation any time during the year. In seasons of extremely high rainfall, grass may have such a high moisture content that maximum milk production in lactating ewes and growth rate in lambs will not be realized. Supplementation under these conditions is seldom economical.

   b. Energy, protein, phosphorus and vitamin A are the nutrients that may be limiting for range animals. One of the first requirements to be met for range sheep is energy. They frequently travel long distances to acquire feed and water. In addition, they must maintain body temperatures during the winter without shelter. When energy-supplying carbohydrates and fats in the forage are inadequate, the animal will use protein for energy. This will further aggravate any protein deficiency already present in the basal diet.

Protein feeds such as cottonseed meal and soybean meal are better supplements when some forage is available than energy supplements such as corn and barley. This is true even when energy is substantially low in the sheep's diet. Supplements such as corn and barley have a tendency to reduce the digestibility of cellulose and other carbohydrates (energy furnishing constituents) of the range forage and therefore do not increase substantially the overall energy intake. The protein supplements (soybean meal and cottonseed meal) actually increase the digestibility of most nutritional constituents in the range forage and thereby enhance the nutritive value of the range feed. In addition, the protein supplements can be used for meeting both the energy and protein requirements of the animals. When practically no consumable forage is left on the range, hay and grain feeding is more suitable.

The supplement chosen will be influenced by availability, transportation cost, feeding cost and form of the supplement. Soybean meal, cottonseed meal, sunflower meal, corn, milo, barley, grass hay, alfalfa hay and alfalfa pellets are feeds often used to supplement range flocks. Urea is generally of little value in range supplements.

 c. Supplements are commonly fed as loose hay, whole corn, cake or pellets so they can be fed on the ground or snow. This method is not to be recommended when the practice can be avoided as many diseases are spread through the contamination of feed and water. One infected ewe may transmit disease to a large portion of the flock resulting in the loss of ewes and lambs. Daily feeding or feeding at less frequent intervals in feed troughs constructed to prevent contamination is recommended. Protein-molasses-grain blocks may be used. Supplemental feed intake is affected by availability of range forage and is regulated to some extent by the hardness of block and frequency of feeding. Blocks are more expensive, but labor requirements for feeding are low. Protein supplements, grain or both mixed with 15% to 25% salt can be fed. However, water intake must increase and required supplementation is often associated with periods of limited water supply.

G. *Pregnancy Disease (Lambing Paralysis or Ketosis)*

This is a nutritional condition. It is caused by lack of usable carbohydrates (sugars, starches). It usually affects older ewes, particularly those carrying twins or triplets, but occasionally affects those carrying singles.

Most cases occur in poor-condition sheep on inadequate diets, such as straw, corn fodder or poor pasture with too little or no grain. The increasing demands of the growing fetus finally becomes too great for the readily available supply of carbohydrates and pregnancy disease results.

Symptoms are progressive—loss of appetite, lagging behind the flock, labored breathing, impaired vision, staggering and finally paralysis. The urine and breath have an abundance of ketone bodies, which give the same symptoms of ketosis (sweet smell) as in cattle. Management entails two phases, prevention and treatment. Prevention has been discussed in feeding the ewe an adequate grain ration during the nutritionally critical period.

Pregnancy disease can be treated satisfactorily if caught in time—before the ewe is paralyzed. Generally, one will exhibit the condition before it is noticed in others. If more ewes show symptoms, supplement ¼ to ½ pound of blackstrap molasses or corn sugar (dextrose) to the grain mixture. Drenching the ewes with these materials is another method.

Administration of dextrose, molasses or propylene glycol to a ewe in a coma should be by stomach tube two to three times a day. Sheep in a coma cannot swallow. Injection of sugars into the veins is costly, but is a method of treatment. Prevention is the cheapest in the long run.

H. *Feeding the Lactating Ewe*

 1. Some general principles can provide the guidelines to feeding lactating ewes. These principles include:

  a. The nutrient requirements are two to three times greater during lactation than during maintenance.

  b. Ewes with twin lambs produce 20% to 40% more milk than those with singles, therefore, have greater nutrient requirements.

  c. Milk production in ewes peaks early (2 to 3 weeks after lambing) and is maintained at a high level until about the eighth week, then declines rapidly. Ewes produce 3 to 6 + lb milk daily.

d. Milk provides virtually the total source of nutrients for lambs the first month and a significant proportion for the first 2 months.

e. Lambs can be weaned successfully at 8 weeks (perhaps earlier). It is more efficient to feed the lamb directly than to feed the ewe to feed the lamb beyond 8 to 10 weeks.

2. It is not necessary to feed the ewe very much for a day or so after lambing. It is important to have plenty of fresh water and a light feed during this period. By the third day the lactating ewe may be brought up to a regular ration (as-fed basis) such as

a. 2½–3 lb of each hay and ground ear corn

b. 4–5½ lb hay plus 1 to 1½ lb shelled corn

c. 8 to 10 lb corn silage plus ¾ to 1½ lb shelled corn + ½ lb soybean meal

d. 10 lb alfalfa haylage plus 1½ lb shelled corn

The lower amount is typically fed to ewes with single lambs and the higher amount to ewes with twin lambs.

3. Daily feed intakes can be increased by feeding at more frequent intervals. A rule of thumb: if the ewe is nursing one lamb, feed once a day. If she is nursing two lambs, twice a day, etc. Splitting the amount fed per day into more frequent feedings reduces the potential for acidosis when high levels of concentrate are fed.

Suggested formulations of ewe lactation rations fed as complete feeds are shown in Table 10–10.

4. The mineral requirements for lactating ewes are greater than during other periods. A greater intake of feed will furnish some of the increased needs, perhaps all of them. But as insurance, free choice feeding of a complete mineral is suggested. (See more complete discussion on mineral in "feeding during late gestation" section.)

5. Under most midwest conditions ewes lambing in the winter would be the only group to be fed harvested feed during lactation. Ewes lambing in late April and May would normally be on pasture. The pasture should be as productive as it will ever be during May and June and should provide most of the ewes' requirements unless it is overstocked. Ewes on pasture can be fed supplemental grain if needed. Ewes can be trained to return to drylot for grain feeding. If pasture is short, lambs can be kept in drylot, and ewes kept with them during the day, then turned to pasture evenings and nights.

6. Fall lambing ewes (September and October) in most flocks would lamb on pasture. A pasture should have been set aside for these fall lambing ewes. If this has been done and with typical fall rains, the pasture ought to provide most of the requirements during the first 4 to 6 weeks of lactation. The good manager will observe pasture and ewe and lamb conditions and begin supplemental grain feeding at the first indication ewes are losing weight or lambs are slowing down in growth rate and losing bloom. Drylotting the lambs and bringing the ewes to the lambs for nursing during the day may be advisable. If this is done, lambs would be creep fed in drylot.

II. Feeding of Lambs

A. *New Life*

1. The shepherd's first job when a lamb is born is to make sure it nurses. A lamb must nurse within 1 hour or its chances of living are limited. Most lambs nurse within 30 minutes after birth. The first milk is an important meal for the lamb. Colostrum, the first milk, is not only a rich food for the offspring, but plays an important role in disease resistance. The transfer of antibodies through the umbilical cord does not occur in sheep and cattle as it does in many other animals. Colostrum furnishes these antibodies. It is also important as a laxative to clean out the fecal matter accumulated during fetal life.

The lamb must consume at least 6 to 8 oz of colostrum. If none is available from the mother, then provide the lamb with colostrum from another ewe. Frozen ewe or cow colostrum, warmed to body temperature and bottle fed, is an adequate alternative.

2. A large portion of lamb death losses are due to starvation during the first week after birth. A number of factors may suggest that lambs be reared on milk replacer.

a. Lambs that are orphans due to death of the mother, one side of udder nonfunctional on ewe with twins, mismothered, etc.

b. The excess of two in sets of triplets or more or an obviously weak lamb in a set of twins.

Ewes nursing twin or triplet lambs require more daily feed that ewes nursing singles; thus, daily feed intake can be increased by feeding at more frequent intervals.

    c. One of a set of twins from a ewe lamb in most instances.

    d. Any lamb that shows symptoms of progressive weakness during the first week after birth where this can be traced to inadequate milk supply.

3. The decision to switch lambs to milk replacer should be made as soon after birth as possible (<24 hrs). The sooner they are removed from their mother, the easier they will be to train. Lambs usually require assistance the first day in learning to nurse the nipple. Place lambs in a warm, dry enclosed area with other lambs on milk replacer. The lambs should not be able to see or hear their dams when weaned onto a milk replacer. Inject the lambs with the following when they are placed in the nursery.

    a. Vitamin A, D, E.

    b. Se, in low Se areas.

    c. Combiotic.

4. There are a number of commercially prepared lamb milk replacers on the market.

    a. Best results have been obtained with milk replacers containing 25% to 30% fat, 20% to 25% milk protein and 30% to 35% lactose.

    b. The dry replacer should be diluted with water to a minimum of 17% to 20% d.m. (1¾ to 2 lb milk replacer per gallon of water).

    c. Mix replacers thoroughly before feeding. Mix the dry powder in warm water and cool immediately to about 33° F. This helps to eliminate the problem of ingredient separation during storage.

    d. Make certain the milk replacer contains a high level of antibiotics to avoid scours and other digestive disorders.

5. Method of feeding milk replacers

    a. Use one of the multiple nipple pails for self feeding the milk replacer.

    b. Research has shown that self fed lambs should be given cold milk rather than warm milk. Cold milk does not sour as quickly and lambs consume only a small amount of cold milk replacer solution at each nursing, but nurse more often. This reduces digestive problems. The one problem with feeding cold milk is getting the lambs to start nursing adequate amounts. Warm milk may be used the first week and cold milk thereafter. In a free-choice system, each lamb will consume ½ to ¾ lb of milk replacer powder in solution daily (2 to 4 pints of liquid milk).

    c. If lambs are hand fed they can safely be given warm milk and research results have not shown any advantage for feeding more often than twice daily after lambs are one week old.

    d. Utensils must be kept clean and milk not allowed to spoil.

    e. Lambs should be provided constant access to fresh water and high-quality, palatable solid feed to accustom them to eating dry feed and to stimulate rumen development.

f. Because milk replacer is expensive, the liquid-feeding period should be as short, as possible. Lambs can be successfully weaned from milk replacer at 3 weeks of age, although the lambs lose some weight the first week before resuming normal gains. Delaying weaning to 4 weeks of age reduces the growth check and results in lamb weights approximately 2 to 3 lb heavier at 70 days of age. The extra weight is not sufficient, however, to offset the extra costs of the 6 to 8 lb of milk replacer required for the extra week of feeding. Postweaning rations, until lambs reach 60 lb, should be high in protein (16% to 22% crude protein) and energy.

B. *Feeding of Market Lambs*

1. Early lambs

    a. Healthy baby lambs show interest in dry feed at 10 days of age, suggesting they can be started on dry feed very early. This is called the creep feeding period.

    Creep feeding involves providing a small pen with an opening (7 to 10 in.) which allows the lambs to enter and exit at will, but keeps ewes out. A lamb-sized feed bunk is needed in the creep.

    The creep pen should be in the most desirable location, in an area that is well lit, warm, and near ewe traffic and resting place. Some producers add light to the creep area. Above all, provide a highly nutritious and palatable feed.

    Commercially prepared, scientifically formulated, complete lamb creep feeds are available. These are generally in complete pellet form and in 50-lb paper bags. This makes creep feeding simple, but relatively costly. However, lambs are small and don't eat much during this period.

    These complete pelleted creep feeds should be 14% to 16% protein of a vegetable meal origin.

    Lambs must like the feed at first taste. Feeds especially palatable to lambs (in addition to commercial feeds) include cracked, shelled corn; bran; rolled oats; molasses; soybean meal and high quality, leafy alfalfa hay.

    Simple mixtures of 80% to 85% cracked, shelled corn, a combination of equal parts corn and oats and 15% to 20% soybean meal or a commercial 32% to 36% lamb supplement can be fed in lamb grain bunks. Feed high quality, leafy alfalfa hay in separate hay racks. Make fresh feed available in clean bunks at least twice daily. Give only the quantity lambs will clean up between each feeding.

    When feeding simple cracked corn and hay-based creep rations, a complete lamb supplement is top dressed on the corn. The supplement provides added protein, calcium, vitamin E, selenium and an antibiotic. Feed the amount recommended by the supplement manufacturer.

    Lambs should consume the cracked corn and supplement in the proportion fed. Observe closely to see if the supplement is being eaten or refused in relation to the corn. Lambs have the capability of sorting feeds.

    Lambs hand fed cracked corn, supplement and hay may be more subject to overeating (enterotoxemia) than those fed complete pellets or complete ground mixed rations. Vaccination at 3 to 4 weeks of age and again 12 to 14 days later will help guard against enterotoxemia losses. But vaccination will not prevent digestive upsets and lambs going off feed. Careful feeding and gradually changing amounts and kinds of feed offered will avoid those problems.

    Lambs being creep fed while nursing also need clean, fresh water and salt. Minerals are not needed when feeding a well-formulated diet. High-phosphorus mineral mixtures are undesirable because of the possibility of too much phosphorus to calcium intake, which predisposes wether lambs to the formation of calculi in the urinary system.

    b. Lambs weighing 25 lb can be weaned at 25 to 30 days of age as is done with lambs artificially reared. Lambs raised by ewes in peak milk production are seldom weaned. Lambs normally are not weaned before 7 to 8 weeks old and weighing more than 40 lb.

    To reduce weaning stress, a lamb should eat about a pound of palatable, nutritious dry feed daily. And it is best if lambs at weaning remain in familiar surroundings and no change is made in their diet for about a week.

The very early weaned 25-lb lambs required an 18% to 19% protein, high-energy, well-fortified diet until they reach about 50 lb. At this weight, they can be fed as the typically weaned lamb.

Research data indicate that for the maximum gain and feed conversion, the protein level for 50-lb lambs is about 10% to 16% on an as-fed, air-dry feed basis. Data are not available to indicate when or at what weight this level can be reduced. Most crossbred lambs will continue rapid lean tissue growth to 85, perhaps 100 lb. At these weights, the protein level could be reduced to 13% to 14%. This would reduce ration costs without lowering lamb performance significantly. Suggested starter rations for lambs to be fed as complete feed are shown in Table 10–11.

In the interest of simplicity, some lamb producers and feed manufacturers will formulate one lamb ration and use it as the creep feed and for the growing-finishing lamb throughout the feeding period. A protein level of 14% to 15% is commonly chosen. This level is somewhat low for the young lamb and higher than needed by the lamb approaching 100- to 110-lb market weight. The protein level of 14% to 15% is a reasonable compromise if one formulation is fed from weaning to market.

Vegetable meal should be used as the source of supplemental protein to lambs up to 70 to 75 lb. After this weight, a supplement containing urea (NPN) could be used if the supplement cost can be lowered.

A simple ration of shelled corn, long alfalfa hay and supplement can be fed to growing-finishing lambs. To make this work well, the following lamb and feed bunk management practices need to be followed.

  (1) Start the lambs on a complete feed, either in pellet or complete ground-mixed form, and make sure they eat it readily.
  (2) Vaccinate for enterotoxemia twice; the last time when the ration chance is initiated.
  (3) Change to the simplified ration gradually over a 7 to 10 day period by top dressing the shelled corn and supplement on the starter ration in the feed bunks. During these 7 to 10 days, the amount of corn-supplement is increased and the starter ration decreased. Hay is restricted to about ½ pound per lamb per day, fed in separate bunks.
  (4) Feed lambs twice daily at regular times. The shelled corn is put in feed bunks and the supplement top dressed or premixed with shelled corn in correct proportion.
  (5) Feed high quality alfalfa hay twice daily following the feeding of the corn and supplement. Restrict hay to the amount the lambs will clean up daily, but feed suggested proportion or amount. (Hay can be fed in the same bunks as the concentrate portion if the bunks are designed for grain and hay.)
  (6) At least 12 linear inches of feed bunk space are needed per lamb. Make sure **all** lambs come to the troughs at **each** feeding time.
  (7) Reduce the concentrate feed fed by one half for one or two feedings if the lambs do not clean up the feed from one feeding to the next. (Total feed consumption will increase gradually, but some daily reductions may be noted; some are weather related.)
  (8) Observe the consumption of the corn and supplement carefully. Supplement should not be left uneaten in the bunks. Ideally, the supplement (probably in pellet form) would not be sorted by the lambs but consumed as fed. Change to a more palatable supplement if lambs sort and leave pellets.
  (9) For spring and summer feeding, place feed bunks in the shade.
  (10) Be sure plenty of fresh, clean water is available. Salt should also be provided free choice. Minerals are not needed.
 c. Lamb feeding methods can be classified as self-feeding, hand feeding, or pasture feeding. Self-feeding is popular in feedlots utilizing pelleted feeds or with high labor costs. Suggested growing-finishing rations for lambs to be fed as complete feeds either self or hand fed are shown in Tables 10–12 and 10–13.
  (1) Self-fed rations generally contain between 60% and 85% concentrate with the remainder as roughage. In recent years, grain as a source of energy has become less costly in relation to hay. Grain feeding reduces labor and finance costs, and overeating disease is less of a problem due to effective vaccination programs. The current practice is to feed rations containing more grain.

(2) Hand-fed rations are fed twice daily on a regular basis. This method has the advantage that those lambs which are sick or off feed can be identified readily. This is the usual method of choice for feeding silage.

(3) Pasture can be used for the entire fattening period or for the early part and then the lambs can be placed in the feedlot for finishing. To provide good gains on pasture, the plants must be palatable and nutritious. Generally the most nutritious are the legumes or a mixture of legumes and grasses (to reduce bloat problems). Pasture gains may be less expensive per pound, but it takes longer to fatten the lambs.

(4) Regardless of feeding method used, care should be taken in changing from high forage to high grain diets. If such changes are not made gradually, problems with acidosis, diarrhea and enterotoxemia can result.

d. While research does not clearly show the need for vitamin additions to rations for early lambs, it has become common practice to fortify these rations with vitamins A, D and E. The level of fortification generally recommended is as follows.

Vitamin A: 1,000,000–5,000,000 IU/ton. The more dehydrated alfalfa or alfalfa included in the ration, the lower the level of fortification required.

Vitamin D: 200,000–625,000 IU/ton. Lambs kept indoors in strict confinement would need the high level of fortification. Natural feedstuffs and sun are uncertain sources of vitamin D. Therefore, all creep rations for winter lambs should probably be fortified with at least the average level.

Vitamin E: 10,000–35,000 IU/ton. In areas where white muscle or vitamin E + selenium deficiency problems are known to exist, fortification at the higher level would be indicated. Fortification at the lower level would be included as insurance against a possible deficiency in the natural feedstuffs or a carryover deficiency in the newborn lamb.

e. Available information generally support a recommendation that a dietary nitrogen to sulfur ratio of 10:1 be maintained in sheep diets. A diet containing 13% CP (2.1% N) would require 0.21% S. Most grains contain 0.10% to 0.15% S so it is conceivable that lambs on high-grain diets could lack sulfur. This would be particularly true if a portion of the protein in the diet was from urea.

f. The calcium to phosphorus ratio in wether lamb diets should be at least 2:1 to minimize problems with urinary calculi.

2. Late lambs

Good pasture is the key to successful production of late lambs. The top lambs can be marketed off the pasture. The rest can be vaccinated for enterotoxemia at the time of the first sort, allowed to graze and then finished in cornfields or drylot. A drylot ration of 1½ to 2 lb of shelled corn (after the lambs have been brought up to this level), 1½ lb of legume hay, plus ¼ to ½ lb of protein supplement daily per lamb will finish them for market. Suggested mixtures of lamb finishing rations to be fed as complete feeds are shown in Table 10–13.

C. *Overeating Disease in Nursing Lambs (Enterotoxemia)*

This disease usually affects the largest, fastest-gaining and greediest lamb. It is caused by the toxins produced from excessive numbers of the organism known as *Clostridium perfringens* type D. The treatment for young lambs under 2 months of age is antitoxin. Antitoxin gives an immediate immunity lasting 2 to 3 weeks. It is used to stop death losses in lambs following an outbreak of entertoxemia and may also be used to immunize feedlot lambs on a short-term feeding basis for up to 3 weeks.

Vaccination with bacterin or toxoid will provide a more long-term prevention program. The vaccination program is effective for 5–6 months. Some producers will vaccinate the pregnant ewe about a month before lambing. Lambs to be early-weaned should then be vaccinated twice (2 to 3 weeks apart) prior to weaning. Older feeder lambs that are transported to the finishing area should be vaccinated twice during the first 2 weeks after arrival. There may be a reaction in the tissues at the site of injection. Because it may persist for at least 30 days, it is inadvisable to use bacterin if animals are intended for slaughter within that time. Bacterin should not be injected into lambs under 3 weeks of age.

The antitoxin or bacterin is injected by aseptic means usually under the skin of the fleece-free area back of the place where the foreleg joins the body. Injection there helps to prevent unnecessary lameness.

If nursing lambs come down with the disease, it is recommended that they be taken off the enriched feeds or pasture for a day or two, until they can eliminate the excess toxins.

D. *Feeding Replacement Ewe Lambs*

Nutrients supplied to replacement ewe lambs, postweaning, will vary with the lambs' age at their first breeding. More and more producers are replacing with selected ewe lambs rather than yearlings. Breeding ewes as lambs (7 to 8 months of age) to lamb at about a year of age has several advantages: (1) gets ewes into production about a year earlier, thereby reducing maintenance cost, (2) shortens generation interval, resulting in more rapid gain from selection, (3) increases lifetime production and (4) identifies ewes that are more productive. If they are to be bred to lamb when they are 12 to 14 months of age, nutrition between weaning and breeding must be on a high plane.

Most early-born ewe lambs (January- through February-born) will have been fed relatively high energy creep and starter rations. After identifying the selected ewe lambs at about 90 to 120 days of age (80 to 90 lb) body weight, remove them from the market lamb finishing lot and off the high-energy finishing type ration. Replacement ewe lambs should not get fat nor have the udder infiltrated with fat cells. This could happen to ewe lambs left in the finishing lot too long.

Provide a growing, developing diet for the selected ewe lambs such as those suggested in Table 10–14. Drylotting until mid-September will permit better control of diet and ewe development; after this the lambs could be fed grain on pasture. Through the fall and winter these lambs should be fed 1 to 1½ lb grain plus good pasture, hay or both.

The summer-fall feeding program for replacement ewe lambs is very different than for the mature ewes in the flock. Because of this, the ewe lambs would be best managed as a separate flock.

E. *Feeding Rams*

Replacement ram lambs (about 130 lb) or yearlings (about 220 lb) need, respectively 5 to 5½ lb or 6½ to 7 lb of feed daily to gain at the recommended daily rate of 0.3 to 0.4 pound. Suggested rations for replacement rams are shown in Table 10–15. Summer pasture alone is not adequate. Good pasture could take the place of about half the hay. Continue to feed grain during the breeding period.

Mature rams should be provided a maintenance ration (pasture) throughout the year except during a 30- to 45-day period prior to breeding. During this time they should be gaining approximately 0.3 to 0.5 lb daily. This can be accomplished by feeding dehydrated alfalfa pellets or a combination of hay and grain with no more than ⅓ of the ration composed of grain. The ingredients of the ram ration should be similar to what he will be eating when put in with the ewes so the ram will not go off feed and be ineffective as a breeder.

III. Miscellaneous

A. *Urea*

Urea and other forms of nonprotein nitrogen can be used to replace up to ⅓ of the protein equivalent of the sheep ration (about 1.5% urea would be the maximum amount usable in a growing-finishing lamb diet). If urea is used, increased attention may be required in mixing procedures and during the adjustment or early stages of the feeding period. Urea should not be used in creep rations, in range sheep rations, or in lamb rations containing low or limited energy.

The metabolizable protein system of protein evaluation for cattle and sheep presented in Chapter 8 provides a better understanding of the utilization of urea in sheep feedlot rations. The usefulness of urea depends upon the amount of fermentable energy present in a feed and the amount of ammonia formed from protein degradation by bacterial fermentation of that feed in the rumen.

B. *Additives and Implants*

The inclusion of aureomycin (chlortetracycline) or terramycin (oxytetracycline) in creep rations for suckling lambs and lamb-finishing rations improves gain and feed efficiency. The greatest response generally occurs under conditions of stress. Either antibiotic is particularly effective for protecting against low level disease infections and also offering some protection against enterotoxemia.

Food and Drug Administration clearances are as follows (*Feed Additive Compendium*, 1992).
1. Aureomycin (chlortetracycline)
   a. 20 g/ton to aid in preventing enterotoxemia.
   b. 20 to 50 g/ton to stimulate gains and improve feed efficiency.

2. Terramycin (oxytetracycline)
   a. 25 mg/hd/day to aid in preventing enterotoxemia.
   b. 10 to 20 g/ton to stimulate gains and improve feed efficiency.
   c. 50 to 100 g/ton as an aid in the prevention or treatment of bacterial diarrhea.
3. Lasalocid (Bovatec)
   a. 20 to 30 g/ton complete feed.
   b. For prevention of coccidiosis.
4. Ralgro
   Results from implanting lambs with Ralgro (12 mg) are rather inconsistent. A summary of University of Minnesota trials involved 987 feeder lambs gathered over a 7-year period and 230 suckling lambs from three separate trials. This summary revealed a 3.6% improvement in weight gain among Ralgro implanted suckling lambs (both sexes) and 5.8% improvement in weight gain among Ralgro implanted feeder lambs (both sexes).
5. Ammonium chloride or sulfate
   The addition of 0.5% ammonium chloride or sulfate to diets can be used to prevent problems with urinary calculi in wethers. Ammonium chloride is slightly more effective for this purpose than ammonium sulfate; however, the addition of 0.5% ammonium sulfate also adds 0.12% S to the diet which would be an important consideration in those diets requiring supplemental sulfur.

IV. Nutrient Requirements of Sheep

The nutrient requirements for sheep presented in Tables 10–1 to 10–8 were adapted from the *Nutrient Requirements of Sheep,* Publication ISBN 0-309-03596-1, Committee on Animal Nutrition, National Academy of Sciences—National Research Council, Washington, D.C., 1985. These values are expressed on a dry-matter basis.

TABLE 10–1. Daily Nutrient Requirements of Sheep

Body Weight (kg)	(lb)	Weight Change/Day (g)	(lb)	Dry Matter per Animal (kg)	(lb)	(% body weight)	Energy[b] TDN (kg)	(lb)	DE (Mcal)	ME (Mcal)	Crude protein (g)	(lb)	Ca (g)	P (g)	Vitamin A Activity (IU)	Vitamin E Activity (IU)
Ewes[c]																
Maintenance																
50	110	10	0.02	1.0	2.2	2.0	0.55	1.2	2.4	2.0	95	0.21	2.0	1.8	2,350	15
60	132	10	0.02	1.1	2.4	1.8	0.61	1.3	2.7	2.2	104	0.23	2.3	2.1	2,820	16
70	154	10	0.02	1.2	2.6	1.7	0.66	1.5	2.9	2.4	113	0.25	2.5	2.4	3,290	18
80	176	10	0.02	1.3	2.9	1.6	0.72	1.6	3.2	2.6	122	0.27	2.7	2.8	3,760	20
90	198	10	0.02	1.4	3.1	1.5	0.78	1.7	3.4	2.8	131	0.29	2.9	3.1	4,230	21
Flushing—2 Weeks prebreeding and first 3 weeks of breeding																
50	110	100	0.22	1.6	3.5	3.2	0.94	2.1	4.1	3.4	150	0.33	5.3	2.6	2,350	24
60	132	100	0.22	1.7	3.7	2.8	1.00	2.2	4.4	3.6	157	0.34	5.5	2.9	2,820	26
70	154	100	0.22	1.8	4.0	2.6	1.06	2.3	4.7	3.8	164	0.36	5.7	3.2	3,290	27
80	176	100	0.22	1.9	4.2	2.4	1.12	2.5	4.9	4.0	171	0.38	5.9	3.6	3,760	28
90	198	100	0.22	2.0	4.4	2.2	1.18	2.6	5.1	4.2	177	0.39	6.1	3.9	4,230	30
Nonlactating—First 15 weeks gestation																
50	110	30	0.07	1.2	2.6	2.4	0.67	1.5	3.0	2.4	112	0.25	2.9	2.1	2,350	18
60	132	30	0.07	1.3	2.9	2.2	0.72	1.6	3.2	2.6	121	0.27	3.2	2.5	2,820	20
70	154	30	0.07	1.4	3.1	2.0	0.77	1.7	3.4	2.8	130	0.29	3.5	2.9	3,290	21
80	176	30	0.07	1.5	3.3	1.9	0.82	1.8	3.6	3.0	139	0.31	3.8	3.3	3,760	22
90	198	30	0.07	1.6	3.5	1.8	0.87	1.9	3.8	3.2	148	0.33	4.1	3.6	4,230	24
Last 4 weeks gestation (130-150% lambing rate expected) or last 4-6 weeks lactation suckling singles[d]																
50	110	180 (45)	0.40 (0.10)	1.6	3.5	3.2	0.94	2.1	4.1	3.4	175	0.38	5.9	4.8	4,250	24
60	132	180 (45)	0.40 (0.10)	1.7	3.7	2.8	1.00	2.2	4.4	3.6	184	0.40	6.0	5.2	5,100	26
70	154	180 (45)	0.40 (0.10)	1.8	4.0	2.6	1.06	2.3	4.7	3.8	193	0.42	6.2	5.6	5,950	27
80	176	180 (45)	0.40 (0.10)	1.9	4.2	2.4	1.12	2.4	4.9	4.0	202	0.44	6.3	6.1	6,800	28
90	198	180 (45)	0.40 (0.10)	2.0	4.4	2.2	1.18	2.5	5.1	4.2	212	0.47	6.4	6.5	7,650	30
Last 4 weeks gestation (180-225% lambing rate expected)																
50	110	225	0.50	1.7	3.7	3.4	1.10	2.4	4.8	4.0	196	0.43	6.2	3.4	4,250	26
60	132	225	0.50	1.8	4.0	3.0	1.17	2.6	5.1	4.2	205	0.45	6.9	4.0	5,100	27
70	154	225	0.50	1.9	4.2	2.7	1.24	2.8	5.4	4.4	214	0.47	7.6	4.5	5,950	28
80	176	225	0.50	2.0	4.4	2.5	1.30	2.9	5.7	4.7	223	0.49	8.3	5.1	6,800	30
90	198	225	0.50	2.1	4.6	2.3	1.37	3.0	6.0	5.0	232	0.51	8.9	5.7	7,650	32
First 6-8 weeks lactation suckling singles or last 4-6 weeks lactation suckling twins[d]																
50	110	−25 (90)	−0.06 (0.20)	2.1	4.6	4.2	1.36	3.0	6.0	4.9	304	0.67	8.9	6.1	4,250	32
60	132	−25 (90)	−0.06 (0.20)	2.3	5.1	3.8	1.50	3.3	6.6	5.4	319	0.70	9.1	6.6	5,100	34
70	154	−25 (90)	−0.06 (0.20)	2.5	5.5	3.6	1.63	3.6	7.2	5.9	334	0.73	9.3	7.0	5,950	38
80	176	−25 (90)	−0.06 (0.20)	2.6	5.7	3.2	1.69	3.7	7.4	6.1	344	0.76	9.5	7.4	6,800	39
90	198	−25 (90)	−0.06 (0.20)	2.7	5.9	3.0	1.75	3.8	7.6	6.3	353	0.78	9.6	7.8	7,650	40

TABLE 10–1 (continued). Daily Nutrient Requirements of Sheep

Body Weight		Weight Change/Day		Dry Matter per Animal			Nutrients per Animal				Crude protein		Ca	P	Vitamin A Activity	Vitamin E Activity
							Energy[b]		DE (Mcal)	ME (Mcal)			(g)	(g)	(IU)	(IU)
							TDN									
(kg)	(lb)	(g)	(lb)	(kg)	(lb)	(% body weight)	(kg)	(lb)	(Mcal)	(Mcal)	(g)	(lb)				
First 6-8 weeks lactation suckling twins																
50	110	−60	−0.13	2.4	5.3	4.8	1.56	3.4	6.9	5.6	389	0.86	10.5	7.3	5,000	36
60	132	−60	−0.13	2.6	5.7	4.3	1.69	3.7	7.4	6.1	405	0.89	10.7	7.7	6,000	39
70	154	−60	−0.13	2.8	6.2	4.0	1.82	4.0	8.0	6.6	420	0.92	11.0	8.1	7,000	42
80	176	−60	−0.13	3.0	6.6	3.8	1.95	4.3	8.6	7.0	435	0.96	11.2	8.6	8,000	45
90	198	−60	−0.13	3.2	7.0	3.6	2.08	4.6	9.2	7.5	450	0.99	11.4	9.0	9,000	48
Ewe lambs																
Nonlactating—First 15 weeks gestation																
40	88	160	0.35	1.4	3.1	3.5	0.83	1.8	3.6	3.0	156	0.34	5.5	3.0	1,880	21
50	110	135	0.30	1.5	3.3	3.0	0.88	1.9	3.9	3.2	159	0.35	5.2	3.1	2,350	22
60	132	135	0.30	1.6	3.5	2.7	0.94	2.0	4.1	3.4	161	0.35	5.5	3.4	2,820	24
70	154	125	0.28	1.7	3.7	2.4	1.00	2.2	4.4	3.6	164	0.36	5.5	3.7	3,290	26
Last 4 weeks gestation (100-120% lambing rate expected)																
40	88	180	0.40	1.5	3.3	3.8	0.94	2.1	4.1	3.4	187	0.41	6.4	3.1	3,400	22
50	110	160	0.35	1.6	3.5	3.2	1.00	2.2	4.4	3.6	189	0.42	6.3	3.4	4,250	24
60	132	160	0.35	1.7	3.7	2.8	1.07	2.4	4.7	3.9	192	0.42	6.6	3.8	5,100	26
70	154	150	0.33	1.8	4.0	2.6	1.14	2.5	5.0	4.1	194	0.43	6.8	4.2	5,950	27
Last 4 weeks gestation (130-175% lambing rate expected)																
40	88	225	0.50	1.5	3.3	3.8	0.99	2.2	4.4	3.6	202	0.44	7.4	3.5	3,400	22
50	110	225	0.50	1.6	3.5	3.2	1.06	2.3	4.7	3.8	204	0.45	7.8	3.9	4,250	24
60	132	225	0.50	1.7	3.7	2.8	1.12	2.5	4.9	4.0	207	0.46	8.1	4.3	5,100	26
70	154	215	0.47	1.8	4.0	2.6	1.14	2.5	5.0	4.1	210	0.46	8.2	4.7	5,950	27
First 6-8 weeks lactation suckling singles (wean by 8 weeks)																
40	88	−50	−0.11	1.7	3.7	4.2	1.12	2.5	4.9	4.0	257	0.56	6.0	4.3	3,400	26
50	110	−50	−0.11	2.1	4.6	4.2	1.39	3.1	6.1	5.0	282	0.62	6.5	4.7	4,250	32
60	132	−50	−0.11	2.3	5.1	3.8	1.52	3.4	6.7	5.5	295	0.65	6.8	5.1	5,100	34
70	154	−50	−0.11	2.5	5.5	3.6	1.65	3.6	7.3	6.0	301	0.68	7.1	5.6	5,450	38
First 6-8 weeks lactation suckling twins (wean by 8 weeks)																
40	88	−100	−0.22	2.1	4.6	5.2	1.45	3.2	6.4	5.2	306	0.67	8.4	5.6	4,000	32
50	110	−100	−0.22	2.3	5.1	4.6	1.59	3.5	7.0	5.7	321	0.71	8.7	6.0	5,000	34
60	132	−100	−0.22	2.5	5.5	4.2	1.72	3.8	7.6	6.2	336	0.74	9.0	6.4	6,000	38
70	154	−100	−0.22	2.7	6.0	3.9	1.85	4.1	8.1	6.6	351	0.77	9.3	6.9	7,000	40
Replacement ewe lambs[e]																
30	66	227	0.50	1.2	2.6	4.0	0.78	1.7	3.4	2.8	185	0.41	6.4	2.6	1,410	18
40	88	182	0.40	1.4	3.1	3.5	0.91	2.0	4.0	3.3	176	0.39	5.9	2.6	1,880	21
50	110	120	0.26	1.5	3.3	3.0	0.88	1.9	3.9	3.2	136	0.30	4.8	2.4	2,350	22
60	132	100	0.22	1.5	3.3	2.5	0.88	1.9	3.9	3.2	134	0.30	4.5	2.5	2,820	22
70	154	100	0.22	1.5	3.3	2.1	0.88	1.9	3.9	3.2	132	0.29	4.6	2.8	3,290	22

Replacement ram lambs[e]

40	88	0.73	1.8	4.0	4.5	1.1	2.5	5.0	4.1	243	0.54	7.8	3.7	1,880	24
60	132	0.70	2.4	5.3	4.0	1.5	3.4	6.7	5.5	263	0.58	8.4	4.2	2,820	26
80	176	0.64	2.8	6.2	3.5	1.8	3.9	7.8	6.4	268	0.59	8.5	4.6	3,760	28
100	220	0.55	3.0	6.6	3.0	1.9	4.2	8.4	6.9	264	0.58	8.2	4.8	4,700	30

Lambs finishing—4 to 7 months old[f]

30	66	0.65	1.3	2.9	4.3	0.94	2.1	4.1	3.4	191	0.42	6.6	3.2	1,410	20
40	88	0.60	1.6	3.5	4.0	1.22	2.7	5.4	4.4	185	0.41	6.6	3.3	1,880	24
50	110	0.45	1.6	3.5	3.2	1.23	2.7	5.4	4.4	160	0.35	5.6	3.0	2,350	24

Early weaned lambs—Moderate growth potential[f]

10	22	0.44	0.5	1.1	5.0	0.40	0.9	1.8	1.4	127	0.38	4.0	1.9	470	10
20	44	0.55	1.0	2.2	5.0	0.80	1.8	3.5	2.9	167	0.37	5.4	2.5	940	20
30	66	0.66	1.3	2.9	4.3	1.00	2.2	4.4	3.6	191	0.42	6.7	3.2	1,410	20
40	88	0.76	1.5	3.3	3.8	1.16	2.6	5.1	4.2	202	0.44	7.7	3.9	1,880	22
50	110	0.66	1.5	3.3	3.0	1.16	2.6	5.1	4.2	181	0.40	7.0	3.8	2,350	22

Early weaned lambs—Rapid growth potential[f]

10	22	0.55	0.6	1.3	6.0	0.48	1.1	2.1	1.7	157	0.35	4.9	2.2	470	12
20	44	0.66	1.2	2.6	6.0	0.92	2.0	4.0	3.3	205	0.45	6.5	2.9	940	24
30	66	0.72	1.4	3.1	4.7	1.10	2.4	4.8	4.0	216	0.48	7.2	3.4	1,410	21
40	88	0.88	1.5	3.3	3.8	1.14	2.5	5.0	4.1	234	0.51	8.6	4.3	1,880	22
50	110	0.94	1.7	3.7	3.4	1.29	2.8	5.7	4.7	240	0.53	9.4	4.8	2,350	25
60	132	0.77	1.7	3.7	2.8	1.29	2.8	5.7	4.7	240	0.53	8.2	4.5	2,820	25

[a]To convert dry matter to an as-fed basis, divide dry matter values by the percentage of dry matter in the particular feed.

[b]One kilogram TDN (total digestible nutrients) = 4.4 Mcal DE (digestible energy); ME (metabolizable energy) = 82% of DE. Because of rounding errors, values in Table 1 and Table 2 may differ.

[c]Values are applicable for ewes in moderate condition. Fat ewes should be fed according to the next lower weight category and thin ewes at the next higher weight category. Once desired or moderate weight condition is attained, use that weight category through all production stages.

[d]Values in parentheses are for ewes suckling lambs the last 4-6 weeks of lactation.

[e]Lambs intended for breeding; thus, maximum weight gains and finish are of secondary importance.

[f]Maximum weight gains expected.

TABLE 10–2. Nutrient Concentration in Diets for Sheep (Expressed on 100% Dry Matter Basis[a])

Body Weight (kg)	(lb)	Weight Change/Day (g)	(lb)	Energy[b] TDN[c] (%)	DE (Mcal/kg)	ME (Mcal/kg)	Example Diet Proportions Concentrate %	Forage %	Crude Protein (%)	Cal-cium (%)	Phos-phorus (%)	Vitamin A Activity (IU/kg)	Vitamin E Activity (IU/kg)
Ewes[d]													
Maintenance													
70	154	10	0.02	55	2.4	2.0	0	100	9.4	0.20	0.20	2,742	15
Flushing—2 weeks prebreeding and first 3 weeks of breeding													
70	154	100	0.22	59	2.6	2.1	15	85	9.1	0.32	0.18	1,828	15
Nonlactating—First 15 weeks gestation													
70	154	30	0.07	55	2.4	2.0	0	100	9.3	0.25	0.20	2,350	15
Last 4 weeks gestation (130-150% lambing rate expected) or last 4-6 weeks lactation suckling singles[e]													
70	154	180 (0.45)	0.40 (0.10)	59	2.6	2.1	15	85	10.7	0.35	0.23	3,306	15
Last 4 weeks gestation (180-225% lambing rate expected)													
70	154	225	0.50	65	2.9	2.3	35	65	11.3	0.40	0.24	3,132	15
First 6-8 weeks lactation suckling singles or last 4-6 weeks lactation suckling twins[e]													
70	154	−25(90)	−0.06 (0.20)	65	2.9	2.4	35	65	13.4	0.32	0.26	2,380	15
First 6-8 weeks lactation suckling twins													
70	154	−60	−0.13	65	2.9	2.4	35	65	15.0	0.39	0.29	2,500	15
Ewe Lambs													
Nonlactating—First 15 weeks gestation													
55	121	135	0.30	59	2.6	2.1	15	85	10.6	0.35	0.22	1,668	15
Last 4 weeks gestation (100-120% lambing rate expected)													
55	121	160	0.35	63	2.8	2.3	30	70	11.8	0.39	0.22	2,833	15
Last 4 weeks gestation (130-175% lambing rate expected)													
55	121	225	0.50	66	2.9	2.4	40	60	12.8	0.48	0.25	2,833	15
First 6-8 weeks lactation suckling singles (wean by 8 weeks)													
55	121	−50	0.22	66	2.9	2.4	40	60	13.1	0.30	0.22	2,125	15
First 6-8 weeks lactation suckling twins (wean by 8 weeks)													
55	121	−100	−0.22	69	3.0	2.5	50	50	13.7	0.37	0.26	2,292	15
Replacement Ewe Lambs[f]													
30	66	227	0.50	65	2.9	2.4	35	65	12.8	0.53	0.22	1,175	15
40	88	182	0.40	65	2.9	2.4	35	65	10.2	0.42	0.18	1,343	15
50-70	110-154	115	0.25	59	2.6	2.1	15	85	9.1	0.31	0.17	1,567	15
Replacement Ram Lambs[f]													
40	88	330	0.73	63	2.8	2.3	30	70	13.5	0.43	0.21	1,175	15
60	132	320	0.70	63	2.8	2.3	30	70	11.0	0.35	0.18	1,659	15
80-100	176-220	270	0.60	63	2.8	2.3	30	70	9.6	0.30	0.16	1,979	15
Lambs Finishing—4 to 7 months old[g]													
30	66	295	0.65	72	3.2	2.5	60	40	14.7	0.51	0.24	1,085	15
40	88	275	0.60	76	3.3	2.7	75	25	11.6	0.42	0.21	1,175	15
50	110	205	0.45	77	3.4	2.8	80	20	10.0	0.35	0.19	1,469	15
Early Weaned Lambs—Moderate and rapid growth potential[g]													
10	22	250	0.55	80	3.5	2.9	90	10	26.2	0.82	0.38	940	20
20	44	300	0.66	78	3.4	2.8	85	15	16.9	0.54	0.24	940	20
30	66	325	0.72	78	3.3	2.7	85	15	15.1	0.51	0.24	1,085	15
40-60	88-132	400	0.88	78	3.3	2.7	85	15	14.5	0.55	0.28	1,253	15

[a]Values in Table 2 are calculated from daily requirements in Table 1 divided by DM intake. The exception, vitamin E daily requirements /head, are calculated from vitamin E/kg diet × DM intake.

[b]One kilogram TDN = 4.4 Mcal DE (digestible energy); ME (metabolizable energy) = 82% of DE. Because of rounding errors, values in Table 1 and Table 2 may differ.

[c]TDN calculated on following basis: hay DM, 55% TDN and on as-fed basis 50% TDN; grain DM, 83% TDN and on as-fed basis 75% TDN.

[d]Values are for ewes in moderate condition. Fat ewes should be fed according to the next lower weight category and thin ewes at the next higher weight category. Once desired or moderate weight condition is attained, use that weight category through all production stages.

[e]Values in parentheses are for ewes suckling lambs the last 4-6 weeks of lactation.

[f]Lambs intended for breeding; thus, maximum weight gains and finish are of secondary importance.

[g]Maximum weight gains expected.

TABLE 10–3. Net Energy Requirements for Lambs of Small, Medium, and Large Mature Weight Genotypes[a] (kcal/d)

Body Weight (kg)[b]:	10	20	25	30	35	40	45	50
NE$_m$ Requirements[c]:	315	530	626	718	806	891	973	1053
Daily Gain (g)[b]								
NE$_g$ Requirements								
Small mature weight lambs[d]								
100	178	300	354	406	456	504	551	596
150	267	450	532	610	684	756	826	894
200	357	600	708	812	912	1,008	1,102	1,192
250	446	750	886	1,016	1,140	1,261	1,377	1,490
300	535	900	1,064	1,219	1,368	1,513	1,652	1,788
Medium mature weight lambs[e]								
100	155	261	309	354	397	439	480	519
150	233	392	463	531	596	658	719	778
200	310	522	618	708	794	878	960	1,038
250	388	653	771	884	993	1,097	1,199	1,297
300	466	784	926	1,062	1,191	1,316	1,438	1,557
350	543	914	1,080	1,238	1,390	1,536	1,678	1,816
400	621	1,044	1,234	1,415	1,589	1,756	1,918	2,076
Large mature weight lambs[f]								
100	132	221	262	300	337	372	407	439
150	197	332	392	450	505	558	610	660
200	263	442	524	600	674	744	813	880
250	329	553	654	750	842	930	1,016	1,099
300	394	663	785	900	1,010	1,116	1,220	1,320
350	461	775	916	1,050	1,179	1,303	1,423	1,540
400	526	885	1,046	1,200	1,347	1,489	1,626	1,760
450	592	996	1,177	1,350	1,515	1,675	1,830	1,980

[a]Approximate mature ram weights of 95 kg, 115 kg, and 135 kg, respectively.
[b]Weights and gains include fill.
[c]NE$_m$ = 56 kcal $\cdot$ W$^{0.75}$ $\cdot$ d^{-1}.
[d]NE$_g$ = 317 kcal $\cdot$ W$^{0.75}$ $\cdot$ LWG, kg $\cdot$ d^{-1}.
[e]NE$_g$ = 276 kcal $\cdot$ W$^{0.75}$ $\cdot$ LWG, kg $\cdot$ d^{-1}.
[f]NE$_g$ = 234 kcal $\cdot$ W$^{0.75}$ $\cdot$ LWG, kg $\cdot$ d^{-1}.

TABLE 10–4. NE$_{preg}$ (NE$_y$) Requirements of Ewes Carrying Different Numbers of Fetuses at Various Stages of Gestation

Number of Fetuses Being Carried	Stage of Gestation (days)[a]					
	100	%[b]	120	%[b]	140	%[b]
	NE$_{preg}$ Required (kcal/day)					
1	70	100	145	100	260	100
2	125	178	265	183	440	169
3	170	243	345	238	570	219

[a]For gravid uterus (plus contents) and mammary gland development only.
[b]As a percentage of a single fetus's requirement.

TABLE 10–5. Crude Protein Requirements for Lambs of Small, Medium and Large Mature Weight Genotypes[a] (g/d)

Body Weight (kg)[b]:	10	20	25	30	35	40	45	50
Daily Gain (g)[b]								
Small mature weight lambs								
100	84	112	122	127	131	136	135	134
150	103	121	137	140	144	147	145	143
200	123	145	152	154	156	158	154	151
250	142	162	167	168	168	169	164	159
300	162	178	182	181	180	180	174	168
Medium mature weight lambs								
100	85	114	125	130	135	140	139	139
150	106	132	141	145	149	153	151	149
200	127	150	158	160	163	166	163	160
250	147	167	174	175	177	179	175	171
300	168	185	191	191	191	191	186	181
350	188	203	207	206	205	204	198	192
400	209	221	224	221	219	217	210	202
Large mature weight lambs								
100	94	128	134	139	145	144	150	156
150	115	147	152	156	160	159	164	169
200	136	166	170	173	176	174	178	182
250	157	186	188	190	192	189	192	195
300	179	205	206	207	208	204	206	208
350	200	224	224	224	224	219	220	221
400	221	243	242	241	240	234	234	234
450	242	262	260	256	256	249	248	248

[a]Approximate mature ram weights of 95 kg, 115 kg, and 135 kg, respectively.
[b]Weights and gains include fill.

TABLE 10–6. Macromineral Requirements of Sheep (Percentage of Diet Dry Matter)[a]

Nutrient	Requirement
Sodium	0.09-0.18
Chlorine	—
Calcium	0.20-0.82
Phosphorus	0.16-0.38
Magnesium	0.12-0.18
Potassium	0.50-0.80
Sulfur	0.14-0.26

[a]Values are estimates based on experimental data.

TABLE 10–7. Micromineral Requirements of Sheep and Maximum Tolerable Levels (ppm, mg/kg of Diet Dry Matter)[a]

Nutrient	Requirement	Maximum Tolerable Level[b]
Iodine	0.10-0.80[c]	50
Iron	30-50	500
Copper	7-11[d]	25[e]
Molybdenum	0.5	10[e]
Cobalt	0.1-0.2	10
Manganese	20-40	1,000
Zinc	20-33	750
Selenium	0.1-0.2	2
Fluorine	—	60-150

[a]Values are estimates based on experimental data.
[b]NRC (1980).
[c]High level for pregnancy and lactation in diets not containing goitrogens; should be increased if diets contain goitrogens.
[d]Requirement when dietary Mo concentrations are <1 mg/kg DM. See text for requirements under other circumstances.
[e]Lower levels may be toxic under some circumstances. See text.

TABLE 10–8. Vitamin E Requirements of Growing-Finishing Lambs and Suggested Levels of Feed Fortification to Provide 100% of Requirements

Body Weight		α-Tocopheryl Acetate		Feed Intake per Lamb		Amount of Vitamin E Added to Concentrate			Amount of Vitamin E Added to Protein Supplement[b]		
(kg)	(lb)	(mg/lamb/day)[a]	(mg/kg diet)	(kg)	(lb)	(mg/kg)	(mg/lb)	(mg/ton)	(mg/kg)	(mg/lb)	(mg/ton)
10	22	5.0	20	0.23	0.50	20	9.1	18,200	133	60	120,000
20	44	10.0	20	0.45	1.00	20	9.1	18,200	133	60	120,000
30	66	15.0	15	0.96	2.10	15	6.8	13,600	100	45	90,000
40	88	20.0	15	1.30	2.86	15	6.8	13,600	100	45	90,000
50	110	25.0	15	1.60	3.50	15	6.8	13,600	100	45	90,000

[a]Rounded values based on approximate diet intake containing recommended vitamin E levels.
[b]Assumes the concentrate diet contains 15 percent protein supplement.

TABLE 10–9. Ewe Gestation Rations Fed as Complete Feeds, % (as-fed Basis)[1,2,3]

Ingredient	IFN	1	2	3	4	5	6
Corn, gr.	4–02–931	12.5	—	12.5	—	—	7.5
Gr. ear corn	4–02–849	—	32.5	—	—	—	—
Alfalfa hay (good)	1–00–063	—	62.5	—	—	—	—
Alfalfa hay (stemmy)	1–00–068	80.0	—	—	—	—	—
Gr. cobs	1–02–782	—	—	72.5	—	—	—
Corn silage	3–02–823	—	—	—	96.25	—	—
Oat silage	3–03–298	—	—	—	—	97.5	—
Alfalfa haylage	3–08–150	—	—	—	—	—	92.5
Molasses (liquid)	4–04–696	7.5	5.0	6.25	—	—	—
Urea (281)		—	—	1.25	—	—	—
Supplement[4]		—	—	7.5	3.75	2.5	—
		100.0	100.0	100.0	100.0	100.0	100.0

Calculated Analysis (%):

1. *As-fed basis*

Crude protein		13.8	13.4	9.6	4.0	3.7	8.1
TDN		54.7	58.8	51.4	25.3	19.4	29.7
Calcium		0.92	0.83	0.35	0.20	0.17	0.53
Phosphorus		0.22	0.22	0.21	0.13	0.11	0.14
Dry matter		89.2	88.7	88.6	36.2	32.0	46.7

2. *Dry matter basis*

Crude protein		15.5	15.1	10.8	11.2	11.7	17.3
TDN		61.3	66.2	58.0	70.0	60.8	63.5
Calcium		1.03	0.94	0.40	0.55	0.54	1.13
Phosphorus		0.24	0.25	0.23	0.37	0.35	0.30

1. These rations can be hand-fed in the amounts to keep ewes gaining weight and maintain their condition during pregnancy.

2. Ration 3 must be completely ground and thoroughly mixed since it contains supplemental urea.

3. Free choice a mineral that is rich in phosphorus for rations 1, 2, and 3. Feed iodized salt free choice.

4. The following is an example supplement formula used in the rations. There are many good commercial supplements which are similar and may be substituted.

Supplement Ingredient	IFN	%
Soybean meal	5–04–604	50
Wheat bran	4–05–190	30
Molasses, liquid	4–04–696	5
Urea (281% CPE)		3
Limestone, gr.	6–02–632	4
Dicalcium phosphate	6–01–080	6
Trace mineralized salt		1
Vitamin premix[a]		1
		100

Calculated analysis of supplement:

	As-fed basis	Dry matter basis
Crude protein, %	36.0	40.0
TDN, %	60.0	67.0
Calcium, %	3.0	3.3
Phosphorus, %	1.8	2.0

a. Vitamin premix supplies the following: Vitamin A, 2,000,000 IU; vitamin D, 200,000 IU; vitamin E, 20,000 IU

TABLE 10–10. Ewe Lactation Rations Fed as Complete Feeds—First 8 to 10 Weeks, Percent (as-fed Basis)[1, 2]

Ingredient	IFN	1	2	3	4	5	6	7
Corn, gr.	4–02–931	20.00	25.00	—	34.0	4.0	13.0	15.0
Gr. ear corn	4–02–849	—	—	42.5	—	—	—	—
Alfalfa hay (good)	1–00–063	75.0	—	50.0	—	—	—	—
Alfalfa hay (stemmy)	1–00–068	—	67.5	—	—	—	—	—
Gr. cobs	1–02–782	—	—	—	45.0	—	—	—
Corn silage	3–02–823	—	—	—	—	92.5	—	—
Oat silage	3–03–298	—	—	—	—	—	85.0	—
Alfalfa haylage	3–08–150	—	—	—	—	—	—	85.0
Molasses (liquid)	4–04–696	5.0	7.5	5.0	5.0	—	—	—
Urea		—	—	—	1.0	0.5	0.5	—
Supplement[3]		—	—	2.5	15.0	2.5	1.0	—
Limestone	6–02–632	—	—	—	—	0.5	0.5	—
		100.0	100.0	100.0	100.0	100.0	100.0	100.0

Calculated Analysis (%):

1. *As-fed basis*

		1	2	3	4	5	6	7
Crude protein		14.7	13.0	12.9	12.7	5.2	5.4	8.1
TDN		58.5	58.7	60.9	60.3	27.0	26.9	33.9
Calcium		0.97	0.79	0.76	0.54	0.34	0.30	0.49
Phosphorus		0.23	0.23	0.26	0.39	0.12	0.12	0.15
Dry matter		89.4	88.8	88.2	88.3	38.3	39.2	50.0

2. *Dry matter basis*

		1	2	3	4	5	6	7
Crude protein		16.5	14.6	14.7	14.4	13.7	13.7	16.2
TDN		65.5	66.2	69.0	68.4	70.5	68.8	67.9
Calcium		1.09	0.89	0.86	0.62	0.90	0.77	0.98
Phosphorus		0.25	0.26	0.30	0.44	0.32	0.30	0.31

1. These rations can be hand-fed in the amounts to keep ewes milking well during lactation. Some weight and condition loss is to be expected.

2. Allow ewes access to a mineral mixture comparable to dicalcium phosphate and salt.

3. See example supplement formula shown in the footnote of Table 10–9 for ewe gestation rations.

TABLE 10–11. Starter Rations for Lambs (Start to 55 lb Body Weight) to Be Fed as Complete Feeds[1, 2, 3]

Ingredient	IFN	1	2	3
Corn, gr.	4–02–931	53.25	—	35.00
Gr. ear corn	4–02–849	—	60.00	—
Oats	4–03–309	—	—	20.00
Alfalfa hay	1–00–063	25.00	—	25.00
Alfalfa, dehy	1–00–023	—	15.00	—
Soybean meal	5–04–604	15.00	18.25	13.50
Molasses, liquid	4–04–696	5.00	5.00	5.00
Limestone	6–02–632	0.75	0.75	0.50
Salt, iodized		0.50	0.50	0.50
Vitamin premix[4]		0.50	0.50	0.50
Antibiotic		+	+	+
		100.00	100.00	100.00
Calculated Analysis (%):				
1) *As-fed basis*				
Crude protein		16.0	15.8	16.1
TDN		71.6	68.4	69.1
Calcium		0.68	0.61	0.60
Phosphorus		0.32	0.31	0.32
Dry matter		88.1	87.5	88.5
2) *Dry matter basis*				
Crude protein		18.2	18.1	18.2
TDN		81.3	78.2	78.1
Calcium		0.77	0.70	0.67
Phosphorus		0.36	0.35	0.36

1. All rations are formulated on an as-fed basis and should be ground and mixed thoroughly. All rations contain adequate minerals with the possible exception of selenium. Free-choice salt may be provided.

2. Hand-feed small amounts of the complete rations at least twice daily to lambs when 2 weeks of age. Clean troughs before each feeding so that feed is palatable to lambs.

3. When lambs are eating well, the complete ground mixed rations may be self-fed. They may also be hand-fed if kept before the lambs at all times in bunks or trough.

4. Vitamin premix should provide the following per kg of ration: vitamin A, 2000 IU; vitamin D, 200 IU; vitamin E, 20 IU.

TABLE 10–12. Grower Rations for Lambs
(55 to 80 lb Body Weight) to Be Fed as Complete Feeds[1,2]

Ingredient	IFN	1	2	3
Corn, gr.	4–02–931	56.0	53.0	36.5
Oats	4–03–309	—	—	22.0
Alfalfa hay	1–00–063	25.0	25.0	25.0
Soybean meal	5–04–604	12.5	—	10.0
Complete supplement[3]		—	16.5	—
Molasses, liquid	4–04–696	5.0	5.0	5.0
Limestone	6–02–632	0.5	—	0.5
Salt, iodized		0.5	0.5	0.5
Vitamin premix[4]		0.5	—	0.5
Antibiotic		+	+	+
		100.0	100.0	100.0
Calculated Analysis (%):				
1) *As-fed basis:*				
Crude protein		15.1	15.1	14.8
TDN		71.9	69.5	69.0
Calcium		0.58	0.85	0.59
Phosphorus		0.31	0.51	0.31
Dry matter		88.0	88.1	88.4
2) *Dry matter basis:*				
Crude protein		17.2	17.1	16.8
TDN		81.7	78.9	78.0
Calcium		0.66	0.97	0.66
Phosphorus		0.35	0.58	0.35

1. All rations are formulated on an as-fed basis and should be ground and mixed thoroughly. All rations contain adequate minerals with the possible exception of selenium. Free-choice salt may be provided.

2. Rations may be self-fed or also hand-fed if kept before the lambs at all times in bunks or trough.

3. See example supplement formula shown in the footnote of Table 10–9 for ewe gestation rations.

4. Vitamin premix should provide the following per kg of ration; vitamin A, 2000 IU; vitamin D, 200 IU; vitamin E, 20 IU.

TABLE 10–13. Finishing Rations for Lambs (80 lb to Market Weight) to Be Fed as Complete Feeds (%)[1,2,3]

Ingredient	IFN	Medium Energy			Low Energy		
		1	2	3	4	5	6
Corn, gr.	4–02–931	59.50	—	24.0	45.0	—	8.0
Gr. ear corn	4–02–849	—	74.2	—	—	60.5	—
Corn silage	3–02–823	—	—	69.0	—	—	85.0
Alfalfa hay	1–00–063	40.00	13.0	—	54.5	30.0	—
Soybean meal	5–04–604	—	6.5	5.5	—	4.0	5.5
Molasses, liquid	4–04–696	—	5.0	—	—	5.0	—
Limestone	6–02–632	—	0.8	0.5	—	—	0.5
Salt, iodized		0.5	0.5	0.5	0.5	0.5	0.5
Vitamin premix[4]		—	—	0.5	—	—	0.5
Antibiotic		+	+	+	+	+	+
		100.0	100.0	100.0	100.0	100.0	100.0

Calculated Analysis (%):
1) As-fed basis

Crude protein		12.0	11.2	6.5	13.2	11.9	5.6
TDN		69.8	67.9	40.6	65.5	64.9	31.3
Calcium		0.51	0.54	0.27	0.68	0.46	0.28
Phosphorus		0.26	0.26	0.16	0.25	0.24	0.12
Dry matter		88.8	86.8	50.9	89.4	87.4	42.4

2) Dry matter basis

Crude protein		13.5	12.9	12.9	14.8	13.6	13.2
TDN		78.6	78.1	79.8	73.3	74.3	73.9
Calcium		0.57	0.62	0.53	0.77	0.52	0.66
Phosphorus		0.29	0.30	0.31	0.28	0.28	0.29

1. These rations are formulated on an as-fed basis and can be fed once daily in troughs or bunks if there is capacity for a day's feed. Rations can also be self-fed. Feed salt free-choice.

2. Lambs should be brought up to full feed gradually and must not be allowed to be without feed even for short periods of time.

3. Complete grinding and mixing to prepare a uniform finished feed is important.

4. Vitamin premix should provide the following per kg of ration: vitamin A, 2000 IU; vitamin D, 200 IU; vitamin E, 20 IU.

TABLE 10–14. Rations for Replacement Ewe Lambs and Yearlings to Be Fed as Complete Feeds (%)[1,2,3]

Ingredient	IFN	Ewes weighing 90 pounds				Ewes weighing 130 pounds			
		1	2	3	4	1	2	3	4
Corn, gr.	4–02–931	15.0	—	15.0	10.0	6.0	—	2.5	—
Oats	4–03–309	15.0	—	—	—	6.0	—	—	—
Gr. ear corn	4–02–849	—	30.0	—	—	—	11.0	—	—
Orchardgrass hay	1–03–438	70.0	70.0	—	—	88.0	89.0	—	—
Alfalfa haylage	3–08–150	—	—	85.0	—	—	—	97.5	—
Corn silage	3–02–823	—	—	—	85.0	—	—	—	96.5
Supplement[4]		—	—	—	5.0	—	—	—	3.5
		100.0	100.0	100.0	100.0	100.0	100.0	100.0	100.0

Calculated Analysis (%):
1) *As-fed basis*

Crude protein		10.4	9.7	8.1	5.0	10.5	10.2	8.0	4.0
TDN		58.7	57.5	33.9	31.6	54.5	53.8	26.8	25.3
Calcium		0.27	0.28	0.49	0.23	0.33	0.34	0.56	0.19
Phosphorus		0.26	0.23	0.15	0.18	0.24	0.23	0.13	0.13
Dry matter		89.2	88.7	50.0	42.2	89.4	89.3	44.5	36.1

2) *Dry matter basis*

Crude protein		11.7	10.9	16.2	12.0	11.7	11.4	18.0	11.0
TDN		65.9	64.9	67.9	74.9	60.9	60.3	60.3	70.1
Calcium		0.31	0.31	0.98	0.54	0.37	0.38	1.25	0.53
Phosphorus		0.29	0.26	0.31	0.42	0.27	0.26	0.30	0.36

1. All rations are formulated on an as-fed basis and are calculated to be nutritionally adequate. However, flockowners must observe the ewes closely to adjust daily intake for animal condition.

2. These rations are designed to be hand-fed at least once daily. Suggested daily amounts are presented in the NRC requirement table.

3. Free choice feed a good complete mineral mixture along with salt.

4. See example supplement formula shown in the footnote of Table 10–9 for ewe gestation rations.

TABLE 10–15. Rations for Replacement Ram Lambs and Yearlings to Be Fed as Complete Feeds (%)[1, 2, 3]

Ingredient	IFN	Lambs—130 lb		Yearlings—220 lb	
		1	2	1	2
Oats	4–03–309	36	—	30	—
Gr. ear corn	4–02–849	—	35	—	25
Molasses, liquid	4–04–696	—	5	—	5
Alfalfa hay	1–00–063	64	60	70	70
		100	100	100	100
Calculated Analysis (%):					
1) *As-fed basis*					
Crude protein		15.1	13.1	15.4	14.0
TDN		58.1	59.2	57.1	57.3
Calcium		0.82	0.80	0.89	0.92
Phosphorus		0.26	0.22	0.26	0.22
Dry matter		90.3	88.6	90.5	89.0
2) *Dry matter basis*					
Crude protein		16.7	14.8	17.0	15.8
TDN		64.3	66.9	63.1	64.4
Calcium		0.91	0.91	0.99	1.03
Phosphorus		0.29	0.25	0.28	0.24

1. All rations are formulated on an as-fed basis and should be fed at a rate which would keep rams growing and gaining but not fattening; approximately 5.0–5.5 lb would be required for lambs and 6.5–7.0 lb for yearlings. When rams are used for breeding, slightly higher daily intakes will be required.

2. These same rations would be nutritionally adequate for mature rams if fed at rates to maintain normal body condition and weight.

3. Feed salt free-choice.

Study Questions and Problems

Chapter 10

I. True or False Questions

1. The percent protein content of the ration for pregnant ewes is similar to that of pregnant beef cows.
2. Pregnancy disease is more likely to occur in pregnant ewes carrying singles than in ewes carrying triplet lambs.
3. Enterotoxemia (overeating disease) in young nursing lambs can be reduced by injecting bacterin into the lambs after they are 3 to 4 weeks of age.
4. For sheep, an animal unit is generally considered as 4 to 6 ewes and their lambs.
5. Less labor and facilities are required for the early lambing program compared to that needed for the late lambing program.
6. A ewe and her lamb(s) in drylot facilities need about 4 bushel of grain and 1600 lb of hay a year.
7. Grains included in sheep diets should be finely ground.
8. Lambing paralysis or ketosis will generally occur within the first 2 weeks after lambing.
9. To reduce problems of enterotoxemia, feedlot lambs should be vaccinated at least 10 days to 2 weeks after being placed on a full feed.
10. Urinary calculi may be a problem in feedlot wether lambs fed high concentrate diets.

II. Multiple Choice (select the single best answer)

1. A postlambing ration for ewes includes good forage and minerals plus __?__ lb of grain per head daily, depending on number of lambs she is nursing.
   a. 0
   b. ¼ to 1
   c. 1 to 3
   d. 3 to 5
   e. Over 5
2. Creep feeding of early lambs starts at
   a. Birth.
   b. About 10 days after birth.
   c. About 10 days after weaning.
   d. When lambs weigh approximately 50 lb.
   e. Not recommended.
3. A recommended ration for dry, nonpregnant ewes is
   a. Average pasture.
   b. Excellent pasture.
   c. Good pasture plus hay.
   d. Good pasture plus 1 lb corn per day.
   e. Full fed corn silage.
4. During the last third of pregnancy, ewes should be fed hay and minerals and
   a. Start with ¼ lb and increase to 1 lb grain per day.
   b. Nothing else if hay is of excellent quality.
   c. 2 to 3 lb of grain per day.
   d. 4 to 5 lb of grain per day.
   e. 2% of the body weight as grain.
5. Enterotoxemia in pregnant ewes fed in drylot is best prevented by
   a. Vaccination.
   b. Drench with molasses.
   c. Additional roughage in the diet.
   d. Antibiotics.
   e. Generally not a problem.
6. Voluntary intake of forages by ewes in late gestation (last third) is usually __?__ compared to intake in early gestation.
   a. Higher
   b. The same
   c. Lower

445

7. Pregnancy disease in sheep is analogous to which disease in dairy cattle?
   a. Ketosis
   b. Milk fever
   c. Fatty liver syndrome
   d. Displaced abomasum
   e. None of the above

8. Which of the following is not true of the late lambing system?
   a. Roughages provide most of the feed needed.
   b. Lambing facilities need not be as good.
   c. Lambs can be marketed from pasture with a minimum of grain feeding.
   d. Parasite problems are generally less serious.
   e. Answers a and c

9. Enterotoxemia is caused by the toxins produced in the digestive tract from excessive numbers of the organism known as
   a. *Escherichia coli.*
   b. *Aspergillus flavus.*
   c. *Fusarium graminearum.*
   d. *Clostridium perfringens.*
   e. none of the above.

10. Which of the following feedstuffs should be supplemented to the feedlot wether diet to help reduce problems with urinary calculi?
    a. Ground corn
    b. Salt
    c. Dicalcium phosphate
    d. Vitamin trace mineral premix
    e. Ground limestone

III. *Matching or Identification*

   Match the appropriate answer to the following:

a. Bloat	f. Urinary calculi
b. Enterotoxemia	g. Estrogens
c. Copper	h. Acidosis
d. Ketosis	i. Calcium
e. White muscle	j. Sodium

   1. Overeating disease.
   2. May occur when sheep graze alfalfa.
   3. Pregnancy disease.
   4. Feed supplemental calcium source.
   5. Feed supplemental vitamin E.
   6. Mineral highly toxic to sheep.
   7. *Colstridium perfringens.*
   8. Present in certain legumes and may influence reproduction.
   9. Caused by a lack of usable carbohydrates.
   10. Prevented by vaccination with bacterin or toxoid.
   11. May occur in wether lambs fed diets high in phosphorus.
   12. Claim for feeding aureomycin.
   13. Administration of dextrose or propylene glycol to the ewe.
   14. Rumen disorder associated with lambs fed high-grain diets.
   15. Usually affects older ewes, particularly those pregnant with twins or triplets.

IV. Short Answer Questions

1. Why might ewes develop enterotoxemia when grazing corn plant residue?
2. List advantages and disadvantages of the early lambing system.
3. Discuss advantages and disadvantages of utilizing alfalfa pasture as a forage source in sheep production.
4. Describe ways of preventing lambing paralysis.
5. Name two feed additives for sheep diets and why they may be used.

V. Problems

1. A 150-lb nonlactating ewe during the first 15 weeks of gestation requires 1.7 lb TDN and 0.3 lb crude protein (dry matter basis). Using one or more of the following feeds, prepare a suitable daily ration for this ewe.

   a) Feedstuffs available (dry matter basis)

	CP, %	TDN, %
Alfalfa hay (90% DM)	17.0	56
Bromegrass hay (91% DM)	9.7	55
Corn grain (89% DM)	10.9	92

   b) Daily ration (as-fed basis)

2. Feedlot lambs weighing 30-kg (66-lb) have the following daily requirements (dry matter basis)

Crude protein	—196 g
TDN	—1020 g
Calcium	—5.0 g
Phosphorus	—3.3 g

   Assume each lamb consumes 1.4 kg of dry matter daily. Formulate a daily ration using some combination of the following feedstuffs (dry matter basis).

Ingredients	CP %	TDN %	Ca %	P %
Orchardgrass hay	11.2	58	.39	.35
Corn grain	10.9	92	.03	.29
Soybean meal	49.9	88	.34	.70
Limestone	—	—	34.00	—

Notes

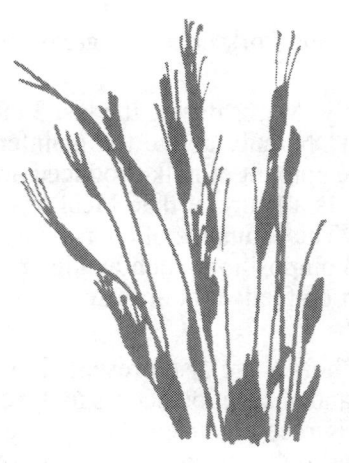

11
Nutrition and Feeding of Horses

Chapter Goals

- Discuss nutrient and energy requirements related to function of the horse.
- Review feedstuffs commonly used in horse diets.
- Outline and discuss a life-cycle horse feeding program.
- Identify nutritionally related diseases or disorders in horses.
- Formulate a horse diet.

Feeding horses for show or performance purposes is more complicated than feeding any other farm animal. Unfortunately, there is limited experimental information on the nutritional needs of the horse. Therefore, one still has to rely quite heavily on the practical results obtained by successful horse producers and on research data from other animals.

This section will present some suggestions on nutrient requirements and on the use of feedstuffs to meet the nutrient requirements of light horses. These suggestions are based on recommendations by the National Research Council (*Nutrient Requirements of Horses,* Publication ISBN 03989-4, Committee on Animal Nutrition, National Academy of Sciences—National Research Council, Washington, D.C., 1989) and from various articles reviewed by this author.

I. Nutrient Requirements

Opinions are divided as to the energy, protein, mineral and vitamin needs of horses. Many of the recommendations made herein were arrived at by extrapolating from the requirements of other farm animals. These suggested levels of nutrients can be used as a guide until more exacting information can be obtained from research studies.

A. *Energy*

The energy requirement of horses for different functions expressed either in terms of pounds of total digestible nutrients (TDN) or as Megacalories (Mcal) of digestible energy (DE) is presented in Table 11–1.

The digestible energy requirement for maintenance of horses has been computed by several workers and is reported to be approximately $1.4 + 0.03$ BW (kg body weight) Mcal daily. Energy for growth would depend on the stage of the growth cycle and rate of average daily gain. The following formulas are suggested for computing DE needs (Mcal/day) for growth.

Weanlings	$1.4 + 0.03$ BW $+ \; 9$ ADG (kg)
Yearlings	$1.4 + 0.03$ BW $+ 16$ ADG (kg)
Long Yearlings	$1.4 + 0.03$ BW $+ 18$ ADG (kg)
2-year-olds	$1.4 + 0.03$ BW $+ 20$ ADG (kg)

The energy requirement for work depends not only on the type of work but the speed and the terrain over which the work is done. The National Research Council (NRC) recommends DE needs per hour of work (above maintenance) to be from 4 Mcal to as high as 16 Mcal depending on the weight of the horse and degree of activity. It is quite apparent that work creates a great demand for additional energy over maintenance.

Pregnancy does not increase the energy requirements appreciably. Mares during the last 3 months of gestation require approximately 2 Mcal or about 1.0 lb of TDN daily above the maintenance requirement. Energy requirements for lactation depend upon the amount of milk produced and the composition of the milk. The National Research Council suggests a figure of 0.36 Mcal or 0.18 lb of TDN per lb of milk as the energy requirements for lactation. The estimates of energy furnished by hay or feeds may be obtained by consulting tables of feed composition, such as illustrated in Table 3–2, using values listed on feed tags or from analysis run on feeds that are used.

B. *Protein*

The recommended protein requirements for different classes of horses are also presented in Table 11–1. It is currently not known how much amino acid synthesis and absorption occurs in the cecum of the horse. Therefore, until more information is known, it is recommended that high quality protein rations, adequate in amino acids, be fed to horses. This should especially be the case with the young growing horse whose cecum may not have much bacterial synthesis during early life. As the horse reaches mature size, its protein needs and importance of protein quality decreases. There probably is no need for supplemental protein for older animals since natural feeds used to meet their energy requirement will also meet their protein requirement. A possible exception is the horse being maintained on poor quality roughage.

Approximately 1 lb of crude protein per 1000 lb of body weight has been suggested as a maintenance level. The estimated protein requirements for pregnancy and lactation, suggested by NRC, are based on what is produced. The mare requires approximately 0.35 lb of total protein per day over her maintenance to maintain and grow the developing fetus (last 3 months of gestation). During lactation, the mare requires, in addition to maintenance, approximately 0.04 lb of protein for each pound of milk produced.

Estimates of crude protein furnished by hays or feeds may be obtained from consulting tables of composition, such as illustrated in Table 3–2, using values listed on feed tags or from analysis run on the feeds that are used.

C. *Minerals*

Minerals are needed to develop sound feet and legs. Developing good strong, sound bones, however, requires more than just calcium and phosphorus. It also involves other minerals and nutrients. Unfortunately, very little is known about these needs. The suggestions contained in Tables 11–1 and 11–2 can be used as a tentative guide until more exacting information can be obtained from horse research studies.

1. Calcium and phosphorus

   To obtain proper calcium and phosphorus utilization, a number of conditions must be met. First, an adequate level of both calcium and phosphorus must be fed. Second, a suitable ratio between them must exist. Third, a sufficient amount of vitamin D must be available. The ratio of calcium to phosphorus will vary depending on the weight and age of the horse and the level of calcium and phosphorus in the diet. During early growth, this ratio approaches 1.5 to 1.0. The ratio of calcium to phosphorus for the mature horse should be approximately 2.0 to 1.0. Studies would indicate that the mature horse needs at least 0.3% of the ration as available calcium and phosphorus. The calcium level might need to be 0.8% for the young horse and the phosphorus level 0.6%. These levels would then be decreased to 0.6% calcium and 0.5% phosphorus as the horse approaches maturity and the stress of training or racing is over.

2. Salt

   Salt serves as both a nutrient and a condiment in the ration. As a nutrient, it contains the mineral elements sodium and chlorine. As a condiment it stimulates the secretion of saliva and makes feed more palatable. The horse needs approximately 85 g (0.2 lb) of salt daily. Salt should be supplied as approximately 0.5% of the total ration and also extra salt if they need it.

3. Trace minerals

   There is enough evidence to indicate that some trace mineral deficiencies exist in some parts of all the major livestock areas of the world. The possibility of a deficiency problem can be taken care of by feeding trace mineral mixtures or trace mineralized salt even though the experimental data are lacking to indicate a definite need. The cost of adding the trace minerals to a salt is very small. The average salt consumption of all horses, young and old, is about 24 lb salt yearly. If trace mineralized salt cost 1¢ more per pound ($20 more per ton) over plain salt, the cost of feeding trace minerals to each horse yearly is 24¢. This is very low-cost insurance. Table 11–2 shows suggested levels for the trace minerals iron, copper, iodine, zinc, manganese, cobalt, selenium and sulfur. Magnesium and potassium are also essential, but known deficiencies have never been recorded, and it is thought that plenty is supplied by commonly fed feedstuffs.

D. *Vitamins*

Although only a limited amount of information is available concerning the dietary vitamin needs of the horse, vitamins are necessary in horse rations to permit proper growth, development, health and reproduction. Also, some of these needs may be accentuated during stress and strain to which many modern horses are subjected. Suggested supplemental vitamin levels for horses are presented in Table 11–3. Nature has provided the horse with the ability to produce most of its needs for vitamins D, K, C, and probably most of its normal requirements for all the B-vitamins.

1. Vitamin A

Vitamin A is one which must be supplemented under certain conditions. When green grazing or good quality green hay is abundant, the vitamin is not a problem as the horse is able to convert the carotene from these products into vitamin A. However, if horses are fed feeds low in carotene, especially in the winter, supplementation is advisable. The average horse can use approximately 7500 to 10,000 international units of vitamin A per day, and three to four times that level per day is thought to be necessary for reproduction and lactation.

2. Vitamin D

Almost all feeds are low in vitamin D activity. In many areas where sunlight is lacking or erratic, or where horses are kept inside a great deal, vitamin D supplementation is used. Vitamin D is necessary for both the absorption and deposition of calcium and phosphorus as bone in the body. In contrast to vitamin A the body has little ability to store much vitamin D. As far as is known, both D_2 and D_3 have the same value for the horse. A good rule of thumb to follow is to use vitamin D at 10% to 15% of the level of the vitamin A in the ration.

3. Vitamin E

Dr. DC Dodd of New Zealand reported a muscle degeneration and a yellowish-brown fat disease in foals, which responded to several days of injections of vitamin E. Injections of vitamin E and selenium are also used for the prevention and cure of the "tying-up" syndrome that often occurs in light horses in intensive training. These reports indicate that vitamin E can be of considerable importance in horse nutrition under certain conditions. More studies are needed to verify these preliminary reports and to give more exact information. A suggested level of vitamin E might be 1 percent of the vitamin A level of the diet.

4. Vitamins C (ascorbic acid) and K

The horse is able to synthesize adequate amounts of vitamins C and K and there is no evidence to indicate that additional levels are needed in a horse ration.

5. B vitamins

At the present time there is little information available on the supplemental needs of the B complex vitamins for the horse. Many recommendations for supplementation of horse rations with B vitamins, such as those presented in Table 11–3, are based on those required for the pig. It is not known that these levels of B vitamins are needed, however, they can be used as a guide in csae one wishes to try supplementation of horse rations. Levels which are adequate for the pig should also meet the requirements of the horse, but they are probably higher than its needs since it is possible that the horse will synthesize more of these vitamins than the pig.

II. Feeds for Horses

In general, feedstuffs for horse diets may be classified either as forage or concentrate materials. Concentrates could contain feedstuffs used for energy, protein, minerals or other unidentified growth factors.

A. *Forages*

Normally, hay or its pasture equivalent will be fed to horses at the rate of 1 to 2 lb/100 lb of body weight per day; therefore, hay quality and the nutrients it furnishes should be given a great deal

451

Pasture provides top quality feed, exercise and a natural environment for developing horses.

of consideration in feeding horses. Undoubtedly, obtaining good hay, storing it and feeding it are major management problems. Some general factors about feeding hay should be considered. Moldy or dusty hay may cause colic and heaves in horses. Large amounts of very poor quality hay should not be fed because it may be so poorly digested that it will not pass the digestive tract and impaction and colic will result. Very high quality clover or small grain hay may be so readily digested that when fed with a high grain feed, a loose feces or a colic condition may result. When very high quality hay is fed with grain, it may be necessary to feed a poorer quality grass hay mixed with it.

Bromegrass, orchardgrass, timothy and Bermudagrass make excellent hay for horses. They are palatable and usually less dusty and less likely to be moldy than legume hays. The legumes are higher in nutrient content than grasses and may be fed by themselves or in combination with the grass hays. Legume hays are heavier and more difficult to cure properly, and are thus more prone to mold and dustiness. Alfalfa hay is more laxative than grass hays and may cause loose feces. This does not seem to be detrimental to the horse.

Most grasses can be pastured by the horse. Legume-grass mixtures make excellent high quality pastures when available. Rotational grazing and/or clipping are important management practices as horses are selective and tend to graze the youngest and most tender grasses. Sorghum and sorghum-Sudan hybrids are not recommended for horses, as they have been linked to cystitis problems in mares.

When plenty of high quality pasture or hay is available, the only horses likely to need grain are rapidly developing weanlings and yearlings, mares that are lactating and are to be bred back and show and performance horses. Again, it is important to emphasize that an excellent pasture is one of the most important feeds on a horse farm. Pasture provides top quality feed, exercise and a natural environment for developing horses.

B. *Concentrates*

Grain or supplemental feed should be fed to make up deficiencies in the energy, protein, minerals and vitamins furnished by the hay. A complete grain mix with other supplements added may be fed.

   1. Grains to use

      a. Oats—still the most widely used and the most popular grain for horses. Heavy, bright or clean oats which contain a small percentage of hull are preferred for horses. Oats are lower in energy content than other grains but will cause less trouble with stomach compaction. Dusty oats should never be used since they may cause colic. It is best to roll or crush oats for horses with poor teeth or for young foals.

      b. Corn—also widely used for horses. It should be cracked, coarsely ground or preferably rolled. Corn is higher in energy than oats and can be used to increase the energy content of the ration. Most horse producers prefer to mix corn with oats, and use it at lower levels than oats.

c. Barley—used some in the United States (especially in the West), and is quite popular in some foreign countries. Since barley is quite hard, it should be coarsely ground or preferably rolled. Most horseowners prefer to mix it with oats in about equal parts when using it.

d. Wheat or milo—wheat or milo (sorghum grain) are not used very much for horses. When ground they become rather doughy and tend to ball up with moisture. If used, they should be rolled and mixed at a low level with bulky feed such as oats or wheat bran.

e. Wheat bran—very valuable for its mild laxative effect and for its bulky nature. It is generally used at levels of 5% to 15% of the ration.

2. Protein supplements

a. Linseed meal—still a popular protein supplement for horse feeding. Most horseowners feel that it contains something which produces bloom and luster in the hair coat. Many prefer to use a pelleted linseed meal because some of the linseed meals may be too dusty because of being too low in fat after solvent extraction. Linseed meal is usually more expensive and contains inferior amino acid composition compared to soybean meal.

b. Soybean meal—also used quite extensively for horses and may be substituted on an equal protein basis for linseed meal. Soybean meal contains higher quality protein and is generally more economical.

c. Milk protein—dried whey or commercial supplements containing high amounts of milk products are often used in the starting ration for foals because of their content of excellent quality protein. Many commercially prepared supplements of this type are available and a decision on which to use should be based on the cost of the nutrients they furnish.

3. Other miscellaneous feedstuffs

a. Molasses—the use of 5% to 10% of cane molasses sweetens the feed and makes it more palatable. This is a good idea when one is trying to get a little extra feed consumption by the horses. Liquid molasses also tends to condition the feed, prevent separation and reduce its dustiness.

b. Dehydrated alfalfa meal—often used at 5% to 10% of a horse ration. It is a good source of vitamins, minerals, protein and unidentified factors. High quality alfalfa meal is the closest substitute for green pasture and will be particularly valuable during periods of the year when the pastures are dormant.

c. The mineral needs of the horse may be supplied by a number of different mineral feeds. Sodium and chlorine needs are easily met by the addition of salt to the horse ration whether it be plain, iodized, or trace mineralized salt. Calcium and phosphorus needs may be supplied to the horse by limestone, dicalcium phosphate, steamed bone meal or defluorinated rock phosphate. The remaining trace mineral requirements for the horse are most easily met by commercial trace mineral premixes or trace mineralized salt.

d. Fats and oils—animal fats and vegetable oils can be used as highly concentrated energy sources for horses. Vegetable oils are generally more palatable than animal fats. These products can also be used as a source of linoleic acid (an essential fatty acid) and are valued for their ability to reduce feed dustiness and put a bloom on the animal haircoat. Up to 5% to 10% fat may be added to the ration of horses subjected to intense and prolonged exercise and stress, such as endurance trials, to increase energy intake and improve performance.

C. *Pelleting of Feeds*

The grain ration or the hay plus grain may be made into a pellet. The use of pellets may be especially useful in creep feeds and rations for weanlings where there is a tendency for the horses to separate out the fine particles made up of the protein and mineral additions to the rations. Complete pelleted rations containing hay as well as grain should contain 60% to 70% coarsely ground hay to decrease problems with colic. It may also be necessry to feed small amounts of unprocessed hay to prevent wood chewing and mane and tail chewing when groups of animals are penned together.

D. *Antibiotics*

There is practically no information available on the value of antibiotics for horses. The addition of an antibiotic may especially be helpful for young foals which suffer set backs because of infections, digestive troubles, lack of milk, poor weather or other stress factors. Presently, the Food and Drug Administration allows the use of 85 mg chlortetracycline per head daily for horses up to one year of age for stimulating growth and improving feed efficiency (*Feed Additive Compendium*, 1992).

III. Feeding the Horse

A. *Feeding Suckling Foals and Weanlings*

The first year of life is a very critical period in the development of horses. Normally, the mare will do an excellent job of taking care of the nutritional requirements of the foal for the first 3 to 4 months of its life provided she is well fed. After this time, it is necessary for the foal to obtain a significant amount of nutrients from other sources. This is an excellent time to familiarize the foal with dry feed of the type it will be receiving after weaning. Concentrate mixes that may be fed prior to weaning and immediately thereafter are shown in Table 11-4. These are high quality rations furnishing large amounts of protein, minerals and vitamins. The foals should be consuming six to eight pounds of concentrate per day at weaning time or 5 to 6 months of age. During the suckling phase, this may be accomplished by furnishing a creep separate from the mares. At weaning time, the foals should be put into a pen or in box stalls in pairs and continued on 6 to 8 lb of concentrate per head per day along with about 1 lb of hay per 100 lb of body weight. The concentrate feed is increased up to a maximum of 10 lb per day adjusting this to what the foals will voluntarily clean up, or in the case of some foals, restricting it to a level which will not cause digestive upset. It is essential at this stage to have the foals as free of internal parasites as possible by having them wormed by the time they are 2 to 3 months of age and repeating this at 30 to 60 day intervals. Also, at the time they are still with the mares, they should be given treatments associated with their health program so that this need not interfere at weaning time. Switch to a weanling ration shown in Table 11-4 when the foals weigh around 500 lb or in the case of heavier weanlings, they are switched to the weanling ration after they are well started on feed after weaning. The weanling ration has slightly lower protein quality and is less expensive. There would be no problem continuing the previous ration if expense was not a consideration. Continue the weanlings on a feeding program of 1 lb of hay per 100 lb of body weight daily plus sufficient concentrate to furnish energy to obtain the desired growth rates based upon the maintenance and energy requirements for the weights and growth rates desired using the values listed in Table 11-1. Quarter horses are expected to weigh 800 lb or more at 12 months of age under this feeding program. This growth rate necessitates about 2 lb average daily gain from birth to 12 month of age.

B. *Feeding Yearlings and 2-Year-Olds*

At 1 year of age or about 800 lb, change to a yearling ration shown in Table 11-4 which is a general purpose herd ration. At this age and stage of development, it is possible to rely on hay and pasture more extensively; however, it is very important to assure an adequate supply of protein, vitamins and minerals so that the horses will continue the maturity of their tissue and bone structure. The total feed fed is based upon requirements for maintenance and growth that is desired. The desirable development and appearance of the horse should be set by the experienced eye of the horseowner. A set of scales is a valuable asset in evaluating this development. As the horse goes into training, it is again best to base your feeding practices on maintaining the desired condition and development and growth of the horse. The work of training will increase the energy requirements approximately at the rate shown in Table 11-1. However, since the intensiveness and amount of work involved in the training will be so variable, it is necessary to use a great deal of judgment in feeding horses at this stage. The horse should be fed as an individual to maintain the desired condition and weight.

C. *Feeding the Mature, Dry Mare or Gelding*

The mature animal which is idle or ridden as infrequently as once or twice per week can be maintained on hay or pasture alone. A calcium-phosphorus and salt supplement should be furnished to such horses. Of course, a combination of hay and grain may also be used. Oats or a simple sweet feed fed with good hay will be adequate.

D. *Reproduction*

1. Gestation

The recommended maintenance requirement for energy and protein should be adequate for conception if the mare is not also lactating at breeding time. Feeding mares a maintenance diet during early gestation appears reasonable because fetal growth rate is extremely low and protein utilization is probably highly efficient. However, during the last 90 days of the gestation period, about 60% of the weight of the fetus develops. An increased energy and nutrient intake is necessary to support this rapid development. The mare should gain about 10% of her normal body weight during the last 3 months of gestation. Suggested rations for the pregnant mare are shown in Table 11–4.

   2. Lactation

     Use the same type of ration fed to yearlings and two-year-olds for the lactating mare (Table 11–4). In the case of the mare which has just foaled, is lactating, and is being prepared for rebreeding, we have another situation of critical nutritional stress in the horse. It seems logical that the mare should be in a top state of health including being well nourished to accomplish all these functions. The poor reproductive record of horses may indicate that this is frequently not the case. At a time of nutritional stress in a horse, they may use nutrients from stores in the body as well as nutrients derived from their feed. A good feeding program should provide nutrients for both sources. The mares should be in good flesh prior to foaling so that her reserves of nutrients are built up, and then she should be gotten on a high level of feeding as soon after foaling as possible. The actual level of feeding is based on her total energy requirement for maintenance plus lactation with an additional amount of feed which should allow her to begin gaining weight and get in the best possible breeding condition

     A few days before foaling, provide the mare a bulky ration to reduce problems with constipation. Allow 7 to 10 days after foaling to bring the mare to a full feed providing 1 to 1½ lb concentrate per 100 lb body weight together with a quantity of hay or pasture equivalent within the same range. Peak lactation occurs between 2 to 3 months postfoaling. An 1100-lb mare may produce over 35 lb milk daily at peak lactation.

     Good success may be obtained with this approach with normal mares; however, mares which have previously foundered or are prone to colic will have to be fed at a lower level. This high level of feeding is continued following breeding and into lactation until the mare begins to gain excess weight.

   3. Stallions are fed the same ration year-round. The amount fed is based on their requirements for energy for maintenance with an additional allowance during the breeding season according to his activity to maintain him in a good breeding condition.

E. *Feeding the Working or Performance Horse*

   1. Includes horses being trained for various performance events or horses used for roping, cutting, jumping etc.; these horses are generally ridden every day and have relatively high energy expenditures.

   2. Exercise or work increases the need for energy but creates little or no increase in demand by the horse for protein, phosphorus, calcium, most trace minerals and most fat-soluble vitamins.

   3. Equal quantities of hay and concentrate work well for most animals, but hard-working horses may need more concentrate than forage to satisfy energy requirements. In no case should they receive less than 1 lb forage/100 lb b.w. daily. Suggested rations for performance horses are shown in Table 11–4.

   4. At high levels of feed intake, multiple feedings may be necessary to minimize digestive disturbances. An animal receiving 7 lb or less concentrate can be fed once daily, but those receiving 8 to 14 lb should be fed twice daily and those receiving more than 14 lb should be fed three times daily.

   5. Following periods of heavy exercise, small quantities (a few swallows) of water should be allowed at 5- to 10-minute intervals until thirst is quenched. This will prevent digestive disturbances and possible founder from excess water intake.

IV. General

A. Additional and more detailed nutrient requirements for horses may be obtained by referring to the recommendations outlined by the National Research Council.

B. Horses should be kept reasonably free of diseases and internal parasites or else little benefit will be realized from good nutritional practices. Worming is first carried out at 8 to 10 weeks of age and is continued at 2-month intervals until the foals are 1 year old, when they are put on the same schedule as the mares, i.e., three or four times a year, with the last worming following the first frost. Frequent fecal checks should be made from the pens. Anthelmintics may be administered through the feed or given directly by means of a stomach tube. It is important to be as sanitary with stomach tubes as with hypodermic needles, soaking them in disinfectant solution between individual cases.

Certain diseases or disorders in horses can be related directly to feeding practices. Colic, founder, heaves and azoturia can be related to the use of improper feeds or improper feeding, although each may be caused by other factors as well. Some nutritionally related diseases are presented in Table 11–5. The good manager will be aware of horses more susceptible to these conditions based on past history and will modify their feeding program accordingly.

There is a need for individual feeding and for a constant study of the peculiarities of needs for each individual horse. This is necessary because, at best, any recommendation made can only serve as a guide and the horse producer must prepare to make adjustments in the feeding program when necessary. In a sense, horses are similar to humans and will vary considerably in likes and dislikes for certain feeds and level of feed intake. Proper attention to these small details will mean the difference between developing a champion or just another horse.

C. The following general hints and feeding rules may be made.
1. Start with a healthy horse.
2. Avoid using moldy, spoiled or otherwise damaged feedstuffs.
3. Consumption:

Body Weight, Lb	Diet Intake per Day, % of Body Weight
Under 600	2.25–2.75
600–1000	2.00–2.25
Over 1000	<2.00

4. Minimum roughage in diet: 1 lb per 100 lb body weight. Normal diets 1 to 2 lb/100 pounds body weight. Use pasture whenever available. Optimum pasture acreage for average-size light horse is 3 acres.
5. Corn silage may make up one-third to half of the dry-matter roughage portion of diet (10 to 15 lb, as-fed). Silage should be of the best quality and not frozen.
6. Feed grain if horses need additional energy. Feed grain by weight and nutrient content rather than by volume. When practical, give a small feeding of hay prior to each grain feeding.
7. Composition of pelleted diet (if only feed fed): 60% to 70% coarsely ground hay or its equivalent, 30% to 40% concentrate. Suggested pellet size is ½ in.
8. Rate of feeding vegetable meal supplement: usually ½ to 1 lb/head daily, rarely 2 lb.
9. Establish a daily schedule for feeding and stick to it. No grain should be left from one feeding to the next, and all edible forage should be cleaned up at the end of each day.
10. Adjust each horse's ration to individual needs. Feed a balanced ration and then feed according to the condition of the horse. Don't get your horse too fat.
11. Make any changes in the ration gradually. It will be best to make rations changes over a week to 10-day period. Changing feed too quickly may throw your horse off feed.
12. Feed horses after cooling out. Divide ration into two or three equal portions daily.
13. Provide clean, fresh water free choice for idle horses. Encourage horses to drink before feeding rather than after. Water horses after cooling out.
14. Follow a strict disease and parasite control program.
15. Provide the horse with adequate exercise.
16. Pay proper attention to nutrition and management details.

V. Steps Involved in Formulating a Horse Ration

A. *Problem*

Balance a daily ration for an 1100-lb mature quarter horse being moderately worked each day (fast trotting, cantering, some jumping). Assume the horse will consume 1% to 2% of its body weight as forage plus an amount of concentrate (grain) needed to meet energy needs. Work the entire problem on an as-fed basis using the horse ration worksheet following this example problem. A sample worksheet is also found in the appendix section at the back of the textbook.

B. *Establish Nutrient Requirements (Table 11–1)*

Record on the worksheet (A 1–3) the requirements for maintenance and for moderate work. Energy is the primary dietary constituent required above maintenance levels for work. The energy requirement for work is met by feeding greater amounts of grain or a concentrate mix. This practice generally provides sufficient amounts of any other nutrient which may be needed.

C. *Determine feeds available and their composition (as-fed basis) from Table 3–2.*
1. Forage ration—feed good quality orchardgrass hay (IFN 1-03-438).
2. Concentrate (if needed).
  a. Oat grain (IFN 4-03-309).
  b. Soybean meal (IFN 5-04-604).
3. Assume salt (plain or trace mineralized) is fed free-choice.

D. *Establish Feeding Levels*
1. Forage—horse rations should not contain less than 1% nor more than 2% of the body weight as forage (air-dry basis). For this problem, assume the horse will be fed orchardgrass hay at 1.75% of body weight, or 19.25 lb hay daily (1100 lb × .0175 = 19.25 lb).
2. Concentrate (grain)—to meet DE needs
  a. Determine the additional nutrients needed from the concentrate (worksheet C.)

	DE	CP	Ca	P
	Mcal	lb	g	g
Daily requirements	24.6	1.4	20	14
From forage	14.8	2.0	32	20
Still need from concentrate	9.8	0	0	0

  b. CP, Ca and P are adequate from the forage but additional energy must be supplied from concentrate. Thus, oat grain will be the only feed needed. From worksheet D., it is calculated that we will feed 7.26 lb of oats (9.8 Mcal needed ÷ 1.35 Mcal/lb in oats = 7.26 lb).
  c. Since the horse in this example requires no additional CP, ignore worksheet E. When other horse formulations would require protein (in addition to energy), a concentrate mixture using a grain and protein feed may be calculated using algebraic or Pearson square solution.

E. *Overall Composition of the Daily Ration*
1. The combination of 19.25 lb orchardgrass hay and 7.26 lb oats satisfied the DE requirements of this horse and provided excess amounts of CP, Ca and P.
2. The final ration amounts to approximately 2.4% of the body weight as daily air-dry feed intake. This figure will vary with the degree of work expended by the horse.

HORSE RATION WORKSHEET

Animal Description <u>Mature horse at moderate work</u> Body Weight <u>1100</u> lb

	(a) Digestible energy Mcal	(b) Crude Protein lb	(c) Calcium g	(d) Phos. g
A. *Daily Requirements* (based on horse activity)				
1. <u>Maintenance</u>	16.4	1.4	20.0	14.0
2. <u>Moderate work</u>	8.2	—	—	—
3. Total Daily Requirements	24.6	1.4	20.0	14.0
B. *Forage Ration* (between 1% to 2% of body weight)				
1. <u>19.25</u> lb of <u>Orchardgrass hay</u>	14.8	2.0	32.0	20.0
2. _____ lb of _____				
3. Total Nutrients from Forage	14.8	2.0	32.0	20.0
C. Remaining, to be supplied from concentrate ("Total Daily Requirements" less "Total Nutrients from Forage"). (A.3. − B.3.)	9.8	0	0	0

D. Amount of Concentrate Required. Assume a typical concentrate will contain approximately 1.35 Mcal/lb DE (air-dry basis) (C.a. ÷ 1.35 = lb)

<u>9.8</u> Mcal DE needed ÷ <u>1.35</u> Mcal/lb
DE = <u>7.26</u> lb

E. Percent Crude Protein Needed in Concentrate Mix. Divide CP to be supplied from concentrate (C.b.) by concentrate required (D.) times 100 = %CP in concentrate mix. (C.b. ÷ D. × 100 = %CP)

<u>0</u> lb CP needed ÷ _____ lb mix
× 100 = _____ % CP

F. *Concentrate*				
1. <u>7.26</u> lb of <u>Oat grain</u>	9.8	0.8	2.6	11.2
2. _____ lb of _____				
3. _____ lb of _____				
4. Total Nutrients from Conc.	9.8	0.8	2.6	11.2
G. Total Ration Nutrients (B.3. + F.4.)	24.6	2.8	34.6	31.2

TABLE 11–1. Nutrient Requirements of Horses, 1100-lb Mature Body Weight (b.w.)[1]

	Typical body weight lb	Dig. energy Mcal per 100 lb b.w.	Dig. energy Mcal Daily	TDN, lb per 100 lb b.w.	TDN, lb Daily	Crude protein %	Crude protein Daily lb	Calcium %	Calcium Daily g	Phosphorus %	Phosphorus Daily g	Vitamin A, IU[4] per lb diet	Vitamin A, IU[4] per lb Daily	Typical Daily Feed (air dry) lb
Mature horses, maintenance	1100	1.5	16.4	0.75	8.2	8.5–10.0	1.4	0.30	20	0.20	14	1500	30,000	19
Mature working horses (per hour above maint.)						8.5–10.0	2	0.30	2	0.20	2	1500	2	2
• light work		0.4	4.1	0.18	2.0									
• moderate work		0.7	8.2	0.37	4.1									
• intense work		1.5	16.4	0.75	8.2									
Gestation (first 8 mo.)		1.5	16.4	0.75	8.2	8.5–10.0	1.4	0.30	23	0.20	14	1500	30,000	19
Gestation (last 3 mo.)		1.7	18.8	0.85	9.4	10.0–11.0	1.7	0.50	36	0.35	23	1500	30,000	19
Lactation (first 3 mo.)		2.6	28.3	1.28	14.1	13.0–14.0	3.0	0.50	56	0.35	42	1250	30,000	26
Lactation (after 3 mo.)		2.2	24.3	1.11	12.2	12.0–13.0	2.4	0.45	41	0.30	34	1250	30,000	24
Foal (2–5 mo.)	300–400	3.6	13.7	1.80	6.8	16.0–18.0[3]	1.6[3]	0.80	33	0.60	20	900	10,000	11
Weanlings (5–12 mo.)	500–700	2.8	16.4	1.40	8.2	14.0–16.0[3]	1.7[3]	0.70	34	0.50	25	1100	14,000	13
Yearling (12–24 mo.)	700–900	2.5	19.7	1.22	9.8	11.0–13.5	1.7	0.50	31	0.40	22	900	14,000	16
2 yr. old: show and performance	1000–1100	1.8	17.9	0.90	9.0	10.0–11.0	1.4	0.45	25	0.35	17	1600	30,000	18

1. For each 100 pounds above or below 1100 mature body weight, add or subtract 7% to this amount.
2. Daily amount will vary with level of dietary intake.
3. High-quality protein recommended (minimum of 0.65% dietary lysine).
4. The following vitamins may be included in certain diets: Vitamin D, 125 IU/lb. diet; Vitamin B, complex mixture.

TABLE 11–2. Suggested Levels of Minerals to Use in Horse Rations

Mineral element	Per Cent	PPM
Calcium*	0.3–1.0	—
Phosphorus*	0.2–0.8	—
Salt	0.2–0.5	—
Iron	—	50
Copper	—	10
Iodine	—	0.1
Zinc	—	40
Manganese	—	40
Cobalt	—	0.1
Selenium	—	0.1
Sulfur	—	0.15

*Level dependent upon function or use of horse. See Table 11–1.

TABLE 11–3. Suggested Daily Supplemental Vitamins*

Vitamin	Early Growth	Training and Heavy Use	Stress or Breeding Season
Vitamin A, IU	5000	10,000	20,000
Vitamin D, IU	500	1,000	2,000
Vitamin E, IU	50	100	200
Thiamine, mg	8–12	20–24	25–35
Riboflavin, mg	10–20	30–40	50–60
Niacin, mg	20–30	100	150–180
Pyridoxine, mg	3–6	10–12	12–18
Pantothenic acid, mg	10–12	24–48	50–70
Choline, mg	100–300	400–600	700–800
Folic acid, mg	4–6	10–12	15–30
Vitamin B_{12}, mcg	25–30	110–120	170–180

*Variations in feeding rate due to size, growth, rate or degree of use will generally correspond to the suggested range in supplementation level.

TABLE 11–4. Suggested Rations for Horses

Following are rations suggested for use by horse owners. The daily ration should consist of forage (hay or pasture) plus a concentrate (grain) mixture, as needed, depending on the size, condition and function of the horse (column A). Total daily consumption of these ingredients (air-dry basis) will range between 2% and 3% of the body weight (column B). Forage can be grass, grass-legume mixture, or legume and may average low (10%) to high (17%) in crude protein content. Six suggested concentrate (grain) mixtures are shown in column C with three to be fed with a low-protein forage (mixtures 1, 2 and 3) and three to be fed with a high-protein forage (mixtures 4, 5, and 6). Each set of three concentrate mixtures are of similar nutrient composition when used as directed. Note footnotes to column C for additional information.

In addition to providing adequate feed, owners should see that horses are healthy and free from internal parasites, that feeds are high quality and mold free, and that changes in feed be made gradually.

| A. Kind of Horse | B. Daily Allowance | C. Suggested concentrate (grain) mixtures[a, b, c] fed with different quality forages (as-fed basis). With all mixtures and for all classes and ages of horses, provide free access in separate containers to plain loose salt and a mineral mixture containing equal parts of trace mineralized salt, dicalcium phosphate, and steamed bone meal. |

	Feed with low-protein forage[d]			Feed with high-protein forage[d]		
Ingredients	1	2	3	4	5	6
Oats, crimped	55.0	38.0	35.0	78.0	55.0	46.0
Corn, rolled	—	15.0	15.0	—	21.0	20.4
Soybean meal, sol.	—	—	25.7	—	—	10.0
Dried milk by-product[e]	10.0	10.0	10.0	10.0	10.0	10.0
Complete supplement[f]	35.0	37.0	—	10.0	12.0	—
Alfalfa meal	—	—	5.0	—	—	5.0
Molasses (liquid)	—	—	5.0	—	—	5.0
Limestone	—	—	0.6	—	—	—
Dicalcium phosphate	—	—	2.2	2.0	2.0	2.6
Trace mineral salt[g]	—	—	.5	—	—	.5
Vitamin premix[h]	—	—	.5	—	—	.5
	100.0	100.0	100.0	100.0	100.0	100.0

Calculated nutrient analysis

	1	2	3	4	5	6
Protein, %	21.7	21.7	21.7	15.9	15.8	16.2
Calcium, %	1.10	1.14	1.04	.89	.93	.87
Phosphorus, %	.87	.89	.87	.90	.92	.88
Dig. energy, Mcal/lb	1.33	1.36	1.36	1.34	1.38	1.35
TDN, %	66.8	68.1	68.3	67.4	69.3	68.0

Foals, creep and starting (weighing 100 to 450 lb)

At 1 to 4 months:
½ to ¾ lb grain per 100 lb body wt

At 5 to 6 months:
1 to 1¼ lb grain per 100 lb body wt, together with a quantity of hay within same range for each period

461

A. Kind of Horse **B. Daily Allowance**

C. Suggested concentrate (grain) mixtures[a,b,c] **fed with different quality forages (as-fed basis). With all mixtures and for all classes and ages of horses, provide free access in separate containers to plain loose salt and a mineral mixture containing equal parts of trace mineralized salt, dicalcium phosphate, and steamed bone meal.**

Weanlings
(450 to 750 lb)

1 to 1½ lb grain per 100 lb body wt, together with a quantity of hay within same range

Ingredients	Feed with low-protein forage[d]			Feed with high-protein forage[d]		
	1	2	3	4	5	6
Oats crimped	70.0	45.0	40.5	80.0	50.0	41.0
Corn, rolled	—	25.0	25.0	—	28.0	30.0
Soybean meal, sol.	—	—	21.0	—	—	16.0
Complete supplement[f]	30.0	30.0	—	20.0	22.0	—
Alfalfa meal	—	—	5.0	—	—	5.0
Molasses (liquid)	—	—	5.0	—	—	5.0
Limestone	—	—	.5	—	—	.2
Dicalcium phosphate	—	—	2.0	—	—	1.8
Trace mineral salt[g]	—	—	.5	—	—	.5
Vitamin premix[h]	—	—	.5	—	—	.5
	100.0	100.0	100.0	100.0	100.0	100.0

Calculated nutrient analysis

	1	2	3	4	5	6
Protein, %	18.5	17.7	17.7	16.2	15.8	15.9
Calcium, %	.85	.83	.83	.59	.63	.66
Phosphorus, %	.74	.72	.74	.60	.62	.69
Dig. energy, Mcal/lb	1.30	1.35	1.34	1.32	1.36	1.35
TDN, %	65.3	67.8	67.4	66.1	68.7	67.9

C. Suggested concentrate (grain) mixtures[a,b,c] fed with different quality forages (as-fed basis). With all mixtures and for all classes and ages of horses, provide free access in separate containers to plain loose salt and a mineral mixture containing equal parts of trace mineralized salt, dicalcium phosphate, and steamed bone meal.

A. Kind of Horse	B. Daily Allowance
Yearlings (700 to 1000 lb)	½ to 1 lb grain and 1 to 1½ lb hay per 100 lb body wt
Performance (900 to 1400 lb)	½ to 1¾ lb grain and 1 to 1½ lb hay per 100 lb body wt, depending on weight of horse and degree of work expended
Pregnant mares (900 to 1400 lb)	First half: 1½ to 2 lb hay per 100 lb body wt. Last half: ½ to 1 lb grain and 1 to 1½ lb hay per 100 lb body wt.
Lactating mares	1 to 1½ lb grain per 100 lb body wt, together with a quantity of hay within same range
Stallions in breeding season (900 to 1400 lb)	¾ to 1¼ lb grain per 100 lb body weight, together with a quantity of hay within same range

Ingredients	Feed with low-protein forage[d]			Feed with high-protein forage[d]		
	1	2	3	4	5	6
Oats, crimped	83.0	50.0	46.8	93.0	55.0	50.0
Corn, rolled	—	30.0	27.0	—	35.0	30.0
Soybean meal, sol.	—	—	13.5	—	—	8.0
Complete supplement[f]	17.0	20.0	—	7.0	10.0	—
Alfalfa meal	—	—	5.0	—	—	5.0
Molasses (liquid)	—	—	5.0	—	—	5.0
Limestone	—	—	.3	—	—	—
Dicalcium phosphate	—	—	1.4	—	—	1.0
Trace mineral salt[g]	—	—	.5	—	—	.5
Vitamin premix[h]	—	—	.5	—	—	.5
	100.0	100.0	100.0	100.0	100.0	100.0
Calculated nutrient analysis						
Protein, %	15.5	15.3	15.2	13.3	13.0	13.3
Calcium, %	.51	.57	.60	.26	.31	.39
Phosphorus, %	.56	.59	.60	.43	.45	.51
Dig. energy, Mcal/lb	1.32	1.37	1.34	1.34	1.40	1.35
TDN, %	66.3	69.0	67.6	67.1	70.3	68.0

TABLE 11–4 (continued). Suggested Rations for Horses

A. Kind of Horse	B. Daily Allowance	C. Suggested concentrate mixtures[a, b, c]
Mature idle horses; stallions, mares, and geldings (900 to 1400 lb)	1½ to 1¾ lb hay per 100 lb body weight	(With grass hay, add ½ to ¾ lb of a high-protein supplement[f] daily)

[a]Grains should be rolled or cracked to increase their bulkiness.
[b]Five to ten percent wheat bran could be substituted for part of the oats or corn.
[c]Five percent linseed meal could be substituted for part of the soybean meal.
[d]Assume low-protein forage averages 10% CP and high-protein forage averages 17% CP (air-dry basis).
[e]Dried skim milk or any similar milk-based product may be used as a source of high quality protein for the foal (25% to 30% crude protein).
[f]Any complete supplement containing 30% to 35% crude protein together with approximately 2.5% calcium and 1.5% phosphorus is suggested.

Sample supplement formula

Ingredient	IFN	Amount lb.		Calculated analyses	
Soybean meal (44% CP)	5-04-604	70.0		Crude protein, %	34.0
Molasses (liquid)	4-04-696	6.4		Calcium, %	2.6
Alfalfa meal	1-00-023	10.0		Phosphorus, %	1.6
Limestone	6-02-632	2.4		Digestible energy, Mcal/lb.	1.2
Dicalcium phosphate	6-01-080	6.2		TDN, %	60.0
Trace mineral salt[g]		2.5		Lysine, %	2.0
Vitamin premix[h]		2.5		Dry matter, %	90.1
		100.0			

[g]Any commercial trace mineralized salt for animal use is satisfactory. The mixture should contain approximately the following:

Zinc (Zn)	0.35%
Manganese (Mn)	0.28%
Iron (Fe)	0.175%
Copper (Cu)	0.035%
Iodine (I)	0.008%
Cobalt (Co)	0.005%

An example trace mineralized salt mixture:

Salt (NaCl)	97.39 lb.
Zinc sulfate (ZnSo₄)	0.86
Manganese Sulfate (MnSo₄)	0.77
Ferrous sulfate, heptahydrate (FeSO₄ · 7H₂O)	0.87
Copper sulfate (CuSO₄)	0.09
Ethylenediamine dihydriodide (EDDI)	0.01
Cobalt carbonate (CoCO₃)	0.01
	100.00 lb.

[h]The vitamin mixture should provide approximately the following minimum daily requirements per pound of final concentrate (grain) ration: vitamin A, 2000 I.U.; vitamin D, 200 I.U.; vitamin E, 20 I.U.; thiamine, 4.0 mg.; niacin, 4.0 mg.; riboflavin, 3.0 mg.; pantothenic acid, 3.0 mg.; and vitamin B₁₂, 5.0 mcg.

An example composition for 1.0 pound of a vitamin premix:

Vitamin A	400,000 I.U.
Vitamin D	40,000 I.U.
Vitamin E	4,000 I.U.
Thiamine	800 mg.
Niacin	800 mg.
Riboflavin	600 mg.
Panthothenic acid	600 mg.
Vitamin B₁₂	1 mg.
Carrier (finely ground corn or soybean meal)	?
	1.0 lb.

TABLE 11-5. Nutritional Diseases of Horses

Disease	Symptoms	Cause	Prevention	Treatment
Azoturia (Monday Morning Sickness) Excess urea or other nitrogen compounds in the urine.	Excessive sweating, nervousness, abdominal distress, stiff gait, reluctant to move. Coffee-colored urine.	True cause of Azoturia is not known. Characteristically it occurs in horses on high grain ration which have rested 1 or 2 days without feed intake reduced, then returned to work. There is evidence that selenium or tocopherol might be involved.	Reduce feed when horses are idle, exercise daily or turn in pasture.	Secure services of veterinarian (urgent).
Colic (Abdominal Pain)	Acute abdominal pain. Profuse sweating. Distended abdomen. Violent kicking and rolling.	Improper management, feeding, watering, working. Parasitism.	Proper feeding, watering and working.	Call veterinarian. Keep animal moving until help arrives.
Heaves	Nostrils will be dilated. Coughing and difficulty in breathing. Rapid inspiration, forced expiration with heave line along rib border.	Essential cause unknown, damaged or moldy feed usually incriminated. Excessive intestinal fermentation. Often follows respiratory disease such as distemper and pneumonia.	Feed good clean dust-free feed; complete pelleted feed. Moisture on feed helpful.	Mild cases progress to severe cases so all should be regarded as serious. Coughing aggravates problem so control the cough. Call veterinarian. Also follow prevention procedures.
Laminitis (Founder)	Lameness, particularly fore feet. Condition may be acute, or chronic. Acute cases are visibly sick. Show colic, perspiration, reluctance to move; hoof hot to touch; chronic cases show deep vertical cracks in hoof and heavy horizontal ridging. Hoof is excessively dry and brittle.	Injury to blood supply to horses hoof. May be caused by overloading the stomach with grain, lush pasture or water. Poor husbandry practices.	Prevent by providing adequate ration and observing good husbandry practices.	Call veterinarian. Application of cold substances to the feet is helpful first aid measure. Acute cases require urgent attention.
Moonblindness (Periodic Ophthalmia)	Hazy vision in one or both eyes which may affect animals intermittently.	Evidence points to lack of riboflavin in the diet. Micro-organisms have also been indicated.	Feed ration which contains at least 40 milligrams of riboflavin per day. Green grass and green leafy hay are excellent sources of riboflavin.	Call veterinarian. Follow prevention procedure.
Rickets	Enlargement of the ends of the long bones, particularly around hock and knee joints. Animals may show pain when moving about.	Imbalance of calcium, phosphorus and/or Vitamin D.	Supply the correct ratio of these three important nutrients.	Same as prevention. If disease is too far advanced animal may not recover fully.
Urinary Calculi (Water Belly)	Frequent attempts to urinate, dribbling or stoppage of the urine. Urine is frequently blood-tinged. Males only usually affected.	Not completely understood. Calcium-phosphorus ratio, high potassium intake appears to increase incidence. Lack of salt.	Incidence appears to be reduced by proper calcium-phosphorus ratio and inclusion of 1 to 3 % salt in the concentrate ration.	Call veterinarian. Also follow prevention procedure.

Study Questions and Problems
Chapter 11

I. True or False Questions

1. Protein requirements for the horse are greatly increased with work.
2. Moldy or dusty hay may cause colic and heaves in horses.
3. Grass hays fed to horses are usually less dusty and less likely to be moldy than legume hays.
4. Wheat or milo should be ground fine when fed to horses.
5. Wheat bran is very valuable in horse rations for its high energy and amino acid content.
6. Milk proteins are often used in mare gestation rations because of their excellent quality protein.
7. An 1100-lb gelding will consume up to about 22 lb of dry matter daily.
8. Urinary calculi may be a common problem in lactating mares fed high grain diets.
9. The stomach of the horse is small in comparison to other species and feed passes through it rapidly.
10. Volatile fatty acids are produced in the cecum of the horse through microbial fermentation and can be used by the horse to meet its protein needs.

II. Multiple Choice (select the one best answer)

1. Lameness and injury to the blood supply to horses hooves is called
   a. Bloat.
   b. Colic.
   c. Rickets.
   d. Laminitis.
   e. Heaves.
2. Perhaps the greatest nutritional need by horses undergoing work is
   a. Protein.
   b. Vitamins.
   c. Minerals.
   d. Energy.
   e. All of the above.
3. Which amino acid is most limiting in the diet of the developing foal?
   a. Lysine
   b. Methionine
   c. Tryptophan
   d. Arginine
   e. Valine
4. Which of the following would be considered the "cleanest" hay for feeding horses?
   a. Bromegrass
   b. Alfalfa
   c. Red clover
   d. Sudangrass
   e. All about the same
5. Foal diets often contain __?__ because of the content of excellent quality protein.
   a. What bran
   b. Dried whey
   c. Rolled oats
   d. Linseed meal
   e. Dehydrated alfalfa
6. Which of the following antibiotics is approved as a feed additive for stimulating growth and improving feed efficiency in horses under one year of age?
   a. Chlortetracycline
   b. Oxytetracyline
   c. Tylosin
   d. Lyncomycin
   e. None of the above

7. Which of the following feeds would be best to provide bulk in the diet at foaling?
   a. Rolled oats
   b. Wheat bran
   c. Timothy hay
   d. Flaked corn
   e. Dried whey
8. An 1100-lb gelding should consume a minimum of ___?___ lb of roughage daily.
   a. 5.5
   b. 11.0
   c. 22.0
   d. 33.0
   e. 44.0
9. Mature horse maintenance diets should contain about ___?___ % crude protein.
   a. 5% to 8%
   b. 8% to 10%
   c. 11% to 13%
   d. 14% to 16%
   e. Over 16%
10. Horses should be fed small quantities of feed at several feedings because of the relatively small capacity of the
   a. Stomach.
   b. Small intestine.
   c. Cecum.
   d. Large intestine.
   e. Answers a and c.

III. Matching or Identification

Match the following statements with the most correct feedstuff for horses.

a. Dried skim milk	f. Salt
b. Sudangrass pasture	g. Wheat bran
c. Linseed meal	h. Milo
d. Meat meal tankage	i. Molasses
e. Dehydrated alfalfa meal	j. Timothy hay

1. Not palatable for horses.
2. May cause cystitis or be toxic to horses.
3. A bulky and laxative feed.
4. A traditionally "clean" hay fed to horses.
5. Excellent source of amino acids for foal diet.
6. Close substitute for green pasture.
7. Often fed "free-choice."
8. Used to "sweeten" the horse feed.
9. Should be coursely processed and mixed with a bulky feed.
10. Produces bloom and luster in the hair coat.

IV. Short Answer

1. Why have timothy hay and oats been popular as horse feeds?
2. What is the most desirable manner to prepare grains for horse diets?
3. During which phase(s) of the horse life cycle is protein nutrition most important?
4. What are some advantages and disadvantages in feeding alfalfa hay to horses?
5. What are advantages and disadvantages of pelleting horse diets?

V. Problems

1. A working quarter horse requires 23 Mcal of digestible energy daily. If the horse is consuming 16 lb of bromegrass hay (0.70 Mcal/lb) per day, how may pounds of oats (1.3 Mcal/lb) must be supplemented to meet the energy needs?

2. Formulate 100 lb of concentrate mix for weanling horses to contain 16% crude protein and 1.4 Mcal/lb digestible energy. Use some combination of the following three feeds.

   Oats   (11% CP, 1.3 Mcal/lb DE)
   Corn   (9% CP, 1.5 Mcal/lb DE)
   Soybean meal   (45% CP, 1.5 Mcal/lb DE)

3. An 1100-lb mature gelding undergoing 2 hours of light work daily has the following nutrient requirements.

   Crude protein   2 lb
   Digestible energy 22 Mcal
   Calcium   30 g
   Phosphorus   20 g

   Formulate a suitable daily ration for this horse using some combination of the following feeds.

Feedstuff	CP, %	DE Mcal/lb	Ca, %	P, %
Orchardgrass hay	10	0.8	.35	.30
Corn grain	9	1.5	.03	.26
Soybean meal	45	1.5	.30	.63
Dicalcium phosphate	0	0	22.0	19.0
Limestone	0	0	34.0	0

Notes

Notes

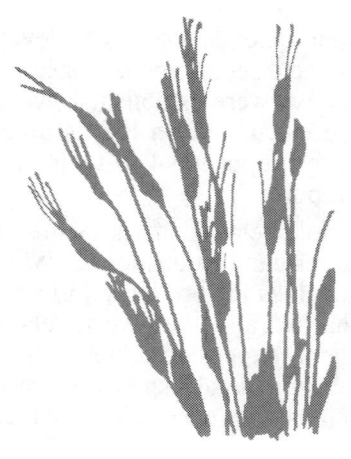

12
Poultry Nutrition and Feeding

Chapter Goals

- Outline and discuss a life cycle chicken and turkey feeding program for meat and egg production.
- Identify specific nutrient needs or additives used within the life cycle.
- Discuss the feeds commonly used in poultry diets and the types of formulations within each part of the poultry life cycle (chickens and turkeys).
- Briefly outline specific aspects of feeding ducks and geese.
- Formulate a diet (or diets) for poultry (either chickens or turkeys).

The following material concerning poultry nutrition and feeding was compiled from recommendations by the National Research Council (*Nutrient Requirements of Poultry*, Publication ISBN 0-309-03486-8, Committee on Animal Nutrition, National Academy of Sciences—National Research Council, Washington, D.C., 1984). Additional and more detailed information may be obtained from publications of various land grant universities and colleges.

Nutrient requirements and feed allowances of chickens and turkeys are given in Tables 12–1 to 12–8. These tables present the requirements in accordance with customary practice, in terms of percentages or units per kilogram of feed. A more valid mode of presentation, from the physiological standpoint, would be units per bird per day, but such values almost inevitably have to be converted to units per unit weight of feed because poultry are generally fed in groups, ad libitum.

I. Nutrients for Poultry

A. *Energy, Protein and Amino Acids*

It can be argued that it is impossible to set an energy requirement in terms of kilocalories per kilogram of diet because birds adjust their feed intake to achieve the necessary daily ration of energy. On the other hand, the protein requirement is one of the most important criteria upon which any feed formula is based, and if it is to be specified, the dietary energy level must be specified because it is essential to maintain the proper ratio of protein to energy in poultry diets. Some variability in optimal protein:energy ratios must be recognized. Some combinations of fats and carbohydrates have a protein-sparing effect. Also, protein:energy ratios may be altered deliberately in order to influence fat deposition. When protein levels are low in relation to energy and with ad libitum feeding, fat deposition is markedly increased; with higher protein levels, less fat is deposited. Increasing the protein level greater than that required for maximum growth rate reduces fat deposition still further.

It is evident that the protein requirement can be defined accurately only in relation to energy concentration, degree of fat deposition and a limited range of nutrient combinations, using those purified and practical feedstuffs that have been subjected to experimental study. For practical purposes in farm operations and feed manufacturing, sufficient work has been done with growing chicks to define, with reasonable accuracy, the minimum protein requirement for maximum growth rate in relation to energy level. Similar estimates can be made from more limited data for laying hens and growing turkeys.

The physiological relationship between levels of energy and protein extends also to the levels of the essential amino acids. In tables 12–1, 12–2 and 12–3 the amino acid requirements for broilers, for replacement pullets, for laying and breeding hens, and for turkeys were established by direct experimentation. Some of the figures in the other columns were calculated on the assumption that amino acid requirements of growing chickens and turkeys are proportional to their protein requirements. This assumption is supported by several research reports.

The figures in each column of tables 12–1 and 12–2 are closely interrelated. If ratios are more important than absolute values, the validity of any given set of figures can be questioned. Why not for example, increase by 10% all energy, protein and amino acid values for broilers? In fact, this can be done very successfully in the case of broilers, except that to raise the levels, relatively expensive fat and protein supplements must be used; to lower them, expensive diluents must be used. For layers and growing pullets, there are other deterrents to a general increase in nutrient concentration. Such an increase leads to excessive fat deposition in layers, causes feather-picking and risks cannibalism in pullets and layers.

Ideally, the diet should provide energy, total protein and the 13 essential amino acids in the amounts specified in Tables 12–1 and 12–2. Obviously, this is too much to expect because of the limited number of protein supplements, each of which has its own fixed ratio of essential amino acids. In the United States grains and soybean meal are the major sources of energy and protein for most poultry feeds. In such feeds methionine is the first amino acid to become limiting. When a considerable amount of the soybean meal is replaced by another plant protein supplement such as cottonseed meal, lysine is likely to become the first limiting.

There are two important relationships between individual amino acids and vitamins. Methionine can spare choline to some degree and tryptophan can spare niacin. But making either of these substitutions is poor management because the two vitamins can be provided more economically than the two amino acids.

It is important to note that there are feeding programs designed to restrict energy, protein and amino acids to levels that support less than maximal weight gains. For example, with pullets of modern broiler strains, which not only grow at a rapid rate but also mature sexually at an early age, it is necessary both to retard growth and to delay the onset of sexual maturity if egg production and the production of viable chicks are to be optimal. There are several feeding programs used to retard growth.

1. From the beginning of the seventh or ninth week, limit the total feed allowed per bird per day. (The allowance usually accepted is 70% of the feed that would be consumed by a bird fed on an *ad libitum* basis.) Continue this program until the birds are placed on the standard laying mash at the beginning of the twenty-third week.
2. Using the same pullet developer formula, feed the birds on the "skip-a-day program" from the seventh or ninth week to the twenty-third week. This feeding program allows the pullets all the feed they will consume on one day and only 2 lb of grain per 100 birds on the alternate day.
3. Use a diet containing only 10% to 10.5% protein. Feed the pullets this protein level from the beginning of the seventh or ninth week until they are placed on the laying mash.
4. Use a diet that contains 0.4% to 0.45% lysine and 0.6% to 0.7% arginine after the seventh or ninth week. The protein content of this diet is in the range of 12.5% to 13%. Apparently, this diet is effective because the imbalance of amino acids, other than the two mentioned, depresses appetite.

The programs listed above are all being used in flocks of growing pullets.

B. *Vitamins*

Vitamin requirements are given in Tables 12–1, 12–2 and 12–3 in terms of milligrams per kilogram of diet, except in the case of vitamins A, D and E, which are given in International Units (IU) or International Chick Units (ICU).

Requirements for vitamin A are expressed in United States Pharmacopeia (USP) units or IU per kg of feed. The international standards for vitamin A activity based on vitamin A and beta-carotene are as follows: 1 IU of vitamin A = 1 USP unit = vitamin A activity of 0.300 μg of crystalline vitamin A alcohol corresponding to 0.344 μg of vitamin A acetate or 0.550 μg of vitamin A palmitate. Beta carotene is the standard for provitamin A. One IU of vitamin A activity is equivalent to 0.6

μg of beta carotene; 1 mg of beta-carotene = 1,667 IU of vitamin A. International standards for vitamin A are based on the utilization of vitamin A and beta carotene by the rat. In the chicken as in the rat, 0.6 μg of beta carotene is equivalent to one USP unit of vitamin A, except when the carotene intake provides vitamin A activity greatly in excess of the requirements.

Requirements for vitamin D are expressed in ICU, which are based on the activity of vitamin D_3. Birds use vitamin D_3 from fish oils and irradiated animal sterol effectively, but they do not use vitamin D_2 from irradiated ergosterol as efficiently as do rats and other mammals. One ICU of vitamin D_3 is defined as the vitamin D activity of 0.025 μg of pure vitamin D_3. The potency of an unknown sample, if it is to be expressed in ICU, must be measured with chicks rather than with rats because of the different response of the two species to vitamin D_2. Vitamin D_3 produced by irradiation of 7-dehydrocholesterol may be more efficiently utilized by growing chicks and turkey poults than vitamin D_3 from fish oils. This difference in the efficacy of vitamin D_3 from fish oils and irradiated animal sterols is particularly apparent when the dietary level of inorganic phosphorus is low. Turkeys are especially sensitive to the type of phosphorus in the diet in relation to the type of Vitamin D. These factors were considered in arriving at the listed requirements for Vitamin D, which are based on suggested minimum quantities of inorganic phosphorus and calcium. With the suggested amounts of inorganic phosphorus and calcium, there should be no difference in the efficacy of vitamin D_3 derived from fish oils and that derived from irradiated animal sterols.

Requirements for vitamin E are highly variable depending on the level of fat in the diet, the level of selenium, and the presence or absence of antioxidants other than vitamin E. Chicks have been successfully reared on purified low-fat diets that do not contain any detectable Vitamin E. The requirement figure specified for chicks and poults applies to diets containing the amounts of selenium and energy indicated in Tables 12–1, 12–2, and 12–3. One IU of vitamin E is equivalent to 1 mg of synthetic DL-α-tocopheryl acetate.

Formulators of poultry feeds have recently shown concern about dietary levels of many vitamins formerly thought to be amply distributed in feeds and feed ingredients. For example, folacin and biotin are now added as supplements to some turkey diets to prevent vitamin deficiency.

On the basis of recent work, it is possible to state a niacin requirement for laying and breeding hens. The requirement is so low that it will always be exceeded by natural feed ingredients.

The growing chicken can use betaine interchangeably with choline to meet its need for methylating agents, but betaine cannot replace choline in preventing perosis. Betaine is widely distributed in practical feedstuffs and may be important in sparing choline. Likewise, vitamin B_{12} has been shown to reduce the requirement of the chick for choline. The figures for choline requirement given in the tables are applicable to diets that contain the specified levels of vitamin B_{12}.

The type of carbohydrate in the diet influences the requirement levels of some water-soluble vitamins. This effect has been investigated in some detail with folacin, and it is possible to specify the requirement level for folacin in chick diets based on either starch or sugar.

C. *Minerals*

The calcium requirement of laying hens is one of the more difficult to define. The problem cannot be solved by simply adding a generous amount because excess calcium interfers with utilization of phosphorus, magnesium, manganese and zinc, and it is likely to decrease the palatability of the diet. The figure of 3.25% is believed correct for conditions usually prevailing in the United States. However, hens that are subjected daily to a temperature of 32° C (90° F) or more for several weeks are likely to require a calcium level of 3.5% to 3.75%. There is no reason to believe that growing pullets fed calcium in amounts that are much too large or much too small are likely to experience difficulties in calcium metabolism after the onset of egg production.

The need for a minimum level of inorganic phosphorus is indicated in the NRC Poultry publication. This need is based on the generally greater availability of inorganic phosphorus than of phytin phosphorus. Inorganic phosphorus supplements have different degrees of availability, and it is important that the minimum level of phosphorus be provided in readily available form.

D. *Unidentified Nutrients*

This discussion specifies the young chick's quantitative requirements for 39 nutrients, including metabolizable energy. It appears, however, that this list is still incomplete. There is evidence of four distinct unidentified growth factors in dried whey, marine and packinghouse byproducts, distiller's solubles, and green forages, and of at least one unidentified hatchability factor in fish solubles and green forages.

E. *Additives*

Consider previous discussion in Chapter 6.
1. Antibiotics—since 1950 several antibiotics have become important additives in broiler and market turkey feeds because they increase growth rate and efficiency of feed utilization. Also, egg production is frequently improved by dietary supplementation with antibiotics. Depending on the antibiotic used, subtherapeutic levels of 2 to 10 mg/kg of feed are needed to be effective.
2. Antioxidants—compounds used to prevent oxidative rancidity in fat. BHT, BHA or ethoxyquin (see also Chapter 6) are used at a level of 0.25 lb per ton of feed to stabilize feed fats against rancidity.
3. Grit—refers to hard insoluble or soluble particles which remain trapped in the thick-muscled gizzard to facilitate grinding of feed materials ingested by the bird. Oyster, clam or coquina shells, limestone, gravel, pebbles or granite products are used for grit. It is definitely needed when birds consume whole grains or coarse, fibrous feedstuffs. When mash or finely ground feeds are used, the value of grit is diminished.
4. Xanthophylls—xanthophylls or carotenoid additives are products which produce a deep yellow color in the beak, skin, shanks, feet, fat, and egg yolks of poultry. Many consumers believe that a deep yellow color of broiler skin/shanks and egg yolks is indicative of top quality. Thus, many producers add ingredients that contain xanthophylls to poultry rations. Alfalfa meal, yellow corn, and corn gluten meal are sources that are commonly used. Common dried algae and marigold petal meal are even richer sources of xanthophyll.

II. Types of Poultry Diets

Various types of poultry formulations are made for use in commercial feeding operations. Several of the major types are briefly outlined as follows.

A. *Starters*
1. Generally fed first 2 to 3 weeks to chickens and 2 to 4 weeks to turkeys.
2. Contain increased levels of all essential nutrients; contain approximately 23% to 28% crude protein for broiler chickens and turkeys, respectively. Leghorn-type pullet chicks are usually started on a diet containing around 18% CP and remain on this diet up to 6 weeks of age.
3. Include high levels of antibiotics to reduce early mortality and initiate more rapid early growth.
4. Contain suitable coccidiostat added at the level recommended by the manufacturer.

B. *Broiler Diets*
1. Fed as a complete feed to meat-type birds.
2. Similar to starter diets with the exception of higher energy and crude protein levels. Vitamin additions are also higher to meet the added requirements for growth under the stress conditions encountered in the average broiler operation.
3. May contain from 3% to 5% added fat to increase caloric content provided the protein content is adjusted to compensate for the fat addition and an optimum protein:calorie ratio maintained.
4. Fortified with antibiotics and should contain a coccidiostat.
5. Generally of two types:
   a. Broiler-grower diets designed to be fed 3 to 6 weeks of age; contain 20% to 22% crude protein.
   b. Broiler-finisher diets (withdrawal diets) designed to be fed from 6 weeks to market age; contain 18% to 20% crude protein.
6. May be fed in crumbles or pelleted form.

C. *Turkey Growing and Finishing Diets*
1. Fed as complete feed to meat-type birds.
2. Similar to broiler diets.
3. General types
   a. Birds 4–8 wk of age—25% to 26% crude protein.
   b. Birds 8–12 wk of age—22% to 23% crude protein.
   c. Birds 12–16 wk of age—19% to 20% crude protein.

      d. Birds 16–20 wk of age—16% to 17% crude protein.

      e. Birds over 20 wk of age—14% to 15% crude protein.

D. *Growing Rations and Developers for Leghorn-type Chickens*

    1. Designed to be fed to the replacement stock from 6 weeks to sexual maturity.

    2. May be of two types:

      a. Complete feed, mash or pelleted. Generally lower in protein than starter diets (14% to 16% CP for 6 to 14 weeks; 12% to 14% CP for 14 to 20 weeks).

      b. A mash concentrate designed to be fed with varying amounts of grain in order to meet the nutrient requirements of the birds.

E. *Laying Diets for Leghorn-type Chickens*

    1. Formulated to be fed to mature hens during egg production.

    2. Should contain about 14% to 16% crude protein.

    3. May be of two types.

      a. Complete feed, mash or pelleted.

      b. Laying mash concentrate to be fed with specified levels of grain or mixed with grain and soybean meal.

    4. The salt content of these feeds may be reduced in an effort to reduce the incidence of wet droppings.

    5. Cage fatigue is considered by some to be the result of a calcium deficiency. If this condition occurs, the calcium content of the diet may need to be increased or a calcium source could be provided on a free choice basis.

    6. Hens kept for the production of hatching eggs, should be fed a more highly fortified feed (especially vitamins) than hens kept merely for commercial egg production. Unless the levels of various nutrients are kept high, normal hatchability cannot be maintained.

    7. Increased levels of protein and vitamins may also be fed during periods of stress or slumps in egg production. A higher level of an antibiotic or a combination of antibiotics may also be included during these periods.

F. *Scratch Grains*

    1. Various grain mixtures which may be used in the concentrate and grain system of feeding.

    2. Usually fed from the 8th week to maturity.

III. Feeds for Poultry

The major ingredients utilized in poultry rations may be categorized as energy feedstuffs, protein supplements, mineral supplements and vitamin supplements.

A. *Selecting Feedstuffs*

Several factors should be considered in selecting feeds for use in poultry rations.

    1. Nutrient availability

      a. Fiber content.

      b. Fat content.

      c. Amino acid balance.

    2. Palatability

      a. Moisture content.

      b. Contaminants.

      c. Feed preparation (whole vs ground).

      d. Color or light reflections from grit.

    3. Content of growth inhibitors or undesirable chemicals or pigments

    4. Cost and market availability of feedstuffs

B. *Energy Feeds*

    1. Grains—corn is the most important grain used in poultry diets; milo, wheat, barley and oats are also used extensively, however, may be inferior to corn in relative efficiency.

    2. Grain by-products—include a great array of milling by-products (for example, the corn gluten and bran feeds and the wheat processing byproducts); and brewery byproducts which may contain unidentified growth factors.

3. Molasses—used as a good source of energy but does have an adverse laxative effect and should be limited to not more than 2% of the diet.
4. Vegetable and animal fats—used extensively as energy sources; fats also reduce feed dustiness; increase palatability, and improve texture and appearance of the feed.

C. *Protein Supplements*
   1. Plant protein supplements
      a. Soybean meal—the most widely used protein supplement in poultry diets because of its ability to furnish the essential amino acids, high digestibility of protein, and freedom from toxic or undesirable substances.
      b. Cottonseed meal—care should be exercised when cottonseed meal is used in poultry diets especially for laying birds. Gossypol found in cottonseed meal produces a mottling and greenish cast to egg yolks. Cyclopropenoic fatty acids, also present in cottonseed meal, impart a pink color to egg whites. Because of these compounds, cottonseed meal is generally not used in laying rations. Cottonseed meal may be used to replace up to 50% of the soybean meal in grower poultry diets.
      c. Linseed meal—can be used in limited amounts, but may depress growth and cause diarrhea if fed at high levels. Linseed meal should not exceed 3% to 5% of the poultry diet.
      d. Alfalfa meal and corn gluten meal are used extensively, both for their high content of carotenoids-compounds that impart a deep yellow pigmentation to the skin. Both should be limited to not more than 10% of the poultry diet.
   2. Animal protein supplements
      a. Fish meals—frequently used at 2% to 5% of the poultry diet as a source of unidentified factors in addition to their favorable amino acid profile. However, since fish meals are high in fat, they tend to create a fishy flavor in meat and eggs when used in larger amounts. The high cost of fish meal may also restrict its use.
      b. Meat products (animal byproducts, poultry meal, blood meal, hydrolyzed poultry feather)—often are economically priced so that they may replace an equal amount of soybean meal protein up to about 10% of the diet. These feeds are also excellent sources of calcium and phosphorus.

D. *Mineral Supplements*
   1. Calcium—common supplements used in poultry diets are ground limestone, crushed oyster shells or oystershell flour, bone meal and dicalcium phosphate.
   2. Phosphorus—common supplements used in poultry diets are bone meal, dicalcium phosphate, deflourinated rock phosphate, monosodium phosphate, and rock phosphate.
   3. Salt—commonly added to poultry diets at a 0.2% to 0.5% level. Too much salt will result in increased water consumption and wet droppings.

E. *Vitamin Supplements*
   In past years a wide variety of feedstuffs were added to poultry diets primarily for their vitamin content; in modern poultry formulation, special vitamin premixes often represent the common approach to providing vitamin needs for poultry.

F. *Feed Preparation*
   1. Most poultry feeds are prepared in one of the following forms.
      a. Mash—grind medium-fine.
      b. Pellets—composed of mash feeds that are pelleted.
      c. Crumbles—produced by rolling pellets.
   2. Pellets or crumbles cost slightly more than the same ration in mash form, but have the following advantages.
      a. Reduced feed wastage.
      b. Reduced sorting.
      c. Readily adapted to automatic equipment.
      d. Less feeder and storage space.
      e. Possible improved palatability.

Caged layers in a light and temperature controlled environment. Note feed trough and egg collecting rack in front of each row of cages.

IV. Feeding Programs

A. *Broilers*

1. Broiler chicks are fed *ad libitum* for 42 to 56 days to an average weight of 4 to 5 pounds.
2. Feed conversion is usually about 2.0 and feed is the largest cost item, representing 60% to 75% of total broiler production cost.
3. Suggested broiler diets for a 3 stage feeding program (starter, grower and finisher) are shown in Table 12–9. Protein requirements vary with age, thus multiple stage diets are usually employed. In a three-stage program the starter feeds should be used for the first 2 to 3 weeks, the grower for about 2 weeks, and the finisher for the remainder of the feeding period.
4. Broiler diets often contain a coccidiostat for the control of coccidiosis and an antibiotic as a growth stimulant.

B. *Replacement Pullets*

1. Suggested all mash rations for replacement pullets are shown in Table 12–10.
2. The replacement pullet feeding program is generally divided into three stages.
   a. Starter—fed from 0 to 6 weeks of age; contains 18% to 20% crude protein and about 3000 kcal/kg ME.
   b. Grower—fed from 6 to 12 weeks of age; contains 14% to 16% crude protein and about 3000 kcal/kg ME.
   c. Developer—fed from about 12 weeks of age until lay (approximately 20 weeks); contains 12% to 14% crude protein and about 3000 kcal/kg ME.
3. The Leghorn-type pullets are seldom restricted-fed during the growing period because varying the lighting during growth from 6 to 20 weeks of age can be used to control feed consumption and sexual development. However, pullets of heavy breeds tend to accumulate excessive amounts of body fat; so, it is common practice to restrict the feed intake of these birds.
   a. A most effective feed restriction program is to daily feed a controlled level of a well-balanced ration. This program requires adequate feeder space and a rapid even distribution of the diet so that all birds can eat at the same time.
   b. An alternate control feeding technique is to use a skip-a-day feeding program. This program complimented by adequate feeder and water space will produce a more uniform flock. The selected daily feed quantity is doubled and fed on alternate days. On the nonfeed day, 2 lb oats per 100 birds or similar scratch grain may be used by scattering on the litter.
4. During the 20-week growing period, rate of feed consumption by each pullet will increse tenfold; from less than 0.02 lb per day to approximately 0.20 lb per day when she begins to lay eggs.
5. When pullets start into egg production, their feed intake should increase. Sometimes it is necessary to reduce the energy concentration of the feed at 18 to 19 weeks of age in order to incresae the pullet feed consumption.

6. When the pullet flock is laying about five eggs per 1000 birds, the birds should be placed on a "prelay" program in which the feed contains about 2% or more calcium. By the time the pullets are up to 5% egg production, they should be placed on a regular layer feed program.

7. Feed represents about 60% of the cost of raising replacement pullets.

C. *Feeding Layers*

1. Table 12–11 provides an example of a complete diet for laying hens. This diet was formulated for White Leghorn type laying hens kept for market egg production. Increased levels of vitamins (A, D, E, riboflavin, pantothenic acid, niacin and B_{12}) and the trace minerals manganese and zinc would be required if eggs are to be used for hatching by the breeding flock.

2. Many research reports have shown that the White Leghorn laying hen requires approximately 18 g of protein per bird daily to support optimum egg production. Thus, hens receiving a 15% protein ration (Table 12–11) must consume about 25 to 26 lb of feed per 100 birds per day in order to obtain 18 g of protein per hen.

3. Methionine is the first limiting amino acid in the laying hen diet based on corn and soybean meal. Adequate dietary methionine can be provided by increasing the protein level in the ration, or by using feeds high in methionine or its analog to the diet. In recent years, the latter approach has been most economical.

4. Calcium is important for proper egg shell formation. The calcium requirement will vary with the age of the bird, environmental temperature, rate of lay and egg size. A general recommendation for laying hens is a daily calcium intake of 3.4 grams. After 40 weeks of age, this intake should be increased to 3.8 grams. The practice of supplying one half to two thirds of the calcium as large particle size limestone or oyster shell has been reported to improve egg shell quality.

5. It is also necessary to assure that the phosphorus level in the diet is not excessive since excess phosphorus tends to inhibit calcium absorption from the GI tract. A level of 0.3% to 0.4% available phosphorus, which is equivalent to about 0.5% to 0.6% total phosphorus, is adequate.

6. The laying hen diet also needs to be adequate in vitamin D_3, since this nutrient is directly involved with calcium metabolism.

7. Feeding grit to chickens helps them grind coarse feedstuffs. The result is a slight improvement in feed efficiency. Grit can be fed in special feeders approximately every 3 weeks, mixed in a complete feed at the rate of 0.25% of the diet; or sprinkled on top of the feed at a rate of 5 lb per 1000 laying hens every week. When mash or finely ground feeds are used, the value of grit is diminished.

8. A relatively common feeding concept for laying hens is that of phase feeding. The main objective of phase feeding is to reduce the waste of nutrients caused by feeding more than a bird actually needs under different sets of conditions. Pullets that are coming into egg production should be fed a protein level varying from 17% to 19%. This may be lowered to 15% to 16% after 3 to 4 months of lay, or when the pullet has attained the adult weight. Decreasing the protein level below 15% is not advocated in practical large scale operations.

Another area where phase feeding of laying hens is indicated is that of varying the protein with temperature and season. The feed intake of laying hens decreases as the temperature increases above 85° to 90° F. It has been fairly well established that the chicken eats to satisfy its energy requirement. If the energy requirement is met with a small amount of feed, a deficiency of protein and of essential amino acids may become apparent in decreased egg production and lowered egg size. It may be necessary to increase the protein level to 18% or possibly 20% where daily temperature exceeds 100° F for an extended period of time.

9. When young pullet flocks appear to have stopped their production increase or have plateaued (leveled off for several consecutive days), they sometimes will respond when "challenged" with additional feed amounts. Challenge the flock (to lay more eggs) with approximately 2 additional pounds of feed per 100 birds added to their daily feed allowance. If the flock does not respond with increased production by the fourth day, return to the amount fed prior to the challenge period. This technique can be repeated as often as necessary, depending on the flock response and if production and other conditions warrant it.

10. Pullet flocks that have peaked in egg production and have begun a gradual decline in lay sometimes will produce more efficiently on less feed. When the flock has passed peak and is showing a normal decrease (4% to 6%), the following feeding procedure may be used.

a. Challenge the flock to produce more efficiently by reducing the daily feed allowance by ½ lb per 100 birds for a period of 3 to 4 days. If this causes an abnormal drop in egg production, return immediately to the prior feed amount.

b. As production continues to decline normally, this challenge technique may be repeated as often as necessary (depending on flock response) and if production and other conditions warrant it.

c. Caution: challenge feeding should not be attempted when an abnormal drop in production occurs, caused by mismanagement of feeding or lighting, weather, disease or other stress.

11. In the final analysis, the objective of feeding laying hens is to produce a dozen eggs of good quality at the lowest possible feed cost. With lightweight layers, the poultry producer should aim for a feed efficiency of 3.5 to 4.0 lb, or less, of feed per dozen eggs.

D. *Feeding Turkeys*

1. Turkeys fed for market purposes

a. The all-mash diets suggested in Table 12–12 are designed for continuous feeding. As can be observed from this schedule, diets are changed frequently to adjust to the specific needs of the birds and to minimize feed costs.

b. Turkeys grow faster than chickens; hence, they have relatively higher feed and protein requirements. Protein requirement levels decrease by age (28% with starting poults to 14% for mature birds), whereas, energy requirement levels tend to increase during the growing stage ranging from 2900 to 3300 kcal/kg of diet.

c. Upon arrival, poults should be encouraged to consume feed and water as soon as possible. The use of colored feed, or the placing of brightly colored marbles in the feed and waters, may help.

d. An antibiotic and a coccidiostat or histostat are desirable in the early starter diets but their use in diets past 8 to 12 weeks of age is optional.

e. Young turkeys use feed efficiently requiring between 2.75 to 3.25 lb of feed per pound of live turkey produced.

f. Males (toms) are usually marketed at 18 to 20 weeks of age and at liveweights of 23 to 35 lb. Younger toms are often sold as oven-ready dressed birds; older toms are generally further processed or used in the restaurant trade. Females (hens) are commonly marketed at 14 to 16 weeks of age and at about 14 lb liveweight.

2. Turkey holding and breeding program

a. Table 12–13 gives examples of diets for breeder candidates and breeder hens.

b. The holding diet would be fed from approximately 16 to 18 weeks of age to the commencement of the lighting program (or about 2 weeks prior to egg production) which usually occurs about 30 weeks of age; at that point they would be fed the breeder diet.

c. The holding diet contains substantially less energy than the starter and grower diets. This type of diet retards sexual maturity and may result in some desirable effects upon later reproductive performance.

E. *Feeding Ducks and Geese*

1. Rations for ducks and geese may for the most part be patterned after those discussed for broilers, making necessary adjustments for either the vitamin and/or mineral levels as indicated in the appropriate nutrient requirement tables.

2. It is recommended that ducks and geese be fed pelleted rations because they are easier to consume and tend to minimize feed wastage. Use ⅛-in. pellets for starter rations, and 3/16-in. pellets for older ducks or geese.

3. Ducklings should be fed a starter diet (22% CP) for the first 2 weeks of age, followed by a finisher-type diet (15% to 16% CP) to market age. Ducks should be ready for market between 7½ and 8 weeks of age.

4. Ducks have a higher requirement for niacin than do chickens, thus, supplemental niacin should be provided if chicken rations are substituted.

5. Potential breeding ducks should be placed on a breeder-holding diet (15% to 16% CP) from about 8 weeks of age until the breeding season begins, without their getting too fat. It is recommended that birds fed holding rations be limited to about ½ lb per bird per day.

6. Goslings should be fed a starter-type feed (22% CP) for the first 3 weeks of life, then finished by one of the following programs.
    a. Foraging the farm or pasture until about 15 weeks of age; then free-choice feeding of a high-energy finishing ration and marketed at about 18 weeks of age.
    b. Foraging and limit-fed prepared feed throughout the growing period; then, marketed at about 14 weeks of age following liberal feeding of a high-energy finishing ration.
    c. The goslings are full-fed in confinement and marketed at about 10 weeks of age.
7. Geese used for breeding purposes should be provided with a holding ration beginning 6 to 8 weeks before the breeding season, followed by a breeding ration formulated for the intensive production of eggs.
8. As noted with ducks, supplemental niacin may be needed in geese formulations.

V. Formulating Poultry Diets

A. The steps involved in formulating a poultry diet are similar to the procedure described for swine diet formulation in Chapter 7. A similar ration worksheet can be used and a sample copy is found in the appendix of this textbook.

B. *Feed Ingredients and Additives*
    1. Since corn and soybean meal usually are the most plentiful and lowest-cost sources of energy and well-balanced protein, most poultry feeds in the United States are based upon these feeds as the major ingredients.
    2. Fish meals and meat meals are good sources of protein, amino acids, and also contain bone, a source of highly available calcium and phosphorus. These feeds help considerably to balance the amino acids in soybean meal (methionine and lysine) and may be desirable to add at a level of 2% to 5% of the diet if they are reasonably priced.
    3. The major minerals required are calcium and phosphorus. It is important that sufficient phosphorus be provided in available form. Only 30% to 40% of the phosphorus in plant products is nonphytin phosphorus and considered available to poultry.

    Salt is added to most poultry rations at a 0.2% to 0.5% level. Too much salt will result in increased water consumption and wet droppings.
    4. Supplemental fats (animal tallow or vegetable oil) may increase energy utilization in adult chickens in association with a decreased rate of food passage through the GI tract. Furthermore, because the heat increment of feeding of fats is less per unit weight than that of carbohydrates, the substitution of fat for a portion of the dietary carbohydrate (up to 5% of the diet) may enhance energy utilization by reducing the heat increment of feeding.

Poultry confinement building with automatic feeders and waterers. Broilers and turkeys are commonly finished to market weight under confinement conditions.

5. Since it is desirable in the United States and many othe countries to produce broilers of a fairly high degree of yellow pigmentation, it is advisable to use as much yellow corn as possible, plus good sources of xanthophyll such as alfalfa meal or corn gluten meal. Xanthophylls constitute a group of naturally occurring oxycarotenoid pigments that are responsible for the yellow coloration in the shanks, feet, skin, and egg yolks of poultry.
6. Modern poultry feeds commonly contain one or more nonnutritive additives. These additives are used for a variety of reasons. They are not nutrients, but some of them improve production under certain circumstances. Others prevent rancidity in the feed. There is no evidence of a nutritional deficiency when they are omitted from a ration. Among such additives are the following.
   a. Antibiotics for growth-stimulating and disease control programs.
   b. Arsenicals and nitrofurans for improved performance.
   c. Antiparasitic drugs.
   d. Antioxidants.
   e. Antifungal agents.
   f. Grit for improved digestion.

TABLE 12–1. Nutrient Requirements of Leghorn-Type Chickens as Percentages or as Milligrams or Units per Kilogram of Diet

Energy Base kcal ME/kg Diet[a] ⟶		Growing			Laying		Breeding
		0–6 Weeks 2,900	6–14 Weeks 2,900	14–20 Weeks 2,900	2,900	Daily Intake per Hen (mg)[b]	2,900
Protein	%	18	15	12	14.5	16,000	14.5
Arginine	%	1.00	0.83	0.67	0.68	750	0.68
Glycine and Serine	%	0.70	0.58	0.47	0.50	550	0.50
Histidine	%	0.26	0.22	0.17	0.16	180	0.16
Isoleucine	%	0.60	0.50	0.40	0.50	550	0.50
Leucine	%	1.00	0.83	0.67	0.73	800	0.73
Lysine	%	0.85	0.60	0.45	0.64	700	0.64
Methionine + cystine	%	0.60	0.50	0.40	0.55	600	0.55
Methionine	%	0.30	0.25	0.20	0.32	350	0.32
Phenylalanine + tyrosine	%	1.00	0.83	0.67	0.80	880	0.80
Phenylalanine	%	0.54	0.45	0.36	0.40	440	0.40
Threonine	%	0.68	0.57	0.37	0.45	500	0.45
Tryptophan	%	0.17	0.14	0.11	0.14	150	0.14
Valine	%	0.62	0.52	0.41	0.55	600	0.55
Linoleic acid	%	1.00	1.00	1.00	1.00	1,100	1.00
Calcium	%	0.80	0.70	0.60	3.40	3,750	3.40
Phosphorus, available	%	0.40	0.35	0.30	0.32	350	0.32
Potassium	%	0.40	0.30	0.25	0.15	165	0.15
Sodium	%	0.15	0.15	0.15	0.15	165	0.15
Chlorine	%	0.15	0.12	0.12	0.15	165	0.15
Magnesium	mg	600	500	400	500	55	500
Manganese	mg	60	30	30	30	3.30	60
Zinc	mg	40	35	35	50	5.50	65
Iron	mg	80	60	60	50	5.50	60
Copper	mg	8	6	6	6	0.88	8
Iodine	mg	0.35	0.35	0.35	0.30	0.03	0.30
Selenium	mg	0.15	0.10	0.10	0.10	0.01	0.10
Vitamin A	IU	1,500	1,500	1,500	4,000	440	4,000
Vitamin D	ICU	200	200	200	500	55	500
Vitamin E	IU	10	5	5	5	0.55	10
Vitamin K	mg	0.50	0.50	0.50	0.50	0.055	0.50
Riboflavin	mg	3.60	1.80	1.80	2.20	0.242	3.80
Pantothenic acid	mg	10.0	10.0	10.0	2.20	0.242	10.0
Niacin	mg	27.0	11.0	11.0	10.0	1.10	10.0
Vitamin B_{12}	mg	0.009	0.003	0.003	0.004	0.00044	0.004
Choline	mg	1,300	900	500	?	?	?
Biotin	mg	0.15	0.10	0.10	0.10	0.011	0.15
Folacin	mg	0.55	0.25	0.25	0.25	0.0275	0.35
Thiamin	mg	1.8	1.3	1.3	0.80	0.088	0.80
Pyridoxine	mg	3.0	3.0	3.0	3.0	0.33	4.50

[a]These are typical dietary energy concentrations.

[b]Assumes an average daily intake of 110 g of feed/hen daily.

TABLE 12–2. Nutrient Requirements of Broilers as Percentages or as Milligrams or Units per Kilogram of Diet

Energy Base kcal ME/kg Diet[a] ⟶		Weeks 0–3 3,200	Weeks 3–6 3,200	Weeks 6–8 3,200
Protein	%	23.0	20.0	18.0
Arginine	%	1.44	1.20	1.00
Glycine + Serine	%	1.50	1.00	0.70
Histidine	%	0.35	0.30	0.26
Isoleucine	%	0.80	0.70	0.60
Leucine	%	1.35	1.18	1.00
Lysine	%	1.20	1.00	0.85
Methionine + Cystine	%	0.93	0.72	0.60
Methionine	%	0.50	0.38	0.32
Phenylalanine + Tyrosine	%	1.34	1.17	1.00
Phenylalanine	%	0.72	0.63	0.54
Threonine	%	0.80	0.74	0.68
Tryptophan	%	0.23	0.18	0.17
Valine	%	0.82	0.72	0.62
Linoleic acid	%	1.00	1.00	1.00
Calcium	%	1.00	0.90	0.80
Phosphorus, available	%	0.45	0.40	0.35
Potassium	%	0.40	0.35	0.30
Sodium	%	0.15	0.15	0.15
Chlorine	%	0.15	0.15	0.15
Magnesium	mg	600	600	600
Manganese	mg	60.0	60.0	60.0
Zinc	mg	40.0	40.0	40.0
Iron	mg	80.0	80.0	80.0
Copper	mg	8.0	8.0	8.0
Iodine	mg	0.35	0.35	0.35
Selenium	mg	0.15	0.15	0.15
Vitamin A	IU	1,500	1,500	1,500
Vitamin D	ICU	200	200	200
Vitamin E	IU	10	10	10
Vitamin K	mg	0.50	0.50	0.50
Riboflavin	mg	3.60	3.60	3.60
Pantothenic acid	mg	10.0	10.0	10.0
Niacin	mg	27.0	27.0	11.0
Vitamin B_{12}	mg	0.009	0.009	0.003
Choline	mg	1,300	850	500
Biotin	mg	0.15	0.15	0.10
Folacin	mg	0.55	0.55	0.25
Thiamin	mg	1.80	1.80	1.80
Pyridoxine	mg	3.0	3.0	2.5

[a]These are typical dietary energy concentrations.

TABLE 12-3. Nutrient Requirements of Turkeys as Percentages or as Milligrams or Units per Kilogram of Feed

		M: 0–4 F: 0–4 2,800	4–8 4–8 2,900	8–12 8–11 3,000	12–16 11–14 3,100	16–20 14–17 3,200	20–24 17–20 3,300	Holding 2,900	Breeding Hens 2,900
Energy Base kcal ME/kg Diet[a] →									
Protein	%	28	26	22	19	16.5	14	12	14
Arginine	%	1.6	1.5	1.25	1.1	0.95	0.8	0.6	0.6
Glycine + serine	%	1.0	0.9	0.8	0.7	0.6	0.5	0.4	0.5
Histidine	%	0.58	0.54	0.46	0.39	0.35	0.29	0.25	0.3
Isoleucine	%	1.1	1.0	0.85	0.75	0.65	0.55	0.45	0.5
Leucine	%	1.9	1.75	1.5	1.3	1.1	0.95	0.5	0.5
Lysine	%	1.6	1.5	1.3	1.0	0.8	0.65	0.5	0.6
Methionine + cystine	%	1.05	0.9	0.75	0.65	0.55	0.45	0.4	0.4
Methionine	%	0.53	0.45	0.38	0.33	0.28	0.23	0.2	0.2
Phenylalanine + tyrosine	%	1.8	1.65	1.4	1.2	1.05	0.9	0.8	1.0
Phenylalanine	%	1.0	0.9	0.8	0.7	0.6	0.5	0.4	0.55
Threonine	%	1.0	0.93	0.79	0.68	0.59	0.5	0.4	0.45
Tryptophan	%	0.26	0.24	0.2	0.18	0.15	0.13	0.1	0.13
Valine	%	1.2	1.1	0.94	0.8	0.7	0.6	0.5	0.58
Linoleic acid	%	1.0	1.0	0.8	0.8	0.8	0.8	0.8	1.0
Calcium	%	1.2	1.0	0.85	0.75	0.65	0.55	0.5	2.25
Phosphorus, available	%	0.6	0.5	0.42	0.38	0.32	0.28	0.25	0.35
Potassium	%	0.7	0.6	0.5	0.5	0.4	0.4	0.4	0.6
Sodium	%	0.17	0.15	0.12	0.12	0.12	0.12	0.12	0.15
Chlorine	%	0.15	0.14	0.14	0.12	0.12	0.12	0.12	0.12
Magnesium	mg	600	600	600	600	600	600	600	600
Manganese	mg	60	60	60	60	60	60	60	60
Zinc	mg	75	65	50	40	40	40	40	65
Iron	mg	80	60	60	60	50	50	50	60
Copper	mg	8	8	6	6	6	6	6	8
Iodine	mg	0.4	0.4	0.4	0.4	0.4	0.4	0.4	0.4
Selenium	mg	0.2	0.2	0.2	0.2	0.2	0.2	0.2	0.2
Vitamin A	IU	4,000	4,000	4,000	4,000	4,000	4,000	4,000	4,000
Vitamin D[b]	ICU	900	900	900	900	900	900	900	900
Vitamin E	IU	12	12	10	10	10	10	10	25
Vitamin K	mg	1.0	1.0	0.8	0.8	0.8	0.8	0.8	1.0
Riboflavin	mg	3.6	3.6	3.0	3.0	2.5	2.5	2.5	4.0
Pantothenic acid	mg	11.0	11.0	9.0	9.0	9.0	9.0	9.0	16.0
Niacin	mg	70.0	70.0	50.0	50.0	40.0	40.0	40.0	30.0
Vitamin B_{12}	mg	0.003	0.003	0.003	0.003	0.003	0.003	0.003	0.003
Choline	mg	1,900	1,600	1,300	1,100	950	800	800	1,000
Biotin	mg	0.2	0.2	0.15	0.125	0.100	0.100	0.100	0.15
Folacin	mg	1.0	1.0	0.8	0.8	0.7	0.7	0.7	1.0
Thiamin	mg	2.0	2.0	2.0	2.00	2.0	2.0	2.0	2.0
Pyridoxine	mg	4.5	4.5	3.5	3.5	3.0	3.0	3.0	4.0

[a] These are typical ME concentrations for corn-soya diets. Different ME values may be appropriate if other ingredients predominate.

[b] These concentrations of vitamin D are satisfactory when the dietary concentrations of calcium and available phosphorus conform with those in this table.

TABLE 12-4. Body Weights and Feed Requirements of Leghorn-Type Pullets and Hens

Age (weeks)	Body Weight[a] (g)	Feed Consumption[b] (g/week)	Typical Egg Production (hen-day %)
0	35	45	—
2	135	90	—
4	270	180	—
6	450	260	—
8	620	325	—
10	790	385	—
12	950	430	—
14	1,060	460	—
16	1,160	460	—
18	1,260	460	—
20	1,360	460	—
22	1,425	525	10
24	1,500	595	38
26	1,575	665	64
30	1,725	770	88
40	1,815	770	80
50	1,870	765	74
60	1,900	755	68
70	1,900	740	62

[a]Pullets and hens of Leghorn-type strains are generally fed ad libitum but are occasionally control-fed to limit body weights. Values shown are typical but will vary with strain differences, season, and lighting. Specific breeder guidelines should be consulted for desired schedules of weights and feed consumption.

[b]Based on diets containing 2,900 ME kcal/kg. Consumption will vary depending upon the caloric density of the diet, environmental temperature, and rate of production.

TABLE 12-5. Body Weights and Feed Requirements of Broilers[a]

Age (weeks)	Body Weights (g)		Weekly Feed Consumption (g)		Cumulative Feed Consumption (g)		Weekly Energy Consumption (ME kcal/bird)		Cumulative Energy Consumption (ME kcal/bird)	
	M	F	M	F	M	F	M	F	M	F
1	130	120	120	110	120	110	385	350	385	350
2	320	300	260	240	380	350	830	770	1,215	1,120
3	560	515	390	355	770	705	1,250	1,135	2,465	2,255
4	860	790	535	500	1,305	1,205	1,710	1,600	4,175	3,855
5	1,250	1,110	740	645	2,045	1,850	2,370	2,065	6,545	5,920
6	1,690	1,430	980	800	3,025	2,650	3,135	2,560	9,680	8,480
7	2,100	1,745	1,095	910	4,120	3,560	3,505	2,910	13,185	11,390
8	2,520	2,060	1,210	970	5,330	4,530	3,870	3,105	17,055	14,495
9	2,925	2,350	1,320	1,010	6,650	5,540	4,225	3,230	21,280	17,725

[a]Typical for broilers fed well-balanced diets containing 3,200 ME kcal/kg.

TABLE 12–6. Typical Body Weights and Feed Allowances for Male and Female Meat-Type Chickens (Replacement Stock)[a]

Age (weeks)	Male Body Weight[b] (g)	Male Feed Consumption[c] (g/week)	Female Body Weight[b] (g)	Female Feed Consumption[c] (g/week)	Typical Egg Production (hen-day %)
0	40	100	40	75	—
2	250	250	225	225	—
4	545	350–385	455	315–330	—
6	795	390–425	660	330–350	—
8	1,020	405–475	840	350–400	—
10	1,250	475–550	1,000	385–445	—
12	1,480	540–625	1,180	425–480	—
14	1,700	575–700	1,360	460–550	—
16	1,930	625–765	1,550	495–600	—
18	2,150	665–825	1,730	525–670	—
20	2,400	—[d]	1,930	570–730	—
22	2,640	—	2,110	635–795	10
24	3,200	—	2,450	800–925	15
26	3,540	—	2,730	950–1,050	30
28	3,750	—	2,880	1,078–1,141	56
30	3,900	—	3,000	1,078–1,141	75
32	4,090	—	3,090	1,078–1,141	80
34	4,220	—	3,130	1,078–1,141	78
36	4,340	—	3,160	1,078–1,141	76
38	4,450	—	3,180	1,071–1,134	73
40	4,540	—	3,180	1,064–1,127	72

[a]Broiler-breeder strains must be grown on a controlled feeding program to limit weight. Values shown are typical but will vary according to strain. Specific breeder guidelines should be consulted for desired schedule of weights and feed allotments.

[b]Values are typical for fall-hatched chicks. Spring-hatched chicks will have decreasing natural daylight during the time of sexual maturity and usually need to be heavier to attain sexual maturity at the desired age.

[c]Adjust as required to maintain desired body weight.

[d]Males and females intermingled.

TABLE 12–7. Growth Rate, Feed and Energy Consumption of Large-Type Turkeys

Age (weeks)	Body Weight (kg) M	Body Weight (kg) F	Feed Consumption per Week (kg) M	Feed Consumption per Week (kg) F	Cumulative Feed Consumption (kg) M	Cumulative Feed Consumption (kg) F	ME Consumption per Week (Mcal) M	ME Consumption per Week (Mcal) F
1	0.11	0.11	0.10	0.10	0.10	0.10	0.30	0.30
2	0.27	0.24	0.20	0.17	0.30	0.27	0.60	0.50
3	0.58	0.47	0.45	0.39	0.75	0.66	1.1	0.80
4	1.0	0.70	0.61	0.46	1.36	1.12	1.7	1.2
5	1.5	1.1	0.70	0.60	2.06	1.72	2.3	1.6
6	2.0	1.6	0.86	0.76	2.92	2.48	2.9	2.1
7	2.6	2.1	1.08	0.89	4.00	3.37	3.5	2.6
8	3.3	2.6	1.30	1.04	5.30	4.41	4.1	3.1
9	4.0	3.1	1.51	1.18	6.81	5.59	4.8	3.6
10	4.7	3.7	1.78	1.34	8.59	6.93	5.2	4.1
11	5.5	4.3	1.99	1.47	10.58	8.40	5.7	4.6
12	6.3	4.8	2.25	1.59	12.83	9.99	6.3	5.1
13	7.1	5.3	2.51	1.70	15.34	11.69	7.1	5.5
14	8.0	5.8	2.66	1.75	18.00	13.44	7.8	5.8
15	8.8	6.3	2.89	1.82	20.89	15.26	8.4	6.1
16	9.7	6.7	3.05	1.92	23.94	17.18	8.8	6.4
17	10.5	7.1	3.13	2.03	27.03	19.21	9.6	6.7
18	11.3	7.5	3.27	2.07	30.34	21.28	10.2	6.9
19	12.1	7.8	3.43	2.15	33.77	23.43	10.9	7.1
20	12.8	8.1	3.60	2.23	37.37	25.66	11.6	7.3
21	13.5	—	3.71	—	41.08	—	12.5	—
22	14.2	—	3.82	—	44.90	—	12.9	—
23	14.8	—	3.94	—	48.84	—	13.2	—
24	15.4	—	4.05	—	52.89	—	13.5	—

TABLE 12–8. Body Weights and Feed Consumption of Large-Type Turkeys During Holding and Breeding Periods[a]

Age (weeks)	Hens Weight (kg)	Hens Egg Production (%)	Hens Feed /Day (g)	Toms Weight (kg)	Toms Feed /Day (g)
20	7.0	—	200	12.0	400
25	8.0	—	215	13.5	420
30	9.0	Start light stimulation	230	16.0	440
35	9.5	66	260	17.0	450
40	9.3	63	255	18.0	460
45	9.1	60	250	18.2	480
50	9.0	50	240	18.5	500
55	9.0	40	230	18.8	510
60	9.0	35	220	19.0	520

[a]These values are based on experimental data involving "in season" egg production (i.e., November through July) of commercial stock. It is estimated that summer breeders would produce 70–90 percent as many eggs and consume 60–80 percent as much feed, respectively, as "in season" breeders.

TABLE 12–9. Suggested Broiler Diets

Ingredient	IFN	Starter 0–3 wks	Grower 3–6 wks	Finisher 6–9 wks
		%	%	%
Corn[1]	4–02–931	55.10	62.75	67.50
Soybean meal	5–04–612	29.00	25.00	20.25
Meat & bone meal	5–08–388	5.00	4.00	4.00
Fish meal, menhaden	5–02–009	2.00	—	—
Alfalfa meal, dehy	1–00–023	2.00	2.00	2.00
Limestone	6–02–632	0.25	0.70	0.60
Dicalcium phosphate	6–01–080	0.30	0.70	0.80
Stabilized fat	4–00–409	5.00	3.50	3.50
Trace mineralized salt		0.25	0.25	0.25
Vitamin premix[2]		1.00	1.00	1.00
Methionine		0.10	0.10	0.10
Antibiotic & coccidiostat[3]		+	+	+
		100.00	100.00	100.00

Calculated Analysis:

Metabolizable energy, kcal/kg		3175	3150	3200
Crude protein, %		23.2	20.2	18.3
Lysine, %		1.31	1.1	0.96
Calcium, %		0.87	0.92	0.89
Phosphorus, Total, %		0.71	0.68	0.68
Phosphorus, available, %		0.48	0.44	0.46

1. Up to 50% of the corn can be replaced by grain sorghum.
2. Vitamin premix should supply the following per kilogram of diet: vitamin A, 7000 IU; vitamin D_3, 1500 ICU; vitamin E, 10 IU; vitamin K (menadione), 0.5 mg; riboflavin, 3.6 mg; pantothenic acid, 10 mg; niacin, 27 mg; choline, 1300 mg; vitamin B_{12}, 0.009 mg.
3. Use at manufacturers recommended levels.

TABLE 12–10. All-mash Rations for Replacement Pullets

Ingredients	IFN	Starter 0–6 wks	Grower 6–14 wks	Developer 14–20 wks
		%	%	%
Corn	4–02–931	61.25	59.00	61.60
Wheat middlings	4–05–205	4.00	3.00	6.80
Oats	4–03–309	8.00	18.00	20.00
Alfalfa meal, dehy	1–00–023	2.50	2.50	2.50
Fish meal, menhaden	5–02–009	2.50	—	—
Meat & bone meal	5–00–388	2.50	—	—
Soybean meal	5–04–612	16.30	14.40	6.00
Dicalcium phosphate	6–01–080	1.00	1.00	1.00
Limestone	6–02–632	0.50	0.75	0.75
Salt		0.25	0.25	0.25
Trace mineral mix[1]		0.10	0.10	0.10
Vitamin mix[2]		1.00	1.00	1.00
Methionine		0.10	—	—
Antibiotic & coccidiostat[3]		+	+	+
		100.00	100.00	100.00

Calculated Analyses:

Metabolizable energy (kcal/kg)		2950	2935	2950
Protein, (%)		18.3	15.3	12.3
Lysine, (%)		0.96	0.74	0.52
Calcium, (%)		0.88	0.60	0.58
Phosphorus, total (%)		0.74	0.54	0.54
Phosphorus, available (%)		0.50	0.31	0.23

1. Trace mineral mix should supply the following per kilogram of diet: copper, 4 mg; iodine, 0.35 mg; iron, 80 mg; magnesium, 600 mg; manganese, 55 mg; selenium, 0.1 mg; zinc, 40 mg.
2. Vitamin mix should supply the following per kilogram of diet: vitamin A, 7000 IU; vitamin D3, 1500 ICU; vitamin E, 10 IU; vitamin K (menadione), 0.5 mg; riboflavin, 3.6 mg; pantothenic acid, 10 mg; choline, 1300 mg; vitamin B_{12}, 0.009 mg.
3. Use at manufacturers recommended levels.

TABLE 12–11. All-mash Ration for Layers

Ingredients	IFN	%
Corn	4–02–931	71.25
Soybean meal	5–04–612	13.00
Meat & bone meal	5–00–388	4.00
Alfalfa meal, dehy	1–00–023	2.00
Dicalcium phosphate	6–01–080	0.60
Limestone	6–02–632	7.75
Salt		0.25
Trace mineral mix[1]		0.10
Vitamin mix[2]		1.00
Methionine		0.05
Antibiotic & coccidiostat[3]		+
		100.00

Calculated Analyses:

Metabolizable energy (kcal/kg)	2875
Protein, (%)	15.0
Lysine, (%)	0.74
Calcium, (%)	3.48
Phosphorus, total (%)	0.62
Phosphorus, available (%)	0.39

1. Trace mineral mix should supply the following per kilogram of diet: copper, 3 mg. iodine, 0.3 mg; iron, 50 mg; magnesium, 500 mg; manganese, 25 mg; selenium, 0.1 mg; zinc, 50 mg.
2. Vitamin mix should supply the following per kilogram of diet; vitamin A, 4000 IU; vitamin D_3, 1200 ICU; vitamin E, 5 IU; vitamin K (menadione), 0.5 mg; riboflavin, 2.2 mg; pantothenic acid, 2.2 mg; niacin, 10 mg; choline, 500 mg; vitamin B_{12}, 0.003 mg.
3. Use at manufacturers recommended levels.

TABLE 12–12. Suggested Turkey Rations[a]

Ingredients	IFN	0–4 wks, %	5–8 wks, %	9–12 wks, %	13–16 wks, %	17–20 wks, %	20+ wks, %
Corn	4–02–931	48.15	51.00	62.30	70.50	75.95	80.65
Soybean meal	5–04–612	35.65	37.00	26.60	18.90	13.75	10.00
Meat and bone meal	5–00–388	7.50	7.50	7.50	7.50	6.45	4.00
Fish meal, menhaden	5–02–009	4.00	—	—	—	—	—
Alfalfa meal, dehy	1–00–023	2.00	—	—	—	—	—
Stabilized fat	4–00–409	1.00	2.0	1.50	1.50	2.00	2.50
Dicalcium phosphate	6–01–080	—	0.35	0.50	0.20	0.50	1.20
Limestone	6–02–632	0.25	0.70	0.15	—	—	0.30
Salt		0.25	0.25	0.25	0.25	0.25	0.25
Trace mineral mix[2]		0.10	0.10	0.10	0.10	0.10	0.10
DL-Methionine		0.10	0.10	0.10	0.05	—	—
Vitamin premix[3]		1.00	1.00	1.00	1.00	1.00	1.00
Antibiotic & coccidiostat[4]		+	+	+	+	+	+
		100.00	100.00	100.00	100.00	100.00	100.00

Calculated Analyses:

Metabolizable energy (kcal/kg)		2900	2970	3060	3150	3225	3300
Protein, (%)		28.4	26.5	22.4	19.3	16.7	14.0
Lysine, (%)		1.66	1.51	1.22	1.01	0.83	0.66
Calcium, (%)		1.17	1.19	1.00	0.86	0.80	0.82
Phosphorus, total (%)		0.86	0.82	0.82	0.73	0.72	0.72
Phosphorus, available (%)		0.61	0.57	0.58	0.51	0.52	0.52

1. All mash diets designed to be fed ad libitum.
2. Trace mineral mix should supply the following per kilogram of diet: copper, 6 mg; iodine, 0.4 mg; iron, 60 mg; magnesium, 500 mg; manganese, 55 mg; selenium, 0.1 mg; zinc, 75 mg.
3. Vitamin premix should supply the following per kilogram of diet: vitamin A, 8000 IU; vitamin D_3, 1500 ICU; vitamin E, 12 IU; vitamin K (menadione), 1 mg; riboflavin, 3.6 mg; pantothenic acid, 11 mg; niacin, 70 mg; choline, 1900 mg; folic acid, 0.4 mg; biotin, 0.05 mg; vitamin B_{12}, 0.003 mg.
4. Use at manufacturers recommended levels.

TABLE 12–13

TABLE 12–13. Suggested Turkey Breeder Complete Mash Rations

Ingredients	IFN	Holding %	Breeding %
Corn	4–02–931	58.20	71.00
Oats	4–03–309	25.00	—
Soybean meal	5–04–612	6.00	11.10
Fish meal, menhaden	5–02–009	2.00	4.00
Alfalfa meal, dehy	1–00–023	5.00	4.00
Dicalcium phosphate	6–01–080	2.00	2.00
Limestone	6–02–632	0.40	4.50
Salt		0.30	0.30
Trace mineral mix[1]		0.10	0.10
Vitamin premix[2]		1.00	1.00
Antibiotic & coccidiostat[3]		+	+
		100.00	100.00
Calculated Analyses:			
Metabolizable energy (kcal/kg)		2900	2950
Protein, (%)		13.1	15.1
Lysine, (%)		0.59	0.78
Calcium, (%)		0.80	2.41
Phosphorus, total (%)		0.73	0.79
Phosphorus, available (%)		0.53	0.59

1. Trace mineral mix should supply the following per kilogram of diet: copper, 6 mg; iodine, 0.4 mg; iron, 60 mg; magnesium, 500 mg; manganese, 35 mg; selenium, 0.1 mg; zinc, 65 mg.
2. Vitamin premix should supply the following per kilogram of diet: vitamin A, 8000 IU; vitamin D_3, 1500 ICU; vitamin E, 25 IU; vitamin K (menadione), 1 mg; riboflavin, 4 mg; pantothenic acid, 16 mg; niacin, 30 mg; choline, 1000 mg; vitamin B_{12}, 0.003 mg.
3. Use at manufacturers recommended levels.

TABLE 12–14. Nutrient Requirements of Pekin Ducks as Percentages or as Milligrams or Units per Kilogram of Diet*

Energy Base kcal ME/kg Diet† →		Staring (0–2 weeks) 2,900	Growing (2–7 weeks) 2,900	Breeding 2,900
Protein	%	22.0	16.0	15.0
Arginine	%	1.1	1.0	—
Lysine	%	1.1	0.9	0.7
Methionine + cystine	%	0.8	0.6	0.55
Calcium	%	0.65	0.6	2.75
Phosphorus, available	%	0.40	0.35	0.35
Sodium	%	0.15	0.15	0.15
Chlorine	%	0.12	0.12	0.12
Magnesium	mg	500	500	500
Manganese	mg	40.0	40.0	25.0
Zinc	mg	60.0	60.0	60.0
Selenium	mg	0.14	0.14	0.14
Vitamin A	IU	4,000	4,000	4,000
Vitamin D	ICU	220	220	500
Vitamin K	mg	0.4	0.4	0.4
Riboflavin	mg	4.0	4.0	4.0
Pantothenic acid	mg	11.0	11.0	10.0
Niacin	mg	55.0	55.0	40.0
Pyridoxine	mg	2.6	2.6	3.0

*For nutrients not listed, see requirements for chickens as a guide.
†These are typical dietary energy concentrations.

TABLE 12–15. Nutrient Requirements of Geese as Percentages or as Milligrams or Units per Kilogram of Diet*

Energy Base kcal ME/kg Diet† →		Staring (0–6 weeks) 2900	Growing (After 6 weeks) 2900	Breeding 2900
Protein	%	22.0	15.0	15.0
Lysine	%	0.9	0.6	0.6
Methionine + cystine	%	0.75	—	—
Calcium	%	0.8	0.6	2.25
Phosphorus, available	%	0.4	0.3	0.3
Vitamin A	IU	1,500	1,500	4,000
Vitamin D	ICU	200	200	200
Riboflavin	mg	4.0	2.5	4.0
Pantothenic acid	mg	15.0	—	—
Niacin	mg	55.0	35.0	20.0

*For nutrients not listed, see requirements for chickens as a guide.
†These are typical dietary energy concentrations.

TABLE 12–16. Suggested Duck Rations

Ingredient	IFN	Starter 0–2 Weeks		Grower 2–4 Weeks		Grower 4 Weeks–Market		Holding	Breeder	
		No. 1 (%)	No. 2 (%)	No. 3 (%)	No. 4 (%)	No. 5 (%)	No. 6 (%)	No. 7 (%)	No. 8 (%)	No. 9 (%)
Ground corn	4-02-931	44.0	46.0	52.8	27.8	60.4	31.8	49.0	54.7	56.6
Ground wheat	4-05-211	—	—	—	27.8	—	31.8	—	—	—
Ground barley	4-00-549	10.0	10.0	10.0	10.0	10.0	10.0	10.0	10.0	10.0
Wheat shorts	4-05-201	5.0	5.0	5.0	5.0	5.0	5.0	20.0	5.0	5.0
Wheat middlings	4-05-205	5.0	5.0	5.0	5.0	5.0	5.0		5.0	5.0
Alfalfa meal, dehydrated	1-00-023	2.0	2.0	—	—	—	—	7.5	2.0	2.0
Meat and bone meal	5-00-388	—	2.0	—	—	—	—	—	—	2.0
Fish meal	5-02-009	—	2.0	—	—	—	—	—	—	2.0
Soybean meal	5-04-612	30.5	25.5	23.5	20.6	16.0	12.8	9.1	14.3	9.4
Limestone	6-02-632	1.0	0.6	1.0	1.1	1.1	1.1	1.8	6.4	6.0
Dicalcium phosphate	6-01-080	1.0	0.4	1.2	1.2	1.0	1.0	1.1	1.1	0.5
Salt (iodized)		0.5	0.5	0.5	0.5	0.5	0.5	0.5	0.5	0.5
Vitamin/mineral premix		1.0	1.0	1.0	1.0	1.0	1.0	1.0	1.0	1.0
TOTAL		100.0	100.0	100.0	100.0	100.0	100.0	100.0	100.0	100.0
Calculated Analyses										
Metabolizable energy, kcal/kg		2770	2810	2865	2790	2940	2860	2690	2730	2775
Crude protein (%)		22.00	22.00	19.00	19.00	16.00	16.00	14.50	15.00	15.00
Lysine (%)		1.20	1.20	0.96	0.91	0.75	0.69	0.67	0.70	0.70
Calcuim (%)		0.73	0.73	0.73	0.77	0.71	0.71	1.10	2.75	2.77
Total phosphorus (%)		0.64	0.66	0.65	0.66	0.59	0.60	0.64	0.60	0.61
Available phosphorus (%)		0.44	0.45	0.44	0.45	0.40	0.40	0.44	0.41	0.42

TABLE 12-17. Suggested Geese Rations

Ingredient	IFN	Starter 0–2 Weeks		Grower 2–4 Weeks		Grower 4 Weeks–Market		Holding	Breeder	
		No. 1 (%)	No. 2 (%)	No. 3 (%)	No. 4 (%)	No. 5 (%)	No. 6 (%)	No. 7 (%)	No. 8 (%)	No. 9 (%)
Ground corn	4-02-931	43.9	23.6	55.6	29.7	62.9	33.2	51.2	56.4	30.2
Ground wheat	4-05-211	—	23.6	—	29.0	—	33.2	—	—	30.2
Ground barley	4-00-549	10.0	10.0	10.0	10.0	10.0	10.0	10.0	10.0	10.0
Wheat shorts	4-05-201	5.0	5.0	5.0	5.0	5.0	5.0	—	5.0	5.0
Wheat middlings	4-05-205	5.0	5.0	5.0	5.0	5.0	5.0	20.0	5.0	5.0
Alfalfa meal, dehydrated	1-00-023	2.0	2.0	—	—	—	—	7.5	2.0	2.0
Meat and bone meal	5-00-388	—	2.0	—	—	—	—	—	—	2.0
Soybean meal	5-04-612	30.4	25.9	21.0	18.0	13.6	10.1	6.8	14.0	8.7
Limestone	6-02-632	1.2	0.9	1.0	1.0	1.0	1.0	2.0	5.1	4.9
Dicalcium phosphate	6-01-080	1.0	0.5	0.9	0.8	1.0	1.0	1.0	1.0	0.5
Salt (iodized)		0.5	0.5	0.5	0.5	0.5	0.5	0.5	0.5	0.5
Vitamin/mineral premix		1.0	1.0	1.0	1.0	1.0	1.0	1.0	1.0	1.0
TOTAL		100.0	100.0	100.0	100.0	100.0	100.0	100.0	100.0	100.0
Calculated Analyses										
Metabolizable energy, kcal/kg		2762	2720	2900	2830	2970	2880	2710	2785	2725
Crude protein (%)		22.00	22.00	18.00	18.00	15.00	15.00	13.50	15.00	15.00
Lysine (%)		1.17	1.12	0.89	0.84	0.69	0.62	0.60	0.70	0.64
Calcuim (%)		0.81	0.78	0.66	0.64	0.66	0.66	1.14	2.25	2.25
Total phosphorus (%)		0.64	0.64	0.59	0.58	0.58	0.59	0.61	0.58	0.58
Available phosphorus (%)		0.44	0.44	0.40	0.40	0.40	0.40	0.42	0.40	0.40

Study Questions and Problems

Chapter 12

I. True or False Questions

1. Poultry requirements for vitamin D are expressed in ICU, which are based on the activity of vitamin D_2.
2. Requirements for vitamin E are dependent on the level of fat and selenium in the poultry diet.
3. The dietary calcium requirement of laying hen rations may exceed 3%.
4. Cage fatigue in laying hens may be caused by dietary calcium deficiency.
5. Gossypol found in linseed meal produces a mottling and greenish cast to egg yolks.
6. Broiler chicks are fed for approximately 6 weeks to an average weight of more than 3.5 lb.
7. Lysine is the first limiting amino acid in the laying hen diet based on corn and soybean meal.
8. Turkeys grow faster than chickens, so they have relatively higher feed and protein requirements.

II. Multiple Choice (select the one best answer)

1. With lightweight layers, the poultry producer should aim for a feed efficiency of __?__ lbs or less of feed per dozen eggs.
   a. 1.5 to 2.0
   b. 2.0 to 2.5
   c. 2.5 to 3.0
   d. 3.0 to 3.5
   e. 3.5 to 4.0
2. A yellow pigmentation to the skin of broilers may be imparted from the carotenoid compounds found in
   a. Soybean meal.
   b. Alfalfa meal.
   c. Milo grain.
   d. Trace mineralized salt.
   e. Vitamin A.
3. Cyclopropenoic fatty acids found in __?__ may impart a pink color to egg whites when fed to hens.
   a. Sorghum grain
   b. Alfalfa meal
   c. Fish meal
   d. Cottonseed meal
   e. Linseed meal
4. Hens kept for the production of hatching eggs should be fed a more highly fortified feed, especially
   a. Protein.
   b. Vitamins.
   c. Minerals.
   d. Salt.
   e. Fat.
5. Which of the following is *not* efficiently utilized by poultry?
   a. Beta carotene
   b. DL-α-tocopheryl acetate
   c. Irradiated ergosterol
   d. Menadione
   e. Irradiated 7-dehydrocholesterol
6. In addition to the essential amino acids required by the pig, broilers also require dietary
   a. Glycine and proline.
   b. Glycine and alanine.
   c. Proline and asparagine.
   d. Serine and glutamine.
   e. Aspartic acid and hydroxyproline.
7. Poultry diets may contain __?__ as a source of unidentified growth factors.
   a. Corn grain
   b. Linseed meal

c. Brewery byproducts
d. Molasses
e. Hydrolyzed poultry feathers
8. In addition to crude protein, meat products supplemented to a poultry diet are an excellent source of
   a. Vitamins A and D.
   b. Unidentified growth factors.
   c. Trace minerals.
   d. Phosphorus and calcium.
   e. Soluble carbohydrates.
9. Pellets or crumbles have the following advantage(s) over the same poultry diet fed in the meal form.
   a. Reduced feed wastage.
   b. Improved nutrient availability.
   c. Lower ration cost.
   d. All of the above.
   e. Only answers a and c.

III. Matching or Identification

Match the following statements with the appropriate letter.

a. 3.5	f. 5
b. 14–16	g. 60
c. 26	h. 18
d. 85–90	i. 6–8
e. .25–.5	j. 120

1. Age at which broilers are marketed, weeks.
2. % crude protein in laying hen diet.
3. Fish meal should be limtied to ___?___ % of the poultry diet.
4. % salt added to poultry diet.
5. Approximate grams of protein per bird daily to support optimum egg production.
6. % calcium in laying hen diet.
7. Feed intake of laying hens decreases as environmental temperature increases above ___?___ ° F.
8. Feed represents about ___?___ % of the cost of raising replacement pullets.
9. Body weight at which broilers are marketed, lb.
10. Approximate lb of feed required per lb of live turkey produced.

IV. Short Answer

1. Which form of vitamin D is utilized most effectively by poultry?
2. Which mineral is extremely important in the nutrition of the laying hens? Why?
3. Why is it important to maintain a minimum level of inorganic P in poultry diets?
4. What are advantages and disadvantages to the use of fish meal in poultry diets?
5. How does a pullet developing diet differ from a broiler finishing diet?

V. Problems

1. Calculate the total and available phosphorus content (%) of the following broiler diet.

Feed (% Total P)	Amount, lb
Corn grain (.26)	1100
Soybean meal (.63)	580
Fish meal (2.9)	50
Meat and bone meal (5.1)	100
Alfalfa meal (.23)	45
Limestone (0)	5
Dicalcium phosphate (18.7)	6
Fat (0)	100
Vitamin-mineral mix (0)	14
	2000

2. Assume the following feedstuff composition.

Feedstuff	CP, %	Ca, %	P, %	Lysine, %
Sorghum grain	11.0	.04	.30	.23
Commercial supplement	40.0	4.0	2.0	—
Gr. limestone	0	34.0	0	0

a. Using the above feedstuffs, formulate 2000 lb of laying hen diet to meet the following requirements.

Crude protein—15%
Calcium—3.25%
Phosphorus—0.50%

b. So your previous formulation meets the lysine requirement of laying hens, 0.60%, calculate what percent lysine must be present in the commercial supplement to complement the level already contributed by the grain.

c. How many grams of manganese oxide (77% Mn) must be added to each ton (2000 lb) of commercial supplement in order to supply the entire manganese requirement of the birds in your previous final diet? The Mn requirement is 25 mg/kg diet. Ignore any Mn in the grain.

3. The following diet is being formulated for growing turkeys.

70.5 lb	Gr. corn (9% CP; 1550 kcal/lb ME)	
19.0 lb	Soybean meal (48% CP; 1100 kcal/lb ME)	
7.5 lb	Meat and bone meal (50% CP; 950 kcal/lb ME)	
3.0 lb	Vitamin-mineral mix	
100.0 lb		

a. Calculate the protein/calorie ratio (g CP/1000 kcal ME) in the above formulation.

b. If 2.5% fat (0% CP; 3500 kcal/lb ME) were included in the previous formulation, (substituted for corn grain) recalculate a diet to contain the same protein/calorie ratio as in part a.

Notes

Notes

13
Nutrition and Feeding of Dogs and Cats

Chapter Goals

- Identify specific nutrient needs within the dog and cat life cycle.
- Highlight unique nutritional characteristics of dogs and cats.
- Describe the different types of commercial dog and cat foods and understand how to interpret container labeling.

I. Introduction[1]

Nutrition of dogs and cats can be maintained at high levels through the use of good quality commercially produced diets available in most food stores. Every pet owner can achieve a top level of nutrition for their animal without becoming a nutrition specialist. One can select from the excellent pet food available and regulate the intake to achieve optimum nutrient intake, economically and conveniently, and still maintain a balanced dietary consumption. A nutritional background will help the pet owner to better understand their pet's nutritional requirements and the problems likely to be encountered during all stages of life; from conception to death, in health and in disease. Information on nutrition is relatively abundant from manufacturers of pet foods in both published literature and advertising. The primary purpose of this section is to point out some fundamentals of dog and cat nutrition and elaborate on deviations from the normal to permit a better understanding of all facets of pet nutrition.

II. Nutrient Requirements

Estimations of the nutritional requirements of dogs and cats are complicated by the wide variation in size, performance during maintenance, reproduction, growth, old age, physical exertion, environmental and psychological stress and temperament found in the "normal" population. Hard and fast rules are impossible to make; feeding is an individual matter and requirements serve only as guides. Diets can be suggested using estimated nutrient requirements of the pet based on body weight, but the final determination must be based on individual response to that diet, i.e., weight gain or loss, physical condition, energy etc. Although both dogs and cats are, by nature, carnivorous, it has been amply demonstrated that vegetable and cereal products are acceptable when used in proper combinations.

A. *Water*

Good quality water should be available at all times, with the exception of animals experiencing persistent, profuse vomiting or waking up from general anesthesia. It should always be clean and fresh—cats often mistake water pan for bedpan and defecate in it, so water should frequently be checked. The water needed may be drunk, or it may be present in the food consumed. With increased amounts present in the food, the amount the animal will drink is decreased. A resting dog or cat with no increased water loss or need due to environmental temperature or humidity, lactation, growth or disease will require approximately 1 ml of water per kcal of metabolizable energy every 24 hours.

---

1. This chapter was prepared by Dr Johnny Hoskins, DVM, PhD, School of Veterinary Medicine, Louisiana State University, Baton Rouge, LA 70803.

Dog's Body Weight (kg)	Water (ml) and Metabolizable Energy (kcal) per kg body wt
3	100
6	85
10	75
20	60
30	55
50	50
75	45
Cat (avg size)	65

A dog gets about 25% of his total requirements from drinking water, but a cat gets only 10% this way. Voluntary water intake will usually range from two- to three-times the amount of food dry matter consumed. Requirements markedly increase under conditions of exercise, fever, kidney disease, diarrhea, increased salt intake, diabetes mellitus or insipidus, etc. During lactation, hot weather, or severe exertion, water intake may reach four- or more times the food dry matter consumed. Decreased water intake may be caused by poor availability, poor quality, or water which is too warm or too cold.

B. *Protein*

In protein evaluation, the first consideration is completeness—are all 10 essential amino acids present? Some natural sources where this is true are milk, eggs, meat, soybeans. Some incomplete ones are gelatin, flour and wheat. The second consideration is the biological value (BV) which is the percentage of protein (or nitrogen) absorbed and retained; not excreted in the urine or feces and, therefore, presumably utilized by the body. The BV of the protein in some commonly used foods are egg 98^+%, fish meal 92%, milk 92%, liver 79%, beef 78%, casein 78%, soybean meal 67%, meat scrap 50%, whole wheat 48%, tankage 47%, whole corn 45% and gelatin 0% because it does not contain any tryptophan. Most commercially prepared pet foods contain a mixture of animal tissues, soybean meal, and cereal grains which form a mixture that yields a BV of 70% or more, thus, capable of supporting the protein functions of the body. As dogs and cats get older, feeding excessive protein tends to affect the kidneys, and this is especially important with reduced kidney function in the older animal—the commercial prescription foods are much better. The third consideration of protein evaluation is the digestibility. Protein digestibility in good quality pet foods average 80% in dry diets, 84% in soft-moist and canned ration diets, and 90% in canned meat diets. Some other typical figures of digestibility are

Fish meal 99%	Meat Scraps 75% to 86%
Liver meal 88%	Soybean meal 86%

The digestibility of a diet designed for dogs and cats is related to the temperature and length of time to which it is submitted to heating during processing. During heating, proteins associated with aldose sugars are altered to form N-glycosides, compounds which are insensitive to the proteolytic enzymes of the intestinal tract, and thus pass out in the feces. This perhaps accounts for the disappointingly poor biological value of some processed pet foods, even though the starting materials were known to have had a high reference protein value. Some products used for protein are beef by-products meal, cheese meal, fish meal, soybean meal, and dried skim milk. The proteins are the most expensive item in dog and cat foods. Inexpensive proteins of low biological value such as gelatin and collagen or those contained in tankage, low quality meat scraps, and cereal wastes are usually present in poor quality pet foods. The dietary crude protein content for adult dogs with maintenance function should be about 18% to 25% on a dry matter basis. Diets of growing kittens should contain about 35% protein of high biological value, and adults at maintenance at least 25%, both calculated on a dry matter basis.

The cat has a substantially higher protein requirement than the dog. The metabolic reason for the high protein requirement of the cat is the high activity of the amino acid catabolic enzymes in the liver. It has been shown that cats are very sensitive to a deficiency of the amino acid arginine. The feeding of a single meal that is devoid in arginine will result in hyperammonemia in less than an hour.

The cat also has a higher requirement for S amino acids relative to other mammals. The presence of high dietary S amino acid levels may be required to support the cat's thick hair coat, which is high in the S amino acid cysteine.

The amino sulfonic acid, taurine, is uniquely important in the nutrition of the cat. Taurine is synthesized from the sulfur-containing amino acids methionine and cystine sufficiently to meet the requirements of the dog. Endogenous taurine biosynthesis in cat tissue is limited and, therefore, cats fed a taurine-devoid diet are unable to maintain sufficient body concentrations. Taurine is present in bile as taurocholic acid and in high concentrations in the retina and olfactory bulb. The cat, unlike the dog, exclusively conjugates cholic acid with taurine and is unable to alternate between taurine and glycine conjugations in the production of bile. Consequently, in the cat there is a decrease in conjugated bile acids, and central retinal degeneration develops. Typically, reduced visual acuity, without total loss of vision, has been seen in older kittens and adult cats. Therefore, the cat has a continual dietary requirement for taurine to replace losses which occur in the feces. As the fiber content in the cat's diet increases, these losses increase necessitating an increased amount of taurine in the diet. Taurine requirement also increases with decreasing intake of other sulfur-containing amino acids. Based on the amount of fiber and sulfur-containing amino acids present in most cat foods, 500 mg of taurine/kg of diet dry matter is recommended. Taurine is most commonly found in natural form in animal tissues, with none to limited amounts found in higher plants.

C. *Fat*

Fats and oils have an important effect on palatability. Oils contain a high percentage of short chain or unsaturated fatty acids, while fats contain a low percentage of these fatty acids. A small increase in the amount of fat or oil in a particular diet may materially improve its acceptability to the dog or cat. Fats and oils also supply essential fatty acids. Linoleic acid, the main essential fatty acid, is the major unsaturated fatty acid in most vegetable oils and makes up 15 to 25% of poultry and pork fat but less than 5% of beef tallow and butter fat. At least 0.17% of the unsaturated fatty acid arachidonic acid is also needed in the cat's diet. Arachidonic acid is synthesized from linoleic acid by animals other than cats, is present only in animal fats, and is not present in plant products of any type. Linoleic acid is converted to linolenic acid by cats and dogs; and, therefore, neither require linolenic acid in the diet. Essential fatty acids should constitute at least 1% of the diet dry matter or 2% of caloric intake. Their sources are lard, suet, corn oil, cottonseed oil, olive oil, soybean oil and safflower oil. Inadequate fat in the diet, in addition to causing a fatty acid deficiency, may result in an energy deficiency resulting in poor growth, poor physical performance, poor reproduction, and weight loss. An essential fatty acid deficiency causes a dry lusterless hair coat and scaly skin, impairs wound healing, and changes the lipid film on the skin which predisposes to skin infection.

The absorption of vitamins A, D, E and K is associated with the absorption of fats. Rancid fat destroys vitamins A and E. Extended use of rancid fat can cause hair loss, rash, loss of appetite, constipation progressing to diarrhea, even death. Use of an antioxidant to retard rancidity, therefore, is necessary in commercial dog and cat food.

The danger of excess dietary fat is important in dogs and cats because this excess energy will be converted and stored within the body which leads to obesity. From 25% to 50% of the daily caloric need of the dog or cat can be supplied by fats. Most commercial pet foods for dogs will contain 5 to 40% fat (dry matter basis) while that for cats will contain 10% to 30% fat on a dry matter basis. Most fats present in pet foods are greater than 90% digestible and are frequently the most digestible nutrient in the diet. High amounts of fat are very advantageous during periods of high caloric need such as growth, lactation, and physical exertion. However, if the amount of dietary fat exceeds the animal's ability to assimilate it, fatty stools occur.

D. *Carbohydrate (CHO)*

Limited information is available concerning the carbohydrate requirement of the dog or cat. A dog can utilize up to 65% to 70% dietary carbohydrate while cats utilize only about 35% to 40%.

A large part of most commercial dog food is composed of carbohydrate which usually supplies the most inexpensive source of energy. When properly prepared, carbohydrates are well-utilized by normal dogs. Some starches, like those from potatoes, oats, and corn, are poorly digested unless first subjected to cooking or the heat used in processing pet foods. High dietary levels of raw starch can lead to diarrhea. Occasionally milk upsets a dog or cat. The explanation offered for this is that some animals do not have the enzyme present to digest the carbohydrate lactose found in milk, namely, lactase, or if the enzyme is present, it is not at a high enough level. Fresh cow's milk has 5% lactose whereas dried skim milk has 50% lactose. A ration with 10% or more dried skim milk may lead to diarrhea. If an adult dog or cat has an unplainable diarrhea and is being fed milk, take it off the milk, and the diarrhea will often spontaneously resolve.

While carbohydrates are not essential for the cat, they provide a cheaper source of calories than other foods. It is best if starch is cooked in the form of biscuit meal, cooked potato, parboiled rice, etc. None of these foods have a good acceptance on their own to the cat. Some sources of carbohydrates are barley, corn, wheat, potatoes, oatmeal, pabulum, corn meal, and cream of wheat.

Carbohydrates ingested in excess of the animal's energy needs are stored in the body as glycogen or fat for later utilization as energy. The only effect of excess usable carbohydrate intake is obesity. A carbohydrate deficiency without a caloric deficiency is difficult to create because of the body's ability to utilize protein for glucose production and fat and protein for energy. The digestibility of usable carbohydrates in average quality commercial dog food is about 85%.

E. Fiber

Fiber is also a part of the carbohydrate portion of the diet. Fiber has several functions in the digestive tract. Some is digested and used for energy. Other fibers absorb water and produce a more voluminous stool than do nonfibrous diets. This is helpful in stimulating and maintaining intestinal action, especially in senile or inactive animals. Fiber aids in the prevention of constipation and other intestinal problems. Most foods have 1–8% fiber. Reducing diets for dogs or cats contain up to 32% fiber. The sources are cellulose, hemicellulose, lignin, cereal grain hulls, bran, beet pulp, agar, and string beans.

F. Gross Energy

Animals expend energy in almost every form of body activity and this energy is obtained from either oxidation of food or from destruction of their own bodily sources of energy. The utilization of dietary energy is influenced by the kinds or type of food consumed, the recency of eating, the quantity of food stored in the body and the amount of exercise.

Carbohydrates and fats provide the largest sources of dietary energy, especially for dogs. Proteins are also an important source of energy for cats. The caloric requirements are higher on a per pound basis in smaller animals than larger and in growing dogs and cats compared to mature adults. From 70% to 90% of dietary dry matter intake is used for energy production.

1. Caloric determination of average quality commercial pet foods for dogs and cats—when food is oxidized completely in a bomb calorimeter, the total combustible energy released as heat is known as gross energy. Gross energy values for carbohydrates, fat, and protein average 4.2, 9.4, and 4.4 kcal/g, respectively. However, not all of the gross energy contained in food is available for support of metabolism. Some is lost during digestion and appears in the feces. The differences between the gross energy consumed and that in the feces is known as apparent digestible energy. Of the apparent digestible energy, a small loss appears in the urine. The differences between the apparent digestible energy and that in the urine is known as metabolizable energy. The regulation of commercially prepared pet foods has designated that the energy content of pet foods be reported as metabolizable energy. However, giving the energy content of a particular pet food is voluntary and not required by regulation. The estimated metabolizable energy concentration (also known as caloric density) of a pet food in which the amount of metabolizable energy provided in the food has not been volunteered may be calculated as follows:

   a. Assume the values below.

Commercial Food Analysis		Reference Values
Moisture	7%	Fat = 9.4 kcal/g
Fat	9%	CHO = 4.2 kcal/g
Crude protein	24%	CP = 4.4 kcal/g
Fiber	3%	Fat = 92.5% digestible
Ash	15%	CP = 80% digestible
Total	58%	CHO = 85% digestible

CHO = 100 − 58 = 42% CHO
Fat = 9% = 9g/100 × 9.4 kcal/g × .925 dig. = 78.2 kcal/100g
CHO = 42% = 42g/100 × 4.2 kcal/g × .85 dig. = 149.9 kcal/100g
CP = 24% = 24g/100 × 4.4 kcal/g × .80 dig. = 84.5 kcal/100g

TOTAL 312.6 kcal/100g

1 lb = 454g
312.6 kcal/100g × 4.54 = 1419 kcal/lb of food

General comments—When the digestibility of fat, CHO and CP nutrients in a pet food isn't known, these are the best reference values to use in calculating the metabolizable energy (caloric density) provided by that diet for dogs and cats. These values and, therefore, the metabolizable energy provided are higher in above average and lower in below average quality pet foods. If the ash content is not available, the percent ash may be estimated as 9% in dry, 6% in soft-moist, and 2.5% in canned foods.

   b. Short method of estimation
      (1) Dry food = 1 lb—figure 75% digestibility;
         thus, 454 g × .75 = 340 g digested.

         Use fudge factor of 4.3 kcal/g
         340 × 4.3 = 1462 kcal/lb
      (2) Canned food—take 454 g × 25% to get solid part, as the rest is water. Assume 95% digestibility,
         thus, 454 g × .25 = 113 g × .95 = 108 g digested.
         Again use fudge factor of 4.3 kcal/g:
         108 g × 4.3 = 464 kcal/lb

2. Caloric requirements for dogs—these vary with age and activity of the individual animal. A rule of thumb is as follows: X = body wt in lbs
   a. 5 to 65 lb dog needs: X (33–1/4X) = kcal/day needed
   b. More than 65 lb: 18X = kcal/day needed.

3. Requirement calculation for dogs—this has a number of variables and the following increases are of kcal above maintenance.
   a. Age variable
      (1) Add 10% for dogs 1 year of age.
      (2) Add 30% for dogs 6 months of age.
      (3) Add 60% for dogs 3 months of age.
   b. Activity variable
      (1) Add 25% for moderate activity.
      (2) Add 60% for heavy activity—hunting dog.
   c. Fever variable—add 7% per degree above normal (102°F)
   d. Pregnancy and lactation variables
      (1) From conception to parturition increase 20%.
      (2) At parturition—increase 25%.
      (3) 2nd week of lactation—increase 50%.
      (4) 3rd week of lactation—increase 75%.
      (5) 4th week of lactation—increase 100%.
   e. Sample problems
      (1) 20 lb adult dog
         20(33 − 5) = 560 kcal
      (2) 20 lb dog 6 months old
         20(33 − 5) = 560 kcal + .30(560) = 560 + 168 = 728 kcal

4. Caloric requirements of cats
   a. Energy requirements of the cat may be summarized as
      (1) Weanling—115 kcal/lb.
      (2) Young adult—65 to 70 kcal/lb.
      (3) Adult (depending on activity)—27 to 70 kcal/lb.
   b. Cats should obtain 25% to 35% of their dietary calories from protein. They will do a poor job of obtaining calories from carbohydrate or fat unless in the presence of good quality proteins. Thus, most cat nutritionist favor a high-fat, high-protein diet.

G. *Minerals*

Some of the essential ones are Ca, P, Mg, S, Na, Cl, K, Fe and the elements C, O, H and N. The essential trace minerals are Cu, Mn, I, Zn and Co.

1. Calcium and phosphorus

    The proper ratio of these is about 1.2:1 (Ca:P). If deficiency occurs one sees rickets, osteomalacia, and loose teeth. Young and old dogs alike need additional Ca at these ages, so one needs to feed a diet appropriate for their age and activity or must supplement to correct Ca or P in their diet. The common sources of Ca are bone meal, skim milk, and alfalfa leaf meal. Sources of P are bone meal and meat scraps. Vitamin D is necessary for proper utilization of Ca and P.

2. Magnesium

    Mg appears to function in a number of processes in intermediary metabolism, in particular as an activator of the enzyme, phosphatase. Mg is closely associated with Ca and P in the body in both distribution and metabolism. Mg is an essential constituent of bone and teeth. Deficiencies of Mg cause hyperirritability and convulsions (tetany) in dogs while Mg deficiency has not been described in cats.

    High dietary magnesium levels in the cat diet may play a role in the development of a condition known as feline urologic syndrome (FUS). The magnesium salt, struvite, is precipitated in the urine forming urinary calculi and obstruction, especially in male cats. However, magnesium intake is not singularly responsible for the development of this disease. The low water intake of the cat, producing a very concentrated urine, is probably a contributory factor. On a meat diet, carnivores have an acid urine. When plant products are introduced into the diet, the urine may become alkaline. Struvite is insoluble at a pH of greater the 6.6 and precipitates out, forming urinary calculi. The inclusion of high amounts of cereal grains and low amounts of meat products in cat food may, therefore, also be a contributing factor to the development of FUS in cats.

3. Sulfur

    A natural deficiency is very unlikely, as a natural source is protein foods. Some amino acids contain S in their structure, namely, cystine, cysteine, and methionine.

4. Sodium and chlorine

    Although required by all animals, there is little evidence that serious deficiencies will occur when natural foodstuffs are fed. The diet should contain 1.0 to 1.5% NaCl.

5. Potassium

    Deficiency doesn't occur naturally in dogs and cats, most foods given are good sources. Potassium maintains proper neuromuscular irritability and good muscle tone. The deficiency signs that have been produced are anorexia, weakness, lethargy and decreased muscle tone, which may cause head drooping, ataxia and ascending paralysis.

6. Iron

    Iron is needed for proper hemoglobin formation. Some natural sources are liver and egg yolk; milk is notoriously low in Fe. Very seldom is additional iron supplementation necessary in normal, healthy animals. A strict vegetarian diet will cause deficiency.

7. Trace minerals

    a. Copper—important in hemoglobin formation as is iron. Good sources are dried skim milk, liver, and egg yolk. The requirement is approximately 10% of the dietary iron level.

    b. Cobalt—this is also needed for hemoglobin formation. A deficiency causes anemia, debility, and weight loss. Natural deficiencies seldom occur in dogs and cats.

    c. Manganese—deficiency causes reproductive failure—delayed puberty, irregular ovulation, weak or stillborn animals. Some sources are grains, soybeans, fish meal, limestone and rock phosphate.

    d. Zinc—a deficiency causes generalized thinning of hair coat, scaly dermatitis, parakeratosis, anorexia, weight loss and slow growth. Zn is found in insulin and pancreatic and duodenal digestive juice. Zn is in all common foods, especially vegetables. Colostrum is very high in Zn.

    e. Iodine—a deficiency produces poor hair coats, anemia, hypothyroidism, and abortions (be suspicious if young aborted near term are hairless). Sources are iodized salt, KI or NaI and occasionally tincture of iodine. Most dog and cat foods contain adequate iodine. Fish scraps are high in iodine.

H. *Vitamins*

The exact requirements for some vitamins are not known because of the variance of diet compositions, physiological status of the animal, presence of disease etc. The average well-balanced dog or cat diet probably contains adequate vitamins for most practical conditions. The water-soluble vitamins are sparingly stored in the animal body. B-vitamin requirements are higher in the cat (nearly five times higher than dog).

1. Vitamin A

Vitamin A is important for good vision, epithelial health and bone growth. A deficiency may cause loss of appetite, poor growth, excessive shedding, lowered resistance of mucous membranes leading to weak and infected eyes; also animals are predisposed to upper respiratory tract infections. If vitamin A is deficient, it is possible for dogs to become deaf. Natural sources are fish, fish liver oils, beef liver, kidney, green vegetables, milk, cheese, butter and eggs. Beta carotene is not utilized by the cat for a source of vitamin A. Cats require preformed vitamin A in their diet since they lack the ability to effectively convert beta carotene to vitamin A. Liver, cod-liver oil, and retinyl acetate or palmitate are satisfactory sources of vitamin A for the cat.

Signs of hypervitaminosis A are seen more frequently than deficiency. Signs are anorexia, weight loss, dystrophic calcification of bone, hypersensitivity, and neck or joint stiffness.

2. Vitamin D

Vitamin D is involved in the metabolism of Ca and P, and probably of Mg. Both ergocalciferol (vitamin D_2) and cholecalciferol (vitamin D_3) have been used successfully in the diet for growth and reproduction of dogs and cats. However, vitamin D_3 is generally more effective in dogs and cats than is D_2.

Vitamin D is also called the antirachitic vitamin. Often see a deficiency in German shepherds, Labradors, and other large dogs with a corresponding presence of rickets. A deficiency causes rickets in young and osteomalacia in adults. Another manifestation is irregular development of teeth with delayed eruption patterns. Sources are sunlight, irradiated yeast, fish liver oils and egg yolks. Signs of hypervitaminosis D are anorexia, nausea, fatigue, renal damage, soft tissue calcification, hypercalemia, diarrhea, dehydration and death.

3. Vitamin E

Vitamin E deficiency causes reproduction and lactation failures. Steatitis (yellow fat disease) in cats is also caused by deficiency of vitamin E. Vitamin E deficiency may be associated with muscular dystrophy. Natural sources are corn oil, wheat germ oil, fish, meat and egg yolk. Vitamin E is easily destroyed by heat processing especially in the presence of oxygen and by the interaction with other constituents such as mineral salts during shelf life. Vitamin E in cat diets high in unsaturated fatty acids is also easily destroyed. No hypervitaminosis E has been observed.

4. Vitamin K

Vitamin K deficiency causes hemorrhage due to reduced prothrombin and prolonged clotting times. Causes of vitamin K deficiency are liver diseases and rodent poisoning. This latter is most common in dogs and cats. The absolute requirements are unknown. Dogs and cats can synthesize enough in their own intestine if no liver disease or poisoning is present. Menadione (synthetic vitamin K_1) is used therapeutically to treat vitamin K deficiency. Natural sources of vitamin K are egg yolk, liver, green leaves and dried alfalfa.

5. Thiamine

Thiamine deficiency causes anorexia, weight loss, dehydration, paralysis and convulsions. Overheating or heat during processing will quite easily destroy thiamine. About two thirds of it is destroyed when commercial dog or cat food is cooked. It must be added from separate sources to overcome this loss. Natural sources include organ meats, especially the liver (raw best), wheat germ and dried brewer's yeast. Green vegetables and table scraps are also a good source of thiamine.

High dietary carbohydrate levels, such as potatoes and bread, increase the dog's requirement for thiamine. The need for thiamine is decreased slightly when higher levels of fat are added. Thus, composition of the diet influences the vitamin requirement. Some pet owners have fed raw, ground fish to dogs. Raw fish has transmitted tapeworms to dogs but has probably caused more trouble by creating a thiamine deficiency. Most freshwater raw fish, such as carp, contain an enzyme, thiaminase, that destroys the thiamine. Cooking destroys thiaminase and makes the fish completely harmless in this respect. This is a good example of an antivitamin in animal nutrition.

6. Riboflavin

   Riboflavin deficiency causes dry scaly skin, erythema of hindlegs and chest, muscular weakness in hindquarters, anemia and sudden death. Natural sources are liver, wheat germ, leafy vegetables, and fresh milk.

7. Niacin (nicotinic acid)

   Niacin deficiency causes poor appetite, diarrhea, and weight loss. In the advanced cases dogs may develop pellagra or "black tongue" characterized by oral ulcers and necrosis of the tongue. Natural sources are liver, leafy vegetables, eggs and milk.

   The quantitative requirement for niacin is influenced by dietary tryptophan which can be metabolically converted to niacin. Much of the niacin in cereals is in a bound form and unavailable or only partly available to the animal. The niacin content of cereals should probably be ignored when calculating the contribution of natural ingredients to the dietary niacin supply. While the niacin requirement of dogs and many animal species can be satisfied by the administration of the amino acid tryptophan, it has been established that tryptophan cannot substitute for preformed niacin in the diet of the cat.

8. Pyridoxine (vitamin B_6)

   Pyridoxine deficiency causes microcytic hypochromic anemia and retarded growth—this latter is probably due to interference with protein metabolism. Results as a deficiency from over cooking of dog or cat food or food source. Natural sources are wheat germ, milk, meat, fish, yeast and liver.

9. Lesser vitamins

   a. Pantothenic acid—deficiency causes anorexia, retarded growth, gastroenteritis, fatty livers and depigmentation or discoloration of hair. Natural sources are liver, yeast and molasses. Pantothenic acid is not easily destroyed by heat, thus there is little chance of a deficiency under normal circumstances.

   b. Choline—deficiency causes fatty liver. Choline is necessary in fat metabolism (synthesis of lecithin) and transportation within the body. Natural sources are liver, yeast and soybean oil.

   c. Biotin—a deficiency causes posterior paralysis and seborrhea. Raw egg whites contain a substance called avidin which is an anti-vitamin and will tie up biotin. Hence, don't feed a diet of all raw egg whites. Natural sources are liver and yeast.

   d. Folic acid—a deficiency causes hypoplasia of the bone marrow, macrocytic anemia and glossitis. Natural sources are yeast and liver.

   e. Vitamin B_{12}—is of primary importance in red blood cell formation. It is essential to prevent and cure anemia and necessary for the growth and development of healthy young pets. Dogs in heavy work or training need vitamin B_{12} to facilitate the development of ample erythrocytes capable of carrying oxygen from the lungs to the muscle. Usually it is supplied in adequate amounts due to synthesis by the intestine, plus the average diet has necessary requirements. If the gut is abnormal, however, then either inadequate synthesis or absorption may occur. Natural sources are liver, milk, cheese, eggs and lean meat.

10. Vitamin C (ascorbic acid)

    Vitamin C is also called the antiscurvy vitamin. A deficiency causes delayed wound healing and scurvy type lesions and symptoms—cutaneous or subcutaneous hemorrhage around the teeth or gums. Teeth may loosen, and bloody diarrhea and tender joints may occur. The normal sources are synthesized by the animal. A deficiency may occur when there is a digestive or intestinal problem of some sort. Fresh fruits or vegetables are essential in diets of man, primates and guinea pigs, who mostly experience vitamin C deficiency.

III. General Nutritional Considerations for Feeding Dogs and Cats
    A. *Variables in Individual Requirements*

Some factors that influence dietary needs of dogs and cats are
1. Size of the animal.
2. Age of the animal.
3. Work performed.
4. Temperament of the animal.
5. Environmental factors (temperature, humidity, air movement, stress).
6. Hormonal balance (spayed, castrated).
7. Health (kidney disease, cardiac problems etc.).

    B. *Types of Commercial Foods and Table Scraps*
1. Dry food

These foods commonly contain whole or dehulled cereal grains, cereal byproducts, soybean products, animal products, milk products, fat and oils and mineral and vitamin supplements. These are blended together in amounts pre-calculated to supply a nutritious protein and proper protein-to-calorie ratio. The cereal (carbohydrate) sources are heat-treated to dextrinize the starches and improve their digestibility. Sufficient amounts of fat are added to increase the caloric density and vitamins and minerals are accurately measured and blended throughout the meat and cereal mixture. Most mixtures contain about 6% to 10% moisture and the average caloric value is 1500 to 1600 kcal per pound or 300 to 400 metabolizable kcal/8 oz. cup. There are three main types of dry foods.

    a. Dry meals—dry food may be pelleted or pelleted and then crumbled to a uniform particle size. These may be fat-coated, which increases their caloric density and enhances their palatability. Meal type of dog foods have been replaced by expanded products. There are currently no nationally marketed meal type dog foods (Gaines Meal is an expanded dog food).

    b. Kibbles—cereal grains and dried meat scraps along with dairy products, vitamins and minerals are ground together into a flour, blended with water and formed into a dough. This may be baked on a large sheet and then crumbled or "kibbled" into uniform-sized fragments. The only national brand of kibbles is Ken-L-Ration Biskit (Quaker Oats).

    c. Expanded dry foods—raw grains, meat meal, vegetables, dairy products, vitamins and minerals are mixed with steam inside a blending pressure cooker allowing the ingredients to be cooked while being whipped into a homogeneous mixture. This is then pushed through a die and expanded with steam and air into small porous nuggets. The nuggets are hardened by passing through heated air streams. The dried, hardened nugget is usually passed through a spray chamber and coated with a liquid fat, carbohydrate or milk product. This coating process provides additional energy or palatability to the nugget. Products which are designed to be expanded must be formulated to contain at least 40% carbohydrate or the expansion process won't work. Prior to packaging, a substance called digest also is sprayed on most cat foods and some dog foods to increase their palatability. This process of spraying substances on the outside of the product is called enrobing. Digest is a liquid substance produced by the controlled enzymatic degradation of a variety of animal tissues. The animal tissues most commonly used are ground poultry byproducts including viscera, heads and feet, fish, liver and beef lungs. It is often used to justify a different flavor designation for pet foods. Most dry pet foods available are of the expanded type.

2. Semimoist foods

These foods generally contain a mixture of soybean meal, corn syrup, fresh meat or meat byproducts, animal fat, vitamins, and minerals together with preservatives and humectants. The moisture content is about 23% to 40%. These are commonly packaged with cellophane or foil in portion controlled servings and because of the preservatives and humectants, can be stored unrefrigerated. Semimoist foods are often shaped and colored to resemble meat chunks or hamburger patties. They are nutritionally complete plus being quite palatable. There are about 250 kcal in a 3-oz patty, 500 kcal in a 6-oz packet. Many semimoist foods, especially cat foods, contain acid as an ingredient. Phosphoric, hydrochloric and malic acids are the most commonly used acids in semimoist foods and lowers the pH to retard bacterial growth and spoilage. Sugars, corn syrup and salts elevate the soluble solids in the product and actually bind the water so it is unavailable to bacteria and fungi. Propylene glycol, an added ingredient, is hygroscopic and binds moisture in the product to keep the food pliable and prevent drying. A major advantage of the semimoist form is that more ingredients, including fresh animal tissues, can be used in their formulation. However, to minimize cost, similar ingredients frequently are used in both dry and semimoist pet foods.

3. Canned foods

   Fresh, wet ingredients are sealed into containers (generally cans) to prevent any recontamination and then subjected to a heat-sterilization process to destroy any microorganisms of spoilage already in the food. Types of canned foods include the following:

   a. Ration-type canned foods—ground fresh meat and meat byproducts along with fat, water, and cereal ingredients are blended to make a complete balanced diet. These diets generally contain 75% water and the better grades contain 500 kcal/lb.

   b. Gourmet or meat-type canned foods—the gourmet products may appear to contain a substantial amount of skeletal muscle (meat) but actually contain a variety of animal byproducts and textured vegetable protein which is composed of extruded soy flour mixed with red or brown coloring in order to make it look like chunks of meat or liver. Unless the purchaser is familiar with the appearance of textured vegetable protein chunks, it is easily mistaken for meat chunks.

   The high protein content of gourmet type canned foods requires the animal to use protein as its major energy source. Because of their high protein and fat content and high palatability, they are excellent to feed when food intake is decreased because of anorexia from any cause (except renal failure) and when protein requirements are increased such as for extensive wound healing and protein losing nephropathy or enteropathy. Cat foods, particularly the canned gourmet foods, are extremely palatable; and, therefore, are a good diet to try to induce voluntary food consumption in either the anorectic dog or cat.

   Gourmet cat foods are sold in 3-, 6-, or 6.5-oz. cans, but ration-type cat foods are sold in 12- to 22 oz. cans. The gourmet cat foods are composed primarily of animal tissues such as shrimp, tuna, kidney, liver, and chicken and numerous combinations. Because these foods are highly palatable, cats frequently become addicted to a specific ingredient. Some of the gourmet cat foods, even nationally available brands produced by reputable companies, are not nutritionally complete. For example, one contains less than 0.2% calcium and more than 0.6% phosphorus in its dry matter. Thus, it is deficient in calcium and contains a Ca:P ratio of less than 1:3. If fed for any length of time, it would cause nutritional secondary hyperparathyroidism and extensive bone demineralization.

4. Table scraps

   Table scraps are frequently quite palatable to dogs but generally not nutritionally balanced. Most table scraps are fats and carbohydrates, yielding lots of calories and little else. As a consequence, the dog obtains a sizeable portion of its daily caloric need from the useless scraps and loses his appetite entirely for the commercial food. Spicy foods should not be given to any animal.

5. Why additives are used in pet foods

   While some additives to pet foods do not provide any diet nourishment to the animal, they are very beneficial. Most additives are included in pet foods to enhance the keeping quality of the product and prevent deterioration following processing, assuring a better quality product at the time of consumption. Additives commonly used in pet foods are emulsifiers or surface active agents, antioxidants, antimicrobial agents, flavors or flavor potentiators, and color additives.

   Emulsifiers keep water and fat from separating. Color additives contribute to the physical appearance of pet foods. Antioxidants prevent fat from becoming rancid, and antimicrobial agents prevent spoilage. Flavors are used to enhance the acceptability by the animal or justify the marketing of a separate flavored product. Digest as a flavor potentiator is the greatest single factor enhancing dry food palatability for cats and to a lesser extent for dogs. Digest is a liquid substance produced by the controlled enzymatic degradation of a variety of animal tissues such as ground poultry byproducts including viscera, heads and feet, fish, liver and beef lungs. Proteolytic enzymes present in or added to the tissues partially digest the protein. Degradation is stopped by the addition of acid which creates a pH unfavorable for enzymatic action. The resulting liquid, called digest, is sprayed on the exterior of dry foods generally in an amount that constitutes 3% to 5% of the food.

C. *Feeding Methods*

There are three methods of feeding for dogs and cats—free-choice, *ad libitum,* or self-feeding; time restricted meal-feeding; and food restricted meal-feeding. Free-choice feeding is that situation in which more food than the animal will consume is always available; thus, the animal can eat as much as it wants and whenever it chooses. With time restricted meal-feeding, the animal is given more food than it will consume within a specified period of time, generally 5 to 30 minutes; whereas, with food restricted meal-feeding the animal is given a specific but less amount of food than it would eat if the amount fed were not restricted. Both types of meal-feeding are repeated at a specific frequency such as once or twice a day. Some people use only one, and others a combination of these three feeding methods; such as free-choice feeding a dry or soft-moist food and meal-feeding a canned food or specific food(s) such as meat, table scraps etc.

The method of feeding will be somewhat determined by the type of food used. Dry foods can be successfully self-fed (*ad libitum*) to most dogs and cats, however, some will overeat and become obese or have digestive disturbances. Dry food is placed in a container with water in another container. The dry food's abrasive action on the teeth help keep them scaled and clean. Gum exercise is also provided by the chewing of the dry food. When dogs and cats tend to overeat on *ad libitum* feeding, the next easiest way of feeding is to employ either the time restricted or food restricted meal feeding method.

For the dog or cat that consumes inadequate quantities of dry food, one should determine if this is attributable to some problem such as sore gums or lips or bad teeth. Should this occur then moistening the food to soften it will increase consumption by that animal. The addition of 20% lean meat to the moistened diet will further increase consumption.

Canned foods, fresh foods, and moistened dry foods should be opened or prepared fresh daily and not exposed to the air for more than 10 to 12 hours in summer. Since pet foods are also nutritious for most bacteria they can "spoil" if left out too long in warm temperatures. A better recommendation is a regular feeding time where the owner may have a check on the animal's appetite each day. Uneaten food should not be in front of them for more than 30 minutes, especially during warm weather.

Avoid between-meal snacks and table scraps. They can easily become excessive and may result in an inadequate or unbalanced diet, obesity, digestive disturbances, and development of a finicky eater or food beggar. If occasional scraps are fed, avoid sweets, rich gravies, and bones. Poultry bones, chopped bones, or small bones may lodge in the animal's mouth or gastrointestinal tract. Large bones may result in broken teeth, although they do help clean the teeth. Table scraps or any single food item should not constitute more than 25% of the animal's ration. Although most adult dogs eat rapidly and voraciously, many dogs are inhibited-type eaters and prefer to be left alone while eating.

Most cats like to eat alone and without distractions or worry of competition. If more than one cat is fed, they should have separate bowls (as well as litter pans), and their bowls should be separated. If dogs are present, the cat should be fed at a different time and place than the dog. Many cats, particularly females, intact or neutered, are erratic eaters, consuming little food for one or several days and then consuming a large amount the following day. If the diet is changed during the time when little food is being consumed, one may mistakenly assume that food consumption was decreased because the cat did not find the previous food palatable. This belief is strengthened if food consumption increases when the diet is changed. However, these fluctuations in food consumption may have nothing to do with the palatability of the diets, but instead may simply be due to the cat's erratic eating behavior or the novelty of a new diet.

Regardless of the method of feeding used for cats, it is best to feed a ration type of cat food and to feed on a regular schedule. A ration type of cat food is one that contains several protein and calorie source foods. Avoid between-meal snacks and table scraps. They may create obesity. Any single food item, such as tuna fish or cat food consisting of primarily or exclusively a single food item such as some of the gourmet cat foods, should not constitute more than 25% of the animal's diet. Many cats fed a diet consisting primarily of a single food item will develop a fixed food preference and may refuse to eat anything else. Even if the food is nutritionally complete and adequate, a fixed food preference is undesirable. It makes it difficult to change the cat's diet as may be necessary for different situations or the management of many feline diseases. A fixed food preference, however, is easily prevented by feeding a ration type cat food which provides a variety of different food items.

D. *How Much to Feed*

The amount of food to be given to the dog or cat is determined by

1. Trial and error.
2. Caloric requirement.
3. Rules of thumb.

   Most household pets consume about ⅓ to ½ ounce of dry matter food per pound of body weight when they are inactive (at maintenance). Puppies will consume approximately three times this quantity during the fast growth period. Hardworking and lactating dogs will consume up to three times maintenance levels. When canned diets are fed, approximately three times as much by weight is needed as when dry foods are fed. Since commercial pet foods vary in density, feeding on a weight basis is far more satisfactory. Quantities consumed by individual dogs vary with individual utilization and two related dogs of the same strain may require different levels of food intake to maintain their body condition.

   Heavy exercise increases the nutritional requirements. A good dog may lose up to 20 lbs during the hunting season. Cold weather will increase the requirement of food. The size of the animal must be considered—a small dog will require more food per lb than will a large dog. Nervousness is another factor. Purebred breeds have a tendency to be more nervous, need more food, but they are always thin and often have a diarrhea problem. Spayed and castrated animals require one third to one half as much food than they needed originally because of less natural exercise.

   Then how much should a dog be fed? Requirements normally include

	% of Body Weight Daily as Dry Food
Maintenance	2½
Gestation—1st 6 weeks	2½
last 3 weeks	3
Lactation—1st 3 weeks	3
3–6 weeks	5–7½
Medium work	5
Extremely hard work	7½

Added for outside housing during cold weather—20% additional food; subtracted for extremely hot weather—20% less food.

E. *Feeding Programs*

1. Feeding during pregnancy and lactation

   The primary goal in feeding during pregnancy and lactation is to provide the bitch or queen and young with a nutritionally balanced diet. The type of food and method of feeding to meet the nutritional needs of the cat are quite different from those of the dog.

   A canine reproduction or growth diet can be fed to the bitch throughout pregnancy but is needed especially during the last 3 to 4 weeks of pregnancy and during lactation. The diet on a dry matter basis should be at least 80% digestible and contain at least 25% protein, 17% fat, 1750 kcal of metabolizable energy per pound, less than 5% fiber, and from 1% to 1.8% calcium and 0.8% to 1.6% phosphorus (with the amount of calcium exceeding phosphorus).

   A feline reproduction and growth diet can be fed throughout pregnancy but is needed particularly during the last 3 weeks of pregnancy and during lactation. The diet on a dry matter basis should be at least 80% digestible and contain at least 35% protein, 17% fat, 1800 kcal of metabolizable energy per pound, and from 1% to 1.8% calcium and 0.8% to 1.6% phosphorus.

   The bitch or queen should not be given any supplements (meat, milk, calcium, phosphorus, or vitamins) or fed anything other than a good quality diet meeting the specifications outlined above. The bitch or queen at optimum body weight at breeding should be fed the same amount of food needed for maintenance during the first 5 to 6 weeks of pregnancy. Thereafter, the amount fed should be gradually increased so that the dam is receiving 15% to 25% more calories by the time of parturition. During pregnancy, feed free-choice or twice a day.

Throughout lactation, feed at least three times a day or free-choice to maintain optimum body weight of the dam. A dam is fed 1.5 times the amount of food needed for maintenance the first week of lactation, two times maintenance for the 2nd week, and three times maintenance for the third week to weaning. Encourage the young to begin eating solid food at 3 weeks of age to assist the dam in maintaining her optimum body weight during peak lactation (third through sixth week).

2. Feeding and raising young dogs and cats
   a. Orphan puppies and kittens
      1) Environment—a separate quarter is recommended for each young dog or cat so they don't bother one another. The environmental temperature for the first 7 days of life should be kept at 85° to 90°F; 80°F for the next 2 to 3 weeks; and 75°F by the fourth week. Bedding should be cleaned daily to prevent skin rash. Disposable baby diapers make good bedding.
      2) Milk replacement—replacement feeding with a prototype of nutritive substance formulated to meet the optimum requirements of the young is needed. Various modifications of homemade and commercially prepared formulas simulating mother's milk have been used with good success.
      3) Homemade prepared formulas for milk replacement
         (a) Evaporated milk, reconstituted to contain 20% total solids and supplemented with 8.0 g of dicalcium phosphate per liter of formula.
         (b) Fresh whole cow's milk: 1 pint (473 ml)
             Single cream (minimum 18% fat): 6 fl oz (182 g)
             One egg yolk: 15 g
             Cod liver oil: 2–3 drops (0.3 g)
             Sterilized bonemeal: 1 teaspoonful (4 g)
             Citric acid: 1/2 teaspoonful (2 g)
         (c) Evaporated milk and sterile water, 4:1 by volume
             Dicalcium phosphate: 5 g per liter
             Multiple infant vitamins: 1 drop
         (d) Fresh whole cow's milk: 8 fl oz (240 ml)
             One egg yolk: 15 g
             Multiple infant vitamins: 1 drop
             Corn oil: 5 ml (4.75 g)
             Cod liver oil: 2–3 drops (0.3 g)
      4) Commercially prepared formulas for milk replacement
         (a) Esbilac, powder or SPF-lac, liquid (Borden Company) for puppies.
         (b) Unilac, liquid & powder (Upjohn Company) for puppies.
         (c) KMR, liquid (Borden Company) for kittens.
         (d) Havolac Food Supplement, liquid in 13 fl oz can (Haver Company) for puppies and kittens.
         (e) Veta-Lac, powder (Vet-A Mix, Inc) for puppies and kittens.
      5) Supplemental feeding in preparation for weaning
         (a) One part dry puppy food blended with 3 parts water.
         (b) Two parts canned puppy food blended with 1 part water.
         (c) One part dry kitten food blended with 1 part water.
         (d) Two parts canned kitten food blended with 1 part water.
      6) Feeding amount and frequency—these formulas generally provide 1 to 1.24 kcal of metabolizable energy per ml of formula. Feed 1 ml/7.5 g body weight during first week; 1 ml/ 6.0 g during second week; 1 ml/5.0 g in third week; and 1 ml/4.5 g in fourth week. This provides a food intake of about 10% to 20% of body weight per day at birth which gradually increases to 20% to 25%. Underfeed slightly during the first several weeks of life to avoid digestive disturbances. Feed three to four times daily. After feeding, the abdomen is slightly enlarged but not overdistended. Start with less than the prescribed amount of formula for the first several feedings and increase the amount gradually to the recommended feeding by the second to third day. Continue to increase the formula as weight gains and favorable responses to feeding are observed.

7) Methods of feeding—keep all equipment scrupulously clean; do not prepare more formula than is required to feed for a 48-hour period; divide the formula into portions that are required for each feeding and store in refrigerator (+4 degrees C); and before feeding, warm the formula to about 100°F or near body temperature.

(a) Nipple bottle feeding—nipple bottles made especially for feeding orphan puppies or kittens are preferred. When feeding, hold the bottle so that the young does not ingest air. Enlarge the hole in the nipple so that when the bottle is inverted, milk slowly oozes from the nipple. When feeding, squeeze a drop of milk onto the tip of the nipple and then insert it into the animal's mouth. Never squeeze milk out of the bottle while the nipple is in the mouth; doing so may result in laryngotracheal aspiration, pneumonia, and death.

(b) Tube feeding—tube feeding is the easiest, cleanest, fastest, safest, and most preferred way to feed the orphan puppy or kitten. Fill a syringe with formula, fit with an 8 French, 12 in. human infant rubber feeding tube and deliver appropriate quantities of milk directly into the stomach. Mark the feeding tube clearly to indicate the depth of insertion to ensure gastric delivery, i.e., once weekly, measure the tube from the last rib to the tip of the noise. When feeding slowly inject the correct quantity of formula over a 2-minute period to allow for slow gastric dilation. Regurgitation rarely occurs, but if it does, withdraw the tube and interrupt feeding until the next scheduled meal. For the first three weeks of life, after each feeding burp the puppy or kitten and swab the genital area with moistened cotton to stimulate defecation and micturition.

(c) Supplemental feeding—to encourage consumption of solid food by the young, mix enough water with the solid food to make a thick mushy gruel. Smear some of the gruel on the animal's lips being careful not to get any in their nose. In licking the gruel off their lips, they will generally get started eating the solid food and progress rapidly to eating from a bowl. Once they are eating from a bowl, gradually decrease the amount of water mixed with the food until only the solid food is fed three times a day.

b. Weanling Puppies and Kittens

Feed the weanlings a diet containing 100 to 115 kcal/lb 3 to 4 times daily. Wean at 4 to 7 weeks of age (5 and one half to 6 weeks is the average) and allow 7–10 days for the weaning process. The dam will often start to wean on her own due to the irritation caused by the presence of animals' teeth and toenails. Take the dam from the young in the daytime for the first few days, putting her back with the young at night. Gradually take her away for longer periods so she will finally wean them permanently.

c. Older Puppies and Kittens

Feeding times—for first 3 months, feed them three times a day. At 6 months of age they should be fed twice a day. Dogs and cats that are 8 months to 1 year and older may be fed once daily. You may feed them twice a day at this time if they aren't fed too much at a time.

F. *Feeding and Caring for Aging Dogs and Cats*

Early detection of nutritional disturbances and proper nutritional management will slow or prevent the progression of organ system failure and possibly slow the aging process.

Maintaining good oral hygiene is important in ensuring adequate food intake and utilization. Brushing the teeth daily with a toothbrush and dentifrice for the dog or wiping the teeth daily with a soft rag for the dog or cat is helpful. Periodic mechanical cleaning of the teeth to remove tartar or extensive plaque assists in maintaining intact mucous membranes of the mouth.

The quantity of food fed should satisfy hunger but not result in unnecessary abdominal distension and discomfort. Feed small meals at least twice a day (on a regular schedule) of a palatable, highly digestible diet. The total quantity of food fed each day should not exceed that required to supply the recommended daily metabolizable energy needed for maintenance of optimum body weight.

To a normal aged dog feed a diet on a dry matter basis that is at least 80% digestible and contains at least 14% to 21% protein, 10% fat, 1,700 kcal of metabolizable energy per pound, less than 4% fiber, and from 0.5% to 0.8% calcium, 0.4% to 0.7% phosphorus, and 0.2% to 0.4% sodium, and be of good quality.

To a normal aged cat feed a diet on a dry matter basis that is at least 80% digestible and contains at least 25% to 35% protein, 15% fat, 1700 kcal of metabolizable energy per pound, less than 4% fiber, and from 0.5% to 0.8% calcium, 0.4% to 0.7% phosphorus, 0.2% to 0.4% sodium, less than 0.10% magnesium, and be of good quality.

Older dogs and cats may have a reduced appetite and digestive-absorption ability resulting in a loss of body weight. These animals should be fed palatable high-caloric diets at frequent intervals. Conditions which decrease food intake because of decreasing senses of smell and taste occur in aged dogs and cats so that it may be necessary to feed a commercially prepared canned or moistened dry food warmed or one with a strong aroma. Altering the nature of the food eaten is helpful in maintaining food intake, such as warming the food prior to feeding to volatilize the odors of wet food for cats (to a lesser degree for dogs) or adding water to a commercially prepared dry food for most dogs. Many cats become addicted to particular shapes of commercially prepared dry food, so ensure they receive the shape of food they are accustomed.

Good quality water should be readily available. The amount of water needed is at least 1 ml of water per kcal of metabolizable energy. Also of importance to the aged dog or cat is adequate physical activity to maintain muscle tone, to enhance circulation, and to improve waste elimination.

Aged animals with disease or dysfunction of organ systems, obesity, or symptoms for which treatment includes nutritional management should be fed special diets such as the commercially prepared prescription diets.

IV. Nutritional Problems

A. *Obesity*

Obesity is the most common nutritional problem occurring in dogs and cats in our affluent society; far exceeding all deficiency diseases combined. Dogs and cats are considered obese when they are 10–15% above their optimum body weight. Obesity is more common in female than male dogs up to 12 years of age and is about twice as high in neutered dogs of both sexes. Beagles, cocker spaniels, collies, dachshunds, and Labradors have the highest incidence of obesity. Obesity in cats is equally common in both sexes with higher incidence in older neutered cats. Domestic shorthairs have the highest incidence of obesity.

General health and longevity are impaired because of locomotion problems, respiratory difficulties, cardiovascular disease, hypertension, reduced hepatic function, decreased reproductive ability, heat intolerance, altered resistance to both viral and bacterial diseases, and increased dystocia, irritability, dermatoses, surgical risk and difficulty, thrombembolic disease, constipation, flatulence, bone/joint disease and diabetes mellitus.

The causes of obesity in dogs and cats are multifactorial. Factors such as endocrine imbalances and abnormal responsive taste that interfere with internal body signals are rare as compared to those that interfere with external signals. The factors most commonly interfering with external signals include constant and easy access to highly palatable energy-dense foods in excessive quantities. People overfeed dogs and cats for many reasons, including (1) treating them like people, (2) cultivating and then indulging an animal's taste for highly palatable foods, (3) force of habit—feeding several times daily or feeding the same quantity regardless of the animal's need, (4) interpreting hunger and eating as a sign of health, (5) guilt feelings—because they have left the animal alone etc. and (6) ignoring calories from treats and other food consumed outside of meal time. Coupled with inactivity, these are the most common causes of obesity.

In addition, social pressures such as competition between animals and a desire to please contribute to overeating. Feeding two dogs together frequently results in more food intake by both than would occur if they were fed separately. This may occur because the dog's natural instinct is to consume as much food as quickly as possible before other animals get it. In some instances the dog may eat because he is rewarded for doing so: a treat is offered, the dog eats it and is rewarded by being petted, noticed or spoken to. In turn, the owner feeds the dog because they think it pleases the dog and shows the dog how much it is loved.

An increased number of fat cells, which greatly predisposes to obesity, may also be induced by the development of obesity during growth. The majority of animals that have obese parents or become obese during growth have a problem with obesity throughout life. Therefore, preventing obesity during growth is very important in preventing obesity as an adult.

A weight reduction program should be instituted for all dogs and cats that are more than 15% above their optimum weight so as to (1) decrease health problems, (2) reduce future health care costs, (3) improve appearance and (4) increase the animal's enjoyment and length of life. There are five possible methods to manage obesity—psychological, exercise, dietary, pharmacological and surgical. Drugs and surgery are of no value. The psychological aspect of weight reduction includes the very necessary duty of convincing all associated with the animal of the necessity for weight reduction and teaching them how to accomplish and maintain weight loss. Exercise is quite helpful, not only in increasing energy expenditure but also because an increase in exercise from sedentary activity reduces appetite and food intake.

Methods for weight reduction program for the obese animal include
      (1) Decrease the regular commercial diet by 50% of that necessary for the maintenance of initial (obese) body weight
      (2) Feed the regular commercial diet at 60% for dogs and 66% for cats of that necessary for maintenance of optimum body weight
      (3) Feed a nutritionally complete and balanced high fiber–low calorie reducing diet, e.g. Prescription Diet r/d (Hill's Pet Products), Prescription Diet w/d (Hill's Pet Products), or prepare a similar homemade recipe.
      (4) Feed at least three times a day with the amount fed restricted to feeding times.
      (5) Keep palatable water available at all times.
      (6) Exclude all table scraps, snacks, sweets, etc.
      (7) Avoid total fasting or starvation for quick weight reduction.

If weight loss is not occurring at a satisfactory rate with the weight reduction program, the likely cause is the continuing consumption of excess calories in the form of snacks, table scraps and sweets; such a reducing diet is not calorie restrictive enough. If the animal is only receiving the prescribed amount of low caloric diet, then decrease the amount prescribed by 20% and continue monitoring the body weight. Weight loss will occur subsequently if continued amounts of food are reduced.

When optimum body weight is reached, the choice of feeding more of the reducing diet or switching to a maintenance diet has to be made. Frequently, the weight reducing diet can be fed free-choice. The amount of maintenance diet fed must be restricted to prevent recurrence of obesity. To prevent recurrence of obesity, reweigh the animal every 3 to 4 weeks, maintain a regular weight chart, and adjust the amount fed to maintain optimum body weight. If the animal begins to regain weight, decrease the amount fed by 10% or switch back to the low-calorie reducing diet.

B. *Nutritional Problems in the Cat from Eating Commercial Dog Foods*

Malnutrition of the adult cat occurs when all the basic nutritional requirements are not being met. A common cause of malnutrition in the adult cat results from the continuous ingestion of commercially prepared, cereal-based dog foods.

The cat's nutritional requirements are quite different from those of the dog. The reasons that dog foods cause nutritional problems in the cat are several. The cat has a much higher protein requirement than the dog. Little dietary arginine is needed by the mature dog but the cat will die within hours after consuming an arginine-free diet. Cats require taurine in the diet but dogs do not. Inadequate taurine results in central retinal degeneration and blindness. Cats cannot convert the fatty acid linoleic acid to arachidonic acid as can the dog, therefore, must consume preformed arachidonic acid. If they do not, they develop a dry lusterless hair coat or, if severe deprivation is present, emaciation and spots of moist dermatitis develop. Cats cannot convert beta carotene, present in plants, to vitamin A as can the dog, and therefore, must consume preformed vitamin A which is present only in animal tissues. Cats cannot convert the amino acid tryptophan to the B vitamin niacin as can dogs and, therefore, require more niacin in the diet.

Primary treatment is to feed a nutritionally balanced commercially prepared or homemade diet formulated for cats. Cats in general should not be fed any single food item or cat food consisting of a single food item, such as some of the gourmet cat foods, at more than 25% of the cat's total food intake.

V. Nutrient Requirement Tables

The nutrient requirements and example formulas for cats in tables 13–1 through 13–5 were taken from the *Nutrient Requirements for Cats,* Publication ISBN 0-309-03682-8, Subcommittee on Laboratory Animal Nutrition, National Academy of Sciences—National Research Council, Washington, D.C., 1986.

The nutrient requirements estimated daily food intake and example formulations for dogs in tables 13–7 through 13–12 were taken from the *Nutrient Requirements of Dogs,* Publication ISBN 0-309-03496-5, Subcommittee on Dog Nutrition, National Academy of Sciences—National Research Council, Washington, D.C., 1985.

TABLE 13–1. Daily Metabolizable Energy Intakes Observed for Cats

| Kitten[a] (age, weeks) | Body Weight (kg) | | | Expected Bodyweight (gain g/day) | | | |
	Male	Female	(kcal/kg BW)	Male	Female	Adult[b]	(kcal/kg BW)
10	1.1	0.9	250	20	14	inactive[b]	70
20	2.5	1.9	130	14	11	active[b]	80
30	3.5	2.7	100	7	4	gestation[c]	100
40	4.0	3.0	80	—	—		

Lactation[d] (kcal/kg BW)

Litter size	1	2	3	4	5	6
Lactation week						
1	60	76	92	108	124	124
2	66	83	100	117	134	134
3	72	94	116	138	160	160
4	78	106	134	162	190	190
5	84	117	150	183	217	250
6	90	136	182	228	274	320

[a] The kitten allowances are adapted from Miller and Allison (1958).

[b] The allowances for adult cats refer to in vivo ME data and apply to cats of 50 weeks of age or greater. The value for inactive cats is adapted from Kendall et al. (1983); that for active cats is based on several studies.

[c] The allowances for gestation are based on work by Scott (1968) and Loveridge (1986b).

[d] The allowances for lactation are adapted from Loveridge (1986b) based on voluntary intakes for queens of mean BW 3.8 kg after parturition. Values from about week 4 onward include energy intake by kittens.

NOTE: The intakes other than for normal adult cats were based on studies where ME intakes were calculated using Atwater factors of 4 kcal/g for dietary protein and carbohydrate and 9 kcal/g for dietary fat. Such ME allowances are likely to be some 20 percent higher compared with those based on determined in vivo ME (Kendall et al., 1985) depending on food type.

TABLE 13–2. Minimum Requirements for Growing Kittens (Units per kg of Diet, Dry Basis)[a]

Nutrient	Unit	Amount	Nutrient	Unit	Amount
Fat[b]			Sodium	mg	500
Linoleic acid	g	5	Chloride	g	1.9
Arachidonic acid	mg	200	Iron	mg	80
Protein[c] (N × 6.25)	g	240	Copper	mg	5
Arginine	g	10	Iodine	μg	350
Histidine	g	3	Zinc	mg	50
Isoleucine	g	5	Manganese	mg	5
Leucine	g	12	Selenium	μg	100
Lysine	g	8	Vitamins		
Methionine plus cystine	g	7.5	Vitamin A (retinol)	mg	1 (3333 IU)
(total sulfur amino acids)			Vitamin D (cholecalciferol)	μg	12.5 (500 IU)
Methionine	g	4	Vitamin E[e] (α-tocopherol)	mg	30 (30 IU)
Phenylalanine plus tyrosine	g	8.5	Vitamin K[f] (phylloquinone)	μg	100
Phenylalanine	g	4	Thiamin	mg	5
Taurine	mg	400	Riboflavin	mg	4
Threonine	g	7	Vitamin B_6 (pyridoxine)	mg	4
Tryptophan	g	1.5	Niacin	mg	40
Valine	g	6	Pantothenic acid	mg	5
Minerals			Folacin (folic acid[f])	μg	800
Calcium	g	8	Biotin[f]	μg	70
Phosphorus	g	6	Vitamin B_{12} (cyanocobalamin)	μg	20
Magnesium	mg	400	Choline[g]	g	2.4
Potassium[d]	g	4	Myo-inositol[h]	—	—

[a]Based on a diet with an ME concentration of 5.0 kcal/g dry matter fed to 10- to 20-week-old kittens. If dietary energy density is greater or lesser, it is assumed that these requirements should be increased or decreased proportionately. Nutrient requirement levels have been selected based on the most appropriate optimal response (i.e., growth, nitrogen retention, metabolite concentration or excretion, lack of abnormal clinical signs, etc.) of kittens fed a purified diet. Some of these requirements are known adequate amounts rather than minimum requirements. Since diet processing (such as extruding or retorting) may destroy or impair the availability of some nutrients, and since some nutrients, especially the trace minerals, are less available from some natural feedstuffs than from purified diets, increased amounts of these nutrients should be included to ensure that the minimum requirements are met. The minimum requirements presented in this table assume availabilities similar to those present in purified diets.

[b]No requirement for fat is known apart from the need for essential fatty acids and as a carrier of fat-soluble vitamins. Some fat normally enhances the palatability of the diet.

[c]Assuming that all the minimum essential amino acid requirements are met.

[d]The minimum potassium requirement increases with protein intake.

[e]This minimum should be adequate for a moderate to low-fat diet. It may be expected to increase three- to four-fold with a high PUFA diet, especially when fish oil is present.

[f]These vitamins may not be required in the diet unless antimicrobial agents or antivitamin compounds are present in the diet.

[g]Choline is not essential in the diet but if this quantity of choline is not present the methionine requirement should be increased to provide the same quantity of methyl groups.

[h]A dietary requirement for myo-inositol has not been demonstrated for the cat. However, almost all published studies in which purified diets have been used have included myo-inositol at 150 to 200 mg/kg diet and no studies have tested a myo-inositol-free diet.

NOTE: The minimum requirements of all the nutrients are not known for the adult cat at maintenance. It is known that these levels of nutrients are adequate and that protein and methionine can be reduced to 140 and 3 g/kg diet, respectively. It is likely that the minimum requirements of all the other nutrients are also lower for maintenance than for the growing kitten.

The minimum requirements of all the nutrients are not known for reproduction for the adult male or female cat. It is known that with the following modifications the Nutrient Allowances as recommended in the 1978 NRC report are adequate for gestation and lactation (in units/kg purified diet, note these recommendations are based on 4.0 kcal/g dry diet): arachidonate, 200 mg; zinc, 40 mg; vitamin A, 5500 IU; and taurine, 500. It is probable that the minimum requirements for growing kittens in this table would satisfy all requirements for reproduction if the following were modified as shown: vitamin A, 6000 IU/kg diet, and taurine, 500 mg/kg diet.

TABLE 13–3. Proximate Composition, Apparent Digestibility, and Digestible and Metabolizable Energy of Some Commercial Cat Diets

	Dry Type		Semimoist	Canned	
Number of Determinations	1[a]	29[b]	2[a]	2[a]	28[b]
Composition (percent)					
Dry matter	93.3	91.7	71.3	28.0	20.8
Crude protein	35.3	33.0	25.1	10.8	9.6
Crude fat	9.7	9.3	9.8	4.0	4.6
Crude fiber	2.7	—	0.7	0.6	—
Nitrogen-free extract	38.2	41.1	22.3	10.1	4.2
Ash	7.4	10.5	6.5	2.6	2.3
Propylene glycol	—	—	7.0	—	—
Gross energy (kcal/g)	4.55	4.22	3.65	1.44	1.10
Apparent digestibility (percent)					
Crude protein	78.3	76.0	83.2	80.1	80.5
Crude fat	85.3	72.3	93.9	88.6	76.1
Nitrogen-free extract	78.8	75.4	80.6	75.1	70.2
Gross energy	80.6	75.4	86.3	80.5	78.9
Digestible energy (kcal/g)	3.67	3.18	3.15	1.16	0.87
Metabolizable energy (kcal/g)	3.54	2.94	2.95	1.11	0.79
CME_1 (kcal/g)[c]	3.81	3.80	3.11	1.19	0.97
CME_2 (kcal/g)[d]	3.25	3.02	2.70	1.09	1.79
CME_3 (kcal/g)[e]	3.24	2.92	—	1.00	0.81
CME_4 (kcal/g)[f]	3.42	2.95	2.96	1.08	0.80

[a]Data from Norvell (1976). Each diet was fed to 16 adult cats housed in individual metabolism cages. All products were fed for a 7-day adjustment period and a 6-day collection period.

[b]Data from Kendall et al. (1985). Each diet was fed to 6 adult cats in individual metabolism cages. Experimental protocol was as for Norvell (1976), except collection periods were 7 days and fat was measured in food and feces by the acid ether extraction method.

[c]Calculated ME_1 values were calculated as follows (dry-type diet):

	Nutrient/ Diet	ME (kcal/g)	ME (kcal)
Crude protein	0.353	× 4	= 1.41
Crude fat	0.097	× 9	= 0.87
Nitrogen-free extract	0.382	× 4	= 1.53
			3.81

Propylene glycol in the semimoist diets was assigned an ME value of 4.7 kcal/g (Weil et al., 1971). The use of CME_1 overestimated in vivo ME by 7 to 29 percent on average, depending on food type.

[d]Calculated ME_2 values were calculated as follows (dry-type diet):

	Nutrient/ Diet	Apparent Digestibility Coefficient	E (kcal/g)	ME (kcal)
Crude protein	0.353	× 0.783	× 4.4	= 1.22
Crude fat	0.097	× 0.853	× 9.4	= 0.78
Nitrogen-free extract	0.382	× 0.788	× 4.15	= 1.25
				3.25

The E value for crude protein was derived by subtraction of energy loss in urine (1.25 kcal/g) from the average gross energy of crude protein (5.65 kcal/g). The CME_2 values for the other diets were estimated in the same way, except that propylene glycol in the semimoist foods was assigned an ME value of 4.7 kcal/g. The CME_2 values ranged from 92 to 103 percent on average of those measured in vivo.

[e]Calculated ME_3 values were calculated using regression equations (Kendall et al., 1985).

Dry-type diets:
CME_3 = 0.99 gross energy − 1.26 (kcal/g), R^2 = 0.53; RSD = 0.15 (kcal/g)

Canned-type diets:
CME_3 = crude protein × 3.9 + ether extract × 7.7 + nitrogen-free extract × 3.0 − 0.05. R^2 = 0.94, RSD = 0.06 (kcal/g)

[f]Calculated ME_4 values were calculated as follows:

CME_4 = food DE − 0.9 kcal/g protein digested, e.g., for 29 dry-type diets,
CME_4 = 3.18 − 0.251 × 0.9 = 2.95 kcal/g.

The E value of urinary urea is based on 151.6 kcal/mole urea equivalent to 5.4 kcal/g N or 0.86 kcal/g crude protein. The CME_4 values ranged on average from 97 to 101 percent of in vivo determined ME values.

TABLE 13–4. Estimated Daily Food Allowances for Cats

Cat	Weight[a] of Cat (kg)	Dry Type[b] (g/kg body wt.)	(g/cat)	Semimoist[c] (g/kg body wt.)	(g/cat)	Canned[d] (g/kg body wt.)	(g/cat)
Kitten							
10 weeks	0.9–1.1	78	70–86	83	75–91	227	204–250
20 weeks	1.9–2.5	41	78–103	43	82–108	118	224–295
30 weeks	2.5–3.8	31	78–118	33	83–125	91	228–346
40 weeks	2.9–3.8	25	73–95	27	78–103	73	212–277
Adult[e]							
Inactive	2.2–4.5	22	48–90	23	51–99	64	141–288
Active	2.2–4.5	25	55–113	27	59–122	73	160–329
Gestation	2.5–4.0	31	78–124	33	83–132	91	228–364
Lactation[f]	2.2–4.0	78	172–312	83	182–332	227	499–908

[a]Derived from Figure 1 by setting lower limit at mean weight of females minus one standard deviation and the upper limit at mean weight of males plus one standard deviation.

[b]Dry matter, 90 percent; ME, 3.2 kcal/g.

[c]Dry matter, 70 percent; ME, 3.0 kcal/g.

[d]Dry matter, 25 percent; ME, 1.1 kcal/g.

[e]Fifty weeks of age or older.

[f]Queens nursing 4-5 kittens in week 6 of lactation.

TABLE 13–5. Formulas of "Practical" and Semipurified Cat Diets

Carvalho da Silva (1950) Item	%	Fox *et al.* (1973) Item	%	Morris (1976) Item	%	Teeter *et al.* (1977a) Item	%
Crude casein	10	Casein	32.1	Soy protein[c]	33.0	Amino acid[f] mix	28.1
Raw beef muscle	20	Sucrose	37.2	Corn starch	20.0	Turkey fat	25.0
Steamed, deboned, and	20	Lard	12.6	Sucrose	15.7	Corn starch	20.71
eviscerated sardines		Turkey fat	11.7	Turkey fat	25.0	Sucrose	18.59
Rolled oat groats,	20	Salts IV[a]	3.8	D,L-methionine[c]	1.0	Glista salts[g]	5.37
slightly cooked		Vitamin mix[b]	2.6	Mineral mix[d]	4.0	Choline Cl	0.33
Potatoes, cooked	15		——	Vitamin mix[e]	1.0	Vitamin mix[h]	0.3
and mashed			100.0	Choline Cl	0.3	NaHCO₃	1.5
Lard	10				——	Taurine	0.1
Steamed bonemeal	2				100.0	Vitamin E[i]	+
Cod liver oil	3					Ethoxyquin[j]	+
	100						100.0

[a]Salts IV (%: CaCO₃, 30.0; K₂HPO₄, 32.2; CaHPO₄·2H₂O, 7.5; MgSO₄·7H₂O, 10.2; NaCl, 16.7; Fe(C₆H₅O₇)₂·6H₂O, 2.75; KI, 0.08; MnSO₄·4H₂O, 0.50; ZnCl₂, 0.025; CuSO₄·5H₂O, 0.03 (Hegsted *et al.*, 1941).

[b]Vitamins per kg diet: vitamin A, 20,000 IU; vitamin D, 2,200 IU; α-tocopherol, 110 mg; ascorbic acid, 1,000 mg; inositol, 110 mg; choline Cl, 1,650 mg; menadione, 50 mg; ρ-aminobenzoic acid, 110 mg; niacin, 100 mg; riboflavin, 22 mg; pyridoxine·HCl, 22 mg; thiamin·HCl, 22 mg; Ca pantothenate, 66 mg; folic acid, 2 mg; biotin, 440 μg; vitamin B₁₂, 30 μg.

[c]An amino acid diet has been prepared by substituting the following amino acid mix (%, total 34.7) of Hardy *et al.* (1977) for soy protein and D,L-methione:L-His·HCl·H₂O, 1.2; L-Ile, 1.8; L-Leu, 2.4; L-Lys·HCl, 2.80; L-Met, 1.1; L-Cys Cys, 0.8; L-Phe, 1.5; L-Tyr, 1.0; L-Thr, 1.4; L-Try, 0.4; L-Val, 1.8; L-Arg·HCl, 2.0; L-Asparagine, 2.0; L-Ser, 1.0; L-Pro, 2.0; Gly, 2.0; L-Glu, 6.0; L-Ala, 1.0; Na acetate, 2.5.

[d]Mineral mix (%): CaHPO₄, 47.4; K₂HPO₄, 10.0; CaCO₃, 13.0; MgSO₄, 5.04; KCl, 7.5; NaCl, 13.9; MnSO₄·H₂O, 0.48; ZnSO₄·7H₂O, 0.56; CuSO₄·5H₂O, 0.10; Fe₆H₅O₇·H₂O, 0.60; pentacalcium orthoperiodate (29% I), 0.018; SnCl₂·2H₂O, 0.01; Na₂SeO₃, 0.003; (NH₄)₆Mo₇O₄·4H₂O, 0.005; CrCl₃·6H₂O, 0.032; NiCl₂·6H₂O, 0.030; NaF, 0.032; NH₄VO₃·4H₂O, 0.0025; α-cellulose, 1.31.

[e]Vitamins per kg diet: retinylacetate, 20,000 IU; cholecalciferol, 2,000 IU; D,L-α-tocopheryl acetate, 160 IU; menaquinone, 15 mg; thiamin·HCl, 25 mg; riboflavin, 10 mg; pyridoxine, 10 mg; nicotinic acid, 100 mg; D-Ca pantothenate, 20 mg; myo-inositol, 200 mg; folic acid, 10 mg; biotin, 1 mg; cyanocobalamin, 50 μg; ascorbic acid, 200 mg; taurine (an amino sulfonic acid), 100 mg.

[f]Amino acid mix (%, total 28.55): L-Arg·HCl, 2.0; L-His·HCl·H₂O, 1.2; L-Iso, 1.8; L-Leu, 2.4; L-Lys·HCl, 2.8; L-Met, 0.45; L-Cys, 0.6; L-Phe, 1.5; L-Thr, 1.4; L-Try, 0.4; L-Val, 1.8; L-Tyr, 1.0; L-Asparagine, 1.6; L-Ser, 0.8; L-Pro, 1.6; Gly, 1.6; L-Glu, 4.8; L-Ala, 0.8.

[g]Glista salts (%, total 5.37): CaCO₃, 0.3; Ca₃(PO₄)₂, 2.8; K₂HPO₄, 0.9; NaCl, 0.88; MgSO₄·7H₂O, 0.35; MnSO₄·H₂O, 0.065; ferric citrate, 0.05; ZnCO₃, 0.01; CuSO₄·5H₂O, 0.002; H₃BO₃, 0.0009; Na₂MoO₄·2H₂O, 0.0009; KI, 0.004; CoSO₄·7H₂O, 0.0001; Na₂SeO₃, 0.00002 (Molitoris and Baker, 1976).

[h]Vitamins per kg diet: thiamin·HCl, 150 mg; niacin, 150 mg; riboflavin, 24 mg; Ca-pantothenate, 30 mg; vitamin B₁₂, 0.03 mg; pyridoxine·HCl, 9 mg; biotin, 0.9 mg; folic acid, 6 mg; inositol, 150 mg; ρ-aminobenzoic acid, 3 mg; menadione, 7.5 mg; ascorbic acid, 375 mg; retinyl acetate, 15,000 IU; cholecalciferol, 900 ICU.

[i]D,L-α-tocopheryl acetate, 30 mg/kg diet.

[j]Ethoxyquin, 125 mg/kg diet.

TABLE 13-6. Amount of Food Needed Daily and Growth Rate of the Cat*

Weight (lbs)			Dry** (cups)	Soft-Moist (1.5 oz pkgs)	Canned (ounces)
Mature, Non-Lactating Cat:					
6			0.7	1.8	7
8			0.9	2.5	9
10			1.1	3	12
12			1.3	3.5	14
14			1.5	4	16

Last 3 weeks of Gestation: 1.25 times more.

Peak Lactation: 2–3 times more.

Nursing Kittens: Should weight 90–100g at birth and gain this amount weekly from birth to 5–6 months of age. Encourage solid food consumption beginning at 3–4 weeks of age.

Weaned Kittens:

Weight (Lbs)	Age (Mo.)†				
	Female	Male			
1	1	1	0.4	1	4
2	2	2	0.6	1.6	6.5
3	3	3	0.8	2.1	8
4	4.5	4	0.9	2.5	9
5	6	5	1.0	2.7	10
6	8	6.5	1.0	2.7	10
7	10	8	1.0	2.7	10
10		10	1.2	3.1	13

*The amount needed by any individual cat may vary as much as 100% from the amounts given. Thus, unless obesity is a problem don't restrict the cat to these amounts; feed all they will eat. Amounts given are based on the energy content of average quality cat foods.

**An 8 oz cup (volume) holding 3 oz (weight) of cat food.

†The age at which these weights are reached varies. Generally the bigger the parents and the more calories consumed (in a nutritionally well-balanced diet), the larger the kitten's size. The amount of food needed/lb body wt decreases with increasing age. For example, the 3 lb kitten that is 2 months old needs more, and the 3 lb kitten that is 4 months old needs less, than the amounts given for the 3 lb kitten that is 3 months old.

Small Animal Clinical Nutrition
Lewis LD and Morris ML, Jr (eds)
Mark Morris Associates
Topeka, Kansas

TABLE 13–7. Minimum Nutrient Requirements of Dogs for Growth and Maintenance (amounts per kg of body weight per day)[a]

Nutrient	Unit	Growth[b]	Adult Maintenance[c]
Fat	g	2.7	1.0
Linoleic acid	mg	540	200
Protein[d]			
Arginine	mg	274	21
Histidine	mg	98	22
Isoleucine	mg	196	48
Leucine	mg	318	84
Lysine	mg	280	50
Methionine-cystine	mg	212	30
Phenylalanine-tyrosine	mg	390	86
Threonine	mg	254	44
Tryptophan	mg	82	13
Valine	mg	210	60
Dispensable amino acids	mg	3,414	1,266
Minerals			
Calcium	mg	320	119
Phosphorus	mg	240	89
Potassium	mg	240	89
Sodium	mg	30	11
Chloride	mg	46	17
Magnesium	mg	22	8.2
Iron	mg	1.74	0.65
Copper	mg	0.16	0.06
Manganese	mg	0.28	0.10
Zinc	mg	1.94	0.72
Iodine	mg	0.032	0.012
Selenium	μg	6.0	2.2
Vitamins			
A	IU	202	75
D	IU	22	8
E[e]	IU	1.2	0.5
K[f]			
Thiamin	μg	54	20
Riboflavin	μg	100	50
Pantothenic acid	μg	400	200
Niacin	μg	450	225
Pyridoxine	μg	60	22
Folic acid	μg	8	4
Biotin[f]			
B$_{12}$	μg	1.0	0.5
Choline	mg	50	25

[a] Needs for other physiological states have not been determined.

[b] Average 3-kg-BW growing Beagle puppy consuming 600 kcal ME/day.

[c] Average 10-kg-BW adult dog consuming 742 kcal ME/day.

[d] Quantity sufficient to supply minimum amounts of available indispensable and dispensable amino acids specified below.

[e] Requirement depends on intake of PUFA and other antioxidants. A fivefold increase may be required under conditions of high PUFA intake.

[f] Dogs have a metabolic requirement, but a dietary requirement was not demonstrated when natural ingredients were fed.

TABLE 13–8. Required Minimum Concentrations of Available Nutrients in Dog Food Formulated for Growth

Nutrient	Per 1,000 kcal ME	Dry Basis (3.67 kcal ME/g)
Protein[a]		
Indispensable amino acids		
Arginine	1.37 g	0.50%
Histidine	0.49 g	0.18%
Isoleucine	0.98 g	0.36%
Leucine	1.59 g	0.58%
Lysine	1.40 g	0.51%
Methionine-cystine	1.06 g	0.39%
Phenylalanine-tyrosine	1.95 g	0.72%
Threonine	1.27 g	0.47%
Tryptophan	0.41 g	0.15%
Valine	1.05 g	0.39%
Dispensable amino acids	17.07 g	6.26%
Fat	13.6 g	5.0%
Linoleic acid	2.7 g	1.0%
Minerals		
Calcium	1.6 g	0.59%
Phosphorus	1.2 g	0.44%
Potassium	1.2 g	0.44%
Sodium	0.15 g	0.06%
Chloride	0.23 g	0.09%
Magnesium	0.11 g	0.04%
Iron	8.7 mg	31.9 mg/kg
Copper	0.8 mg	2.9 mg/kg
Manganese	1.4 mg	5.1 mg/kg
Zinc[b]	9.7 mg	35.6 mg/kg
Iodine	0.16 mg	0.59 mg/kg
Selenium	0.03 mg	0.11 mg/kg
Vitamins		
A	1,011 IU	3,710 IU/kg
D	110 IU	404 IU/kg
E[c]	6.1 IU	22 IU/kg
K[d]	—	—
Thiamin[e]	0.27 mg	1.0 mg/kg
Riboflavin	0.68 mg	2.5 mg/kg
Pantothenic acid	2.7 mg	9.9 mg/kg
Niacin	3 mg	11.0 mg/kg
Pyridoxine	0.3 mg	1.1 mg/kg
Folic acid	0.054 mg	0.2 mg/kg
Biotin[d]	—	—
Vitamin B$_{12}$	7 μg	26 μg/kg
Choline	340 mg	1.25 g/kg

[a] Quantities sufficient to supply the minimum amounts of available indispensable and dispensable amino acids as specified below. Compounding practical foods from natural ingredients (protein digestibility ± 70%) may require quantities representing an increase of 40% or greater than the sum of the amino acids listed below, depending upon ingredients used and processing procedures.

[b] In commercial foods with natural ingredients resulting in elevated calcium and phytate content, borderline deficiencies were reported from feeding foods with less than 90 mg zinc per kg (Sanecki et al., 1982).

[c] A fivefold increase may be required for foods of high PUFA content.

[d] Dogs have a metabolic requirement, but a dietary requirement was not demonstrated when foods from natural ingredients were fed.

[e] Overages must be considered to cover losses in processing and storage.

TABLE 13-9. Factors for Consideration in Formulation of Dog Foods from Natural Ingredients[a]

Nutrient	Factors for Consideration
Fat	Degree of unsaturation, antioxidants, vitamin E
Carbohydrate	Fiber, lactose, reducing sugars, processing, stage-of-life cycle
Protein	Energy content, digestibility, amino acid balance, processing, antinutrients, antitryptic factors
Amino acids	Availability; heat treatment in presence of reducing sugars reduces availability, especially of lysine; requirement for individual amino acids increases with increased dietary nitrogen.
Minerals	Ratios, source, availability
Calcium	Phytates, ligands, vitamin D
Phosphorus	Phytates, calcium, plant-animal
Sodium, potassium, chloride	High availability
Zinc	Phytates, calcium, plant-animal, fiber
Copper	Phytates, zinc
Iron	Source, availability, plant-animal
Vitamins	Processing, lipid content, source
A	Oxidation, toxicity
D	Toxicity, calcium level
E	PUFA, selenium
B_1	Losses in processing and storage, product pH, storage time and temperature, thiaminases
B_2	UV light
B_6 (Pyridoxine)	Protein level in diet
Niacin	Tryptophan, low availability of plant sources
Folate	Processing losses
B_{12}	Plant versus animal proteins
Choline	Methionine, folate, vitamin B_{12}, availability, fat

[a] See text discussion for details relative to individual nutrients.

TABLE 13-10. Calculated Metabolizable Protein and Metabolizable Energy Requirements of Dogs in Various Physiological States[a]

Physiological State	Protein Requirement (g metabolizable protein $W_{kg}^{0.67}$ per day)	Metabolizable Energy Requirement (kcal per $W_{kg}^{0.67}$ per day)
Weaning		
Start (3 weeks)	8.1	400
Finish (6 weeks)	6.5	375
Early growth	6.0	353
Half grown	3.8	225
Adult (average)	1.5	132–159
Pregnancy, late	5.7	225
Lactation	12.4	560

[a] Adapted from Payne (1965). Calculated metabolizable protein equals food nitrogen minus fecal and urine N (retained N) × 6.25. Calculated metabolizable energy estimates were based on 4 kcal/g of dietary carbohydrate and protein and 9 kcal/g of dietary fat. These requirements are presumed to apply in a thermoneutral environment at moderate levels of activity.

TABLE 13-11. Recommended Energy Needs of Adult Dogs at Maintenance (kcal ME/day)[a]

Body Weight (kg)	NRC (1974) ($132\,W_{kg}^{0.75}$)	Thonney (1983) ($100\,W_{kg}^{0.88}$)[b]	Thonney (1983) ($144 + 62.2\,W_{kg}$)[b]
1	132	100	207
3	301	262	331
5	441	412	455
10	742	758	766
20	1,248	1,396	1,388
30	1,692	1,995	2,010
40[c]	2,099	2,569	2,632
50[c]	2,482	3,127	3,254
60[c]	2,846	3,671	3,876

[a] Intended to apply in a thermoneutral environment at moderate activity.

[b] The contributions by Professor M. L. Thonney, Cornell University, to the development of these data are gratefully acknowledged, as is the assistance of Dr. C. A. Banta, Allen Products; Dr. Hanson Lee, Quaker Oats; and Dr. Lloyd Miller, Carnation, for supplying data on individual dogs.

[c] Data based on feeding records are needed for dogs in these weight categories.

TABLE 13–12. Examples of Three Types of Commercial Foods (Percent)[a]

	Dry[b]	Semimoist[b]	Canned[b]
Corn	49.1	—	—
Corn gluten feed	19.0	—	—
Meat and bone meal	19.0	—	—
Meat and meat by-products	—	32.8	65–80
Poultry and poultry by-products	—	—	10–20
Soybean meal	7.5	—	—
Soybean flakes, bran flakes	—	32.3	—
Textured soy protein, soy flour	—	—	10–20
Soluble carbohydrates	—	21.0	—
Animal fat	4.5	1.0	—
Mineral mix[c]	0.8	3.3	0.5
Vitamin mix[c]	0.1	0.3	0.2
Antimycotic and emulsifier	—	3.8	—
Propylene glycol	—	3.0	—
Dried skimmed milk	—	2.5	—

[a] For examples of foods from semipurified and purified sources, refer to papers listed in References under Protein and Amino Acids, and Vitamins.

[b] Courtesy of M. C. Stillions, Agway, Inc.; Gaines Nutrition Center, Gaines Foods, Inc.; and C. A. Banta, Alpo Petfoods, Inc.

[c] Quantities to meet NRC requirements with sufficient overages to compensate for lack of availability and/or losses due to processing and storage.

TABLE 13–13. Nutrient Content of Commercial Pet Foods

Form of Food	Water (%)	Metab. Energy (kcal/oz as fed)	In Food Dry Matter (%)					
			Digestibility	Protein	Fat	Ca	P	Sodium
Dog Foods:								
Dry	6–10[b]	95[a]	75	22–31	7–13	1–3	1–2	0.3–0.8
Soft-Moist	23–38	85	90	25–33	9–15	1–3	1–2	0.3–1.0
Canned Ration[c]	68–78	33	85	30–40	10–25	1–3	1–2	0.8–1.2
Canned Gourmet[d]	70–78	37	85	45–60	16–37	1–3	1–2	0.8–0.9
Cat Foods:								
Dry	6–10	100[a] (97–106)	80	34 (30–40)	12 (9–15)	1.6 (1.1–2.1)	1.1 (0.9–1.4)	0.6 (0.3–0.9)
Soft—Moist	30–40	75 (60–90)	90	36 (33–39)	16.5 (12–23)	2.4 (1.4–4.1)	2.2 (1.7–3.2)	0.85 (0.6–1.5)
Canned Ration[c]	70–78	29 (23–41)	85	41 (31–56)	14 (3–30)	2.0 (1.0–4.3)	1.4 (0.8–2.6)	0.6 (0.2–0.9)
Canned Gourmet[d]	72–78	35 (29–46)	90	53 (39–78)	27 (11–40)	1.5 (0.2–3.5)	1.5 (0.7–4.0)	0.9 (0.4–4.2)

[a] An 8 oz. (by volume) measuring cup holds 3–3.5 oz. (by wt.) of most dry foods.

[b] Average and/or (range) for those available in grocery stores.

[c] Multiple ingredient type food. These are generally in 13 oz. or larger cans.

[d] Foods consisting exclusively, or primarily, of a single ingredient. These cat foods are in 3 to 6.5 oz cans

Small Animal Clinical Nutrition
Lewis LD and Moris ML, Jr (eds)
Mark Morris Associates
Topeka. Kansas

Study Questions and Problems

Chapter 13

I. True or False Questions

1. The dietary crude protein content for adult dogs should be more than 30% on a dry matter basis.
2. The amino sulfonic acid, taurine, is uniquely important in the nutrition of the dog.
3. Fats and oils have an important effect on palatability of dog and cat diets.
4. Diets for growing kittens should contain about 33% protein of high biological value.
5. Corn oil is more easily absorbed through the digestive tract of dogs than other types of fats or oils.
6. High dietary levels of raw starch can lead to diarrhea in dogs.
7. Dietary fiber may cause constipation and other intestinal problems in the dog.
8. Cats should obtain 25% to 35% of their dietary calories from protein.
9. B-vitamin requirements are higher in the cat than the dog.
10. Cats have a specific requirement for the vitamin A precursor, carotene.
11. Most raw fish contain an enzyme, avidin, that destroys the thiamine in a dog formula.
12. The amino acid tryptophan can substitute for preformed niacin in the diets of the cat.
13. The caloric density of dry pet foods may be increased by spraying with a liquid fat.
14. In general, table scraps offer the dog a nutritionally balanced diet.
15. An ''all meat'' diet is probably the best natural nutritionally balanced diet that can be fed the dog or cat.

II. Multiple Choice (select the one best answer)

1. Which of the following contains protein of the highest biological value for the cat?
   a. Soybean meal
   b. Extruded corn grain
   c. Organ meat
   d. Wheat germ meal
   e. All about the same
2. Which of the following is uniquely important in nutrition of the cat?
   a. Lysine
   b. Taurine
   c. Methionine
   d. Cystine
   e. Trypsin
3. Which of the following fat sources is often used in the diet of show dogs to produce more ''bloom'' and sheen to the hair coat?
   a. Lard
   b. Suet
   c. Olive oil
   d. Corn oil
   e. Tallow
4. Which of the following nutrients may be damaged or rendered less available after being submitted to heating during processing of dog and cat foods?
   a. Fat
   b. Carbohydrates
   c. Minerals
   d. Proteins
   e. None are affected
5. Dog diets containing high levels of raw starch may result in
   a. Improved palatability.
   b. Diarrhea.
   c. Improved digestability.
   d. Reduced palatability.
   e. Obesity.

6. Digestive disturbances and diarrhea may occur in some dogs fed diets containing milk because the dog may lack the enzyme
   a. Sucrase.
   b. Lactase.
   c. Maltase.
   d. Trypsin.
   e. Lipase.
7. Which of the following is not well utilized by the cat?
   a. Cholecalciferol
   b. Ergocalciferol
   c. Beta carotene
   d. α-tocopherol
   e. Thiamine
8. Which vitamin will be rendered unavailable by a substance found in raw egg whites?
   a. Folic acid
   b. Biotin
   c. Thiamine
   d. Riboflavin
   e. Pyridoxine
9. Dog foods which have been blended, formed into dough, and baked are called
   a. Dry meals.
   b. Kibbles.
   c. Expanded dry foods.
   d. Chunk-style foods.
   e. None of the above.
10. The consumption of __?__ dog or cat food helps keep the teeth clean and gums healthy.
   a. Dry
   b. Frozen
   c. Soft-moist
   d. Canned
   e. Table scrap

III. Matching or Identification

Match the appropriate answer with the following statements.

a. Corn oil	f. Taurine
b. Carbohydrate	g. Egg whites
c. Dried skim milk	h. Fiber
d. Linoleic	i. Protein
e. Tryptophan	j. Raw fish

1. High dietary levels may cause diarrhea in dogs.
2. As dogs get older, diets high in this nutrient tend to overwork the kidneys.
3. A deficiency in cats may cause a decrease in conjugated bile acids and central retinal degeneration development.
4. Essential fatty acid required by dogs and cats.
5. Highly digestible type of dietary fat.
6. Dietary constituent that is helpful in stimulating and maintaining intestinal action.
7. Nutrient that contains approximately 4.2 kcal/g.
8. Most cat diets contain higher levels of this nutrient than do dog diets.
9. Amino acid which cannot substitute for preformed niacin in the diet of the cat.
10. Feed containing an antivitamin that ties up biotin.

IV. Short Answer Questions

1. What are the advantages of heating feedstuffs in the processing of dog and cat foods? Are there any harmful effects from heating?
2. What advantages can fiber contribute in the dog diet?
3. How does kibbled dog food differ from expanded dry dog food?
4. Describe the processing of "chunk-style" dog foods.
5. Why should you not feed dry dog food to cats?

Notes

Notes

SECTION V

TABLES

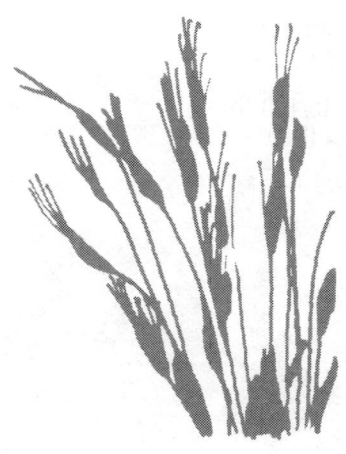

14
Tables of Equivalents

WEIGHT AND VOLUME

1 kilogram (kg)	= 2.2 lb	1 gallon of	
1 kilogram (kg)	= 1000 g	Water	= 8.34 pounds
1 gram (g)	= 1000 mg	Molasses	= 11.7 pounds
1 milligram (mg)	= 1000 mcg	Fat	= 7.5 pounds
1 microgram (mcg)	= 1 gamma (γ) or 1 ug	Milk	= 8.6 pounds
1 pound	= 453.6 g (454 g)	1 fluid ounce (US) = 29.573 milliliters (ml)	
1 pound	= 16 ounces	1 liquid pint (US) = 0.473 liter (l)	
1 ounce	= 28.4 g	1 liquid quart (US) = 0.946 liter	
1 ton (short)	= 2000 pounds or 907.2 kg	1 gallon (US) = 3.785 liter	

WEIGHTS PER BUSHEL (IN LB)

1 bushel = 1.244 cu ft

Alfalfa meal, dehydrated	19	Oats	32
Barley	48	Rye	56
Beet pulp, dried	19	Sorghum grain (milo)	56
Corn, shelled	56	Soybeans, whole	60
Corn, whole ear	70	Soybean, meal	48–50
Corn, ground ear	45	Wheat, whole	60
Cottonseed	32	Wheat, bran	16

ENERGY

1 calorie = Amount of heat required to raise 1 gm of water from 14.5 to 15.5°C
1 calorie = 4.184 joules
1 large calorie (kcal) = 1000 calories
1 mega calorie (Mcal) = 1000 kcal = 1 therm
1 g TDN = 4.41 kcal DE or 1 lb TDN = 2000 kcal DE

CONVERSION FACTORS

Percent	=	Parts per hundred (gm/100 gm; lb/100 lb)
1 gallon H_2O	=	231 cu in or .13368 cu ft or 8.34 lb
1 cu ft H_2O	=	62.43 lb; 7.48 gal; .028 tons
1 gram	=	.03527 oz.
1 BTU	=	.252 kcal
Protein, %	=	Percent N × 6.25
Phosphorus, %	=	Percent P_2O_5 × .4367
Centigrade	=	(F − 32) × 5/9
Fahrenheit	=	(C × 9/5) + 32
1 Bushel (Bu)	=	1.244 cu. ft
1 cu ft	=	0.804 bushel
ppm	=	1 mcg/gm
	=	1 mg/kg
	=	mg/lb × 2.2
	=	mg/lb ÷ 0.454
	=	percentage × 10,000
	=	gm/ton × 1.1

CONVERSION FACTORS USED IN RATION CALCULATION

Convert	to	by	Example
gram (g)	kilogram (kg)	move decimal 3 places to left	2000 g = 2.0 kg
gram (g)	milligram (mg)	move decimal 3 places to right	2.0 g = 2000 mg
gram (g)	pound (lb)	divide by 454	658 g ÷ 454 = 1.45 lb
gram/pound	percent (%)	divide by 4.54	90 g/lb ÷ 4.54 = 19.8%
gram/ton	percent	multiply by 11, move decimal 5 places to left	45 g/ton × 11 = 495 = 0.00495%
kilogram (kg)	gram	move decimal 3 places to right	5.5 kg = 5500 g
milligram (mg)	gram	move decimal 3 places to left	95 mg = 0.095 g
percent	gram/pound	multiply by 4.54	25 × 4.54 = 113.5 g/lb
percent	parts/million (ppm)	move decimal 4 places to right	0.025% = 250 ppm
percent	gram/ton	divide by 11, move decimal 5 places to right	0.011 ÷ 11 = 0.001 = 100 g/ton
pound	gram	multiply by 454	0.5 lb × 454 = 227 g
ppm	percent	move decimal 4 places to left	100 ppm = 0.01%

FIGURING GRAIN, SILAGE AND HAY STORAGE CAPACITIES

I. No. 2 Shelled Corn (15.5% Moisture) or Similar Air-Dry Grain

A. Capacity of a cylinder (circular bin), cu ft =
radius^2 × height × 3.1416

B. Capacity of a cone, cu ft =
radius^2 × height × 3.1416 ÷ 3

C. Capacity of a square or rectangular bin, cu ft =
length × width × height

D. For capacity of high-moisture shelled corn or ground ear corn, multiply bushels of air-dry corn by the following appropriate factors:

Moisture %	Shelled Corn	Ground Ear Corn
15.5	1.000	0.644
20.0	0.962	0.610
25.0	0.919	0.573
30.0	0.868	0.543

II. Silo Capacities

A. Trench or bunker silo (tractor packed)

1. Silage wt.:
   Corn—35 lb/cu ft
   Sorghum—35 lb/cu ft
   Grass—40 lb/cu ft
2. Silo volume, cu ft:

$$\frac{\text{Top width} + \text{bottom width}}{2} \times \text{height} \times \text{length}$$

3. Silo capacity, tons:

$$\text{Corn or sorghum silage} = \frac{\text{Silo volume, cu ft}}{57} = \text{tons}$$

$$\text{Grass silage} = \frac{\text{Silo volume, cu ft}}{50} = \text{tons}$$

B. *Estimated tonnage capacity* of tower silos for corn or grass-legume silage (68–70% moisture)*

Silo height (ft)	Height of settled silage (ft)	Lb/ cu ft	Inside diameter of silo (ft) 10	12	14	16	18	20	22	24	26	28	30
			(tons)										
20	20	44	33	48	66	86	—	—	—	—	—	—	—
30	29	49	56	80	109	143	180	223	270	321	377	—	—
40	38	51	77	110	150	196	248	307	371	442	520	600	690
50	47	53	—	—	193	252	320	394	477	580	668	773	886
60	56	55	—	—	—	—	392	483	585	697	818	947	1087
70	65	56	—	—	—	—	—	574	694	827	970	1125	1290

*Capacity according to depth after settling.
(a) Silo cpacity in cubic feet = radius^2 × height × 3.1416
(b) Cubic feet of settled silage = radius^2 × height of settled silage × 3.1416
(c) Silo capacity in tons of settled silage $= \frac{\text{(b)} \times \text{pounds per cubic foot of silage}}{2{,}000}$
(d) Pounds per cubic foot of unsettled silage varies from 35 to 45.
(e) Factors that cause variation from the estimated capacities include: moisture content, length of cut, speed of filling, extent of refilling, use of distributor in silo.

III. Baled Hay

Approximate bale size = 14 in. × 18 in. × 36 in.
Approximate bale volume = 5.26 cu ft
Approximate weight per cu ft = 11.4 lb
Approximate bale weight = 60 lb

IV. Hay Weight in a Stack

In order to estimate the tonnage of hay in a stack or in a barn, it is necessary (1) to calculate the volume of hay, and (2) to know the number of cubic feet per ton of hay. The following values can be used as cubic feet per ton of stacked hay:

Feed	Settled 1 to 2 Months (cu ft)	Settled Over 3 Months (cu ft)
Alfalfa	485	470
Clover	510	495
Hay, chopped	225	210
Straw, loose	1000	600–1000
Timothy	640	625
Wild hay	600	450

When using the previous values, many factors—other than kind of hay, form, and period of settling—affect the density of hay in a stack or in a barn, including moisture content at haying time, and texture and foreign material.

TABLE OF TANK CAPACITY

To find the number of gallons of water in a tank, measure the depth of water in inches and multiply by the gallons per inch of depth.

Round Tanks			Round End Tanks		
Diameter in Feet	Gal per Inch Depth		Width	Length	Gal per Inch Depth
3	4.4		2	4	4.5
3 1/2	6.0		2	5	5.7
4	7.8		2	6	7.0
4 1/2	9.9		2	7	8.2
5	12.2		2	8	9.5
5 1/2	14.8		2	10	11.9
6	17.6		3	5	8.1
6 1/2	20.7		3	6	10.0
7	24.0		3	7	11.9
7 1/2	27.5		3	8	13.8
8	31.3		3	10	17.5
9	39.7		3	12	21.2
10	49.0		3	14	25.0
			3 1/2	8	15.8
			3 1/2	10	20.2
			3 1/2	12	24.5
			3 1/2	14	28.9

Notes

Notes

SECTION VI

GLOSSARY, APPENDICES AND INDEX

Glossary of Animal Nutrition

Abomasum the fourth stomach compartment of a ruminant. Sometimes called the true or glandular stomach.

Abortion the expulsion of a nonviable, immature fetus.

Abscess a localized collection of pus in the tissues of the body, often accompanied by swelling and inflammation and often caused by bacteria.

Absorption the passage of nutrients or other substances to the blood, lymph and cells as from the alimentary tract (e.g., digested foods) or from the tissues.

Acidification the act of reducing the pH of a substance or solution; increasing the acidity.

Acidosis a condition in ruminants where excess acid from the rumen is absorbed into the blood stream causing a lowered blood bicarbonte level. Acidosis in ruminants is caused by excessive ingestion of fermentable carbohydrates (grains) which results in production of large quantities of lactate and volatile fatty acids and reduced rumen pH. Outward signs of acidosis are anorexia, diarrhea, mucous in feces, dehydration, incoordination and, sometimes, death.

Additive an ingredient or combination of ingredients added to the basic feed mix or parts thereof to fulfill a specific need of certain essential nutrients and/or medicines. Usually used in microquantities and requires careful handling and mixing.

ADF acid detergent fiber. Fiber extracted with acidic detergent in a technique employed to help appraise the quality of forages.

ADG average daily gain. The average amount of daily live weight increase as applied to farm animals.

Adipose consisting of, resembling, or having relation to fat.

Ad Libitum (ad. lib.) at pleasure. Commonly used to express the availability of feed on a free-choice basis.

Adrenal Gland an endocrine gland, located near the kidney, which secretes hormones needed for nutrient utilization.

Aerial Part the above ground part of a plant.

Aerobic organisms or tissues living or functioning in the presence of free oxygen.

Afterbirth the placenta and allied membranes expelled from the uterus following parturition.

Agglomerate a process that compacts a material into a dense cluster or mass.

Air-Dry (Approximately 90% dry matter) this refers to feed that is dried by means of natural air movement, usually in the open. Many feeds are fed in the air-dry state.

Albumen a water soluble protein found in blood, milk, eggs, and other animal substances.

Algae single celled plants which synthesize proteins by the use of sunlight. If used as an animal feedstuff, contain about 50% protein on a dry matter basis.

Alimentary Tract the food passage in an animal from mouth to anus. Same as digestive tract.

Alkali (Alkaline) any of various bases, the hydroxides of alkali metals and ammonium, which neutralize acids to form salts and turn red litmus paper blue.

Ambient Temperature the prevailing or surrounding temperature about us.

Amino Acid nitrogen-containing compounds which serve as building blocks from which more complex proteins are formed. Contain both an amino (NH_2) group and a carboxyl (COOH) group.

Ammoniated combined or impregnated with ammonia or an ammonium compound.

Amylase any of several starch splitting enzymes.

Anabolism constructive metabolism in which simple substances are converted into more complex substances by living cells.

Anaerobic living or functioning in the abscence of air or molecular oxygen.

Analogue something having analogy (or similarity) to something else.

Anasarca (Dropsy) generalized edema.

Anemic a quantitative deficiency of hemoglobin, often accompanied by a reduced number of red blood cells, and causing weakness and breathlessness.

Animal Protein Factor (APF) a term formerly used to describe an unidentified growth factor essential for swine and poultry and present in protein feeds of animal origin. It is now known to be the same as vitamin B_{12}.

Anorexia lack or loss of the appetite for food.

Anoxia oxygen deficiency.

Antacid a substance that counteracts or neutralizes acidity.

Anthelmintic a product which destroys or expells internal parasites.

Antibiotic a chemical substance produced by living microorganisms that when used in proper dilution can inhibit the growth of, and even destroy, other microorganisms.

Antibody any of various substances existing in the blood or developed in immunization which counteract disease.

Antigen any substance which when injected into animal tissues will stimulate the production of antibodies.

Antioxidant any substance inhibiting oxidation. Antioxidants are used to prevent rancidity in livestock feeds.

Antivitamin any substance which inhibits the normal function of a vitamin.

Anus the opening at the lower end of the alimentary tract, through which the solid refuse of digestion is excreted.

Appetite a desire for food or drink.

Artificially Dried dried, or dehydrated, by other than natural means, usually by heat from fossil fuel sources.

As-Fed this refers to feed as consumed by the animal. It may range from 0% to 100% dry matter.

Ash the mineral matter of a feed. The residue that remains after complete burning of the organic matter.

Aspirated the removal of chaff, dust or other light material by use of air.

Assay determination of the potency or purity of a substance or the amount of any particular constituent found in a mixture.

Atom the smallest unitary constituent of a chemical element, composed of a more or less complex aggregate of protons, neutrons, and electrons, whose number and arrangement determine the element.

Atrophy a wasting away of the body or of an organ or part as from defective nutrition, disease, or other cause.

Available Nutrient a nutrient which can be digested, absorbed, and/or used in the body.

Average Daily Gain see ADG.

Avidin a substance found in the white of eggs which prevents the action of biotin; an antivitamin.

Avoirdupois Weight and Measures a system of weights in British and U.S. use that differs from the metric system.

Bacteria microscopic, single-cell organisms. Some are beneficial, others are capable of causing disease.

Bactericide a product that destroys bacteria.

Balanced Ration one which provides an animal with the proper amounts and proportions of all the required nutrients for 24 hours.

Baled Hay forage that has been cured (partially dried) and has been compressed into a bale (round or rectangular) to save space in storage and aid in handling.

Barn-Cured a process in which forage material is dried with forced ventilation in an enclosure.

Basal Diet a diet common to all groups of experimental animals to which the experimental substance(s) is added.

Basal Metabolic Rate (BMR)   the heat produced by an animal during complete rest following fasting, when using just enough energy to maintain vital cellular activity, respiration and circulation, the measured value of which is called the BMR. Basal conditions include thermoneutral environment, resting, postabsorbitive state (digestive processes are quiescent), consciousness, quiescence and sexual repose. BMR is determined in man 14 to 18 hours after eating and when at absolute rest. It is measured by means of a calorimeter and is expressed in calories per square meter of body surface.

Beans   seeds of leguminous plants.

Bile   a bitter yellow or greenish liquid secreted by the liver and aiding in digestion of fats.

Bioassay   determination of the relative effective potency of a substance (as a vitamin, drug or hormone) by comparing its effect on a test organism with that of a standard prepration.

Biochemistry   the chemistry of living organisms, animals and plants.

Biological   pertaining to the science of live or living matter in all its forms and phenomena.

Biological Value   a measure of the efficiency with which a feed protein furnishes the proper amount and proportions of the essential amino acids to the animal. A protein which has a high biological value is said to be of good quality.

Biscuits   refers to a physical form of animal feed from shaped and baked dough.

Blending   a process in which two or more feed ingredients are mingled or combined together. It does not imply a uniformity of distribution.

Bloat   a distention of the rume or paunch or of the large colon by gases by fermentation.

Blocks   refers to a physical form in which agglomerated feed is compressed into a solid mass cohesive enough to hold its form and weighing over 2 pounds, generally weighing 30 to 50 pounds.

Bloom   a term commonly used to describe the beauty and freshness of an animal as evidenced by a glossy hair coat and attractive appearance.

Bolted   a process by which materials are separated by means of a bolting cloth as flour from bran.

Bolting Feed   animals that ingest or eat too rapidly or greedily are said to be bolting their feed.

Bolus   regurgitated feed that has been chewed and is ready to be swallowed (also known as the cud); is also a large pill for dosing animals.

Bomb Calorimeter   an instrument for measuring the gross energy (heat of combustion) of any material.

Bone Meal   animal bones that were steamed under pressure and then ground. It is used as a fertilizer and as a mineral supplement for feeding farm livestock.

Bran   the pericarp or seed coat of grain which is removed during processing and used as animal feed.

Bricks   refers to the physical form of agglomerated feed, other than pellets, compressed into a solid mass cohesive enough to hold its form and weighing less than two pounds.

British Thermal Unit (BTU)   The amount of heat required to raise the temperature of 1 lb of water 1° Fahrenheit. Equivalent to 252 calories.

Brix   a term used to indicate a sugar (sucrose) content of molasses. It is expressed in degrees and was originally used to indicate the percentage by weight of sugar in sucrose solutions, with each degree brix being equal to one percent sucrose.

Brouse   the small stems, leaves, and/or flowers and fruits of shrubs, trees or woody vines as feed for ruminant animals.

Buffer   a substance in a solution that makes the degree of acidity resistant to change when an acid or base is added.

Bushel   a unit of dry measure equivalent to 2150.42 cubic inches (approximately 1.25 cubic feet).

By-Product   the secondary products produced in addition to the principal product from plant and animal processing and from industrial manufacturing.

Cake   the physical form of the mass resulting from the pressing of seeds, meat, or fish in order to remove oils, fats, or other liquids.

Calcification   the process by which organic tissue becomes hardened by a deposit of calcium salts.

Calculi an occurrence in which mineral deposits appear in the urinary tract.

Calorie the amount of energy as heat required to raise the temperature of one gram of water from 14.5 to 15.5° Centigrade.

Calorimeter an instrument for measuring heat (or energy value).

Canned a term applied to a feed which has been processed, packaged, sealed, and sterilized for preservation in cans or similar containers.

Cannibalism the act of one animal pecking at or eating on another animal, such as fowl pecking at or eating on another fowl or one pig biting the tail of another.

Cannula a tube for insertion into the body used to connect internal structures with the outside of the animal. Liquids and other materials may be introduced into the body or removed from the body from a cannula.

Carcass the body of a slaughtered animal after removal of the viscera and usually the head, skin, and lower leg.

Carcinogenic cancer producing.

Cardiovascular pertaining to the heart and blood vessels.

Carotene a yellowish-orange colored hydrocarbon compound found in many plants that is a precursor of vitamin A.

Carrier an edible material to which ingredients (generally micronutrients) are added to facilitate uniform incorporation of the latter into feeds.

Carrying Capacity the number of animal units an area will support on a year-round basis. This includes the land grazed plus the land necessary to produce winter feed.

Casein a protein precipitated from milk by rennin and/or acid, and forming the basis of cheese.

Catalyst a substance which accelerates the rate of a chemical reaction without itself being permanently affected by the reaction.

Catobolism the breaking down of complex substances into more simple compounds by living cells. Destructive metabolism.

Cecum an intestinal pouch or pocket located at the junction of the small and large intestine. Also caecum.

Cereal a plant in the grass family yielding an edible grain for use by humans and animals.

Chaff glumes, husks, or other seed covering together with other plant parts separated from seed in threshing or processing.

Charcoal dark-colored porous forms of carbon made from the organic parts of vegetable or animal substances by their incomplete combustion.

Chelate derived from the Greek word meaning "claw." It refers to a ring-like structure which is formed between an organic molecule and a metal ion, the latter being held within the organic molecule as if by a claw.

Chipped refers to a processing method in which a feed material is cut or broken into fragments; also meaning prepared into small, thin slices.

Chlorophyll the green coloring substance of leaves and plants, associated with the production of carbohydrates by photosynthesis.

Cholesterol a fatlike substance found in bile, blood, brain tissue and several of the body glands. It is important in metabolism and is a precursor of several hormones.

Chopped a process by which feed material is reduced in particle size by cutting with knives or other edged instruments.

Cleaned the process of removing material by such methods as scalping, aspirating, magnetic separation, or by any other method.

Cleanings the chaff, weed seeds, dust and other foreign matter fremoved from the cereal grains.

Coagulated material changed from a fluid into a thickened mass; curdled, clotted, or congealed.

Cobs the fiberous inner portion of the ear of corn from which the kernals have been removed.

Coenzyme a substance, usually containing a vitamin, which is required by certain enzymes to produce their reactions.

Collagen   the protein contained in a white, papery transparent type of connective tissue that yields gelatin on boiling.

Colostrum   the milk secreted by mammalian females during the first few days of lactation, which is high in antibodies and laxative.

Compaction   tightly packed feed in the stomach and intestines of an animal causing digestive disturbances and/or constipation.

Compensatory Growth   accelerated animal growth following a period of limited feed intake.

Complete Ration   a nutritionally adequate feed mixture for animals other than man; by specific formula is compounded to be fed as the sole ration and is capable of maintaining life and/or promoting production without any additional substance being consumed except water.

Concentrate   a broad classification of feedstuffs which are high in energy and low in crude fiber. Also, a concentrated source of one or more nutrients used to enhance the nutritional adequacy of a supplement mix.

Condensed   a process in which feed material is reduced to denser form by removal of moisture.

Condition   refers to the amount of flesh (body weight), quality of hair coat, and general health of animals. Also refers to a process of having achieved predetermined moisture characteristics and/or temperature of ingredients prior to further processing.

Cooked   a process in which feed material is heated in the presence of moisture to alter chemical and/or physical characteristics or to sterilize.

Cooled   a process in which feed temperature is reduced by air movement, usually accompanied by a simultaneous drying action.

Coprophagy   the ingestion of fecal material.

Cracked   a process in which feed particle size is reduced by combined breaking and crushing action.

Creep   a feeder or an enclosure used for supplemental feeding of nursing young, which excludes their dams.

Cribber   a horse that has the vice of biting or setting the teeth against some object, such as a manger, while sucking air.

Crimped   a process by which grain is rolled by use of corregated rollers. The grain may be tempered or conditioned before crimping, and may be cooled afterward.

Crop Residue   the forage remaining after harvesting a seed or grain crop, such as corn stalks and husklage, sorghum stalks, soybean refuse, small grain straws, and legume and grass seed straws.

Crumbles   pelleted feed reduced to granular form.

Cubing   refers to the process of compressing long or coarsely cut hay into cubes about 1¼ in. square and 2 in. long, with a bulk density of 30 to 32 pounds per cubic foot.

CUD   a portion of feed (bolus) which a ruminating animal returns from the stomach to the mouth to chew a second time.

Culture   nutrient medium innoculated with specific microorganisms which may be in a live or dormant condition.

Curd   the semisolid mass consisting of primarily casein obtained from milk by coagulation with an acid or the enzyme rennin.

Cure   a process to prepare for keeping or use, or to preserve. The process may be by drying, use of chemical preservatives, smoking, salting or by use of other processes and/or materials for preserving.

Cystitis   inflammation of the urinary bladder.

D-Activated   plant or animal sterol fractions that have been vitamin D–activated by ultra-violet light or by other means.

Debeaking   the removal of part of the beak of chickens and poults with an electric debeaker to prevent cannibalism.

Decortification   removal of the bark, husk, hull or shell from a plant, seed, or root. Also refers to the removal of portions of the cortical substance of a structure or organ, as in the brain, kidney and lung.

Defecation the voiding of fecal material from the rectum.

Deficiency Disease a disease resulting from the lack of one or more basic nutrients.

Defluorinated processed in some manner that the fluorine content is reduced to a level that is nontoxic under normal use.

Degermed a process in which the embryo of seeds is wholly or partially separated from the starch endosperm.

Dehulled a process in which the outer covering is removed from grains or other seeds.

Dehydrate a process of removing most or all moisture from a substance for the purpose of preservation, primarily through artificial drying.

Dermatitis an inflammation of the skin.

Desiccate to dry completely.

Dicoumarol an anticoagulant compound that can cause internal hemorrhages when consumed by animals. This compound may be found in spoiled sweetclover hay or produced synthetically.

Diet feed ingredient or mixture of ingredients including water, which is consumed by animals.

Digested a process in which material is subjected to prolonged heat and moisture, or to chemicals or enzymes with a resultant change or decomposition of the physical or chemical nature of that substance.

Digestive Tract the passage from the mouth to the anus through which feed passes following consumption as it is subjected to various digestive processes.

Diluent an edible substance used to mix with and reduce the concentration of nutrients and/or additives to make them more acceptable to animals, safer to use, and more capable of being mixed uniformly in a feed. It may also be a carrier.

Diuretic a drug or agent used to increase the flow of urine.

Dough Stage stage of maturity at which the seeds of a plant are immature and soft.

Dormant stage of maturity in which plants have ceased growing and have cured on the stem. Applied to most non-growing range plants after the seeds have formed.

Dried materials from which water or other liquid has been removed.

Drug **(as defined by FDA as applied to feed)** a substance (a) intended for use in the diagnosis, cure, mitigation, treatment or prevention of disease in man or other animals or (b) a substance other than food intended to affect the structure or any function of the body of man or other animals.

Drylot a relatively small area in which animals are confined indefinitely as opposed to being allowed to have free access to pasture.

Dry Matter the moisture-free content of feeds.

Dry-Milled a process in which grain is tempered with a small amount of water or steam to facilitate the separation of the various component parts of the kernel in the absence of any significant amount of free water.

Dry Period the period of time between lactations (when a female is not secreting milk).

Dry-Rendered a process in which animal tissue residues are cooked in open steamjacketed vessels until the water has evaporated. Fat is removed by draining and pressing the solid residue.

Dry Rolling a process in which feed material is compressed between rollers without added steam or other moisture.

Dung the excrement or feces (manure) of animals and birds.

Dust fine, dry pulverized particles of matter usually resulting from the cleaning or grinding of grain.

Early Bloom a period of maturity in forages between initiation of bloom to stage at which one-tenth of the plants are in bloom.

Early Leaf stage of maturity at which the plant reaches one-third of its growth before blooming.

Ears the fruiting heads part of the corn plant, including only the cob and grain.

Edema swelling of a part or all of the body due to the accumulation of excess water.

Emaciated an excessively thin condition of the animal body.

Emulsify to disperse a suspension of fine particles or globules of one liquid into another liquid.

Endocrine pertaining to the endocrine glands or their secretions that pass directly into the blood or lymph instead of into a duct (internal secretions).

Endogenous originating from within the animal body.

Endosperm the carbohydrate (or starchy) portion of seed.

Ensilage the product of acid fermentation of green forage crops that has been pressed and stored under anerobic conditions in a container called a silo. Also called silage.

Ensiled the process in which aerial parts of plants have been preserved by ensiling. Normally the original material is finely cut and blown into an airtight chamber as a silo, where it is pressed to exclude air and where it undergoes an acid fermentation that retards spoilage.

Enteritis an inflammation of the intestines.

Enzymatic related to an enzyme.

Enzyme any of the various complex organic substances originating from living cells and capable of producing by catalytic action certain chemical changes, as digestion in organic substances.

Epithelial refers to those tissues or cells that form the outermost layer of the skin and other membranes.

Ergot a disease of rye and other cereals, due to a fungus which replaces the grain by a long, hard, hornlike, dark-colored body.

Eructation the act of belching gas from the the stomach.

Essential Amino Acid an amino acid that is needed for animal metabolism but cannot be synthesized by the animal in the amount needed and thus, must be present in the diet of that animal.

Essential Fatty Acid a fatty acid that cannot be synthesized in the body, or that cannot be made in sufficient quantities for the body's needs.

Estrus recurrent, restricted period of sexual receptivity in female mammals. Marked by intense sexual urge. Also called the period of heat.

Etiology study or theory of the causation of any disease.

Evaporated a process in which material is reduced to a denser form; concentrated as by evaporation or distillation.

Eviscerated having had all the organs in the great cavity of the body removed.

Excreta the products of excretion—primarily feces and urine.

Exogenous originating from outside of the organism.

Expanded a feed process in which the material is subjected to moisture, pressure, and temperature to gelatinize the starch portion. When completed, its volume is increased, due to abrupt reduction in pressure.

Expeller Process a process for the mechanical extraction of oil from seeds, involving the use of a screw press.

Experiment a procedure (test or trial) for the purpose of discovering something unknown or testing a principle, supposition, etc.

Extracted, Mechanical a process in which fat or oil is removed from materials by heat and mechanical pressure. Similar terms: expeller extracted, hydraulic extracted, "old process."

Extruded a process by which feed has been pressed, pushed, or protruded through orifices (die) under pressure.

Fasting to abstain from all food.

Fat Soluble refers to a material soluble in fats and fat solvents but generally not soluble in water.

Fattening the deposition of energy in the form of fat within the animal body tissues.

FDA Food and Drug Administration, a federal regulatory agency.

Feathers the light, horny epidermal outgrowths that form the external covering of birds.

Feces the waste matter discharged from the intestines through the anus.

Feed (Feedstuff) any material of natural or artificial origin, fed to animals for the purpose of sustaining them.

Feed Additive nonnutritive products that improve the rate and/or efficiency of gain of animals, prevent certain diseases, or preserve feed.

Feed Efficiency (Feed conversion) the ratio expressing the number of units of feed required for one unit of production (meat, milk, eggs) by an animal.

Feed Grade feed materials suitable for animals, but not for human consumption.

Feed Grain any of several seeds from cereal plants commonly used for livestock or poultry feed.

Feeder's Margin the difference between the cost per body weight unit of feeder animals and the selling price per body weight unit of the same animals when finished.

Feedlot an area of land on which animals are fed or finished for market.

Fermentation enzymatic decomposition by various microorganisms especially of carbohydrates as used in the production of alcohol, bread, vinegar and other food or industrial materials.

Fermentation Products products formed as a result of an enzymatic transformation of organic substances.

Fermented a process in which materials are acted upon by yeast, molds or bacteria in a controlled aerobic or anaerobic process in the manufacturer of such products as alcohols, acids, vitamins of the B-complex group or antibiotics.

Fetus the unborn offspring of an animal.

Fibrous high in content of cellulose and/or lignin.

Fill a term designating the fullness of the digestive tract of an animal. With market animals, the fill refers to the amount of feed and water consumed upon their arrival at the market and prior to selling.

Fines any material which will pass through a screen whose openings are immediately smaller than the specified minimum crumble size or pellet diameter.

Finish to fatten a slaughter animal, also the degree of fatness of such an animal.

Fistula a narrow passage or duct from some part of the body to another part or to the exterior—sometimes surgically inserted. Also see cannula.

Flaked see rolled.

Flora the plant life of a given region or locality. In nutrition, it generally refers to the bacteria present in the digestive tract.

Flour soft, finely ground and bolted meal obtained from the milling of cereal grains, other seeds, or products. It consists essentially of the starch and gluten of the endosperm.

Flush the practice of feeding females more generously approximately two weeks before breeding.

Fodder the fresh or cured plant, containing all the ears or seed heads, if any, grown primarily for forage. It has been applied more specifically to corn and sorghum.

Food when used in reference to animals, is synonymous with feed.

Forage vegetative plants in a fresh, dried, or ensiled state which are fed to livestock (pasture, hay, silage or green chop).

Formula Feed two or more ingredients proportioned, mixed and processed according to specifications.

Fortify nutritionally, to add one or more nutrients to a feed.

Founder a condition of indigestion or overloaded stomach in the animal due to overeating. May also be the crippled condition of an animal afflicted with laminitis.

Free Choice a feeding system by which animals are given unlimited access to the separate components or groups of components constituting the diet.

Fresh refers to the green, newly produced or gathered form of feed material; not stored, cured or preserved. Also denotes a cow that has recently given birth to a calf.

Full Bloom a stage of maturity when two-third's or more of the plants are in bloom.

Full-Feed a term indicating that animals are being provided as much feed as they will consume safely without going off the feed.

Fungi any one of a class of plants of a low order of development, including mushroom, toadstools, molds and yeasts and that contain no chlorophyll. They may get their nourishment from either dead or living organic matter.

Fungicide any substance used to kill fungi.

Gallbladder the pear-shaped reservoir for the bile on the under surface of the liver.

Gastric pertaining to the stomach.

Gastric Juice a clear liquid secreted by the wall of the stomach. It contains hydrochloric acid and the enzymes rennin, pepsin, and gastric lipase.

Gastritis inflammation of the stomach

Gastroenteritis inflammation of the stomach and intestines.

Gastrointestinal pertaining to the stomach and intestine.

Gelatinized a feed process in which the starch granules are completely ruptured by a combination of moisture, heat and pressure and in some instances by mechanical shear.

Germ embryo of a seed.

Gestation pregnancy. The period from conception to birth.

Glucogenesis the formation of glucose by the breakdown of glycogen.

Gluconeogenesis the formation of glucose and glycogen from nonglucose matter.

Gluten the tough, viscid, nitrogenous substance remaining when the flour of wheat or other grain is washed to remove the starch.

Glycogenesis the formation or synthesis of glycogen.

Glycogenolysis the splitting up of glycogen in the body tissue to glucose.

Glycolysis the decomposition of sugars and metabolism to lactic acid in animals or pyruvic acid in enzymatic reactions.

Goiter enlargement of the thyroid gland, causing a swelling in the front part of the neck. Sometimes caused by an iodine deficiency.

Goiterogenic producing or tending to produce goiter.

Gossypol a phenolic, yellow pigment found in cottonseed, which is toxic to swine and certain other nonruminents and which cause discoloration of egg yolks during cold storage.

Grain seed from cereal plants.

Grain Screening the small imperfect grains, weed seeds, and other foreign material having feeding value that is separated in cleaning grain by a screen.

GRAS abbreviation for the phase "generally recognized as safe." A substance which is generally recognized as safe by experts qualified to evaluate the safety of the substance for its intended use.

Gravid pregnant.

Graze to feed on standing vegetation, as by livestock or wild animals.

Green Chop (Soilage) forages harvested (cut and chopped) in the field and hauled to livestock for consumption.

Grind to reduce to small segments or particle size by impact, shearing, or attrition.

Grits coarsely ground grain from which the bran and germ have been removed, usually screened to a uniform particle size.

Groats grain from which the hulls have been removed.

Ground a process in which feed material has been reduced in particle size by impact, shearing or attrition.

Growth may be defined as the increase in size of the muscles, bones, internal organs and other parts of the body as contrasted to fattening or fat deposition.

Gruel a feed prepared by mixing ground ingredients with hot or cold water.

Hay   the aerial portion of grass or herbage especially cut and cured for animal feeding.

Hay Belly   a term used in reference to a horse exhibiting a distended barrel due to excessive consumption of bulky rations, such as grass, hay or straw.

Haylage   low moisture forage stored as silage (40% to 60% moisture). Legume and grass crops are cut and wilted in the field to a lower moisture level than normal for silage, but the crop is not sufficiently dry for baling. It is commonly stored in a silo.

Heads   the seed or grain-containing portions of a plant.

Heat Labile   unstable to heat.

Heat-Processed   a process in which feed material is subjected to a method of preparation involving the use of elevated temperatures with or without pressure.

Heat Rendered   a process in which feed is melted, extracted, or clarified through use of heat. Usually, water and fat are removed.

Hemoglobin   the oxygen-carrying red pigment of the red corpuscles.

Hemorrhage   a copious escape of blood from the vessels.

Hepatitis   inflammation of the liver.

Herb   a flowering plant whose stem above ground does not become woody and persistant.

High-Lysine Corn (Opaque 2)   a variety of corn that is higher in lysine and tryptophan content than normal corn; hence, it has a better balance of the amino acids for monogastric animals.

Hives   any of various eruptive (swelling) diseases of the skin. Can be caused by feed allergies.

Homogenized   a process in which particles are broken down into evenly distributed globules small enough to remain emulsified for long periods of time.

Hormone   a discreet chemical substance secreting into the body fluids by an endocrine gland, which as a specific effect on the activities of other organs.

Hulls   outer covering of grain or other seeds.

Husks   leaves enveloping an ear of corn; or the outer coverings of kernels or seeds especially when dry or membranous.

Hydraulic Process   a process for the mechanical extraction of oil from seeds, involving the use of a hydraulic press. Sometimes referred to as "old process."

Hydrogenation   to combine or treat any unsaturated compound with hydrogen.

Hydrolysis   a process in which complex molecules are split to simpler units by chemical reaction with water, usually by catalysts.

Hydroponics   the cultivation of plants by placing the roots in liquid nutrient solutions rather than in soil. These include the growing of grains sprouted in chambers to provide a source of green feed at times when it is not possible to produce it normally.

Hygroscopicity   the tendency for a substance to absorb or attract moisture from the air.

Hyperthyroidism   a condition due to excessive functional activity of the thyroid gland and characterized by increased basal metabolism.

Hypertrophied   enlargement or overgrowth of an organ or part due to increased size of its constituent cells.

Hypervitaminosis   an abnormal condition due to an excessive of one or more vitamins.

Hypocalcemia   below normal concentration of ionic calcium in blood, resulting in convulsions, as in tetany or milk fever.

Hypoglycemia   a reduction in concentration of blood glucose below normal.

Hypomagnesemia   an abnormally low magnesium content of the blood.

Impaction   see compaction.

Implant   a substance that is inserted into the body tissue for the purpose of growth promotion or controlling some physiological function.

Inert   without active properties.

Ingest to put or take food into the body through the mouth.

Ingesta food and drink taken into the stomach.

Ingestion that act of taking food, medicines etc., into the body as into the stomach.

Ingredient a component part or constituent of any combination or mixture making up a feed.

Inoculation introduction of microorganisms or other substances into tissues of living plants and animals.

Inorganic pertaining to compounds not containing carbon.

Insulin a protein hormone formed in the pancreas and secreted into the blood, where it regulates carbohydrate (sugar) metabolism.

International Chick Unit (ICU) the vitamin D requirements of poultry are generally expressed as ICU, based on using the chick as the assay animal because of the unequal activities of vitamin D_2 and vitamin D_3 for this species.

International Unit (IU) a standard unit of potency of a biologic agent (e.g., a vitamin, hormone, antibiotic, antitoxin) as defined by the International Conference for Unification of Formulae. Potency is based on bioassay that produces a particular effect agreed on internationally. Also called a USP unit.

Intestinal Juice a clear liquid secreted by the glands in the wall of the small intestine. It contains the enzymes lactase, maltase, and sucrase, and several peptidases.

Intestinal Tract the small and large intestine.

Intramuscular located or occurring within a muscle.

Intraperitoneal situated within the peritoneal cavity.

Intravenous in, into, or from within a vein or veins.

Intrinsic Factor a chemical substance secreted by the walls of the stomach which is necessary for the absorption of vitamin B_{12}. A deficiency of this factor may lead to a deficiency of vitamin B_{12}.

In Vitro "in a glass." Occurring in an artificial environment, as in a test tube.

In Vivo occurring in the living body.

Iodize to treat with iodine or an iodide.

Irradiation a process in which a material is treated, prepared, or altered by exposure to a specific radiation. Yeast contains considerable ergosterol, which, when exposed to ultraviolet light, produces vitamin D.

Joule a unit of work or energy, as well as the concept of heat (4.184j = 1 calorie).

Juice the aqueous substance obtainable from biological tissue by pressing or filtering with or without addition of water.

Keratin a sulphur containing protein which is the principle constituent of epidermis, hair, nails, horny tissues and the organic matrix of the enamel of the teeth.

Kernal the whole grain of a cereal. For other species, dehulled seed.

Ketonuria the presence of ketone bodies in the urine, as with ketosis in high producing dairy cows.

Ketosis (Acetonemia) a condition characterized by an elevated concentration of ketone bodies in body tissues and fluids. It is more common among high producing dairy cows in a negative energy balance.

Kibbled cracked or crushed baked dough, or extruded feed that has been cooked prior to or during the extrusion process.

Kwashiorkor a nutritional disease of childen produced by a severe protein deficiency, with characteristic changes in pigmentation of the skin and hair, edema, skin lesions, anemia and apathy.

Lactation the secretion or formation of milk.

Laminitis an inflammation of the sensitive laminae under the horny wall of the hoof. Characterized by ridges running around the hoof, often associated with over feeding. All feet may be affected, but the front feet are most susceptible.

Lard   rendered fat of swine.

Late Bloom   a stage of maturity in plants when blossoms begin to dry and fall.

Laxative   a medicine or agent that will induce bowel movements and relieve constipation.

Leached   the condition of a product following subjection of the material to the action of percolating water or other liquid.

Lead Feeding   in dairy cattle operations, lead feeding is a practice of increasing grain fed to cows to a level equal to 1.0% to 1.5% of body weight beginning about 3 weeks prior to the predicted calving date. Following calving, grain is increased until a cow reaches maximum milk production or maximum feed intake, whichever comes first. This concept is aimed at finding the potential of cows to secrete milk. Hence, an abundance of high energy feed, balanced for the needs of high milk production, is provided cows above what might apparently be needed. Also called challenge feeding.

Leaves   lateral outgrowths of stems that constitute part of the forage of a plant, typically a flattened green blade, and primarily functions in photosynthesis.

Legume   a plant member of the leguminosae family (alfalfa, clovers and similar crops), with the character of forming nitrogen-fixing nodules on its roots, in this way making possible the use of atmospheric nitrogen.

Lesion   any pathological or unhealthy change in the tissue or loss of function of a part of the body.

Lignin   a complex strengthening material in the thickened cell walls of plants. It is practically indigestible.

Limited Feeding   a feeding system in which animals are fed less than they would like to eat. Animals are given sufficient feed to maintain weight and growth, but not enough for their potential production or finishing.

Limiting Amino Acid   the essential amino acid of a protein which shows the greatest percentage deficit in comparison with an animal's amino acid requirements.

Lipolysis   the decomposition or splitting up of fat to yield glycerol and fatty acids.

Liquid Protein Supplements   high-protein containing products (primarily nonprotein nitrogen) that usually contain molasses, urea, and some water, with added vitamins and trace minerals. Fed to ruminant animals.

Liver Abscesses   single or multiple inflammations on the liver, observed at slaughter. Usually the abscesses consist of localized collections of pus in a cavity formed by the disintegration of tissues. At slaughter, those livers affected with abscesses are condemned for human food. Liver abscesses are often found in feedlot cattle consuming high energy rations.

Liveweight   weight of an animal when alive or on foot.

Lymph   a transparent, slightly yellow liquid of alkaline reaction, found in the lymphatic vessels.

Lyophilization   the evaporation of water from a frozen product with the aid of high vacuum. Also called freeze drying.

Malignant   a cancerous growth, tending to go from bad to worse, as distinguished from a benign growth.

Malnutrition   imperfect or lack of proper nutrition.

Malt   sprouted and steamed whole grain from which the radicle has been removed.

Mammary Gland   the milk secreting gland in mammals.

Manure   the refuse from animal pens consisting of excreta with or without litter or bedding.

Mash   a mixture of feed ingredients in meal form. Also called mash feed.

Mastitis   an inflammation of the mammary gland (or glands).

Meadow   an area covered with grasses or legumes grown primarily for hay.

Meal   the physical form of an ingredient which has been ground or otherwise reduced in particle size.

Meat Analogs   food material usually prepared from vegetable protein to resemble specific meats in flavor, texture, and color.

Meconium   a dark green excrement material accumulated in the intestines during fetal development.

Medicated Feed   any feed which contains drug ingredients intended or represented for the cure, mitigation, treatment, or prevention of diseases of animals other than man or which contains drug ingredients intended to affect the structure of any function of the body of animals other than man.

Metabolic Body Size (Metabolic Weight)   the weight of an animal raised to the three-quarter power (Wt.$^{0.75}$).

Methane (CH$_4$)   methane and carbon dioxide are the two primary waste gases produced by rumen fermentation. The production of methane results in a loss of energy. It is produced by the reduction of carbon dioxide by hydrogen.

Metritis   an inflammation of the uterus.

Microbe   the same as microorganism.

Microflora   the overall microbial life present in a given region, such as the bacteria and protozoa population of the rumen.

Microingredient   any ration component, such as vitamins, minerals, antibiotics, drugs and other materials, normally required in small amounts and measured in milligrams, micrograms or parts per million.

Microorganism   a minute living organism, usually microscopic, such as bacteria and protozoa.

Mid-Bloom   a stage of maturity in plants during which one-tenth to two-thirds of the plants are in bloom.

Milk   total lacteal secretion from the mammary gland.

Milk Fat   the lipid components of milk.

Milk Stage   a period of maturity in plants after bloom when the seeds begin to form.

Mill Byproduct   a secondary feed product obtained in addition to the principle product in milling practice.

Mill-Dust   fine feed particles of undetermined origin resulting from handling and processing feed and feed ingredients.

Mill Run   the state in which a material comes from the mill, ungraded and usually uninspected.

Mixing   to combine by agitation, two or more materials to a specific degree of dispersion.

Moisture   a term used to indicate the water content of a feed, expressed as a percentage.

Molasses   the thick, viscous by-product resulting from refined sugar production or the concentrated partially dehydrated juices from fruits.

Mycotoxin   a fungus or bacterial toxin produced by molds during growth. Sometimes present in feed materials.

National Research Council (NRC)   a division of the National Academy of Sciences established in 1916 to promote the effective utilization of scientific and technical resources. Periodically, this private, nonprofit organization of scientists publishes bulletins giving nutrient requirements and allowances of domestic animals, copies of which are available on a charge basis through the National Academy of Sciences, National Research Council, 2101 Constitution Avenue, N.W., Washington, D.C., 20418.

Native Grass   a grass species indigenous to an area; not introduced from another environment or area.

NDF   neutral detergent fiber. A chemical analysis which separates forage material into cell contents and cell wall components.

Necrosis   death of a cell or a group of cells which is in contact with living tissue.

Neonate   a new born animal.

Nephritis   inflammation of the kidney.

Neuritis   inflammation of a nerve.

New Process   pertains to the extraction of oil from seeds. Same as solvent process.

Nitrate Poisoning   a condition sometimes resulting when animals ingest nitrates (NO$_3$) bacteria convert to nitrites (NO$_2$). The nitrites compete with oxygen, tying up the oxygen-carrying mechanism in the blood, causing the animal to suffocate.

Nitrogen Balance   a nutritional state in the animal determined from the nitrogen in the feed intake minus the nitrogen in the feces, minus the nitrogen in the urine.

Nitrogen Fixation   conversion of atmosperic nitrogen to organic nitrogen compounds by microbial activity.

Nodule   a tubercle, particularly such as are formed on legume roots by the symbiotic nitrogen-fixing bacteria of the genus *Rhizobium*.

Nonessential Amino Acid   any of the several amino acids that are required for the nutrition of the animal but which can be synthesized in adequate quantities by the animal from other amino acids.

Nonprotein Nitrogen (NPN)   nitrogen originating from other than an amino acid source but may be used by bacteria in the rumen to synthesize protein. NPN sources include compounds like urea and anhydrous ammonia, which are used in feed formulation for ruminants.

Nonruminant   a simple-stomached animal that does not ruminate. Examples are swine, poultry, horses, dogs, cats, and humans.

Nutrient Allowances   a term referring to the nutrient recommendations for animals that allow for the many variations in determining that animal's nutritive needs. These variations may include losses during storage and processing, age and size of the animal, degree of activity, stress, management, health and the kind, quality, and amount of feed consumed.

Nutrient/Calorie Ratio   an expression of nutrients as weight per unit of energy. For example, it is suggested that the protein to calorie ratio should be expressed as grams of protein per 1000 kcal metabolizable energy (g protein/1000 kcal ME).

Nutrient Requirements   values which refer to satisfying the animal's minimum nutrient needs, without margins of safety, for maintenance, growth, reproduction, lactation, and work. To meet these nutritive requirements, the animals must receive sufficient feed to furnish the necessary quantity of energy, protein, minerals, and vitamins.

Nutrients   any feed constituent or group of feed constituents that aid in the support of life.

Nutrition   the science encompassing a series of processes which food or feed is taken in and absorbed into the body of an organism which serves for purposes of growth, work, maintenance and repair of the vital process.

Obese   an excessive accumulation of fat in the body.

Offal   material left as a byproduct from the preparation of some specific products, less valuable portions and the by-products of milk.

Off Feed   having ceased eating with a healthy and normal appetite.

Oil   a substance composed chiefly of triglycerides of fatty acids, and liquid at room temperature.

Old Process   refers to the extraction of oil from seeds. Same as hydraulic or expeller process.

Olfactory   pertaining to the sense of smell.

Open   a term commonly used of female farm mammals to indicate a nonpregnant status.

Oral   pertaining to the mouth.

Organic   pertaining to substances derived from living organisms or all other compounds containing carbon.

Orts   fragments of feed which an animal refuses to eat.

Osmosis   the tendency of a fluid to pass through a semipermeable membrane into a solution where its concentration is lower, thus equalizing the condition on either side of the membrane.

Ossification   the formation of bone or of a bony substance.

Osteitis   inflammation of a bone.

Osteomalacia   a condition marked by increasing softness of the bones, so that they become flexible and brittle leading to deformities. It occurs chiefly in adults and is due to calcium, phosphorus and/or vitamin D deficiency.

Osteoporosis   abnormal porousness or rarefication of bone by the enlargement of its canals or the formation of abnormal spaces. Generally a result of calcium, phosphorus and/or vitamin D deficiency.

Overfinishing   excess finishing or fatness indicated by patchy fat deposits, especially over the shoulders, back and thighs.

Oxidation   the combining of oxygen with another element to form one or more new substances. Chemically, it consists in the increase of positive charges on an atom or the loss of negative charges. The animal combines carbon from feedstuffs with inhaled oxygen to produce carbon dioxide, energy, water and heat.

Oxytocin   the hormone that controls milk letdown.

Paddock   a small fenced field used for grazing purposes.

Palatability the relative attractiveness of a feed to the animal. Could include such factors as appearance, odor, taste, texture, temperature, and in some cases, auditory properties of the feed.

Pancreas a large, enlongated gland behind the stomach that secretes juice into the upper small intestine via the pancreatic duct.

Pancreatic Juice a thick, transparent liquid secreted by the pancreas into the upper small intestine. It contains the enzymes pancreatic amylase, pancreatic lipase and trypsin; also the hormone insulin.

Papillae small nipple-shaped elevations or projections located on the inner surface of the rumen wall.

Parakeratosis any abnormality of the outermost or horny layer of the skin, especially a condition caused by edema between the cells which prevents the formation of keratin.

Paralysis loss or impairment of power of voluntary motion.

Parathyroid any one of four small glands situated beside the thyroid gland, concerned chiefly with maintaining normal calcium in the body.

Parturition the act or process of giving birth to an animal.

Pasture a fenced area of domesticated forages, usually improved, on which animals are grazed.

Pathogen any disease-producing microorganism or material.

Pearled dehulled grains reduced by machine brushing into smaller smooth particles.

Peelings outer layers or coverings that have been removed by stripping or tearing.

Pellet Binders materials that enhance the firmness of pellets.

Pellets a physical form of agglomerated feed formed by compacting and forcing through die openings by a mechanical process.

Permeable capable of being penetrated.

Perspiration the functional excretion of sweat or moisture from the pores of the skin.

pH the pH scale is a measure of acidity or alkalinity of a solution. Values range from 0 (most acid) to 14 (most alkaline), with neutrality at pH 7.

Phagocyte any cell that ingests microorganisms of other cells and substances.

Phase Feeding denotes changing or varying the animal diet to adjust for any of several factors which may influence overall productivity. These may include temperature, stage production, season, body weight, age, economy or availability of feedstuffs.

Photosynthesis the process by which carbohydrates are produced by green plants from carbon dioxide and water using sunlight as a source of energy and with the aid of a catlyst such as chlorophyll.

Physiological pertaining to the functions of the body and organs.

Pica a depraved appetite characterized by a craving for unnatural articles of food (dirt, sand, hair, feces etc.).

Plasma the fluid portion of the blood in which the corpuscles are suspended.

Polioencephalomalacia (PEM) the softening of the gray matter of the nervous system. Has occurred in feedlot cattle fed high-concentrate diets. Cattle may exhibit blindness and/or muscular tremors (especially of the head), progressing to convulsions and death after one to several days in a coma. Animals tend to press on fixed objects with their heads. PEM is attributed to a wide variety of causes but is shown to respond to large, intravenous doses of thiamin.

Polished having a smooth surface produced by a mechanical process, usually by friction.

Polydipsia excessive thirst.

Polyneuritis inflammation of many nerves at once.

Polyunsaturated a compound, often a fatty acid, that contains more than one double carbon bonding.

Pomace pulp from fruit.

Postpartum occurring after the birth of young.

Pot-Bellied a term referring to an animal that has developed an abnormally large abdomen.

Poultry Litter material used on the floor of poultry houses along with the excreta that accumulates therein.

Prairie a level or rolling area of treeless land naturally covered with grass, under which fertile soils usually have developed.

Prebloom   a stage of maturity in plants occurring as the last third of growth before blooming.

Preconditioning   refers to a schedule of practices preparing the feeder calf for feedlot adaptation before it leaves the production sight.

Precursor   a compound that can be used by the body to form another compound; for example, carotene is a precursor of vitamin A.

Pregnant   the condition of having a developing fetus in the body; gravid.

Premix   a uniform mixture of one or more microingredients with diluent and/or carrier. Premixes are used to facilitate uniform dispersion of the microingredients into a larger mix.

Preservatives   any material(s) added to the forage crop at ensiling or haying time, with claims made to improve the preservation of nutrients, nutritive value, and/or palatibility of the feed.

Pressed   a process by which material is compacted or molded by pressure; also meaning having fat, oil or juices extracted under pressure.

Pressure Cooker   air tight container for the cooking of feed at high temperatures under steam pressure.

Presswater   the aqueous extract of fish or meat free from the fats and/or oils. Presswater is the result of hydrolic pressing of the fish or meat followed by separation of the oils either by centrifuging or other means.

Protein   any of a large class of naturally-occurring complex combinations of amino acids, always containing carbon, hydrogen, oxygen and nitrogen, and sometimes phosphorus and sulphur. Proteins are essential parts of all living matter and in the feed rations of animals.

Protein Bumps   a condition in some horses which appear to be allergic to certain proteins or to excesses of specific amino acids.

Protein Equivalent   a term indicating the total nitrogenous contribution of a substance in comparison with the nitrogen content of protein (usually plant). For example, the nonprotein nitrogen compound, urea, contains approximately 45% nitrogen and has a protein equivalent of 281% (45% N × 6.25).

Protoplasm   the essential protein substance of living cells.

Provitamin   a hypothetical precursor of vitamin, especially a substance from which the animal organism can form vitamin. Provitamin A is carotene. Ergosterol has been spoken of as provitamin D.

Prussic Acid (Hydrocyanic acid)   a toxin or poison produced as a glucoside by several plant species, especially sorghums.

Puberty   the age at which the reproductive organs of an animal become functionally operative and secondary sex characteristics develop.

Pulp   the solid residue remaining after extraction of juices from fruits, roots, or stems. Also called pomace.

Purgative   a laxative compound.

Purified Diet   a mixture of the known essential dietary nutrients in a pure form that is fed to experimental (test) animals in nutrition studies.

Pus   a liquid product of inflammation consisting of leukocytes, lymph, bacteria, dead tissue cells and fluid derived from their decomposition.

Putrification   the microbial decomposition of proteins.

Quality   a term referring to the desirability and/or acceptance of an animal or food product.

Quality of Protein   refers to the amount and ratio of essential amino acids present in a protein. A protein is said to be of good quality when it contains all the essential amino acids in proper proportions and amounts needed by specific animals.

Rancid   a term used to describe fats that have undergone partial decomposition and have an unpleasant, stale smell or taste.

Range (Rangeland)   large, naturally vegetated areas of relatively low productivity, mostly unfenced, grazed by livestock and game mammals. In general, range areas are too dry, poorly drained, rough, elevated or othewise unsuited to farming.

Range Cubes   large pellets designed to be fed on the ground.

Rate of Passage   the time taken from the ingestion of feed materials until excretion as feces.

Ration(s)   an amount of feed consumed by an animal in 24 hours.

Raw   food in its natural or crude state not having been subjected to heat in the course of preparation as food.

Refuse   damaged, defective or superfluous edible material produced during or left over from a manufacturing or industrial process.

Residue   part remaining after the removal of a portion of its original constituents.

Resorption   a return of the nutritive components of a partially developed fetus and fetal membrane to the system of the mother.

Respiration   the act or function or breathing.

Rhizobia   species of bacteria that live in symbiotic relation with legume plants within nodules on their roots. They carry on the fixation of atmospheric nitrogen in forms that are used as nutrients by the host legumes.

Rolled   having changed the shape and/or size of particles by compressing between rollers. It may entail tempering or conditioning.

Roots   subterranean parts of plants.

Roughage   feeds that are relatively bulky, high in crude fiber and low in total digestible nutrients. Opposite of concentrate.

Rumen Flora   the microorganisms of the rumen.

Rumenitis   inflammation of the rumen wall. Most frequently found in ruminants ingesting high grain diets which result in reduced rumen pH. A weakness or small break in rumen epithelium allows bacterial entry into the portal circulation which can subsequently cause liver abscesses.

Ruminate (Rumination)   the act of regurgitating previously consumed feed and chewing a soft mass of coarse feed particles, called a bolus or cud. This process is important to stimulate additional salivation and to further reduce feed particle size to allow passage of the material from the rumen.

Saliva   a clear, alkaline, somewhat viscid, digestive fluid secreted by the salivary glands into the mouth.

Salmonella   a pathogenic, diarrhea-producing organism sometimes present in contaminated feeds.

Saponification   the formation of soap and glycerol from the reaction of fat with alkali.

Satiety   to satisfy to the full. May refer to satisfaction of appetitle.

Saturated Fat   a completely hydrogenated fat in which each carbon is associated with the maximum number of hydrogens and there are no double bonds.

Scours   a persistant diarrhea in animals.

Scratch   a poultry feed term referring to whole, cracked or coarsely cut grain.

Screened   a process in which feedstuffs have been separated into various size particles by passing over or through screens.

Seed   the fertilized and ripened ovule of a plant.

Self-Fed   a feeding system where animals have continuous free access to some or all components of a ration, either individually or as mixtures.

Self-Feeder   a feed container from which animals can eat at will.

Separating   classification of particle by size, shape and/or density.

Sequester   to combine with and hold. For example, sequestering compounds can combine with calcium and prevent its entry into certain reactions.

Serum   the colorless fluid portion of blood remaining after clotting and removal of corpuscles. It differs from plasma in that fibrinogen has been removed.

Shells   the hard, fibrous, or calcareous covering of a plant or animal product, such as, nut, egg, oyster.

Shoots   the immature aerial parts of plants; stems with leaves and other appendages in contrast to the roots.

Shorts   fine particles of bran, germ, flour or offal from the tail of the mill from commercial flour milling.

Shrinkage   a term used to indicate the amount of body weight loss when animals are exposed to stressful conditions such as being transported, severe weather, or feed shortage.

Sifted   a process in which feed materials are passed through wire sieves to separate particles in different sizes. The separation of finer materials then would be done by screening.

Silage   see ensilage.

Silo   an airtight structure in which forage materials are fermented and stored. See ensilage.

Slotted Floors   floors in an animal pen with slots through which the feces and urine pass to a storage area below or nearby.

Sodium Bentonite   a fine grade of clay used as a pellet binder.

Sodium Bicarbonate   a chemical compound which is known to function as a buffer.

Soilage   see green chop.

Solubles   liquid containing dissolved substances obtained from processing animal or plant materials. It may contain some fine suspended solids.

Solvent Extracted   a process for the extraction of oil from seeds involving the use of an organic solvent.

Specific Dynamic Action (SDA)   a term denoting the extra heat produced by the animal, over and above the basal heat production, as a result of feed ingestion.

Specific Gravity   a technique to determine the relative amounts of fat and lean body mass comprising body weight. The measurement of specific gravity involves comparing the body weight under water (corrected for residual air in the lungs) to the body weight in air. As the proportion of body fat increases, the specific gravity decreases, since fat is lighter per unit of volume than lean body mass. The lower the specific gravity, the greater the proportion of fat in the body.

$$\text{Specific gravity} = \frac{\text{Body wt in air}}{\text{Body wt in air} - \text{Body wt under water}}$$

Specific Heat   the heat absorbing capacity of a substance in relation to that of water.

Spore   the reproductive element of one of the lower organisms, such as protozoa.

Spray Dehydrated   a process in which material has been dried by spraying on the surface of a heated drum. It is recovered by scraping from the drum.

Stabilized   a condition in which a material has been made more resistant to change by the addition of another particular substance.

Stalk(s)   the main stem of a herbaceous plant; often with its dependent parts as leaves, twigs, and fruit.

Starch   a white, granular polymer of plant origin. The principal part of seed endosperm.

Steamed   a process in which ingredients have been treated with steam to alter physical and/or chemical properties.

Steep-Extracted   a process in which feed materials are soaked in water or other liquid (as in the wet milling of corn) to remove soluble materials.

Steepwater   water containing soluble materials extracted by steep extraction.

Stem   the coarse, aerial parts of plants which serve as supporting structures for leaves, buds, fruit etc.

Sterility   inability to produce young.

Stickwater   the aqueous extract of cooked meat or fish free from the oil. Stickwater contains the aqueous cell solution, the soluble glue proteins and the water condensed from steam used in processing.

Stillage   the mash from fermentation of grains after removal of alcohol by distillation.

Stock Cattle (Stocker)   usually, young steers or cows that are light and thin and lack finish.

Stocking Rate   a pasture management term pertaining to animal numbers in relation to carrying capacity of a unit or area of the pasture.

Stool   the fecal discharge from the digestive tract.

Stover   the mature, cured stalks and leaves of corn after the ears, or sorghum after the heads have been harvested.

Straw   the plant residue remaining after separation of the seeds in threshing. It includes chaff.

Stress any adverse condition or influence which tends to disrupt the normal, steady functioning of the body and its parts. The greater the stress, the more exacting the nutrient requirements.

Stubble the lower portion of herbaceous plants remaining after the top portion has been harvested either artificially or by grazing animals.

Subcutaneous situated or occurring beneath the skin.

Substrate a term applied to the substance upon which a ferment or enzyme acts.

Succulence a condition of plants characterized by tenderness, juiciness, and freshness, making them appetizing to animals.

Suckle to nurse at the breast or mammary glands.

Sun-Cured a process in which feed material is dried by exposure in open air to the direct rays of the sun.

Supplement a feed used with another feed to improve the nutritive balance or performance of the total mix. Supplements are usually high in protein, minerals, vitamins, antibiotics or a combination of part of all of these. They are commonly combined with basal feeds to produce a complete feed.

Sweet Feed refers to a horse feed which is characterized by its sweetness due to the addition of molasses.

Symbiosis the living together or close association of two dissimilar organisms, with a resulting mutual benefit.

Synthetics pertaining to artificially produced materials that may be similar to natural materials.

Tallow the fat extracted from adipose tissue of cattle and sheep.

Tankage a protein supplement made from the residues from animal tissues including bones and exclusive of hair, hooves, horns, and contents of the digestive tract.

TDN see Total Digestible Nutrients.

Tempered a process of having achieved pre-determined moisture characteristics and/or temperature of ingredients rior to further processing. Also called conditioned.

Tetany a condition in animals in which there are localized, spasmodic, muscular contractions.

Tether to tie an animal with a rope or chain to allow grazing but prevent straying.

Therapeutic pertaining to the art of healing or treatment of disease.

Thermal refers to heat.

Thrombosis the obstruction of a blood vessel by coagulation of the blood.

Tiller a branch or shoot originating at a basal node in grasses.

Toasted a process by which feed materials are browned, dried, or parched by exposure to a fire, or to gas or electric heat.

Tonic a term formerly used for a class of medical preparations believed to have the power of restoring normal tone to tissue. Also denotes a drug, medicine, or feed designed to stimulate the appetite.

Total Digestible Nutrient (TDN) a term used to express the energy value of a feedstuff or feed mixture. It is determined by the summation of the digestible protein, digestible nitrogen-free extract, digestible crude fiber, and digestible fat times 2.25.

Toxemia a general intoxication or poisoning due to the absorption of bacterial products (toxins) formed at a local source of infection.

Toxic pertaining to a poison.

Toxin the poison produced by certain microorganisms.

Trace Element a chemical element used in minute amounts by an organism and considered essential to its physiology.

Trace Mineral a mineral nutrient required by animals in micro amounts only (measured in milligrams per pound or smaller units).

Trauma a wound or injury.

Tubers   the part of plants characterized by short, thickened, fleshy stems or terminal portions of stems or rhizomes that are usually formed underground, their minute scaled leaves each with a bud capable under suitable conditions of developing into a new plant, and constitutes the resting stage of various plants.

Twigs   small shoots or branches, usually without leaves, or portions of stems of variable length or size.

Udder   the encased group of mammary glands provided with teats or nipples, as in a cow, ewe, mare or sow. Also called bag.

Underfeeding   a term referring to not providing the animal sufficient dietary energy. Usually results in lowered production, the degree related to the extent of underfeeding and the length of time it exists.

Unthriftiness   lack of vigor, poor growth of development; the quality or state of being unthrifty in animals.

Urea   a white, crystalline, water-soluble substance with the formula $CO(NH_2)_2$. Used as a nonprotein nitrogen compound in ruminant diets. It is made synthetically by combining ammonia and carbon dioxide.

Urease   an enzyme which breaks down urea to yield carbon dioxide and ammonia. It is found in considerable concentrations in the soybean and the jackbean, and is produced by certain microorganisms in the rumen.

Uremia   the presence of urinary constituents in the blood, and the toxic condition produced thereby.

Urine   the fluid secreted by the kidneys, stored in the bladder, and discharged by the urethra.

USDA   United States Department of Agriculture.

USP (United States Pharmacopoeia)   a unit of measure or potency of biologicals that usually coincides with an international unit. Also see International Unit.

Vaccination   artificial immunization. To inoculate with a mildly toxic preparation of bacteria or a virus of a specific disease to prevent or lessen the effect of that disease.

Vaccine   a suspension of attenuated or killed microorganisms (bacteria or viruses), administered for the prevention, improvement or treatment of infectious diseases.

Vascular   pertaining to the blood vessels of the body.

Veal   a calf fed for early slaughter (usually less than 3 months of age).

Vegetative   a term used to designate the stem and leaf development, in contrast to flowers and seed development.

VFA   see Volatile Fatty Acids.

Villi   small fingerlike projections attached to the interior side of the wall of the small intestine.

Vines   a part of any plant whose stems require support or lie on the ground.

Viscera   all the organs in the great cavity of the body, excluding contents of the intestinal tract.

Viscosity   a property of a fluid which resists change in the shape or arrangement of its elements during flow.

Vitamin   any of a group of feed constituents essential in small quantities to maintain life but not themselves supplying energy. They may be water soluble or fat soluble.

Void   to cast out or evacuate as waste matter, feces and/or urine.

Volatile Fatty Acids (VFA)   any one of several volatile organic acids found especially in rumen or large intestine contents and/or silage. Acetic, propionic and butyric acids are ordinarily the most prevalent.

Vomiting   the forcible explusion of the stomach contents through the mouth.

Wafers   a physical form of agglomerated feed based on fibrous ingredients in which the finished form usually has a diameter or cross section measurement greater than its length.

Weaning   the stopping of young animals from nursing or suckling their mothers.

Weanling   a recently weaned animal.

Wet-Milled   a process in which feed material is steeped in water with or without sulphur dioxide to soften the kernel in order to facilitate the separation of the various component parts.

Wet Rendered   a process in which material is cooked with steam under pressure in closed tanks.

Whey the watery part of milk separated from the curd.

Wilted a physical form of a forage becoming limp as a result of water loss.

Wind Sucker see cribber.

Wood Chewing an abnormal behavior in horses characterized by chewing on wood (mangers, fences etc.)

Wort the liquid portion of malted grain. It is a solution of malt sugar and other soluble extracts from malted mash.

Yearling refers to a male or a female farm animal (especially cattle and horses) during the first year of its life.

Appendix Table I. Nutrient Composition of Feedstuffs Commonly Fed to Swine and Poultry

Feedstuff (IFN)	D.M. %	CP %	Lys %	Thr %	Tryp %	Meth %	Ca %	P %	Fat %	ME Swin kcal/kg	ME ChKn kcal/kg
Alfalfa meal, dehydrated	91.8	17.4	.8	.7	.3	.3	1.38	.23	2.8	1322	1508
(1-00-023)	100.0	18.9	.9	.8	.4	.3	1.51	.25	3.0	1441	1643
Barley, grain	88.6	11.5	.4	.4	.2	.2	.05	.34	1.8	2937	2621
(4-00-549)	100.0	13.0	.4	.4	.2	.2	.05	.38	2.0	3316	2959
Beet pulp, dehy	91.0	8.9	.5	.4	.1	—	.62	.09	.5	2704	—
(4-00-669)	100.0	9.8	.6	.4	.1	—	.68	.10	.6	2971	—
Blood meal, spray dr.	92.6	87.7	7.6	3.8	1.0	1.1	.22	.20	.7	1942	2770
(5-00-381)	100.0	94.7	8.2	4.1	1.1	1.2	.24	.21	.8	2097	2991
Corn, grain	87.3	8.8	.3	.3	.1	.2	.02	.29	3.9	3359	3425
(4-02-931)	100.0	10.1	.3	.4	.1	.2	.02	.33	4.5	3846	3921
Corn, gluten feed	89.9	22.9	.6	.8	.2	.4	.32	.75	2.2	2478	1730
(5-02-903)	100.0	25.5	.7	.9	.2	.4	.35	.83	2.5	2755	1924
Corn, hominy feed	90.2	10.3	.4	.4	.1	.2	.05	.51	6.5	3381	2894
(4-02-887)	100.0	11.4	.4	.4	.1	.2	.05	.57	7.2	3748	3208
Cottonseed meal, sol	91.0	41.3	1.7	1.4	.5	.6	.17	1.11	1.5	2359	1950
(5-01-621)	100.0	45.4	1.9	1.5	.6	.6	.18	1.22	1.7	2592	2142
Fat, animal, stab.	99.1	0.0	0.0	0.0	0.0	0.0	0.0	0.0	99.1	7973	7677
(4-00-409)	100.0	0.0	0.0	0.0	0.0	0.0	0.0	0.0	100.0	8044	7744
Fish meal, menhad.	91.7	62.2	4.7	2.5	.6	1.7	5.01	2.87	9.8	2602	2851
(5-02-009)	100.0	67.9	5.2	2.8	.7	1.9	5.46	3.14	10.7	2839	3110
Meat meal	93.8	50.6	3.0	1.7	.3	.7	8.61	4.58	8.9	2224	2087
(5-00-385)	100.0	54.0	3.2	1.8	.4	.7	9.17	4.88	9.5	2371	2225
Meat & bone meal	93.3	50.2	2.8	1.6	.3	.7	10.0	4.96	10.4	2266	2017
5-00-388)	100.0	53.8	3.0	1.8	.3	.7	10.71	5.32	11.2	2428	2161
Molasses, cane, liq.	74.3	4.3	—	—	—	—	.74	.08	.2	2193	1902
(4-04-696)	100.0	5.8	—	—	—	—	1.00	.10	.2	2950	2560
Oats, grain	89.2	11.8	.4	.4	.2	.2	.08	.34	4.6	2626	2543
(4-03-309)	100.0	13.3	.4	.4	.2	.2	.09	.38	5.2	2943	2850
Oat, groats	89.6	15.5	.6	.4	.2	.3	.08	.42	6.1	2928	3251
(4-03-331)	100.0	17.3	.6	.5	.2	.3	.09	.47	6.8	3269	3630
Poultry meat meal	93.8	61.1	3.1	2.1	.5	1.1	3.95	2.06	13.1	2869	2865
(5-03-798)	100.0	65.1	3.3	2.2	.5	1.2	4.21	2.19	13.9	3058	3054
Rye, grain	87.5	12.0	.4	.3	.1	.2	.06	.32	1.5	2911	2652
(4-04-047)	100.0	13.7	.5	.4	.1	.2	.07	.36	1.7	3327	3031
Skim milk, dehy	94.1	33.4	2.5	1.6	.4	.9	1.28	1.02	1.0	3591	2537
(5-01-175)	100.0	35.5	2.7	1.7	.5	1.0	1.36	1.09	1.0	3815	2695
Sorghum grain (milo)	88.5	10.0	.2	.3	.1	.2	.04	.30	2.8	3247	3226
(4-04-444)	100.0	11.3	.3	.4	.1	.2	.05	.34	3.2	3669	3645
Soybean meal, w/o hull, sol	89.9	49.3	3.1	1.9	.7	.7	.26	.64	1.0	3361	2473
(5-04-612)	100.0	54.8	3.4	2.1	.8	.8	.29	.71	1.1	3737	2750
Soybean meal, sol	89.6	45.7	2.8	1.8	.6	.6	.30	.69	1.2	3006	2331
(5-04-604)	100.0	51.0	3.2	2.0	.7	.6	.34	.77	1.3	3355	2602
Soybeans, full-fat cooked	92.6	36.6	2.2	1.4	.5	.5	.26	.61	18.7	3642	—
(5-04-597)	100.0	39.5	2.4	1.5	.6	.5	.28	.66	20.2	3933	—
Sunflower meal, sol	92.5	45.2	1.7	1.6	.6	.8	.42	.94	2.7	2636	2073
(5-04-739)	100.0	48.9	1.8	1.8	.6	.9	.45	1.02	2.9	2851	2242
Wheat, bran	89.0	15.4	.5	.5	.2	.2	.13	1.13	3.8	2372	1226
(4-05-190)	100.0	17.4	.6	.6	.2	.3	.14	1.27	4.3	2664	1377
Wheat, grain	89.0	13.0	.4	.4	.1	.2	.05	.35	1.7	3267	3069
(4-05-211)	100.0	14.7	.4	.5	.2	.2	.05	.39	2.0	3670	3449
Wheat, middlings	88.9	16.4	.7	.6	.2	.2	.13	.89	4.2	2712	2075
(4-05-205)	100.0	18.4	.8	.6	.2	.2	.14	1.00	4.7	3049	2333
Whey, dehy	93.3	13.1	.9	.9	.2	.2	.85	.76	.7	3104	1941
(4-01-182)	100.0	14.0	1.0	1.0	.2	.2	.92	.81	.8	3328	2081

Appendix Table II. Nutrient Composition of Feedstuffs Commonly Fed to Ruminant Animals and Horses

Feedstuff (IFN)	D.M. %	CP %	ADF %	CA %	P %	TDN Catl %	TDN Shp %	NEm Catl M/kg	NEg Catl M/kg	NEl Catl M/kg	DE Hors M/kg
Alfalfa, dehydrated	91.8	17.4	31.5	1.38	.23	55.6	55.0	1.27	.73	1.25	2.16
(1-00-023)	100.0	18.9	34.3	1.51	.25	60.6	60.0	1.38	.80	1.37	2.36
Alfalfa, hay, early-bim	90.5	18.0	29.1	1.48	.19	51.8	53.2	1.33	.80	1.21	2.24
(1-00-059)	100.0	19.9	32.1	1.63	.21	57.2	58.8	1.47	.88	1.33	2.48
Alfalfa, hay, mid-bim	91.0	17.0	33.6	1.24	.22	52.1	52.2	1.17	.65	1.15	2.07
(1-00-063)	100.0	18.7	36.9	1.37	.24	57.2	57.4	1.28	.71	1.27	2.28
Alfalfa, silage	43.4	8.0	15.1	.57	.13	23.0	25.4	.46	.22	.51	—
(3-08-150)	100.0	18.4	34.7	1.32	.31	53.0	58.6	1.07	.52	1.18	—
Alfalfa-grass, hay	91.4	14.5	37.5	1.34	.22	51.8	49.3	.96	.46	.99	1.90
(1-08-331)	100.0	15.9	41.0	1.47	.25	56.7	53.9	1.05	.50	1.08	2.08
Barley, grain	88.6	11.5	7.7	.05	.34	73.7	76.0	1.73	1.16	1.61	3.26
(4-00-549)	100.0	13.0	8.7	.05	.38	83.2	85.9	1.95	1.31	2.05	3.68
Brome, smooth, hay	89.6	12.4	33.4	.34	.24	53.5	53.4	1.19	.67	1.17	1.71
(1-00-947)	100.0	13.9	37.3	.38	.26	59.7	59.6	1.33	.75	1.30	1.91
Corn-and-cob meal	86.5	7.8	—	.06	.24	72.7	71.4	1.90	1.31	1.68	2.85
(4-02-849)	100.0	9.0	—	.07	.27	84.0	82.5	2.20	1.52	1.94	3.29
Corn, silage	34.1	2.8	9.7	.09	.07	23.9	24.0	.56	.35	.54	—
(3-02-823)	100.0	8.1	28.3	.27	.20	70.0	70.5	1.63	1.03	1.60	—
Corn, stover	85.2	5.4	—	.49	.08	51.3	49.9	1.12	.63	1.15	1.38
(1-02-776)	100.0	6.4	—	.57	.10	60.2	58.5	1.32	.74	1.35	1.62
Corn, grain	87.3	8.8	5.3	.02	.29	78.4	82.3	1.95	1.35	1.84	3.38
(4-02-931)	100.0	10.1	6.1	.02	.33	89.8	94.2	2.24	1.55	2.11	3.87
Clover, red, hay	88.4	13.0	36.2	1.22	.22	51.6	52.0	1.11	.61	1.16	1.96
(1-01-415)	100.0	14.7	41.0	1.38	.24	58.4	58.9	1.26	.69	1.31	2.22
Cottonseed, hulls	90.4	3.8	59.0	.13	.08	40.6	43.8	.55	.08	.86	1.71
(1-01-599)	100.0	4.2	65.3	.15	.09	44.9	48.5	.61	.08	.95	1.89
Cottonseed, meal, sol	91.0	41.3	18.2	.17	1.11	71.0	64.9	1.64	1.07	1.61	2.74
(5-01-621)	100.0	45.4	20.0	.18	1.22	78.0	71.3	1.80	1.17	1.77	3.01
Linseed meal, mech	90.8	34.5	—	.41	.87	74.3	75.6	1.79	1.20	1.71	2.74
(5-02-045)	100.0	38.0	—	.45	.96	81.8	83.2	1.97	1.32	1.88	3.04
Molasses, cane, liq.	74.3	4.3	.3	.74	.08	61.1	58.7	1.70	1.18	1.41	2.60
(4-04-696)	100.0	5.8	.4	1.00	.10	82.2	79.0	2.28	1.58	1.89	3.50
Oats, grain	89.2	11.8	14.2	.08	.34	68.9	68.5	1.77	1.19	1.55	2.95
(4-03-309)	100.0	13.3	15.9	.09	.38	77.3	76.7	1.98	1.33	1.74	3.30
Oat, hay	90.7	8.6	34.8	.29	.23	54.3	48.9	1.26	.73	1.22	1.75
(1-03-280)	100.0	9.5	38.4	.32	.25	59.9	53.9	1.39	.81	1.35	1.92
Orchard grass, hay	89.6	10.5	34.0	.37	.23	51.5	51.6	1.10	.59	1.15	1.69
(1-03-438)	100.0	11.8	37.9	.41	.26	57.4	57.6	1.22	.66	1.29	1.89
Ryegrass, per., hay	86.6	8.4	—	—	.24	48.2	51.6	1.01	.52	1.08	1.54
(1-04-077)	100.0	9.7	—	—	.27	55.7	59.6	1.16	.60	1.24	1.77
Sorghum, grain (milo)	88.5	10.0	3.5	.04	.30	75.0	77.7	1.65	1.09	1.62	—
(4-04-444)	100.0	11.3	4.0	.05	.34	84.8	87.8	1.87	1.23	1.83	—
Sorghum, Sudan grass, hay	91.1	10.9	37.2	.47	.28	50.9	49.1	1.07	.55	1.14	1.74
(1-04-480)	100.0	12.0	40.8	.51	.31	55.8	53.8	1.17	.61	1.25	1.91
Soybean meal, sol	89.6	45.7	—	.30	.69	75.0	78.8	1.84	1.25	1.73	3.15
(5-04-604)	100.0	51.0	—	.34	.77	83.7	87.9	2.05	1.39	1.93	3.52
Sunflower meal, sol	92.5	45.2	—	.42	.94	60.1	69.0	1.36	.82	1.36	2.59
(5-04-739)	100.0	48.9	—	.45	1.02	65.0	74.6	1.47	.88	1.47	2.80
Timothy hay, early-bim	89.1	9.6	31.4	.45	.25	52.7	50.4	1.14	.63	1.18	1.83
(1-04-882)	100.0	10.8	35.2	.51	.29	59.1	56.5	1.28	.71	1.33	2.06
Trefoil, Birdsfoot, hay	90.6	13.9	—	1.54	.21	55.3	52.3	1.22	.69	1.25	1.99
(1-05-044)	100.0	15.3	—	1.70	.23	61.0	57.8	1.34	.77	1.37	2.20
Urea, 45% N, 281% CPE	97.0	276.9	—	.09	0.0	0.0	0.0	0.0	0.0	0.0	0.0
(5-05-070)	100.0	285.4	—	.09	0.0	0.0	0.0	0.0	0.0	0.0	0.0

NAME _____ SEAT NO. _____ SECTION _____

SWINE RATION WORKSHEET

Ration _____

Calculated Analysis

Ingredient	kg.	Protein %	Calcium %	Total Phosphorus %	Vit. A units/kg	Vit. D units/kg	Riboflavin mg/kg	Niacin mg/kg	Pantothenic acid mg/kg	Vit. B₁₂ mcg/kg	Metabolizable energy kcal/kg
Nutritive Requirements											
Amount Required		kg	kg	kg	IU	IU	mg	mg	mg	mcg	kcal
TOTALS (kg, IU, mg, mcg)											
Calculated Analysis: % or amt. per kg Feed											
DIFFERENCE FROM NEEDS											

565

Appendix III (continued).

SWINE RATION WORKSHEET (CONTINUED)

					Calculated Analysis						
								mg/kg			
Ingredient	kg.	Lysine %	Thre-onine %	Trypto-phan %	Cu	Fe	I	Mg	Mn	Se	Zn
Nutritive Requirements											
Amount Required		kg	kg	kg	mg	mg	mg	mg	mg	mg	mg
TOTALS (kg, IU, mg, mcg)											
Calculated Analysis: % or amt. per kg Feed											
DIFFERENCE FROM NEEDS											

Appendix IV

NAME _____ SEAT NO. _____ SECTION _____

METABOLIZABLE PROTEIN FORMULATION WORKSHEET FOR CATTLE AND SHEEP
(When Individual Amino Acid Requirements Are Not Computed)

1. Kind of Animal _____ Body Weight _____ kg
   Daily Dry Matter Consumption: _____ % of Body Weight = _____ kg

2. Daily Ration Ingredients Available in Satisfying Energy Needs:

Ration Ingredient	Amount, %	D.M. kg	NE_m Mcal	NE_g Mcal	NE_L Mcal
TOTAL		(a)	(b)	(c)	(d)
	Mcal/kg =		(e)	(f)	(h)
			$(b \div a)$	$(c \div a)$	$(d \div a)$

3. Daily Net Energy Requirement from NRC Tables:

   NE_m required, Mcal   (i) _____ (from tables)

   Total kg. ration D.M.   (a) _____ (a)

   Kg. ration to supply NE_m (j) _____ $(i \div e)$

   Kg. ration left for NE_g or NE_L (k) _____ $(a - j)$

   NE_g or NE_L available, Mcal   (m) _____ (k x f or h)

   Expected daily gain or milk prod., kg   _____ (m compared to tables)

4. MP Required for Expected Daily Gain or Milk Production (n) _____ g. (from tables)

5. MP and UFP Supplied by Above Daily Ration:

Ration Ingredient	Amount, D.M., kg	MP, g	UFP, g

6. MP (from satisfaction of positive UFP value) to come from negative UFP feeds [total negative UFP (not exceeding the positive UFP)] _____ x 2.225 = _____ g from neg. UFP feeds

   Total MP = (o) _____
   Net positive UFP remaining (p) _____

7. Supplemental MP Needs:
   a. From urea: MP needed (n − o) = (q) _____ g ÷ 2.225 = _____ g urea*
      *Note: If total g of urea needed exceed the net positive UFP remaining (p), urea *cannot* be fed in order to achieve predicted gain or milk production. Thus, feed natural protein.

   b. From natural protein source: _____ containing _____ g /kg MP and _____ UFP
      If there is a net positive UFP remaining (p), then this must be multiplied by 2.225 to predict the amount of MP which will be realized from the negative UFP value of the natural protein source to be added. Thus:

      (1) Net pos. UFP remaining (p) _____ x 2.225 = (r) _____ g MP from pos. UFP
      (2) MP needed (q) _____ g − (r) _____ = (s) _____ g MP from natural protein source
      (3) (s) _____ ÷ _____ (MP value of natural protein) = _____ kg. natural protein

8. Final Supplemented Ration:

Ingredient	kg	%
TOTAL		100%

567

Appendix V

NAME _____ SEAT NO. _____ SECTION _____

DAIRY RATION WORKSHEET

	(a) Crude Protein kg	(b) Net Energy for Lactation Mcal	(c) Calcium gm	(d) Phosphorus gm
A. *Daily Requirements for Dairy Animal* (dry-matter basis)				
1. For maintenance, cow weighing _____ kg	_____	_____	_____	_____
2. For maintenance + pregnancy (last 2 months), cow weighing _____ kg	_____	_____	_____	_____
3. For milk, _____ kg per day testing _____ % fat	======	======	======	======
4. Total Daily Requirements	_____	_____	_____	_____
B. *Forage Ration* (dry-matter basis)				
1. _____ kg of _____	_____	_____	_____	_____
2. _____ kg of _____	_____	_____	_____	_____
3. _____ kg of _____	======	======	======	======
4. Total Nutrients from Forage	_____	_____	_____	_____
C. *Remaining,* to be supplied from concentrate mix ("Total Daily Requirements" less "Total Nutrients from Forage") ($A_4 - B_4$)	_____	_____	_____	_____

D. *Amount of Concentrate Mix Required*
A good dairy concentrate will contain approximately 1.8–2.0 Mcal/kg NE_L on a dry matter basis.

_____ Mcal NE_L needed ÷ _____ Mcal/kg NE_L =

_____ kg

E. *Percent Crude Protein Needed in Concentrate Mix*
Divide Crude Protein to be supplied from concentrates ($C_{(a)}$) by concentrates required (D) times 100 = % Crude Protein needed in concentrate mix. ($C_{(a)}$) ÷ D × 100 = %

_____ kg CP needed ÷ _____ kg mix × 100 =

_____ % CP

568

NAME _____ SEAT NO. _____ SECTION _____

HORSE RATION WORKSHEET

Animal Description _____ Body Weight _____ lb

	(a) Digestible energy Mcal	(b) Crude Protein lb	(c) Calcium g	(d) Phos. g
A. *Daily Requirements* (based on horse activity)				
1. _____	_____	_____	_____	_____
2. _____	_____	_____	_____	_____
3. Total Daily Requirements	_____	_____	_____	_____
B. *Forage Ration* (between 1% to 2% of body weight)				
1. _____ lb of _____	_____	_____	_____	_____
2. _____ lb of _____	_____	_____	_____	_____
3. Total Nutrients from Forage	_____	_____	_____	_____
C. Remaining, to be supplied from concentrate ("Total Daily Requirements" less "Total Nutrients from Forage"). (A.3. − B.3.)	_____	_____	_____	_____

D. Amount of Concentrate Required. Assume a typical concentrate will contain approximately 1.35 Mcal/lb DE (air-dry basis) (C.a. ÷ 1.35 = lb)

_____ Mcal DE needed ÷ _____ Mcal/lb
DE = _____ lb

E. Percent Crude Protein Needed in Concentrate Mix. Divide CP to be supplied from concentrate (C.b.) by concentrate required (D.) times 100 = %CP in concentrate mix. (C.b. ÷ D. × 100 = %CP)

_____ lb CP needed ÷ _____ lb mix
× 100 = _____ % CP

F. *Concentrate*				
1. _____ lb of _____	_____	_____	_____	_____
2. _____ lb of _____	_____	_____	_____	_____
3. _____ lb of _____	_____	_____	_____	_____
4. Total Nutrients from Conc.	_____	_____	_____	_____
G. Total Ration Nutrients (B.3. + F.4.)	_____	_____	_____	_____

F. Concentrate Mixture

Dry-Matter Basis					
Ingredient	Amount kg	Crude Protein kg	NE$_L$ Mcal	Ca gm	P gm
Totals (kg, Mcal, g)					
Calculated Analysis: % or amt per kg					

G. Overall Composition of Daily Ration

Dry-Matter Basis					
Ingredient	Amount kg	Crude Protein kg	NE$_L$ Mcal	Ca gm	P gm
Totals (kg, Mcal, g)					
Calculated Analysis: % or amt per kg					
Difference from needs (A$_4$)					

NAME _____ SEAT NO. _____ SECTION _____

POULTRY RATION WORKSHEET

Ration _____

Ingredient	kg.	Calculated Analysis										
		Protein %	Calcium %	Phosphorus %	Vit. A units/kg	Vit. D₃ units/kg	Riboflavin mg/kg	Niacin mg/kg	Pantothenic acid mg/kg	Vit. B₁₂ mcg/kg	Choline mg/kg
Nutritive Requirements												
Amount Required		kg	kg	kg	IU	ICU	mg	mg	mg	mcg	mg
TOTALS (kg, IU, mg, mcg)												
Calculated Analysis: % or amt. per kg Feed												
DIFFERENCE FROM NEEDS												

POULTRY RATION WORKSHEET (CONTINUED)

Ingredient	kg.	Metabol energy kcal/kg	Lysine %	Methionine %	Cu	Fe	I	Mg	Mn	Se	Zn
								mg/kg			
Nutritive Requirements											
Amount Required		kcal	kg	kg	mg	mg	mg	mg	mg	mg	mg
TOTALS (kg, IU, mg, mcg)											
Calculated Analysis: % or amt. per kg Feed											
DIFFERENCE FROM NEEDS											

Calculated Analysis

Index

573